9th International Conference on Thermal Equipments, Renewable Energy and Rural Development (TE-RE-RD 2020)

E3S Web of Conferences Volume 180 (2020)

Constanta, Romania
26 – 27 June 2020

Editors:

Gabriel P. Negreanu
Valentin Vlăduţ
Mohand Tazerout

ISBN: 978-1-7138-1421-4

Printed from e-media with permission by:

Curran Associates, Inc.
57 Morehouse Lane
Red Hook, NY 12571

proceedings.com

Some format issues inherent in the e-media version may also appear in this print version.

For additional information, please contact EDP Sciences – Web of Conferences at the address below.

EDP Sciences – Web of Conferences
17, Avenue du Hoggar
Parc d'Activité de Courtabœuf
BP 112
F-91944 Les Ulis Cedex A
France

Phone: +33 (0) 1 69 18 75 75
Fax: +33 (0) 1 69 28 84 91

contact-edps@webofconferences.org

Additional copies of this publication are available from:

Curran Associates, Inc.
57 Morehouse Lane
Red Hook, NY 12571 USA
Phone: 845-758-0400
Fax: 845-758-2633
Email: curran@proceedings.com
Web: www.proceedings.com

TABLE OF CONTENTS

PREFACE

The 9-th International Conference on Thermal Equipment, Renewable Energy and Rural Development (TE-RE-RD 2019 - http://www.tererd.pub.ro/) was organized Online (powered by Microsoft Teams) in the period 26-27 June 2020, due to the restrictions caused by the COVID-19 pandemics.

TE-RE-RD is an International pluridisciplinary Conference aimed to be a forum for information exchange between scientists from academic, research and development communities around the world on topics related to thermal machines, equipments and technologies, equipments, technologies and policies for rural development, renewable energy sources conversion and other technical domains.

Another target of the Conference is to offer to the young people involved in research activities (PhD, Master, and regular students) the opportunity to disseminate their work results and learn from experimented scientists.

We also welcome people and entities from industry, agriculture, administration, trade, politics and media who want to cooperate with academic and R&D organizations in the fields of the Conference.

The main Organizers, University POLITEHNICA of Bucharest (Faculty of Mechanical Engineering and Mechatronics and Faculty of Biotechnical Systems Engineering), National Institute of Research – Development for Machines and Installations Designed to Agriculture and Food Industry and the Romanian Society of Agricultural Mechanical Engineers produce every year since 2012 an itinerant Conference in academic cities or touristic resorts. This year, the chosen partner was Constanța Maritime University, who offered its halls for the sessions and a part of his staff for the event success. Unfortunately, the social distance restrictions lead the Organizers to move the sessions online.

The main topics of the Conference were:

- Processes in thermal equipment
- Steam and hot water generators
- Steam and gas turbines
- Internal combustion engines
- Environmental protection
- Solar, wind, geothermal, biomass
- Building efficiency
- Equipment for rural development
- Politics of rural development
- Agricultural engineering
- Energy of rural space

For this year the Organizers continued to use the Conference Management Toolkit provided by Microsoft in order to manage the paper submissions and reviewing process. We received

106 papers distributed to 108 reviewers. 84 papers were accepted to be presented and 22 papers were rejected or retired by the authors after revision process.

The Conference was developed on 4 tracks:

- 17 papers in the section Thermal Equipments
- 20 papers in the section Renewable Energy
- 27 papers in the section Rural Development
- 20 papers in the section Miscellaneous

Besides the representatives of the Organizers, the Conference also welcomed researchers from University Angel Kanchev of Ruse (Bulgaria), Technical University of Varna (Bulgaria), Technical University of Sofia (Bulgaria), University of Telecommunications and Post Sofia (Bulgaria), Almaty University of Power Engineering and Telecommunications (Kazahstan), Satpayev University (Kazahstan), University of Food Technologies of Plovdiv (Bulgaria), Moscow State University of Civil Engineering (Russia), RUDN University (Russia), Al - Furat Al- Awsat Technical University (Irak), University “Dunărea de Jos” of Galați (Romania), University of Craiova (Romania), Naval Academy Mircea cel Batran Constanța (România), "Ion Mincu" University of Architecture and Urbanism (Romania), INCD Comoti (Romania), INCD ICEMENERG (Romania), INOE2000-IHP Bucharest (Romania), Agricultural Research and Development Secuieni (Romania), Etudes et Productions Schlumberger, etc.

We hope that next edition the number and the value of papers will increase, the current participants will come back and new authors will attend.

See you again in 2021,

Prof. Gabriel Negreanu

Dr. Valentin Vlăduț

Prof. Mohand Tazerout

Editors of the Conference Proceedings TE-RE-RD 2020

https://doi.org/10.1051/e3sconf/202018000002

OPENING ADDRESS

Dear participants to the 9th International Conference on Thermal Equipment, Renewable Energy and Rural Development (TE-RE-RD 2020),

Dear researchers, ladies and gentleman

On behalf of the Organizing Committee, allow me to welcome you to this event, organized each year by a consortia of Universities, Research Institute and NGO.

This year, the host of the event should have been Constanța Maritime University, that generously offered its support in order to contribute to the event success.

But, in full effort to organize the Conference, to complete the research works, to evaluate it, the pandemic of COVID-19 suddenly appeared, temporarily closing many activities, among which the academic and research ones. In this unfavorable climate to classical scientific events, the organizers decided to move the presentations of the works online, based on the experience gained in the few months of remote teaching activities.

We are using today four parallel channels of Microsoft Teams app, corresponding to the four Conference Tracks, and I hope that the connections will keep stable and will allow the voice and image streaming during all the works period.

Ladies and gentlemen,

We will keep in this edition of our Conference the agreement signed with the prestigious Publishing House - E3S Web of Conferences stipulating that accepted and presented papers will be published and indexed in Web of Sciences, Scopus and other important international scientific databases.

Finally, I wish to thank to all participants to the Conference for their presence here, and especially to the members of the Scientific and Organizing Committees, who worked hard to make it possible.

I also hope that we will be able to organize the 2021 edition in the "face-to-face" format, and I hope that happens in Constanța, as planned for this year.

Prof. Lucian Mihăescu

Chairman of TE-RE-RD 2020

Statement of Peer review

In submitting conference proceedings to *Web of Conferences*, the editors of the proceedings certify to the Publisher that

1. They adhere to its **Policy on Publishing Integrity** in order to safeguard good scientific practice in publishing.
2. All articles have been subjected to peer review administered by the proceedings editors.
3. Reviews have been conducted by expert referees, who have been requested to provide unbiased and constructive comments aimed, whenever possible, at improving the work.
4. Proceedings editors have taken all reasonable steps to ensure the quality of the materials they publish and their decision to accept or reject a paper for publication has been based only on the merits of the work and the relevance to the journal.

Title, date and place of the conference

9-th International Conference on Thermal Equipment, Renewable Energy and Rural Development TE-RE-RD 2020

26-27 June 2020 Online powered by Microsoft Teams

Proceedings editor(s): Prof. Gabriel Negreanu, Dr. Valentin Vlăduţ, Prof. Mohand Tazerout

Date and editor's signature 08.07.2020

17, avenue du Hoggar - PA de Courtabœuf – BP 112 - 91944 Les Ulis Cedex A (France)
Tél. : 33 (0)1 69 18 75 75 - Fax : 33(0)1 69 07 45 17 – www.edpsciences.org

E3S Web of Conferences **180**, 01001 (2020) https://doi.org/10.1051/e3sconf/202018001001
TE-RE-RD 2020

Seismic qualification by analysis of a nuclear class horizontal pressure vessel

Liviu-Constantin Stan[1,*], *Ioan* Călimănescu[2] and *Catalin* Faitar [1]

[1] Constanta Maritime University, Faculty of Naval Electromechanics, Department Electromechanics, 104 Mircea cel Batran Street, 900663 Constanta, Romania
[2] Nuclear Power Plant Cernavoda-Romania

Abstract*:* The seismic qualification for the structural strength it is an issue for the assessment of the seismic safety of the Nuclear Power Plants. The qualification for seismic events in nuclear industry [4] includes the assessment of the structural integrity and fitness for operability during and after an earthquake. Such qualifications are done for various safety and non-safety equipment among which the pressure vessels are of a paramount importance. The seismic qualification is done either directly on prototypes or scaled prototypes (mock ups) or via analytical methods like analysis with finite elements. In any case the model should accurately represent the actual performance of the component or structure when it is subjected to the prescribed effects. The goal of this article is to underline the procedure to be followed for nuclear industry horizontal pressure vessels in order to obtain sound and credible results for equipment seismic qualification by using the unique analysis features of ANSYS 19. A seismic qualification analysis is carried out for the horizontal pressure vessel. The seismic event is overloading the anchoring system beyond the ultimate strength of the material. The zone where these big stresses fields are calculated is at the anchors level. The vessel as designed failed to be qualified for the seism excitations imposed to the model. In order to have the vessel qualified the anchoring way should be redesigned. By comparing the overall calculated equivalent stresses with the tensile yield strength we may pull the conclusion that the original structure design will fail. Instead the new redesigned anchoring system with 8 extra-anchors will stand.

* Corresponding author: liviustan14@yahoo.com

1 Introduction

The seismic qualification for the structural strength it is an issue for the assessment of the seismic safety of the Nuclear Power Plants.

The qualification for seismic events in nuclear industry [4] includes the assessment of the structural integrity and fitness for operability during and after an earthquake. Such qualifications are done for various safety and non-safety equipment among which the pressure vessels are of a paramount importance.

The seismic qualification is done either directly on prototypes or scaled prototypes (mock ups) or via analytical methods like analysis with finite elements. In any case the model should accurately represent the actual performance of the component or structure when it is subjected to the prescribed effects.

Seismic qualification by analysis should be used for items without a functional safety requirement that are unique or that are of a size or scale to preclude their qualification by testing. Civil engineering structures, tanks, pressure vessels, distribution systems and large items of equipment are usually qualified by analytical methods[5].

In some countries the nuclear class horizontal cylindrical vessels resistant to temperature and internal pressure are designed according to ASME Section III design code, which often result in pressure vessels with wall thicknesses calculated for all the loading scenarios including the seismic one, are governed by the calculation of circumferential stress[6].

But their supports do not always are checked for seismic loads during their design, especially the anchoring system. In seismic response of a structure such a pressure vessel, the saddle supports and the anchoring system to the floor, is mandatory to be checked for resistance when it is called for seismic qualification[7,8].

For numerical analysis the latest advances in simulation technology of ANSYS 19 will be deployed. The main module to be used will be the Response Spectrum meant to perform calculations over the maximum response of a structure when subjected to a transient load (variable in time). The maximum response can be computed as the mode shape times a certain scale factor [1]. Multiple maximum responses then may be combined to give the total response of the vessel. This is an alternative to the more cumbersome full time history.

By knowing the natural frequencies and the mode shapes of the vessel we can simply determine what the displacement would be for a 1-DOF oscillator, if it were subjected to the same transient loading, and scale the response by the appropriate amount. If there is more than one load, in our case one for any direction X, Y, Z, each will have its own spectrum. A modal analysis must first be completed to determine the natural frequencies, mode shapes, and participation factors for each mode. For each natural frequency, the spectrum value can be determined by a simple look-up from the response-spectrum table. The mode coefficients can be determined from the participation factors, depending on the type of spectrum input. Participation factors measure the amount of mass moving in each direction for a unit displacement. The response (displacement, velocity or acceleration) for each mode can then be computed from the frequency, mode coefficient, and mode shape. Finally the mode combinations are computed ant they take the form:

$$[R] = \left(\sum_{i=1}^{n} \sum_{j=1}^{n} \epsilon_{ij} \, [R]_i [R]_j\right)^{1/2} \qquad (1)$$

where *[R]* is the total modal response and $[R]_i[R]_j$ is the product of modes *i* and *j*. The modal correlation coefficientsϵ_{ij}, are uniquely defined, depending on the method chosen for evaluating the correlation coefficient.

The goal of this article is to underline the procedure to be followed for a nuclear industry horizontal pressure vessel in order to obtain sound and credible results for equipment seismic qualification by using the unique analysis features of ANSYS 19. A seismic qualification analysis is carried out for the horizontal pressure vessel as shown in the Figure 1.

2 Materials and methods

The solving strategy is following a series of fixed stages in order to have the right results.

Firstly the CAD geometry of the vessel will be generated by using SpaceClaim as seen in Figure 1:

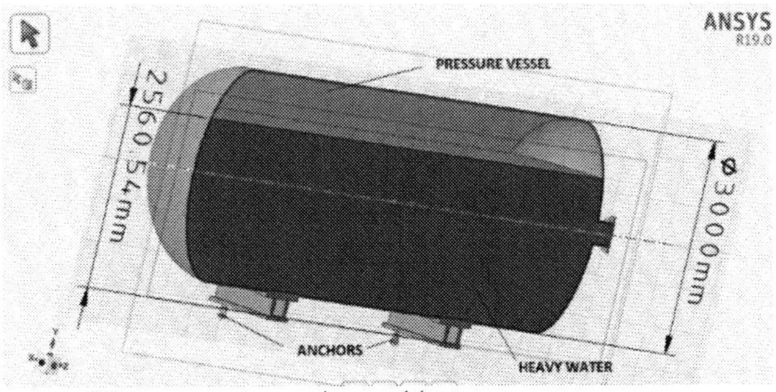

Fig.1. The CAD geometry of the pressure vessel on subject

The geometry is somehow simplified targeting only the major structural features and neglecting small details as instrumentation or name plates. The external diameter of the vessel is 3000 mm and its length is about 7500 mm. It is sitting on the floor on two saddles fixed via M56 anchors in concrete. The material of the vessel is stainless steel and the stored fluid is radioactive heavy water. The thickness of the vessel wall is 20 mm.

The material imposed is the stainless steel with density of 29,314 kg/m^3, Young modulus of 1.93e11 Pa, Poisson ratio of 0.31, tensile yield strength of 2.07e8 Pa and the ultimate strength of 5.86e8 Pa.

The normal density of the stainless steel material is 7750 kg/m$^{3.}$ But inside the material definition the density has the value of 29,314 kg/m$^{3.}$ The explanation of this paradox is in the follows.

The basic idea is that we need the mechanical dynamic system of the empty vessel to account for the participation of the heavy water inside. The simplest way is to modify the density of the steel material in order to consider the effects of the water mass and the difference between the centres of mass of the empty vessel and the filled one. For the seismic qualification the vessel is considered partially filled with heavy water as seen in Figure 1. No sloshing effect is considered, the effects of such a phenomenon are negligible in comparison with the seismic effects.

Firstly we will consider the masses of the vessel empty and of the contained water. The volume of the steel for the vessel is 2.2 m^3 and the volume of contained water is 41.06 m^3. Taking into account the densities of the steel and the heavy water the mass of the steel is 17,050 kg and the mass of the water is 41,142 kg. The total mass is 58,192 kg. By dividing the total mass to the steel volume the new density of the steel to mimic the mass of water should be 26,450 kg/m^3.

Secondly (see Figure 2) let's analyse the position of centre of mass for the empty vessel and the vessel filled with heavy water. The empty vessel has the coordinates of the mass centre (-4.65, -411.7, 13.29) mm and the filled one (-4.85, -253.26, 2.99) mm. Knowing that the distance from the saddle to vessel axis is 1870 mm then the overturn moment arm of the empty vessel is 1459 mm and of the filled vessel is 1617 mm (in the Oy direction-the vertical direction). We need to have an equal overturn moment and since the acting forces are proportional to the masses which are proportional to the densities, the final steel density is given by the equation (26,450x1617)/1459=29,314 kg/m^3. This new density for the steel is not affecting the mechanical properties of the system but is mimicking the presence of the water inside the vessel. Thus the new mechanical dynamic model is a good approximation of the real model.

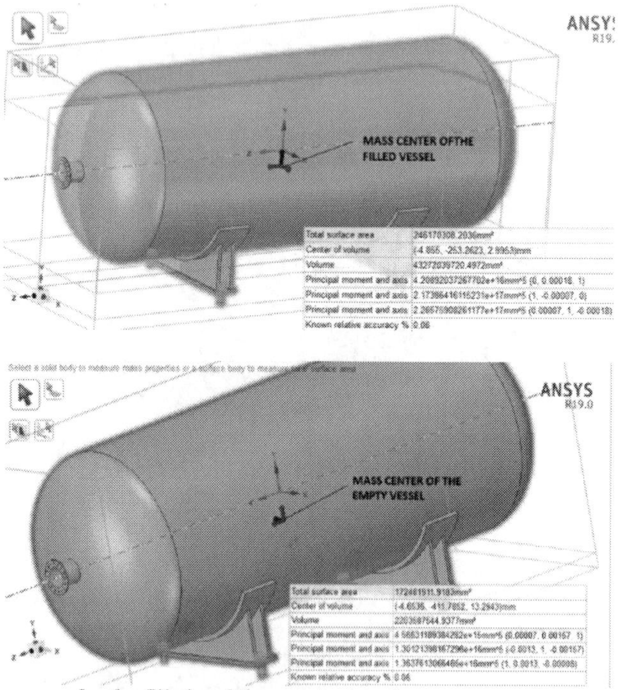

Fig.2. The centres of masses for the filled and the empty vessel

The connections are left bonded type as default and the finite elements mesh is shown in the Figure 3:

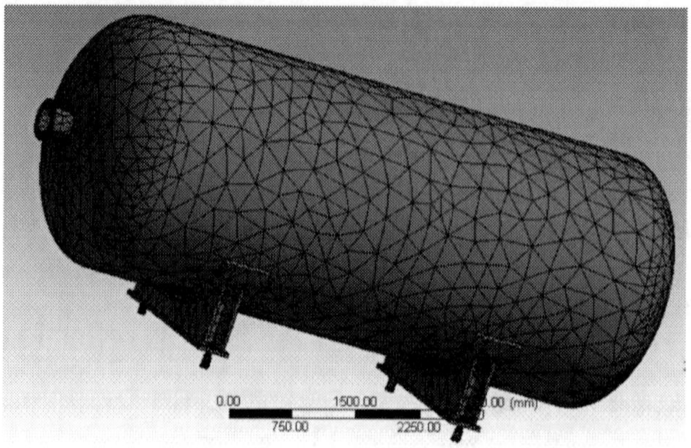

Fig.3. The finite elements mesh

There will be 20995 finite elements with 40709 nodes with an acceptable element quality.

There will be some boundary conditions to be imposed for the model.

For the static structural module the loads are as seen in the Figure 4.

The four anchors transmitting the load from the vessel to the floor will be set as fixed supports. Thru the same fixed supports the seismic excitation will be imposed from the building to the vessel.

The standard earth gravity will be imposed for the entire model and is placed in the centre of gravity of the structure. It is acting in the negative Oy direction of the model.

The contact surfaces between the saddles and the floor are modelled as compression only supports. A 0.3 MPa pressure of the fluid is imposed on the inner surfaces of the vessel. The large deflection is on and the solver to be used is the direct one.

Fig.4. The boundary conditions

Once the deformations and the stresses are calculated inside the structural module, these results are imported inside the modal module in order to calculate the natural frequencies and the mode shapes. The natural frequencies will be calculated only for 0-33 Hz interval that is meaningful in the seismic assessment for the nuclear industry. The value of 33 Hz is known as the cut-off frequency as well.

The modal shapes will be furthermore used inside the Response spectrum modules for calculating the loaded structure response to any seismic excitation coming for any of the three directions. Therefore the response spectrums are to be defined for the three directions. These response spectrums are defined for any building and any elevation from the design stage of the plant and they are well known. They are given in g's meaning multiples of the gravitational acceleration of the floor. Therefore a scale factor of 9.81 m/s^2 will be imposed once the values are tabulated.

For Ox axis of the vessel which is horizontally perpendicular on the axis the graph is looking as in Figure 5:

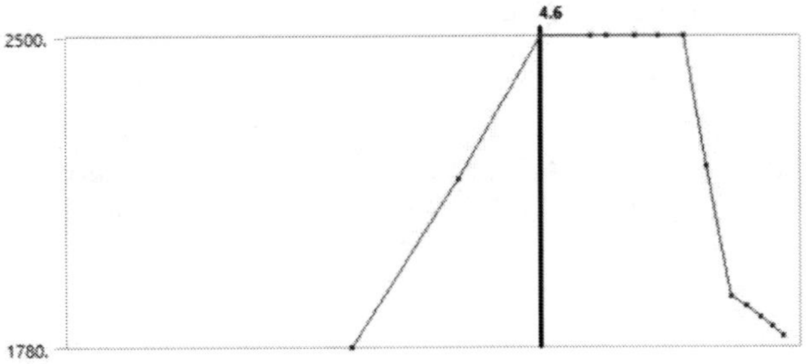

Fig.5. Response spectrum for Ox direction

For Oz axis of the vessel which is horizontally parallel to axis the graph is looking as in Figure 6:

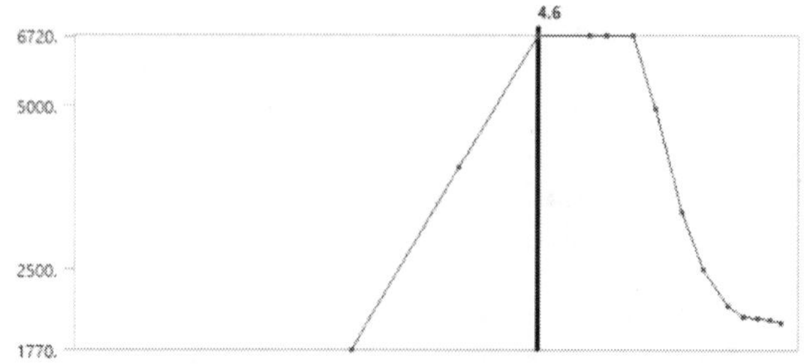

Fig.6. Response spectrum for Oz direction

For Oy axis of the vessel which is vertically perpendicular to the axis the graph is looking as in Figure 7:

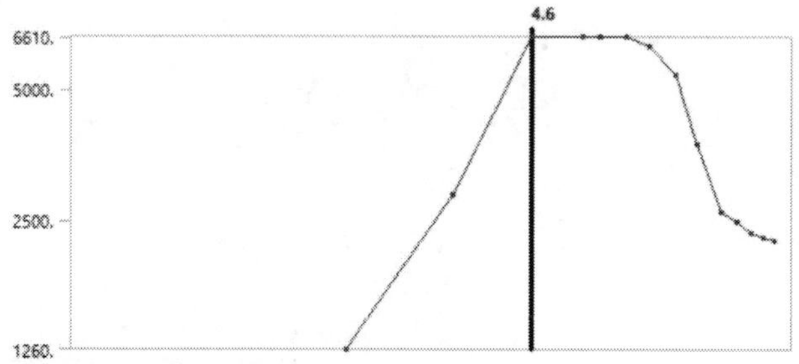

Fig.7. Response spectrum for Oy direction

Care should be taken in response spectrum definition since the crane might be placed in between elevations thus a linear interpolation of the values should be acceptable. Furthermore the Regulatory for nuclear industry in most of the countries and following the Fukushima accident requested for increasing of these values with a certain amount, usually 0.4g.

All these modules will calculate the total deformations and stresses for the structure in three different directions. In order to combine these results between themselves and with the structural results a last module will be used: the Design assessment one. This module will combine all the above calculated results in one overall result.

3 Results and discussion

3.1 The structural module results

After running the model some important results were retrieved in the structural model. The total deformations are shown in the Figure 8. Since the loads are acting in the vertical direction the middle section of the vessel will deform in this direction with an amount of 0.62 mm. This is in accordance with any simple engineering educated guess.

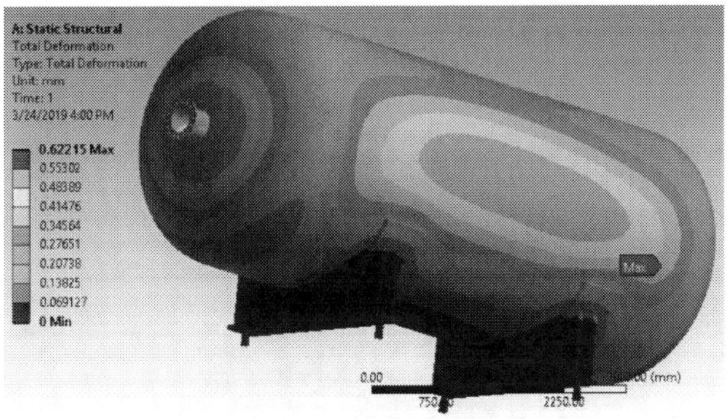

Fig.8. The total deformation calculated with the structural module

In the Figure 9 the von Mises equivalent stress is shown.

The maximum pressure of 3 bar and the weight of the structure is developing mild stresses inside the structure. The maximum stress of 44 MPa is calculated for the nozzle neck.

Fig.9. The equivalent stresses calculated with the structural module

3.2 The Modal Module

This modal module is calculating the mode shapes and natural frequencies for the loaded structure in the interval 0-33 Hz of interest. Actually four frequencies were calculated for 10.4, 23.2, 24.4 and 27.1 Hz. For the first frequency the total deformation is given in the Figure 10:

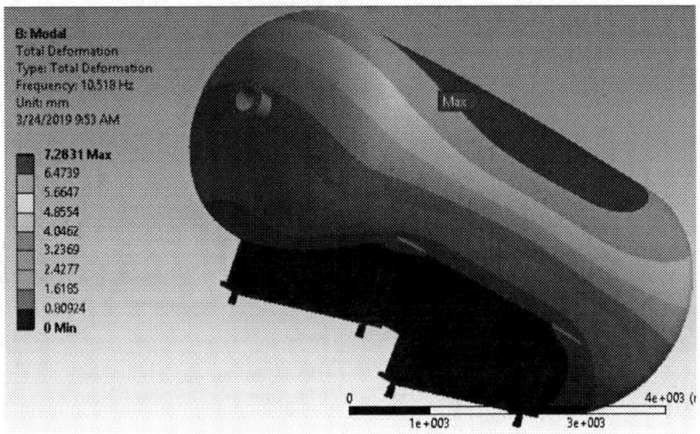

Fig.10. The total deformation calculated in modal module

The first mode shape is perpendicular to the axis in the vertical direction.

The mass participation factors for the first natural frequency of 10.4 Hz are given in Table 1:

Table 1 Participation factors

DIRECTION	PARTIC. FACTOR	RATIO EFF. MASS TO TOTAL MASS
X	5.72	0.49
Y	0.04	0.25e-4
Z	-0.03	0.17e-4
ROTX	7.92	0.16e-6
ROTY	23.8	0.15e-5
ROTZ	-4921	0.16

By analysing the participation factors one might the conclusion that the crane is very prone to rotate over Oz which is the vessel axis.

3.3 The Response Spectrum module

- **Ox direction**

In this direction the earthquake excitation will make the vessel to deform as seen in Figure 11 below.

The seismic excitation is imposing a rather big deformation of 7.5 mm. The maximum calculated equivalent stresses are big as well as in Figure 12. The maximum stress is on one of the anchors and it is 310 MPa. Even at this stage we may pull the conclusion that the seismic qualification of the vessel failed.

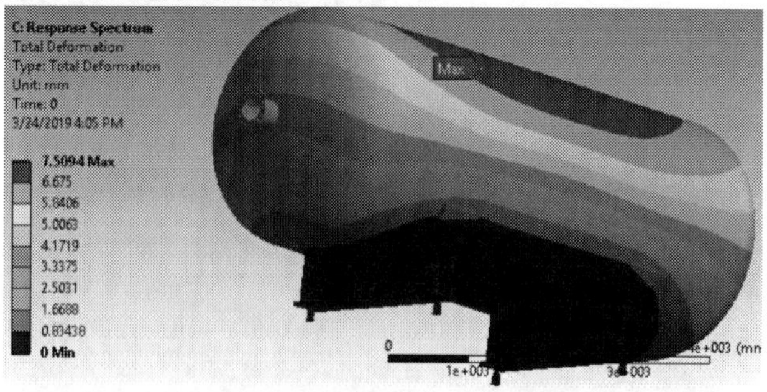

Fig.11. Total deformation for Ox excitation

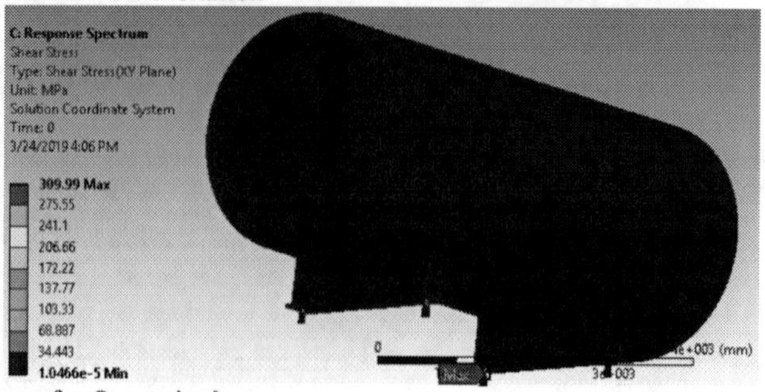

Fig.12. Equivalent stress for Ox excitation

- **Oz direction**

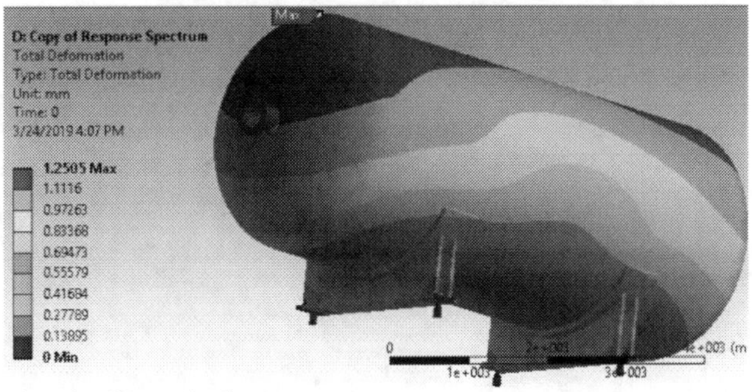

Fig.13. Total deformation for Oz excitation

The seismic excitation is imposing a relatively small deformation on the structure no more than 1.25 mm. The stresses are rather big on the anchor level, 561 MPa, as in Figure 14:

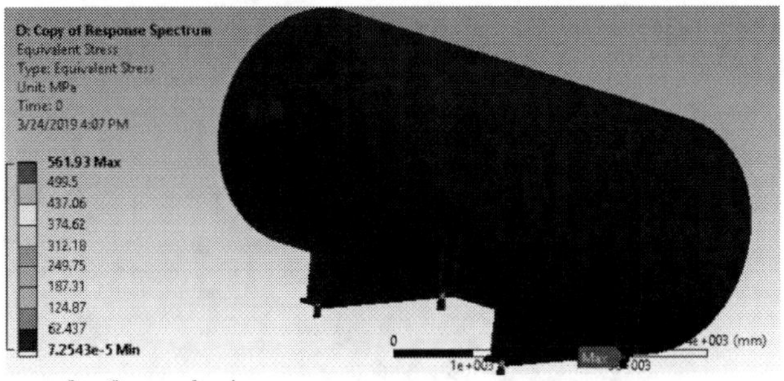

Fig.14. Equivalent stress for Oz excitation

- **Oy direction**

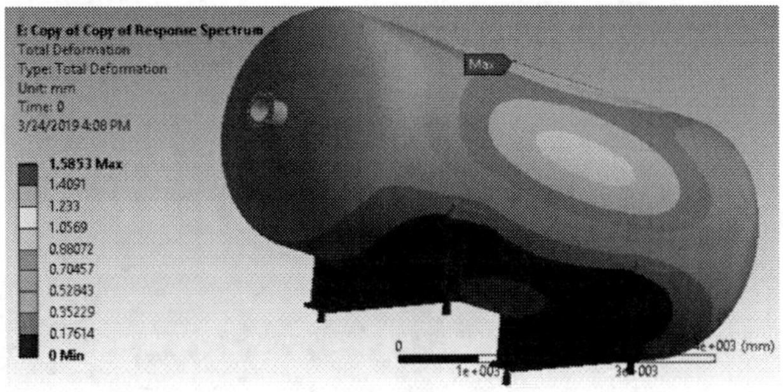

Fig.15. Total deformation for Oy excitation

The seismic excitation is imposing a small deformation on the structure no more than 1.58 mm.

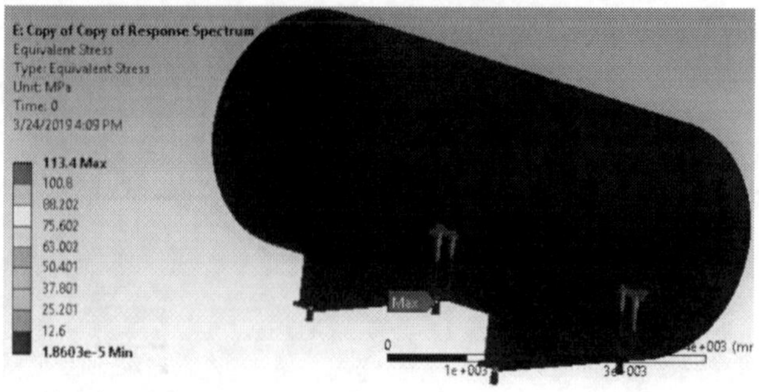

Fig.16. Equivalent stress for Oy excitation

The stresses developed in Oy direction are big with a maximum of 113 MPa somewhere on anchors (Figure 16).

3.4 The Design assessment module

This last module is taking all the results from Structural and response spectrum modules and is consolidating everything into one single structure state by combining them.

Regarding the total deformation the calculated overall result is given in Figure 17:

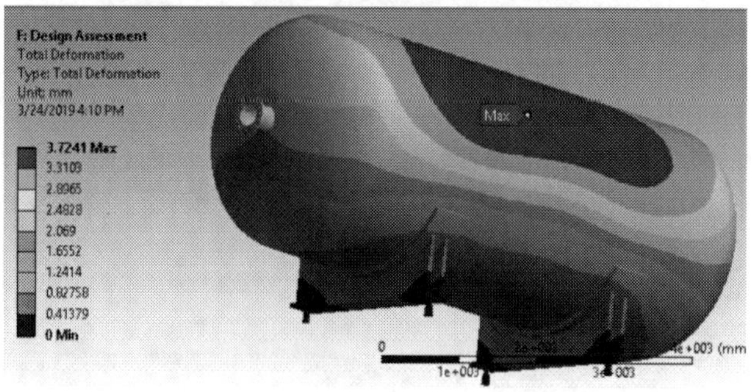

Fig.17. The overall total deformation

The overall total deformation is showing that the vessel structure suffers a bending and rotation over Oz axis yet the value rather big, 3.6 mm deformation is recorded for the upper side vessel.

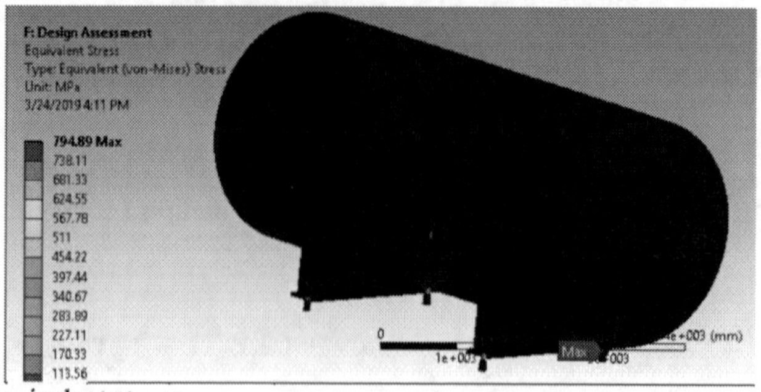

Fig.18. The overall equivalent stress

The overall stress in Figure 18 shows that the peak stress of 794 MPa calculated is bigger than the ultimate strength of the stainless steel material of 586 MPa. This means the anchors are breaking following a seismic event.

The procedure above is adding algebraically with the rule of 100-40-40 the maximum responses and is allowed by US NRC Regulatory Guide 1.92. It is an alternative for SRSS method of combining the results. That is the bigger result in the Oz direction is taken 100% and the other 2 in Oy and Ox direction are taken in the algebraic addition as 40%. As per US NRC Regulatory Guide 1.92 [3]:

"The 100-40-40 percent rule was originally proposed as a simple way to estimate the maximum expected response of a structure subject to three-directional seismic loading for response spectrum analysis, and is the only alternative method for spatial combination which may be used. The results of the 100-40-40 spatial combination have been compared with the SRSS spatial combination. Generally, they indicate that the 100-40-40 combination method produces higher estimates of maximum response than the SRSS combination method by as much as 16 percent, while the maximum under-prediction is 1 percent."

3.5 Anchoring system redesigned

Since the performance of the anchoring system of the vessel seems to render unsatisfactory results, the system was redesigned by adding other 8 anchors as in the Figure 19:

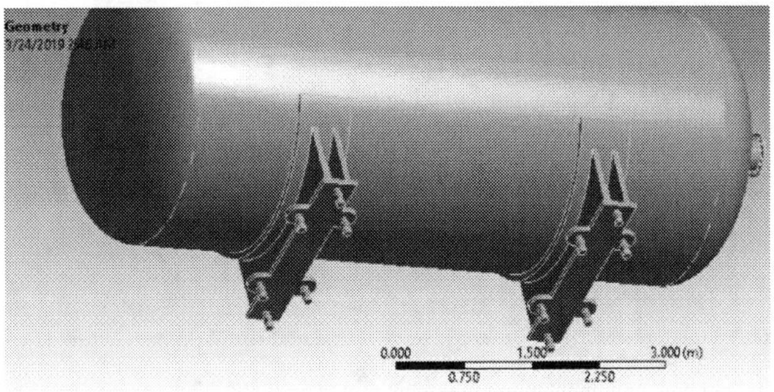

Fig.19. The anchoring system redesigned

The new model was solved by going through all the stages described above and the overall equivalent stress is shown in the Figure 20.

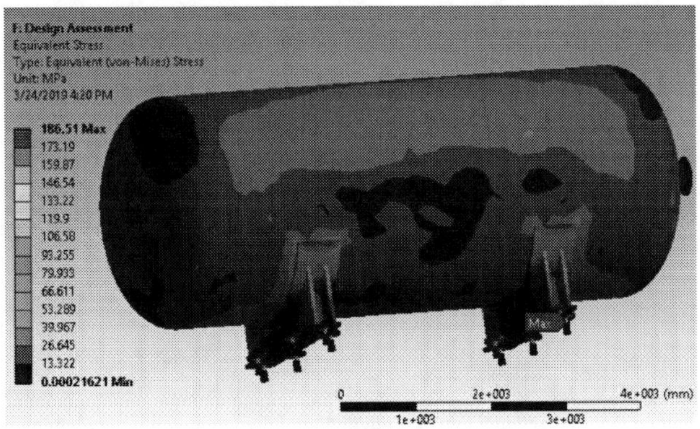

Fig.20. The overall equivalent stress for the anchoring system redesigned

The maximum stress is now recorded for a single stud and has the value of 186 MPa which is lower than the yield strength of 207 MPA. The safety factor is 1.11.

4 Conclusions

This study is meant to show how a complex structure as a pressure vessel might be treated in order to have meaningful results for a decision whether or not such a structure is able to withstand the loads generated by a seismic event.

The seismic event is overloading the anchoring system beyond the ultimate strength of the material. The zone where these big stresses fields are calculated is at the anchors level. The vessel as designed failed to be qualified for the seism excitations given above. In order to have the vessel qualified the anchoring system should be redesigned.

By comparing the overall calculated equivalent stresses with the tensile yield strength, we may pull the conclusion that the original structure design will fail. Instead, the new redesigned anchoring system with 8 extra-anchors will stand.

References

1. *ANSYS* **19** documentation
2. Seismic Design Classification, Regulatory Guide 1.29, U.S. Nuclear Regulatory Commission, (2007).
3. US NRC Regulatory Guide 1.92, Combining Modal Responses and Spatial Components in Seismic Response Analysis, Rev.2, (2006).
4. Safety Standards Series No.NS-G-1.6, Seismic Design and Qualification for Nuclear Power Plants Safety Guide
5. EPRI, Recommended Piping Seismic Adequacy Criteria Based on Performance During and after Earthquakes, Report NP-5617, January 1988
6. US NUCLEAR REGULATORY COMMISSION, Seismic fragility test of a 6-inch diameter pipe system, NUREG/CR-4859 (1988)
7. SEISMIC QUALIFICATION UTILITY GROUP (SQUG), Generic Implementation Procedure (GIP) for Seismic Verification of Nuclear Plant Equipment, February 1992
8. EPRI, NP-7500-SL, The October 17, 1989 Loma Prieta Earthquake: Effects on selected power and industrial facilities, September 1991

Researches on the delimitation between the water aeration and oxygenation processes

Rasha Cusma[1], *Nicolae* Băran[1], *Nicolae Vlad* Sima[1], *Mihaela* Petroșel[1], *Remus* Voicu[1], *Mariana Mirela* Stoican (Prisecaru)[1] and *Mihaela* Constantin[1,*]

[1] University Politehnica of Bucharest, Faculty of Mechanical Engineering and Mechatronics, Splaiul Independenței, no. 313, Bucharest, Romania

Abstract. Considering that in certain papers the aeration processes are confused with the oxygenation processes, a distinction between them must be made. The paper presents modern water aeration and oxygenation installations. The experimental results obtained for the two categories of installations are highlighted. Theoretical and experimental researches aim to increase the concentration of dissolved oxygen in water; this can be done in two ways: I - by introducing atmospheric air into the water; II - by introducing a gaseous mixture consisting of atmospheric air and oxygen. Version II is better, the results being shown in Figure 11.

1 Introduction

In order to increase the dissolved oxygen content in water, there are several methods and procedures that do not differentiate between the notion of "water aeration" and " water oxygenation"; there are papers in which the two expressions are confused [1].

Water aeration is achieved by introducing atmospheric air into the water as follows [2] [3]:

1. By mechanical aeration;
2. By pneumatic aeration.

- In the case of mechanical aeration, the air is introduced into the water under the action of rotors, blades, etc. performing surface aeration.

- In the case of pneumatic aeration, the air is sent to diffusers made of plastics, ceramics, etc.; these diffusers are placed below the water level, thus aerating the water with air bubbles.

The aeration equipment's are classified according to the following criteria [4] [5]:

❖ By the way of obtaining the interfacial contact surface:

• equipment's that sprays water in the air and equipment in cascade;

• equipment's that disperses the gas in the water (deep mechanical aerators, etc.);

• mixed equipment's - spray the water in the form of drops and entrain the air through the jet effect upon mass re-entry of the water from the basin (surface mechanical aerators).

❖ By the movement of the active body of the aeration equipment:

---

* Corresponding author: i.mihaelaconstantin@gmail.com

• static equipment's (static aerators, ejectors, etc.);
• dynamic equipment's (surface or depth mechanical aerators).
❖ Depending on the type of gas used for aeration:
• equipment's that disperses air into water (deep mechanical aerators, pneumatic aerators, ejectors, etc.);
• equipment's that disperses pure oxygen in water (pneumatic type);
• equipment's with ozone or ozone enriched air in water (such as fluid jet pumps).
❖ By the constructive solution:
• pneumatic equipment's with porous diffusers, static aerators, etc.;
• surface mechanical equipment's, medium or high depth with rotor, brush, etc. ;
• mixed equipment's.
❖ By the immersion mode of the dispersion device:
• surface equipment's (mechanical surface aerators with rotor or brush);
• medium depth equipment's - the dispersion device is located at a depth of 1 ÷ 2 m (INKA type pneumatic devices, medium depth mechanical aerators, etc.);
• deep water equipment's - the dispersion device is located at about 3 ÷ 120 m (pneumatic, injector, mechanical deep water equipment's, etc.).

2 Boundary between water aeration and water oxygenation installations

This delimitation refers to the nature and composition of the gas introduced into the water.

In table 1 a clear distinction is made between the two processes (two installations), namely [1], [2], [3]:

I - Aeration installations;

II - Oxygenation installations.

Table 1. Classification of pneumatic aeration installations.

Installation type	Nature of the gas	Gas source	Volumetric gas composition	Advantages	Disadvantages
I. Installation for water aeration	Air	Air compressor	21% O_2 + 79% N_2	Source available for free	Electricity consumption
II. Installations for water oxygenation	Air + oxygen	Air compressor + O_2 from a cylinder	Air + (20% ÷ 80%) O_2	Reduction of water oxygenation time	Cost O_2 from a cylinder
		Oxygen concentrators	95% O_2 + 5% N_2	Reduction of water oxygenation time	Electricity consumption
		Ozone from O_3 generators	Air + O_3	Reduction of water oxygenation time	Electricity consumption

It is useful to sort the expressions, as follows:

I. Water aeration refers only to the introduction of atmospheric air into water (21% O_2 + 79% N_2);

II. Water oxygenation refers to the introduction of gaseous mixtures in which the oxygen exceeds 21%.

Obviously I + II have the same purpose, namely to increase the dissolved oxygen concentration in water. When increasing the percentage of O_2 introduced into water by more than 21%, the time in which water oxygenation takes place ($C_0 \rightarrow C_s$) is reduced.

https://doi.org/10.1051/e3sconf/202018001002

3 The installation of water aeration

Fine air bubbles generators (FBG) will be used to introduce air into water. From the literature it is known that the rate of oxygen transfer to water increases with decreasing air bubble diameter; the air bubble diameter of is a function of the orifice diameter in the perforated plate of the FBG.

The gas bubbles (atmospheric air or a mixture of gases) can be classified according to the data in figure 1.

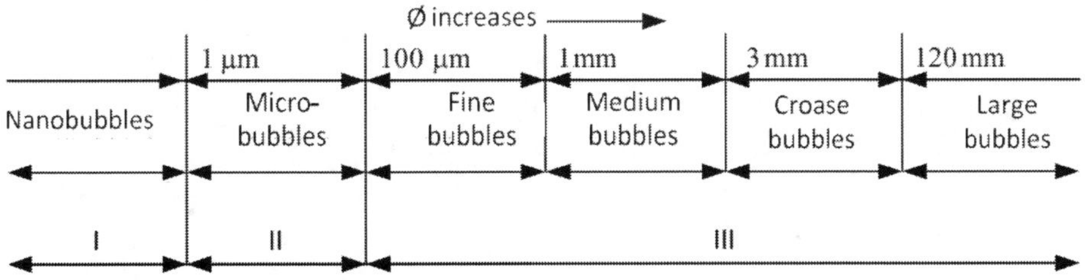

Fig. 1. Classification of gas bubbles according to their diameter
I - the area where the gas bubbles can be observed under the microscope;
II - the area where gas bubbles can be observed with difficulty;
III - the area where gas bubbles can be observed with the naked eye.

Figure 2 shows a FBG that has a rectangular shape orifices plate. Taking into account the dimensions of the water tank and the height of the water layer, a section of air outlet in water of $1.2 \cdot 10^{-6}$ m^2 [6] [7] has been chosen, resulting in a number of orifices of Ø 0.1 mm equal to:

$$n = \frac{A}{\frac{\pi \cdot d_0^2}{4}} = \frac{1.2 \cdot 10^{-6}}{\frac{\pi \cdot \left(0.1 \cdot 10^{-3}\right)^2}{4}} = 152 \ \text{orifices} \quad (1)$$

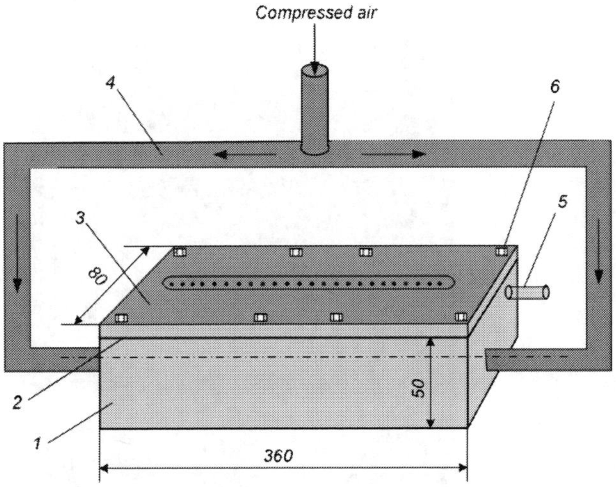

Fig. 2. FBG with rectangular orifices plate [4]
1- compressed air tank; 2 - sealing system; 3 - orifices plate; 4 - compressed air line; 5 - connection to the manometer; 6 - elements for fastening the orifices plate

The plate (3) contains 152 orifices (Ø 0.1 mm) made using a KERN Micro machine; the constructive solution of this FBG is an original and very efficient constructive solution.

The framing of this FBG in the experimental installation is shown in figure 3.

During the measurements, the following data were kept constant:

- Air flow rate: $\dot{V}$ = 0.6 m^3 / h;
- Air pressure at the entrance of the FBG: p = 573 mmH_2O;
- Hydrostatic load: H = 500 mmH_2O;
- Water temperature 24 °C for which from [3]: C_s = 8.4 mg /dm^3, C_o =5.84 mg /dm^3.

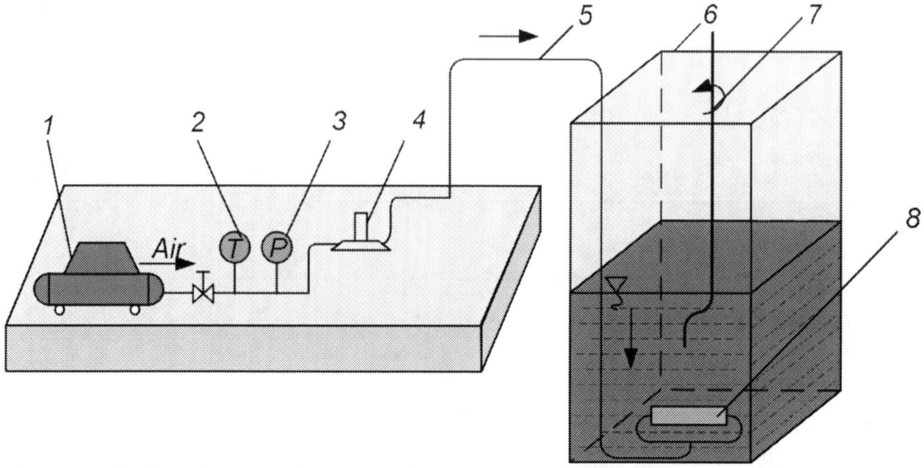

Fig. 3. Installation scheme for the introduction of atmospheric air into water [4]: 1-electrocompressor; 2– digital thermometer; 3- digital manometer; 4– rotameter; 5- air transport pipeline; 6– oxygen sensor; 7– water tank; 8– fine bubble generator.

Figure 4 shows the functioning of the studied FBG.

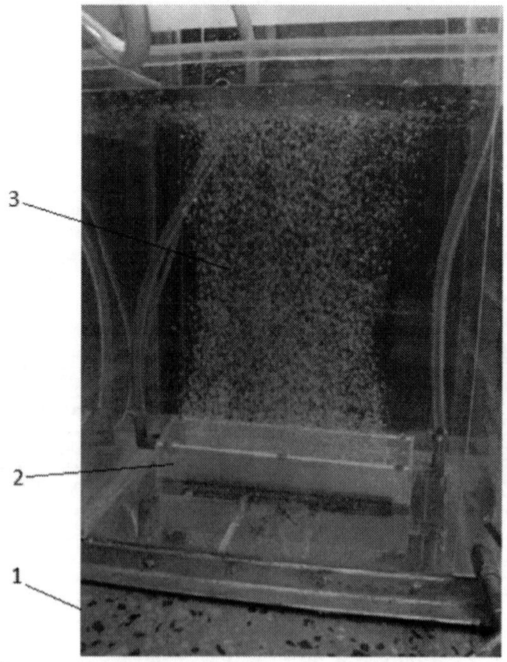

Fig. 4. FBG in operation.
1 - water tank; 2 - FBG; 3 - plane jet with air bubbles.

Figure 4 show that the air coming out through the orifices in the straight line creates a bubble curtain, similar to a plane jet of air. The increase of the dissolved oxygen concentration in the water depends on the rate of its increase, given by the relation [8] [9] [10]:

$$\frac{\mathrm{d}C}{\mathrm{d}\tau} = ak_L\left(C_s - C\right) \qquad (2)$$

Where:
ak_L - the volumetric mass transfer coefficient [s^{-1}];
C_s - the mass concentration of oxygen at saturation in liquid phase [kg / m^3];
C - the current mass concentration of oxygen in the liquid phase [kg / m^3].
The relation (2) was numerically integrated and a calculation program was performed, the results being presented in figure 5, curve 1.

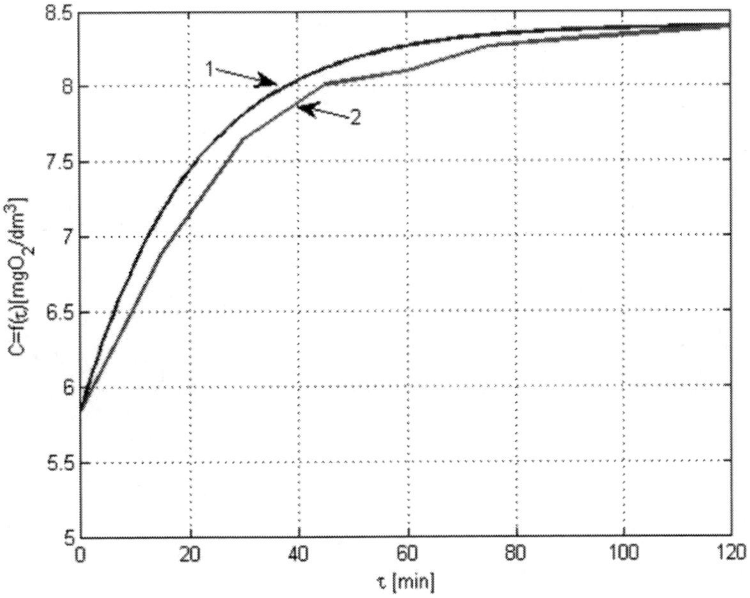

Fig. 5. $C_{O2} = f(\tau)$ for atmospheric air
1 - curve drawn on the basis of theoretical data;
2 - curve drawn on the basis of experimental data.

In the experimental researches, the dissolved oxygen concentration in the water was measured from 15 in 15 minutes resulting curve 2 (figure 5). Between the two curves (1) and (2) in figure 5 a good coincidence is observed.

4 Water oxygenation installations

The operating scheme and the experimental results for three water oxygenation installations are presented.

4.1. Installation that uses a mixture of atmospheric air and oxygen from a cylinder.
4.2. Installation that uses low nitrogen air.
4.3. Installation that uses a mixture of air and ozone.

All the three installations 4.1, 4.2, 4.3 were built in the laboratory of the Department of Thermotechnics, Engines, Thermal and Refrigeration Equipment's.

4.1 Installation that uses a mixture of atmospheric air and oxygen from a cylinder

The working parameters of the installation are:

- air flow: $\dot{V} = 600$ dm^3 / h;
- air pressure: p =573 mmH_2O;
- hydrostatic load: H = 500 mmH_2O; the hydrostatic load is: p =ρ g H [N/m^2], where H comprises the height at which the water is to be pumped and the pressure losses on the hydraulic circuit.
- working time: two hours.

For atmospheric air there is rotameter (4) in figure 6, and rotameter (4 ') for the oxygen from the cylinder.

Operating variants:

$\dot{V}_{O_2} = 150$ dm^3 / h, i.e. 25% of $\dot{V}$;

$\dot{V}_{O_2} = 300$ dm^3 / h, i.e. 50% of $\dot{V}$;

$\dot{V}_{O_2} = 450$ dm^3 / h, i.e. 75% of $\dot{V}$;

$\dot{V}_{O_2} = 600$ dm^3 / h, i.e. 100% of $\dot{V}$.

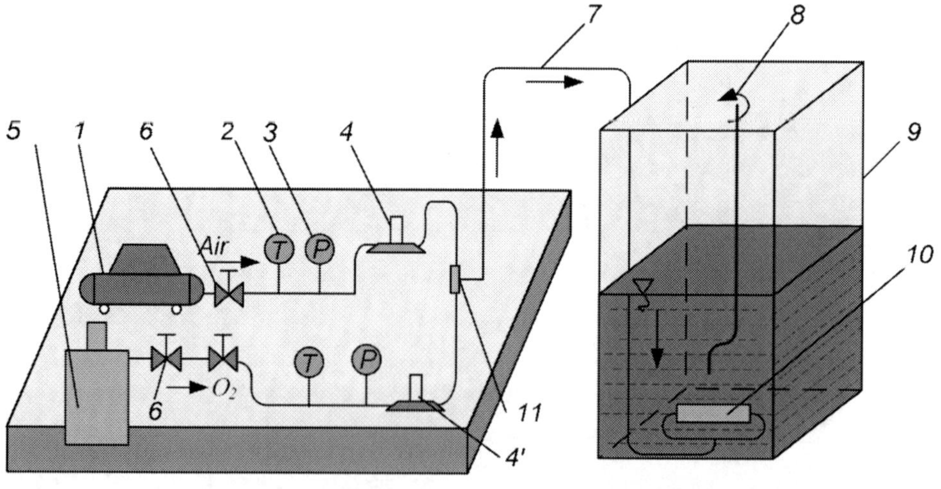

Fig. 6. Scheme of the installation for introducing a gas mixture: air + oxygen from the cylinder.
1- electrocompressor; 2- thermometer; 3- manometer; 4,4'- rotameter;
5- oxygen cylinder; 6- pressure reducer; 7- air duct + oxygen;
8- probe for measuring the concentration of oxygen, dissolved in water; 9- water tank;
10- fine bubble generator (FBG); 11- T-joint.

Figure 7 shows an overview of the experimental installation.

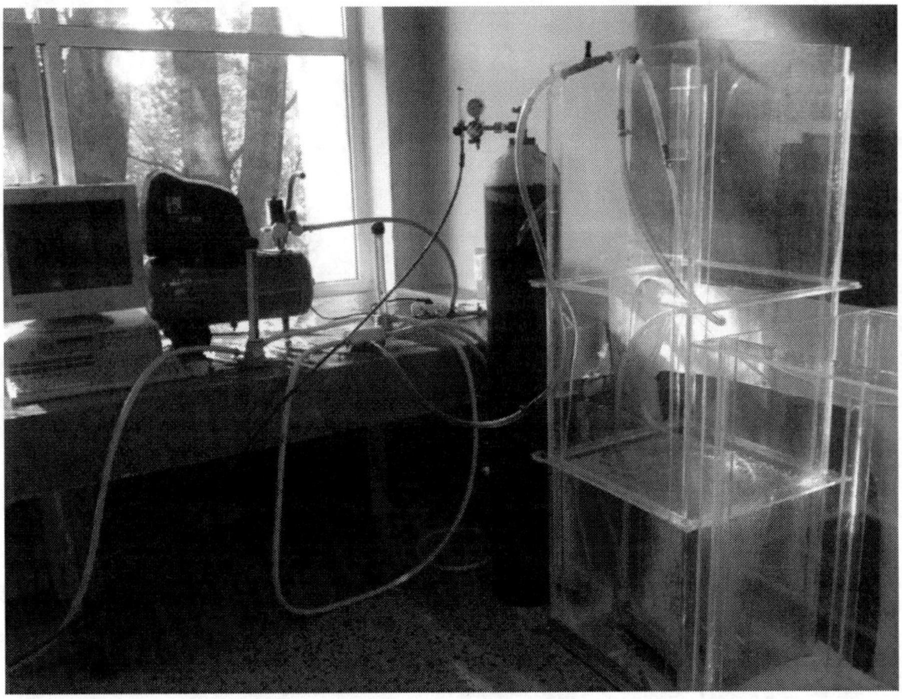

Fig. 7. Overview of the experimental installation for the introduction of a mixture (air + oxygen).

In the centre of the photo one can see the oxygen cylinder, and on the right side there is the transparent plexiglass water tank.

4.2 Installation that uses low nitrogen air

Figure 8 shows two oxygen concentrators each delivering 300 dm^3/h of gas with the volumetric composition $O_2 = 95\%$, $N_2 = 5\%$.

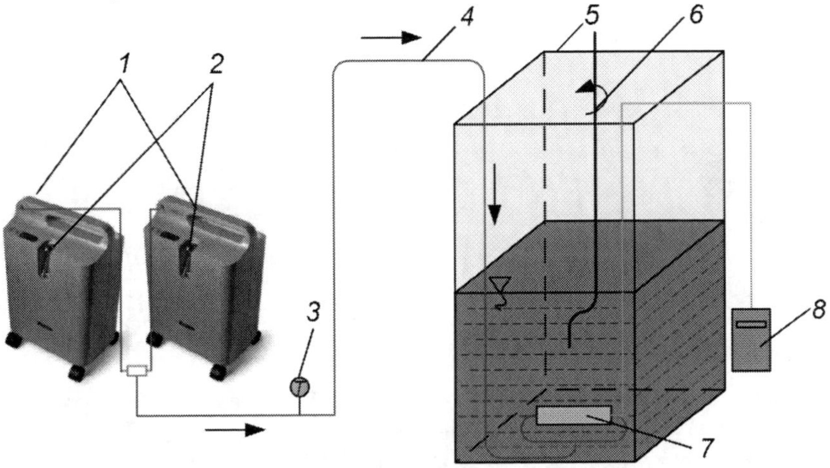

Fig. 8. FBG supplied with air with low nitrogen content
1- oxygen concentrators; 2- rotameter; 3- thermometer; 4- gas pipe (95% O_2 + 5% N_2); 5- water tank; 6- oxygen sensor; 7- FBG; 8- manometer.

In figure 9, on the left side, two oxygen concentrators (1) are observed, each device being equipped with a rotameter.

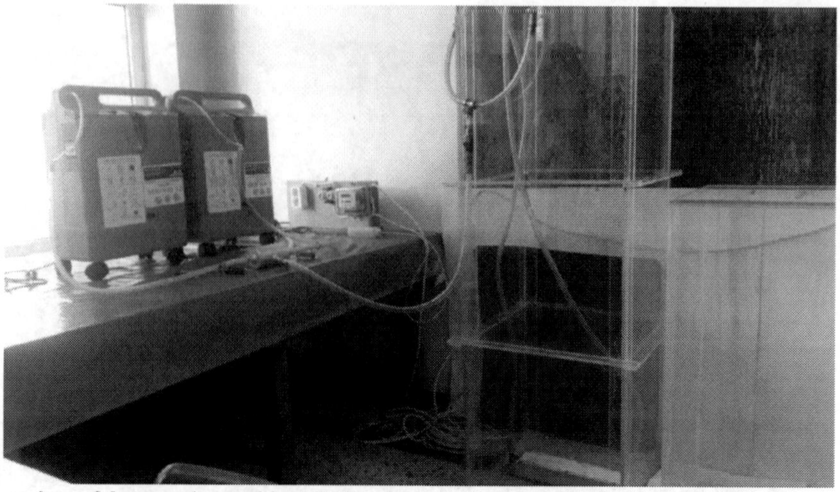

Fig. 9. Overview of the experimental installation for low nitrogen content air supply [5].

The atmospheric air extracted from the atmosphere is compressed and sent to the zeolite filters where nitrogen is retained; upon exiting the unit, a gas containing 95% oxygen is obtained.

4.3 Installation that uses a mixture of air and ozone

Figure 10 shows the ozone generator (4); at the exit from it the produced ozone is mixed with air and enters the rotameter (7) and then into the FBG (11) [11] [12].

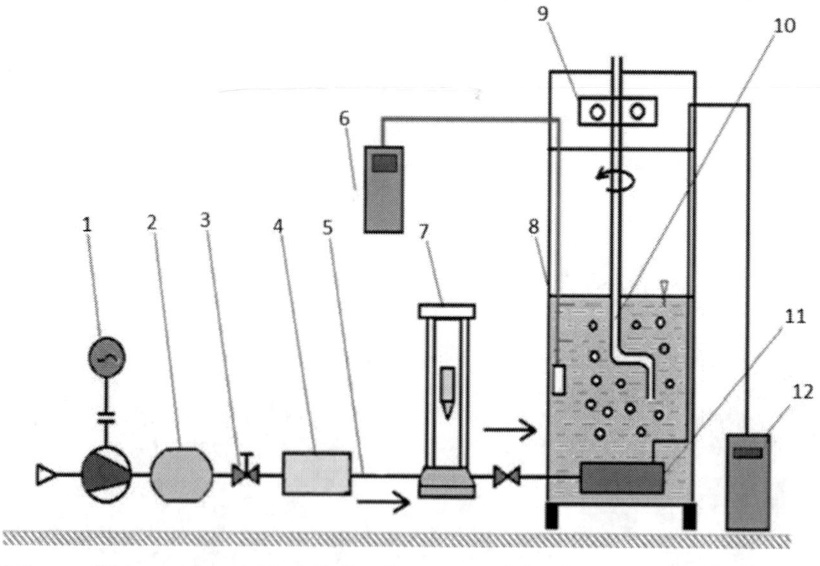

Fig. 10. Scheme of the experimental installation for water oxygenation researches, by introducing into the water tank a gas mixture containing atmospheric air and ozone
1- electrocompressor; 2- compressed air tank; 3- pressure reducer; 4- ozone generator; 5- pipe; 6- thermometer with digital indication; 7- rotameter; 8- water tank; 9- mechanism for the rotation of the oxygen probe in water; 10- oxygen probe; 11- fine bubble generator; 12- manometer with digital indication.

The compressed air supplied by the compressor (1) passes through the ozone generator (4), through the rotameter (7) and enters the FBG (11).

Depending on the water temperature (t = 23.7 °C) in [3] the saturation concentration values are found for:

- Atmospheric air: C_s = 8.4 mg / dm^3;
- Air + 25% O_2: C_s = 10.73 mg / dm^3;
- Air + 50% O_2: C_s = 21.46 mg / dm^3;
- Air + 75% O_2: C_s = 32.21 mg / dm^3;
- Air + 100% O_2: C_s = 43.00 mg / dm^3.

The value of the dissolved oxygen concentration in water was initially 5.84 mg / dm^3.

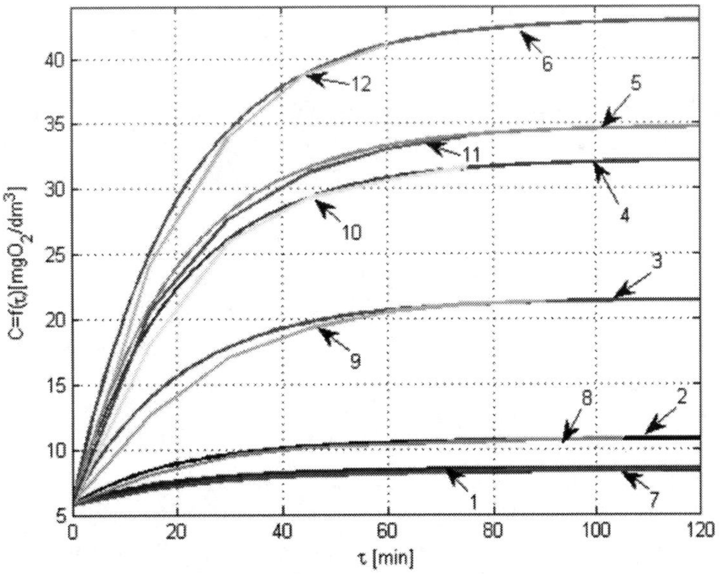

Fig. 11. Increase of dissolved oxygen concentration in water for:
1/7 - atmospheric air; 2/8 - air + oxygen (150 dm^3 / h); 3/9 - air + oxygen (300 dm^3 / h); 4/10 - air + oxygen (450 dm^3 / h); 5/11 - low nitrogen air; 6/12 - air + oxygen (600 dm^3 / h).
1, 2, 3, 4, 5, 6 - curve drawn on the basis of theoretical data;
7, 8, 9, 10, 11, 12 - curve drawn on the basis of experimental data

Figure 10 shows the change of dissolved oxygen concentration in water as a function of time.

The curves 1, 2, 3, 4, 5 and 6 are plotted on the basis of theoretical data, and the curves 1 ', 2', 3 ', 4', 5 'and 6' are plotted on the basis of experimental data.

5 Conclusions

a) The fine bubbles generator with 0.1 mm orifices is original construction and very efficient solution in the process of water aeration and oxygenation.

b) The results obtained in a theoretical way (figure 5 and figure 11) coincide well with the experimental data.

c) The value of dissolved oxygen concentration in water, at saturation, is reached most rapidly when pure oxygen is introduced into water (curve 6).

d) Future researches will follow the use of microtechnologies and nanotechnologies in water aeration processes.

References

1. S. Stoianovici, D. Robescu, D. Stamatoiu, Calculul şi construcţia echipamentelor de oxigenare a apelor, Editura CERES, Bucureşti, 1985.
2. D. L. Robescu, S. Lanyi, A. Verestoy, D .Robescu, Modelarea şi simularea proceselor de epurare, Ed. Tehnică Bucureşti, 2004.
3. G. Oprina, I. Pincovschi, Ghe. Băran, Hidro-Gazo-Dinamica Sistemelor de aerare echipate cu generatoare de bule, Editura POLITEHNICA PRESS, Bucureşti, 2009.
4. Elena Beatrice Tănase, Influenţa compoziţiei gazului insuflat în apă asupra conţinutului de oxIgen dizolvat, Teză de doctorat, Universitatea POLITEHNICA din Bucureşti, 2017.
5. Al. Pătulea, Influenţa parametrilor funcţionali şi a arhitecturii generatoarelor de bule fine asupra eficienţei instalaţiilor de aerare, Teză de doctorat, Universitatea POLITEHNICA din Bucureşti, 2012.
6. Ionela Mihaela Căluşaru, N. Băran, Al. Pătulea, The influence of the constructive solution of fine bubble generators on the concentration of oxygen dissolved in water, Advanced Materials Research, Trans Tech Publications, Switzerland, vols. 538-541, 2012, pp .2304-2310.
7. Ionela Mihaela Constantin, N. Băran, Elena Beatrice Tănase, A New Solution for Water Oxygenation, International Journal of Innovative Research in Advanced Engineering (IJIRAE), Vol. 2, Issue 7, 2015, pp. 49-52.
8. Ionela Mihaela Căluşaru-Constantin, Elena Beatrice Tănase, N. Băran an, Rasha Mlisan-Cusma, Researches Regarding the Modification of Dissolved Oxygen Concentration in Water, IJISET - International Journal of Innovative Science, Engineering & Technology, Vol. 1 Issue 6, 2014, pp. 228-231.
9. Elena Beatrice Tănase, N. Băran, Rasha Mlisan, An Efficient Solution for Water Oxygenation, Asian Engineering Review Vol. 1, No. 3, 36-40, 2014.
10. Droste Ronald L. Theory and Practice of Water and Wastewater Treatment, 1996. ISBN 978-0-471-12444-3 - John Wiley & Sons Inc.
11. A. G. Roza, Cercetări privind utilizarea ozonului la oxigenarea apelor, Teză de doctorat, Universitatea POLITEHNICA din Bucureşti, 2019.
12. Roza Giovani Albertino, Albu Nicoleta Dorina, Donţu Octavian, Moga Corina, Băran Nicolae, Constantin Mihaela, „Researches on the use of ozone generators for wastewater treatment", Proceedings of the International Conference of Mechatronics and Cyber-Mix Mechatronics – 2019, pp 11-21.

The efficiency analysis of the SH-25A ball drum mill when grinding industrial products of fossil fuels

Saule Kamarova[1] , *Saule* Abildinova[1] , *Angel* Terziev[2,*], and *Aliya* Elemanova[1]

[1]Department of Industrial Heat Technology, Almaty University of Power Engineering and Telecommunications, Almaty, Kazakhstan
[2]Department of Power Engineering and Power Machines, Technical University of Sofia, 8 Kliment Ohridski blvd., 1000 Sofia, Bulgaria

Abstract. The working characteristics of a laboratory ball drum mill during grinding of Karaganda coal products were studied in order to apply the results obtained to industrial installations. The performed analysis shows that the most complicated and energy-intensive stage of preparation of coal dust is the grinding of fuel. The operational characteristics of the mill productivity, consumed electric power and specific energy consumption in terms of the relative rotational speed of the drum as well as the optimal values of the relative velocity determining the qualitative grinding of the fuel have been determined. The performance of the processed drum mill reaches its maximum at a relative speed of rotation of the drum 0.71 ensuring a waterfall mode of grinding of fuel related to the high-speed mode of a coal-grinding mill. The estimated specific energy consumption for fuel grinding varies in the range from 0.11 kWh/kg up to 0.23 kWh/kg at fuel milling coefficient of Kl = 1.2.

1. Introduction

The most complicated and energy-intensive stage of preparation of coal dust is the grinding of fuel. For reliable, economical and safe operation of the dust preparation equipment, it is necessary to focus to the entire system for preparation and grinding of fuel at thermal power plants. The process of fuel grinding must be implemented in such way to ensure: uniform fuel supply to all boiler burners; the required amount of coal fuel, grounded to a required fineness and dried to the required humidity [1].

The dust preparation system equipped with ball drum mills (BDM) is widely used in many thermal power plants (TPP) for grinding different types of solid fuel. The main disadvantage of the BDM is the significant electricity consumption, which at the TPP level is about a quarter of the total electricity consumption for own needs. Hence the total electricity production decreases. The stated above shows that the solid fuel requires the most complicated and expensive preparation before combustion. Therefore, methods for

* Corresponding author: aterziev@tu-sofia.bg

improving the technology for coal preparation and further combustion are currently very relevant [2].

The purpose of the study is to improve the efficiency of the BDM because of their widespread use in many industries.

2. General information

The focus of the study is the TPP-2 station which is the structural subdivision of OJSC ArcelorMittalTemirtau (AMT) of the city of Temirtau, Karaganda region. An individual closed pulverization system with an intermediate bunker equipped with a ball drum mill currently operates at TPP-2. The installation includes two ball drum mills SH-25A with a total capacity of 34.3 t/h when grinding industrial Karaganda coals [3].

At present, the efficiency of the installed in TPP-2 of OJSC AMT equipment is below the average level for such type of systems. The degree of deterioration of the main equipment is more than 60% while about 50% of electrical grid failures and emergencies often occur during operation. Analysis of the dust preparation system for TPP-2 shows: increased specific energy consumption for dust preparation from 25 kWh up to 35 kWh per 1 ton of dust produced; significant wear of the metal (ball units) during operation (about 400 g of metal is spent when preparing 1 ton of dust); degraded grinding properties and insufficient drying capacity of the mill; technological processes of grinding are poorly automated [4].

Based on the above, improvements of the pulverizing system of BDM at TPP-2 must be searched in the optimization of the operating modes of the mills and the automation of the grinding process. Also determining the optimum speed of the mill during grinding is crucial for increasing the mill efficiency [5].

3. Experimental study

For the purpose of the study an experimental equipment of BDM grinding industrial products was built. The results obtained should be successfully used to improve the efficiency of the industrial facilities. The test bench consists of a laboratory non-ventilated BDM equipped with a frequency converter and energy meter. The block-scheme of the laboratory equipment is presented in Fig.1.

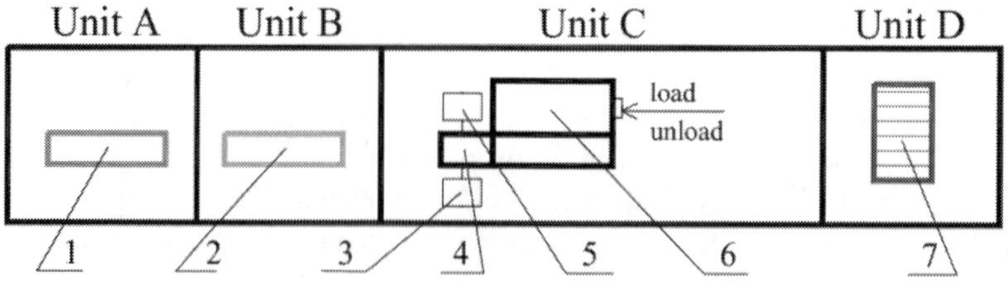

1-drying chamber, 2-scale CAS MWP-600 type, 3-electricity meter (multimeter) IMS F-1 type, 4-electric drive, 5-pulse-frequency control point adjustment unit (frequency transformer vector FTV1-HH), 6 - porcelain ball drum grinder (active zone: internal diameter - 210 mm, depth - 190 mm), 7-shaker screen AS-200 type.
Unit A - place for measuring humidity and particle size distribution at the inlet; Unit B - place for measuring of weight coal characteristics at the inlet and outlet; Unit C - input coal grinding unit; Unit D is the Screen analysis of the processed dust.

Fig. 1. Block-scheme of the process of grinding industrial products

The experimental study shows that the number of revolutions of the mill is determined by the diameter of the drum. At slower speed of the mill, the angle of lifting of the balls is low hence the performance is low. At high speeds of rotation of the mill, centrifugal forces exceed the gravitational forces of the balls. At a critical speed, the centrifugal force acting on the balls will catch up with their weight, and the balls, pressed by the centrifugal force to the wall of the drum will rotate with the drum. The minimum peripheral speed of the inner surface, at which the critical number of revolutions of the BDM occurs, is determined by the Eq. 1 [6]:

$$n_{cr} = \frac{42.4}{D^{1/2}}, \text{rpm} \tag{1}$$

where n_{cr} - range of speeds, min^{-1}; D - diameter of a BDM, m

The working number of revolutions of BDM is about 75% of the critical number of revolutions:

$$n = \frac{32}{D^{1/2}}, \text{rpm} \tag{2}$$

Estimated working speed of the mill:

- The critical and working number of revolutions of a laboratory mill are: 92 rpm; 71rpm.
- The degree of filling the drum with grinding balls also affects the performance of the mill and the efficiency of grinding. The degree of filling of the drum with balls is according the methodology [7]:

$$\phi_b = \frac{G_b}{\gamma_{bd} \cdot V_i}, \tag{3}$$

where G_b is the mass of balls loading the mill, which is function of filling degree of mill with balls and their diameter, kg; γ_{bd} - bulk density of balls, kg/m^3; V_i - internal volume of the drum, m^3.

The degree of filling of the drum with balls is ϕ_b =0.2.

Study of the efficiency of the ball drum mills SH-25A was performed with sample of Karaganda industrial coals. The Karaganda coals was collected from the boiler room of TPP-2 of AMT OJSC. The composition of the fuel is as follow: S^p=0.8%; C^p =53.7%; H^p =3.3%; N^p=0.8%; O^p=4.8. Wp=9.0%; A^p=27.6%. summarized information for the thermal characteristics of the Karaganda colas are presented in Table 1.

Table 1. Thermal properties of the proposed fuel.

Coal brand	Heat of combustion Q_c^H, MJ / kg	Content of volatiles V^v,%	Grinding coefficient K_1,%	Ash melting temperature		
				Initiating deformation	Softening	Liquid-fusible state
Karaganda coal	20.222	28	1.2	1 400	1 450	1 470

The study of the grinding process includes the following steps: drying the coal in a convective dryer; weighing the grinding amount of coal; direct grinding to determine the energy performance of the mill; determining the fractional composition of the resulting coal dust via sieve analysis. Sieving analysis of coal dust is performed with vibrating screens with various diameters 1000, 500, 200, 90, 71 μm

The experimental equipment is presented in Figure 2.

To remove extra moisture, the test sample of coal was pre-dried in a convective dryer to a moisture content of $W^c = 4\%$. Analytical moisture (W^c) is determined for a laboratory sample of coal where particle size is from 0 up to 3 mm with a mass of 500 g.

The laboratory ball drum mill (Fig. 2) was filled with metal balls 28 mm in diameter (with a total mass of 5 kg) and 15 mm in diameter (with a total mass of 2 kg).

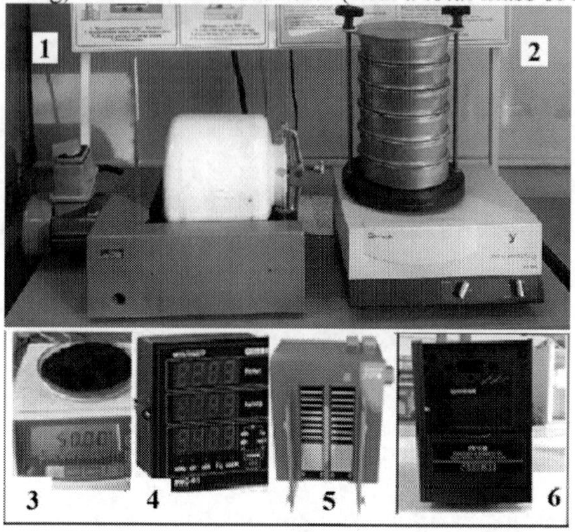

1-ball drum mill; 2 - Retsch AS200 vibrating screen; 3- electronic scales CAS MWP-600; 4- IC F-1 multimeter; 5- drying chamber; 6- vector frequency converter frequency converter B1-XX.

Fig. 2. Overview of the grinding and measuring equipment using during the experimental study

The performance of an industrial BDM generally depends on the properties of the coal being milled, more precisely on the strength, grinding ability, size of pieces entering the mill (up to 50 mm), moisture content of materials (up to 0.5%), comminution degree, degree of filling BDM with coals and balls. Experimentally, the performance of mill can be expressed with the following expression [7]:

$$B = k \cdot D^{2.4} \cdot n^{0.8} \cdot \phi^{0.6} \cdot L \text{, kg/h} \tag{4}$$

where B - is the mill productivity, kg/h; k is the correction factor for the fineness of the grinding (with a residue on a sieve of 90 μm from 2 up to 20%, the value of k varies from 0.5 to 1.5, respectively), here it is accepted k = 1; D - internal diameter, m; L- is the length of the mill, m; ϕ - filling factor of mill; n - mill rotation frequency, rpm.

The consumed electrical power of the laboratory mill depends on the working speed of the engine:

$$N_{ep} = 0.214 \cdot D^2 \cdot n \text{, kWt} \tag{5}$$

The mill efficiency will be either specified as specific performance or specific consumption of electricity for grinding [1]:

$$E = \frac{N_{ep}}{B},, \text{ kWh/kg} \quad (6)$$

Table 2 presents the results from the experimental study when grinding the Karaganda coals in terms of relative rotation speed.

Table 2. The performance characteristics of experimental BDM

Relative rotational speed	Power, kWt	Performance, kg/h	Specific electricity consumption, kWh/kg
0.59	0.32	2.86	0.11
0.65	0.55	2.89	0.19
0.7	0.59	2.92	0.20
0.71	0.65	3.0	0.22
0.81	0.67	2.93	0. 23
0.86	0.61	2.87	0.21
0.92	0.58	2.84	0.20
0.97	0.5	2.78	0.18
1.03	0. 4	2.75	0.15

The efficiency of the mechanical mode of operation of a ball drum mill is determined by two main parameters [8]:

•relative speed of rotation of the drum $\psi = \frac{n}{n_{cr}}$;

•coefficient of relative filling of the mill with grinding balls $\phi = \frac{V_B}{V}$.

On Figure 3 are presented the results of grinding the test coal in a laboratory non-ventilated BDM with D = 210mm, relative length $\frac{L}{D}$=0.9 and the wave-like form of cover at various mill revolutions of the mill and the degree of filling with balls $\varphi_b = 0.2$ and the fineness of dust $R_{90} = 22\%$.

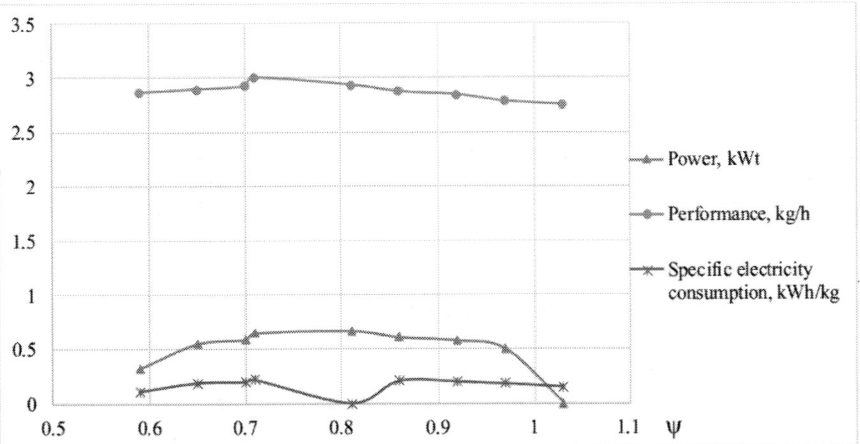

Fig. 3. Performance, power and specific power consumption curves at different operational modes of mill drum

The change in performance of the mill in terms of relative rotational speed of the drum is uneven, where the maximum change is at value ψ = 0.71. Further increase in the relative speed causes grinding bodies (balls) to fall in the center of the drum because of a decreased separation angle. In this case the grinding bodies do not reach the fuel but only bare surface, i.e. fuel grinding is not in place. The potential energy of the falling balls is not spent efficiently for a heating, as metal wear and losses in mechanical transmission occurs.

Power and specific energy consumption with an increase in the relative speed of drum rotation in terms of performance slightly changes.

The following motion modes of the grinding bodies inside the rotating drum can be distinguished:

- cascade, with rolling grinding bodies without flight;
- mixed, partly with rolling milling bodies and partly with flight;
- waterfall, with the predominant flight of grinding media, including subcritical, when approaching centrifugation or with centrifugation of the outer grinding layers;
- supercritical or "flywheel", with centrifuging of the entire volume of grinding bodies.

High-quality grinding of fuel corresponds to the waterfall regime [8]. The angle of separation of the grinding body is shown along the section of the drum in Figure 4. With a faster rotation of the drum, the balls under the influence of centrifugal force are pressed against the wall, rise and then fall down with a parabolic trajectory (Figure 4a).

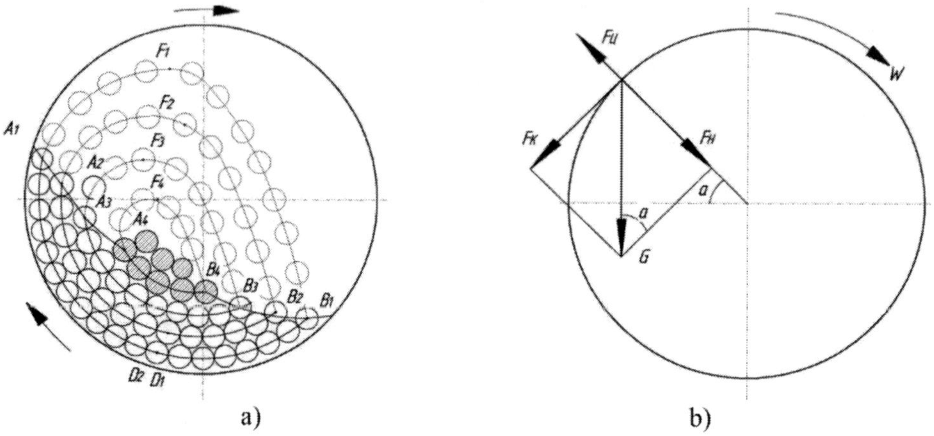

Fig. 4. Motion of grinding bodies inside rotating drum, corresponding to high-quality frame of fuel and its characteristics: a) movement of the balls inside the BDM; b) Applied force scheme.

The relative speed of rotation of the drum has a maximum at a value of $\varphi = 0.71$ ensuring the transition of most layers of grinding bodies from a circular trajectory $B_1\ D_1\ A_1$ at point A1 to a parabolic $A_1\ F_1\ B_1$, where the grinding body makes a free fall onto the fuel layer in a parabolic way and makes circular movement again at the point of fall.

The quality of milled coal has a large, and sometimes decisive effect on the efficiency of grinding and incineration, as well as, the operation of individual elements of the dust preparation plant, the furnace, and the boiler unit as a whole. Therefore, the amount of supplied milled coal should be given special attention.

Larger dust particles lead to higher productivity of any type of mill and lower specific energy consumption for grinding and transport, lower specific wear of the metal and consumption of lubricant, etc. Of course, coarse-gained dust production is one of the easiest ways to increase grinding efficiency. However, it should be considered that the fineness of the dust has a very strong effect on the combustion process in the furnace, on the stability of ignition, the efficiency of combustion and slugging of heating surfaces [1].

Two different parameter affecting the efficiency are specified:

- The greater the yield of volatiles in the fuel, and the lower its ash content, the coarser the dust can be;
- The greater yield of volatiles in the fuel turns the fuel in combustible gas when heated, the easier it is to ignite this fuel and to maintain stable combustion process.

The quality of the final dust, ensuring effective combustion is determined primarily by the fineness of its grinding i.e. surface which is usually evaluated by particle size on the basis of sieve analysis of dust samples.

The reliability of dust ignition is directly related to the total residue of dust on a sieve with cell sizes of 90 μm (R90), and its complete burnout due to the presence of bigger fractions in dust characterized by the total residue on a sieve with cell sizes of 200 μm (R200), which determines heat losses associated with mechanical underburning of fuel [9].

Sieved analysis was performed with an average portion of dust in the amount of 50 g (Fig. 1). According to the results of sieve analysis R_{90}=22.08%, and the magnitude of the total balance amounting to R_{200}=4.2%.

The process of ignition and combustion of coal dust with the same fineness of grinding is largely affected by the amount of volatile substances.

A correlation between the fineness of grinding and the amount of volatiles, providing a stable ignition of dust and its complete burnout, was proposed in [10]:

$$R_{90} = \alpha + \beta V^{v}, \quad (7)$$

where for coals, based on empirical data the values are determined: α = 4; β = 0.7. Following Eq. 7, the grinding fineness for Karaganda coals amounts to $R_{90} = 23.6$. The experimental study shows $R_{90} = 22.08$ which is within 6% error. This proves a positive statement about correlation of important characteristics of the processed fuel, fineness of grinding and the amount of volatiles in the dry ash-free mass of fuel.

When transferring the experimental data obtained to an existing coal-grinding installation, the coefficient of grinding ability and specific energy consumption for fuel grinding should be considered as the most significant thermotechnical characteristics.

As follows from table 2, the specific energy consumption for fuel grinding varies in the range from 0.11 kW h/kg up to 0.23 kWh/kg. The values for the specific energy consumption obtained are consistent with the current standards for BDM - 0.15 up to 0.25 kW/kg of coal with a fuel milling coefficient of K_l = 1.4. According to the norms of dust-control plants [1], the selection of mills for grinding industrial Karaganda coals is allowed, when $K_l \geq 1.1$.

The results of the sieve analysis confirm the qualitative grinding of the fuel and correspond to the standard values of the fineness of the dust recommended [11].

4. Conclusion

The performed experimental study concerns the operational characteristics of BDM for grinding of Karaganda coal products. The results obtained can be used to improve the energy efficiency of large industrial grinding BDM. The following conclusions have been made:

- The relation between productivity, power and specific energy consumption in terms of relative rotational speed of the mill drum is determined;
- The performance of the drum mill reaches its maximum at a relative speed of rotation of the drum 0.71 ensuring a waterfall mode of grinding of fuel related to the high-speed mode of a coal-grinding mill.
- When transferring the experimental data obtained to an existing coal-grinding installation, the coefficient of grinding ability and specific energy consumption for fuel grinding should be considered as the most significant thermotechnical characteristics.

- The specific energy consumption for fuel grinding varies in the range from 0.11 kW h/kg up to 0.23 kWh/kg at fuel milling coefficient of $K_l = 1.2$.

The development of a methodology for thermodynamic analysis of coal dust preparation, as well as the development of a mathematical model for the operation of a dust preparation plant with BDM-25A with improved technical and economic indicators, is ahead.

References

1. Levit G T, Dust preparation at thermal power plants. Energoatomizdat, p. 384, 1990
2. Mingalieva A, Zatsarinnaya Yu and Vachagina E.K, Analysis of the system for the preparation of fuel pulverized CPP, Energy issues, 2005, №1-2, p. 22-31.
3. Prothrombin index 102-47-11. JSC "ArcelorMittalTemirtau" CHP-2, Production and technological instruction manual for the pulverization system with ball mills of the type BDM-320/570, Temirtau 2011, p.19.
4. Abildinova S, Kamarova S, Optimization of electricity costs for dust preparation in Sh-25A Sharpen Mills CHP-2 OJSC ArcelorMittalTemirtau, Energy: Management, quality and energy efficiency, Proceedings of the IX International Scientific and Practical Conference, March 11-12, 2019. Blagoveshchensk: AmSU Publishing House, 2019, p.370-376.
5. Sokolova N, Sokolov N, Calculation and design of dust preparation installations for boiler units (Regulatory materials) / Under the general editorship of N. Sokolova, M. Kisselgof. ONTI CKTI, 197, p. 312
6. Savchenko V, Enamelling technology and equipment for enamelling workshops: studies for technical schools / V.I. Savchenko; by ed. L.D. Svirsky. Kharkov: Metallurgizdat, 1961, p.388
7. Pivniak G, Grinding, energy and technology. M: Ore and metals, 2007, p. 238
8. Khakhalev P, Improving the design of the stepped lining and study of the grinding process in a ball-type drum mill, BSTU them, V. G. Shukhov, Belgorod, 2017, p.177
9. Butko A, Educational Complex. Fuel and its use, Minsk: BSU, 2019, p.70
10. Lavrantev A. Y., Tolchinski A. Y., Improving the methods of calculating dust preparation plants, Electrical stations Journal, 2003, Vol. 12, p.p. 45-58
11. Zhukov E, Calculation and design of dust preparation systems: a tutorial E. B. Zhukov; Alt. state tech. un-t them. I. Polzunova, Barnaul: Izd-vo AltGTU, 2015, p. 123

E3S Web of Conferences **180**, 01004 (2020) https://doi.org/10.1051/e3sconf/202018001004
TE-RE-RD 2020

Condensing economizers for large scale steam boilers

Iliya Iliev[1], *Angel* Terziev[2*], and *Hristo* Beloev[1]

[1]University of Ruse, Department of Thermotechnics, Hydraulics and Ecology, 8 Studentska Str., 7017 Ruse, Bulgaria
[2]Department of Power Engineering and Power Machines, Technical University of Sofia, 8 Kliment Ohridski blvd., 1000 Sofia, Bulgaria

Abstract. An energy analysis and assessment of the feasibility of the development and implementation of a condensing economizer (CE) for steam generator TGM-96A has been made. The condensing economizer is characterized with a very high heating capacity of 21 to 23.5 Gcal/h and will cool the fuel gases from 123°C down to 47°C. The utilized heat will be used to heat three water loops: two loops with District Heating Network water (DHN) and one with raw water from the Dnieper River. The expected improvement in the efficiency of the steam generator is 10.55%. Complete thermal calculations of the condensing economizer have been carried out along with a definition by flow and temperature of the heat carriers. Based on the calculations of the heat and mass transfer processes, the thermal capacity of the economizer under the conditions of partial condensation of water vapor from the flue gases was estimated. The analysis presents an objective assessment of the project's investments, as well as other financial indicators and environmental benefits that give the investor the opportunity to put the project into operation.

1. Introduction

The efficient use of fuels in large energy boilers is predominantly related to the utilization of wasted heat from the exhaust gases. One of the ways to increase the energy efficiency is to use state-of-the-art condensing economizers to utilize the heat from the exhaust gases of the steam generators [1, 2, 3]. Common practice shows that at this stage deeper heat utilization from boilers burning natural gases, methane, and propane butane is more promising [2,3]. The efficiency of the boilers burning natural gas is in the range of 82-94% [4], estimated by the lower heating value (LHV). At temperatures between 110-220°C for the exhaust gases, the heat losses are estimated at 6-12% [1, 4]. In recent years, the so-called condensing boilers, in which the gas is cooled at temperatures below the dew point (58-60°C), have gained popularity [1, 2, 9]. In these boilers both the physical heat of the gases and the latent heat of condensation are utilized, which allows the fuel efficiency to be

* Corresponding author: aterziev@tu-sofia.bg

significantly increased by up to 15%. For example, in [5] it is shown that when the gas is cooled down to 0°C, the amount of heat during condensation is 11.9% relative to the LHV of the natural gas. However, due to the absence of a source with such a low temperature, the gases are cooled to 20-40°C, which to some extent reduces the effect of using condensation heat.

A number of studies [6] have found that the condensation of water vapors from flue gases dramatically increases the cost of each cooled degree. The most important factor in this cost increase is the initial moisture in the flue gases; the higher the moisture content, the higher the cost. To increase the efficiency of the conventional boilers by 1%, a reduction of gas temperature by 15-20°C is needed. The same share of efficiency with condensing boilers can be achieved with cooling the gases by 2-3°C only. The aforementioned studies [2, 6] prove that when the exhaust gas is cooled from 150°C to 40°C the fuel economy reaches 10-12%. Moreover, the heat loss of the exhaust gases, related to the upper LHV, is 2-3%, with efficiency reaching 96-97%. If the heat balance of the steam generator is made by the lower combustion heat, the efficiency reaches 105-107% [1, 2, 3, 10].

2. Subject of this energy analysis

CHP-5 is the largest thermal power station in Ukraine with an installed thermal capacity of 1874 Gcal/h and installed electric power of 700 MW. The plant has four units, two power units of 100 MW and two of 250 MW. The plant has two vertical water-tube steam generators, model TGM-96A; two co-current vortex steam boilers, model TGMP-314A; three water heating water-tube boilers, model PTVM-180; and two hot water boilers, model KVGM-180. The main fuel for the steam generators is natural gas, and fuel oil, M100, is used as a standby fuel.

CHP-5 provides domestic hot water to the 850,000 inhabitants of Kyiv, as well as heat energy to 3 million square meters of industrial buildings.

The heat energy is supplied through 6 main pipelines with diameters from 900 mm to 1200 mm.

For both thermal and electricity production 800,000 m^3 of natural gas, with a calorific value of 34,330 kJ/Nm^3 is consumed annually.

The construction of CHP-5 started in 1960. The first and second power blocks were put into operation in 1971-72 with the third and fourth in 1974-1976.

The subject of this energy analysis are the two TGM-96A steam generators. The gross thermal efficiency of the steam generators, according to their load, is within η = 92% to 94% at exhaust gases temperatures of 123°C to 125°C.

The project's goal is to assess the opportunity for condensing the flue gases below the dew point cooling temperature, using a Condensing Economizer (CE) thereby increasing the efficiency of the steam generators by 8% to 10% - and also reducing natural gas consumption and CO_2 and NO_x emissions while saving significant quantities of heat in CHP-5's regeneration cycle and the heat distribution network.

2.1. Analysis of the Parameters for Determining the Degree of Flue Gas Cooling

Today, indirect surface-type heat exchangers, using water as their coolant, are the most common. Their critical advantage is the provision of optimal price, performance and technological solution.

The key factor, that influences the residual heat of the flue gases utilization in this type of regenerative heat exchanger is the quantity and temperature of the cooling fluid (usually water) [7]. The secondary factors affecting the performance and design characteristics include: type of heat exchanger; pitch and arrangement of the heat exchanger's tubes; velocities of the gases and the water; pressure drop in water and gas flow; composition of the exhaust gases (including excess air ratio and air humidity); and heat exchange surface area/type [8, 9,10].

2.2. Boilers' Capacity and Quantity of Exhaust Gases

Based on the data provided by KTE:

- Boiler #1 operates about 2,648 hours per year at an average capacity of 206 Gcal/h (240 MW);
- Boiler #2 operates about 5,204 hours at an average capacity of 230 Gcal/h (268 MW).

Both boilers do not operate simultaneously with the CE – it operates with either boiler and will be switched from boiler to boiler as appropriate.

During three of the winter months, the average capacity of one boiler is about 259 Gcal/h, which is when the conditions for condensing mode are the best. Therefore, when determining the amount of exhaust gases required for calculating the size of the CE it is assumed that the maximum boiler capacity is 260 Gcal/h (302 MW).

Taking into consideration the quantity and the composition of the fuel and the combustion products, and the excess air ratio (which must not exceed 1.4), the amount of exhaust gas is assumed to be 411,000 Nm^3/h. The analysis uses this method of estimating exhaust gas volume instead of calculating the theoretical amount of gas at full boiler capacity.

2.3. Exhaust Gas Temperature

The temperature of the exhaust gases at different time periods and loads ranges is between 110°C and 130°C. This analysis assumes an average exhaust gas temperature of 123°C.

2.4. Quantity and Temperature of Water Flows for Gas Cooling

A CE has one to several water inlets and outlets, with water flow rates and temperatures sufficient to cool the exhaust gases below the dew point of the water vapor in the flue gases at the respective partial pressure.

At CHP-5 it is possible to use three types of water streams: district heating net (DHN) water; additional water to refill the district heating system; or fresh water from the Dnieper river for the CE's technological needs.

2.5. District Heating Network (DHN) Water

Using water from the DHN's return water is a significant advantage for the CE as there is sufficient water at a temperature close to the condensation temperature of the water vapor in the flue gases.

It is possible to build the CE with two or three independent water loops and outlets in the exhaust flue gases, using DHN water with a flow rate of 2000 m^3/h to 3000 m^3/h. The circulation pump with a required flow rate and head will be used to transport the water. Another possibility is to use the head of the network connected pumps, installed before the boiler system, for running the water through the economizer. (The different strategies will be considered during the CE's design process.)

The DHN return water temperature ranges between 52°C and 55°C in the summer months - June, July and August. During the winter season the water temperature is between 46°C and 50°C. Experience shows that when water temperature rises from 53°C to 55°C the outlet gas temperature is usually higher than 56°C to 58°C (when sufficient cooling water flow rate is present). At this temperature, the CE partially condenses the exhaust gases. Therefore, during the 4 summer months, the inlet water does not ensure a stable condensing process in the CE. During the winter period, the network water in the CE can utilize a power of 10 to 15 Gcal/h, and in summer 8 to 10 Gcal/h. The distance between the proposed CE site and the heating network is about 250m.

2.6. Water Added to Compensate the Water Losses from the District Heating System (Makeup Water)

The CE could use the district heating system's makeup water. The chemically treated and purified makeup water, with a flow rate of 400÷500 m^3/h and a temperature of about 30°C, could be used by the CE for providing year-round condensation of water vapor in the exhaust gases at almost all loads of the heating networks and the boilers. If this water source is used, it is recommended that the makeup water is heated by 10°C before entering the deaerator and added back to the heating network.

Adding water with such properties is an advantage ensuring condensation at the CE and saving about 8 Gcal/h to 10 Gcal/h of thermal energy output.

Circulation would be provided by a pump with suitable head and flow rate. The distance from the CE's proposed site to the makeup water source is about 80m to 100m.

2.7. Fresh Water from the Dnieper River for Technological Needs of the Plant

The Dnieper River's waters could also serve as the CE's cooling water. Water circulation is supported by a pump with suitable pressure and flow rates. It is very likely that the river water would need to be treated. The distance to the water source (i.e., the Dnieper River) is about 80 - 100 m.

Table 1. Average annual data of water loops of condensing economizer

Element	Unit	Value
Average flow of DHN water of nearest network, G_{avDHN}	m^3/h	17,614
Minimum flow of DHW (not accident) of nearest network, G_{minDHN}	m^3/h	15,009
Maximum flow of DHN water (more than along a day), G_{maxDHN}	m^3/h	29,000
Average temperature of return network, t_{ret}	°C	50.0
Number of hours with temperature > t_{ret}	h/year	4,104
Distance from network to CE, L	m	≈250
Diameter of the network pipes at the point of connections, d	m	0.9
Operation time of network, h_{net}	h/year	8,760
Pressure of return network, P_{net}	MPa	0.33
FEEDWATER FOR DHN		
Average feedwater flow, G_F	m^3/h	512
Feed water temperature before heating, t_H	°C	20÷30
Distance between water source and CE	m	≈80
RAW RIVER WATER		
Temperature of raw water after condenser, t_{RW}	°C	15÷25
Average raw water flow, G_{RW}	m^3/h	665

Table 1 presents the average annual flow and temperature data for the different possible water sources.

3. Condensation Conditions

The easiest and most straightforward way to significantly reduce CHP-5's fuel use is deep flue gas cooling to and below, the dew point temperature of water vapor in the flue gases and further use the latent heat of condensation.

The maximum amount of heat utilized from the flue gases is represented by the amount of heat released under full condensation (at a temperature below the dew point t=0°C and moisture content of d=0 g/kg dry gas) related to the lower heat value (LHV) of the fuel. The utilized latent heat is 11.9% more, related to the LHV – depending on the amount of air in the flue gases and its moisture content, the partial pressure of the gas mixture, the temperature of the gases, the characteristics of the cooling surface, etc. – this percentage may vary. There is no precise theoretical relationship used for determining the percentage of heat recovery in industrial-scale condensation in pipe economizers due to the significant physical, construction/design and other factors that affect the process.

However, a number of studies have shown that the condensation of water vapor in flue gases significantly depends on the excess air ratio, which changes the condensation temperature [8]. High excess air ratios may compromise the economizer's performance by limiting condensation and collecting a reduced share of the latent heat. By increasing the excess air ratio (α), flue gas condensation is hindered, in turn requiring deeper cooling of the flue gases. Therefore, an excess air ratio of $\alpha < 1.4$ before and after the economizer is recommended, to allow the process of water vapor condensation in the flue gases to start at a temperature higher than 53.5°C (Figure 1). Reducing and maintaining the excess air ratio (α) at the lowest possible level is important for operating the CE in an efficient manner.

In conclusion, there are three potential water sources with the needed quantities and at the required temperatures able to cool and condense the flue gases from the boilers. The CE's maximum output is expected to be in the range of 21 to 23.5 Gcal/h (24.5 to 27.3 MW), which represents an increase in efficiency from 9% to 10.5%. Generally, it is possible to achieve an additional increase of about 1% to 2%, but this requires complex engineering and equipment, which are not cost effective.

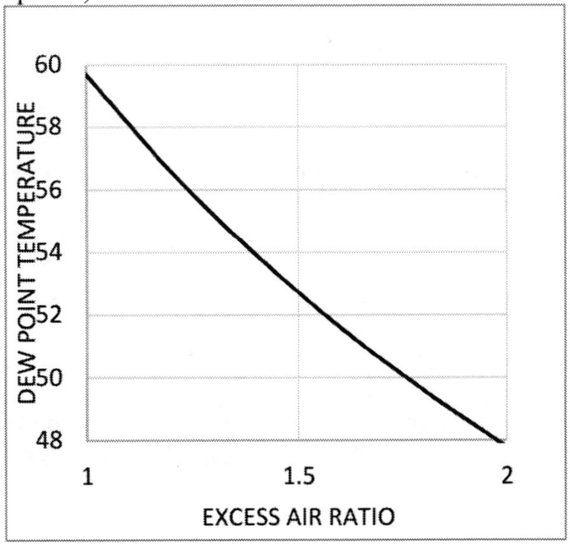

Fig. 1. Dependence between the dew point temperature and the excess air ratio [1]

Fig. 2 shows the general appearance of the CE installed at Kyiv's CHP-6 after the ALSTOM water-heating boiler with a capacity of 180 Gcal/h. The thermal capacity of the CE is 15.6 Gcal/h, saving 8% to 10% of the hot water boiler's fuel consumption.

Fig. 2. The flue condensing economizer (CE) recently installed at CHP-6

This analysis assumes the following will take place at CHP-5: the CE will operate with an average real load on the boiler (see Table 2) of 222 Gcal/h; the boilers will operate with an 8.6% increase in their efficiency (which is both theoretically, reasonably and practically possible), the CE's average annual power value is 21 Gcal/h, and the CE has a stable and reliable operation and maintains its design parameters.

The average annual performance indicators of the CE, which can be switched between the two boilers, are presented in Table 2.

A schematic of CHP-5's condensing economizer is shown in Figure 3. The figure shows the three water inlets and outlets and their temperatures as well as the temperatures of the incoming and outgoing flue gases used in the calculations.

Thermal calculations of the condensing economizer are performed with specialized software. Input data accepted for the analysis of the condensing economizer are presented in table 3.

The parameters that most strongly affect the degree of utilization of the residual heat of the flue gases in the concerned recuperative heat exchanger types are the flow rate and cooling fluid temperature.

The next other significant factors influencing the efficiency and design features are the type, step and pipes arrangement, optimal velocities of gases and water, acceptable resistance losses on the water and gas path, composition of the exhaust gases, incl. excess air coefficient, air humidity, and total heat exchange surface.

Input data analysis shows that when using the condensing economizers (CE) a couple of closed water loops with flow rates and temperature sufficient to cool the gases below flue gas dew point temperature at the respective partial pressure.

Three types of water flows are possible at CHP-5: network water supplied by the district heating network; additional water to fill the district heating system; raw water from the Dnieper River for the power plant's technological needs

The temperature of the returning network water is in the range from 52 to 55 °C during the summer period (June, July and August). During the winter season the temperature is between 46 - 50 °C.

The performed analysis and thermal calculations show that both in a winter and summer mode when the inlet water temperature in CE is in the range of 46 - 56°C and the optimal flow rate of the used network water is lower than 1000 m^3/h, the temperature of the exhaust gases will always be greater than the flue gas dew point temperature (which is 56-58°C), i.e. this loop will operate as an ordinary economizer without condensation of water vapor from the flue gases. In this case the flue gases outlet temperature will also drop to about 75°C.

In terms of the above, the partial condensation of water vapor in the flue gases will take place in the second or third contour of the CE. If main water runs in the second loop again at the same temperature and flow, condensation of water vapor in the flue gases will not occur again, but such a process will instead be initiated only on a part of the pipe wall surface.

Therefore a third loop with make-up water at a flow rate of 420 m^3/h and temperature of 32°C, is considered, while the gases will be cooled to about 45°C, and condensation is expected to reach more than 60% (it is well known that practical results and empirical relations for degree of condensation are not available not only for different constructions but also for different condensation terms).

The results of reaching high levels of condensation can be achieved by using a third loop with filtered water for technological needs from the river at up to 620 m^3/h and a temperature lower than 25 °C.

Table 2. Average annual data for one boiler incorporated with condensing economizer

Element	Unit	Value
Average steam boiler load, G_{st}	t/h	372
Average natural gas consumption, B_F	Nm^3/h	29,154
Average steam boiler output, N_B	Gcal/h	222
Flue gas temperature after steam boiler, t_{FG}	°C	123
Excess air ratio in the burning chamber, α_{ch}	-	1.11
Excess air ratio after boiler, α_{AH}	-	1.4
Boiler efficiency (Gross)	%	92.36
Total operating hours for both boilers	h/year	7,800

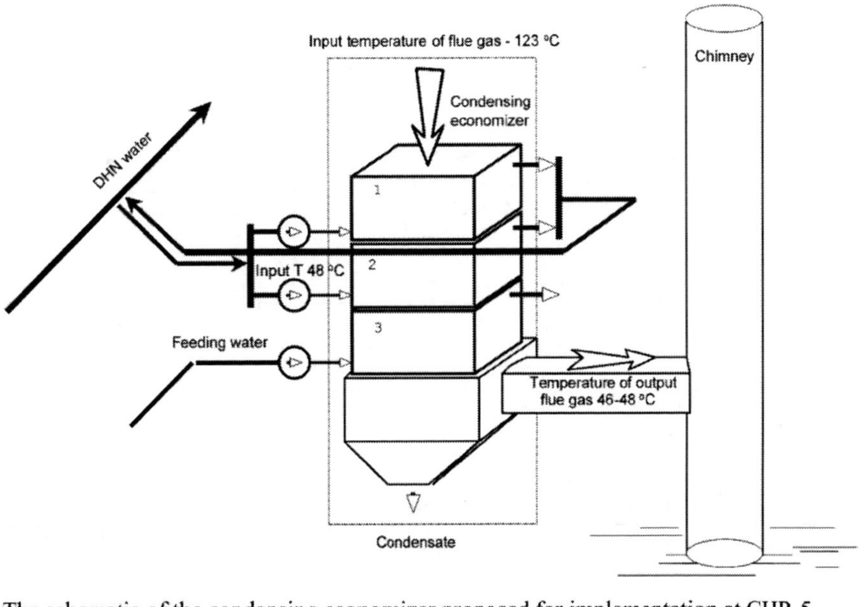

Fig. 3. The schematic of the condensing economizer proposed for implementation at CHP-5

Table 3 presents the input data (composition and calorific value of the fuel, steam generator parameters, etc.) needed for the CE calculations.

Table 4 summarizes the calculation parameters used for the analysis of the CE.

Table 3. Input data for analysis of the condensing economizer

Element	Unit	Value
Low heating value (LHV), Q^r_i	kJ/Nm^3	34,330
Excess air ratio in the burning chamber, α_{ch}	-	1.11
Maximum permissible value of the excess air after the air heater, α_{AH}	-	1.40
Nominal output of superheated steam, D_{sh}	t/h	480
Flue gas temperature at the economizer inlet, υ_{FG}	°C	123
Flue gas enthalpy at economizer inlet, H_{FG}	kJ/Nm^3	2,576.40
Steam pressure in steam preheater, P_{ne}	MPa	11.50
Steam temperature in steam preheater, t_{SP}	°C	540
Steam enthalpy in steam superheater, h_{sh}	kJ/kg	3,460.40
Feed water pressure, P_{FW}	MPa	14.60
Feed water temperature, t_{FW}	°C	230
Feed water enthalpy, h_{FW}	kJ/kg	993.00
Enthalpy at saturation at P_{BSD}; h’	kJ/kg	2,788.65
Pressure in the boiler drum, P_{BSD}	MPa	12.88
Heat loss from outgoing gases, q_2	%	7.49
Heat loss from chemically incomplete combustion, q_3	%	0.0
Heat loss from mechanically incomplete combustion, q_4	%	0.0
Heat loss from radiation to the ambient, q_5	%	0.13
Boiler efficiency	%	92.36

Table 4. Calculation parameters of the condensing economizer

Element	Unit	Value
Cooled flue gases		
Flue gas temperature at the economizer inlet, u_{FG}	°C	123
Flue gas enthalpy at economizer inlet, H_{FG}	kJ/Nm^3	2,576
Flue gas temperature at the economizer outlet, u_{EFG}	°C	47
Flue gas enthalpy at the economizer outlet, H_{EKO}	kJ/Nm^3	888
Maximum boiler capacity for calculating of CE, Q_{cal}	Gcal/h	260
Flue gas flow at average boiler capacity, V_G	Nm^3/h	411,000
Average boiler capacity delivering flue gases to CE (at Q_B=260 Gcal/h), Q_{aver}	Gcal/h	222
Average CE capacity (at average Q_B=222 Gcal/h), Q_{aver}	Gcal/h	21

Working hours	h/year	7,800
Utilized waste heat (Heat energy savings)	Gcal/year	163,323
Reduction of natural gas consumption (Natural gas savings), B''_T	1000 Nm^3/h	23,500
Heat loss from outgoing gases after construction of CE, q''_2	%	-2.26
Boiler efficiency after construction of CE	%	102.1
Boiler efficiency improvement	%	10.55
Heated water in condensing economizer		
Inlet network water temperature entering the economizer, t'_{DHN}	°C	50.0
Outlet network water temperature leaving the economizer, t''_{DHN}	°C	55.7
Network water flow, G_{NHW}	t/sec	0.8
Feed water temperature entering the economizer, t'_{HFW}	°C	28.0
Feed water temperature leaving the economizer, t''_{HFW}	°C	44.0
Feed water flow through economizer, G_{FW}	t/sec	0.14
Inlet temperature of technological water from river Dnieper, t'_{RW}	°C	22.0
Outlet temperature of technological water from river Dnieper, t''_{RW}	°C	30.0
Technological water flow, G_{RW}	t/sec	0.18
Circulating pump power H=22 m H_2O, m=1700 m^3/h – 2 circuits	kW	700
Circulating pump power H=15 m H_2O, m=600 m^3/h – 1 circuit	kW	150
Circulating pump power H=15 m H_2O, m=900 m^3/h – 1 circuit	kW	240
Total electrical capacity of water pumps	kW	1,090
Annual heat losses from network pipes to the CE	Gcal/year	477
Decrease of specific nat. gas consumption after installing of CE	$Nm^3/Gcal$	118.8

It is well known that carbon steel corrodes rapidly when exposed to hot condensing flue gases. When there is a presence of higher concentration of acidic components in the combustion gases, corrosion is accelerated.

Flue gas dew point corrosion occurs when these aggressive acids condense on carbon and stainless steels in the convection sections, flue ducts, and stacks [11]. The amount of contaminants in the fuel is directly correlated with the concentration of acid droplets, and therefore with the degree of corrosion itself. To mitigate these corrosive processes in the proposed CE more resistant materials in the construction of flues are installed.The flue gases will be discharged into the atmosphere via a chimney made of polymer composite materials with a high corrosion resistance and operational lifespan up to 50 years. This will avoid secondary condensation of water vapor in the chimney without applying the very common methodology of mixing condensed flue gases with hot gases before the chimney.

In order to avoid low temperature corrosion over the heating surfaces, ribbed pipes with extruded aluminum ribs are recommended for the CE. The calculations are made on the basis of pipes D=32 mm and wall thickness of 2.3 mm with extruded aluminum ribs 320 pcs/m and a rib height of 15 mm. For the primary loop, where condensation of water vapor does not occur, carbon steel pipes can be used – boiler type st20 /GOST 9567-75; ГОСТ8733; DIN 2391; EN 10305-1/.

For the second and third loop surfaces where the process of water vapor condensation takes place and carbon and nitrogen oxides from the exhaust gases are dissolved in the condensate, the pH of the condensate is lower than 7. This suggests the use of alloy steel pipes with universal application type 304, 304L, 304H, which boast better corrosion resistance to nitric acid with concentrations below 65%, as well as most other organic and inorganic acids.

It is mandatory that stainless-steel pipes resistant to carbonic acid and nitric acid in all are used in all water contours. It is recommended that the CE is designed as a vertical structure to permit a downward flow of the gases. Additionally, when the CE's condensate is to be used for technical needs or the district heating network considerations should be made to ensure that it is collected and neutralized (to a pH>7)..

4. Project benefits

The project's benefits are estimated based on calculations of the CE's energy use and efficiency:

4.1. Heat energy & natural gas savings

After installing the CE, CHP-5 would use the waste heat of the boiler's exhaust gases. The boiler's exhaust gas temperature will be reduced from 123°C to 47°C. DHN water (or heated feeding or raw water) will be running and heating in the CE's heat pipe bundles. Assuming an average boiler power of 222 Gcal/h, an average power of the CE about 21 Gcal/h and 7,800 operational hours per year, the CE will produce 163,800 Gcal/year. Considering the annual heat losses in the pipes to the CE of 477 Gcal/year, the estimated heat savings are 163,323 Gcal/year. 23,500,000 Nm^3/year of natural gas would be needed to produce such an amount of thermal energy.

4.2. Natural gas cost savings

- Natural gas prices in Ukraine are regularly changing and the price change can be significant. Also, there are 2 different natural gas prices, depending on the type of energy (heat or electricity) produced by natural gas: "heat natural gas price" and "electrical natural gas price". KTE staff recommended that this SPFS use the "heat natural gas price", because the CE produces heat that otherwise (when the CE does not work) is produced by a hot-water boiler producing only heat. The data for "heat natural gas price" for 2018, was provided by KTE, and is set in calculation model. This price is: 6,763.51 UAH/1000 Nm^3 (229.51 EUR/1000 Nm^3).
- Extended forecasts for the changes in price of natural gas should be used in a more detailed feasibility study.
- For a natural gas price of 229.51 EUR/1000 Nm^3 ,the annual natural gas cost savings alone would amount to 5,393,454 EUR.

4.3. Additional costs

- The electricity price is 58.36 EUR/MWh based on KTE's average electricity production cost in 2018 (data provided by KTE).
- Electricity consumption of circulation pumps: Three water pumps with a total installed power of 1,090 kW are used for water circulation in the three loops of the CE with 7,800 operational hours per year. The pumps' annual electricity consumption is 8,502 MWh/year. At an electricity price of 58.36 EUR/MWh, the annual expenses are 496,159 EUR/year.
- The costs for chemical reagents needed for neutralizing the condensate: with an average price of 20 EUR/h and with 7,800 operational hours, the annual costs are 156,000 EUR/year.
- Net cost savings, calculated as a difference between the natural gas cost savings and additional costs, are estimated to be 4,741,295 EUR/year.

The CE project's net cost savings: EUR 4,741,295 per year.

The preliminary capital investment costs estimated are EUR 9,739,167 including a contingency of EUR 205,833 or 2.11% taking into consideration the very preliminary status of the project.

This results in a 2.05 years estimated payback period, which is highly promising.

With the installation of the CE at CHP-5, annual CO_2 emissions will drop by 44,791 t/year as a result of the 23,500,000 Nm^3 of natural gas saved each year. However, 8,502 MWh/year, of additional electricity will be used to power the water pumps, which will increase CO_2 emission by 8,723 t/year.

5. Conclusion

The energy analysis and assessment of the feasibility of the development and implementation of a condensing economizer (CE) for steam generator TGM-96A showed that:

- There is considerable potential for the recovery of waste heat from the boiler and boiler efficiency can be increased by 10.55% using a condensing economizer;
- The CE's maximum output is expected to be in the range of 21 to 23.5 Gcal/h (24.5 to 27.3 MW);
- The energy saving measure would result in the following:
  - Electricity consumption of circulation pumps will increase the overall CE electricity consumption: The pumps' annual electricity consumption is 8,502 MWh/year equal to an annual expense of 496,159 EUR/year.
  - Thermal energy savings of 163,323 Gcal/year;
  - Natural gas consumption would be reduced by 23,500,000 m^3/year, amounting to cost savings of 5,393,454 EUR.
  - The proposed energy efficiency measure has a payback period of 2.05 years.
- Implementing the proposed condensing economizer will lead in particular to the reduction of the emissions of:
  - NOx by 52.17 tons/year;
  - Carbon dioxide by 36,068 tons/year.

References

1. I. Z. Aronov, Contact heating of water by natural gas combustion products, St. Petersburg, (1990), (In Russian)
2. Iliev I. Means and methods for waste heat recovery from low-grade gas-steam flows. Monography, Publishing Centre at the Ruse university, (2013), (in Bulgarian)
3. Y.D. Georgiev, I.K. Iliev, G.D. Genchev, P.G. Mushakov, V.V. Bobilov. Investigation of heat and mass transfer in contact condensing economizer with "active" packing. Proceedings of University of Ruse, Vol **38**, Series **5**, (2001) (In Bulgarian)
4. Simplified Pre-Feasibility Study for Installing a Condensing Economizer at CHP-5 in Kyiv, Ukraine, Energy Security Project in Ukraine, (2019)
5. Consider Installing a Condensing Economizer, US Department of energy, Energy efficiency & renewable energy, Steam Tip Sheet #**26A** (2007).
6. U.S. Department of ENERGY, Energy Efficiency & Renewable Energy DOE/GO-102012-3393 (2012) https://www1.eere.energy.gov/manufacturing/tech_assistance /pdfs/steam26a_condensing.pdf
7. G.P. Negreanu, Numerical simulation of a large power steam turbine run-up, Int. J. Energy Technology and Policy, Vol. **6,** Nos. 1/2, (2008)

8. G.P. Negreanu, I. Pîşă, L. Mihaescu, I. Oprea, T. Prisecaru, V. Berbece, Experimental and numerical performance tests of the new 150 MW unit of CHP Paroşeni, The Proceedings of CIEM 2009 (on CD), 4rd International Conference on Energy and EnviroNment, Bucharest, Romania, (2009)
9. C.I. Ioniţă, N. Vatachi, I.V. Ion, The Influence of the Dew Temperature of the Flue Gas on the Steam Boiler Performance, ECOS'98-Efficiency, Cost, Optimization, Simulation And EnviroNmental Aspects of Energy Systems And Processes, Nancy-France, 8-10 July, Vol. II, (1998)
10. K. Uzuneanu, I. Diaconescu - Analysis of New Technologies used for CHP Systems. Proceedings of 2nd International Conference on Energy and EnviroNment Technologies and Equipment (EEETE'13), Brasov, Romania, (2013)
11. https://inspectioneering.com

Consideration regarding the reduction of pollutions emissions by increasing energy efficiency in ship operations

Lucian Mihail Dumitrache[1], *Nicolae* Buzbuchi[1], *Cătălin* Faităr[1*]

[1]Maritime University of Constanta, Faculty of Naval Electromechanics, 104 Mircea cel Batran Street, 900663, Constanta, România

Abstract. The use of experimental investigation techniques in the research of internal combustion engines has contributed substantially to their progress and development over time. The performances obtained in the development of modern engines have been strongly influenced by the emergence and development of the investigation equipment, especially the laser and the modern computing technique. Maritime transport generates about 4% of the total CO2 emissions produced by human activities, which means that its carbon value is almost as large as that of Germany. Emissions from this sector are not yet internationally regulated, but this issue is currently being debated within the IMO and the United Nations Framework Convention on climate change (UNFCCC). Of the strategies to reduce pollutant emissions from ships, identified so far, the technical solutions are the first ones that have been implemented on board. The advantage of using technical solutions onboard ships is the high efficiency of reducing a product or several pollutant products, but the main disadvantage is the high cost of implementing and operating them on board ships.

1 Introduction

Combustion is defined as the transformation of chemical energy (of oxidizable elements contained in fuel) into thermal energy, materialized by a rapid exothermic oxidation reaction, after which a significant amount of energy is released in the form of heat. [1]

Combustion is studied from two points of view: thermodynamic and kinetic. From a thermodynamic point of view, the combustion process is treated globally, in the sense of evaluating the energy of the reactants and the final products, without considering the kinetic mechanism and the intermediate products of combustion.

The carrier of thermal energy resulting from the combustion process consists of the exhaust gases discharged at a temperature that depends on the nature of the fuel, the perfection of the process how the combustion takes place. [1]

From a kinetic point of view, the combustion process highlights the existence of a mechanism consisting of several stages of process accomplishment and the role of active radicals in propagating the flame front. [2] [3]

*Corresponding author: catalinfaitar@yahoo.ro

2 Formating of pollutant emissions

The exhaust gases resulting from the combustion of the hydrocarbons in the combustion chambers of the internal combustion engines result in the elimination of oxygen from the atmosphere and the release of some compounds as follows: [5] [6]

- H_2O and CO_2
- hydrocarbon (as: CH_4, C_2H_2, C_2H_6, C_2H_8, C_6H_6, CH_2, CHO)
- carbon monoxide (CO)
- nitrogen oxides (NO, NO_2, N_2O)
- nitrogen compound NH_3 and HCN
- sulfur gas (SO_2, OCS, CS_2)
- halo-carbonates (CHCl and CH_3Br)
- particles of material.

Pollutant emissions can be divided into primary and secondary pollutant emissions. Primary pollutant emissions are those which, following combustion, are emitted directly into the atmosphere. Secondary pollutant emissions, on the other hand, are formed as a result of reactions between primary pollutant emissions and other compounds found in the atmosphere. [4] [5]

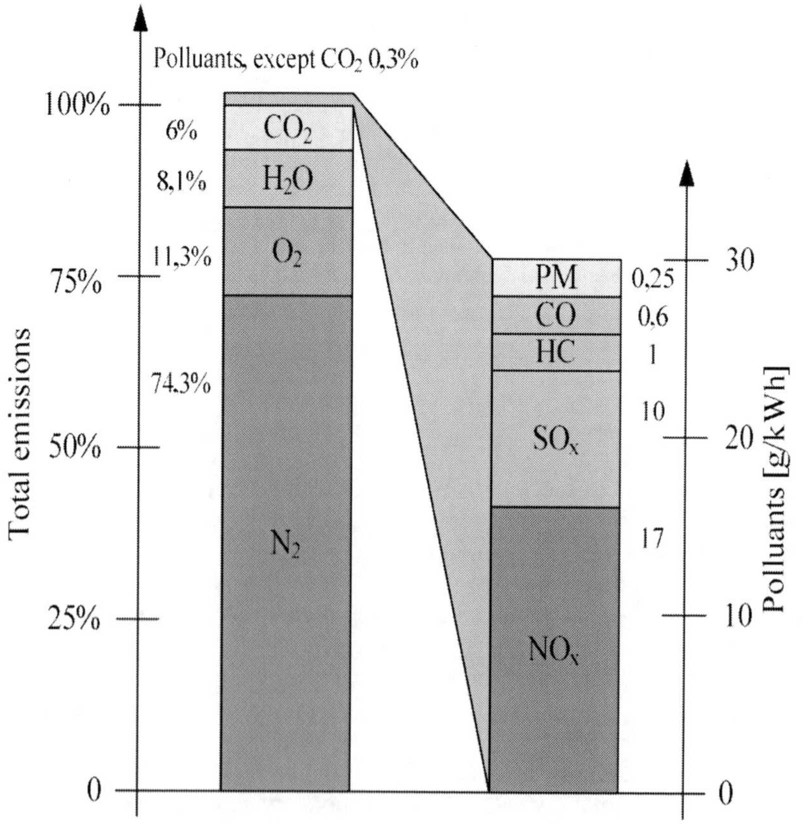

Fig.1. Percentage composition of the exhaust gases of an engine with internal combustion that burns fuel with an average of 3% Sulfur [7]

3 Measurement of the chemical composition of gases

Chemical gas analyzers are used to determine the chemical composition of the gases. Most of the methods are for the determination of the exhaust gas composition, but some systems allow the analysis of the gases inside the cylinder.

A simple instrument for chemical analysis of flue gases is the Orsat analyzer, which, however, could only measure relatively high concentrations (%), without being able to measure the concentrations of unburnt hydrocarbons and nitrogen oxides, which are so much needed today. A sample of flue gas is successively passed through a test tube containing potassium hydroxide, which absorbs carbon dioxide, a test tube with pyrogallic acid, which absorbs oxygen, and a test tube containing chlorate, which absorbs carbon oxide.

From these molecular fractions and the hydrogen-carbon ratio of the fuel, respectively the air-fuel ratio, the chemical balance of the flue gases can be determined. The increased attention lately given to pollutant emissions has led to the emergence and use of a new range of chemical analyzers.

These instruments, compared to the classic Orsat analyzer, have a much higher resolution, the routine measurements being made in ppm, compared to the percentages and being able to measure a much wider range of gases, of which, in particular, those required by the regulations in force for polluting emissions.

3.1 The gas analyzer

The HORIBA EXSA-1500 gas analyzer is an integrated system for measuring pollutant emissions from engine exhaust gases. The appliance is mounted directly on the exhaust pipe and measures the concentrations of pollutant emissions (CO, CO_2, NO / NOx and unburned hydrocarbons) from the undiluted exhaust gases

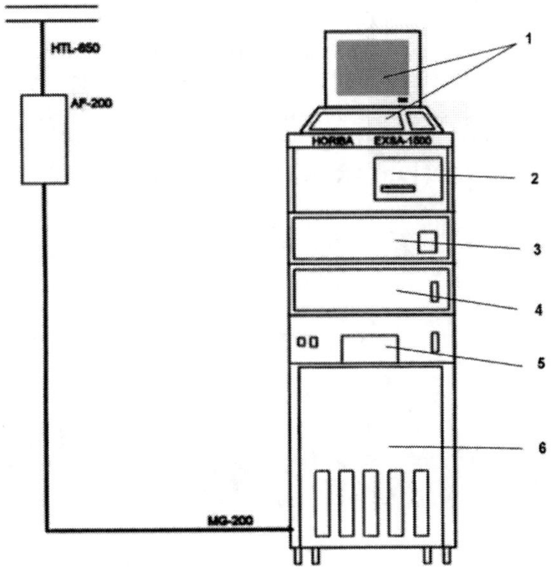

Fig.2. Gas analiser type HORIBA EXAS1500

3.2 The operating principles of the measurement subsystems

The NOX CLA-220UA analyzer is of the chemiluminescence (CLD) type. Chemiluminescence analysis is a precise method of determining the nitrogen oxide content of the exhaust gases, using the luminescence effect characteristic of the reaction between nitrogen monoxide (NO) and ozone (O_3).

The formed NO_2 molecules are partly electronically excited NO_2 and relax by emitting photons in the band 0.6 ... 3. Deactivation is also possible by colliding with other inert molecules M, the likely reactions being:

$$\begin{aligned} & NO + O_3 \rightarrow NO_2 + O \\ & NO + O_3 \rightarrow NO_2^* + O \\ & NO_2^* \rightarrow NO_2 + h\nu \\ & NO_2^* + M \rightarrow NO_2 + M \end{aligned} \tag{1}$$

The intensity of the light emitted is proportional to the concentration of NO and O_3, respectively inversely proportional to the concentration of the substance M:

$$I = I_0 \frac{[NO][O_3]}{[M]}. \tag{2}$$

In chemiluminescence analyzers, the reaction between NO and O_3 takes place in a continuous flow reactor, at a pressure of 5...10 towers. The light signal is sensed by a photomultiplier tube (figure 3). The presence of CO_2 molecules in the gas sample (which plays the role of substance M in reactions 1) causes the measurement result to be altered depending on the CO_2 concentration, by decreasing the signal intensity, a phenomenon known as CO_2 interference.

The analyzer used has an error of up to 3% of the read value if the CO_2 concentration is 12% or less. The analyzer enclosure is thermostated to 450 °C. The ozone required for the reactions is produced in a generator, passing the current of oxygen through an electric arc. The analyzer can also be used to determine the concentration of NO_2 in the mixture if NO_2 is previously converted to NO, which is done in a converter by thermal dissociation, which uses as Mo-C catalyst.

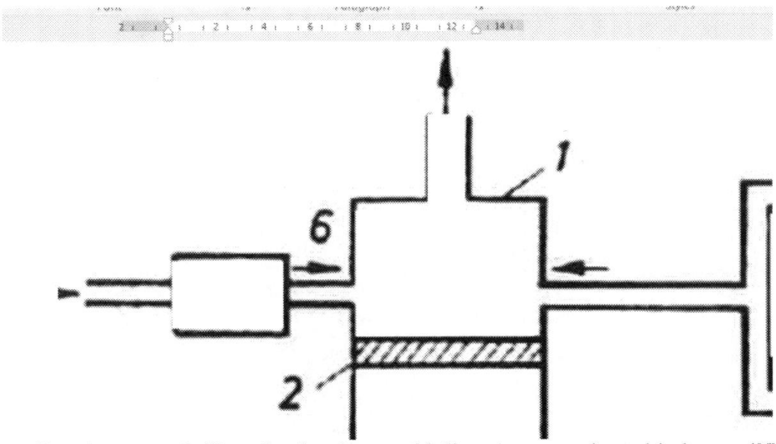

Fig.3. NO_X analiser 1 reactor; 2-filter; 3-tube photomultiplier; 4- connection with the amplification system; 5 oxygen; 6-ozone generator; 7-NO test; 8 convertorNO_2 / NO.

4 Experimental measurement

To underline the development and implementation of the SEEMP, the situation of a tanker with a capacity of 50000 dwt will be considered (figure 4) [7].

Fig. 4. Reference ship [7]

Ship characteristics:

- L_{pp} =183 m
- B = 32,2 m
- V_{max} = 15,7 Knots
- ME - MAN B&W 6S50MC-C, P = 8600 kW, 127 rpm

CO, CO_2, HC, NO_X and SO_X emissions for the main engine and a diesel generator were monitored. The results are presented in Table 1 and Table 2.

Table 1. NOx emissions of the main engine depending on the main engine load

Load [%]	NO_x [ppm]	CO [ppm]	SO_x [ppm]	CO_2 [%]	C [mg/m^3]
100	939	60	63	5	61
80	722	41	68	4	56
60	471	21	66	4	51
40	449	19	40	3	50
20	571	18	19	1	58

Table 2. Generator diesel emissions related to main engine load

Load	NO_x [ppm]	CO [ppm]	SO_x [ppm]	CO_2 [%]	C [mg/m^3]
100	627	178	53	5	32
80	571	188	50	4	33
60	638	173	52	5	32
40	620	177	50	4	32
20	624	168	51	4%	32

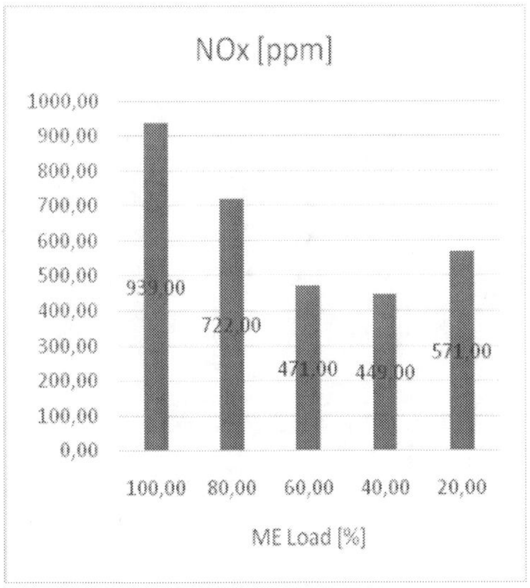

Fig. 5. NO_x emissions – main engine [ppm]

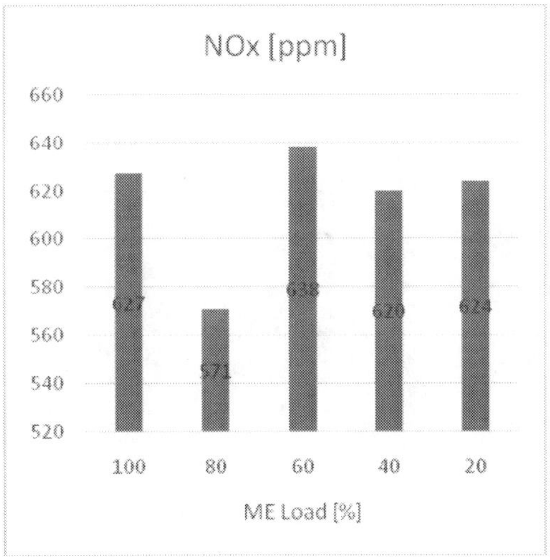

Fig. 6. NO_x emissions – auxiliary engine [ppm]

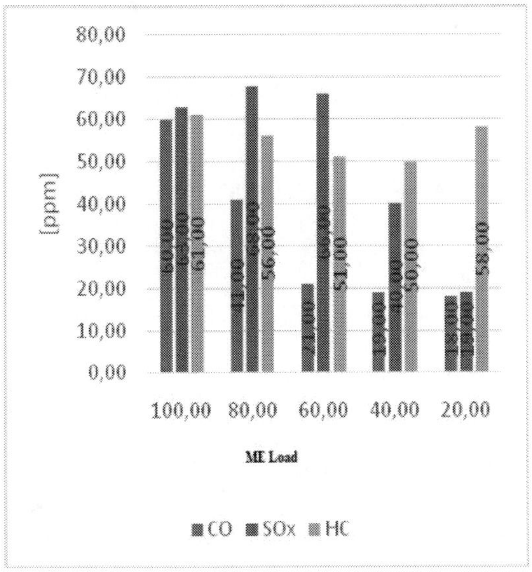

Figure 7. CO, SOx and HC emissions - main engine [ppm]

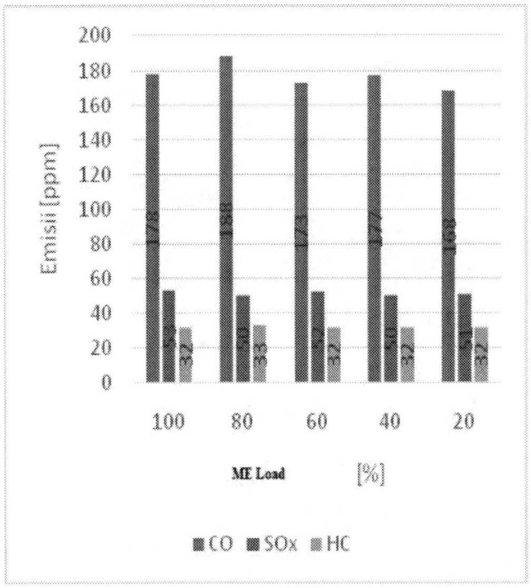

Figure 8. CO, SOx and HC emissions - auxiliary engine [ppm]

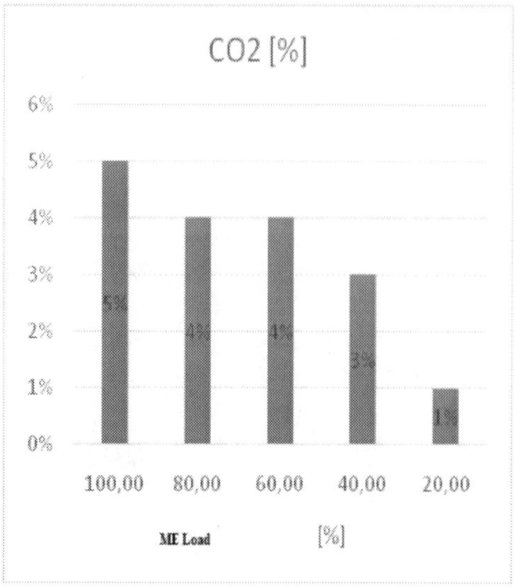

Fig. 9. CO_2 emissions – main engine [%]

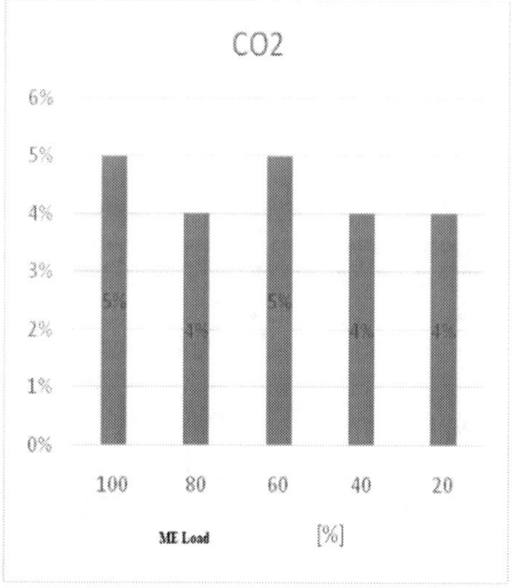

Fig. 10. CO_2 emissions – auxiliary engine [%]

Analyzing Figures 5 - 10, it is found that, with the reduction of the main engine load, NOx emissions have a decreasing tendency, up to a load of 20 %, when an increase of 22 ppm is observed compared to the value recorded for the load of 40 %.

The same emission category has a relatively constant variation for the auxiliary engine. As regards CO, SO_X and HC emissions, there is a clear tendency to decrease these categories of pollutants, for the main engine, except for the emissions of unburnt hydrocarbons, which, due to the low coefficient of excess air at low load, show a marked increase.

5 Conclusions

For the auxiliary engine, due to the loads induced by the various consumers on board, the emission variation does not follow a trend associated with the variation of the main engine load. However, for short intervals, in the transient range around 40 % of the nominal load, at the moment when the turbocompressor motors serving the main engine at low loads come into operation, there is a temporary increase in the level emissions because of diesel generator running. CO_2 emissions are in the same trend as those shown above, with a clear tendency to lower emissions for the main engine, while for the auxiliary engine, emissions vary according to a law independent of the load of the main engine. It is found that by implementing such measures to increase energy efficiency, a reduction in emissions of pollutants and, consequently, a reduction in the consumption of consumed fuel is achieved.

In this way, it can be stated that the operational approach to increase the overall efficiency of the energy system represented by the ship is a viable solution.

„This work is supported by the project ANTREPRENORDOC, in the framework of Human Resources Development Operational Programme 2014-2020, financed from the European Social Fund under the contract number 36355/23.05.2019 HRD OP /380/6/13 – SMIS Code: 123847."

6.References

1. European Communities, *Council Directive of 7 March 1985 on Air Quality standard for Nitrogen Dioxid (85/203/EEC)*, Official Journal of the European Communities, No. L87/1, Luxemburg, (1985)
2. Arcadis Belgium nv/sa, *Directive European Commission, Contract N° 070307/2010/577902/ETU/F1 EC DG Environment Project number 11601*, Final report, (2012)
3. D. Bodolan, *Metode de control şi reducere a emisiilor poluante ale sistemelor energetic navale*, Universitatea Maritimă Constanţa, (2003)
4. http://www.marineinsight.com/news/nichio-maru-the-new-green-car-carrier-from-nissan
5. P. Arsenie, G. Martinas, C. Gheorghe, A. Arsenie A ,*Technologies for the Reduction of Nitrogen Oxides Emissions*, TRANSNAV, Vol 9 Nr 2, p251-256 (2015)
6. ABS *Ship Energy Efficiency Measures Advisory, TX 05/13 5000 13015*
7. Manual de utilizare al simulatorului Transas ERS TechSim, model MAN B&W 6S50MC-C

The amplitude influence on the thermal and hydraulic performances for a wavy air fin in a compact heat exchanger used in agriculture applications

Septimiu Albețel[1,2,*], *Alexandru* Rus[1,2], *Eveline* David[1], and *Vlad* Marțian[2]

[1]Politehnica Timişoara, Mechanical Faculty, 1 M. Viteazu Streeet, Timişoara, Romania
[2]RAAL S.A., Technical Department, 7 Industriei street, Bisțrita, Romania

Abstract. Wavy air fins are used in the construction of aluminum compact heat exchangers which are used in thermal equipment for agriculture, construction, and industrial applications. Since atmospheric air convective heat transfer coefficient is among the lowest of all fluids, there is a special interest in the optimizing of the geometry for this type of heat exchangers. One of the biggest challenges in designing cooling solutions for agriculture applications is the clogging effect on the performances of the heat exchangers. Clogging leads to a critical performance reduction of the compact heat exchanger used for these types of applications. This specific study, which was conducted in RAAL S.A company in collaboration with the University "Politehnica" of Timisoara", checks the influence of the wavy air fin amplitude on the thermal and hydraulic performances and tries to find an optimum solution suitable for these specific projects. So, starting from the standard existing amplitude, different amplitude values were studied analytically. The analytical studies were further confirmed by real life measurements. The scope of the study was to find an optimal value for the amplitude that can further improve the overall performance of the aluminum compact heat exchangers and reduce the clogging effect.

1 Introduction

Wavy air fins such as the one from Fig. 1, are the most commonly means of extending the heat transfer surfaces in aluminum compact heat exchangers cooled by air. The wavy fin has also the role of intensifying the heat rejection by breaking the thermal boundary

* Corresponding author septimiu.albetel@raal.ro

layer inducing turbulence and a better mixing of the air through the heat exchanger. Due to its resistance to damage and its lower degree of clogging than other types of extended surfaces, such as louvered fins, this type of fin is used almost exclusively in the construction of compact heat exchangers which are used in thermal equipment for agriculture, construction, and industrial applications.

Fig. 1. Wavy fin isometric view

The above stated facts taken in conjunction with the fact that the atmospheric air convective heat transfer coefficient is among the lowest of all fluids, the wavy fin was studied since 1950 when Kays and London [1] have presented thermal and hydraulic performance charts for 3 variations of this wavy fin, without presenting analytical equations. Later, Beng and MPhil [2] have presented experimental and analytical data for 3 types similar with those of Kays and London. There have been papers presenting numerical investigation of the lateral vortexes in the low Reynolds Number domain [$10 \leq Re \leq 1000$] such as Zhang et. al. [3], and Metwally et. al. [4]. A variation of the wavy fin was numerically studied by Wang and Chen [5] as a convergent divergent way channel for $Re \leq 700$ and reached the conclusion that the greatest influence in heat transfer increase is due to the first wave the rest of the waves having only a marginal influence. Gadeck [6] studied corugated channels on a much larger Reynolds numbers ($0 \leq Re \leq 7500$) containing all the flow regimes from laminar to turbulent, he then comparred the results with the flow between two plates and reach the conclusion that the corrugated chanels transfers the heat 2 times better. In his paper Ismail [7] numerically studied the wavy and offset fins and proposed criterial relations for f, and j. Dong in his papers [8]–[10] studied the effect of pitch, height and thickness of the wavy fin experimentally and numerically. In recent years there are several papers that have sumarizes and generalizes the criterial relations obtained on the wavy fin from these the one of the Aliabadi et. al. [11] which compares different relations using different fluids, Gum [12] studied a variation of wavy fin by creating cross cuts. Xiao [13] studied the performances of the wavy fin in evaporative cooling conditions, and the one of the Qasem et. al. [14] which generalizes the criterial relations of the performances for the wavy fin found in the literature.

The wavy fin presented in Fig. 1 is different from the rest of the wavy fins from the presented literature by having a straight section at the inlet and outlet of the passage. This straight section was introduced to reduce the pressure drop of this kind of fin. Studies on this specific type of fin can be found in [15], [16] which study the height and the straight section influence on the wavy fin performances.

Even if this kind of fin has a lower clogging factor this is still one of the biggest challenges in designing cooling solutions for agriculture applications. Clogging leads to a critical performance reduction of the compact heat exchanger used for these types of applications.

This paper presents a numerical study of the influence of the wavy air fin amplitude on the thermal and hydraulic performances of a coolant to air compact heat exchanger. The numerical data is validated using experimental data from a production cooler.

2 Experimental Setup

The thermal and hydraulic test bench used for this study can be seen in Fig. 2. It can make acquisitions for different types of fluids (air, water, and oil) in different working conditions.

Fig. 2. Thermal and hydraulic performance test bench at RAAL S. A.

The main components of this test bench, see Fig. 3, are the wind tunnel, the fluid loop, and the tested sample.

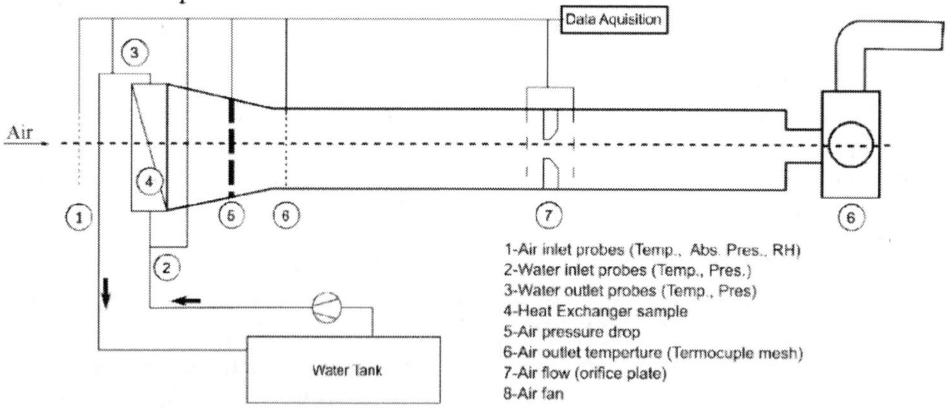

Fig. 3. Test bench functional sketch

The cooling air data is obtained using K-thermocouples, barometric pressure and relative humidity sensors each having an accuracy of 0.2 % of the entire measurement domain and thermocouples for relative humidity [15]. The exit temperature is measured using a mesh of 10 K–thermocouples to evaluate the non-uniformity of the air temperature. All the used thermocouples are pre-calibrated with an accuracy of 0.1 [°C]. The air flow is measured by means of pressure differential device, i.e. orifice plate, in conformity with EN ISO 5167-2 [17].

Fig. 4. Wavy fin dimensions.

For the validation of the numerical study, a test was performed on a sample water cooler that was built using a 9 mm fin height, 8 mm fin pitch and a 1.7 mm fin amplitude, as in Fig. 4. The measurements were made using as hot fluid 100% pure water at an inlet temperature of 88 °C with a maximum deviation of ±2 [°C].For the cooling air flow the measured inlet temperature was 20 [°C] with a maximum deviation of ±2 [°C]. The cooling air frontal area speed ranged from 7 [m/s] up to 13 [m/s].

In order to make the experimental data independent of the test conditions we used the well-known ϵ-NTU methodology [18] to express the performances of the heat exchanger in Nusselt number versus Reynolds Number, see Fig. 5, also the hydraulic performances are express a friction coefficient as defined by Kays and London [1]

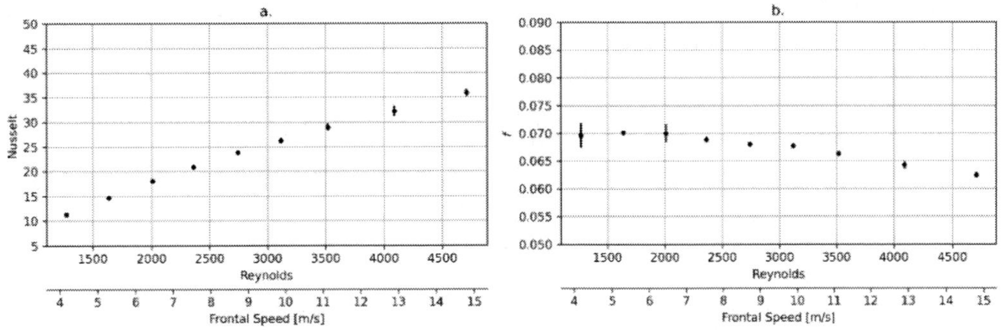

Fig. 5. Experimental dimensionless data. a. Nusselt vs Reynolds, b. friction vs Reynolds

3 Numerical simulation

The numerical simulations were done using the SolidWorks Flow Simulation package [19] which has a Cartesian grid system and a proprietary k-ϵ turbulence model implemented. Firstly, the wavy fin was simplified to reduce the computational time. For the simulation was used only a single fin as in the Fig 5 where we used the periodic boundary condition on the z axis. The water temperature was simulated by imposing a constant temperature on the top and bottom walls. The simulation domain was extended at the inlet and outlet of the fin to accommodate with the simulation software's constrains. At the inlet boundary a constant speed and temperature is applied, and at the exit boundary a constant pressure is applied.

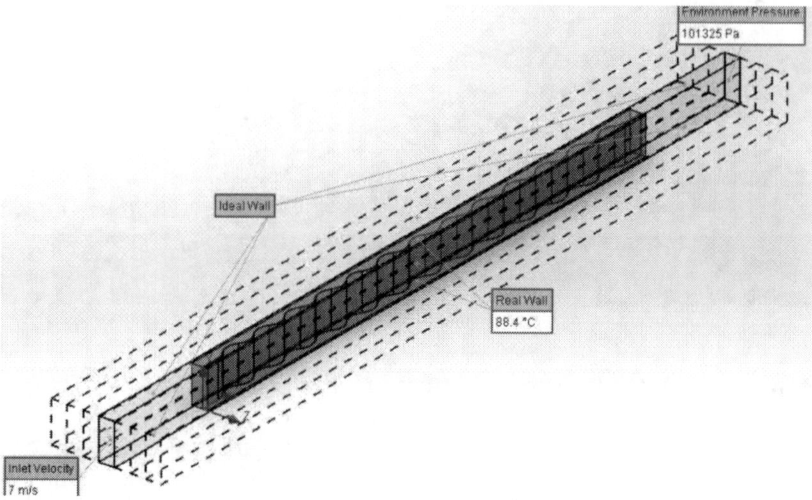

Fig. 6 Simulation domain and boundary conditions

Validation of the numerical model was done in two steps:

Firstly, we had to achieve a mesh for which the results are independent of the discretization. For this we ran the simulations with 4 different meshes discretization.

To assess the mesh convergence, we used 3 physical quantities (pressure difference, temperature difference, and relative difference of the mass flow) plotted against the cell count in Fig 6. As can be seen from the Fig. 7, the mesh with approximately 2×10^6 cells is a good mesh because by doubling the number of the cells for the next mesh we see only a 4% difference in the results, also for this mesh the mass flow error is around 1%

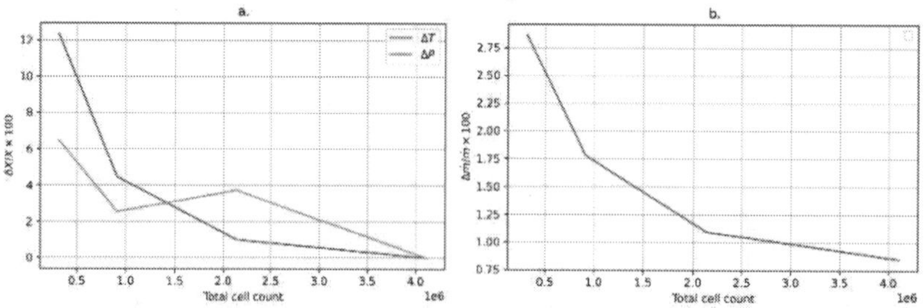

Fig. 7. Mesh independence

To compare the simulation performances, we must transform the data obtained from simulation into dimensionless quantities. As we considered the wall temperature to be constant, we can use logarithmic temperature difference to calculate the Nusselt using the equations below:

$$\Delta T_{mln} = \frac{T_{out} - T_{in}}{\ln\left(\frac{T_w - T_{in}}{T_w - T_{out}}\right)} \#(1)$$

where T_{in} and T_{out} are the inlet and outlet temperatures, T_w is the wall temperature.

The heat rejection can be calculated as:

$$\dot{Q} = \dot{m}\, c_p \Delta T = kA \cdot \Delta T_{mln} \Rightarrow kA = \frac{\dot{Q}}{\Delta T_{mln}} \#(2)$$

From equation (2) we can find the overall heat transfer coefficient kA which can be written in the form of a thermal resistance as follows:

 https://doi.org/10.1051/e3sconf/202018001006

$$\frac{1}{kA} = R_{conv} + R_{cond}$$

$$R_{conv} = \frac{1}{\alpha \cdot \eta \cdot A} \#(3)$$

$$R_{cond} = \frac{\delta}{\lambda_{Al} \cdot A_w}$$

Because we have imposed a constant temperature $T_w = 88.4$ °C we can consider that δ=0 which in equation (3) results that $R_{cond} \rightarrow 0$ and we arrive at the final form in the equation (4) below:

$$\frac{1}{kA} = R_{conv} = \frac{1}{\alpha \cdot \eta \cdot A} \#(4)$$

Considering the fin efficiency (η) as 1 we can calculate Nusselt number with the equation (5) as follows:

$$Nu = \frac{\alpha \cdot D_h}{\lambda_{air}} = \frac{\dot{Q}}{A \cdot \Delta T_{mln}} \cdot \frac{D_h}{\lambda_{air}} \#(5)$$

Calculation of the friction coefficient is straight forward and can be calculated form the equation (6) [1] directly:

$$f = \frac{A_c}{A} \cdot \frac{2\Delta p}{\rho \cdot v^2} \#(6)$$

With these quantities defined we can make a comparison of the simulation with the experimental results. As we can see form the Fig. 8 the difference in thermal performance (Nusselt) is around ± 11 % which it is considered an acceptable deviation. For the friction coefficient the deviation is greater reaching -20%, but this can be accounted if we consider that the simulations were done without considering the roughness of the material, and also because in reality the fins are not perfect which may induce higher pressure drop.

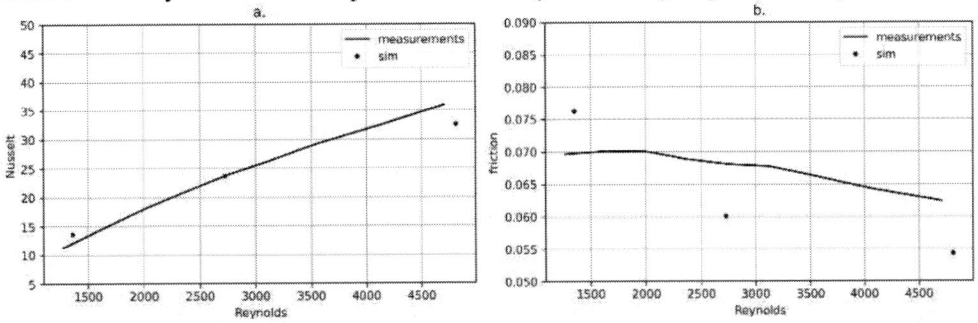

Fig. 8. Comparison of the simulation results with experimental results. a. Nusselt comparison, b. friction comparison

4 Results and discussions

For the current analysis 3 amplitude values (A) were chosen with the results presented in the form of Colburn number (j) and friction coefficient (f) in Fig. 9.

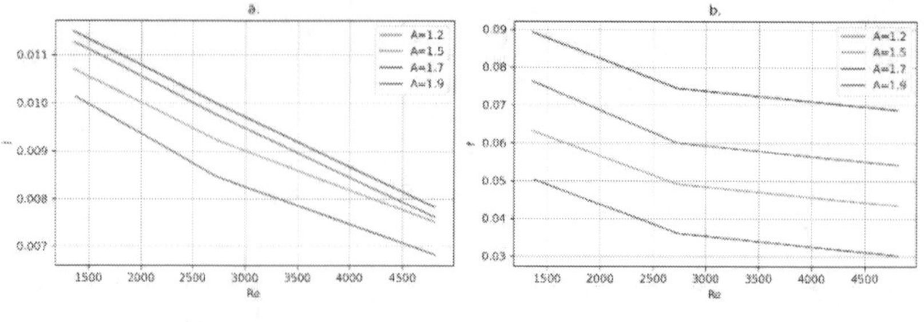

Fig. 9 Colburn and friction coefficient

The Colburn number rises with the amplitude, but we can see that for values of A=1.5, and A=1.9 the difference is small. As a first conclusion that can be drawn is that increasing the amplitude gives us diminishing returns. If we analyze the Fig. 9b, we can see that the friction coefficient has a similar dependency of the amplitude, it rises with the amplitude. To assess the overall performance of the wavy fin we must use another factor which is a combination of the two factors from the Fig. 9. In [20] it is defined a factor called JF which takes into consideration the ratio of the Colburn number and friction coefficient based on a reference performance value, as in equation (7)

$$JF = \frac{j/f^{1/3}}{j_r/f_r^{1/3}} \#(7)$$

In Fig. 10 we can see that as the amplitude of the wavy fin increases the JF factor drops. Since we had experimental values for the A=1.7 wavy fin we had choose this wavy fin as a reference.

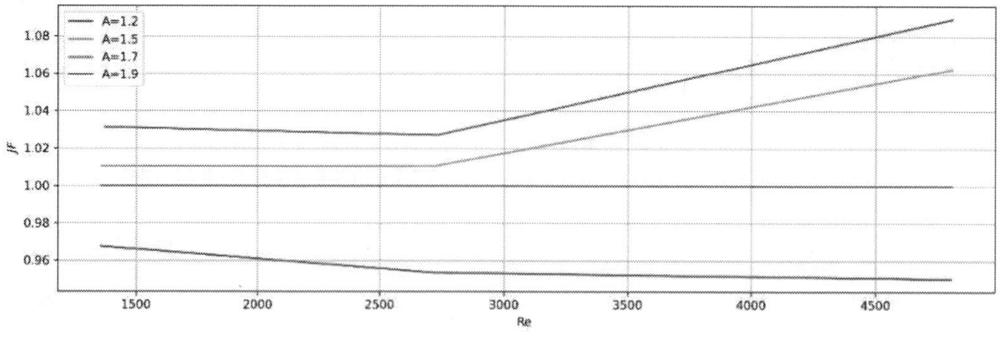

Fig. 10. JF factor comparison

The JF factor is a factor of type the "higher the better" that means that a higher value results in better performances. So, comparing the performances at different amplitudes A and at the same Reynolds number or speed – since the flow section doesn't change – it results that for the tested amplitudes the lower the amplitude the greater overall performances for the fin.

Considering that a heat exchanger in the real applications uses a fan to drive the cooling air through it, a higher value of JF at the same Reynolds number means that there is a reserve in pressure drop and in reality the heat exchanger will work with more air and also the heat rejection will be higher.

Having lower amplitude is good also from the clogging point of view because the passage is easier to traverse by the solid particles in the air. This translates in a longer service of the heat exchanger before cleaning maintenance is needed.

As a future work there arises the need to investigate other amplitudes to see at which amplitude we find an optimum.

References

1. W. Kays and A. London, *Compact heat exchangers*, Third ed. New York: McGraw-Hill, (1984)
2. R. J. G. BEng and T. T. Al-Shemmeri, *Proc. Inst. Mech. Eng. Part A J. Power Energy*, **208**, 225–229, (1994)
3. J. Zhang, J. Kundu, and R. M. Manglik, Int. J. Heat Mass Transf., **47**, 1719–1730, (2004)
4. H. M. Metwally and R. M. Manglik, Int. J. Heat Mass Transf., **47**, 2283–2292, (2004)
5. [5] C. C. Wang and C. K. Chen, Numer. Heat Transf. Part A Appl., **48**, 879–900, (2005)
6. M. Gradeck, B. Hoareau, and M. Lebouché, Int. J. Heat Mass Transf., **48**, 1909–1915, (2005)
7. L. S. Ismail and R. Velraj, Numer. Heat Transf. Part A Appl., **56**, pp. 987–1005, (2009)
8. J. Dong, J. Chen, W. Zhang, and J. Hu, Appl. Therm. Eng., **30**, 1377–1386, (2010)
9. J. Dong, C. Jiangping, C. Zhijiu, Z. Yimin, and Z. Wenfeng, Appl. Therm. Eng., **27**, 2066–2073, (2007)
10. J. Dong, L. Su, Q. Chen, and W. Xu, Appl. Therm. Eng., **51**, 32–39, (2013)
11. M. K. Aliabadi, F. Hormozi, and E. H. Rad, Int. J. Numer. Methods Heat Fluid Flow, **24**, 1086–1108, (2014)
12. G. W. Kim, H. M. Lim, and G. H. Rhee, Int. J. Heat Mass Transf., **96**, 110–117, (2016)
13. L. Xiao, T. Wu, S. Feng, X. Du, and L. Yang, Int. J. Heat Mass Transf., **110**, 383–392, (2017)
14. N. A. A. Qasem and S. M. Zubair, Int. J. Refrig., **97**, 21–30, (2019)
15. V. Marţian, M. Nagi, S. Albeţel, and M. Sucilă, *13th Uk Heat transfer Conference*, UKHTC2013/26, (2013)
16. S. Albeţel, M. Vlad, and M. Nagi, Appl. Mech. Mater., **659**, 283–288, (2014)
17. A. Association Française de Normalisation, "NF EN ISO 5167-2." p. 56, (2003)
18. F. P. Incropera, D. P. DeWitt, T. L. Bergman, and A. S. Lavine, *Fundamentals of Heat and Mass Transfer*, 6th ed. Wiley & sons, (2006).
19. SolidWorks Corp., *Technical reference SOLIDWORKS FLOW SIMULATION 2018*, (2018)
20. J. Yun and K. Lee, Int. J. Heat Mass Transf., **43**, 2529–2539, (2000)

E3S Web of Conferences **180**, 01007 (2020) https://doi.org/10.1051/e3sconf/202018001007
TE-RE-RD 2020

Pipelines Designing By Using Hydraulic Criteria Equations

Gencho Popov[1,*], *Kliment Klimenntov*[1] *and Boris Kostov*[1]

[1]University of Ruse, Department "Heat, Hydraulics and Environmental Engineering", Ruse, Bulgaria

Abstract. This work intend an original methodology, newly established by the authors, to design typical pipe systems by using hydraulic criteria equations, excluding the determination of the relevant Darcy factor. Therefore, it can be eliminated the necessity for searching iterative solutions of such problems. The established equations are easy to use, in terms of mathematical processing, which guarantees their proper application in solving engineering problems. It is considered for the proposed physical relations (equations) concerning the determination of the pipe system resistance characteristics to be highly useful in engineering practice, as for example: in determining a hydraulic machine work regime, used in a system transporting fluids, as well as in performing a synthesis and analysis of such systems or in accomplishing an energy analysis of pipe systems. A comparative analysis between the results found after applying the classical methods of calculations and by using a licensed specialized software product indicates that a highly sufficient (for engineering calculations) overlapping is achieved.

1. Introduction

Performing hydraulic calculations of a pipe system is one of the main tasks in designing a system, used to transport fluids. It is well-known that a major difficulty is to determine the resistance coefficients and more specifically the relevant Darcy friction factors - λ_i . In [1] and [2], the authors present a method for avoiding difficulties in determining this coefficient.

It is also known [3, 4] that the basic tasks concerning the hydraulic estimation (analysis) of a pipe system are to determine the required pressure (pressure drop between the pipe inlet and outlet) p for ensuring a given flow rate Q, or to determine the flow rate Q in case that the pressure p is known.

The method for solving the above problems concerning simple pipes (no branches) with constant diameters d, without determining the friction factor, is

* Corresponding author: gspopov@uni-ruse.bg

presented in [1, 2]. By using the Dimensional Analysis (DA) it is established different criteria equations including three dimensionless complexes (π - criterions), each containing only one of the analyzed parameters – pressure p, flow rate Q or diameter d. These equations, obtained in [2], respectively are:

$$\pi_p = p\,\frac{\rho \Delta^3}{L \mu^2}, \tag{1}$$

$$\pi_Q = Q\,\frac{\rho}{\mu\,\Delta}, \tag{2}$$

$$\pi_d = \frac{\Delta}{d}, \tag{3}$$

where p is the pressure difference between the pipe inlet (start) and outlet (end); Q - the ensured flow rate; L - the referred pipe length; d-the pipe diameter; Δ - the pipe walls roughness; ρ , μ - the fluid density and dynamic viscosity.

For obtaining the following hydraulic criteria equations a methodology, proposed in [2], is used:

$$\pi_p = A \pi_Q^a \pi_d^b \tag{4}$$

$$\pi_Q = B \frac{\pi_p^e}{\pi_d^f}, \tag{5}$$

where the values of the coefficients A and B, as well as the exponents (a, b, e, f), are found by using the pre-cited method.

Using the above equations enables determining the main hydraulic parameters for each "i" pipe. The flow rate Q_i can be estimated by using the flow rate criterion π_Q (4) for the current pipe:

$$Q_i = \pi_{Q,i} \Delta_i \frac{\mu}{\rho}. \tag{6}$$

Analogically, the pressure p_i for each pipe is determined by using the pressure criterion (3):

$$p_i = \pi_{p,i} \frac{L_i}{\Delta_i^3} \frac{\mu^2}{\rho}. \tag{7}$$

A significant detail in accomplishing an analysis of a pipe system, used to transport fluids, is the obtaining of its resistance characteristic, which is often presented in the following way:

$$p = k Q^{\alpha}, \tag{8}$$

where k is the coefficient of the pipe system resistance characteristics, m – is an exponent, which in case of completely rough pipes is equal to two ($\alpha = 2$).

The main purpose of this work is to establish an easily applicable in engineering calculations methodology for accomplishing hydraulic pipe sizing and analysis of typical pipe systems – pipes connected in parallel, in series or combined, by using the criteria equations (4) and (5).

2. Main types of pipe systems transporting fluids

This paragraph represents the application of equations (1) and (2) in performing hydraulic analysis of several types of typical complex pipe systems.

2.1 Pipe systems containing pipes connected in parallel

In case of pipes connected in parallel the following physical relations are valid:

- summing the flow rates ensured in each pipe should be equal to the system total flow rate Q:

$$Q_1 + Q_2 + \ldots + Q_n = Q \; ; \tag{9}$$

- the pressure loss in each of the parallel connected pipes is equal the system total pressure loss:

$$p_1 = p_2 = \ldots = p_n = p \, . \tag{10}$$

2.1.1 Different type of pipes connected in parallel

In general, a pipe system contains different types of pipes. In this regard, the absolute equivalent wall roughness Δ_i for each pipe has a different value.

By using equation (6) for the ensured flow rate in each pipe it is obtained that:

$$\pi_{Q,1}\Delta_1 \frac{\mu}{\rho} + \pi_{Q,2}\Delta_2 \frac{\mu}{\rho} + \ldots + \pi_{Q,n}\Delta_n \frac{\mu}{\rho} = \pi_Q \Delta \frac{\mu}{\rho} \; , \tag{11}$$

where $\pi_{Q,i}$ and Δ_i $(i = 1 \ldots n)$ respectively are the flow rate criterion and wall roughness for each pipe, and π_Q and Δ - represent the same parameters but concerning the common pipe (for example, it may be the pipe just before the splitting or after the collecting point of the analyzed parallel connected pipes).

Indicating that the fluid transported through each pipe has the same properties – $\mu / \rho = const.$, it may be concluded that:

$$\pi_{Q,1}\Delta_1 + \pi_{Q,2}\Delta_2 + \ldots + \pi_{Q,n}\Delta_n = \pi_Q \Delta \, . \tag{12}$$

Equation (7), enabling the estimation of the pressure loss in each of the parallel connected pipes by taking into the account the relevant criterion - $\pi_{p,i}$, is involved in (10). Indicating that $\mu^2 / \rho = const.$ it may be stated:

$$\pi_{p,1} \frac{L_1}{\Delta_1^3} = \pi_{p,2} \frac{L_2}{\Delta_2^3} = \ldots = \pi_{p,n} \frac{L_n}{\Delta_n^3} \; . \tag{13}$$

The criteria equation (1), applied for each pipe, is used to determine the relevant criterions - $\pi_{p,i}$.

For presenting the parameters $\pi_{p,i}$ it is used the criteria equation (4). Indicating that the constant A, given in (4), is the same for each pipe, equation (11) may be presented as it follows:

$$\pi_{Q,1}^{a}\,\pi_{d,1}^{b}\,\frac{L_1}{\Delta_1^3}=\pi_{Q,2}^{a}\,\pi_{d,2}^{b}\,\frac{L_2}{\Delta_2^3}=\ldots=\pi_{Q,n}^{a}\,\pi_{d,n}^{b}\,\frac{L_n}{\Delta_n^3}\,. \tag{14}$$

Using the equations (12) and (14) enables the establishing of the following system of equations:

$$\left|\begin{array}{l}\pi_{Q,2}^{a}\,\pi_{d,2}^{b}\,\dfrac{L_2}{\Delta_2^3}=\pi_{Q,1}^{a}\,\pi_{d,1}^{b}\,\dfrac{L_1}{\Delta_1^3}\\ \pi_{Q,3}^{a}\,\pi_{d,3}^{b}\,\dfrac{L_3}{\Delta_3^3}=\pi_{Q,1}^{a}\,\pi_{d,1}^{b}\,\dfrac{L_1}{\Delta_1^3}\\ \quad\ldots\\ \pi_{Q,n}^{a}\,\pi_{d,n}^{b}\,\dfrac{L_n}{\Delta_n^3}=\pi_{Q,1}^{a}\,\pi_{d,1}^{b}\,\dfrac{L_1}{\Delta_1^3}\\ \pi_{Q,1}\Delta_1+\pi_{Q,2}\Delta_2+\ldots+\pi_{Q,n}\Delta_n=\pi_Q\Delta\end{array}\right.\,. \tag{15}$$

Indicating that a pipe relative roughness criterion is $\pi_d=\dfrac{\Delta}{d}$, it can be concluded:

$$\pi_{d,i}^{b}\,\frac{L_i}{\Delta_i^3}=\Delta_i^{b-3}\left(\frac{L_i}{d_i^b}\right), \tag{16}$$

and respectively the system of equations (15) has to be transformed in the following way:

$$\left|\begin{array}{l} \pi_{Q,2}^{a}\,\Delta_2^{b-3}\dfrac{L_2}{d_2^b}=\pi_{Q,1}^{a}\,\Delta_1^{b-3}\dfrac{L_1}{d_1^b}\\ \pi_{Q,3}^{a}\,\Delta_3^{b-3}\dfrac{L_3}{d_3^b}=\pi_{Q,1}^{a}\,\Delta_1^{b-3}\dfrac{L_1}{d_1^b}\\ \quad\dots\\ \pi_{Q,n}^{a}\,\Delta_n^{b-3}\dfrac{L_n}{d_n^b}=\pi_{Q,1}^{a}\,\Delta_1^{b-3}\dfrac{L_1}{d_1^b}\\ \pi_{Q,1}\Delta_1+\pi_{Q,2}\Delta_2+\ldots+\pi_{Q,n}\Delta_n=\pi_Q\Delta \end{array}\right. . \tag{17}$$

Then, this system of equations may be presented as it follows:

$$\left|\begin{array}{l} \left(\pi_{Q,2}\Delta_2\right)^a\Delta_2^{b-a-3}\dfrac{L_2}{d_2^b}=\left(\pi_{Q,1}\Delta_1\right)^a\Delta_1^{b-a-3}\dfrac{L_1}{d_1^b}\\ \left(\pi_{Q,3}\Delta_3\right)^a\Delta_3^{b-a-3}\dfrac{L_3}{d_3^b}=\left(\pi_{Q,1}\Delta_1\right)^a\Delta_1^{b-a-3}\dfrac{L_1}{d_1^b}\\ \quad\dots\\ \left(\pi_{Q,n}\Delta_n\right)^a\Delta_n^{b-a-3}\dfrac{L_n}{d_n^b}=\left(\pi_{Q,1}\Delta_1\right)^a\Delta_1^{b-a-3}\dfrac{L_1}{d_1^b}\\ \pi_{Q,1}\Delta_1+\pi_{Q,2}\Delta_2+\ldots+\pi_{Q,n}\Delta_n=\pi_Q\Delta \end{array}\right. . \tag{18}$$

After being mathematically transformed the above equation may be presented in the following way:

$$\left|\begin{array}{l} \pi_{Q,2}\Delta_2=\left(\pi_{Q,1}\Delta_1\right)\left(\dfrac{L_1}{L_2}\right)^{\frac{1}{a}}\left(\dfrac{d_2}{d_1}\right)^{\frac{b}{a}}\left(\dfrac{\Delta_1}{\Delta_2}\right)^{\frac{b-a-3}{a}}\\ \pi_{Q,3}\Delta_3=\left(\pi_{Q,1}\Delta_1\right)\left(\dfrac{L_1}{L_3}\right)^{\frac{1}{a}}\left(\dfrac{d_3}{d_1}\right)^{\frac{b}{a}}\left(\dfrac{\Delta_1}{\Delta_3}\right)^{\frac{b-a-3}{a}}\\ \quad\dots\\ \pi_{Q,n}\Delta_n=\left(\pi_{Q,1}\Delta_1\right)\left(\dfrac{L_1}{L_n}\right)^{\frac{1}{a}}\left(\dfrac{d_n}{d_1}\right)^{\frac{b}{a}}\left(\dfrac{\Delta_1}{\Delta_n}\right)^{\frac{b-a-3}{a}}\\ \pi_{Q,1}\Delta_1+\pi_{Q,2}\Delta_2+\ldots+\pi_{Q,n}\Delta_n=\pi_Q\Delta \end{array}\right. . \tag{19}$$

By taking into account equation (6), $\pi_{Q,i}\Delta_i = Q_i \frac{\rho}{\mu}$ and the system of equations (17) may be presented as it follows:

$$\left| \begin{array}{l} Q_2 = Q_1 \left(\frac{L_1}{L_2}\right)^{\frac{1}{a}} \left(\frac{d_2}{d_1}\right)^{\frac{b}{a}} \left(\frac{\Delta_1}{\Delta_2}\right)^{\frac{b-a-3}{a}} \\ Q_3 = Q_1 \left(\frac{L_1}{L_3}\right)^{\frac{1}{a}} \left(\frac{d_3}{d_1}\right)^{\frac{b}{a}} \left(\frac{\Delta_1}{\Delta_3}\right)^{\frac{b-a-3}{a}} \\ \ldots \\ Q_n = Q_1 \left(\frac{L_1}{L_n}\right)^{\frac{1}{a}} \left(\frac{d_n}{d_1}\right)^{\frac{b}{a}} \left(\frac{\Delta_1}{\Delta_n}\right)^{\frac{b-a-3}{a}} \\ Q_1 + Q_2 + \ldots + Q_n = Q \end{array} \right. . \tag{20}$$

For estimating the flow rate Q_1 in the first pipe the following equation is used:

$$Q_1 = \frac{1}{1 + \left(\frac{L_1}{L_2}\right)^{\frac{1}{a}} \left(\frac{d_2}{d_1}\right)^{\frac{b}{a}} \left(\frac{\Delta_1}{\Delta_2}\right)^{\frac{b-a-3}{a}} + \ldots + \left(\frac{L_1}{L_n}\right)^{\frac{1}{a}} \left(\frac{d_n}{d_1}\right)^{\frac{b}{a}} \left(\frac{\Delta_1}{\Delta_n}\right)^{\frac{b-a-3}{a}}} Q \cdot \tag{21}$$

Using the first "n" number of equations in (20) enables the estimation of the other flow rates Q_i .

For example, the flow rate Q_k ensured in the pipe „k" (k = 1…n) is estimated by the following way:

$$Q_k = \frac{1}{\sum_{i=1}^{n} \left(\frac{L_k}{L_i}\right)^{\frac{1}{a}} \left(\frac{d_i}{d_k}\right)^{\frac{b}{a}} \left(\frac{\Delta_k}{\Delta_i}\right)^{\frac{b-a-3}{a}}} Q . \tag{22}$$

The pressure drop between the inlet and outlet (collector points) of the parallel connected pipes is determined by using the system total flow rate Q and the applied method of calculation is described below. First, for analyzing the "kth" pipe the relevant criteria equation $\pi_{pk} = A\pi_{Qk}^{a}\pi_{dk}^{b}$ is applied in the following way:

$$p\frac{\rho \Delta_k^3}{L_k \mu^2} = A\left(Q_k \frac{\rho}{\mu \Delta_k}\right)^a \left(\frac{\Delta_k}{d_k}\right)^b . \tag{23}$$

Then for the pressure drop it may be stated:

$$p=\left(A\,\rho^{a-1}\mu^{2-a}\right)\frac{L_k}{d_k^b}\Delta_k^{b-a-3}\,Q_k^a\,. \tag{24}$$

Taking into account equation (22), concerning the estimation of the flow rate in the "k[th]" pipe, enables the estimation of the pressure drop:

$$p=\left(A\,\rho^{a-1}\mu^{2-a}\right)\frac{\dfrac{L_k}{d_k^b}\Delta_k^{b-a-3}}{\left[\sum_{i=1}^{n}\left(\dfrac{L_k}{L_i}\right)^{\frac{1}{a}}\left(\dfrac{d_i}{d_k}\right)^{\frac{b}{a}}\left(\dfrac{\Delta_k}{\Delta_i}\right)^{\frac{b-a-3}{a}}\right]^a}Q^a\,. \tag{25}$$

Transforming mathematically the above equation enables the establishment of the final equation for estimating the pressure drop in case of parallel connected pipes:

$$p=\left(A\,\rho^{a-1}\mu^{2-a}\right)\frac{1}{\left[\sum_{i=1}^{n}\left(\dfrac{d_i^b}{L_i}\right)^{\frac{1}{a}}\left(\dfrac{1}{\Delta_i}\right)^{\frac{b-a-3}{a}}\right]^a}Q^a\,. \tag{26}$$

Finally, taking into account equation (26) enabling the determination of the resistance coefficient of a pipe system, containing "n" number of parallel connected pipes (the exponent is: $\alpha = a$):

$$k=\left(A\,\rho^{a-1}\mu^{2-a}\right)\frac{1}{\left[\sum_{i=1}^{n}\left(\dfrac{d_i^b}{L_i}\right)^{\frac{1}{a}}\left(\dfrac{1}{\Delta_i}\right)^{\frac{b-a-3}{a}}\right]^a}\,, \tag{27}$$

2.1.2 Same type of pipes connected in parallel

In this case, the absolute equivalent wall roughness will be the same for all the pipes. Therefore:

$$\Delta_1=\Delta_2=\ldots=\Delta_n=\Delta\,. \tag{28}$$

The ratio between the pipes roughness in equation (22) is $\Delta_k/\Delta_i=1$ and therefore the flow rate in the "k[th]" pipe is given in the following way:

$$Q_k=\frac{1}{\sum_{i=1}^{n}\left(\dfrac{L_k}{L_i}\right)^{\frac{1}{a}}\left(\dfrac{d_i}{d_k}\right)^{\frac{b}{a}}}Q\,. \tag{29}$$

The pressure drop is determined by using the following equation:

$$p = \left(A\rho^{a-1}\mu^{2-a} \right)\Delta^{b-a-3} \frac{1}{\left[\sum_{i=1}^{n} \left(\frac{d_i^b}{L_i} \right)^{\frac{1}{a}} \right]^a} Q^a . \tag{30}$$

Using equations (26) and (30) enables for the resistance characteristic $p = f(Q)$ of a system, containing "n" number of parallel connected pipes, to be presented graphically.

2.2 Systems containing pipes that are connected in series

In case of a system containing "m" number of serial connected pipes, having different diameters, the following statements are valid [3, 4]:

- the total pressure loss is equal to the sum of the pressure losses obtained in each separate pipe:

$$p = p_1 + p_2 + ... + p_m ; \tag{31}$$

- the flow rate in each separate pipe is equal to the system total flow rate:

$$Q = Q_1 = Q_2 = ... = Q_m . \tag{32}$$

Taking into account equation (6) and presenting the flow rate Q_i by the relevant flow rate criterion $\pi_{Q,i}$ the above equation may be transformed and presented in the following way:

$$\pi_{Q,1}\Delta_1 = \pi_{Q,2}\Delta_2 = ... = \pi_{Q,m}\Delta_m . \tag{33}$$

For determining the pressure loss in each pipe it is necessary to involve equations (7) and (31):

$$p = \pi_{p1} L_1 \frac{\mu^2}{\rho\Delta_1^3} + \pi_{p2} L_2 \frac{\mu^2}{\rho\Delta_2^3} + ... + \pi_{pm} L_m \frac{\mu^2}{\rho\Delta_m^3} = \frac{\mu^2}{\rho}\left(\pi_{p1}\frac{L_1}{\Delta_1^3} + \pi_{p2}\frac{L_2}{\Delta_2^3} + ... + \pi_{pm}\frac{L_m}{\Delta_m^3} \right) . \tag{34}$$

Using the hydraulic criteria equation (1) for presenting the criterions $\pi_{p,i}$ leads to the equation above being presented in the following way:

$$p = A\frac{\mu^2}{\rho}\left(\pi_{Q,1}^a \pi_{d1}^b \frac{L_1}{\Delta_1^3} + \pi_{Q,2}^a \pi_{d2}^b \frac{L_2}{\Delta_2^3} + ... + \pi_{Q,m}^a \pi_{dm}^b \frac{L_m}{\Delta_m^3} \right), \tag{35}$$

which by replacing the criterions: $\pi_{Q,i} = Q\frac{\rho}{\mu}\frac{1}{\Delta_i}$; $\pi_{di} = \frac{\Delta_i}{d_i}$, and after being mathematically transformed may be presented in the following way:

$$p = \left(A\rho^{a-1}\mu^{2-a} \right)\left(\Delta_1^{b-a-3}\frac{L_1}{d_1^b} + \Delta_2^{b-a-3}\frac{L_2}{d_2^b} + ... + \Delta_m^{b-a-3}\frac{L_m}{d_m^b} \right) Q^a . \tag{36}$$

Equation (36), describing the resistance characteristic of a pipe system containing serial connected pipes, is considered to be more useful after being converted and presented analogically to equation (26):

$$p=\left(A\rho^{a-1}\mu^{2-a}\right)\left(\sum_{i=1}^{m}\Delta_i^{b-a-3}\frac{L_i}{d_i^b}\right)Q^a . \qquad (37)$$

Taking into account equation (8) enables for the coefficient of the system resistance characteristic to be estimated by using the following equation:

$$k=\left(A\rho^{a-1}\mu^{2-a}\right)\left(\sum_{i=1}^{m}\Delta_i^{b-a-3}\frac{L_i}{d_i^b}\right). \qquad (38)$$

The next equation describes the system resistance characteristic in case that it contains of the same type serial connected pipes:

$$p=\left(A\rho^{a-1}\mu^{2-a}\Delta^{b-a-3}\right)\left(\sum_{i=1}^{m}\frac{L_i}{d_i^b}\right)Q^a . \qquad (39)$$

By using equations (37) and (39) the resistance characteristic $p=f(Q)$of the system, containing "m" number of serial connected pipes, may be presented graphically.

2.3 Systems containing combined (parallel and serial) connected pipes

In case of a system containing "m" number of serial connected pipes, including a sector of "n" number of parallel connected pipes, the equation describing the system resistance characteristic represents:

$$p=\left(A\rho^{a-1}\mu^{2-a}\right)\left(\sum_{i=1}^{m}\Delta_i^{b-a-3}\frac{L_i}{d_i^b}+\frac{1}{\left[\sum_{i=1}^{n}\left(\frac{d_i^b}{L_i}\right)^{\frac{1}{a}}\left(\frac{1}{\Delta_i}\right)^{\frac{b-a-3}{a}}\right]^a}\right)Q^a . \qquad (40)$$

In case that the analyzed system contains different types of pipes the coefficient of its resistance characteristic may be estimated in the following way:

$$k=\left(A\rho^{a-1}\mu^{2-a}\right)\left(\sum_{i=1}^{m}\Delta_i^{b-a-3}\frac{L_i}{d_i^b}+\frac{1}{\left[\sum_{i=1}^{n}\left(\frac{d_i^b}{L_i}\right)^{\frac{1}{a}}\left(\frac{1}{\Delta_i}\right)^{\frac{b-a-3}{a}}\right]^a}\right). \tag{41}$$

In case that the analyzed system contains same types of pipes, i.e. $\Delta_1=\Delta_2=\ldots=\Delta_n$, the equation for estimating the system pressure loss, describing its resistance characteristic, represents:

$$p=\left(A\rho^{a-1}\mu^{2-a}\right)\Delta^{b-a-3}\left(\sum_{i=1}^{m}\frac{L_i}{d_i^b}+\frac{1}{\left[\sum_{i=1}^{n}\left(\frac{d_i^b}{L_i}\right)^{\frac{1}{a}}\right]^a}\right)Q^a. \tag{42}$$

Therefore, the coefficient of the system resistance characteristic may be estimated as it follows:

$$k=\left(A\rho^{a-1}\mu^{2-a}\right)\Delta^{b-a-3}\left(\sum_{i=1}^{m}\frac{L_i}{d_i^b}+\frac{1}{\left[\sum_{i=1}^{n}\left(\frac{d_i^b}{L_i}\right)^{\frac{1}{a}}\right]^a}\right). \tag{43}$$

3. Comparative and numerical analysis and results discussion

3.1 Output data

In [2], for determining the exponents by applying equations (1) and (2) the friction factor $\lambda$is estimated by using the equation of Konakov - $\lambda=\left(1.8\log\frac{\mathrm{Re}}{6.9}\right)^{-2}$, characterizing hydraulic smooth pipes, while in [3] - for the other regimes of motion, the equation of Altschul - $\lambda = 0.11\left(\frac{68}{\text{Re}} + \frac{\Delta}{d}\right)^{0,25}$, is used.

A comparative analysis indicates good corresponding between the results found in the previously mentioned works and the results found after applying the proposed methodology, but in case that the friction factor is estimated by using the well-known and worldwide recognized equation of Colebrook-White [1, 4], which requires for the values of the relevant constants and exponents in equations (1) and (2) to be preliminary determined. Table 1 provides information about their values, where the exponents are also presented as simple fractions.

Table 1. Values of the coefficients and exponents belonging to equations (1) and (2)

A	a	b		B	e	f
0.137	1.96	5.18		2.75	0.51	2.64
	45/23	119/23			23/45	119/45

Fig.1 illustrates a scheme of a pipe system, containing combined connected pipes, which is analyzed by applying the methodology, newly established by the authors, as for this aim the required information concerning the system parameters is given in Table 2. The flow rate is considered to be: $Q = 40\ l/s$.

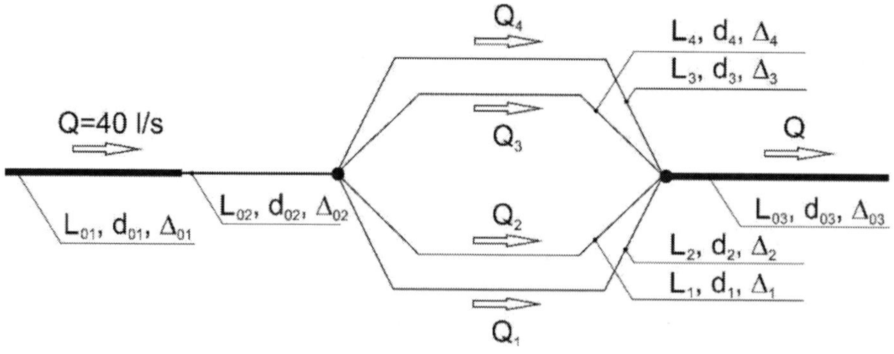

Fig. 1. A scheme of the analysed pipe system.

Table 2. Geometric characteristics of the key elements of the analyzed pipe system

Pipe	Diameter		Pipe referred length		Wall roughness	
	[mm]		[m]		[mm]	
01	d_{01}	160	L_{01}	150	Δ_{01}	0.05
02	d_{02}	150	L_{02}	40	Δ_{02}	0.1

1	d_1	80	L_1	100	Δ_1	0.05
2	d_2	70	L_2	50	Δ_2	0.1
3	d_3	60	L_3	40	Δ_3	0.1
4	d_4	90	L_4	120	Δ_4	0.05
03	d_{03}	160	L_{03}	200	Δ_{03}	0.1

The object of this research respectively is to determine: the flow rate ensured in each of the parallel connected pipes - Q_1, Q_2, Q_3 and Q_4; the system pressure drop p ;the equation of the system resistance characteristic (8) - $p = kQ^{\alpha}$.

3.2 Results found by using the proposed criteria equations

Taking into account the values, given in Table 1, concerning the exponents and constants belonging to equations (1) and (2), the required equations for solving the previously considered tasks are being transformed into the following way:

- Parallel connected pipes

Equation (22), applied in estimating the ensured flow rate Q_k in the „k[th]" pipe ($k = 1...4$), represents:

$$Q_k = \frac{1}{\sum_{i=1}^{n} \left(\frac{L_k}{L_i}\right)^{\frac{23}{45}} \left(\frac{d_i}{d_k}\right)^{\frac{119}{45}} \left(\frac{\Delta_k}{\Delta_i}\right)^{\frac{1}{9}}} Q. \tag{44}$$

By applying equation (26) it may be determined the pressure drop p_p in case of parallel connected pipes:

$$p_p = \left(0.137\, \rho^{\frac{22}{23}} \mu^{\frac{1}{23}}\right) \frac{1}{\left[\sum_{i=1}^{n} \left(\frac{d_i^{\frac{119}{23}}}{L_i}\right)^{\frac{23}{45}} \left(\frac{1}{\Delta_i}\right)^{\frac{1}{9}}\right]^{\frac{45}{23}}} Q^{\frac{45}{23}}. \tag{45}$$

- Serial connected pipes:

The pressure drop p_S in case of serial connected pipes is estimated in accordance with equation (37):

$$p_S = \left(0.137\, \rho^{\frac{22}{23}} \mu^{\frac{1}{23}}\right) \left(\sum_{i=1}^{3} \Delta_i^{\frac{5}{23}} \frac{L_i}{d_i^{\frac{119}{23}}}\right) Q^{\frac{45}{23}}. \tag{46}$$

Equation (40), applied for a system containing combined connected pipes, is used to determine its resistance characteristic (8):

$$p=\left(0.137\,\rho^{\frac{22}{23}}\mu^{\frac{1}{23}}\right)\left(\sum_{i=1}^{3}\Delta_i^{\frac{5}{23}}\frac{L_i}{d_i^{\frac{119}{23}}}+\frac{1}{\left[\sum_{i=1}^{4}\left(\frac{d_i^{\frac{119}{23}}}{L_i}\right)^{\frac{23}{45}}\left(\frac{1}{\Delta_i}\right)^{\frac{1}{9}}\right]^{\frac{45}{23}}}\right)Q^{\frac{45}{23}}. \quad (47)$$

The results found are given in Table 3.

The final equation of the system resistance characteristic represents:

$$p=81134454.5\,Q^{\frac{45}{23}}. \quad (48)$$

3.3 Results found after the specialized software “Pipe Flow Expert” is used

The specialized software product “PipeFlow Expert” [5] has a powerful and reliable computing “engine” containing preliminary added algorithms, based on well-known physical relations that are published in many research works concerning problems in industry, as well as particular algorithms and methodologies established and developed (in addition) by the British company “Daxesoft” Ltd.

For the aims of this current work it is used a licensed version of the above mentioned software product, which is donated by the company “Daxesoft” Ltd. to the University “Angel Kanchev” of Ruse and more specifically to the dept. of “Heat, Hydraulics and Environmental Engineering”.

By involving accurately the output data concerning the analyzed pipe system key elements: lengths, diameters, wall roughness and the ensured flow rate Q, it is possible to design its computational scheme, given in Fig.2.

The numerical computation is accomplished under the following conditions:

- Calculation Engine:
  - Method of Solution: Non-Compressible Flow [Accurate];
  - Method of Calculation: Darcy-Weisbach Equation;
  - Friction Model: Colebrook-White Friction Factors;
- Results found:
  - Found linear solution flow values. Refining...
  - Final solution iterations: 2.
- Verifying pipe losses against node pressures at start and end of pipe.

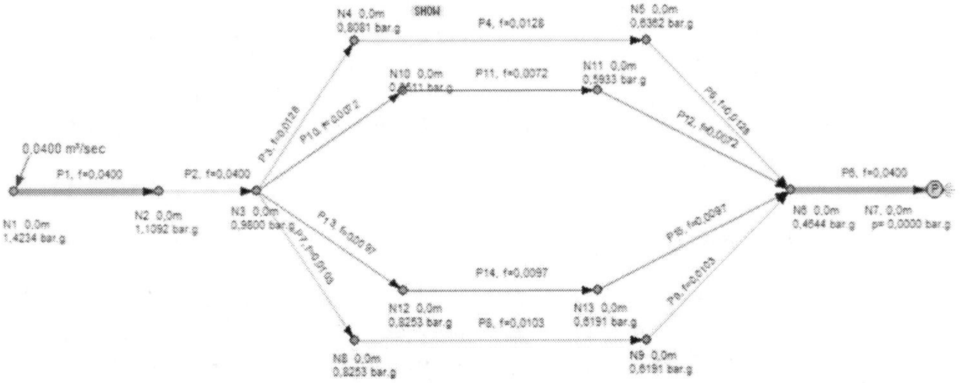

Fig. 2. Computational scheme of the analyzed pipe system designed and solved by using "Pipe Flow Expert".

The final results are given in Table 3.

Table 3. Comparative analysis results

Pipe	*By using criteria equations*				*By using "Pipe Flow Expert"*			
	Flow rate		**Pressure drop**		**Flow rate**		**Pressure drop**	
	[m³/s]		**[Pa]**		**[m³/s]**		**[Pa]**	
01	Q_{01}	0,040	p_{01}	**31807,8**	Q_{01}	0,040	p_{01}	**31420**
02	Q_{02}	0,040	p_{02}	**13770,9**	Q_{02}	0,040	p_{02}	**12920**
1	Q_1	**0,01036**	p_1	**54429,4**	Q_1	**0,0103**	p_1	**51560**
2	Q_2	**0,00960**	p_2	**54429,4**	Q_2	**0,0097**	p_2	**51560**
3	Q_3	**0,00716**	p_3	**54429,4**	Q_3	**0,0072**	p_3	**51560**
4	Q_4	**0,01288**	p_4	**54429,4**	Q_4	**0,0128**	p_4	**51560**
03	Q_{04}	0,040	p_{03}	**49307,6**	Q_{04}	0,040	p_{03}	**46440**
Total system pressure drop				**149316**	***Total system pressure drop***			**142340**

The analysis of the results found clearly indicates sufficient overlapping in using both the selected methods of calculation, since the values concerning the analyzed parameters are practically equal. This is an indirect indicator confirming the accurate using of these newly established equations in engineering calculations concerning pipe systems.

It may be stated that, the accomplishing of a hydraulic analysis of pipe systems by applying the classical methods of calculation leads to the obtaining of inaccuracies that in some cases exceed 5…6 %. The inaccurate determining of the pipe wall roughness has most significant impact on this process, while it is less influenced by the equations used for estimating the friction factor and local resistance coefficients. In this regard, the discrepancy between the results found by using the newly established criteria equations and "Pipe Flow Expert" is considered to be insignificant (negligible) which confirms the great potential in applying this methodology in accomplishing engineering calculations concerning fluid transporting systems.

4 Conclusion

The proposed in this work original methodology, newly established by the authors, to accomplish hydraulic analysis of typical pipe systems contrasts with the classical methods of calculations used as a basis in developing the modern software products. The established equations are characterized by simple mathematical processing enabling them to be easily applicable in finding a solution of many engineering problems. As an important particularity in using the criteria equations (4) and (5) it is considered to be the fact that there is no need to perform all the required steps concerning the accomplishment of such hydraulic estimations, including the determining of the regime of motion (Reynolds number) , selecting an appropriate equation for estimating the friction factor, etc. In addition to that, applying the proposed methodology excludes the necessity of searching for an iterative determination of the ensured flow rate, which in case of a more complex pipe system is one of the most laborious tasks, because most often the final solution is found after assuming that the regimes of motion are fully turbulent. Taking into account the above statements, the methodology, proposed in this work, may be considered as a new approach at accomplishing the hydraulic analysis of pipe systems, used to transport fluids.

The equations, newly established by the authors, concerning the determination of the pipe system resistance characteristics are considered to be especially useful in engineering practice and more specifically: in determining a hydraulic machine work regime, used in a system transporting fluids, as well as in the synthesis and analyzing of such systems or in accomplishing an energy analysis of pipe systems. However, the method derived in this work has the following limitations: cannot be applied for looped pipe networks, neither for branched pipe networks; cannot be applied in case of involving one or several inflows and many outflows (end-users).

Acknowledgements: This paper contains results found after working on a project 2020-FAI-02, funded by the Fund "Scientific research" of the University of "Angel Kanchev" of Ruse, Bulgaria.

References

1. Eugene Adiutori. Why the Fluid Friction Factor should be Abandoned, and the Moody Chart Transformed. The Open Mechanical Engineering Journal, 2009, pp 43-48
2. Popov, Gencho. Hydraulic Criteria Equations for Pipelines Design //PROCEEDINGS of the University "A. Kanchev" of Ruse, volume 50, book 1.2, Ruse, 2011, pp. 54-59

   https://www.researchgate.net/publication/312531977_Kriterialni_uravnenia_za_hidravlicno_presmatane_na_trboprovodi_Hydraulic_Criteria_Equations_for_Pipelines_Design
3. Altshul, A., L. Zhivotobski, L. Ivanov, Hydravlika i aerodinamika. Moskva, 1987
4. White, F. M. Fluid Mechanics. Fourth Edition. McGraw-Hill, 1998.
5. https://www.pipeflow.com/

Effects of CNG quantity on combustion characteristics and emissions of a dual fuelled automotive diesel engine

Silviu Rotaru*[,1], *Constantin* Pana[1], *Nicolae* Negurescu[1], *Alexandru* Cernat[1], *Dinu* Fuiorescu[1] *and Cristian Nikolaos* Nuțu[2]

[1]Dept. of Thermotechnics, Engines, Thermal and Frigorific Equipment, University POLITEHNICA of Bucharest, Bucharest, Romania
[2]Automotive Engineering Department, University POLITEHNICA of Bucharest, Bucharest, Romania

Abstract. The paper reveals some experimental aspects of compressed natural gas (CNG) use in dual fuel mode at an automotive diesel engine. Brake specific energetic consumption, in-cylinder pressure, emissions and variability of indicated mean effective pressure are analysed at operating regime of 2000 rpm and 40% load. Using CNG as an alternative fuel reduces brake specific energetic consumption by 50%, the CO_2 emission by 10% and sets the in-cylinder maximum pressure 13 bar higher comparative to diesel fuel fuelling. The smoke and hydrocarbons emissions and the variability of indicated mean effective pressure are affected by the injection of compressed natural gas into intake manifold: HC emission grows 24 times, the smoke number and the coefficient of variability of IMEP double their values. The use of compressed natural gas at an automotive diesel engine improves its energetic performances and combustion process, having positive effects on CO_2emission and fuel consumption.

Keywords: diesel engine, CNG, combustion, diesel-gas, dual-fuel, COV, IMEP, BSEC.

1. Introduction

The automotive industry experiences nowadays a tremendous pressure as the more aware society demands reduction of emission that have a negative effect on climate safety and on the health of population. Restrictions on CO_2, NO_x, HC and smoke emissions are more and more severe, lots of publications showing the negative effect of diesel emission on humans' health [1, 2, 3]. In this context the diesel engine loses a lot of ground because

[1]Corresponding author: silviu_r.rotaru@yahoo.com

of its high smoke and NO_x emissions level. Even though there exist optimistic opinions about electric propulsion (50% percent for passenger cars by 2040 [4]) and there are ways to facilitate the transition from internal combustion engines to full electric propulsion (mild and full hybridization), the pollution actual issues impulse the engineers to find solutions that could be applied very fast and with real improvements on the current or next generation of diesel engines. Two main directions are taken into consideration related to internal combustion engines: development of more efficient internal combustion engine that have better aftertreatment systems and development of technologies for cleaner alternative fuel use. Due to the higher efficiency of diesel engine comparative to spark ignition engine, researcher all over the world try to reduce its emissions by fuelling it with clean alternative fuels (LPG, LNG, CNG, animal fat, biogas, butanol). The most common technology for use of alternative fuels is the dual-fuel mode when the alternative fuel is injected into the intake manifold [5]; this offers the main advantage of less engine design modification and so the procedure can be applied to the in-use engines. The compressed natural gas (CNG) is an alternative fuel used in the transport industry due to its important features: low carbon content (1 kg of CNG contains 750 g of carbon as opposite to diesel fuel that has over 870g of carbon), lower heating value (49 MJ/kg) and high octane number (130) [6]. There are many studies that reveal diesel engine operation when it is fuelled in dual fuel mode with compressed natural gas. Basavarajappa [7] states that the electronically controlled injection of compressed natural gas has the best results in combustion and gave as optimum an injection timing of 5°before TDC and injection duration 60°CA. Egúsquiza [8] presents better energetic performances and lower emission of a supercharged diesel engine fuelled with compressed natural gas. Imran shows that the NO_x emission level is 53% less for all range of substitution and at all engine loads; the CO_2 emission decreases by 25% at all engine loads due to the lower carbon content of the CNG; the HC emission increases 8 times at medium load and with 25% at high load [9]. Yousefi presents a 32% growth of in-cylinder maximum pressure and lower levels of NO_x at highest substitution ratio; the CO emission rises because of the lack of oxygen as opposite to the conventional fuelling [10]. Mahla [11] presents improvements of the brake thermal efficiency, hydrocarbons and nitrous oxides emissions levels and negative effects on the ignition delay, in-cylinder pressure, smoke and CO emissions levels when using exhaust gases recirculation at a diesel engine fuelled with CNG.

This paper comes in accordance with specialty literature studies mentioned above and has as objective the study of combustion at CNG use at an automotive diesel engine, focusing on in-cylinder pressure, brake specific energetic consumption, emissions level of carbon dioxide, nitrous oxides, smoke, hydrocarbons and cyclic variability of indicated mean effective pressure (IMEP).

2. Research methodology

Figure 1 present the design of the experimental test bed which use a K9K diesel engine (1), designed for Dacia Logan, which is equipped with a CNG fuelling system which incorporates a CNG injector (11) connected to the inlet manifold, flame arrestor (10), mass flow meter for gaseous fuels (9),pressure regulator (8) and CNG reservoir (7) pressurised at 220 bar. The steel pressure tank is tested at pressure around 300 bar. The CNG reducer releases CNG at fueling pressure (2.5 – 3 bar) after two stages of reduction (first stage is at 10 bar). The CNG injector is controlled with an open electronic control unit (ECU) (12). By this ECU the operator has access to the operating parameters (CNG quantity and injection timing) of the fuelling system by using computer (13). The conventional fuelling system is compounded of diesel fuel tank (3), liquid mass flow meter (4), low/high pressure pump system (5) and common rail and injectors (6). The inlet air quantity admitted into the cylinders is measured by volumetric flow meter (14). The engine is loaded by an eddy current dyno (2); the accelerator pedal (20) is actuated by a step-by-step controller (19) driven by electronic unit (18). In-cylinder pressure is measured by a piezoelectric transducer (21); the in-cylinder pressure can be observed on oscilloscope`s (17) screen; the in-cylinder pressure data is stored on computer (16). Emissions level from the exhaust gases are measured with a gas analyser – opacimeter system (15).

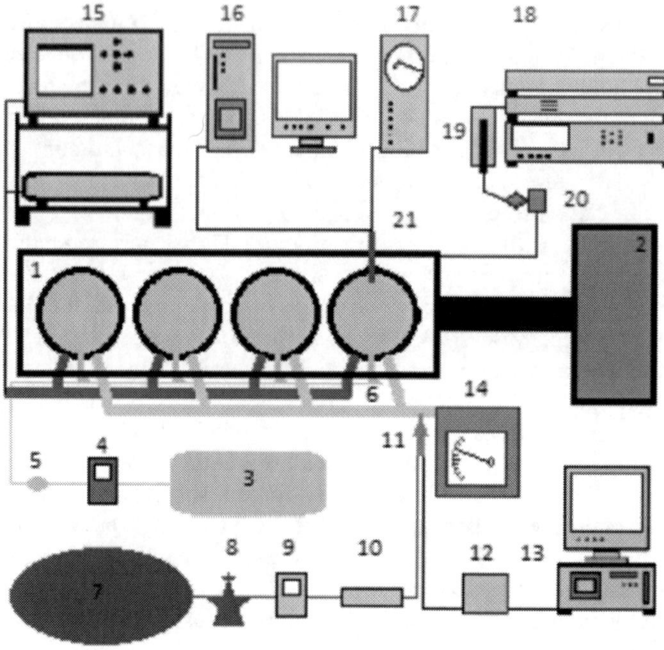

Fig. 1. The research test bed

The paper's objective is to demonstrate that the use of CNG as an alternative fuel at an automotive diesel engine at low loads, determines improvements in fuel performances and emission exhaust.

At the investigated operating regime, 40% engine load and 2000 rev/min, diesel fuel and CNG quantities, volume and temperature of air admitted into the cylinder, the emissions levels for NO_x, CO_2, hydrocarbons and smoke, temperatures of exhaust gases, and 200 in-cylinder pressure diagrams of consecutive combustion cycles had been recorded. Also, temperature in certain points on intake and exhaust manifolds and of cooling liquid have been recorded; atmospheric pressure and temperature were also monitored. All measurement equipment was previously calibrated and before the official experimental investigation the engine ran until it reached the optimal values of the operating parameters. During entire investigating procedure the GNC injection timing was kept constant for all energetic substitution ratios.

During the entire experimental investigation procedure there were several main activities developed:

- engine fuelled only with diesel fuel was set to regime of 2000 rpm and 40% load.
- the diesel fuel quantity was reduced.
- the CNG was injected until the engine power was re-established at the reference.

A set of parameters has been calculated:

-the brake specific energetic consumption (BSEC) [12] in kJ/(kWh) is:

$$BSEC = (C_{hCNG} * LHV_{CNG} + C_{hDiesel} * LHV_{Diesel}) / P_e \quad (1)$$

-the energetic substitution ratio (x_c) is:

$$x_c = (C_{hCNG} * LHV_{CNG}) / (C_{hCNG} * LHV_{CNG} + C_{hDiesel} * LHV_{Diesel}) * 100 \quad (2)$$

where $C_{hCNG/diesel}$ is CNG/diesel consumption, in kg/h, and $LHV_{CNG/diesel}$ is lower heating value for CNG/diesel in kJ/Kg.

The effective power [13] is:

$$P_e = M_e * (\pi * n) / 30 \quad (3)$$

where M_e is the effective torque measured in Nm, n is engine speed in rev/min.

Indicated mean effective pressure (IMEP) represents the indicated mechanical work divided by total displacement of engine and it represents a criterion to evaluate the capability of the engine to convert heat into mechanical work. The IMEP it is calculated using the analytic formula:

$$imep = n * \{\sum_0^{720}[p_i * (V_{(i+1)} - V_i)]\} / Vt \quad (4)$$

where n represents the number of engine's cylinders, i is the position of crankshaft in °CA, p_i/V_i represents the in-cylinder pressure/volume at certain crankshaft position measured in bar/cubic centimetres and V_t total engine's displacement in cubic centimetres.

The coefficient of variability [14]is used to determine the degree of variation for the interest parameters and it is calculated using the formula:

$$COV = \sigma_x / x_{avg} * 100 \quad (5)$$

where σ_x represents the standard deviation and x_{avg} is the average value of all elements belonging to the analysed set.

The standard deviation is calculated using the formula:

$$\sigma_x = (\sum_{i=1}^{n}(x_i - x_{avg})^2 / n)^{1/2} \quad \textbf{(6)}$$

where n is the number of elements in the series, $x_i$represents a specific element of the series and x_{avg} is the average value of all elements belonging to the analyzed series.

3. Results

The most important aspect of this work was to determine the maximum quantity per cycle of CNG that can be injected. For all energetic substitution ratios, the influence of CNG dose on the in-cylinder maximum pressure are presented, figure 2. The maximum pressure p_{max} rises with16 bars at highest energetic substitution ratio.

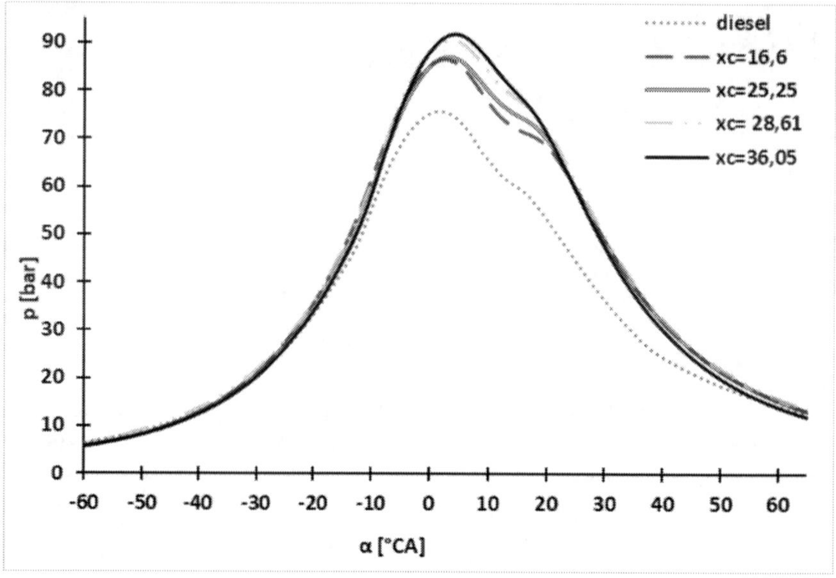

Fig. 2. In-cylinder pressure as function of crankshaft position for different energetic substitution ratios.

As seen in figure 2 and 3 the in-cylinder maximum pressure rises with over 20% at highest energetic substitution ratio. A very important aspect to be highlighted is that the highest in-cylinder pressure attained in the conventional fuelling mode –76 bar – is reached 10 °CA sooner in dual fuel mode and it also lasts for 12 times longer at highest energetic substitution ratio. In standard fuelling mode the pressure peak of 76 bar is recorded for 2 °CA as in dual fuel mode at highest x_c, equal or higher pressure is recorder for 24 °CA. This may have a great impact on NO_x emission level, figure 6, as it determines higher in-cylinder temperature for longer periods. Higher in-cylinder pressure are determined by two very important factor: the lower heating value of CNG is 17% higher than that of diesel fuel; the charge in dual fuel mode has a greater percentage of fuel in gaseous state and this means that the heat used to vaporize the fuel is lower comparative to classic fuelling.

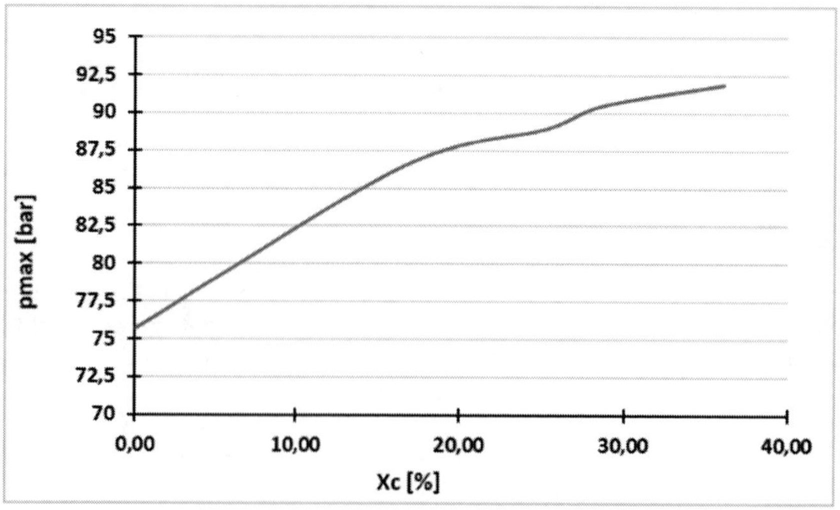

Fig. 3. In-cylinder maximum pressure as function of energetic substitution ratio

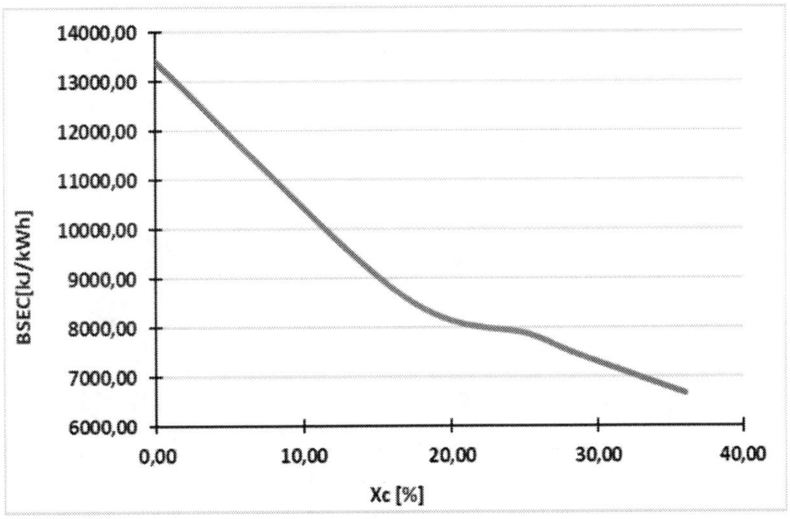

Fig. 4. Break specific energetic consumption as function of energetic substitution ratio

The lower heating value of compressed natural gas (48 MJ/kg) is higher than that of diesel fuel (41,7 MJ/kg). This added to the gaseous state of the alternative fuel determines a drop of the specific energetic consumption (BSEC). At highest energetic substitution ratio BSEC drops by 50% comparative to classic fuelling.

Compressed natural gas has 75% to 77% carbon content and diesel fuel has a content of 82% to 87 % in carbon, fact related to a lower CO_2 emission level in dual-fuel mode. At highest energetic substitution ratio, the carbon dioxide drops by 10%, as figure 5 shows.

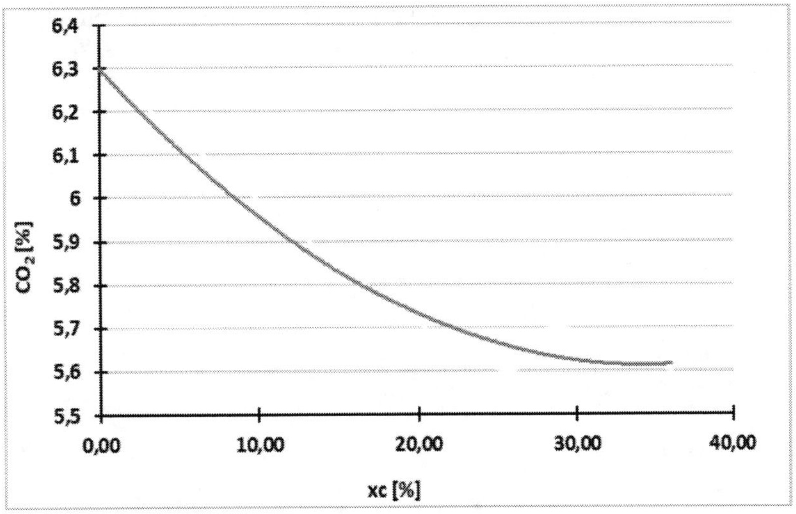

Fig. 5. Carbon dioxide emission as function of energetic substitution ratio

Also, the reduction of the BSEC, figure 4, leads to the reduction of the CO_2 emission level, figure 5.

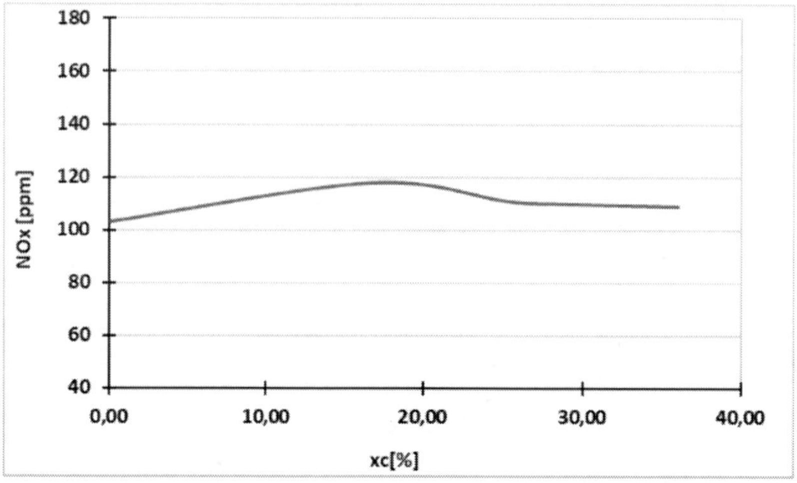

Fig. 6. Nitrous oxides emission as function of energetic substitution ratio

Nitrous oxides are formed through Zeldovich mechanism [13] favoured by in-cylinder high temperature, amount of oxygen and the time spent by nitrogen close to oxygen molecules. At medium engine loads the NO_x is very little influenced by injection of compressed natural gas into the intake manifold. Two out of the three factors mentioned earlier are influenced by CNG injection: higher temperature inside cylinder (the in-cylinder much higher pressure for longer periods and the higher percentage of gaseous premixed fuel determine a higher in-cylinder temperature) determines a growth of the NO_x emissions level. Less oxygen available per cycle (a percentage of air admitted into cylinder in conventional fuelling mode, is replaced by compressed natural gas – in dual-fuel operating mode - determining a lower quantity of oxygen) favours lower NO_x emissions. The two aspects mentioned in this paragraph have opposite effects over NO_x emission and it grows with less than 20 parts per million at highest energetic substitution ratio, the influence of temperature being more significant.

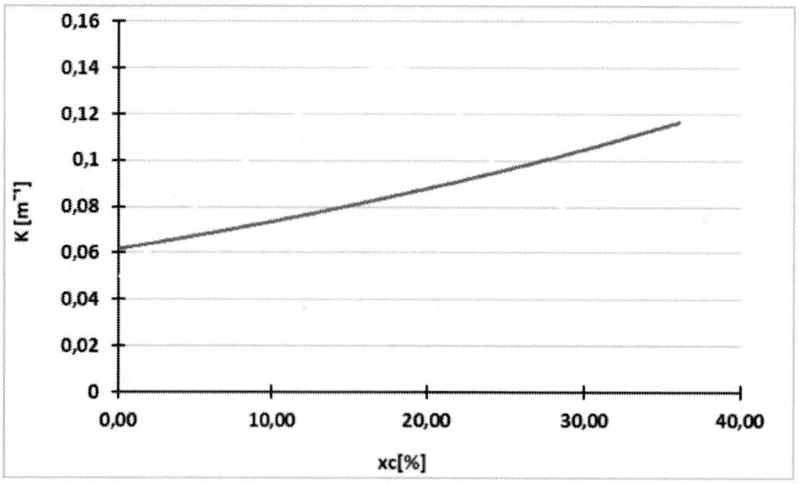

Fig. 7. Smoke number as function of energetic substitution ratio

 https://doi.org/10.1051/e3sconf/202018001008

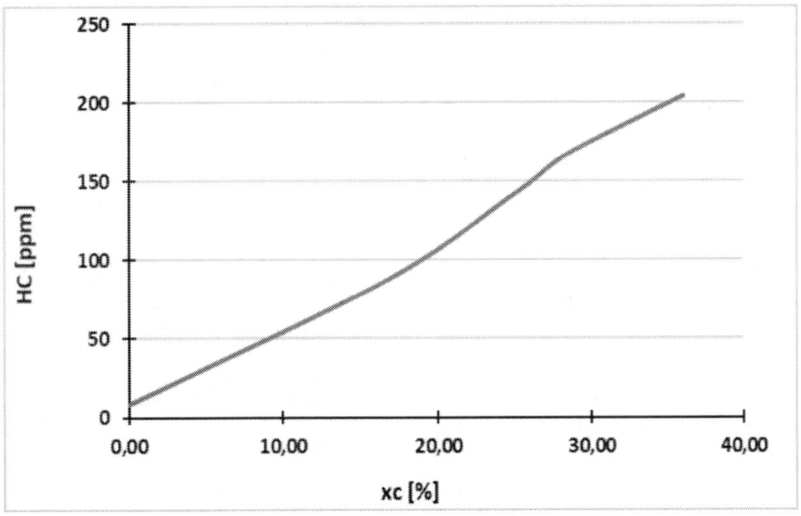

Fig. 8. Hydrocarbons emission as function of energetic substitution ratio

At medium loads, at maximum x_c the smoke number, figure 7, doubles its value at CNG use comparative to classic fuelling. At high temperatures with low intensity of in-cylinder turbulences and in absence of oxygen–determined by the percentage of air replaced by GNC – the hydrocarbons molecules lose the hydrogen atoms (dehydrogenate) and form particle matter that groups in smoke flakes.

Lack of oxygen, flame quenching at cold cylinder wall or in volume of inert gases and gaseous fuel trapped in crevices (piston top land, fire ring and cylinder wall)represent main reasons for the growth of hydrocarbons emission (figure 8). The HC emission at highest energetic substitution ratio is 24 times higher than that of engine running in conventional mode. Also, the reduction of inlet air quantity at CNG use that leads to the increase of smoke emission influences the HC emission level.

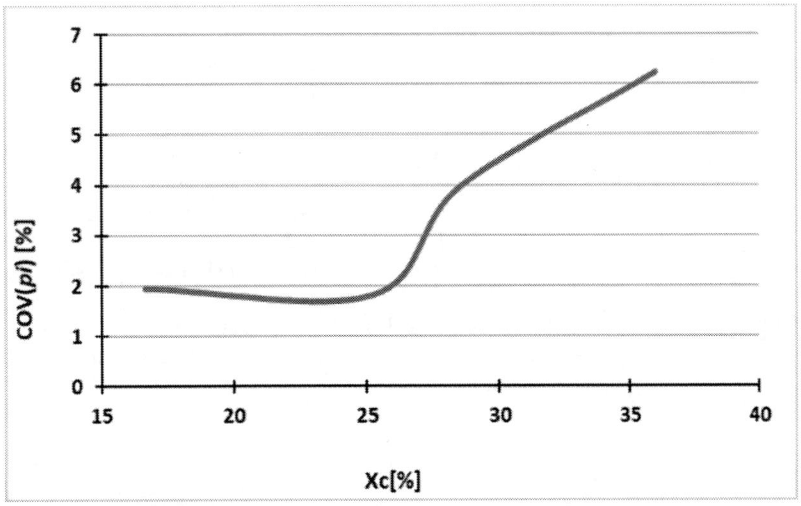

Fig. 9. COV of indicated mean effective pressure as function of energetic substitution ratio

The variation of $(COV)_{IMEP}$ reflects a reduced influence of CNG quantity on cycle to cycle variation of the combustion process, for xc=0% … 25% (figure 9). After this value of energetic substitution ratio, the stability of IMEP is affected, the $(COV)_{IMEP}$ for maximum xc being 3 times higher comparative to $(COV)_{IMEP}$ of xc=0, the irreproducibility of the combustion phase from cycle to cycle being influenced by large CNG doses. The rise of the $(COV)_{IMEP}$ at the increase of CNG cyclic dose is related with the decrease of the inlet air quantity. The maximum value of $(COV)_{IMEP}$ is under 10%, the normal running of the engine being assured at dual fuelling, [13].

The rise tendency registered for HC and smoke emissions levels and for $(COV)_{IMEP}$ may represents a criterion for limitation of the CNG used quantity at xc=36%.

4. Conclusions

The results obtained during the experimental investigation of CNG use to fuel an automotive diesel engine at the regime of 2000 rev/min and40%load allow the formulation of the following conclusions:

- the in-cylinder maximum pressure rises from 76 bar, in conventional fuelling mode, to 92 bar in dual fuel mode;
- CNG gaseous state and a higher heating value determines a 50% reduction in terms of brake specific energetic consumption at highest x_c comparative to efficiency registered for the engine fuelled in conventional mode;
- at the highest energetic substitution ratio, the carbon dioxide emission level is with 10% lower comparative to standard mode, due to a lower carbon content in compressed natural gas;
- the NO_x emission level is affected by the increased pressure and temperature reached during the combustion of the air-CNG mixture, the NO_x emission level slightly increase with 15%; factors like temperature rise and oxygen reduction have opposite effects thus the rise of NO_x emission level is limited;
- the smoke emission level doubles its value as *xc* reaches highest level, mainly due to low quantity of oxygen inside the cylinder;
- the HC emission level grows 24 times when the energetic substitution ratio is at 36%, because of the gaseous fuel trap inside crevices and reduction of inlet air;
- for xc=0…25% the combustion stability is assured, but for xc>25% the COV of IMEP rises from 2% (at only diesel fuel fuelling) to 6.3% at CNG-diesel fuel use at maximum xc, because of a more significant reduction of inlet air quantity at rise of CNG at the maximum value. However, the engine operates normally at dual fuelling since COV of IMEP doesn't exceed 10%.

This work has been funded by the European Social Fund from the Sectorial Operational Programme Human Capital 2014-2020, through the Financial Agreement with the title "Scholarships for entrepreneurial education among doctoral students and postdoctoral researchers (Be Entrepreneur!)", Contract no. 51680/09.07.2019 - SMIS code: 124539.

The authors address special thanks to AVL GmbH Graz Austria for providing the necessary equipment.

Gratitude is also presented to Mr. Ovidiu Pavelescu for providing the CNG fuelling system.

This work was supported by a grant of the Romanian Ministry of Research and Innovation, CCCDI – UEFISCDI, project number PN-III-P1-1.2-PCCDI-2017-0404 / 31PCCDI/2018, within PNCDI III.

References

1. D. T. Silverman, C. M. Samanic, J. H. Lubin, A. E. Blair, P. A. Stewart, R.Vermeulen, J. B. Coble, N. Rothman, P. L. Schleiff, W. D. Travis, R. G. Ziegler, S. Wacholder, M. D. Attfield, J Natl Cancer Inst.; **11**: 855–868(2012).
2. E. S. Schultz,A. Litonjua,E. Melén Curr Allergy Asthma Rep **17** pp 1-17 (2017)
3. Ambient air pollution: A global assessment of exposure and burden of disease ISBN 9789241511353
4. https://about.bnef.com/" BNEF's annual forecast of electric vehicles, shared mobility and road transport to 2040."
5. L. Sun, Y.F. Liu, K. Zeng, R. Yang, Z.H. Hang, Proc. Inst. Mech. Eng. D J. Automob. Eng. **229**235–246 (2015).
6. L.Wei , P. Geng, Fuel Processing Technology, 142 264–278 (2016)
7. Y. H. Basavarajappa, R. B. NagarajInt. J. of A. Eng. and Tech.**4**, 223 – 244, (2015).
8. Egúsquiza J C, Braga S L, Braga, J. of the Braz. Soc. of Mech. Sci. & Eng. **31** pp 142-150 (2009)
9. S. Imran, D.R. Emberson, A. Diez, D.S. Wene, R.J., Crookes T. Korakianitis, Applied Energy, **124**, 354–365 (2014)
10. A.Yousefi, , M. Birouk, , B. Lawler, A. Gharehghani, Eng. Conv. and Mng.**106**, 322–336, (2015)
11. S. K.Mahla, L. M.Das, M. K.Babu, B 2010 Jordan J. of Mech. and Indus. Eng.**4**, 524 – 528, (2010).
12. B. Grunwald „Teoria, calculul şi construcţia motoarelor pentru autovehicule rutiere" Editura Didactică şi Pedagogică (1980).
13. J. B. Heywood, Internal combustion engine fundamentals (New York: McGraw-Hill, Inc) ISBN 0-07-028637-X (1988).
14. E. Ţiţan,Statistică, Teorie şi aplicaţii în sectorul terţiar, Editura STATISTICĂ Meteor Press, Bucureşti, (2003).

Numerical modeling of the thermal conduction process in water-air convector's fins

Zhivko Kolev[1,*], and *Seher* Kadirova [2]

[1]University of Ruse, Department of Heat, Hydraulics and Environmental Engineering, 7017 Ruse, Bulgaria
[2]University of Ruse, Department of Electronics, 7017 Ruse, Bulgaria

Abstract. The paper presents results from numerical modelling of thermal conduction process in ABAQUS. The computing mesh of the model has been verified. Based on the simulation results, the average temperature of the heat exchanger's external surface and the external heat convection coefficient have been determined. The process of thermal conduction in the fins is modelled on the basis of boundary conditions, which values are obtained by simulation of the heat convection process in the convector's pipes. The average heat parameters have been determined relative to the first, middle and last fins, by developing a geometric model in ABAQUS. The computing mesh has been verified by increasing the number of cells and nodes. The external heat convection coefficient is calculated by Newton's law.

1 Introduction

The processes of thermal conduction and the determination of their parameters have been studied in various fields. In [1] the thermal conductivity of building composite materials has been investigated. The thermal conductivity coefficient of the materials has been determined by TCi Mathis Thermal Conductivity Analyzer for non-destructive measurements. The mass specific heat capacity of the materials has been also determined. The paper presents investigation of thermal-physical characteristics of environmentally friendly building materials made by clay, sand and straw. Paper [2] presents determination of the heat capacity, thermal conductivity and thermal diffusivity of linalool - ethanol - water solutions. The purpose is determination of physicochemical and thermal properties of solutions with different ethanol concentrations, in order to describe the stability of the system. Heat transfer in thin plates with temperature dependent conductivity, has been considered in [3]. Due to the small thickness of the plates, two-dimensional modeling has been used. Graphical dependences of the thermal conductivity coefficient on the temperature of the plates are built. Investigation [4] presents numerical study of transient heat transfer process in longitudinal fins with different profiles. In the numerical models, linear or nonlinear functions of the temperature for the heat transfer coefficient and thermal conductivity coefficient have been used.

* Corresponding author: zkolev@uni-ruse.bg

Many papers present investigations of heat transfer processes in finned tube exchangers. In [5] heat transfer and flow characteristics of air over flat finned tubes, have been determined. Non-perforated fins and perforated with different shapes fins, have been investigated. The aim is determination of effective passive techniques for thermal performance enhancement of finned heat exchangers. Paper [6] presents experimental study of heat transfer in cross-flow heat exchanger with smooth tubes or integral finned tubes. The heat flow, heat transfer coefficient and overall heat transfer coefficient, have been determined at different velocities of the water flow in tubes, and different velocities of the air flow, which crosses the tubes. Experimental investigation of spiral finned tube heat exchanger has been considered in [7]. Different fin pitch, fin height, transverse tube pitch, and longitudinal tube pitch, have been investigated. The heat transfer Nusselt number and the flowing resistance Euler number, have been determined at different fluid flowing Reynolds number. Another research presents determination of the heat transfer coefficient of longitudinal finned tubes [8]. The local heat transfer coefficient and the temperature distribution have been determined for various Reynolds numbers of the fluid flow. In [9] a fin and tube heat exchanger has been studied by numerical modeling method. The gas flowing over the fin side is simulated at constant internal tube wall temperature, and the temperature distribution in the fins and the fluid velocity profile, have been determined. Paper [10] presents experimental study of the convective heat transfer process in finned tube counter flow heat exchanger. The convection heat transfer coefficient has been determined for plain tube, two finned tube and three finned tube. The investigation has been accomplished by different compositions of the liquid flowing in the pipes, at different Reynolds numbers. Numerical modeling of cross flow heat exchanger with louvered fins has been considered in [11]. A numerical model has been developed based on the thermal resistance concept and the finite difference method. Investigation [12] presents simulation study of a finned-tube latent heat thermal energy storage system. Numerical method based on heat capacity formulation, has been used for investigation of the influence of the convection heat transfer on the processes in the storage system. In [13] numerical simulation of the heat transfer in finned tube heat recovery unit, is considered. A method based on fluid-solid coupling is used for calculation of some parameters for the heat transfer process between the tube and the fluids. Paper [14] presents numerical study of circular tube bank fin heat exchanger with vortex generators. The fin's temperature distribution and the fluid's velocity distribution have been determined by simplified 3D model. CFD simulation investigation of the heat transfer in the pipes of water-air convector is considered in [15]. Pipe's sections with reduced heat transfer intensity, have been investigated.

The numerical modeling of heat transfer processes is related to determination of the temperature field [15-17]. ABAQUS is used in [15] and [16], while in [17] the used software product is ANSYS FLUENT.

The investigation of heat transfer processes in the heat consumers of heat pump installations is important, in terms of the wide-ranging use of the heat pumps for different needs [15, 18-21]. An experimental reversible water-to-water heat pump installation for air-conditioning is considered in [15] and [18]. The operation of air-to-water heat pump system for building's heating, has been investigated in [19]. In [20] and [21] mechanical vapor recompression heat pump for treatment of industrial wastewater, has been investigated.

For determination of certain parameters, in some cases it is necessary to study a sequence of heat transfer processes. The paper presents investigation of thermal conduction processes in the fins of water-air convector. The study has been accomplished in heating regime of the laboratory heat pump installation. The convector is used as a heat consumer. The temperature field in several fins of the convector's heat exchanger has been modelled, based on the results obtained from numerical modelling of heat transfer process in the convector's pipes [15]. The temperature distribution results in the convector's fins have been used for determination of the heat convection coefficient ($\overline{h_{c-external}}$, W/m^2.K) between the heat exchanger's external surface and the air flow. For this purpose, the average temperature of the surface ($\overline{t_w}$, °C) has been calculated. The calculations are on the base of two types of simulation results. One of them is the temperature field along the internal surface of the pipes. The other one is the temperature results for the heat exchange surface of three fins along the heat exchanger. The investigated fins are the first (№1), the middle (№79) and the last one (№157), respectively.

2 Methodology

2.1 Principal scheme of heat exchanger

Figure 1 presents a principal scheme of the convector's heat exchanger. It is two-pipe staggered finned pipe sheaf. When the heat pump installation operates in heating mode, the heat flow is transferred from the water flow in the pipes to the ambient air. The first row of finned pipes in the direction of air flow has been investigated.

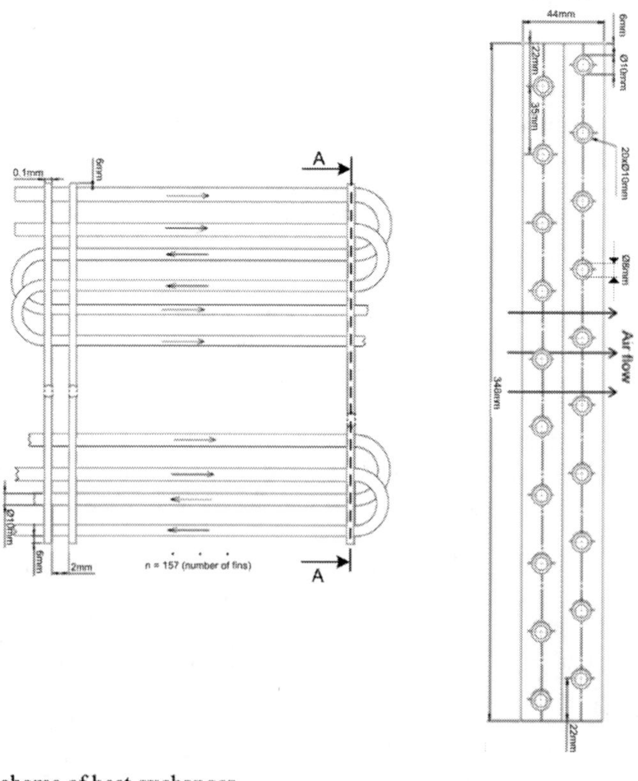

Fig. 1. Principal scheme of heat exchanger.

2.2 Schematic diagram of the simulation model

Figure 2 presents a schematic diagram of the numerical model for simulation of the temperature field in the convector's fins. For this purpose, the heat flux (q, W/m^2) from the heat exchange surface to the environment, and the average temperatures of the fins' internal cylindrical surfaces ($t_{w-1(I)}$ ÷ $t_{w-4(X)}$, °C), have been set as boundary conditions. The temperature values have been determined from the simulation of the heat transfer process in the convector's pipes.

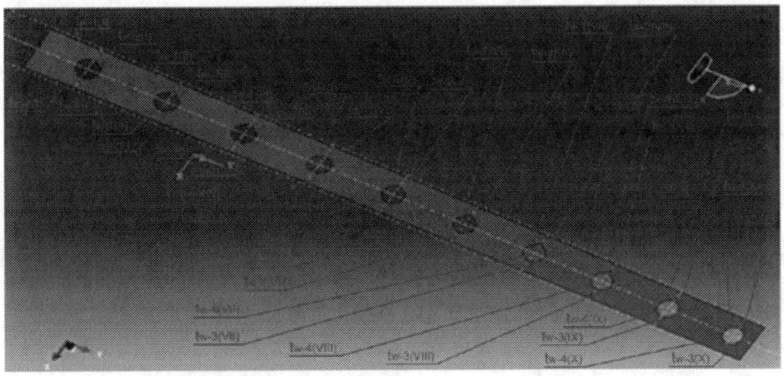

Fig. 2. Schematic diagram of the simulation model.

2.3 Verification of computing mesh

The verification of computing mesh has been performed using two temperatures (t_1 and t_2, °C), defined for computing nodes. They are located on two of the edges of the middle fin (Fig. 3). Since the thermal resistance of the fin's wall is negligible, it is assumed that these are the values of the temperatures of the edges' nodes, located on the fin's opposite side.

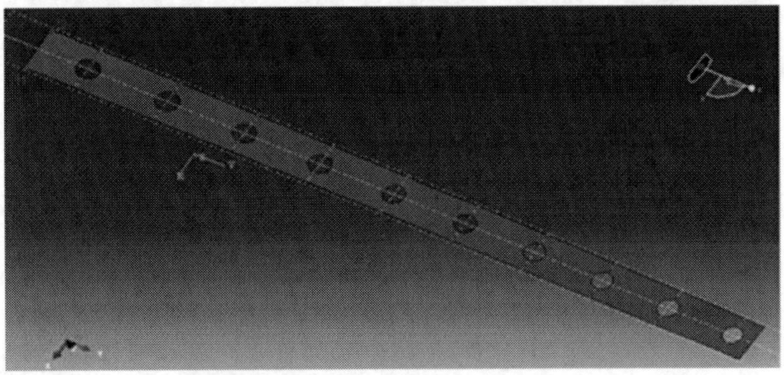

Fig. 3. Temperature points for verification of computing mesh.

2.4 Calculation of the heat convection coefficient

The heat convection coefficient between the heat exchanger's external surface and the air flow has been calculated by (1):

$$\overline{h_{c-external}} = \frac{q}{\overline{t_w} - \overline{t_{heated\ air}}} \tag{1}$$

where $\overline{t_{heated\ air}}$ is the average temperature of the air, heated by the exchanger's external surface, °C. This temperature has been calculated on the base of the basic heat equation, using experimentally determined parameters of the heat pump installation.

3 Results and discussion

3.1 Results from determination of the boundary conditions

Figures 4 - 6 present results obtained for the temperature field in the convector's pipes. They are base for calculation of the average temperature of the external cylindrical surfaces of the pipes between the fins, and the temperatures $t_{w-1(I)} \div t_{w-4(X)}$. The temperatures $t_{w-1(I)} \div t_{w-4(X)}$ are boundary conditions for simulation the thermal conduction process in the three investigated fins.

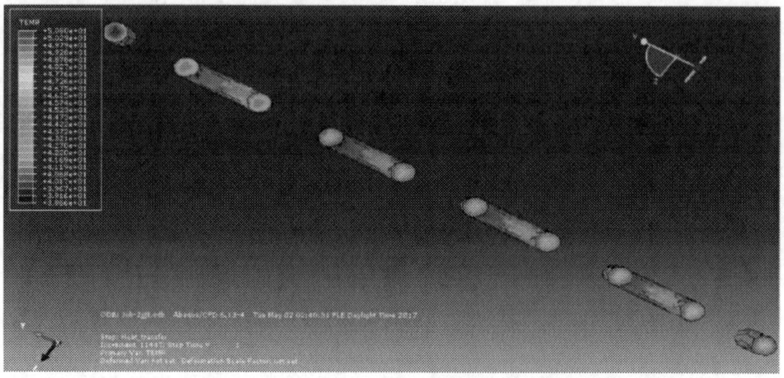

Fig. 4. Temperature field of the fluid – cross-section in the middle of the first fin.

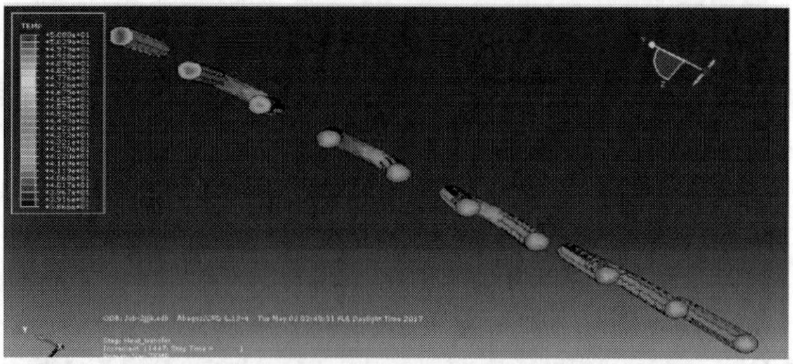

Fig. 5. Temperature field of the fluid – cross-section in the middle of the middle fin.

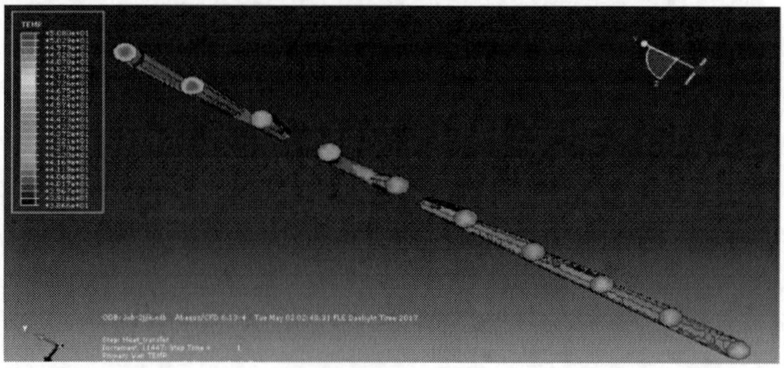

Fig. 6. Temperature field of the fluid – cross-section in the middle of the last fin.

In these simulations, „Tet" shape of the computing cells is set. The approximate size of the cells is 0.005 m. The number of computing cells is 69293, and the number of computing nodes is 18384. The used turbulent model is „Spalart–Allmaras". The obtained temperature field of the fluid presents the heat transfer intensity in different sections of the pipes.

Tables 1 – 3 present the values of the temperatures, used as boundary conditions for simulation the thermal conduction process in the three fins. The first row of the tables gives information about the number of the pipe's cross-sections. The average heat flux q is 1108.79 W/m^2, and is with constant value for all fins. The value of heat flux has been determined by calculations of the heat transfer process in the convector's pipes.

Table 1. Temperature values for simulation of the first fin.

	I	II	III	IV	V	VI	VII	VIII	IX	X
t_{w-1}, °C	48.109	45.511	44.990	44.188	44.271	44.061	44.169	44.162	43.918	44.293
t_{w-2}, °C	47.912	46.242	46.005	44.500	44.723	44.641	45.009	44.533	44.929	44.237
t_{w-3}, °C	47.582	45.714	46.300	44.111	44.879	44.405	44.883	44.321	44.936	44.218
t_{w-4}, °C	47.862	45.392	45.578	44.183	44.275	44.042	44.188	44.123	44.166	44.079

Table 2. Temperature values for simulation of the middle fin.

	I	II	III	IV	V	VI	VII	VIII	IX	X
t_{w-1}, °C	47.559	46.387	45.444	44.411	44.317	44.479	44.685	44.425	44.220	44.207
t_{w-2}, °C	46.516	46.364	45.536	44.767	44.301	44.572	44.292	44.940	44.231	44.059
t_{w-3}, °C	46.943	45.807	45.240	44.514	44.239	44.098	44.169	44.167	44.057	44.099
t_{w-4}, °C	47.569	45.947	44.910	44.505	44.181	44.167	44.462	43.705	43.969	44.197

Table 3. Temperature values for simulation of the last fin.

	I	II	III	IV	V	VI	VII	VIII	IX	X
t_{w-1}, °C	46.879	46.150	44.999	44.139	44.279	43.848	44.111	43.744	44.557	43.883
t_{w-2}, °C	47.174	46.694	45.164	45.207	44.267	43.556	44.344	44.389	44.369	44.718
t_{w-3}, °C	46.859	47.222	45.056	45.338	44.193	44.618	44.392	44.614	44.592	44.908
t_{w-4}, °C	46.498	46.076	45.096	44.553	44.415	43.956	44.373	43.612	44.507	44.745

The presented temperatures are obtained from the average values in the respective computing nodes in the internal surface of the pipes. The temperatures are highest in the first cross-section, because there the temperature of the fluid flowing in the pipe is the highest.

3.2 Results from verification of computing mesh

Table 4 presents the results from verification of computing mesh. The differences in temperature values have been calculated by subtracting of the down row values from the up row values in each column of the table. The selected shape of computing cells is "Hex (sweep)".

Table 4. Results from verification of computing mesh.

Mesh	Cell's size, m	Number of cells	t_1, °C	Difference of t_1, °C	t_2, °C	Difference of t_2, °C
I	0.01000	220	30.1896	-	30.1470	-
II	0.00800	238	30.1365	0.0531	29.9749	0.1721
III	0.00700	258	30.1168	0.0197	29.9467	0.0282
IV	0.00600	282	29.7362	0.3806	29.6725	0.2742
V	0.00500	320	29.7321	0.0041	29.6640	0.0085
VI	0.00450	356	29.5580	0.1741	29.4934	0.1706
VII	0.00400	463	29.3519	0.2061	29.2850	0.2084
VIII	0.00350	517	29.1894	0.1625	29.1871	0.0979
IX	0.00335	613	29.3248	-0.1354	29.2689	-0.0818

From the data presented in the table it can be seen that the tendency is to increase the values of temperatures t_1 and t_2, until reaching of computing mesh IX. The change of temperatures with increasing mesh density is presented. On the base of the obtained results, computing mesh VIII has been selected. In Fig. 7 can be seen the number of cells and their specific shapes. The number of the nodes is 1312.

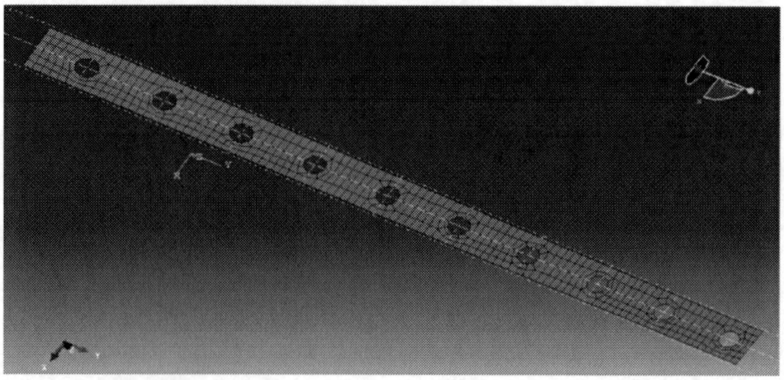

Fig. 7. View of selected computing mesh.

3.3 Simulation results for the temperature distribution in fins

Figures 8 - 10 present the simulation temperature fields obtained for the three fins. The presented temperature distribution expresses the heat transfer intensity between the respective zones of the fins and the air flow. It can be seen that the temperatures of the fins' areas, located at the greatest distance from the pipes, are the lowest.

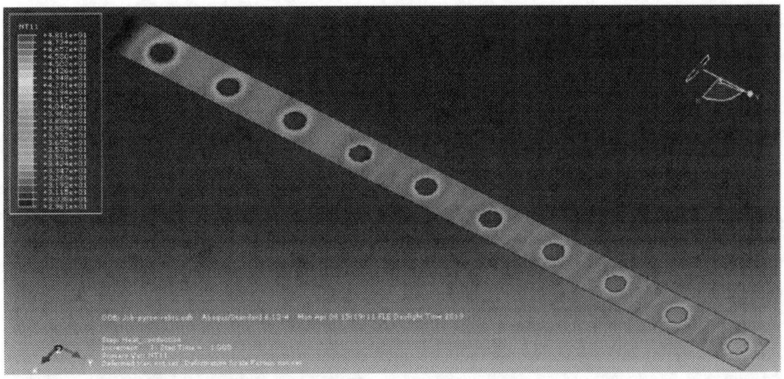

Fig. 8. Veiw of the temperature field in the first fin.

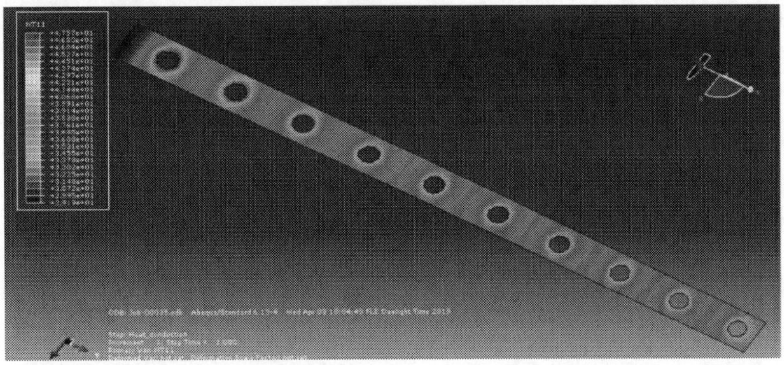

Fig. 9. Veiw of the temperature field in the middle fin.

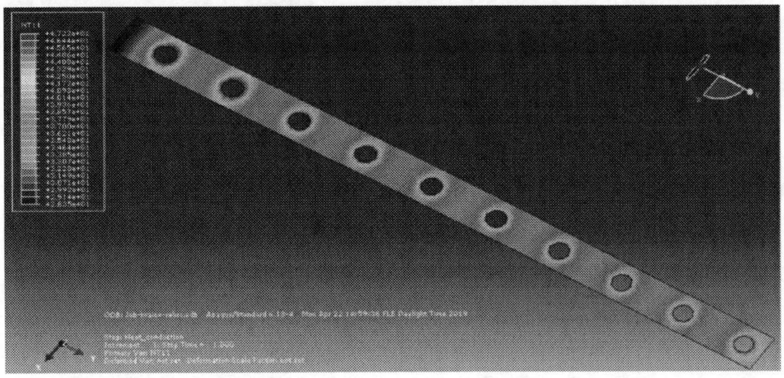

Fig. 10. Veiw of the temperature field in the last fin.

3.4 Results for the average temperatures and heat convection coefficient

Table 5 presents the calculated results for the following parameters: average temperatures of the heat exchange surfaces of the fins, average temperature of the heat exchanger's external surface, and external heat convection coefficient.

Table 5. Results for the calculated parameters.

$\overline{t_{w-first\_fin}}$, °C	$\overline{t_{w-middle\_fin}}$, °C	$\overline{t_{w-last\_fin}}$, °C	$\overline{t_w}$, °C	$\overline{t_{heated\,air}}$, °C	$\overline{h_{c-external}}$, W/(m².K)
38.492	38.319	38.387	39.386	19.662	56.215

The ambient temperature during the investigation has been 14.2 °C.
The relatively small difference in the average temperatures of the studied fins is an indicator of a relatively small difference in the amount of heat they exchange with the air flow crossing them.
For comparison of the results presented in Table 5 with results obtained by another method, the heat convection coefficient of the two-pipe sheaf is calculated, based on a criterion equation. The value obtained is 40.737 W/m^2.K. It should be noted that the data in Table 5 refer to the first row of the sheaf.

4 Conclusions

A methodology for determination of the external heat convection coefficients of finned water-air heat exchangers has been developed. The methodology is consist of numerical modeling of the heat convection processes in the ducts and subsequent numerical modeling of the thermal conduction processes in the fins.
Schematic diagram of the simulation model has been built in ABAQUS. The model has been developed, and the computing mesh has been verified.
The temperature field of the pipes' external heat exchange surface has been obtained by ignoring the thermal resistance of the pipes walls.
The average temperature of the convector's external heat exchange surface has been calculated based on the temperature field of the pipes' external surface in the zones between the fins, as well as based on the fins' temperature field.
The average external heat convection coefficient has been calculated on the base of the average temperature of the heat exchanger's external surface.
The big number of temperatures used as boundary conditions to simulate the process of thermal conductivity in the fins is a prerequisite for accurate determination of the simulation temperature field.
The developed methodology allows the results obtained in the first simulation to be used as boundary conditions for the implementation of the second simulation. Thus, the temperature field in the fins of the heat exchanger can be obtained, given the existing difficulties in measuring the maximum number of temperatures and its experimental building.
The precise determination of the external heat convection coefficient is very important, given the fact that it has a higher impact on the overall heat transfer coefficient of the investigated heat exchanger.

References

1. R. Petkova-Slipets, P. Zlateva, Thermal insulating properties of straw-filled environmentally friendly building materials, De Gruyter, 52 (2017)
2. S. Tasheva, V. Gandova, K. Dobreva, I. Dincheva, V. Prodanova-Stefanova, A. Stoyanova, Studies of physicochemical and thermal properties of linalool-ethanol-water system, Research Journal of pharmaceutical, Biological and Chemical Sceince, **10 (6)**, 220 (2019)
3. R. M. Saldanha da Gama, F. B. de Freitas Rachid, M. L. Martins-Costa, Mathematical modeling of heat transfer problems in thin plates with temperature dependent conductivity, Heat Transfer Research, (2017)
4. N. Namdari, M. Abdi, H. Chaghomi, F. Rahmani, Numerical solution for transient heat transfer in longitudinal fins, International Research Journal of Advanced Engineering and Science, 131 (2018)

5. M. Zaidan, A. Alkumait, T. Ibrahim, Assessment of heat transfer and fluid flow characteristics within finned flat tube, Case Studies in Thermal Engineering **12**, 557 (2018)
6. Z. Kadhim, M. Kassim, A. Hassan, Effect of integral finned tube on heat transfer characteristics for cross flow heat exchanger, International Journal of Computer Applications **(0975 – 8887)**, 139 (2016)
7. He Fa Jiang, Cao Wei Wu, Yan Ping, Experimental investigation of heat transfer and flowing resistance for air flow cross over spiral finned tube heat exchanger, Energy Procedia **17**, 741 (2012)
8. A. Cebula, T. Sobota, Determination of the heat transfer coefficient distribution on the longitudinal finned tubes in staggered arrangement using inverse and CFD method, European Conference on Computational Fluid Dynamics (ECCOMAS CFD), (2006)
9. S. Singh, K. Sørensen, T. Condra, Multiphysics numerical modeling of a fin and tube heat exchanger, Proceedings of the 56th SIMS, 383 (2015)
10. M. S. Baba, M. Bhagvanth Rao, A. V. Sita Rama Raju, Experimental study of convective heat transfer in a finned tube counter flow heat exchanger with Fe3O4 – water nanofluid, International Journal of Mechanical Engineering and Technology (IJMET), 500 (2017)
11. D. Jung, D. N. Assanis, Numerical modeling of cross flow compact heat exchanger with louvered fins using thermal resistance concept, SAE Technical Paper Series, (2006)
12. S. Arena, E. Casti, J. Gasia, L. F. Cabeza, G. Cau, Numerical simulation of a finned-tube LHTES system: influence of the mushy zone constant on a phase change behavior, Energy Procedia **126 (201709)**, 517 (2017)
13. Lin Wei, Guorong Zhu, Zhijiang Jin, Numerical simulation of heat transfer in finned tube of heat recovery unit using fluid-solid coupled method, Advances in Mechanical Engineering, (2015)
14. Shubham Singh, Venkata Krishnan K., Spandana H., Mahesh Kumar N., P. S. Kulkarni, Numerical study of the heat transfer enhancement of circular tube bank fin heat exchanger with vortex generators, 20th Annual CFD Symposium – Bangalore, (2018)
15. Zh. Kolev, S. Kadirova, Numerical modelling of heat transfer in convector's pipes by ABAQUS, IOP Conf. Series: Materials Science and Engineering, 595 (2019)
16. Z. Kolev, S. Kadirova, CFD simulation of forced heat transfer of gas in pipe, E3S Web of Conferences **112**, 01008, TE-RE-RD (2019)
17. E. Dimofte, F. Popescu, I. Ion, Numerical modelling of mixing fluids at different temperatures, Proceedings of TE-RE-RD, 35 (2016)
18. Z. Kolev, S. Kadirova, T. Nenov, Research of reversible heat pump installation for greenhouse heating, INMATEH - Agricultural Engineering, 77 (2017)
19. P. Zlateva, K. Yordanov, Experimental study of heat pump type air-water for heating system performance, E3S Web of Conferences **112**, 01007, TE-RE-RD, (2019).
20. S. Valchev, N. Nenov, Determination of specific energy consumption of mechanical vapor recompression heat pump, Scientific Researches of the Union of Scientists in Bulgaria – Plovdiv, 99 (2016).
21. S. Valchev, N. Nenov, Study of thermodynamic parameters of mechanical heat pump system, Scientific Works of University of Food Technologies, 247 (2016)

The study of the spark ignition engine operation at fuelling with n-butanol-gasoline blends

Cristian Sandu[1*], *Constantin* Pană[1], *Niculae* Negurescu[1], *Alexandru* Cernat[1], *Cristian* Nuțu[2], and *Rareș* Georgescu[1]

[1]Thermotechnics, Engines, Thermical and Frigorific Equipment Department, University Politehnica of Bucharest, Bucharest, Romania

[2]Automotive Engineering Department, University Politehnica of Bucharest, Bucharest, Romania

Abstract. For conventional internal combustion engines alternative fuels such alcohols (ethanol, methanol and butanol) have attracted more attention. This aspect is due to the fact that alcohols have good combustion properties and high oxygen content. Butanol is a viable fuel for blending with conventional fuels such as gasoline or diesel because of its high miscibility with these conventional fuels. The high combustion speed of butanol compared to that of gasoline ensures a shorter burning process thus the engine thermal efficiency can potentially be improved. Moreover, the additional oxygen content of the alcohol n-butanol can potentially improve the combustion process and can lead to a reduction of carbon monoxide and unburnt hydrocarbons emissions level. Utilizing butanol-gasoline blends can provide a good solution for the reduction of greenhouse gases level (CO_2) and pollutants level (CO, HC, and NOx). An experimental study was carried out in a spark ignition engine which was fueled with a blend of n-butanol-gasoline at different volume percentages. The objective of this paper is to determine the effects of butanol on the engine energetic performances and on the emissions (HC, CO and NOx). At first the engine fueled with pure gasoline to set up a reference at the engine load χ=55%, engine speed of n=2500 min-1 and different excess air coefficients (λ). After setting the reference the engine was fueled with butanol-gasoline blend (10% vol. butanol - 90% vol. gasoline) with the same engine adjustments. At butanol use the CO, HC and CO_2 emissions level decreased, but the NOx emission level increased. The butanol can be considered a good alternative fuel for the spark ignition engines without modifications.

1 Introduction

Current worldwide fossil fuel consumption will lead to a depletion of conventional fuels in a matter of decades, thus more researches have been focused on alternative fuels. The great consumption of fossil fuel is also associated with greenhouse gases such as CO_2 and pollutants products, especially for passenger vehicles over the last years. For internal combustion engines, alcohols have gained a lot of attention in recent years. The use of the alcohols like methanol, ethanol or butanol - conventional fuels such as gasoline or diesel

* Corresponding author: sandu_cristian2007@yahoo.com

blends can improve the combustion process in the engine cylinder and reduce the fossil fuels consumption. Due to their better combustion proprieties, alcohols can increase the thermal efficiency for a given engine size and compression ratio, improve of the combustion resulting in lower conventional fuel consumption and thus lower CO_2 emissions [1, 2, 3]. A well-to-wheel analysis carried out by Wu. M. et al, [4], concluded that on a life-cycle basis, the use of the butanol resulted from corn production as fuel could reduce the consumption of gasoline by 39-56% and reduce the greenhouse emissions by up to 48%.

Out of all the alcohols mentioned before, n-butanol is regarded as encouraging solution because of its better properties over the other alcohols such as higher energy density, low viscosity and good blending properties, [5]. When compared to ethanol, butanol is much less corrosive and hygroscopic as makes it more compatible with the fossils fuels distribution infrastructure, [6]. Liu et al. [7, 8] studied the flame natural luminosity and spray of different oxygenated fuels using laser diagnostics. The levels of soot concentration were much lower for n-butanol and more restricted when compared to biodiesel. Another study by Liu et al., [9] studied the effects of oxygenated compounds on engine emissions and thermal efficiency. For this study five different fuels were used for testing such as n-heptane, iso-octane, n-butanol, 2-butanol and methyl octyonate. The study found that oxygenated compounds have little impact on engine pollutants such as nitrogen oxides, carbon monoxide and unburned hydrocarbons. He also noticed that there is no impact on the indicated thermal efficiency. A study by Yuqiang Li et al., [10] analyzed the impact of isopropanol-n-butanol-ethanol gasoline blends on performance and emissions of a spark ignition engine. The blended fuel (IBE30) is made of 30% isopropanol-butanol-ethanol and 70% gasoline. Results with IBE30 show an improvement of the brake thermal efficiency by up to 4,3% for the blended fuels and lower HC (15,1-20,3%), lower CO (4%) and NO_x (3,3 -18,6%) in some cases. The study concluded that IBE30 can be considered as a good alternative to gasoline. J. Serras-Pereira et al., [11] analyzed of the combustion process in a direct injection engine by using ethanol, butanol, iso-octane, gasoline and methane. Results show that ignition is more difficult for butanol in a cold environment (they indicate a high coefficient of variability of IMEP and longer duration of MFB in the early stages of fuel ignition of 0-10%). The initial combustion phase was longer for all alcohols with 5-7°CA. The coefficient of variation of IMEP was higher for some alcohols but for butanol and ethanol did not exceed 8%. The COV for iso-octane has increased beyond 10-20%. The ignition timing increased for all alcohols with 10-15°CA to reach maximum indicate mean effective pressure with minimum variation. Butanol as a standalone fuel was used in a homogenous charge compression ignition engine. Studies were focused on using n-butanol as fuel, engine performances were limited by the reduction of combustion stability at lean fuel mixtures [12]. At low to mid-engine loads, it was possible to obtain low NO_x and soot levels without the use of exhaust gas recirculation, [13]. Brake thermal efficiency is similar to diesel engines. Alasfour measured the efficiency in a single-cylinder engine at fueling with a blend of 30% butanol-gasoline. The study indicates a reduction of 7% engine power at fueling with the blended mix vs pure gasoline, [14]. In another study also carried out by Alasfour [15], nitrogen oxides emissions were measured at different air-fuel ratio (λ). Results indicate a reduction in nitrogen oxides emissions for λ between 0,9 and 1,05 at fueling with the blended fuel. A 9% decrease of NO_x emissions level was also observed. The highest NO_x emissions level was found at a slightly leaner mixture for the blended mix vs gasoline due to the additional oxygen content. Another study, [16] concluded that the increase the spark timing will lead to higher NO_x emissions.

The objective of this paper is to determine the impact of n-butanol on the engine energetic performances and pollutants (HC, CO and NO_x) in a 4 cylinder spark ignition 1,5l type Cielo Nubira A15MF. The engine is turbocharged and was firstly fueled with pure

 https://doi.org/10.1051/e3sconf/202018001010

gasoline to set the reference at the engine load χ=55%, engine speed of n=2500 min^{-1} and different air-fuel ratios (λ). After setting the reference the engine was fueled with n-butanol-gasoline (10% vol. butanol - 90% vol. gasoline) with the same engine adjustments.

2 Experimental environment

The experimental test bench is presented in Fig. 1.

The engine used is a 4 cylinder, turbocharged, 1,5l type Cielo Nubira A15MF. The engine is fueled by gasoline and the engine load of χ=55% and speed of 2500 min^{-1} ; different parameters are recorded such as air and water temperature, intake air temperature, exhaust gases temperature, fuel consumption, air consumption, boost pressure and emissions level. In cylinder pressure is measured across 250 engine cycles and the mean values of these measurements are used to determine of the maximum pressure (p_{max}), maximum pressure rise rate $(dp/d\alpha)_{max}$, heat release rate ($dQ/d\alpha$) and total heat released (Q).

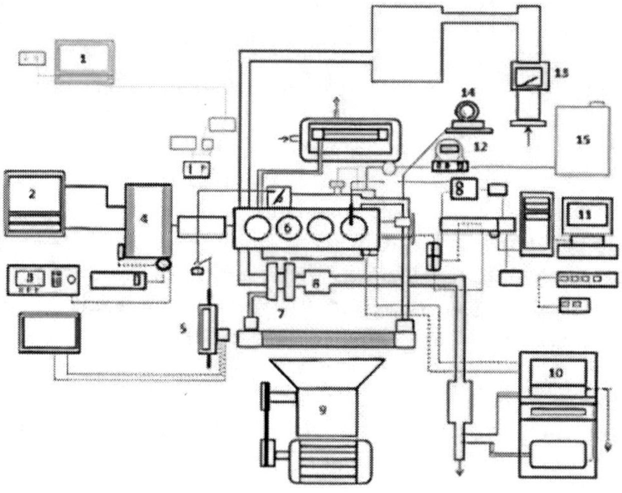

Fig. 1. Test bench schema

The main components of the test bench are: 1 – ECU, 2 – dyno cooling system, 3 – dyno control cabinet, 4 – Eddy current dyno, 5 – throttle body actuator, 6 – A15MF engine, 7 – turbocharger, 8 – catalytic converter, 9 – radiator cooling system, 10 – gas analyzer, 11 – desktop, 12 – fuel mass flow transducer, 13 – air flow transducer, 14 – boost pressure transducer, 15 – fuel reservoir.

3 Results

In the figures 2-11 experimental results are presented while fueling the engine with gasoline and butanol-gasoline blended mix.

In Fig.2 the pressure diagrams for both fuels are represented. Because the spark timing was kept constant at the butanol-gasoline blend use the initial combustion phase duration increased and the combustion is moved after the fuel ignited.

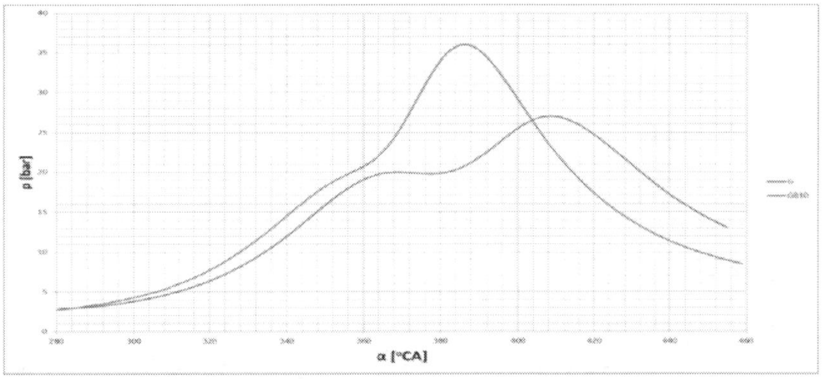

Fig. 2 Pressure diagrams for gasoline and blended mix

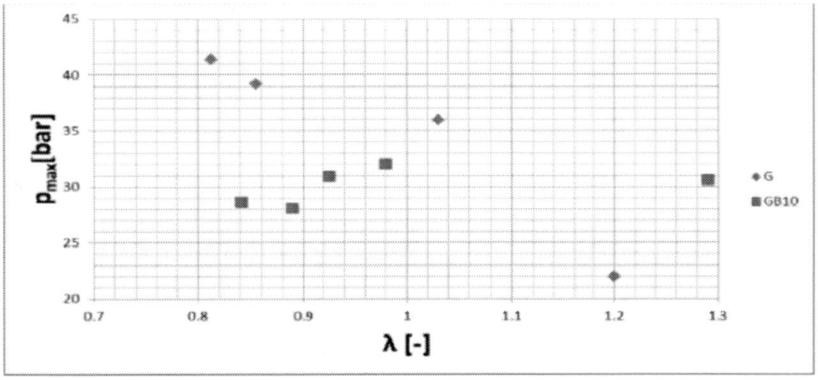

Fig. 3 Maximum cylinder pressure for gasoline and blended mix

The maximum pressure for gasoline and butanol-gasoline blend is presented in the Fig.3. It is observed the higher maximum pressure value for gasoline, with 25%, at rich mixtures comparative to the blended mix because the spark timing was the same in both cases. This difference decreased in the lean mixtures zone due to the combustion improvement at the butanol use. It is possible that by increasing the spark timing, the maximum pressure for butanol-gasoline blend to become higher than gasoline. Same observations and for maximum pressure rise rate, Fig. 4.

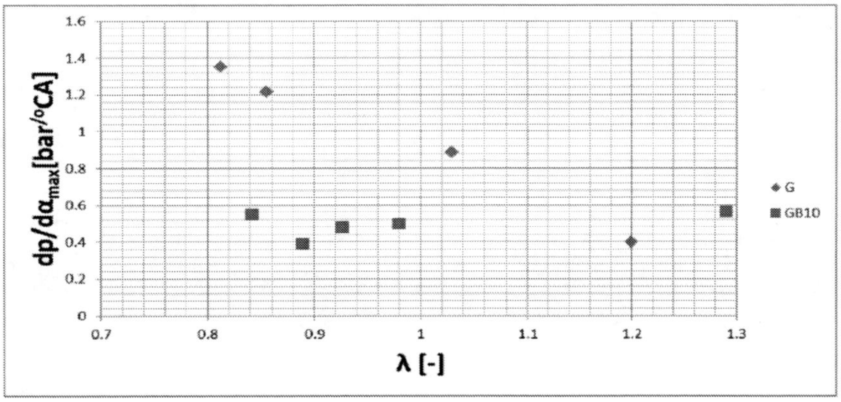

Fig. 4. Maximum pressure rise rate for gasoline and blended mix

 https://doi.org/10.1051/e3sconf/202018001010

The heat release rate is calculated with the relation:

$$\frac{dQ}{d\alpha} = \frac{V}{k-1} \cdot \frac{dp}{d\alpha} + \frac{k}{k-1} \cdot p \cdot \frac{dV}{d\alpha} \tag{1}$$

where:

- $\frac{dU}{d\alpha}$ -the internal energy variation
- $\frac{dL}{d\alpha}$ - the mechanical work variation
- k - adiabatic exponent
- V, p – the volume and pressure

The heat release rate is presented in Fig 5 where a higher value is observed for gasoline. It is also observed the main combustion phase move to the expansion at the butanol-gasoline use.

By integration of the (1) relation the heat release laws were obtained for both fuels, Fig. 6. The burning process duration of the butanol-gasoline is longer comparative to gasoline, because the spark timing was kept constant.

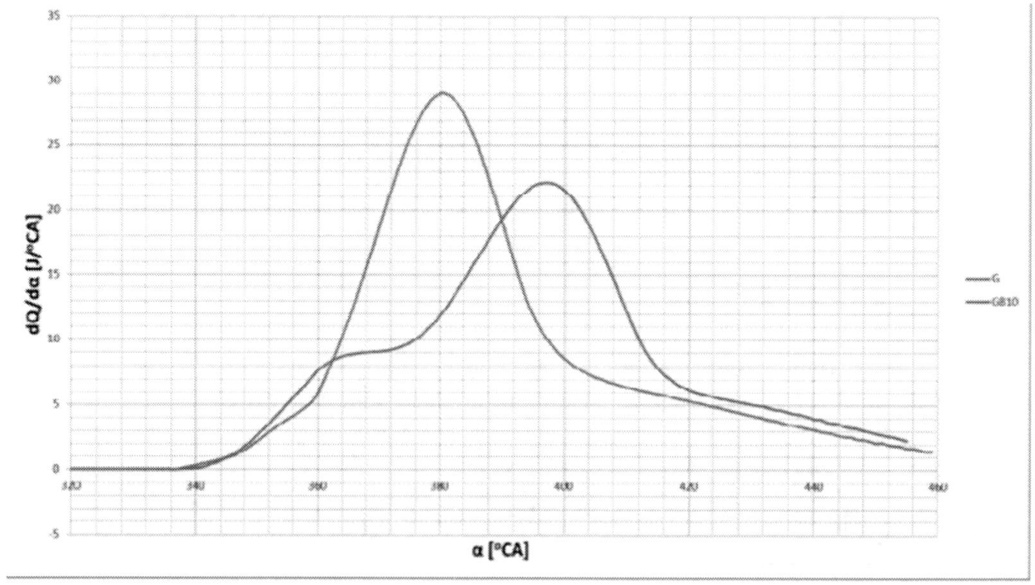

Fig. 5. Heat release rate for gasoline and blended mix

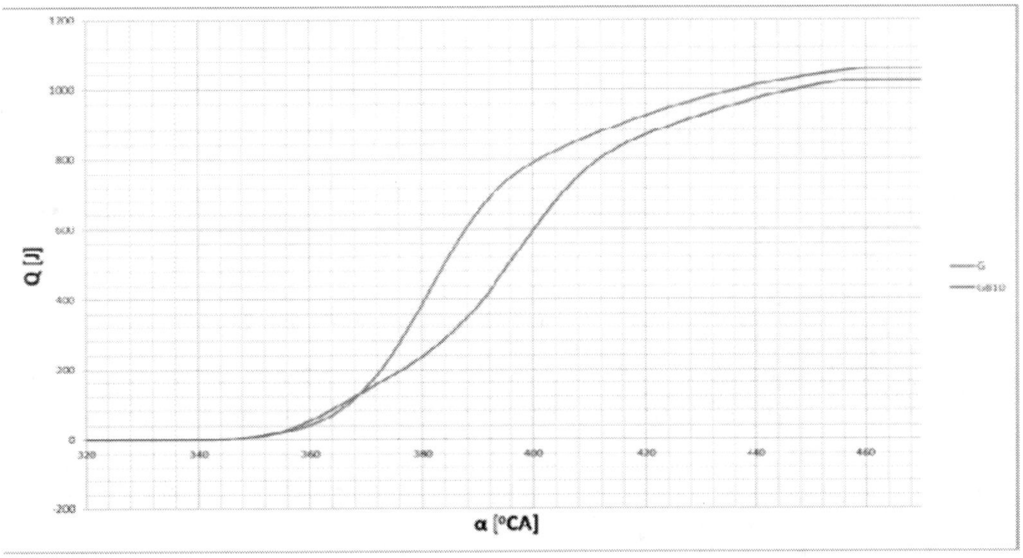

Fig. 6. Heat release laws for gasoline and blended mix

The brake specific energetic consumption (BSEC) for the gasoline-butanol blend is higher than gasoline, Fig.7.The first explanation is that the spark timing has same value both fuels and the initial combustion phase duration increased and the thermal engine efficiency decreased, Fig. 8. The main combustion phase moved to the expansion. The second explanation is the lower caloric heat of the butanol comparative to gasoline.

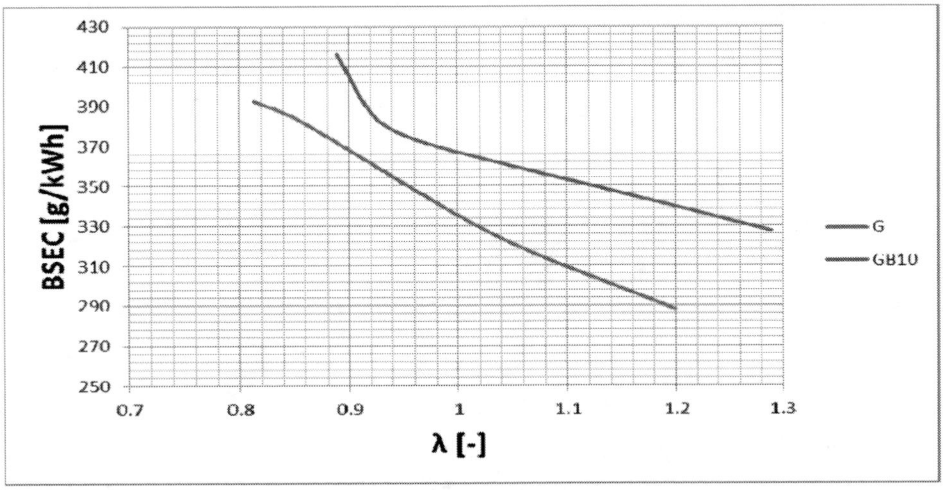

Fig. 7. BSEC for gasoline and butanol-gasoline blend

These negative effects can be eliminated by the spark timing increase.

At the engine fueling with butanol, in the lean mixtures zone the engine operation becomes more stable than gasoline. Such, for λ=1.29 the thermal engine efficiency has the same value as gasoline at λ=1.06, Fig. 8.

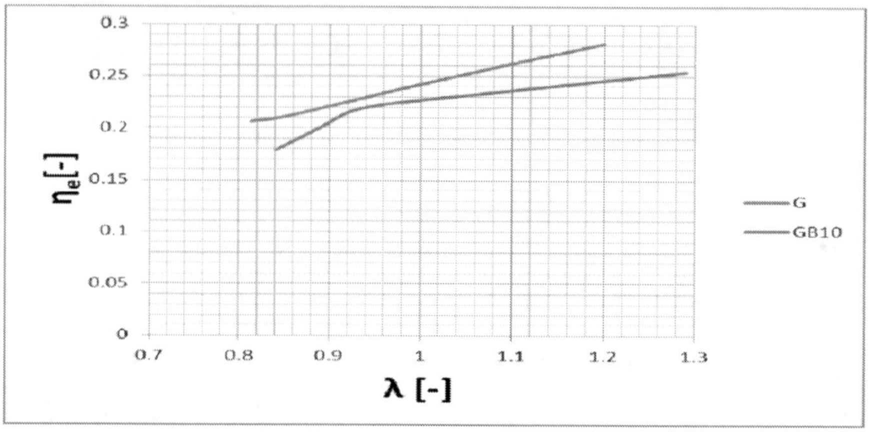

Fig.8. Brake thermal efficiency for gasoline and gasoline- butanol-gasoline

Carbon monoxide is the product of the incomplete combustion, indicating losses of heat. In general, the carbon monoxide level has high values in the rich mixtures domain. In the Fig. 9 it is observed that CO emissions level is higher for gasoline-butanol blend comparative to gasoline at rich mixtures (λ<0,95) due to lower temperature from the engine cylinder and degradation of combustion (butanol has a higher vaporization heat and the gases temperature from the engine cylinder are reduced and the combustion velocity is lower). At stoichiometric mixtures and lean mixtures the CO emissions level is lower at the use of the butanol-gasoline blend with around 1% less than gasoline because of combustion improvement (butanol has a higher content of oxygen than gasoline). Unburnt hydrocarbons (HC) are also a result of incomplete burning process due to air being absent.

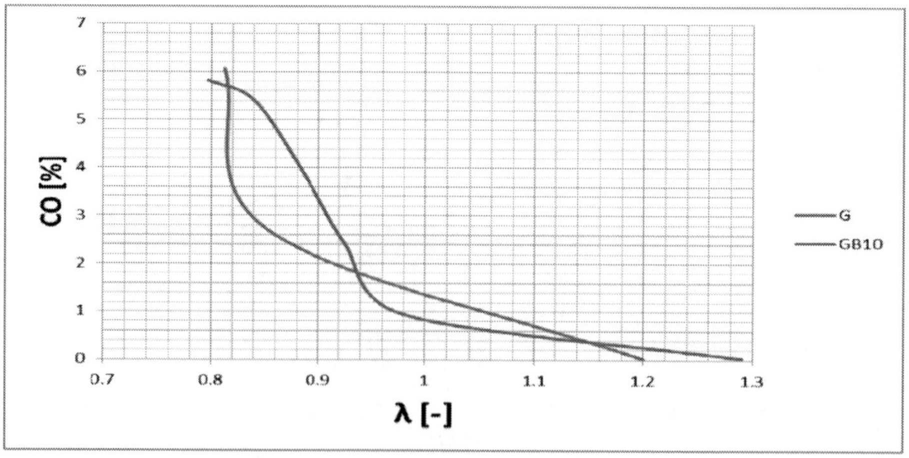

Fig.9. CO emissions level for gasoline vs butanol-gasoline blend

Same as in the case of carbon monoxide, HC emissions level is slightly higher at rich mixtures at the use of the butanol-gasoline blend but they significantly decrease with ~20% at lean mixtures due to the combustion improvement, Fig. 10.

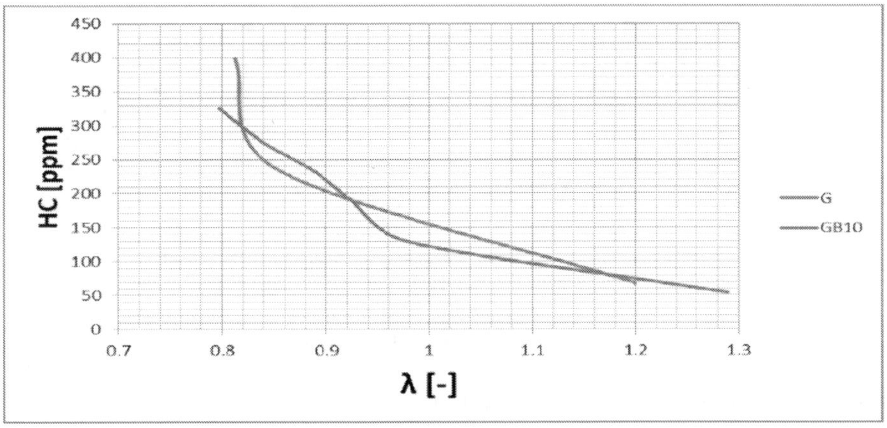

Fig. 10. HC emissions level for gasoline vs butanol-gasoline blend

The nitrogen oxides formation is an endothermic process and depends, in special, on temperature and oxygen content. The variation of nitrogen oxides level is presented in Fig. 11. As shown in this figure, NO_x emissions level for the blended mix is significantly higher than gasoline for all excess air coefficients because of the additional oxygen content of butanol. Thus, NOx emission maximum level at the butanol–gasoline blend use is higher with ~50% than gasoline. At the spark timing increase, necessary adjustments for to move the main combustion phase to TDC, when the combustion duration decreases is possible the NOx emission level will continue to increase. This trend can be reduced by the very lean mixtures use.

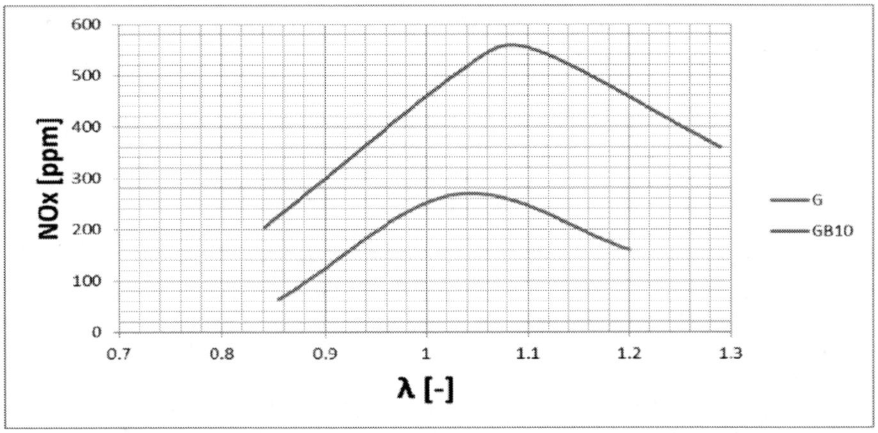

Fig. 11. NO_x emissions level for gasoline vs butanol-gasoline blend

Carbon dioxide is the result of the complete combustion, unlike CO and HC. Hydrocarbons burning with the sufficient amount of air will generate heat producing carbon dioxide and water as a final products of the combustion process. The gasoline-butanol blend improves the combustion process and CO_2 emission level is smaller than gasoline due to lower carbon content, Fig. 12.

In the domain of λ=1,06…1,29, HC, CO and CO_2 emissions levels, at the gasoline-butanol blend use, are lower that gasoline. The results can be improved by the spark timing optimization, Fig. 9, 10 and 12.

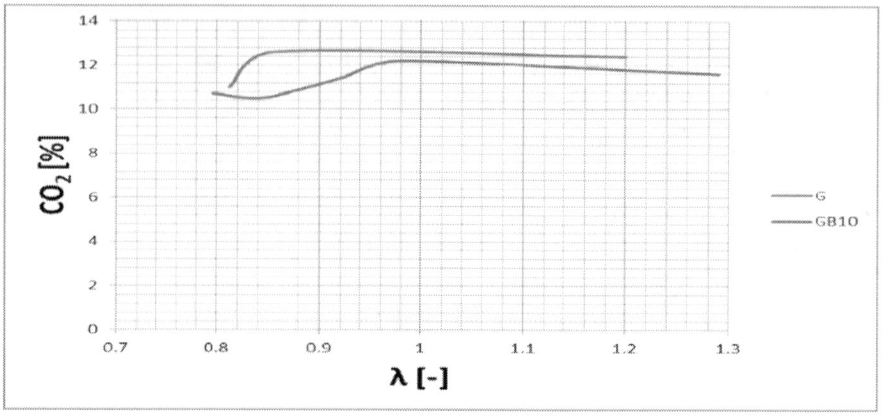

Fig. 12 CO_2 emissions level for gasoline vs butanol-gasoline blend

4 Conclusions

The butanol-gasoline blend use at the spark ignition engines is a good solution for pollutants emissions and greenhouse gases reduction for spark ignition engines.

Therefore, CO emission level decreased with 28% comparative to gasoline due to combustion improvement. Just like CO, HC emission level decreased with 20% comparative to gasoline due to combustion improvement. NOx emissions level at the butanol–gasoline blend use is higher than gasoline. This effect can be reduced by the very lean mixtures use.

Brake specific energetic consumption increased at the butanol-gasoline blend use, but by optimizing the spark timing it can become lower than gasoline.

Maximum pressure and maximum pressure rise rate are smaller at the use of butanol-gasoline blend, with favorable effects for engine strengths.

Butanol can be considered a good alternative solution for gasoline engines without any constructive modifications. The researches will continue for the engine adjustments and butanol fraction in mixture with gasoline optimization.

The authors would like to thank to AVL List GmbH Graz, Austria, for providing the possibility to use the Simulation Software AVL BOOST.
This work has been funded by:
-the European Social Fund from the Sectoral Operational Program Human Capital 2014-2020, through the Financial Agreement with the title "Scholarships for entrepreneurial education among doctoral students and postdoctoral researchers (Be Entrepreneur!)", Contract no. 51680/09.07.2019 - SMIS code: 124539
-a grant of the Romanian Ministry of Research and Innovation, CCCDI – UEFISCDI, project number PN-III-P1-1.2-PCCDI-2017-0404 / 31PCCDI/2018, within PNCDI III

References

1. Gautam M., Martin II D.W. (2000) Combustion characteristics of higher-alcohol/gasoline blends, Proceedings of the IMechE Part A: J. Power and Energy 214, 5, 497-511.Yoon J, Ryu J and Lim K B 2005 A novel reconfigurable ankle rehabilitation robot for various exercises Proceedings of the 2005 IEEE International Conference on Robotics and Automation ICRA 2005 pp 2290-2295

2. Gautam M., Martin II D.W., Carder D. (2000) Emission characteristics of higher-alcohol/gasoline blends, Proceedings of the IMechE Part A: J. Power and Energy 214, 2, 165-182.
3. Yacoub Y., Bata R., Gautam M. (1998) The performance and emission characteristics of C1-C5 alcohol-gasoline blends with matched oxygen content in a single-cylinder spark-ignition engine, Proceedings of the IMechE Part A: J. Power and Energy 212, 5, 363-379.
4. Wu M, Wang M, Liu J, Huo H. Assessment of potential life-cycle energy and greenhouse gas emission effects 293 from using corn-based butanol as a transportation fuel. Biotechnol Progr 2008;24:1204-1214
5. Jin C, Yao M, Liu H, Lee CF, Ji J. Progress in the production and application of n-butanol as a biofuel, Renew Sust Energ Rev 2011;15:4080-4106.
6. Doğan O. The influence of n-butanol/diesel fuel blends utilization on a small diesel engine performance and emissions. Fuel 2011;90:2467–2472.
7. H, Huo M, Liu Y, Wang X, Wang H, Yao M, Lee CF. Time-resolved spray, flame, soot quantitative measurement fueling n-butanol and soybean biodiesel in a constant volume chamber under various ambient temperatures. Fuel 2014;133:317-325.
8. Liu H, Bi X, Huo M, Lee CF, Yao M. Soot Emissions of Various Oxygenated Biofuels in Conventional Diesel Combustion and Low-Temperature Combustion Conditions. Energ Fuel 2012;26:1900-1911.
9. Liu H, Li S, Zheng Z, Xu J, Yao M. Effects of n-butanol, 2-butanol, and methyl octynoate addition to diesel fuel on combustion and emissions over a wide range of exhaust gas recirculation (EGR) rates. Appl Energ 2013; 112:246-256.
10. Yuqiang L, Lei M, Karthik N, Timoty H L, Yilu L, Shengming L, Chia-Fon L, Combustion, performance and emissions characteristics of a spark-ignition engine fueled with isopropanol-n-butanol-ethanol and gasoline blends. Elsevier 2016; 184: 864-872.
11. Serras-Pereira J, Aleiferis P.G, Richardson D. An Analysis of the Combustion Behavior of Ethanol, Butanol, Iso-Octane, Gasoline, and Methane in a Direct-Injection Spark-Ignition Research Engine. Combustion Science and Technology (2013), 185:3, 484-513.
12. He B-Q, Yuan J, Liu M-B, Zhao H. Combustion and emission characteristics of a n-butanol HCCI engine. Fuel 2014;115:758–64.
13. Zheng M, Han X, Asad U, Wang J. Investigation of butanol-fuelled HCCI combustion on a high efficiency diesel engine. Energy Convers Manage 2015;98:215–24.b
14. Alasfour F.N. (1997) Butanol -A single-cylinder engine study:Availability analysis, Appl. Therm. Eng. 17, 6, 537-549.
15. Alasfour F.N. (1998) NOx emission from a spark-ignition engine using 30% iso-butanol – gasoline blend: Part 1: Preheating inlet air, Appl. Therm. Eng. 18, 5, 245-256.d
16. Alasfour F.N. (1998) NOx emission from a spark-ignition engine using 30% iso-butanol – gasoline

Researches on biofuels gasification using the Lurgi process with homogeneous air inlet over the combustion space

Gheorghe Lăzăroiu[1], *Lucian* Mihăescu[2], *Gabriel* Negreanu[2], *Iulia* Simion[1]*, *Dana-Alexandra* Ciupăgeanu[1]
[1]Power Engineering Faculty, University Politehnica of Bucharest, Splaiul Independentei 313, 060042 Bucharest (Romania)
[2] Mechanics and Mechatronics Faculty, University Politehnica of Bucharest, Splaiul Independentei 313, 060042 Bucharest (Romania)

Abstract. The experimental investigations carried out within the present research focus on a simple gasification technology dedicated to biofuels conversion according to the Lurgi procedure. Specifically, an installation with a fixed grill and a homogeneous distribution of the air inlet over the combustion space are considered. In order to provide a thorough background for the experimental research, this paper presents first the challenges related to the air distribution. If for coal gasification the difficulty of the homogeneous penetration of the air inlet within the whole combustion volume is balanced by the possibility of the direct emission of CO valorization, for biomass gasification this factor becomes fundamental. The original contribution of the technology introduced in this paper assumes an improved combustion process for Lurgi-type gas generators. The experimental installation employed has a particular design, enabling a homogeneous distribution of the air inlet over the entire combustion zone, up to the top of the embers layer. This allows achieving a maximum CO_2 content in the flue gas flux, effectively reducing it inside the embers bed. The high calorific value of biofuels used favors developing an efficient combustion process, occurring at high temperatures. Thus, the reduction process of the CO_2 is self-controlled. The experimental installation operates at a slow fuel rate, with discontinued supply and precise airflow control. The quality of the gas obtained is evaluated based on the resulting open flame, analyzing its composition.

Key words: biomass combustion, biomass gasification, energy recovery, homogeneous air inlet, Lurgi process

1. Introduction

The agricultural requirements for energy need an alternative solution for fuel gasification; one of the possible means is by gasification of biofuels, with a special focus on biomass [1]. To this end, it is necessary to develop simple operating methods, both on the fuel supply side and on the air supply side. The novel installation developed explores the possibility of

*Contact person: julia.simion@gmail.com

building a gas generator with a more flexible operation than the original Lurgi approach, thus being suitable for biofuels, such as wood or agricultural waste.

Employing steam in the gasification process becomes very difficult because assumes a complementary production plant [2]. Therefore, biomass energy valorization methods are of great interest currently [3].

Experimental tests and theoretical models developments are depicted in the literature, addressing global efficiency improvement paths, combustion facilities upgrading and burning parameters optimization [4]–[6].

The rest of the paper is organized as follows. Section 2 presents some general considerations of gasification models, depicting underlying equations, elemental composition of wood samples and process parameters. Further, Sections 3 describes the experimental stand and Section 4 discusses the results. Finally, conclusions are drawn in the last section.

2. Gasification of biofuels

Basically, the gasification of solid fuels (including biomass) consists of the thermochemical process converting the organic mass into combustible gases under the action of oxygen [6]. Table 1 presents the reactions that occur during the gasification process.

The first three reactions are the defining ones, representing the formation of CO and CO_2 compounds through the burning of carbon, with the final decomposition of carbon dioxide on the bed of gasifier, with the formation of CO combustible gas. A gasogenous can only work if the three reactions show stability in time. The remaining reactions are usually a consequence of the dynamics of the thermodynamic reactions from the space charged with solid fuel.

Table 1. The reactions of the gasification process [7]

No.	Reaction	Thermal effect $\left[\frac{kJ}{kgK}\right]$
1	$C + O_2 = CO_2$	34096
2	$C + 0{,}5O_2 = CO$	10276
3	$C + CO_2 = 2CO$	-13544
4	$C + (H_2O) = CO + H_2,$*	-9904
5	$CO + 0{,}5O_2 = CO_2$	23800
6	$CO + (H_2O) = CO_2 + H_2,$*	3635
7	$H_2 + 0{,}5O_2 = H_2O, water$	-
8	$S + O_2 = SO_2$	-
9	$CH_4 + 2O_2 = CO_2 + 2(H_2O), steam$	67180
10	$H_2S + 1{,}5O_2 = SO_2 + (H_2O), steam$	-
11	$C + 2H_2 = CH_4$	6285
12	$CO + 3H_2 = CH_4 + H_2O$	17004

* $H_2O - steam$

The main benefit of this process is represented by the wider employment range of the resulting gaseous fuel compared to the initial solid fuel (for instance, energy generation technologies using internal combustion engines or gas turbines cannot use solid fuels). Furthermore, the ecological effect can also be taken into account as a contributory factor [8]. Reactions (1), (2) and (3) are heterogeneous reactions, taking place on the separation surface between the solid phase and the gas phase. The rest of the reactions are homogeneous, which take place between the gas phases. The equilibrium of the CO and CO_2 gas phases depending on the temperature is showed in Table 2. The formation of hydrogen H_2 and methane CH_4 depends on the thermochemical cycle of water vapor. The process can be stimulated or only controlled by steam injection into the gas stream.

Table 2. The equilibrium composition for the reaction between carbon and oxygen

Temperature [°C]	Experimental values [%]		Calculated values [%]	
	CO	CO_2	CO	CO_2
450	0.6	99.4	-	-
550	10.7	89.3	11	89
650	39.8	60.2	39	61
800	93.0	7.0	90	10
925	96.0	4.0	97	3

Scientific literature depicts several gasification models [2], presented in the following:

A. Gasification process using air, oxygen and air enriched with oxygen

The general reaction for gasification of biofuels using air, oxygen and oxygen enriched oxygen can be expressed as in eq. (1) [8]:

$$\begin{aligned} CH_aO_bN_cS_d + wH_2O + n(n_{O_2}O_2 + n_{N_2}N_2) \\ \rightarrow n_{H_2}H_2 + n_{CO}CO + n_{CO_2}CO_2 + n_{H_2O}H_2O + n_{CH_4}CH_4 + n_{N_2}N_2 \\ + n_{NO}NO + n_{NO_2}NO_2 + n_{SO_2}SO_2 \end{aligned} \quad (1)$$

Where: $CH_aO_bN_cS_d$ represents the general formula of the biofuel, for which the detailed composition is given in Table 3; *w* is the moisture content, *n* represents the ratio given by kmoles of agent / kmoles of biofuel entered in the gasifier; n_{O_2} is the number of kmoles of oxygen, $n_{H_2}$is the number of kmoles of hydrogen, n_{CO} is the number of kmoles of carbon monoxide, n_{N_2} is the number of kmoles of nitrogen, n_{H_2O} is the number of kmoles of water, n_{CO_2} is the number of kmoles of carbon dioxide, n_{CH_4} is the number of kmoles of methane, n_{NO} is the number of kmoles of nitrogen oxide, n_{NO_2} is the number of kmoles of nitrogen dioxide, n_{SO_2} is the number of kmoles of sulfur dioxide.

By applying the mass balance for hydrogen, carbon, oxygen, nitrogen and sulfur, the relations between the coefficients n_{H_2} and $n_{SO_2}$are determined. Based on the ultimate analysis of several biomass types (in particular wood, paper, municipal solid waste and sawdust), Table 3 presents their compositions taking into account a single carbon atom. Specifically, the chemical formula for solid biofuels will be:

$CH_{1,44}O_{0,66}$; $CH_{1,55}O_{0,766}N_{0,00593}$; $CH_{1,15}O_{0,518}N_{0,0216}S_{0,0047}$ and $CH_{1,4}O_{0,599}N_{0,004}$.

Table 3. Characteristics of anhydrous fuels [3], [8]–[10]

Biomass	C [%]	H [%]	N [%]	S [%]	O [%]	Ash [%]	Thermal effect $[kJ/kmol]$
Wood	50	6	0	0	44	-	449 568
Paper	43.4	5.8	0.3	0.2	44.3	6	454 864
Municipal solid waste	47.6	6	1.2	0.3	32.9	12	433 034
Sawdust	52	6.07	0.28	0	41.55	0.1	485 294

For modelling purposes, scientific literature introduces a series of coefficients to define anhydrous fuels, as given in Table 4.

Table 4. Coefficients that define anhydrous fuels [2]

Biomass	A	b	c	d
Wood	1.44±0.0047	0.66±0.00079	0	0
Paper	1.59±0.0055	0.766±0.001	0.00593±0.00019	0.0034±0.00017
Municipal solid waste	1.51±0.0052	0.518±0.00072	0.0216±0.00018	0.0047±0.00015
Sawdust	1.4±0.0046	0.599±0.00071	0.004±0.00014	0

The moisture content per unit of biomass can be determined based on the moisture content (MC), as showed in eq. (2) [8]:

$$w = \frac{M_{biomass} \times MC}{M_{H_2O} \times (1 - MC)} \times 100 \quad [\%] \tag{2}$$

where $M_{biomass}$ and M_{H_2O} are the molecular weight of the biofuel and water, MC is the moisture content per mole of biomass. The main reactions that occur in the gasification process are as introduced in the following, according to [6]:

$$C + CO_2 = 2CO \tag{3}$$

$$C + H_2O = CO + H_2 \tag{4}$$

$$C + 2H_2 = CH_4 \tag{5}$$

Based on eq. (3) and (4) it can be obtained by specific operations the following relation, as given in eq. (6):

$$CO + H_2O \leftrightarrow CO_2 + H_2 \tag{6}$$

Table 5 lists the parameters of the gasification process.

Table 5. Input data [4], [5], [11]

Air inlet temperature	298 $[K]$
Biomass inlet temperature	298 $[K]$
Temperature of syngas existing gasifier	1 073 $[K]$
Environmental pressure	1.013 $[bar]$
Environmental temperature	298 $[K]$
Pressure of gasifier	1.013 $[bar]$
Molar composition of environmental air	21% O_2, 79% N_2

B. Gasification process with steam

The general reaction of the gasification process of a biofuel using steam can be expressed by eq. (7):

$$CH_aO_bN_cS_d + wH_2O + m(H_2O) \rightarrow n_{H_2}H_2 + n_{CO}CO + n_{CO_2}CO_2 + n_{H_2O}H_2O + n_{CH_4}CH_4 + n_CC_{(s)} \tag{7}$$

where $CH_aO_bN_cS_d$ is the general formula of the biofuel, w is the moisture content of the biomass and m is the molar ratio between steam and biomass.. In the scientific literature, the steam to biomass mass ratio (STBM), is defined and used as follows [12]

$$STBM = \frac{M_{H_2O} \times m}{\left(M_{biomass} + M_{H_2O} \times w\right)} \tag{8}$$

The dominant chemical reactions that take place in the steam gasification process are [12]:

$$CH_4 + H_2O \leftrightarrow CO + 3H_2 \tag{9}$$

$$CO + H_2O \leftrightarrow CO_2 + H_2 \tag{10}$$

The equilibrium constants for the methane decomposition reaction (9) and the gas-water reaction (10) can be written as follows [12]:

$$K_1 = \frac{n_{CO} n_{H_2}^3}{n_{CH_4} n_{H_2O}} \left(\frac{P/P_{ref}}{n_{tot}}\right) \quad (11)$$

$$K_2 = \frac{n_{H_2} n_{CO_2}}{n_{CO} n_{H_2O}} \left(\frac{P/P_{ref}}{n_{tot}}\right) \quad (12)$$

where P is the gas pressure (bar), P_{ref} is the reference pressure, and the number of moles of the gas elements produced corresponds to the general equation, K_1 and K_2 are equilibrium constants that can be related to the Gibbs function as follows [10]:

$$-\frac{\Delta G_1^0}{\bar{R}T_g} = \ln K_1 \quad (13)$$

$$-\frac{\Delta G_2^0}{\bar{R}T_g} = \ln K_2 \quad (14)$$

where ΔG_1^0 and ΔG_2^0 are modifications of the Gibbs free energy for the methane decomposition reaction and T_g is the gas temperature.

Considering the general equation of steam gasification, the equilibrium equations for carbon, hydrogen and oxygen can be written as:

$$n_{CO} + n_{CO_2} + n_{CH_4} + n_C - 1 = 0 \quad (15)$$

$$2n_{H_2} + 2n_{H_2O} + 4n_{CH_4} - a - 2w - 2m = 0 \quad (16)$$

$$n_{CO} + 2n_{CO_2} + n_{H_2O} - b - w - m = 0 \quad (17)$$

To validate the steam gasification model using wood, the theoretical data reported by [8] can be used for a gasification temperature of 1 073 K and a moisture content of 25%.

The comparison between the percentage component of the gas obtained from the paper [8] with those from the specialized literature is presented in Table 6.

Table 6. Percentage component of gas obtained from steam gasification of biofuels

Components	Hydrogen	Carbon monoxide	Carbon dioxide	Water	Methane	Nitrogen
Results reported by [5]	44.58	23.51	9.44	16.56	0.07	0
Theoretical results reported by [11]	46.57	25.84	10.30	17.11	0.09	0

C. Functional aspects and performance analysis of air gasification

The efficiency of carbon conversion is defined by the relation:

$$\eta = \frac{12(CO + CO_2 + CH_4 + 2.5C_mH_n)}{22.4\left(\frac{298}{273}\right) \times C} \times G_V \times 100 \quad [\%] \quad (18)$$

The efficiency of the gasification is defined by the relation:

$$\eta = \frac{H_{i\,gas}^i \times G_V}{H_{i\,fuel}^i} \times 100 \quad [\%] \quad (19)$$

where $H_{i\,gas}^i \left[\frac{kJ}{Nm^3}\right]$ - is the lower calorific power of the generated gas, and $H_{i\,fuel}^i \left[\frac{kJ}{kg}\right]$ - is the lower calorific power of the fuel. The speed of the reactions depends on the thermal level first, followed by the effect of concentration and pressure [13]. The reaction rate for two substances having the concentrations C_A and C_B is given by eq. (20):

$$W = KC_A^m C_B^n \quad (20)$$

where m and n are the stoichiometric coefficients of the reaction, and K is the reaction rate constant.

In Table 2 is presented the equilibrium composition, for various temperatures, for the basic reduction reaction for CO_2 on the embers bed. The thermochemical processes in the body of a gasifier are shown in Figure 1, highlighting moreover the particularities of the updraft flow and the downdraft flow [14].

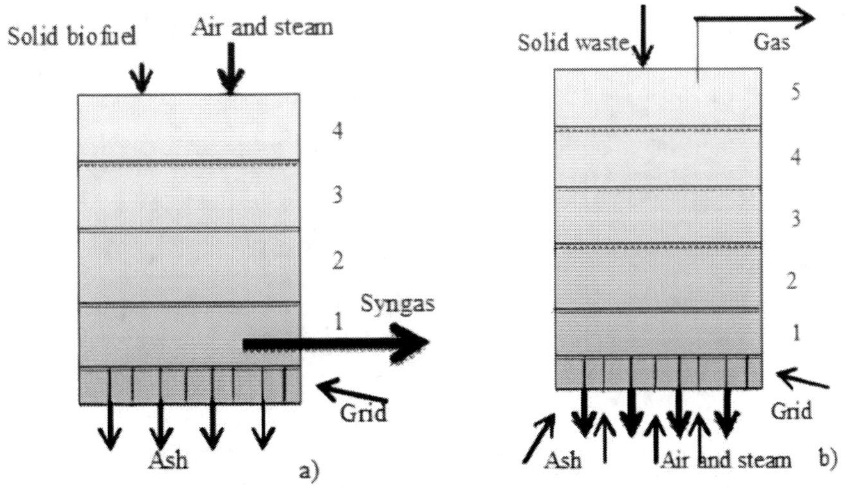

Fig. 1. The thermochemical processes in the body of a gasifier a) updraft; b) downdraft

For updraft flow, zone 1 represents the ash cooling, zone 2 is the combustion zone (temperature range between 1 000 and 1 300 °C), zone 3 is the reduction zone, zone 4 is the preheating process (pyrolysis), with a temperature range within 600-800 °C, and zone 5 is characteristic to the drying process.

For the downdraft flow, the characteristic areas follow the same order (1-heating, 2-pyrolysis, 3-combustion, 4-reduction). The technology is suitable for low humidity biomass because heating and dehydration is not aided by burning of the fuel. Instead, the temperature in the reduction zone is maximum, close to that of combustion in the fuel layer. There have been developed also systems of air supply with transverse flow, derived from the application of the flow for both presented technologies. The system implemented in this research partially feeds with air the pyrolysis area, and continues to be distributed mainly in the combustion zone. The air in the end zone of the pyrolysis process ensures the ignition of the released volatiles and the preparation of the solid phase combustion process, as depicted in Figure 2 and Figure 3 [14], [15].

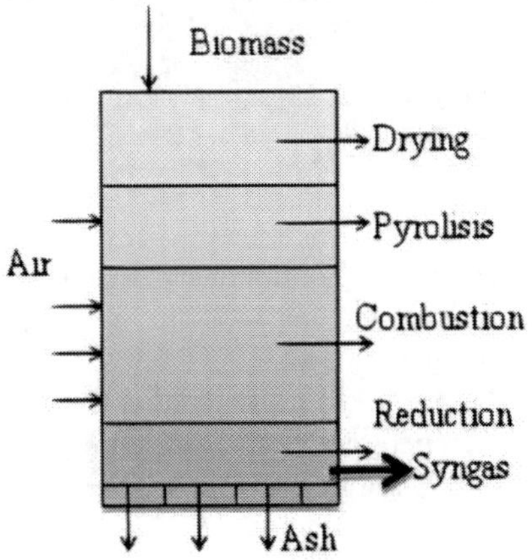

Fig. 2. Installation with homogeneous distribution of air

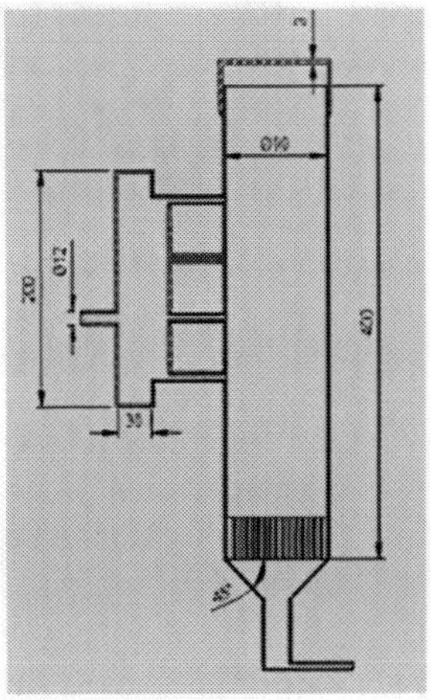

Fig. 3. The experimental installation

3. The experimental installation

The experimental installation is designed to produce gasifier gas using an inlet rate up to 1 kg of biomass per hour. The installation shown in Figure 3 uses a discontinuous supply system. The gaseous fuel passes through the ash, being discharged beneath it.
For the initial phase of the research, wood (beech species) with low humidity (9-14%), calorific power was considered within the range:
$H_i^i = 14\,200 - 14\,700\ \left[\frac{kJ}{kg}\right]$. The theoretical volume of air is $V_a^0 = 3.8 - 3.9\ \left[\frac{m_N^3}{kg}\right]$, while the total volume of combustion gases is evaluated at $V_g^0 = 4.6 - 4.9\ \left[\frac{m_N^3}{kg}\right]$.
Moreover, the theoretical volume of CO_2 is $V_{CO_2}^0 = 0.8 - 0.85\ \left[\frac{m_N^3}{kg}\right]$ and the theoretical volume of steam is $V_{H_2O}^0 = 0.8 - 0.9\ \left[\frac{m_N^3}{kg}\right]$.

On the second phase of the research, the influence of humidity on the gasification process is assessed by employing wood with an average humidity of 37%. .The experimental stand is presented in Figure 4, including an air supply compressor and a two rotameter system to allow the measurement and control of the inlet.

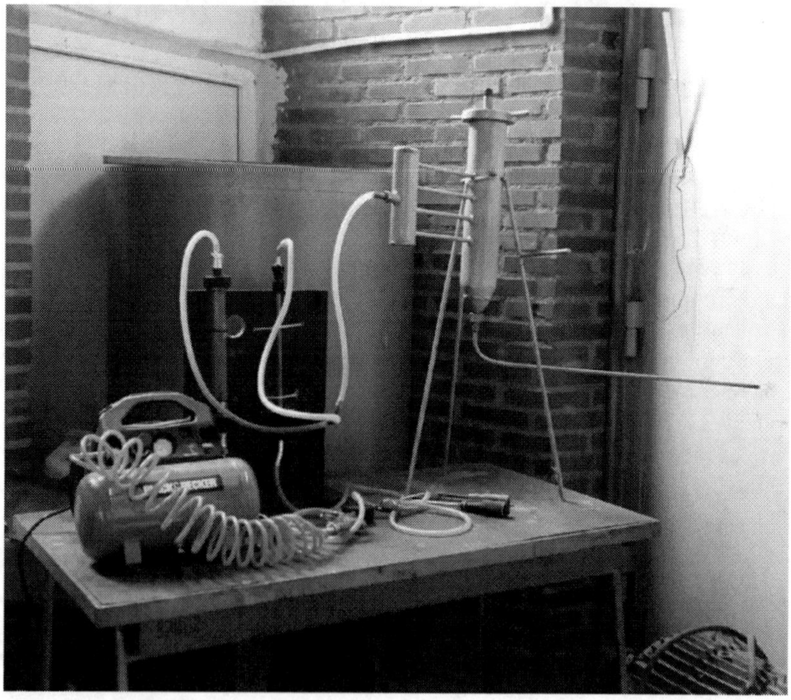

Fig. 4. Experimental stand.

The temperature of the gasifier body was monitored continuously using a contact thermometer or a thermal imaging chamber. The syngas is evacuated in the environment, having the possibility to easily measure its flow, temperature and its composition. Figure 5 shows the combustion of syngas when discharged into the environment and the appearance of the flame is highlighted on the right side. The temperature of the outside of the gasifier body, measured by a thermal vision camera is shown in Figure 6.

Fig. 5. Evacuation of the syngas for the experiment with the dry wood

Fig. 6. Temperature measurement with the thermal imaging chamber

4. Results

The first round of results refer to dry wood, with the following characteristics:

- humidity $W_t^i < 12.2\ \%$;
- lower heating value $H_i^i = 16\ 000 - 16\ 500\ \left[\frac{kJ}{kg}\right]$;
- particles size $0 - 15\ mm$;

For this fuel, the air flow rate was $V_a^0 = 1.2 - 1.5\ \left[\frac{m_N^3}{h}\right]$. Figure 7 pictures the analyzer.

Fig. 7. The measure of syngas composition

According to the measurements results, the average composition of the syngas, results as: $CO = 4.8\%$; $CO_2 = 8.4\%$; $N_2 = 55\%$; $CH_4 + H_2 = 31.8\%$.

The analysis of the syngas was carried out in the combustion section of the installation in order not to take into account the flow of air driven through the environment by the gas jet. For a ratio $\frac{H_2}{CH_4} \cong 4$ it results: $CH_4 = 6.36\%$ and $H_2 = 25.44\%$. The lower heating value is:

$$H_i^i = 0.084 * 12644 + 0.0636 * 35700 + 0.254 * 10760 = 5614.61 \left[\frac{kJ}{m_N^3}\right] \quad [20]$$

The second round of results is obtained for biomass with 37% humidity, it was observed the increase of the participation of carbon oxide and the decrease of the one of hydrogen in the gas produced by gasification. The average composition was: $CO = 5.7\%$; $CO_2 = 9.5\%$; $N_2 = 54\%$; $CH_4 = 7.7\%$; $H_2 = 23.1\%$. For what concerns the lower heating value, it is evaluated at:

$$H_i^i = 0.057 * 12644 + 0.077 * 35700 + 0.231 * 10760 = 5955.1 \left[\frac{kJ}{m_N^3}\right] \quad [21]$$

The reason why the net calorific value is higher in second case than in first one is because the content of methane and carbon oxide is higher in this case due to it's high moisture content.

5. Conclusions

In this work, on the basis of theoretical studies, the potential of gasification of biomass was demonstrated, and through experimental work, was demonstrated the possibility of gasification of wood in the created laboratory facility. More in detail, the process is researched by using only the reduction processes from the final stage with a gasification agent represented only by air.

The pilot plant under study is equipped with discontinuous fuel supply and has a thermal power of $3.3\ kW$ and a combustion under a slight air overpressure.

A cross-sectional distribution over the entire area of chemical combustion reactions of air was used.

During the experimental investigations, the humidity of the wood varies within the range 12.2% to 37%. The gas flow rate was $2.06\ m_N^3/h$ for a wood flow of $0.5\ kg/h$.

This work was supported by a grant from the Romanian Ministry of Research and Innovation, CCCDI—UEFISCDI, project number PN-III-P1-1.2-PCCDI-2017- 0404/31PCCD/2018, within PNCDI.

Nomenclature

n_{H_2} hydrogen moles [kmol]
n_{CO} carbon monoxide moles [kmol]
n_{CO_2} carbon dioxide moles [kmol]
n_{H_2O} water moles [kmol]
n_{CH_4} methane moles [kmol]
n_{N_2} nitrogen moles [kmol]
n_{O_2} oxygen moles [kmol]
n_C carbon moles [kmol]
n_{NO} nitrogen oxide moles [kmol]
n_{NO_2} nitrogen dioxide moles [kmol]
n_{SO_2} sulfur dioxide moles [kmol]
K equilibrium constant
n kmol of oxygen per kmol of biomass
m kmol of steam per kmol of biomass
T temperature [K] and [°C]
P pressure [bar]
ΔG change in Gibbs function
η efficiency [%]
w moisture content [%]
$H_{i\,gas}^i$ lower calorific power of the gas $\left[\frac{kJ}{m_N^3}\right]$
$H_{i\,fuel}^i$ lower calorific power of the fuel $\left[\frac{kJ}{kg}\right]$
G_V gas yield during the whole gasification $\left[\frac{m_N^3}{kg}\right]$
V_a^0 theoretical volume of air $\left[\frac{m_N^3}{kg}\right]$
V_g^0 volume of combustion gases $\left[\frac{m_N^3}{kg}\right]$
$V_{CO_2}^0$ theoretical volume of CO_2 $\left[\frac{m_N^3}{kg}\right]$
$V_{H_2O}^0$ theoretical volume of steam $\left[\frac{m_N^3}{kg}\right]$

References

1. L. Mihaescu *et al.*, "Achievements in energy valorization of cereal straw in boiler manufactured in Romania," in *5th International Conference on Trends in Agricultural Engineering*, pp. 417–421, (2013)
2. E. Shayan, V. Zare, and I. Mirzaee, "Hydrogen production from biomass gasification; a theoretical comparison of using different gasification agents," *Energy Convers. Manag.*, vol. 159, no. January, pp. 30–41, (2018)
3. G. Lazaroiu *et al.*, "Experimental Investigations of Innovative Biomass Energy Harnessing Solutions," *Energies*, vol. 11, no. 12, p. 3469, (2018)
4. C. Colpan, F. Hamdullahpur, I. Dincer, and Y. Yoo, "Effect of gasification agent on the performance of solid oxide fuel cell and biomass gasification systems," *Int. J. Hydrogen Energy*, vol. 35, pp. 5001–9, (2010)
5. Z. Zainal, R. Ali, C. Lean, and K. Seetharamu, "Prediction of performance of a downdraft gasifier using equilibrium modeling for different biomass materials," *Energy Convers. Manag.*, vol. 42, pp. 1499–515, (2001)
6. S. Sansaniwal, K. Pal, M. Rosen, and S. Tyagi, "Recent advances in the development of biomass gasification technology: a comprehensive review," *Renew. Sustain. Energy Rev.*, vol. 72, pp. 368–84, (2017)
7. I. Carabogdan, A. Badea, L. Ionescu, A. Leca, I. Nistor, and I. Cserveny, *Industrial termical installtion*. Bucharest: Editura Tehnica Bucuresti, (1978)
8. E. Gholamian, V. Zare, and S. Mousavi, "Integration of biomass gasification with a solid oxide fuel cell in a combined cooling, heating and power system: a thermodynamic and environmental analysis," *Int. J. Hydrogen Energy*, vol. 41, pp. 20396–406, (2016)
9. E. Gholamian, "Proposal, exergy analysis and optimization of a new biomass-based cogeneration system," *Appl. Therm. Eng.*, vol. 93, pp. 223–35, (2016)
10. C. Altafini, "Prediction of the working parameters of a wood waste gasifier through an equilibrium model," *Energy Convers. Manag.*, vol. 44, pp. 2763–3277, (2003)
11. G. Schuster, G. Loffler, K. Weigl, and H. Hofbauer, "Biomass steam gasification- an extensive parametric modeling study," *Bioresour. Technol.*, vol. 77, pp. 71–9, (2001)
12. E. Balu, U. Lee, and J. Chung, "High temperature steam gasification of woody biomass - a combined experimental and mathematical modeling approach," *Int. J. Hydrogen Energy*, vol. 40, pp. 14104–15, (2015)
13. G. Lazaroiu *et al.*, "Biomass combustion with hydrogen injection for energy applications," *Energy*, vol. 127, pp. 351–357, (2017)
14. I. Pisa, G. Lazaroiu, L. Mihaescu, T. Prisecaru, and G. Negreanu, "Mathematical model and experimental tests of hydrogen diffusion in the porous system of biomass," *Int. J. Green Energy*, (2016)
15. A. D. Bondrea, G. Lazaroiu, L. Mihaescu, and V. Berbece, "The particularities flame to burning bio-fuels," in *International Multidisciplinary Scientific GeoConference Surveying Geology and Mining Ecology Management, SGEM*, (2018)

E3S Web of Conferences **180**, 01012 (2020) https://doi.org/10.1051/e3sconf/202018001012
TE-RE-RD 2020

Determination of hydraulic resistance of the aerothermopressor for gas turbine cyclic air cooling

Dmytro Konovalov[1], *Halina* Kobalava[1], *Mykola* Radchenko[1]*, *Ionut-Cristian* Scurtu[2], *Roman* Radchenko[1]

[1] Admiral Makarov National University of Shipbuilding, Heroes of Ukraine Avenue, 9, Mykolayiv, 54025, Ukraine
[2]. Mircea cel Batran Naval Academy, Romania

Abstract. One of the promising trends to increase the fuel and energy efficiency of gas turbines is contact cooling of cyclic air by using a two-phase jet apparatus – an aerothermopressor. The rational parameters of work processes of the aerothermopressor were studied. The experimental setup was designed to simulate the aerothermopressor operation in the cooling air cycle of the gas turbine and to determine pressure losses in the aerothermopressor flow part. Based on the obtained experimental data, an empirical equation was proposed to determine the hydraulic resistance coefficient of the aerothermopressor flow part, depending on the initial pressure and the amount of water injected. The deviation of the calculated hydraulic resistance coefficient from the experimental ones is ± 25 %. The obtained results can be used in the practice of designing the aerothermopressor for gas turbine cyclic air cooling.

1. Introduction

To bring the process of compressing air in gas turbine compressors closer to isothermal ones the complex schemes with cyclic air cooling are usually used [1, 2]. It results in increasing the fuel and energy efficiency [3, 4]. One of the promising trends is contact cooling of cyclic air [5, 6]. For this purpose the jet technologies can be used [7-10]. To provide efficient evaporative cooling of gas turbine cyclic air without total pressure loss the aerothermopressor are quite useful [11]. Due to evaporative cooling in the aerothermopressor, an effect of thermogasdynamic compression is taken place – gas pressure is increased in the process of instantaneous evaporation of water injected into the gas (air) flow accelerated to a speed close to sound [12, 13]. An effective water dispersed atomization in the air flow occurs in the aerothermopressor. This ensures efficient evaporation of the water dispersed flow in the flow part of the gas turbine compressor bringing the compression process closer to isothermal.

*Corresponding author: nirad50@gmail.com

2. Literature Review

The aerothermopressor allows to compensate the pressure loss and reduce the work of compression in the compressor and to increase the working fluid flow and, accordingly, the gas turbine power output [11]. In addition, the aerothermopressor is compact and structurally and technologically simple to manufacture in comparison with surface air coolers [12, 14, 15].

The working processes in the aerothermopressor are considerably influenced by design factors. The total pressure of the gas flow in the aerothermopressor is significantly affected by pressure losses caused by hydraulic and local resistance in the aerothermopressor flow part (confuser, evaporation chamber and diffuser) [13, 16]. Therefore for designing the aerothermopressors it is important to determine the rational parameters of working processes in their flow part.

The operation of the aerothermopressor on the gas turbine exhaust gases was investigated in [12]. Data on the pressure loss caused by local and hydraulic resistance were obtained for the experimental aerothermopressor (gas flow rate G_g = 11.5 kg/s). It was shown that the total pressure of compressed air and the work of compression correspondingly was decreased by approximately 14 %.

The pressure losses in the aerothermopressor flow part depend primarily on the operating mode nature. The process of accelerating the gas flow to a transonic speed (0.5-0.9 *M*) occurs in the confuser. To the inlet chamber, located before the confuser, liquid is supplied using a nozzle or other special devices [12, 16, 17]. In the evaporation section, these flows interact, and, as a result, droplets are accelerated, crushed, heated and evaporated, and gas is also cooled [13]. There are three main operating modes of the aerothermopressor [12, 13, 17]:

1. The influence of the water droplets resistance prevails over the positive effect of evaporation and determines the gas flow behaviour. The Mach number is increased, the static pressure of the flow is decreased, and the water temperature has risen approaching the saturation temperature.

2. Water evaporation predominates. The Mach number is decreased, and the total and static pressure is increased.

3. Surface friction (in the first two modes was relatively insignificant) becomes the predominant factor. This mode is taken place when the liquid completely evaporates.

All three modes determine the effect of hydraulic resistance on pressure losses. One of the ways to increase the aerothermopressor efficiency is to increase the total pressure by reducing losses due to hydraulic resistance of the flow part. It might be achieved by providing liquid incomplete evaporation - dispersed mode. Incomplete evaporation should provide a greater pressure increase, due to a friction losses decrease in the two-phase flow and it should provide effective fine atomization of water, due to a drop reduction during partial evaporation.

It is known that pressure losses in the two-phase flow (liquid-gas) may be less than pressure losses in the single-phase flow (gas) [18, 19, 20]. When providing liquid injection through the nozzle in an amount more necessary for evaporation, mode 3 is absent (in this mode, surface friction of the single-phase flow predominates). It positively affects the increase in total flow pressure as a result of the thermogasdynamic effect.

To determine pressure losses in the aerothermopressor, classical methods are used. These techniques are based on the analytical determination of the local and friction resistance coefficients of each of the aerothermopressor elements [21, 22]. Pressure losses due to resistance (acceleration or deceleration of the droplet) can be determined by the aerodynamic resistance coefficient of the droplet in the flow ζ_w in accordance with the technique proposed in [21]. Pressure losses in the evaporation chamber ζ_{ch} might be determined by the classical Blasius equation for flow in the channels [22] or by the Lockcart-Martinelli phases slip model [19, 20]. To calculate losses from the total resistance for the confuser and diffuser are used equations for the local loss coefficients ζ_c, ζ_d [22, 23].

Thus, the general equation for determining pressure losses in the aerothermopressor:

$$\Delta P_r = \zeta_{ATP} \cdot \frac{w_{air}^2 \cdot \rho_{air}}{2}, \tag{1}$$

where ζ_{ATP} – total coefficient of hydraulic resistance in the aerothermopressor flow part; w_{air} – average air velocity in the evaporation chamber; ρ_{air} – air density.

The total coefficient of hydraulic resistance is determined by:

$$\zeta_{ATP} = \zeta_w + \zeta_c + \zeta_d + \zeta_{ch} \tag{2}$$

The application of the indicated methods for calculating losses by the elements of the aerothermopressor flow part often gives incorrect results. First of all, it is due to the complexity of accounting for the operating modes of the flow in the aerothermopressor.

Even more difficult is to determinate the losses due to the resistance of the droplet [24] in the confuser and in the evaporation chamber initial part. This is due to the difficulty in determining the initial velocity of the droplet, its diameter, etc. It has been experimentally established that such losses can reach 10–20% [12, 13], but there are no equations for the exact determination of such losses for the aerothermopressor.

Determination of pressure losses in the aerothermopressor flow part experimentally, taking into account water injection, will clarify the design methodology for such jet devices. This will provide an accurate determination of the aerothermopressor effective use as a part of the gas turbine for cooling cyclic air under various operating conditions, as well as in various climatic conditions.

3. Research Methodology

The installation of the aerothermopressor in the gas turbine cyclic air cooling system is proposed in accordance with the scheme (Fig. 1) [11]. This arrangement between the compressors will reduce the operation of the gas turbine high-pressure compressor due to lower initial temperature of the compression process. This will ensure isothermal compression during the evaporation of the finely dispersed mixture in the high-pressure compressor flow part [25].

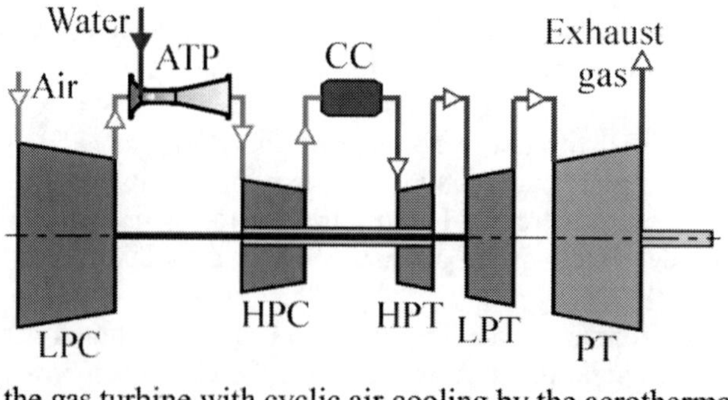

Fig. 1. Scheme of the gas turbine with cyclic air cooling by the aerothermopressor: ATP – aerothermopressor; LPC, HPC – low and high pressure compressors; CC – combustion chamber; HPT, LPT – low and high pressure turbines; PT – power turbine.

To conduct an experimental study of determining pressure losses in the aerothermopressor flow part, an experimental setup was developed (Fig. 2). The experimental setup is designed to simulate the aerothermopressor operation in the cooling cycle air of the gas turbine.

The simulation of the aerothermopressor operation to cooling cyclic air of the gas turbine WR-21 from Rolls Royce (N_e = 25.250 kW, g_e = 0.190 kg/(kW·h), η_e = 41.2%) were considered.

The principle of experimental setup operation (Fig. 3): after cleaning in the air filter (2) (Caterpillar - 4N-0015 CAT), the air was supplied for compression to the screw compressor (3) (Atlas Copco XA 85), and pumped at 0.6 MPa into the air receiver. After deep cleaning in the three-section moisture separator and oil separator (7), the air was heated in the ducted gas heater (11) to t_1 = 50-180 °C. After preparation (approximation to gas turbine cyclic air parameters), the air entered into the experimental aerothermopressor (19). Water for injection came from a distilled water reserve tank (14) and it was injected by a high-pressure pump (18) (STIHL RE 98). Water spray was carried out by nozzles of type FMT. At the pressure of 7.5 MPa, the nozzles provided a water flow rate at the receiving chamber inlet of the aerothermopressor: FMT-43.0 – g_w = 0.0175 kg/s; FMT-100.0 – g_w = 0.0407 kg/s; FMT-120.0 – g_w = 0.0487 kg/s. The spraying angle was 70-90 ° and the average droplets diameter was δ_w = 18–20 μm with a maximum diameter of δ_w = 50 μm.

The experimental aerothermopressor (Fig. 4 a) consists of the following elements: a receiving chamber with a nozzle (Fig. 4 b) and system for injecting water into the flow (Fig. 5 a); confuser (Fig. 5 b); evaporation chamber (Fig. 6 a); diffuser (Fig. 6 b); nozzles for installing temperature and pressure sensors. All elements of the aerothermopressor are removable allowed to carry out studies for the different geometric characteristics (Table 1).

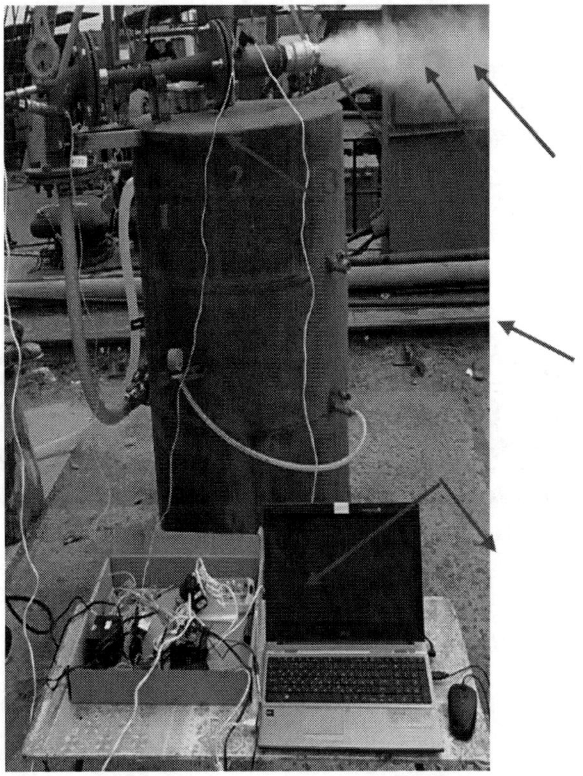

Fig. 2. Photo of the experimental setup:
1 – aerothermopressor receiving chamber with installed temperature and pressure sensors; 2 – aerothermopressor; 3 – pipe insert at the exit of the aerothermopressor with installed

temperature and pressure sensors; 4 – control valve with a device for trapping droplets; 5 – three-section moisture separator for air coming from the compressor module; 6 – computer monitoring and control system.

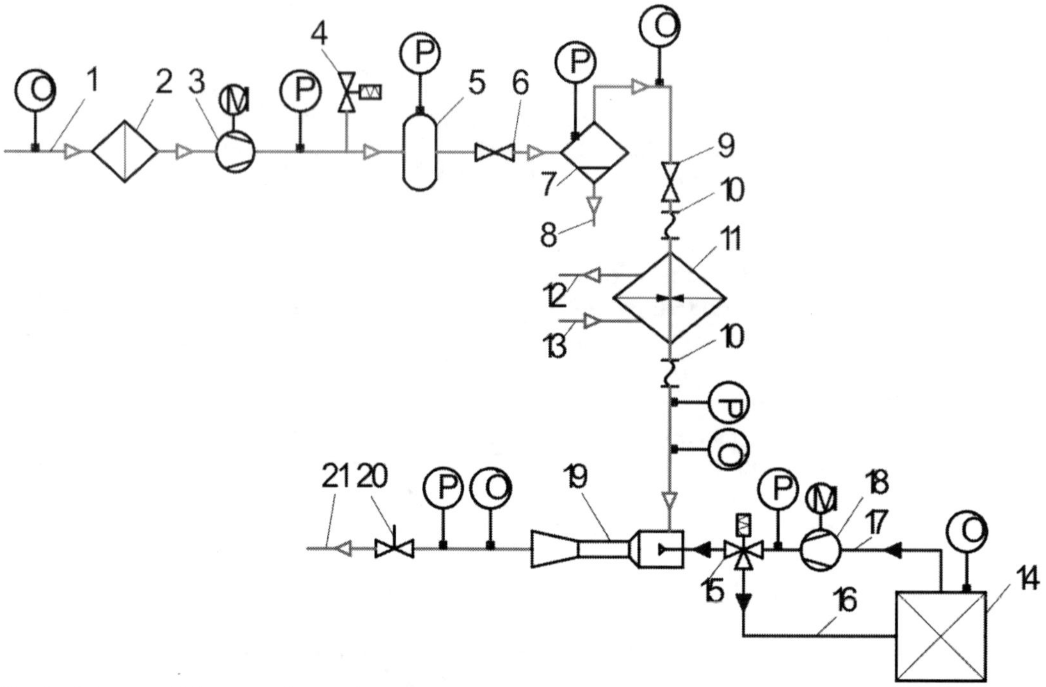

Fig. 3. Scheme of the experimental setup:
1 – inlet pipeline; 2 – air filter; 3 – screw compressor; 4 – safety valve; 5 – air receiver; 6, 9 – control valves; 7 – three-section moisture separator; 8 – drain oil and moisture; 10 – temperature compensator; 11 – gas air heater; 12, 13 – inlet and outlet of hot gas; 14 – water reserve tank; 15 – control valve of the injection system; 16 – return pipeline; 17 – pipeline for supplying water to the injection system; 18 – high pressure pump; 19 – experimental aerothermopressor; 20 – control valve at the outlet; 21 – output pipeline; P – pressure sensors; T – temperature sensors.

(a) (b)

Fig. 4. Experimental aerothermopressor (a), a nozzle for water injection system (b).

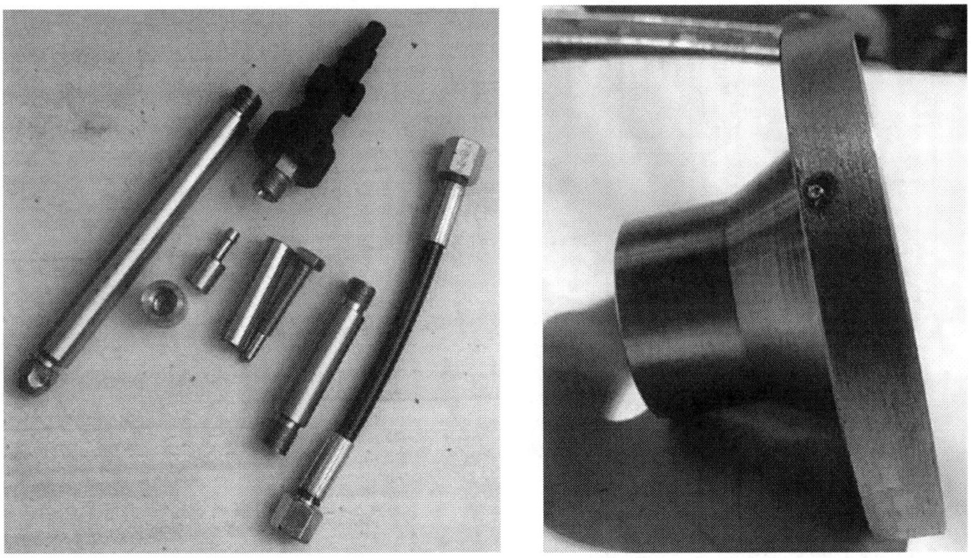

(a) (b)

Fig. 5. Water injection system (a), confuser (b).

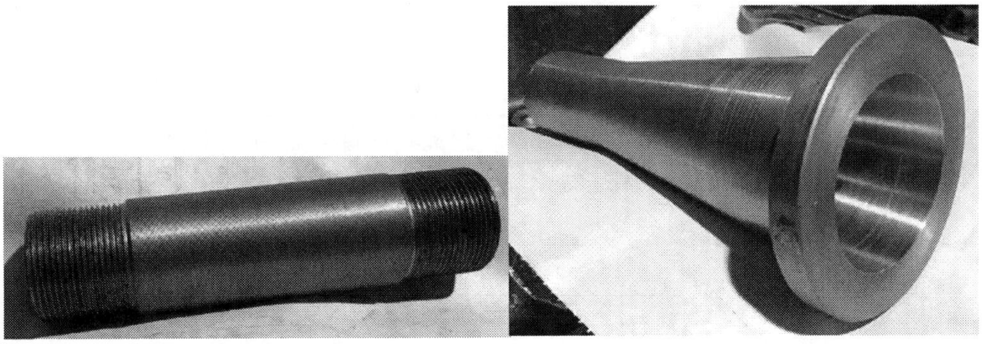

(a) (b)

Fig. 6. Evaporation chamber (relative length (L/D) = 5) (a), diffuser (b).

All temperature, pressure and air flow sensors were connected to a developed computerized monitoring system. The parameters were measured with an interval $\tau = 1$ s. To record the readings of measuring instruments, eight-channel meters I8-TS (temperature measurement) and I8-AT (pressure measurement) were used by RegMik. To collect and organize information about the data, a PI485 / USB RS485 communication interface converter was used, which converts USB interface signals (USB 1.1 and USB 2.0 compatibility) to RS485 / RS-422 / V.11 interface signals (EIA-485, EIA- 422A).

Testing and measuring devices with a list of the parameters they measure, the measurement range, accuracy classes and measurement errors are given in Table 2. The error of the experimental results was determined by the error of the measuring instruments, methodological and systematic errors.

To determine the local resistance coefficients for the diffuser and confuser, classical dependences of fluid dynamics [13, 14] were used. The energy equation (Bernoulli equation for air flow, taking into account mechanical specific losses):

$$\rho g z_1 + p_1 + N_1 \frac{\rho w_1^2}{2} = \rho g z_2 + p_2 + N_2 \frac{\rho w_2^2}{2} + \Delta p_t , \quad (3)$$

where $\rho g z_1$, $\rho g z_2$ – geometric pressure; p_1, p_2 – static pressure; Δp_t – total losses of total pressure; $N_1 \frac{\rho w_1^2}{2}$, $N_2 \frac{\rho w_2^2}{2}$ – dynamic pressure.

Dividing by ρg:

$$z_1 + \frac{p_1}{g\rho} + N_1 \frac{w_1^2}{2g} = z_2 + \frac{p_2}{g\rho} + N_2 \frac{w_2^2}{2g} + \Delta h_t , \quad (4)$$

where Δh_t – total head losses.

The equation for total pressure is:

$$p_t = p_1 + N_1 \frac{\rho w_1^2}{2} \quad (5)$$

Taking into account the fact that $z_1 = z_2$ and Eq. (4), the total head losses Δh_t:

$$\Delta h_t = \frac{p_{t1} - p_{t2}}{g\rho} \quad (6)$$

The local resistance coefficient ζ_{ATP} is:

$$\zeta_{ATP} = \frac{2\Delta p_t}{\rho w^2} \quad (7)$$

Table 1. Geometrical characteristics of the experimental aerothermopressor.

Parameter	Value
Receiving chamber	Diameter D_1 = 65 mm; length L_1 = 200 mm
Confuser	Inlet diameter D_{c1} = 65 mm; outlet diameter D_{c2} = 25 mm; length L_c = 34 mm; convergent angle α = 30°
Evaporation chamber	Diameter D_{ch} = 25 mm; length L_{ch} = 125; 175 mm (relative length l_{ch} = 5; 7)
Diffuser	Inlet diameter D_{d1} = 25 mm; outlet diameter D_{d2} = 65 mm; length L_{ch} = 192 mm; divergent angle β = 6°
Nozzle	The distance between the water exit point to the receiving chamber inlet: L_f = 5; 65; 125 mm (L_f = 125 mm corresponds to the evaporation chamber inlet)

Based on the obtained experimental data, for different values of the initial pressure at the aerothermopressor inlet P_1 and the relative amount of water injected g_w, an empirical dependence was established by approximation to determine the pressure losses coefficient in the aerothermopressor with the flow rate to 0.52 kg/s.

The reliability of the work results is ensured by: the tasks correct formulation of the theoretical and experimental research; confirmation of the adequacy of the mathematical model with satisfactory agreement between the calculated and experimental data; using modern methods of experimental research and analytical modeling.

Table 2. Testing and measuring devices.

Measured value	Name of measuring instruments	Measuring range	Error
Outside air temperature	Multifunction tester for environment parameters 1052 (Chauvin Arnoux)	–20…+80 °C	± 0.4% (± 0.3 °C)
Outside air pressure		0…110 kPa	± 0.2 %
Air pressure on the discharge line, in front of the "Cyclone" air separator	Measuring pressure transducer, model A-10 ("WIKA", Germany)	0…600 kPa	± 0.14 %
Air pressure in front of the aerothermopressor			
Steam-water mixture pressure in the evaporation chamber of the aerothermopressor			
Air pressure in the discharge pipe of the aerothermopressor			
Discharge air temperature	Resistive temperature transducer TSP-103 ("RegMik", Ukraine)	-40…+270 °C	± 0.5 %
Air temperature in front of the aerothermopressor			
Steam-water mixture temperature in the evaporation chamber of the aerothermopressor			
Air temperature in the discharge pipe of the aerothermopressor			
Water pressure in the system (in front of the nozzle)	Technical manometer MT-UHL3 (GOST 15150-82)	0…10 MPa	± 4 %
Reserve tank water temperature	Resistive temperature transducer TSP-103 ("RegMik", Ukraine)	-40…+270 °C	± 0.5 %
Air flow rate	Variable area flow meter type RM-25 GUZ	0…25 m^3/h	± 4.0 %

4. Results

To determine the hydraulic resistance coefficient of the aerothermopressor flow part, a number of experimental measurements of pressure losses ΔP_r were carried out under various conditions (Fig. 7, 8): the pressure at the inlet to the receiving chamber – P_1 = 150; 200; 250; 300 kPa; injected water flow rate relative to the flow rate of the two-phase flow – g_w = 5-35% (up to 0.0487 kg/s). The measurements are showed: at the inlet pressure P_1 = 300 kPa pressure losses are ΔP_r = 55-75 kPa (20-24%), at P_1 = 250 kPa pressure losses are ΔP_r = 40-45 kPa (16-18%), at P_1 = 200 kPa pressure losses are ΔP_r = 22-24 kPa (12-16%). In this case, with an increase in water flow rate g_w, pressure losses increase by 2-4%, and when the initial pressure decreases from 300 to 150 kPa, the effect of the amount of injected water gradually decreases. At P_1 = 150 kPa, pressure losses are ΔP_r = 18-20 kPa (11-12%), but with an increase g_w to 35% pressure losses remain practically unchanged, the effect of water flow rate on pressure losses ceases.

Pressure losses increase ΔP_r with injected water flow rate increase g_w at inlet pressures P_1 = 200-300 kPa is explained by the greater difference between the initial droplet velocity at the nozzle exit and the air velocity – (w_w / w_{air}) = 1.1–1.3. Hence, there is a greater influence of the droplet resistance. It can reach up to 20% of the total pressure loss in the aerothermopressor. At the pressure P_1 = 150 kPa, the droplet and air velocities are almost equalized and the droplet resistance decreases significantly.

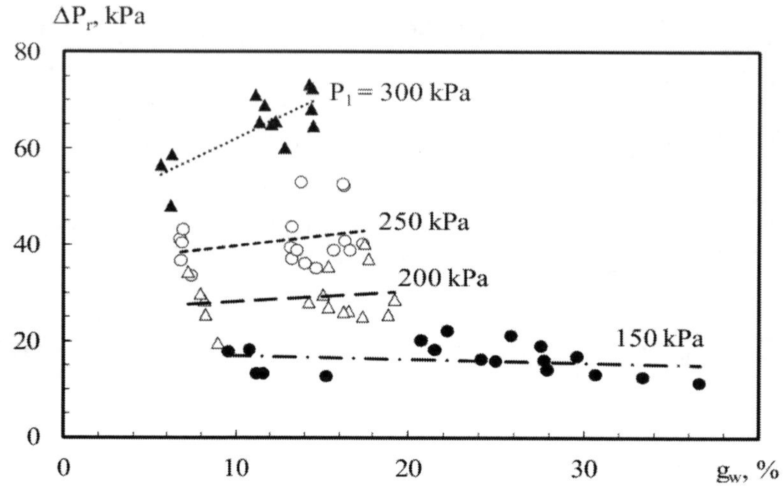

Fig. 7. Experimental dependences of pressure losses ΔP_r on the relative flow rate of the injected water g_w at various initial pressures: P_1 = 150; 200; 250; 300 kPa

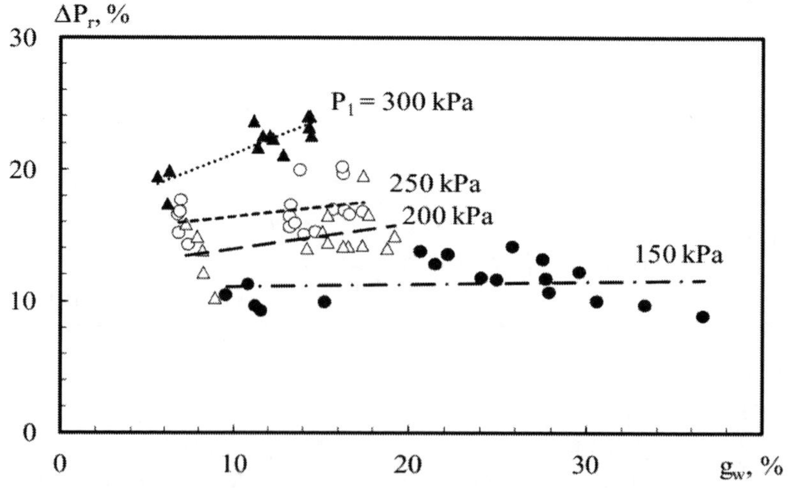

Fig. 8. Experimental dependences of pressure losses $\Delta P_r = 100\cdot(P_1 - P_2)/P_1$ (%) on the relative flow rate of the injected water g_w at various initial pressures: P_1 = 150; 200; 250; 300 kPa.

Determination of the hydraulic resistance coefficient $\zeta_{ATP.E}$ for the aerothermopressor based on the data (Fig. 7, 8, 9): at P_1 = 300 kPa – $\zeta_{ATP.E}$ = 1.15-1.30, at P_1 = 250 kPa – $\zeta_{ATP.E}$ = 0.90-1.00 and at P_1 = 200 kPa – $\zeta_{ATP.E}$ = 0.70-0.75. The tendency of the hydraulic resistance coefficient behavior is the same as for absolute pressure losses ΔP_r, that is, with a water flow rate increase, the hydraulic resistance coefficient $\zeta_{ATP.E}$ increases by 0.05-0.15.

At P_1 = 150 kPa – the value of $\zeta_{ATP.E}$ practically does not change and it is – 0.35-0.40. This behavior is explained by the fact that there are no influences of the *Re* and the relative velocity $M = (w_{air}/w_{sound})$ on the hydraulic resistance coefficient. This indicates there is a self-similar flow regime in the elements of the aerothermopressor flow part. The value of the hydraulic resistance coefficient depends only on the geometric parameters of the channel: confuser convergent angle α and diffuser divergent angle β; the degree of narrowing of confuser n_c, the degree expansion of diffuser n_d; the relative length of the evaporation chamber (caliber) $l_{ch} = (L_{ch}/D_{ch})$.

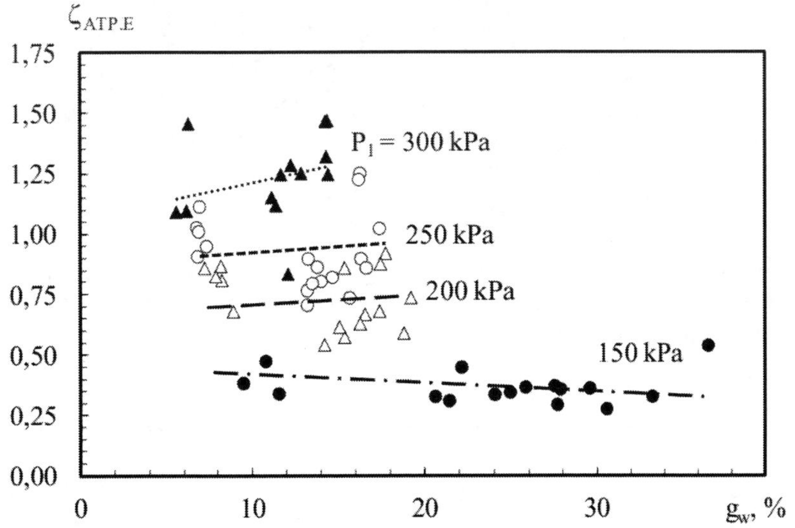

Fig. 9. Experimental dependences of the hydraulic resistance coefficient $\zeta_{ATP.E}$ on the relative flow rate of the injected water g_w at various initial pressures: P_1 = 150; 200; 250; 300 kPa.

Based on the obtained experimental data, the equations of the dependence of the hydraulic resistance coefficient in the aerothermopressor flow part ζ_{ATP} were determined by the approximation method according to a number of equations and depending on the initial pressure at the receiving chamber inlet P_1 and the relative water injected flow rate g_w. In this case, the plane equations were selected (Fig. 10):

$$\zeta_{ATP} = 0.0983 \cdot g_w + 5.8633 \cdot 10^{-6} \cdot P_1 - 0.5004 \quad (8)$$

This equation (regression coefficient – R = 0.8847; R^2 = 0.7827) is obtained for the following characteristics of the aerothermopressor operation: $1.2 \cdot 10^5 < Re < 3.4 \cdot 10^5$; M = 0.25–0.65; g_w = 0.05–0.35; P_1 = 150–300 kPa.

The deviation of the calculated values of the hydraulic resistance coefficient $\zeta_{ATP.C}$ (equation (8)) from those obtained during the experimental study $\zeta_{ATP.E}$ is $\delta_\zeta = \pm 25$ % (Fig. 11). Based on the calculated values $\zeta_{ATP.C}$, the absolute pressure loss in the aerothermopressor $\Delta P_{r.C}$ was determined (equation (1)) and their comparison with experimental $\Delta P_{r.E}$ was carried out. In this case, the deviation of the calculated data with the experimental ones was no more than $\delta_{\Delta P} = \pm 25$ % (Fig. 12). This indicates that the empirical dependence (8) for determining ζ_{ATP} can be applied to the considered conditions when designing aerothermopressor for gas turbine cyclic air cooling. But it should be noted that experimental data were obtained only at air flow rates to 0.52 kg/s, i.e., it is possible to apply the obtained dependence only for the low-flow aerothermopressor with small diameters of the evaporation chamber.

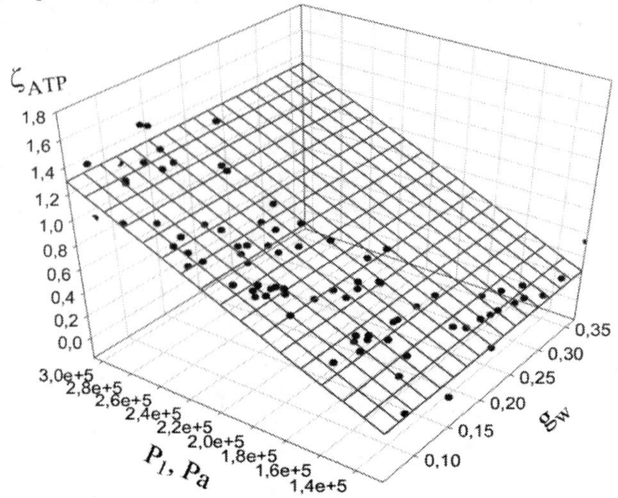

Fig. 10. Experimental dependences of the hydraulic resistance coefficient $\zeta_{ATP.E}$ on the relative flow rate of the injected water g_w at various initial pressures P_1 (distribution over the surface in accordance with equation (8))

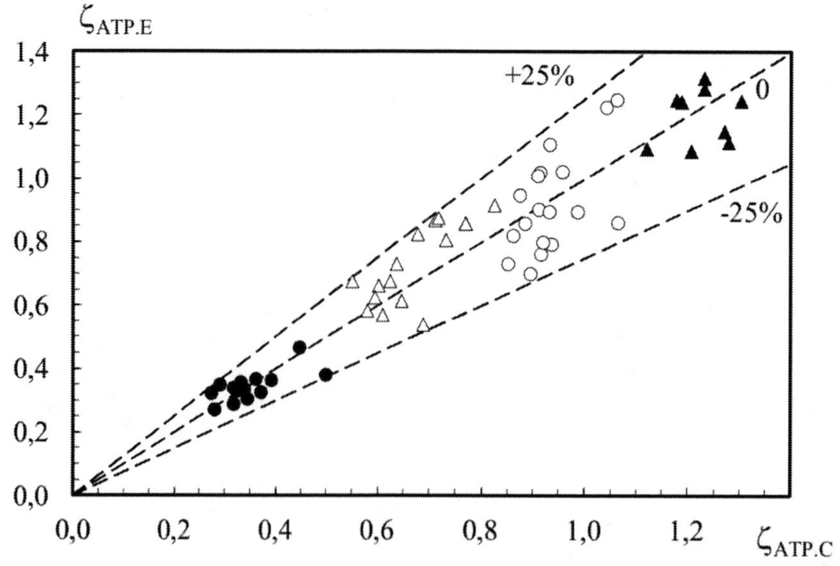

Fig. 11. Comparison of the hydraulic resistance coefficient calculated values $\zeta_{ATP.C}$ in the aerothermopressor flow part (equation (8)) with the experimental $\zeta_{ATP.E}$

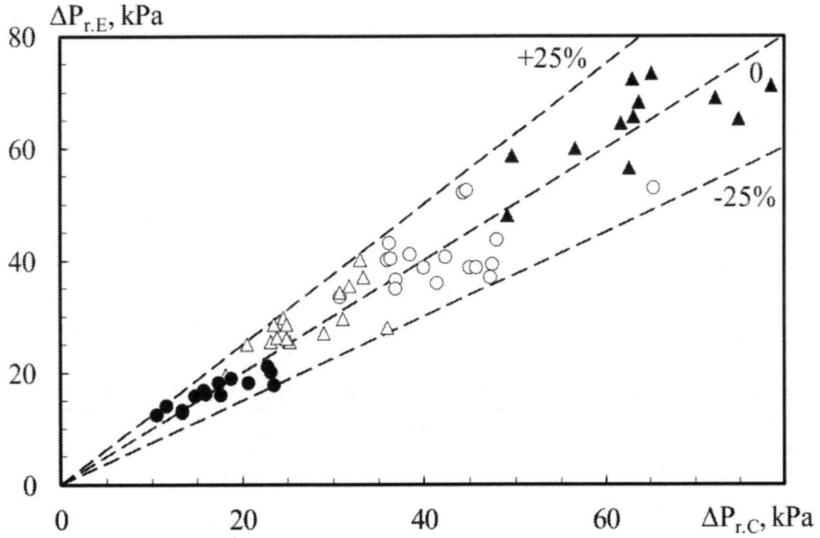

Fig. 12. Comparison of calculated pressure losses values in the aerothermopressor flow part $\Delta P_{r.C}$ with experimental $\Delta P_{r.E}$

5. Conclusions

The data on hydraulic resistance coefficients for the flow part of the low-flow aerothermopressor were determined experimentally. The experimental hydraulic resistance coefficient $\zeta_{ATP.E}$ = 0.35-1.30, depending on the initial pressure P_1, and the relative flow rate g_w of water injected into the inlet chamber has been received. The absolute values of pressure losses were ΔP_r = 18-75 kPa (11-24%).

An empirical equation was proposed to determine the hydraulic resistance coefficient of the aerothermopressor flow part ζ_{ATP}, depending on the amount of water injected. This coefficient takes into account hydraulic losses: local losses in the confuser and diffuser; friction losses in the evaporation chamber; losses due to resistance of water droplets. The deviation of the calculated hydraulic resistance coefficient $\zeta_{ATP.C}$ from the experimental ones $\zeta_{ATP.E}$ is $\delta_\zeta = \pm 25$ %. The deviation of the calculated pressure loss ΔP_r from experimental is $\delta_{\Delta P} = \pm 25$ %. This indicates the possibility of applying this dependence under the conditions: $1.2 \cdot 10^5 < Re < 3.4 \cdot 10^5$; $M = 0.25–0.65$.

The results can be used in the practice of designing the low-flow aerothermopressors for contact air cooling.

References

1. Radchenko, A., Radchenko, M., Konovalov, A., Zubarev, A.: Increasing electrical power out-put and fuel efficiency of gas engines in integrated energy system by absorption chiller scavenge air cooling on the base of monitoring data treatment. In: Proceedings of E3S Web of Conferences 70, 03011, p. 6. HTRSE-2018 (2018).
2. Kumar, T. A., Muzaffarul, H. M., Islam, M.: Effect of ambient temperature on the performance of a combined cycle power plant. Transactions of the Canadian Society for Mechanical Engineering 37(4), 1177–1188 (2013).

3. Eidan, A. A., Alwan, K. J.: Enhancement of the Performance Characteristics for Air-Conditioning System by Using Direct Evaporative Cooling in Hot Climates. Energy Procedia 142, 3998–4003 (2017).
4. Günnur, S. G. et al.: The effect of ambient temperature on electric power generation in natural gas combined cycle power plant-A case study. Energy Reports 4, 682–690 (2018).
5. Jonsson, M., Yan, J.: Humidified gas turbines – a review of proposed and implemented cycles. Energy 30, 1013–1078 (2005).
6. Boyce, M. P.: Gas Turbine Engineering Hand-book. Gulf Publishing Company, Houston, Texas (2002).
7. Elbel, S., Lawrence, N.: Review of recent developments in advanced ejector technology. International Journal of Refrigeration 62, 1-18 (2016).
8. Elbel, S.: Historical and present developments of ejector refrigeration systems with emphasis on transcritical carbon dioxide air-conditioning applications. International Journal of Refrigeration 34(7), 1545-1561 (2011).
9. Elbel, S., Hrnjak, P.: Ejector refrigeration: an overview of historical and present developments with an emphasis on air-conditioning applications (2008).
10. Lawrence, N., Elbel, S.: Theoretical and practical comparison of two-phase ejector refrigeration cycles including First and Second Law analysis. International Journal of Refrigeration 36(4), 1220-1232 (2013).
11. Konovalov, D., Kobalava, H.: Efficiency Analysis of Gas Turbine Plant Cycles with Water Injection by the Aerothermopressor. In: Ivanov V. et al. (eds.) Advances in Design, Simulation and Manufacturing II. DSMIE 2019. Lecture Notes in Mechanical Engineering, pp. 581–591. Springer, Cham (2020).
12. Fowle, A.: An experimental investigation of an aerothermopressor having a gas flow capacity of 25 pounds per second. Massachusetts Institute of Technology, USA (1972).
13. Stepanov, I. R., Chudinov, V. I.: Some Problems of Gas and Liquid Motion in the Channels and Pipelines of Power Plants. The Science Publ., Leningrad (1977).
14. Zhivitsa, V. I.: Intercoolers with thermopressor for two-stage ammonia refrigeration machines. Refrigeration Engineering 5, pp. 18–20 (2002).
15. Radchenko, M., Radchenko, R., Ostapenko, O., Zubarev, A., Hrych, A.: Enhancing the utilization of gas engine module exhaust heat by two-stage chillers for combined electricity, heat and refrigeration. In: Proceedings of 5th International Conference on Systems and Informatics, ICSAI 2018, pp. 240–244. Jiangsu, Nanjing, China (2019).
16. Shapiro, A. H., et al.: The Aerothermopressor – a Device for Improving the Performance of a Gas-Turbine Power Plant. In: Proceedings of the Trans. ASME, pp. 617-653. Cambridge, USA (1956).
17. Erickson, J.: A theoretical and experimental investigation of the aerothermopressor process. Massachusetts Institute of Technology, USA (1958).
18. Radchenko R., Radchenko A., Serbin S., Kantor S., Portnoi B. Gas turbine unite inlet air cooling by using an excessive refrigeration capacity of absorption-ejector chiller in booster air cooler. In: Proceedings of E3S Web of Conferences 70, 03012, 6 p. HTRSE-2018 (2018).
19. Lockhart, R. W., Martinelli, R. C.: Proposed correlation of data for isothermal two-phase, two-component flow in pipes. Chem. Eng. Progr. 45(1), 39–48 (1949).
20. Martinelli, R. C., Nelson, D. B.: Prediction of pressure drop during forced – circulation boiling of water. Trans. ASME 70, 695–702 (1948).
21. Dolinsky, A. A., Ivanitsky, G. K.: Heat and mass transfer and hydrodynamics in vapor-liquid dispersed flow. Naukova Dumka, Kyiv (2008).
22. Kulinchenko, V. R.: Hydraulics, hydraulic machines and hydraulic drive. Tsentr navchalnoi literatury Publ., Kyiv (2006).

23. Konovalov D., Trushliakov, E., Radchenko, M., Kobalava, H., Maksymov, V.: Research of the Aerothermopressor Cooling System of Charge Air of a Marine Internal Combustion Engine Under Variable Climatic Conditions of Operation. In: Tonkonogyi V. et al. (eds.) Advanced Manufacturing Processes. InterPartner 2019. Lecture Notes in Mechanical Engineering, pp. 520-529. Springer, Cham (2020).
24. Lin, A., et al.: Evaluation of Mass Injection Cooling on Flow and Heat Transfer Characteristics for High-Temperature Inlet Air in a MIPCC Engine. International Journal of Heat and Mass Transfer 135, 620–630 (2019).
25. Radchenko, A., Bohdal, L., Zongming, Y., Portnoi, B., Tkachenko, V.: Rational designing of gas turbine inlet air cooling system. In: Tonkonogyi V. et al. (eds.) Advanced Manufacturing Processes. InterPartner 2019. Lecture Notes in Mechanical Engineering, pp. 591-599. Springer, Cham (2020).

Calculation of the main parameters involved in the combustion process of CH_4-H_2 mixtures at different proportions

Razvan Carlanescu[1], *Marius* Enache[1], *Raluca* Maier[1], *Andreea* Alcea[1], *Raluca* Condruz[1], *Cristian* Stoica[1] and *Madalina* Ghilvacs[2,*]

[1]Romanian Research and Development Institute for Gas Turbines COMOTI, 220 D Iuliu Maniu Bd., sector 6, Bucharest, Romania
[2]Automobile Dacia SA, Group Renault, 3G Preciziei Bd., sector 6, Bucharest, Romania

Abstract. By setting clear targets for reducing pollutant emissions, the researchers in the field of combustion are pushed lately to find new alternatives for cleaner combustion. The partial or total transition to other types of fuels, such as hydrogen, involves substantial changes in the combustion process and possible necessary constructive changes. In the study of the combustion of CH_4-H_2 mixtures, both numerically and experimentally, preliminary calculations are required, which will help to easily establish the parameters and working regimes and then to use for verifying the results. This paper aims to find an easier method of calculating these parameters, depending on the percentage of gas in the fuel mixture. The calculated values resulted this way will lead to some logical estimates of important aspects of combustion, such as flame field and temperature variation, related to the variation of the amount of hydrogen in the mixture. The method can be a useful tool in the preliminary design of a combustion chamber for CH_4-H_2 mixtures.

1 Introduction

Starting from the environmental problems form the past decades, the scientific world is lately in a constant search for less pollutant and more efficient fuels. Hydrogen is studied as a possible alternative, since new ways for producing and transporting it developed in the past few years [1]. For example, from wind or solar energy, hydrogen can be produced on site by electrolysis and can be transported with the existing natural gas distribution network and then used for different existing or new applications. In the field of energy, it could be interesting to use this new fuel in gas turbines, at industrial level, in order to produce electricity and heat.

The new resulting gas, as a mixture of hydrogen and natural gas, depending on the proportions, can change significantly the combustion parameters of the final application, due to the different mixtures properties. From this perspective, numerical and experimental investigations are needed. Various studies and experiments were made with results pointing

* Corresponding author: make_madalina@yahoo.com

that above 15-20% H_2, some structural modifications of the equipment are requested [2] [3] [4].

For both numerical simulations and experimental sessions, the preliminary calculations of the input values and nominal regimes are needed. In order to find the main operating parameters of the combustion, formulas and calculations are usually required, based on theory and literature. These calculations and the preliminary results obtained can be later used to define the input and output parameters in the numerical simulations and to establish the operating regimes, air and fuel flows, temperatures, pressures and stability limits in the experiments. The present paper will point below a method to easily calculate the excess air, the calorific value of the fuel mixture, adiabatic temperature and other important parameters in the combustion chamber and the variation of the products in the combustion gases, according to this mixture's different proportions between methane and hydrogen. Hence, the acquired formulas can be used for obtaining useful values that can be used both as starting data and also for verifying the experimental measured results.

2 Excess air - mixture ratio - gas analysis

The combustion process for a C_nH_m, hydrocarbon fuel, as a global stoichiometric reaction using air as an oxidizer, is written as [5] [6]:

$$C_nH_m + \left(n + \frac{m}{4}\right)(O_2 + 3.76N_2) \rightarrow nCO_2 + \frac{m}{2}H_2O + 3.76\left(n + \frac{m}{4}\right)N_2 \tag{1}$$

Although the air contains other compounds in small quantities, beside oxygen (20.95%) and nitrogen (78.08%) and other [5], here is simply considered to be composed only of oxygen (21%) and nitrogen (79%). Thus the molar ratio between the two compounds is considered to be 3.76 in the above relationship.

Therefore, for the combustion of each mole of fuel are required $(n+m/4)(1+3.76)$ moles of air, resulting in $4.76(n+m/4)+m/4$ moles of reaction products.

2.1 Combustion of CH_4

The overall reaction for stoichiometric combustion process in the case of CH_4 is written as follows [7]:

$$CH_4 + 2\,O_2 \rightarrow CO_2 + 2\,H_2O \tag{2}$$

The molar mass of CH_4 is 16.043 g/mol . Thus, the fuel / stoichiometric air ratio is:

$$\left(\frac{\dot{m}_a}{\dot{m}_{f(CH_4)}}\right)_s = \frac{2 * \left(32 + \frac{79}{21} * 28\right)}{16.043} = 17.16 = L_{0\,CH4} \tag{3}$$

To burn 1 kg of CH_4 requires 17.16 kg of air, an amount known in the literature as $Lmin_{CH4}$, or $L_{0\,CH4}$.

2.2 Combustion of H_2

The overall reaction for stoichiometric combustion in the case of H_2 is:

$$H_2 + \frac{1}{2}O_2 \rightarrow H_2O \tag{4}$$

The molar mass of H_2 is 2.016 g/mol. Thus the fuel/stoichiometric air ratio is:

$$\left(\frac{\dot{m}_a}{\dot{m}_{f(H2)}}\right)_s = \frac{0.5*(32+\frac{79}{21}*28)}{2.016} = 34.32 = L_{0\,H2} \tag{5}$$

To burn 1 kg of H_2 requires 34.32 kg of air, an amount known in the literature as $Lmin_{H2}$ or $Lo_{\ H2}$.

2.3 Combustion of CH_4 + H_2 mixtures

The classical stoichiometric method of calculation is adopted, writing the individual equations for CH_4 and H_2 based on the relation (1): [8]

$$CH_4 + 2(O_2 + 3.76N_2) \rightarrow CO_2 + 2H_2O + 2*3.76*N_2 \tag{6}$$

$$H_2 + \frac{1}{2}(O_2 + 3.76N_2) \rightarrow H_2O + 3.76*\frac{1}{2}*N_2 \tag{7}$$

Then the stoichiometric equations are multiplied by the molar fractions of the fuel mixture. For this paper, the following notations and rule are adopted: the molar fraction (or the volume percentage by multiplying by 100) for CH_4 is noted with $\boldsymbol{x}$, and for H_2 it is noted with $\boldsymbol{y}$. The condition is $x + y = 1$; (1 = 100% as percentage). In these conditions:

$$x*[CH_4 + 2(O_2 + 3.76N_2) \rightarrow CO_2 + 2H_2O + 2*3.76*N_2] \tag{8}$$

$$y*[H_2 + 0.5\,(O_2 + 3.76N_2) \rightarrow H_2O + 3.76*0.5*N_2] \tag{9}$$

By mixing the above results in the stoichiometric equation of combustion for the mixture CH_4 - H_2 :

$$\begin{aligned} x*CH_4 + 2x*(O_2 + 3.76N_2) + y*H_2 + 0.5y*(O_2 + 3.76N_2) \rightarrow \\ \rightarrow x*(CO_2 + 2H_2O + 2*3.76*N_2) + y*(H_2O + 3.76 \\ *0.5*N_2) \end{aligned} \tag{10}$$

By transformations the stoichiometric equation is obtained:

$$\begin{aligned} (1-y)*CH_4 + y*H_2 + (2-1.5*y)(O_2 + 3.76*N_2) \\ \rightarrow (1-y)*CO_2 + (2-y)*H_2O + (2-1.5*y)*3.76 \\ *N_2 \end{aligned} \tag{11}$$

From the relation (11), depending on the molar (volumetric) participations, the stoichiometric air is further defined – L_s [Kmol $_{air}$/Kmol $_{fuel}$] ([$m^3_{N\,air}$/ $m^3_{N\,fuel}$])

$$L_s = \frac{(2-1.5y)(1+3.76)}{(1-y)+y} = (2-1.5*y)*4.76 \quad [m^3_{Nair}/\,m^3_{Nfuel}] \tag{12}$$

The minimum theoretical air required [8] [kg_{air}/kg_{fuel}] :

$$L_0 = L_s*\frac{Mair}{Mfuel} = \frac{4.76*(2-1.5*y)*Mair}{Mfuel} \quad [kg_{air}/kg_{fuel}] \tag{13}$$

Where: Mair = 28.84 [kg/kmol] is the molar mass of air;
Mfuel = $(1\text{-}y)*16+y*2$ [kg/kmol] is the molar mass of the fuel mixture

It results:

$$L_0 = \frac{8.5799 * (2 - 1.5 * y)}{1 - 0.875 * y} \qquad [kg_{air}/kg_{fuel}] \tag{14}$$

The excess air is further defined by an excess air coefficient, known also as air–fuel equivalence ratio [5] [9]:

$$\lambda = \frac{\dot{m}_a/\dot{m}_{fuel}}{(\dot{m}_a/\dot{m}_{fuel})_s} = \frac{\dot{m}_a}{\dot{m}_{fuel} * L_0} \tag{15}$$

where: $\dot{m}_a$ - air flow
$\dot{m}_{fuel}$ - fuel mixture flow

Fuel–air equivalence ratio, is defined as:

$$\phi = \frac{\dot{m}_{comb}/\dot{m}_a}{(\dot{m}_{comb}/\dot{m}_a)_s} = \frac{1}{\lambda} \tag{16}$$

Thus, for lean fuel combustion ($\phi \leq 1$; $\lambda \geq 1$) the relation (11), can be redefined, obtaining the global equation in the form:

$$\begin{aligned}(1-y) * CH_4 + y * H_2 + \lambda * (2 - 1.5 * y)(O_2 + 3.76 * N_2) \\ \rightarrow (1-y) * CO_2 + (2-y) * H_2O + \lambda * (2 - 1.5 * y) * 3.76 \\ * N_2 + (\lambda - 1) * (2 - 1.5 * y) * O_2\end{aligned} \tag{17}$$

Starting from the above, the excess air and dosage are defined (required for comparisons with data from the literature, where dosage is used as a basis):

Stoichiometric dosage:

$$f_0 = \frac{1}{L_0} \qquad [kg_{fuel}/kg_{air}] \tag{18}$$

$$\varphi_0 = \frac{1}{L_s} \qquad [m^3_{N fuel}/m^3_{N air}] \tag{19}$$

The table below shows the variations of these parameters, depending on the composition of the mixture:

Table 1. Calculated values for various percentages of H_2 in the mixture

H_2 (y) % vol	H_2 % mass	M_{fuel}	f_0 [kg/kg]	L_0 [kg/kg]	φ_0 [m^3_N/m^3_N]	L_s [m^3_N/m^3_N]
0	0.00%	16	0.058	17.160	0.105	9.520
0.2	3.03%	13.2	0.057	17.680	0.124	8.092
0.4	7.69%	10.4	0.054	18.480	0.150	6.664
0.6	15.79%	7.6	0.050	19.869	0.191	5.236
0.8	33.33%	4.8	0.044	22.880	0.263	3.808
1	100.00%	2	0.029	34.320	0.420	2.380

The flue gas analysis results from the relation (17), depending on the volume fraction of H_2 and on the excess air:

$$CO_2 = \frac{(1-y)}{(3-2y) + \lambda * (2 - 1.5 * y) * 3.76 + (\lambda - 1) * (2 - 1.5 * y)} \tag{20}$$

$$H_2O = \frac{(2-y)}{(3-2y) + \lambda * (2-1.5*y) * 3.76 + (\lambda - 1) * (2-1.5*y)} \quad (21)$$

$$N_2 = \frac{\lambda * (2-1.5*y) * 3.76}{(3-2y) + \lambda * (2-1.5*y) * 3.76 + (\lambda - 1) * (2-1.5*y)} \quad (22)$$

$$O_2 = \frac{(\lambda - 1) * (2-1.5*y)}{(3-2y) + \lambda * (2-1.5*y) * 3.76 + (\lambda - 1) * (2-1.5*y)} \quad (23)$$

Considering that in practice, when using the gas analyzer, water is condensed on the well path, or at the inlet through the special installation in the analyzer, the analysis of dry flue gases, respectively CO_2, is considered as the main reference element:

$$CO_2 = \frac{(1-y)}{(1-y) + \lambda * (2-1.5*y) * 3.76 + (\lambda - 1) * (2-1.5*y)} \quad (24)$$

It results:

$$\lambda = \frac{1 - y + CO_2 - 0.5 * y * CO_2}{4.76 * CO_2 * (2-1.5y)} \quad (25)$$

Or by transforming the relationship (24) :

$$CO_2 = \frac{1}{1 + \lambda * (2 + 0.5 * \frac{y}{1-y}) * 4.76 - \left(2 + 0.5 * \frac{y}{1-y}\right)} \quad (26)$$

$$CO_2 = \frac{1}{4.76 * \lambda * (2 + 0.5 * \xi) + 0.5 * \xi - 1} \quad (27)$$

Where: $y/(1-y) = \xi$ represents the molar ratio H_2/CH_4
In this way, it turns out that:

$$\lambda = \frac{\frac{1}{CO_2} + 1 + 0.5 * \xi}{4.76 * (2 + 0.5 * \xi)} \quad (28)$$

These latter relationships will be used during the experiments to determine the excess air, as an additional, and more accurate, verification of the measured CH_4, H_2 and air flow ratios. Based on the measurements recorded with the gas analyzers, obtaining values for CO_2, CO, O_2, the excess air resulting from the measurements can be calculated and then it can be compared as a verification with the calculated one. There are similar formulas obtained and used by the literature in the calculation of excess air based on the concentrations of the components in the flue gases. [10] [6].

3 Adiabatic temperature - mixture ratio

It starts from the relationship known for defining the efficiency of combustion: [11]

$$\zeta * Hi = (1 + \lambda * L_0) * h_3 - \lambda * L_0 * h_2 \quad (29)$$

where:
- ζ – combustion efficiency [%];
- Hi – lower calorific value (fuel) [kJ/kg];
- h_3 – flue gas enthalpy [kJ/kg];
- h_2 – enthalpy of the air at the entrance to the combustion chamber [kJ/kg];

The enthalpy can be defined by the following relationship [12]:

$$\Delta h = Cp * (T - T_{ref}) \quad (30)$$

It was considered that the reference temperature is $T_{ref} = 0\ ^{o}C$. So, it can be written:

$$\zeta * Hi = (1 + \lambda * L_0) * Cp_{flgas} * T_3 - \lambda * L_0 * Cp_{air} * T_2 \quad (31)$$

It results:

$$T_3 = \frac{\zeta * Hi + \lambda * L_0 * Cp_{air} * T_2}{(1 + \lambda * L_0) * Cp_{flgas}} \quad (32)$$

where:

- Cp_{air} – average specific heat for air [kJ/kg K]; [8]
- Cp_{flgas} - average specific heat for flue gases [kJ/kg K];
- T_3 – the temperature at the outlet of the combustion chamber (flue gases temperature) [K]
- T_2 – the temperature at the inlet of the combustion chamber [K]
- $T_0 = T_{ref}$ – ambient temperature, reference [K]; value is considered $0\ ^{o}C = 273$ K
- λ - air–fuel equivalence ratio (equation (15));
- L_0 – the minimum theoretical air required $[kg_{air}/kg_{fuel}]$ (capitolul 2):
- H_i – lower calorific value $[kJ/m_N^3]$

For the calculation of the calorific value, it is possible to calculate by volume or by mass.

If we start from the **calorific value by volume**, we can write the relationship [6]:

$$Hi_v = (1 - y) * Hi_{vCH4} + y * Hi_{vH2} \quad [kJ/m_N^3] \quad (33)$$

where: Hi_{vCH4}, Hi_{vH2} the volumetric calorific values of CH_4 respectively H_2.

$$Hi_v = \frac{Hi_{vH2} * (\frac{Hi_{vCH4}}{Hi_{vH2}} * (1 - y) + y)}{Mfuel} * Vmol \quad [kJ/kg] \quad (34)$$

where: - Vmol = Molar volume = 22.4 m^3_N/kmol
- Mfuel = (1-y)*16+y*2 [kg/kmol] is the molar mass of the fuel mixture

Values considered [13]: Hi_{vCH4} = 35800 kJ/m^3_N; Hi_{vH2} = 10700 kJ/m^3_N;
It results:

$$Hi_v = \frac{10700 * (3.345 * (1 - y) + y)}{Mfuel} * Vmol \quad [kJ/kg] \quad (35)$$

After transformations it results:

$$Hi_v = 239680 * \frac{[3.345 * (1 - y) + y]}{[(1 - y) * 16 + y * 2]} \quad [kJ/kg] \quad (36)$$

If we start from the **calorific value by mass**, the mass fractions are taken into account and thus the following relation results. [6]:

$$Hi_m = \frac{(1 - y) * 16}{(1 - y) * 16 + y * 2} * Hi_{mCH4} + \frac{y * 2}{(1 - y) * 16 + y * 2} * Hi_{mH2} \quad [kJ/kg] \quad (37)$$

$$Hi_m = Hi_{mCH4} * \frac{\left[(1-y) * 16 + \frac{Hi_{mCH4}}{Hi_{mH2}} * y * 2\right]}{[(1-y) * 16 + y * 2]} \quad [kJ/kg] \tag{38}$$

The following values are considered: Hi_{mCH4}=50000 kJ/kg; Hi_{mH2} = 120000 kJ/kg (the values presented in the literature differ by +/- 0.5%, preferring these values to simplify the relationship).

Therefore:

$$Hi_m = 50000 * \frac{[(1-y) * 16 + 2.4 * y * 2]}{[(1-y) * 16 + y * 2]} \quad [kJ/kg] \tag{39}$$

With the two calculation relations for the calorific values by mass, the differences are:

Table 2. Comparison between the values obtained for the calorific value by mass

	y	0	0.1	0.2	0.4	0.6	0.8	1
Rel (36)	Hi [kJ/kg]	50108	51063	52221	55472	61118	73352	119840
Rel (39)	Hi [kJ/kg]	50000	50959	52121	55385	61053	73333	120000
	Differences [%]	-0.22%	-0.20%	-0.19%	-0.16%	-0.11%	-0.03%	0.13%

Relation (39) was chosen, and the calculations are presented in the following table:

Table 3. Calculated values of calorific value, for different percentages of H_2 in the mixture

Y (molar fraction) (H_2 % by vol)	**0 (0%)**	**0.1 (10%)**	**0.2 (20%)**	**0.4 (40%)**	**0.6 (60%)**	**0.8 (80%)**	**0.9 (90%)**	**1 (100%)**
Hi [kJ/kg]	50000	50959	52121	55385	61053	73333	87059	120000
Mfuel	16	14.6	13.2	10.4	7.6	4.8	3.4	2
Mass fraction H_2 by mass [%]	0 0	0.0137 1.37%	0.0303 3.03%	0.0769 7.69%	0.1579 15.79%	0.3333 33.33%	0.5294 52.94%	1 100%
Hi increment [%]	0%	2%	4%	11%	22%	47%	74%	140%

It can be observed that over 40% H_2 (y=0.4) the rising of the lower calorific value is increasing, which is also observed in the diagram below, in which the slope of the curve changes essentially after this value. This observation correlates with the reports in the literature, in which experiments say that up to 10-15% by volume H_2, there are not special problems for the combustion process [14]. This should be monitored during experiments. In addition to this, the variation of the fuel mixture's flow for CH_4-H_2 will be considered, which will be adjusted with the change of the percentage of H_2, in order to obtain the same constant thermal power.

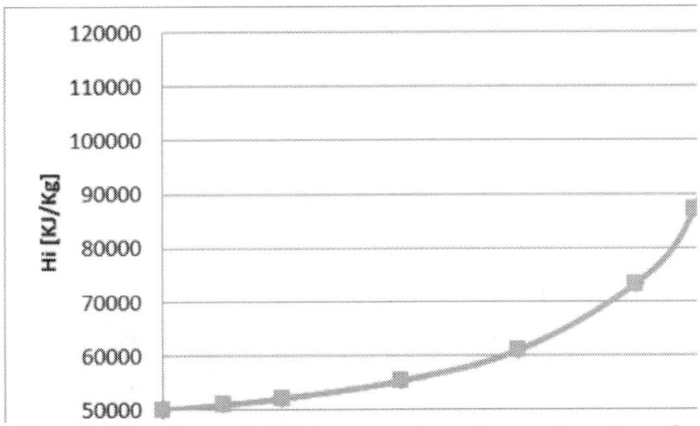

Fig. 1 Variation of the calorific value and of H_2 by mass, depending on the volume fraction of H_2

Finally, for the calculations, the relation (39), will be preferred, which together with the relation (14), by introduction in the relation (32), results:

$$T_3 = \frac{\zeta * 50000 * \frac{[(1-y) * 16 + 2.4 * y * 2]}{[(1-y) * 16 + y * 2]} + \lambda * \frac{8.5799 * (2 - 1.5 * y)}{1 - 0.875 * y} * Cp_{air} * T_2}{\left(1 + \lambda * \frac{8.5799 * (2 - 1.5 * y)}{1 - 0.875 * y}\right) * Cp_{flgas}} \quad (40)$$

To determine the specific heats, the literature and the common methods indicates various ways, or charts to calculate. One example is presented and used in the program and calculation procedures for combustion chambers used in 1968 by NASA [15]. After transformations, the specific heats can be written:

$$Cp_{air} = [\,0.2419 - 0.8181 * 10^{-5} * T + 17.91 * 10^{-9} * T^2 - 2.743 * 10^{-12} * T^3\,] * 4.186 \quad [kJ/kg\,K] \quad (41)$$

$$Cp_{flgas} = \frac{(0.431 + 52.558 * 10^{-5} * T - 38.24 * 10^{-9} * T^2 + 0.25 * 10^{-12} * T^3)}{8.5799 * (2 - 1.5 * y) * \lambda} * \frac{(1 - 0.875 * y)}{8.5799 * (2 - 1.5 * y) * \lambda} \quad (42)$$

The procedure is iterative (2-3 iterations), in which we start from a specific heat value and depending on the resulting temperature, the calculation is repeated with the value of the specific heat corresponding to the temperature resulting in the first iteration. The method is often used in the field, with different calculation relationships [8] [15] [16].

The presented relations do not take into account the thermal dissociation.

In order to make the dissociation correction, we will start from the correction concept according to which the lower calorific value is diminished by the energy lost by dissociation [15] [16]:

$$Hi_{corrected} = Hi - Hi_{dissociation} \quad [kJ/kg] \quad (43)$$

where $Hi_{dissociation}$ is approximated with the relation [15]:

$$Hi_{dissociation} = \frac{3 * 10^{-26} * T_3^{\,7.5}}{f} * 2.326 \quad [kJ/kg] \quad (44)$$

where: f – fuel-air ratio; T_3 – flue gases temperature [K];

After transformations and adaptation to the CH_4-H_2 mixture, the relationship becomes:

 https://doi.org/10.1051/e3sconf/202018001013

$$\mathrm{Hi}_{\mathrm{dissociation}} = \frac{581 * 10^{-26} * T_3{}^{7.5} * (1 - 0.875 * y)}{8.5799 * (2 - 1.5 * y) * \lambda} \qquad [\mathrm{kJ/kg}] \qquad (45)$$

$\mathrm{Hi}_{\mathrm{dissociation}}$ is shown graphically in the next figure below, in values [kJ/kg]:

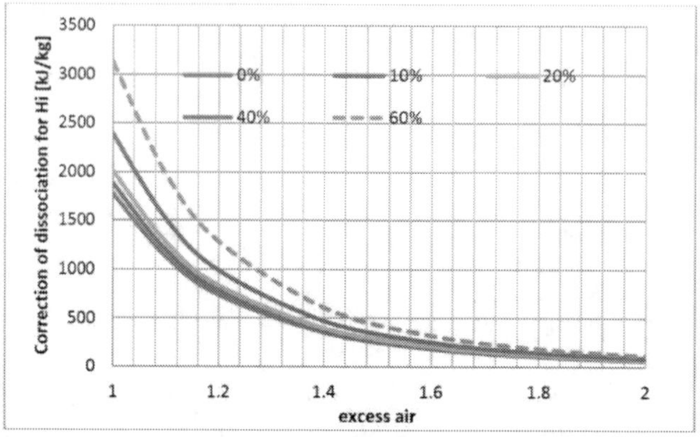

Fig. 2 Corrections of lower calorific value for different values of H_2 %

In the end, the dissociation correction is applied. Calculations can also be done directly taking into account the dissociation correction, the final relationship becoming:

$$T_3 = \frac{\zeta * \left(50000 * \frac{[(1-y) * 16 + 2.4 * y * 2]}{[(1-y) * 16 + y * 2]} - \frac{581 * 10^{-26} * T_3{}^{7.5} * (1 - 0.875 * y)}{8.5799 * (2 - 1.5 * y) * \lambda}\right)}{\left(1 + \lambda * \frac{8.5799 * (2 - 1.5 * y)}{1 - 0.875 * y}\right) * \mathrm{Cp}_{\mathrm{flgas}}}$$

$$+ \frac{\lambda * \frac{8.5799 * (2 - 1.5 * y)}{1 - 0.875 * y} * \mathrm{Cp}_{\mathrm{air}} * T_2}{\left(1 + \lambda * \frac{8.5799 * (2 - 1.5 * y)}{1 - 0.875 * y}\right) * \mathrm{Cp}_{\mathrm{flgas}}} \qquad [\mathrm{K}] \qquad (46)$$

where: T_2 – air temperature; T_3 – adiabatic temperature of combustion (in the equation above, T_3 is iterative, as in the case of specific heats $\mathrm{Cp}_{\mathrm{flgas}}$.

Results of calculations using these relationships are presented in the following graphs:

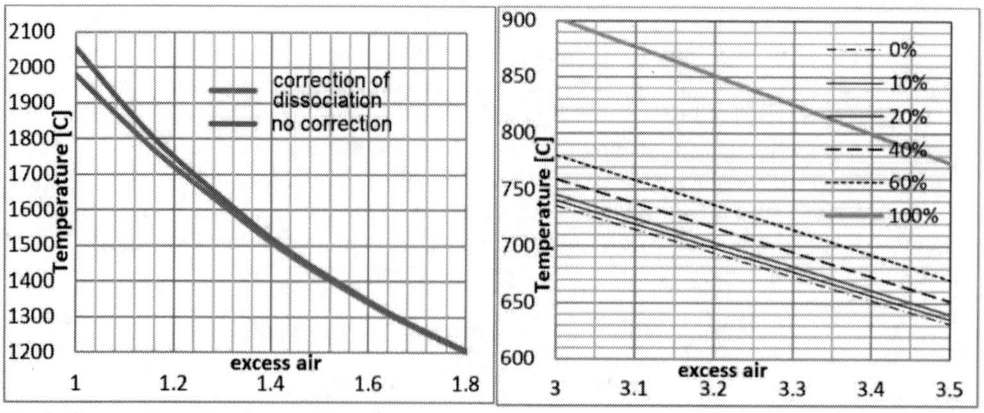

Fig. 3 Corrected combustion temperature
a) for CH_4 100%, b) for the mixture CH_4-H_2, at various % of H_2

Fig. 3b shows the combustion temperatures for the mixtures at values of interest for the experimental sessions.

Verification was performed with data from the known literature [15] [16] [17] [18] [19] [20]. The error margin (T_{calc}-T_{lit})/T_{calc} [%] is in between the limits of 1.32 % and -1.27%, the compared cases being methane and pure hydrogen, at different air temperatures. Some examples are given in the table below.

Table 4. Comparison of calculated values with the literature

	y=0	y=1		Deviation from the literature (T_{calc}-T_{lit})/T_{calc} [%]	
K	2251	2494	calculation		
C	1978	2221			
C	1963	2254	literature	-0.69%	1.32%
C	2004	2210		1.13%	-0.44%
K	2225	-		-1.18%	
K	2223	2483		-1.27%	-0.44%

As it can be seen, the deviations are small and of both positive and negative values, not only in one direction (which would suggest a methodological error).

4 Conclusions

The work here does not aim in this phase to bring special innovations in the field of combustion for CH_4-H_2 mixtures, it starts from known formulas, specific to the physical and chemical processes that take place in the combustion chambers. However, going deeper into the calculation details for the peculiarities of burning CH_4 and H_2 mixtures, new aspects come to light, highlighting specific tendencies that should direct numerical or experimental research for a certain course. For example, a clear picture of the variation in lower calorific value, depending on the variation in the percentage of hydrogen in the fuel mixture, can lead to some logical estimates of important aspects of combustion, such as flame field and temperature variation, related to the variation of the amount of hydrogen.

Moreover, in this paper very useful formulas have been developed for establishing the starting parameters for possible numerical calculations, as well as for establishing operating regimes for experiments. Finally, the data obtained with these calculations can be compared with those obtained by CFD analysis, or from experiments, as a very useful, correct and easy verification.

This work has been funded by the European Social Fund from the Sectorial Operational Programme Human Capital 2014-2020, through the Financial Agreement with the title "*Scholarships for entrepreneurial education among doctoral students and postdoctoral researchers (Be Antreprenor!)*", Contract no. 51680/09.07.2019 - SMIS code: 124539.

5 References

1. E.-I. Koytsoumpa, C. Bergins, B. Buddenberg, S. Wu, O. Sigurbjornsson, K.C. Tran, E. Kakaras, *The Challenge of Energy Storage in Europe: Focus on Power to Fuel,* ASME J. Energy Resour. Technol., **138(4),** p. 042002, 2016
2. M. Aigner, EU-Turbines & GERG Workshop, Brussels, *Burning Natural Gas / Hydrogen Mixtures in DLN Gas Turbines* October 2011

3. R. Carlanescu, T. Prisecaru, M. Prisecaru, I. Soriga, *Swirl Injector for Premixed Combustion of Hydrogen–Methane Mixtures,* Journal of Energy Resources Technology, **140(7),** 2018. Paper No: JERT-16-1506; doi: 10.1115/1.4039267
4. M. Andersson, J. Larfeldt, A. Larsson, B., *Co-Firing With Hydrogen in Industrial Gas Turbines,* Swedish Gas Technology Centre (SGC), Malmo, Sweden, Report No. **2013:256,** 2013
5. A.H. Lefebvre, D.R. Ballal, *Gas Turbine Combustion. Alternative Fuels and Emissions* CRC Press, New York, 2010
6. L. Mihaescu, T. Prisecaru, I. Oprea, *Cazane si Turbine - Notiuni de baza,* Editura MatrixRom Bucuresti, 1999
7. T. Prisecaru, *Simularea Numerica a Proceselor de Arder a Combustibililor Gazosi si Solizi Pulverizati,* Editura Bren, Bucuresti, ISBN 973-8143-76-4, 2001
8. S. McAllister, J-Y. Chen, A. C. Fernandez-Pello, *Fundamentals of Combustion Processes, Chapter 2 - Thermodynamics of Combustion,* Springer, ISBN 978-1-4419-7943-8, 2011
9. C. Carlanescu, I. Manea, C. Ion, S. Sterie, *Turbomotoare - Fenomenologia Producerii Si Controlul Noxelor,* Editura Academiei Tehnice Militare, 1998
10. R. Chiriac, R. Radu, *Cercetarea Experimentala a Masinilor Termice - Indrumar de Laborator,* Universitatea Politehnica Bucuresti, Facultatea de Inginerie Mecanica, Bucuresti, 1997
11. V. Pimsner, C.A. Vasilescu, G.A. Radulescu, *Energetica turbomotoarelor cu ardere interna,* Ed. Academiei Republicii Populare Romania, 1964
12. B. P. Meherwan, *Principles Of Operation And Performance Estimation Of Centrifugal Compressors,* Proceedings of the Twenty-First Turbomachinery Symposium, [Dallas, Texas], 1992
13. M. Fossum, R.V. Beyer, *Co-combustion: Biomass Fuel Gas and Natural Gas,* SINTEF Energy Research, Technical Report TRA **473**, ISBN 82-594-1296-94, 1998
14. R. Carlanescu, T. Prisecaru, R. Kuncser, E. Pop**,** *The optimization of a swirl injector for combustion of hydrogen fuel mixtures flame temperature of hydrogen in combination with gaseous fuels,* U.P.B. Sci. Bull., Series D, **81** (1), ISSN 1454-2358, 2019
15. NASA report, Northern Research and Engineering Corp., *Computer program for the analysis of annular combustors. Volume 1 - Calculation procedures,* Cambridge, MA, United State, NAS3-9402, 1968, https://ntrs.nasa.gov/search.jsp?R=19680008088 (accesed: may 2019)
16. J. Hodge, *Cycle and Performance Estimations,* Butterworth Scientific Publications London, England, 1955
17. S.K. Mishra, R.P. Dahiya, *Adiabatic flame temperature of hydrogen in combination with gaseous fuels,* International Journal of Hydrogen Energy, **14(11),** Pages 839-844, 1989
18. B. Lewis, G. von Elbe, *Combustion, Flames and Explosions of Gases - Third Edition,* Academic Press, ISBN: 978-0-12-446751-4, 1987
19. A. M. Helmenstine, ThoughtCo, , https://www.thoughtco.com /flame-temperatures-table-607307 (accessed: may 2019)
20. R. J. Richard, *North American Combustion Handbook: A Basic Reference on the Art and Science of Industrial Heating with Gaseous and Liquid Fuels, Vol. 1,* North Amer Manufacturing Co, Cleveland, Ohio, U.S.A., 1985

Application of waste oil purification technology on a heating installation of an industrial hall

Mihai Dragne[1], *Elena* Pop[1*], and *Ionel* Pîşă[1]

[1]University Politehnica of Bucharest, Faculty of Mechanical Engineering and Mechatronics, Department of Thermodynamics, Engines, Thermal and Refrigeration Equipment, Splaiul Independentei, 313, Romania;

Abstract The paper proposes the development of an experimental model of boiler using waste oil revitalized with bentonite, for heating an industrial hall. The main objective is to present a new perspective on the use of waste oils, a framework in which it no longer presents a danger to the environment and to human health, as well as their energy use from an economic and ecological point of view. Experiments were performed to determine the adsorption of the impurities from the waste oil. One of the experiments is to compare the combustion of the revitalized waste oil with the combustion of raw waste oil. The results show improvements in the combustion process and decreases of the pollutant emissions for revitalized waste oil. The CO concentration decreased on average by 52% when and the NOx concentration increased by 57% due to the increase of temperature in the furnace in the case of burning purified oil. The experimental installation is adapted on an industrial hall.

1 Introduction

Waste oils are a more attractive raw material for obtaining basic oils than the corresponding fractions in the oil, because the yield is higher and does not require extraction or solvent deparaffining. However, the processing of waste oils is having great difficulties and more and more complex technologies have been developed that take into account the type and degree of impurity of the waste oil [1]. An important benefit of the treatment-revitalization of waste oils is the elimination of a complex source of pollution and the return to the consumption of a combustible material [2]. The application of the research results will contribute to ensuring a sustainable economic development at the project field level. This promotes the exploitation and use of the deposits of natural absorbents (bentonites) and the use of waste energy resources while saving the classical resources. As a result of the investments for the exploitation and use in the project field of the natural deposits of absorbents, there will be an economic development of the area, which will lead to the improvement of the living and health conditions of the local population [3].

2 Experimental determinations

* Corresponding author: elena.pop@upb.ro

2.1. Materials and methods

From the analysis of waste oils (from both spark-ignition engine and compression ignition engine), the initial concentration of metal ions is between 50 and 200 mg / liter, predominantly iron (Fe). The next step was mixing the used oil with bentonite in 4 concentrations, respectively adding 20 g, 40 g, 60 g, 80 g bentonite / liter. The mixture was made with an electric mixer, about 30 minutes, for 3-4 days in a row [4]. After mixing, the impurities adsorbed by bentonite were retained. The following experiment was conducted at the National Center for Micro and Nanomaterials (C.N.M.N.), the structural analysis laboratory within the University Politehnica of Bucharest (U.P.B.). After mixing, the bentonite adsorbed metal ions and by filtration the solid matter was removed. The mixing time was not monitored in these experiments. The solid sample containing bentonite who absorbed metal ions was burned for 2 hours at 815 degrees Celsius. We then performed the elemental analysis of ash using an Agilent 8800 Inductive Coupled Plasma Mass Spectrometry (ICP-MS) equipment (Agilent Technologies, Japan). Prior to the measurements at the ICP-MS, the samples were individually weighed and placed in special tubes. After the addition of 2 ml of HNO3, the samples were digested in a microwave system (Milestone Ethos, FKV, Bergamo, Italy), at 200° C for 35 minutes, maximum power of 1800 W. After cooling, the digestion fluids were diluted with ultrapure water to 25 ml. The measurements were made for the most abundant isotopes of each element.

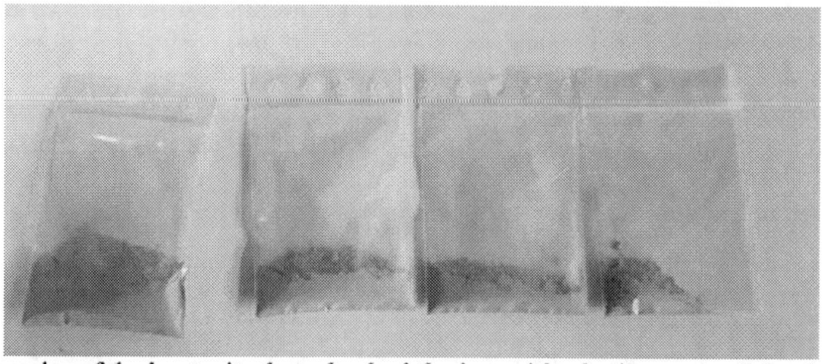

Fig.1 The ash samples of the bentonite that adsorbed the impurities in the waste oil for the four cases considered

2.2 Determination of impurities removed from waste oil using bentonite

From the experiment presented above we obtained the following results presented in the table below:

Table 1 The results from the analysis of the ash of bentonite mixed with the used oil

Sample type	Sample [g]	56 Fe [mg/ g]	63 Cu [mg/ g]	66 Zn [mg/ g]	111 Cd [mg/ g]	208 Pb [mg/ g]
B 20 g	0.1009	63.24	0.077	9.8	0.0012	0.122
B 40 g	0.1005	75.24	0.058	10.5	0.00078	0.178
B 60 g	0.0999	71.16	0.043	9.4	0.00091	0.101
B 80 g	0.1002	71.93	0.046	9.8	0.0011	0.482

It is noticed that the amount of zinc [Zn], copper [Cu], cadmium[Cd] and lead[Pb], are small ratios compared to iron [Fe], which is found in quite large quantities. It turns out that 20g bentonite /liter waste oil is optimal for copper (Cu), zinc (Zn), cadmium (Cd), lead (Pb) removal and 40 g bentonite / liter waste oil for iron (Fe) removal. By eliminating the metal ions, which are found in the highest proportions in the waste oil, we increase the life of the injector nozzle, which will no longer be impacted by these metal ions.

2.3 Tests for combustion processes to a pilot boiler

The experimental tests in the previous research [4] were continued for new results. The experiments were carried out on raw waste oil from the automotive field and revitalized waste oil with bentonite, also from the automotive field, by combustion first on a laboratory pilot boiler of 55 kW. The boiler is equipped with a special burner for liquid which ensures the preheating of the fuel up to 80°C. For the experiments, the case of mixture of 30 g bentonite/liter waste oil was chosen considering as an optimal range value between analyzed cases of 20 g bentonite/liter and 40 g bentonite/liter, in order to have a balance in the adsorption of all metals from used oil.
The characteristics of the measurements regarding the fuel oil waste and the revitalized waste oil combustion are included in the tables below:

Table 2 Measurements of raw used oil pollution by combustion

Sample	O_2	CO	CO_2	NO	NO_2	SO_2	NOx	t gas	λ
	%	ppm	%	ppm	ppm	ppm	ppm	°C	
1	17.3	1628	2.7	11	0	0	11	238	5.68
2	17.3	1722	2.7	14	0	0	14	238	5.68
3	17.2	1813	2.8	15	0	0	15	242	5.53

Table 3 Measurements of used oil revitalized with 30 g bentonite/liter pollution by combustion

Sample	O_2	CO	CO_2	NO	NO_2	SO_2	NOx	t gas	λ
	%	ppm	%	ppm	ppm	ppm	ppm	°C	
1	16.9	1172	3	28	4	0	32	280	5.12
2	16.9	1293	3	27	4	0	31	278.3	5.12
3	17.4	1277	2,6	20	4	0	24	263.3	5.83

Pollutant emissions, flue gas temperature (t gas) and the excess air (λ) at the end of the process has been monitored.

2.4. Experimental installation on an industrial hall

The experimental data obtained in the laboratory, on pilot plant, has been transferred to an industrial installation, having at our disposal a company that deals in the automotive field (repairs and overhauls of road vehicles) [5]. A heating system for the two halls of the company, using the revitalized used engine oil, has been designed. For this purpose, the heat losses, the heat requirement, the fuel flow has been calculated. At a yield of 90%, it has an hourly consumption of 4.2 kg/h of revitalized used oil. For the entire heating period (November – March) approximately 4 tones of revitalized used oil is required.

About a year ago the boiler (Hiton) was purchased in the industrial hall and the first experiment was done using raw waste oil. A year later the experiment was carried out on the same boiler, but this time using 5 liters of waste oil treated with 150 g of bentonite meaning 30 g bentonite/liter waste oil (the same fuel like in laboratory experiment) [6]

The installation on which the experiment was performed is a Hiton HP-145 used oil boiler with 42 kW thermal power. The boiler is equipped with a burner for liquid which ensures the preheating of the fuel up to 40°C. Once the burner is heatedup to 40°C the joints of thermostat placed next to combustion chamber clutch activating the oil inlet pipeand fans.

Fig.2 Industrial hall on which the experiment was carried out

Fig.3 The boiler on which the experiments were performed

Fig.4 Aspect of flame inside the boiler furnace

The results are present in tables below:

Table 4 Polluting emissions by combustion of raw used oil, dated 14.01.2019

Sample	O_2	CO	CO_2	NO	NO_2	SO_2	NOx	t gas	λ
	%	ppm	%	ppm	ppm	ppm	ppm	°C	
1	12.8	169	6.1	33	0	0	33	179	2.56
2	12.9	170	6.1	34	0	0	34	178	2.56
3	12.7	168	6.1	33	0	0	33	180	2.56

Table 5 Polluting emissions by combustion of purified oil treated with 30 g bentonite / liter, 10.03.2020

Sample	O_2	CO	CO_2	NO	NO_2	SO_2	NOx	t gas	λ
	%	ppm	%	ppm	ppm	ppm	ppm	°C	
1	14.1	91	5.4	51	0	0	51	200	3.04
2	13.7	94	5.2	52	0	0	52	196	2.88
3	13.9	76	5.3	51	0	0	51	198	2.96
4	13.6	92	5.0	53	0	0	53	197	2.84

The boiler is always designed to maintain a constant temperature, thus reducing fuel consumption. The gas outlet pipe is kept clean due to its inclination; the ash is collected in the tray which must be emptied periodically. Normally, in 30-40 minutes the cleaning is done.

3.Conclusions

By analyzing the data from the tables above, the following conclusions can be drawn:
- the CO concentration decreased on average by 52% when burning the purified oil;
- the NOx concentration increased by 57% (from 33 ppm to 52 ppm), due to the increase of the temperature in the furnace to the burning of the purified oil;
- the results obtained in the experiments on the pilot boiler at Politehnica University of Bucharest are comparable with the data obtained at industrial hall on boiler Hiton.

The installation on an industrial hall that has as main source of heating a boiler on waste oil is economically and practically advantageous. This type of heating system is perfect for heating large areas of the following sectors:
- Industry sector: industrial constructions, warehouses of goods, production halls;
- Automotive sector: car depots and car workshops;
- Agriculture sector: agricultural buildings, basements, garages, sports and recreation facilities, tents, churches [7].

Waste oil purification technology presented, compared with technologies like REVIVOILD, BLOWDEC, DUNWELL WFE [8, 9, 10, 11], has the advantage of a very low energy consumption, without involving energy of chemical reaction.

This work was supported by a grand of Romanian Ministry of Research and Innovation, CCCDI e UEFISCDI, project number PN-III-P1-1.2-PCCDI-2017-0404/31PCCDI/2018, within PNCDI III.

References

1. C. Stan, C.A.M. Toma, *Some aspects of the regeneration of used motor oil*, Procedia Manufacturing, Volume **22** (2018) 709-713
2. D.I. Osman, S.K. Attia, A.R. Taman, *Recycling of used engine oil by different solvent,* Egyptian Journal of Petroleum 27 (2018) 221-225
3. D. A. Ciupăgeanu, G. Lăzăroiu, M. Tîrşu, *Carbon dioxide emissions reduction by renewable energy employment in Romania*, Proc. of 11th international conference and electromechanical and power systems SIELMEN, (2017)
4. M. Dragne, E. Pop, C. I. Covaliu, E. Matei, I. Pîşă, *Experiments in use of bentonite for energy recovery of used oils*, E3S WEB CONF., Volume **122**, (2019)
5. J. Zhao, Y. Duan, X. Liu, *Study on the policy of replacing coal-fired boilers with gas-fired boilers for central heating based on the 3E system and the TOPSIS method: A case in Tianjin China*, Energy, Volume **189** (2019)
6. S. Fujii, N. Horie, K. Nakaibayashi, Y. Kanematsu, Y. Kikuchi, T. Nakagaki, *Design of zeolite boiler in thermochemical energy storage and transport system utilizing unused heat from sugar mill,* Applied Energy, Vol.**238**, (2019)
7. N. Zhao, B. Li, D. Chen, R. Ahmad, Y. Zhu, G. Li, Z. Yu, J. Li, E. Wang, S. Yun, H. Yoon, Y. Zhou, R. Dong, H. Wang, J. Cao, J. He, X. Ju, *Direct combustion of waste oil in domestic stove by an internal heat re-circulation atomization technology: Emission and performance analysis,* Waste Management, Volume **104** (2020) 20-32
8. F.D. Giovanna, O. Khlebinskaia, A. Lodolo, S. Miertus, *Compendium of used oil regeneration Technologies,* Trieste, (2003)
9. C.T. Pinheiro, R.F. Pais, M.J. Quina, L.M. Gando-Ferreira, *Regeneration of wastelubrificant oil with distinct properties by extraction-flocculation using green solvents*, Journal of cleaner production, Vol. **200** (2018)
10. H. Shabanzade, A. Salem, S. Salem, *Management of adsorbent content in waste motor oil regeneration by spectrophotometrical study and effective acidification in production of nano-porous clay*, Spectrochimica Act Part A: Molecular and Biomolecular Spectroscopy, Volume **202** (2018)
11. S. Salem, A. Salem, A.A. Babaei, *Preparation and characterization of nano porous bentonite for regeneration of semi-treated waste engine oil: Applied aspects for enhanced recovery*, Chemical Engineering Journal, Volume **260** (2015)

Combustion Experiments On A Solid Fuel With Low Sulphur Content

Tudor Prisecaru[1], *Adrian* Adam[2], *Lucian* Mihăescu[1], *Ionel* Pîşă[1,*], *Elena* Pop[1], *Malina* Prisecaru[1], *Viorel* Berbece[1], *Mihai* Dragne[1]

[1] University POLITEHNICA of Bucharest, Romania

[2]ICEMENERG SA, Bucharest, Romania

Abstract. Research was carried out at the experimental installation defined by the 2 MW furnace pilot from the Politehnica University of Bucharest and ICEMENRG institute who was involved for measurements. The purpose of the experimental study is to confirm that the use of a coal with low sulphur content S<0.08% is possible without the use of sulphur-purification installations for industrial purposes. But, the high moisture content of the coal has imposed the problem of its compatibility with a certain milling technology. The pilot plant is equipped with fan mill and pre-drying tower. Under these conditions, this installation allows the grinding of fuels with high moisture content. The experiments focused on the process of ignition, combustion and pollutant emissions, with the exemplification, in particular, of the emission of sulfur dioxide. Also, the outbreak temperatures and the slugging temperature were monitored. For this purpose, the outbreak of the pilot boiler was equipped with all the equipment necessary to follow these parameters. The experiments have shown positive aspects to the burning of this fuel, thus laying the foundations of future tests at industrial installations (first of all, the power plant from Deva is considered).

1. Introduction

The purpose of the experimental study is to confirm that the use of a coal with low sulphur content S<0.08% is possible without the use of sulphur-purification installations for industrial purposes. The industrial use of this fuel can be implemented after the positive response obtained by the research on the pilot furnace [1]. The energy characteristics of the coal made by ICEMENERG-SA, indicate the following values:

$$W_t^i = W_h^i + W_i^i = 5.31 + 29.73 = 35.04\,\% \;; C^i = 43.99\,\% \;; H^i = 3.22\,\% \;;$$

$$A^i = 2.79\,\% \;;\; N^i = 0.65\,\% \;; S_c^i = 0.08\,\% \;;\; O^i = 14.23\,\%,$$

$$order\ no.\ CD\ 797/11.06.2019.$$

*Corresponding author: ionel.pisa@upb.ro

A high humidity, high content, also for oxygen and a very low value for the sulfur in the fuel are noted. The ash is very small. The volatile matter content is high, having the value: $V^i = 32.42\%$. The lower heating value is: $Q_i^i = 15{,}860\,kJ/kg$. The fixed carbon, an element of technical analysis has the value:

$$C_f^i = 100 - \left(A^i + W_t^i + V^i\right) = 29.75\% \quad (1)$$

The ignition capacity is defined by the criterion $k_i = V^i/C_f^i$, having the value $k_i = 1.09$. This higher than one value indicates a high ignition capacity, allowing a less fine grinding. This is a characteristic of coal with high humidity. From the calculation of combustion (stoichiometric) resulted:

- the theoretical air volume: $V_a^0 = 4.24\,Nm3/kg$;
- the theoretical volume of gases: $V_g^0 = 4.95Nm3/kg$.

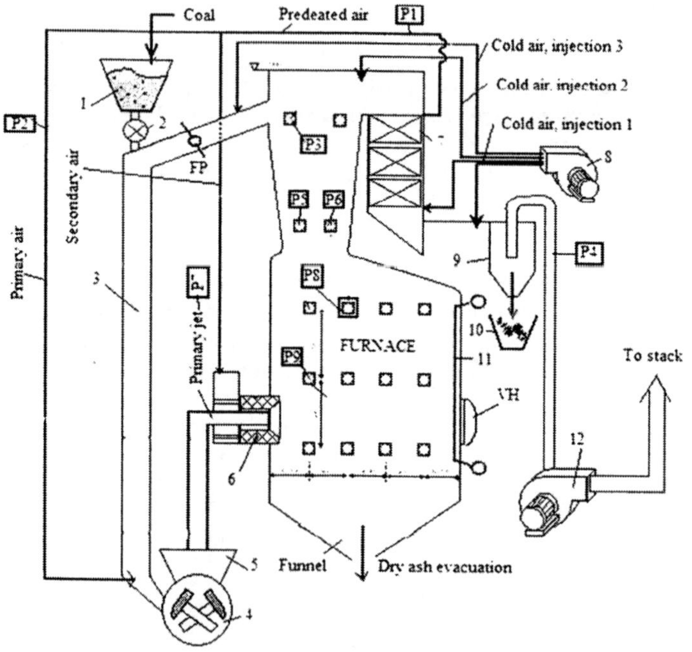

Fig.1 Scheme of 2 MWt installation pilot used to experiments:

1-Raw fuel bunker; 2-Fuel feeder; 3-Pre-drying tower; 4-Mill fan; 5-Dust separator; 6-Straight-flow burner; 7-Tubular air heater; 8-Air fan; 9-Flying ash separator; 10-Ash bunker; 11-Working fluid; 12-Flue gases fan; 13-Flue gases recirculated fan; FP-Flap control; VH-Visit hole; P1, P2, P3…..P9-Measurement points (flow, temperature, pressure).

The experiment carried out on the pilot boiler at UPB (Figure 1) has realized the following elements of similitude with the industrial propose operation expected for this coal [2]:

- milling with fan mill, with a peripheral speed of 85 *m/s*. The fineness of the grinding is adjusted by means of an inertial separator. The fan type mill allows a degree of recirculation of the flue gases before the end of the furnace of *r*=0.1. It's an individual pulverization system with direct blowing of pulverized fuel into furnace;

- the preheating of the combustion air is carried out in the range $250°C \div 300°C$ (in industrial installations the range is between $280°C \div 300°C$);
- the flow of primary air introduced into the mill can be adjusted according to the volatile matter content of the fuel. Usually, it is a dependence between the proportion of air in the form of primary air and the content of volatile fuel material (ideally, these proportions should be equal);
- the temperature of the flue gases recirculated by the mill is in the temperature range above $840°C$;
- there is thermal support for the sustaining of solid fuel combustion, achieved by combustion of natural gas;
- the thermal loading of the furnace volume is within the limits imposed by the combustion stability.

2. Experiments Results

The experiments were performed using a solid fuel consumption B_c that varied in the field $(100 \div 120)\, kg/h, (0.0277 \div 0.033\, kg/s)$ and natural gas consumption (as a thermal support) of $B_g = 16\, Nm3/h\ (0.0044\, Nm3/s)$.

2.1 The study of grinding

For the drying of coal, a primary air flow, $V_{ap} = 120\, Nm3/h\ (0.033\, Nm3/s)$ at $275°C \div 300°C$ was used and a flow of flue gas recirculated at the end of the furnace $V_{gr} = 0.0166\, Nm3/s$ (corresponding to a degree of recirculation of $r = 0.1$).

The calculation of the drying was performed on the basis of the simplified balance presented in Figure 2. The physical sizes of the thermal balance represent:

V_{ap} - primary air flow, $Nm3/s,$; c_a, c_g −the specific heat of the air and the flue gas, respectively, $Nm3/(kg \cdot K)$; t_c, t_a, t_r, t_m − the temperature of the fuel, air, recirculated flue gas and flue gas, respectively, °C; B_c - fuel consumption, *kg/s*; ΔW - moisture evaporating, kg_{water}/kg_{fuel} and c_c - the specific heat of the fuel, *kJ/(kg·K)*.

The following initial values were used in the mathematical model:

$t_c = 20°C,\ t_r = 860°C,\ t_m = 62°C,\ W_t^i = 35.04\%$.

For evaporation of water, the latent heat of water vaporization has been consumed (2,500 *kJ/kg*). From the calculations resulted the heat introduced into the mill, $Q_m = 26.47$ *kW*. Unknown size of the thermal balance equation is the quantity of vaporized water in the heel drying process, ΔW. From calculations resulted: $\Delta W = 0.282$ (28.2%). The moisture of the coal dust at the exit of the mill will be $W^p = W_t^i - \Delta W =$ 35.04 – 28.2 = 6.84 %.

This value is above the hygroscopic value of the fuel ($W_h^i = 5.31\%$). Under these conditions we have a functioning at the limit of avoiding the danger of explosion [3]. The fineness of the coal dust was determined on a sample taken at the exit of the mill, from the path of the burner mill. The fineness of the grinding is shown in the particle size distribution curve of Figure 3. With R_x was noted the total sieve residues for a sieve with the mesh size *x*, expressed in μm. The grinding performed a residue on the standard sieve of $90\mu m$ by 80%. It results in a semi-rough grinding, but sufficient considering that the volatile content is high, which ensures an easy ignition. The milling coefficient had the value for an adsorbed humidity of 30%. This value indicates the existence of low energy consumption for coal milling.

2.2 The study of the combustion process

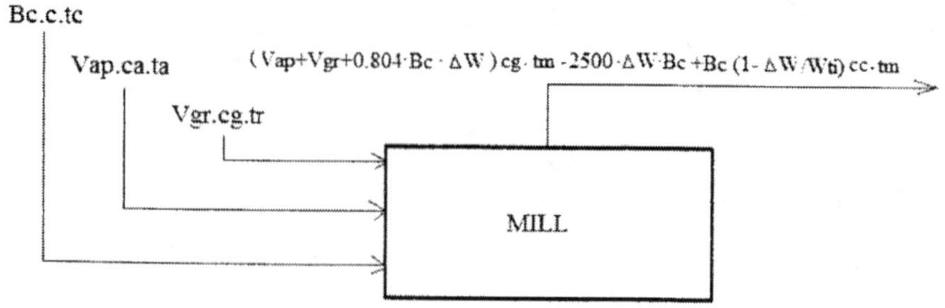

Fig.2 Balance of energy flows from the mill

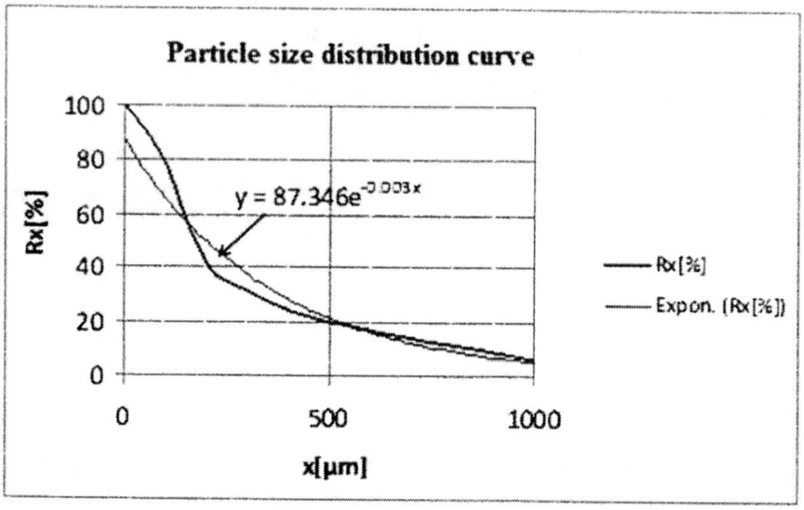

Fig.3 Particle size distribution curve

The combustion conditions in the furnace were carried out by compliance with the global thermal indicators, the main one being the thermal loading of the furnace volume [4]: $q_v = P_t/V_f,\ kW/m^3$, where P_t is thermal power, in kW and V_f is the furnace volume, m^3. For the solid fuel consumption and of natural gas consumption (as support fuel) the thermal power was: $P_t = B_c \cdot Q_i^i + B_g \cdot Q_i^{um}, kW$ where Q_i^i and Q_i^{um} represent the lower heating value of solid fuel and heating value of natural gas, respectively ($Q_i^{um} =$ $35{,}700\ kJ/Nm3$). Thermal power results:
P_t=620 kW. For a volume of the furnace of 10 m^3resulted a thermal loading of the furnace volume $q_v = 62\ kW/m^3$.

The calorific participation of the supporting fuel had the value:

$$q_t = B_g \cdot Q_i^{um}/P_t \cdot 100 = 25\% \tag{2}$$

The temperature at the end of the furnace had an average value of about 980°C. Figure 4 shows the technical operating data for the furnace, recorded by the control panel. The through incomplete combustion, determined by measuring the slag and ash unburned had the value $q_m = 0.55\%$. This value indicates a properly combustion process, validating the chosen combustion technology. The chemical analysis of the ash resulting from the combustion process indicated:

$SiO_2 = 22.52\%;\ Fe_2O_3 = 27.89\%;\ Al_2O_3 = 7.35\%;\ CaO = 28.74\%;\ MgO = 5.89\%$

 https://doi.org/10.1051/e3sconf/202018001015

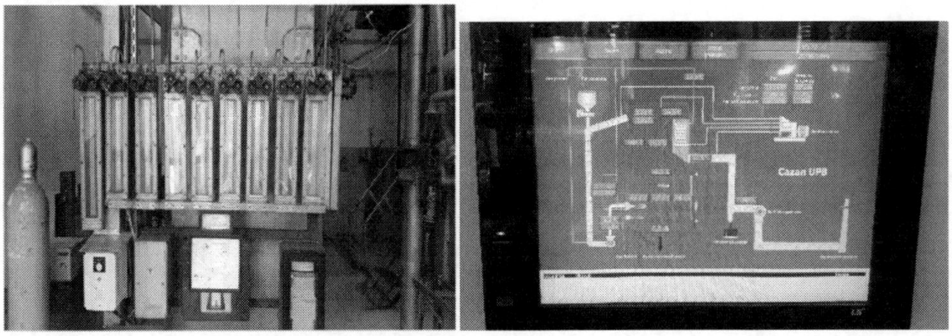

Fig.4 *Technical operating data for the furnace*

The rest of the elements had been participated under 1%. This ash chemical composition indicates average erosion for the milling elements, because the content of SiO_2 it is relatively low. An analysis of the tendency to form adherent slag in the furnace was also imposed. The temperature behavior of the ash depends on its chemical composition, the acid oxides, SiO_2 and Al_2O_3 having a higher temperature resistance then that of the basic ones as Fe_2O_3, CaO, MgO. Fusibility index of the ash has the value:

$$k = (SiO_2 + Al_2O_3)/(Fe_2 + CaO + MgO) = 0.47 \quad (3)$$

This value indicates a melting temperature of less than 1200 °C. Laboratory tests have shown a softening temperature of over 1000 °C. Further research in this field is required.

2.3 Emission of SO_2

The numerical applications for an analytical calculation model [5] led to the determination of a SO_2 maximum concentration of 205.6*mg/Nm3* .Calculations were performed for the real operating mode of the boiler furnace. This analytically determined size represents a maximum possible value for a given operating regime. In the parallel, experimental measurements were made with a device TESTO 350. The emission was in the field: $SO_2 = 137 \div 163\ mg/Nm3$.

Thus, there is a possible classification in the current pollution norms ($200\ mg/Nm3$ maxim), at the industrial operation of the boiler, with this fuel.

3. Conclusions

The complex tests performed on the combustion of a solid fuel with low sulfur content on a pilot boiler, demonstrated the possibility of passing the researches on an industrial scale.

This conclusion resulted from:

- classifying the emission of SO_2 (the main purpose of the tests) within the accepted environmental limits;
- proper grinding with the help of a fast mill, with a good fineness of dust grinding, a drying accentuated with normal energy consumption;
- obtaining a high temperature field and a low value of through incomplete combustion indicates an efficient combustion of this fuel in the experimental furnace.

The data determined during the experimental researches are compatible with those for an industrial boiler, the similarity of the two installations constituting, besides, a general objective.

Referances

1. M. Reznikov, Yu. Lupov, „*Steam boilers of thermal power station*", Mir Publishers, Moscow, (1985);
2. ***, European Commission „*Integrated Pollution Prevention and Control*", Draft March, (2003);
3. V. Ganapathy „*Steam Generators and Waste Heat Biolers*", CCR Press, 2002, p.493, (2015);
4. L. Mihăescu, G. Negreanu, M. Ceclan, I. Oprea, I. Pîşă, M. Prisecaru, M. Georgescu, E. Pop, V. Berbece, „*Sisteme şi echipamente termice pentru producerea de energie*" Ed. Printech, ISBN 978-606-521-872-7, Bucureşti (2012), (in Romanian);
5. I. Pîşă, „*Generatoare de abur. Calculul instalaţiilor anexe*", Ed. Politehnica Press, ISBN 978-606-515-836-8, Bucureşti (2018), (in Romanian).

E3S Web of Conferences **180**, 01016 (2020) https://doi.org/10.1051/e3sconf/202018001016
TE-RE-RD 2020

Structural and Technological Features of an Installation for Recovery of End-of-life Automobile Tires

Violeta Rasheva[1,*], *Georgi* Komitov[2], *Ivan* Binev[3], and *Georgi* Valtchev[1]

[1]Department of Industrial Thermal Engineering, University of Food Technologies, Plovdiv, Bulgaria
[2]Department of Agricultural Mechanization, Agricultural University, Plovdiv, Bulgaria
[3]Department of Energetics, Trakia University St. Zagora, Faculty of Technics and Technology, Bulgaria

Abstract. Automobile tires are one of the generated wastes from the road transport. According to an EU directive after 2006, end-of-life tires are subject to controlled collection and storage by traders. Direct incineration is unacceptable due to high levels of hazardous emissions. In addition to the high rate of depletion of natural resources, and in line with the "sustainable development" model, it is appropriate to create new advanced technologies for the proper utilization of natural resources through the reuse of end-of-life products. The pyrolysis technology of decomposition of raw materials is an innovative technology for production of fuels and metals from the disposal tires. This technology shows a reduced risk to public health and reduced levels of harmful substances in flue gases, has low operating costs and does not require separate waste collection. This paper focuses on the structural features of the pyrolysis installation for used automobile tires recycling. The installation has been implemented into production and has a capacity of 1 ton of end-of-life tires per hour. The result is 400 kg of liquid fuel, 400 kg of carbon and 100 kg of metal waste per hour.

1 Introduction

Global warming is becoming a serious problem. According to the latest research and the need for urgent action to combat climate change, the European Council supports the objective of achieving a neutral climate impact of the EU by 2050, in accordance with the Paris Agreement. The transition to the neutral climate will bring significant opportunities such as potential for economic growth, new business models and markets, new jobs and technological development. However, achieving climate neutrality will require overcoming serious challenges and significant investment. The European Council recognizes the need to ensure energy security and respects the right of Member States to choose the most appropriate technologies [1].

Due to the global policy on CO_2 emissions reduction, governments, municipalities and operators of industrial and commercial installations are looking for ways to use energy more efficiently [2]. A number of scientists have focused their research in this area [3-5].

* Corresponding author: v_rasheva@abv.bg

The daily human activity is related to the generation of waste. The issue of environmental solid waste disposal is still being addressed. Waste is not only a threat to humans, but also a resource through the proper processing of which the use of raw materials for energy generation could be minimized. The policy on waste management in the European Union (EU) aims to reduce their impact on the environment and human health, and to promote the efficient use of resources. In the long term, the goal is to make Europe a recycling society that prevents waste generation and uses inevitable waste as much as possible as resources. The aim is to achieve a higher level of recycling and to minimize the use of additional natural resources. EU waste management policy is regulated in a number of directives and regulations [6-8]. The end-of-life tyres (ELTs) within the EU are managed by three different models as follows: extended producer responsibility (EPR), liberal system (free market) and the tax system (government responsibility, financed through a tax) [9].

In Bulgaria, the Waste management act (WMA) regulates the hierarchy of waste management: prevention, recovery (recycling, reuse and/or recovery of recyclable materials and energy from waste that cannot be prevented) and disposal [10]. According to the WMA, the cost of collecting, transporting, utilizing and disposing of waste should be at the expense of the owners and producers of the waste, which is in accordance with the principles "polluter pays" and "producer responsibility".

The industrialization and growing global population have led to an increase in the number of road vehicles which in turn results in an increase in ELTs [11-14]. Annually, 1 billion tires, 40% of which are heavy truck tires, are dumped at landfills where they pile up due to their non-degradable nature, creating a problem for the environment. At the same rate of production, the problem increases with each passing day, forcing researchers to find alternatives to discarding and for utilizing energy as tires are a good source of carbon fuel [15].

Disposing of used tires is an important environmental issue in terms of the huge spaces turned into dumps and other related problems such as poor biodegradability, easy self-ignition causing long-lasting fires, stagnant water spreading insects and risk of infection [16].

According to the waste hierarchy, tires' recycling is most preferred in terms of waste management. However, the use of recycled tires is not well received. The recovery of energy from ELTs through their burning in cement kilns is becoming unacceptable in view of the increasingly stringent emissions requirements [17].

Tires contain valuable components, such as carbon black, organic oils and steel, but their extraction in a cost effective manner is extremely difficult. Only a few of the ELTs recovery technologies are viable, while others have not been used on an industrial scale yet [18].

Granulation is a recent sector. Rubber granules, obtained from fine ground tires could be used for production of various products. It seems that the use of rubber granules in civil engineering and other applications is a good approach, proven by a number of successful cases. Apart from lightweight, good thermal insulation and good drainage properties of shredded tires, rubber also possesses excellent vibration absorption capability, which has not been fully exploited yet. This property is expected to be useful in reducing vibrations and earthquake protection [16].

However, there is little evidence of occupational exposure and associated pathologies. More research should be carried out on the levels of inhalation exposure to the particles produced during granulation and the atmospheric pollutants in the workshops [19].

Many tire recycling options have been used, such as repairing, grinding and incineration. However, these processes have their limitation and do not fully utilize ELTs as an excellent material and energy recovery [20]. For this reason, the pyrolysis of ELTs has received new attention for the production of valuable products. The pyrolysis process is both environmentally friendly and technologically feasible for the recycling of ELTs [21]. The method involves thermal decomposition of waste rubber in the absence of oxygen, producing oil, carbon, gas, steel and residual products from waste tires. The oil obtained from the ELTs pyrolysis could be used as an alternative diesel or as a source for production of chemicals [22-23].

Currently, many researchers have focused their works on the thermal behaviour of the pyrolysis of ELTs crumb and whole tires. Pyrolysis proved to be a feasible alternative to deal with the problem of scrap tire disposal. The process yielded 46% of oil, 40% of activated carbon and 14% of gases, which could be used as a source of heat in the pyrolysis process. As for the pyrolysis, the higher the temperature and the lower the pressure the higher the fuel oil yield [24]. It could be seen that the waste tire rubber is not completely decomposed at the temperature less than 500^{0}C. The optimum conditions for the process are at temperature of 500 ^{0}C, 80 mesh particle size and 5 L/min carrier gas flow rate to achieve the highest yield of pyrolysis oil [25]. The technology of pyrolysis, based on using the whole tires significantly reduces the disposal costs of expenses incurred for its grinding, according to [26]. Variation in process parameters at small and laboratory pilot scale ensure that the total conversion of rubber tire does not register significant changes. It is found that the three obtained products (gas, liquid and solid fractions) have a good energetic potential: the solid (20 – 32 MJ/kg), the liquid (41 – 43 MJ/kg) and the gas (32– 36 MJ/m^3). Nevertheless, the liquid needs to be refined in order to be used as diesel-like fuel and gases should be treated to remove sulphur compounds [27].

This paper focuses on building of an installation for pyrolysis treatment of end-of-life automobile tires in Bulgaria.

2 Description of the plant

The installation is designed according to the innovative technology for recovery of materials and energy from ELTs [28], the scheme of which is shown in Figure 1.

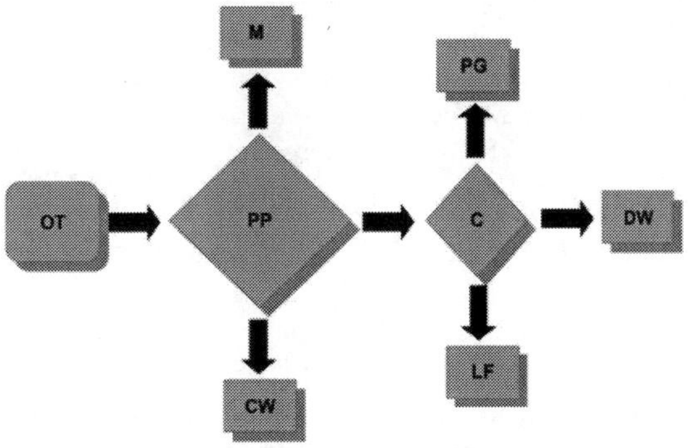

Fig. 1. Flow diagram of ELTs pyrolysis process: OT – old tires; PP - pyrolysis process; M - metal residue; CW - carbon residue; C - condensation; LF - liquid fuel; PG-pyrolysis gas; DW - domestic hot water

The general view of the installation for pyrolysis of ELTs is shown in Figure 2 and has been developed with a view to simplify the design and facilitate the maintenance of the system.

The plant has capacity of 1 ton raw material (ELTs) per hour. Its core modules are a pyrolysis boiler (5), a mixing heat exchanger (8), a pyrolysis gas condenser with a water jacket (9) and a module for cooling the solid carbon obtained from the pyrolysis (25).

The pyrolysis boiler is fed through a system of conveyors. The ELTs are fed to the conveyor 1. The loading is organized either by pouring them directly from the truck on a ramp above the conveyor or by a forklift. The conveyor is inclined so that it can feed the next unit in the system - shredder (2). The function of shredder is to fragment tires to a certain size, suitable for further processing. Fragmentation is necessary in order to reduce the size of tires and to facilitate their decomposition in the pyrolysis. The size of the shredded tires should be determined experimentally according to the capabilities of the pyrolysis boiler and its supply system. The fragmented tires are fed to another inclined conveyor (3), at the end of which there is a rotor valve. This valve is required to supply the specified amount of shredded tires to the pyrolysis boiler and to establish a steady pyrolysis process. The tires from the outlet

 https://doi.org/10.1051/e3sconf/202018001016

of the rotor valve are fed to the auger conveyor (4). The auger speed is adjusted by a worm gearbox attached to the auger. The speed of rotation of the s auger is determined by the speed of the metal conveyor in the pyrolysis boiler and by the open position of the rotor valve.

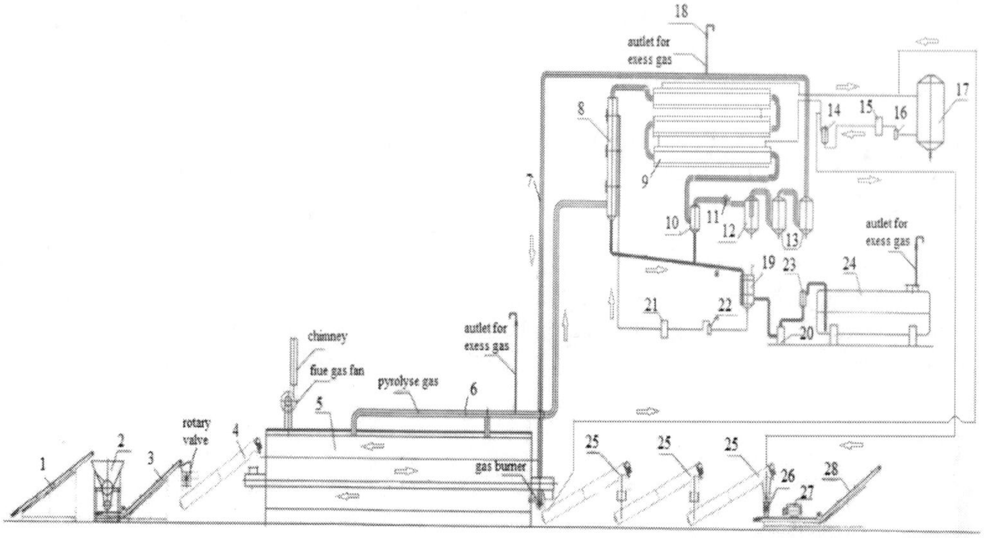

Fig. 2. General view of the plant for utilization of valuable materials from ELTs: 1 - inclined conveyor; 2 - shredder; 3 - inclined conveyor with rotary valve; 4 – auger conveyor; 5 - pyrolysis boiler; 6 – output pipes for pyrolysis gas; 7 - gaseous fuel inlet to the pyrolysis boiler burner; 8 - mixing heat exchanger; 9 - condenser with water jacket; 10 - gas/liquid separator; 11 - fan; 12 - centrifugal separator; 13 - gas receivers; 14 - water filter; 15 - refrigeration machine; 16 - circulation water pump; 17 - water tank; 18 - fire barrier; 19 - intermediate fuel tank; 20 - circulating fuel pump; 21 - cooler for liquid fuel; 22 - fuel pump; 23 - fuel filter; 24 - liquid fuel tank; 25 - auger conveyor with water jacket; 26 - crusher of carbon; 27 - magnetic separator; 28 - outgoing conveyor

The shredded tires are heated to 600^0C in an oxygen-free environment and a vacuum of 0.2 bars in the pyrolysis boiler. The tires are heated by a gas burner fueled by pyrolysis gas obtained in the same installation. The parameters of this gas are flow rate 100 kg/h and a temperature of 40^0C. The flue gas resulting from the combustion moves counter currently to the metal conveyor for tires pieces in the pyrolysis boiler and leave it sucked by an explosion-proof fan through the chimney. As a result of the pyrolysis, the pieces of tires decompose into pyrolysis gas and solid carbon residue. The pyrolysis gas leaves the boiler at a temperature of 600^0C and flow rate 500 kg/h through the pipelines (6), secured by an outlet of excess gas and a fire barrier.

The solid residue is discharged from the boiler through its metal conveyor and fed to an auger conveyor with a water jacket (25), a general view of which is shown in Figure 3. The auger is inclined at an angle of 25^0 for a more convenient transfer of the solid residue and to save area. The solid residue enters the inlet of auger conveyor at a temperature of 600^0C and a flow rate of 500 kg/h and passes through three auger conveyors (25) to be cooled to a temperature of 40^0C.

The solid residue contains carbon and metal particles from the tire frame. Because of that after being cooled, it is fed to the carbon crusher (26) and is crushed into finer parts for better separation of the metal from it. Usually the crushing takes place in roller mills and the size to which the solid residue is ground depends on the distance between the two shafts.

The milled solid residue falls onto a belt conveyor for removal of carbon. It has two sections - horizontal (27) and sloping (28) and after its outlet, the carbon is stored in containers and used as intended. The yield of the solid carbon is 400 kg/h. A magnetic separator positioned on the horizontal section of the conveyor is used to separate the metal particles from the solid residue. The metal particles are extracted from carbon through magnetic induction and stored. The yield of the metal obtained by separation with the magnetic separator is 100 kg/h.

The pyrolysis gas is discharged from the boiler via outlet pipes (6) and enters the mixing heat exchanger (8), a general view of which is shown in fig 4. The gas is cooled there by mixing with the liquid fuel obtained after condensation of the same pyrolysis gas. The liquid fuel with a temperature of 30^{0}C and a mass flow of 1980 kg/h is injected into the hot gas and evaporates, cooling the gas. The supply of liquid fuel to the heat exchanger (8) takes place via three inlets distributed at three levels over its apparatus height for more even spread. The cooled pyrolysis gas with a temperature of 150^{0}C leaves the upper end of the heat exchanger and enters the next module - condenser with a water jacket (9). The general view of this apparatus is shown in Figure 5. Part of the pyrolysis gas cooled by cold water condenses as it passes along the entire length of the condenser and flows to the outlet. A mixture from liquid and gaseous fuel with a temperature of 40^{0}C leaves the condenser outlet, as the ratio between the liquid and the gaseous fuel is 2380:100 kg/h. The cooling fluid in the condenser is water that enters it at a temperature of 30^{0}C and a flow rate of 1692 kg/h and leaves it at a temperature of 70^{0}C.

A gas/liquid separator (10) is installed after the condenser (9). It completely separates the liquid and the gaseous parts of the fuel. The liquid fuel, separated from the lower end of the separator, enters the intermediate fuel tank (19). It is intended to reduce the temperature of the liquid fuel and to distribute it. One part of the liquid fuel with a flow rate of 1980 kg/h is returned to the mixing heat exchanger (8), and the rest part through the fuel pump (21) is fed to the fuel filter (23) and collected in the liquid fuel tank (24). The liquid fuel directed to the heat exchanger (8) is sucked in by the fuel circulation pump and cooled in the fuel cooler (21) by a refrigerating machine to 30^{0}C before entering the mixing heat exchanger (8).

The gaseous fuel is sucked in from the other outlet of the gas/liquid separator by an explosion-proof fan (11) and fed to a centrifugal separator (12) for separation of any liquid impurities and water droplets. After the centrifugal separator, the gaseous fuel is supplied to the gas receivers (13), where it is stored until it is drained for sale or to feed the gas burner of the pyrolysis boiler. The gaseous fuel line between the gas receivers and the gas burner is secured by an outlet of excess gas, equipped with a fire barrier (18).

The circulating water used for cooling the pyrolysis gas in the condenser and solid carbon residue in auger conveyor with water jacket (25) is collected in the water tank (17). The water is then sucked in by the circulation pump (16) and cooled from 70^{0}C to 30^{0}C in a heat exchanger (15) through a refrigerating machine. Then the water passes through a water filter and again enters the water jacket of the condenser. The circulation pump and water filter must be selected so as to provide the required flow rate of water of 2988 kg/h. The same refrigerating machine is used to cool the liquid fuel in the cooler 21 to a temperature of 30^{0}C before being injected into the mixing heat exchanger.

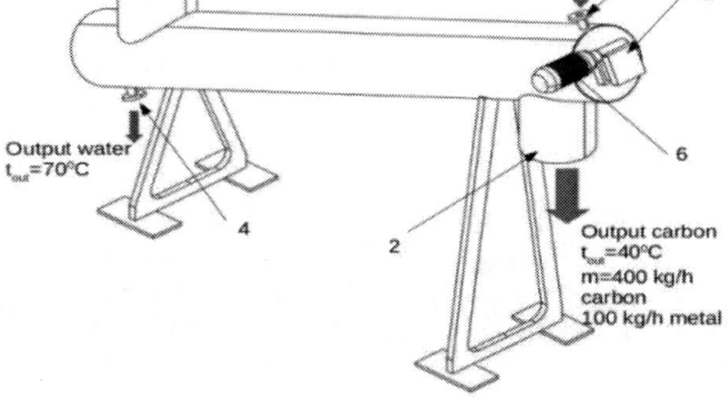

Fig. 3. General view of auger conveyor with a water jacket: 1 - inlet of the solid carbon residue; 2 - outlet of the solid carbon residue; 3 - inlet of cooling water; 4 - outlet of cooling water; 5 - reducer; 6 - electric motor

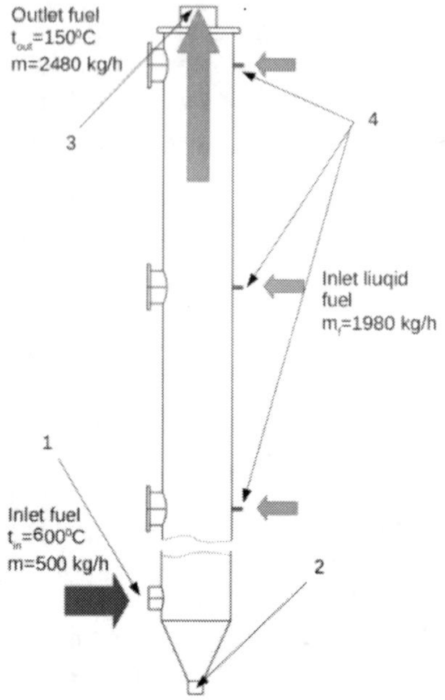

Fig. 4. General view of mixing heat exchanger: 1, 3 - pyrolysis gas inlet and outlet; 2 - drainage outlet; 4 - liquid fuel inlet pipes

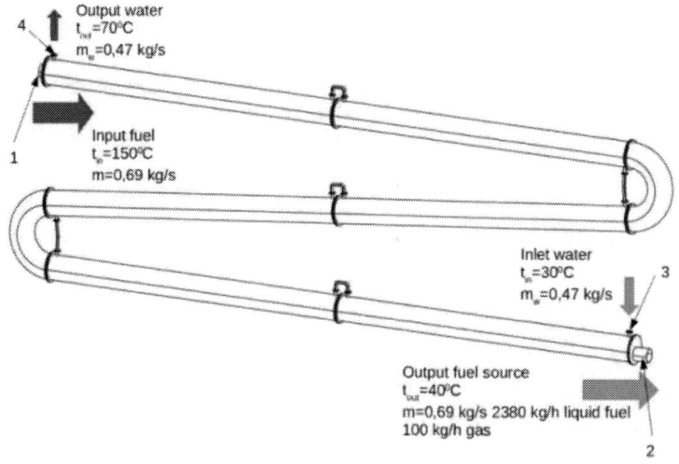

Fig. 5. General view of condenser with water jacket: 1 - pyrolysis gas inlet; 2 - outlet for fuel mixture; 3 - cooling water inlet; 4 - cooling water outlet

3 Materials and energy recovered from ELTs

Material balance: 400 kg of liquid fuel, 100 kg of gaseous fuel, 100 kg of metal and 400 kg of carbon residue are obtained from 1 ton of ELTs treated by pyrolysis in the presented installation.

The gaseous fuel produced as a result of pyrolysis is used as a fuel for the gas burner of the pyrolysis boiler in the same plant. For this reason, it must be ensured that the burner is powered by an external source until gas is generated from the pyrolysis plant.

Liquid fuel has parameters similar to the diesel oil and is suitable for transport. It could be used as additives to other liquid fuels, or as a source for production of various chemical products.

The carbon obtained is applied in industry, such as different fillers for many products, concrete, plastics, etc, or for the production of activated carbon.The metal residue is delivered as secondary raw materials for subsequent recycling processing.

The material and energy flows necessary for the proposed pyrolysis installation are:

Energy balance of mixing heat exchanger is shown by the equation (1)

$$H_g^{in} + H_{oil}^{in} = H_g^{out} + Q_{loss} \tag{1}$$

Enthalpy of pyrolysis gas entering heat exchanger, enthalpy of pyrolysis gas leaving heat exchanger and enthalpy of pyrolysis oil entering heat exchanger are represented by equations (2), (3) and (4) respectively. Q_{loss} is heat loss to the outside air in kW.

$$H_g^{in} = m_g^{in} \; x \; c_g \; x \; T_g^{in} = \frac{500}{3600} \; x \; 1.382 \; x \; 600 = 115.17 \; kW \tag{2}$$

$$H_g^{out} = m_g^{out} \; x \; c_g \; x \; T_g^{out} = \frac{2480}{3600} \; x \; 1.382 \; x \; 150 = 142.81 \; kW \tag{3}$$

$$H_{oil}^{in} = m_{oil}^{in} \; x \; c_{oil} \; x \; T_{oil}^{in} = \frac{1980}{3600} \; x \; 1.848 \; x \; 30 = 30.49 \; kW \tag{4}$$

where m_g^{in}, m_g^{out} and m_{oil}^{in} are mass flows of pyrolysis gas entering heat exchanger, pyrolysis gas leaving heat exchanger and pyrolysis oil entering heat exchanger, kg/s;

c_g and c_{oil} is specific heat capacity of pyrolysis gas and oil respectively, kJ/(kg.K) [29];

T_g^{in} , T_g^{out} and T_{oil}^{in} is temperature of inflow pyrolysis gas, outflow pyrolysis gas and inflow pyrolysis oil, ^{0}C.

The heat flux removed by the pyrolysis gas in condenser - Q_1 is calculated by the energy balance of the pyrolysis gas side in the condenser - equation (5), where enthalpy of pyrolysis gas entering condenser, enthalpy of pyrolysis gas leaving condenser and enthalpy of pyrolysis oil leaving condenser are represented by equations (6), (7) and (8) respectively.

$$H_{g1}^{in} = H_{g1}^{out} + H_{oil1}^{out} + Q_1 \rightarrow Q_1 = 92.4 \; kW \tag{5}$$

$$H_{g1}^{in} = m_{g1}^{in} \; x \; c_g \; x \; T_{g1}^{in} = \frac{2480}{3600} \; x \; 1.382 \; x \; 150 = 142.8 \; kW \tag{6}$$

$$H_{g1}^{out} = m_{g1}^{out} \; x \; c_g \; x \; T_{g1}^{out} = \frac{100}{3600} \; x \; 1.382 \; x \; 40 = 1.53 \; kW \tag{7}$$

$$H_{oil1}^{out} = m_{oil1}^{out} \; x \; c_{oil} \; x \; T_{oil1}^{out} = \frac{2380}{3600} \; x \; 1.848 \; x \; 40 = 48.87 \; kW \tag{8}$$

where m_{g1}^{in}, m_{g1}^{out} and m_{oil1}^{out} are mass flows of pyrolysis gas entering condenser, pyrolysis gas leaving condenser and pyrolysis oil living condenser, kg/s;

T_{g1}^{in} , T_{g1}^{out} and T_{oil1}^{out} is temperature of inflow pyrolysis gas, outflows pyrolysis gas and pyrolysis oil, ^{0}C.

The heat flux absorbed by the cooling water in the condenser – Q_2 is calculated by the energy balance of the water side in the condenser - equation (9), where enthalpy of cooling water entering condenser, enthalpy of cooling water leaving condenser and heat loss to the outside air - $Q_{loss,c}$ are represented by equations (10), (11) and (12) respectively.

$$H_w^{in} + Q_2 = H_w^{out} \rightarrow Q_2 = 78.77 \; kW \tag{9}$$

$$H_w^{in} = m_w^{in} \; x \; c_w \; x \; T_w^{in} = \frac{1692}{3600} \; x \; 4.19 \; x \; 30 = 59.08 \; kW \tag{10}$$

$$H_w^{out} = m_w^{out} \; x \; c_w \; x \; T_w^{out} = \frac{1692}{3600} \; x \; 4.19 \; x \; 70 = 137.85 \; kW \tag{11}$$

$$Q_1 = Q_2 + Q_{loss,c} \tag{12}$$

where m_w^{in} and m_w^{out} are mass flows of cooling water entering and leaving condenser, kg/s;

c_w is specific heat capacity of cooling water, kJ/(kg.K);

T_w^{in} and T_w^{out} is temperature of inflow and outflow cooling water, ^{0}C.

The heat flux removed by the solid carbon residue in cooling heat exchanger - Q_{1r} is calculated by the energy balance of the carbon residue side in the heat exchanger - equation (13), where enthalpy of carbon residue entering heat exchanger, enthalpy of char and metal residue leaving heat exchanger are represented by equations (14), (15) and (16) respectively.

$$H_r^{in} = H_{ch}^{out} + H_m^{out} + Q_{1r} \rightarrow Q_{1r} = 60.88\ kW \tag{13}$$

$$H_r^{in} = m_r^{in}\ x\ c_r\ x\ T_r^{in} = \frac{500}{3600}\ x\ 0.783\ x\ 600 = 65.23\ kW \tag{14}$$

$$H_{ch}^{out} = m_{ch}^{out}\ x\ c_{ch}\ x\ T_{ch}^{out} = \frac{400}{3600}\ x\ 0.853\ x\ 40 = 3.79\ kW \tag{15}$$

$$H_m^{out} = m_m^{out}\ x\ c_m\ x\ T_m^{out} = \frac{100}{3600}\ x\ 0.5\ x\ 40 = 0.56\ kW \tag{16}$$

where m_r^{in}, m_{ch}^{out} and m_m^{out} are mass flows of solid carbon residue entering heat exchanger, char leaving heat exchanger and metal residue living heat exchanger respectively, kg/s;

c_r, c_{ch} and c_m is specific heat capacity of solid carbon residue, char and metal residue respectively, kJ/(kg.K);

T_r^{in} , T_{ch}^{out} and T_m^{out} is temperature of inflow solid carbon residue, outflows char and metal residue respectively, ^{0}C.

The heat flux absorbed by the cooling water in the heat exchanger cooling solid carbon residue – Q_{2r} is calculated by the energy balance of the water side in the heat exchanger - equation (17), where enthalpy of cooling water entering and leaving heat exchanger and heat loss flux to the outside air are represented by equations (18), (19) and (20) respectively.

$$H_{wr}^{in} + Q_{2r} = H_{wr}^{out} \rightarrow Q_{2r} = 60.34\ kW \tag{17}$$

$$H_{wr}^{in} = m_{wr}^{in}\ x\ c_w\ x\ T_{wr}^{in} = \frac{1296}{3600}\ x\ 4.19\ x\ 30 = 45.25\ kW \tag{18}$$

$$H_{wr}^{out} = m_{wr}^{out}\ x\ c_w\ x\ T_{wr}^{out} = \frac{1296}{3600}\ x\ 4.19\ x\ 70 = 105.59\ kW \tag{19}$$

$$Q_{1r} = Q_{2r} + Q_{loss,r} \tag{20}$$

where m_{wr}^{in} and m_{wr}^{out} are mass flows of cooling water entering and leaving heat exchanger respectively, kg/s;

T_w^{in} and T_w^{out} is temperature of inflow and outflow cooling water respectively, ^{0}C.

The heat flux released by the pyrolysis gas combustion in the pyrolysis boiler - Q_{in} is calculated by the equation (21).

$$Q_{in} = m_{b,g}\ x\ Q_{b,g} = \frac{1296}{3600}\ x\ 4.19 = 358,6\ kW \tag{21}$$

where $m_{b,g}$ is mass flow of pyrplysis gas burning in boiler, kg/s;

$Q_{b,g}$ is calorific value of pyrolysi gas, kJ/kg [29];

In general, the following resources are required for the operation of the pyrolysis installation: circulating water with a flow rate of 2988 kg/h and 139 kW of electricity power for operation of cooling water refrigeration machine and circulation pumps and the fan. The heat of the circulating water could be used for space heating or domestic hot water. This will reduce the energy consumption of the refrigerator.

4 Conclusion

The proposed installation for pyrolysis of ELTs has been implemented into real operation. It is a good option for recovery of valuable materials and energy from waste such as ELTs.

Due to decomposition of hazardous tires in pyrolysis process and the use of the resulting pyrolysis gases to produce liquid and gaseous fuels, the system proves to be an environmentally friendly source of heat and energy.

The fuels obtained in the plant are recyclable energy sources and can be used as percentage additives to standard fuels or used directly.

Various products or activated carbon could be made from the obtained carbon residue, and the metal could be used to produce various steels and cast irons.

The operation of this pyrolysis plant leads to reduction in the amount of ELTs in landfills and to protection of the environment and human health.

5 References

1. https://www.consilium.europa.eu/media/41768/12-euco-final-conclusions-en.pdf, European Council meeting (12 December 2019)
2. N. El Bassam, P. Maegaard, and M. L. Schlichting, Current distributed renewable energy rural and urban communities, Distributed Renewable Energies for Off-Grid Communities (2013)
3. G. Lazaroiu, V. Berbece, L. Mihaescu, I. Pisa, G.-P. Negreanu, Biogas production from animal protein waste, an environmental protection technology coupled with energy production, Scientific Works of University of Food Technologies- Plovdiv, Volume **64**, Issue **1**, (2017)
4. S. Dichev, Y. Lambrev, Application and perspectives for use of heat pumps in industry, Scientific Works of University of Food Technologies - Plovdiv, Vol. **XLIII**, Issue **3** (1998) (in Bulgarian)
5. G. Raichkov, M. Valchev, M. Georgieva, Permanent ventilation of rooms and buildings - factor of healthy comfort, J. Heattecnics, No. **6**, Year **5**, Book **1**, TU-Varna, (2014) (in Bulgarian)
6. Council Directive 1999/31/EC on the landfill of waste, Official Journal of the European Communities (1999)
7. Directive 2006/12/EC of the European Parliament and of the Council on waste, Official Journal of the European Union (2006)
8. Directive 2008/98/EC of the European Parliament and of the Council on waste and repealing certain Directives (2008)
9. M.R. Sebola, P.T. Mativenga, J. Pretorius, A benchmark study of waste tyre recycling in south africa to European union practice, Procedia CIRP **69** (2018), 950–955
10. Waste Management Act (2012) https://www.moew.government.bg/static/media/ups/tiny/УООП/Законодателство/WASTE MANAGEMENT ACT_13.pdf
11. T.A. Mallah, A.R. Sahito, Optimization of castor and neem biodiesel blends and development of empirical models to predict its characteristics, Fuel (2020),
12. K. Georgiev, I. Evtimov, R. Ivanov, Analytical study on wearing of pneumatic tires of wheeled machines, Proceedings of Scientific conference with international participation

"Transport, Ecology - Sustainable Development", Varna (2013), 13-17, (in Bulgarian)

13. R. Ivanov, K. Georgiev, G. Kadikyanov, G. Staneva, An experimental research on the wear of truck tire, Poland, SJ Transport Problems, Vol. **10**, Issue **4** (2015), 91-98
14. K. Georgiev, R. Ivanov, G. Kadikianov, G. Staneva, Analytical study of the effect of tire air pressure on tire wear, Proceedings of International Scientific Conference BulTrans, Sozopol, (2015), 113-116
15. R.K. Singh, B. Ruj, A.K. Sadhukhan P. Gupta, Energy recovery from tyre waste pyrolysis: product yield analysis and characterization, Energy Recovery Processes from Wastes, Springer Nature Singapore Pte Ltd., (2020)
16. V. Torretta, E.C. Rada, M. Ragazzi, E. Trulli, I.A. Istrate, L.I. Cioca, Treatment and disposal of tyres: two EU approaches. A review, Waste Management **45** (2015) 152–160
17. H.-Ho Tsang, Uses of scrap rubber tires, Rubber: types, properties and uses, 2010 Nova Science Publishers, Inc. (2019)
18. C. Bulei, M. P. Todor, T. Heput, I. Kiss, Directions for material recovery of used tires and their use in the production of new products intended for the industry of civil construction and pavements, Materials Science and Engineering **294** (2018)
19. B. Savary, R. Vincent, Used tire recycling to produce granulates: Evaluation of occupational exposure to chemical agents, Ann. Occup. Hyg., Vol. **55**, No. **8** (2011), 931–936
20. A. Ayanog˘lu, R. Yumrutas, Production of gasoline and diesel like fuels from waste tire oil by using catalytic pyrolysis, Energy (2016)
21. P.T. Williams, A.J. Brindle, Fluidised bed catalytic pyrolysis of scrap tyres: Influence of catalyst:tyre ratio and catalyst temperature, Waste Manag. Res. **20** (2002), 546–555
22. J.D. Martínez, N. Puy, R. Murillo, T. García, M.V. Navarro, A.M. Mastral, Waste tyre pyrolysis - a review, Renew. Sustain. Energy Rev. (2013)
23. D. Perondi B. Santinon Scopel G. Carvalho Collazzo J. Pessutto Silva M. Leoratto Botomé A. Dettmer M. Godinho A.C Faria Vilela, Characteristics of pyrolysis products from waste tyres and spent foundry sand co-pyrolysis Prog. Rubber, Plast. Recycl. Technol. **32** (2016), 213-240
24. C.F.S. Rombaldo, A.C.L. Lisbôa, M.O.A. Méndez, A. dos R. Coutinho, Effect of Operating Conditions on Scrap Tire Pyrolysis, Mat. Res. vol. **11** no. **3** (2008)
25. D.Ab. Taleb, H.A. Hamid, R.R.R. Deris, M. Zulkifli, N.A. Khalil, A.N.A. Yahaya, Insights into Pyrolysis of Waste Tire in Fixed Bed Reactor: Thermal Behavior, Materials today: Proceedings - February (2020), doi: 10.1016/j.matpr.2020.01.569
26. M. Ryms, K. Januszewicz, W.M. Lewandowski, E. Klugmann-Radziemska, Pyrolysis Process of Whole Waste Tires as a Biomass Energy Recycling, Ecol Chem Eng S., **20(1)** (2013), 93-107
27. S. Kordoghli, M. Paraschiv, R. Kuncser, M. Tazerout, M. Prisecaru, F. Zagrouba, I. Georgescu, Managing the Environmental Hazards of Waste Tires, Journal of Engineering Studies and Research, Volume **20**, No. **4** (2014)
28. G. Komitov, V. Rasheva, I. Binev, I. Kiryakov, Innovation technology for using of disposals automobile tyres, European Journal of technical and natural science, Vol.**4**, Vienna (2016), 20-22
29. D. Chakarova, A. Andreev. Energy and exergy analysis of pyrolysis of waste tires, Proceedings of the **XXI** Scientific Conference EMF'2016, TU-Sofia, (2016), (in Bulgarian)

Some design characteristics of micro steam turbines for agricultural biomass energy conversion

Gabriel-Paul Negreanu[1,*], *Ion* Oprea[1] , and *Viorel* Berbece[1]

[1] University Politehnica of Bucharest, Departament Termotechnics, Engines, Thermal and Refrigeration Equipment, Adress: Splaiul Independenţei, no 313, sector 6, Romania

Abstract. The paper continues the study of reconversion of a 400 kW hot water boiler in a steam generator suitable to valorise the energy content of briquettes and pellets of agricultural biomass. After steam parameters selection (pressure, temperature, mass-flow rate), an overview of main steam machines types (axial, radial, screw, piston engine) is done. Further, a parallel design of most wide-spread ones (Laval, Curtis and radial) were performed, at different rotation speeds, in order to find the best configuration in respect with the flow section dimensions, internal efficiency and power (electrical and thermal) output. The results of the paper could be very useful for the investors in agricultural "waste-to-energy" projects in order to select appropriate technology and equipment.

1 Introduction

During last 15 years, the authors belong to a large research team who studied the biomass by-products (solid, liquid and gaseous) in useful energy (thermal and electrical) to be used locally or remote. One of these researches [1] was focused on mini thermal steam power plant able to give economic value to new energy willow crops developed in our Country.

In a previous recent paper [2] we focused on the feasibility to convert a hot water boiler powered by solid biomass waste into a steam generator, in order to couple it with a condensing or backpressure steam turbine. The purpose of current paper is to select the type of steam engine suitable for this task, based on design feasibility and efficiency estimation.

In a "white paper" edited by a manufacturer [3], is stipulated that the steam microturbines are commonly used to reduce the pressure of process steam, working in parallel to a conventional pressure reducing station with high reliability with a service life of 15-25 years and important cost savings related to the generated electricity.

Other equipment producers [4-5] underline the advantages of the microturbines used in district heating systems such as challenging space limitations, satisfying the reliability requirements, operational simplicity, meeting the sound criteria of equipment rooms; and a

* Corresponding author: gabriel.negreanu@upb.ro

2 to 5 year payback investment. With all the parts into one seamless part, the volume is vastly reduced, allowing the maintenance simple, compact and affordable.

The specification of such axial (Laval and Curtis) micro-turbine-generators are revealed by other manufacturer [6-7]:

- Inlet pressure: 2 – 30 [bar]
- Back Pressure: 0,1 – 15 [bar]
- Turbine speed: 12000 – 14400 [rpm]
- Frequency: 50 / 60 Hz
- Voltage: 380 V / 415V

Kubawara et al [8] present the design characteristics of a micro steam expander based on the screw principle, with outputs of 12-132 kW, with internal efficiencies of 60-70 %.

Another manufacturer [9] gives the range of electric power for its products (50 – 200 kWe) and announces other advantages:

- Efficient operation in the conditions of wet steam
- Excellent ratio power/weight
- Working life min. 100 000 hours
- Full power reaction time in 10 minutes

Finally, Alford et al [10] offer a comprehensive analysis of a microturbine based upon an integral high speed air compressor, optimised by means of numerical modelling, designed, manufactured and tested in a wide range of conditions with good efficiency and endurance results.

2 Design characteristics of axial turbines

2.1 Input data and initial hypothesis

According to Figure 1 that shows the main thermal circuit of the plant, the input parameters revealed in paper [1] are:

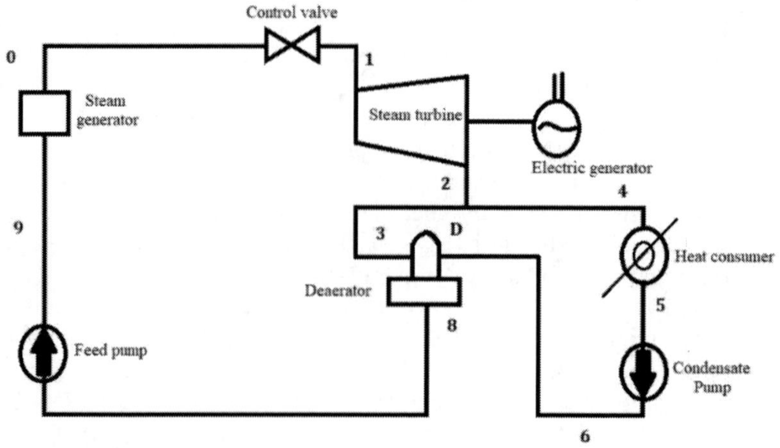

Fig. 1. Thermal circuit of the plant

- Live steam pressure p_0= 8 [bar];
- Live steam vapour content x_0=1 [-] (saturated);

- Exhaust pressure to the consumer p_2= 1.2 [bar];
- Live steam mass flow rate $\dot{m}$=0.35 [kg/s].

The two axial turbines selected for this analyse were the well known one stage Rateau and Curtis types, presented in figure 1

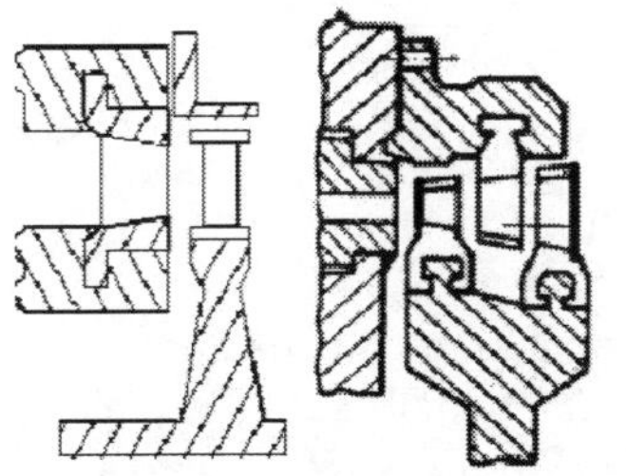

Fig. 2. One stage axial turbines Rateau and Curtis

Due to the very small expansion ratio β=0.16 < 0.546 = β_{cr}, a convergent-divergent nozzle type was selected, in order to obtain high steam velocities. The other common feature is the impulse stage type.

Meanwhile, in order be able to convert efficiently and to have small dimension, the rotation speed should be higher than 50 rps. Thus, three values were selected for this parameter, respectively 100, 150 and 200 rps. It is obvious that the system should be provided with a reducing gearbox or with a frequency convertor in the electric generator.

2.2 Methodology

The designed methodology is inspired from [11] and is currently used in didactic and applied research activities in our University.

First of all, the optimal velocity ratio is calculated:

$$x_{opt} = \frac{u}{c_1} = k_x \cdot \frac{\cos \alpha_1}{2 \cdot i \cdot (1-\rho)} \tag{1}$$

where k_x = 0,8…1;
ρ = 0 (reaction degree)
α_1- effective exit nozzle angle [°]
i – number of mobile rows of blades (1 for Rateau and 2 for Curtis)

Secondly, the absolute nozzle exit velocity can be found:

$$c_1 = \varphi \sqrt{2(H_a + H_0)} \tag{2}$$

where φ – nozzle losses coefficient
H_a – nozzle isentropic drop, equal to the stage isentropic drop H_t [J/kg]
H_0 – input kinetic energy [J/kg]

Combining eq. (1) and (2) the peripheral stage velocity u [m/s] allows to compute the average stage diameter d_1 [m]:

$$d_1 = \frac{u}{\pi \cdot n} \tag{3}$$

Imposing a minimal (related to the acceptable flow losses) and constant (respecting the speed rotation n) nozzle height L_{n1} of 10 [mm] for Rateau and 20 [mm] for Curtis, from flow equation it results the criterion for establishing the optimum rate of admission ε [-], meaning the ratio between the surface of nozzles exit and the whole circular crown surface available:

$$\varepsilon L_{n1} = \frac{\dot{m} \cdot v_1}{\mu \cdot \pi \cdot d_1 \cdot c_1 \cdot \sin\ _1} \quad (4)$$

where μ – filling surface coefficient depending on φ;
v_1 – specific volume at nozzles exit [m^3/kg].

It's now possible to compute the components of velocity diagram for the two selected turbines, in order to assess both main losses (ΔH_m [kJ/kg] related to the nozzles, blades, exit kinetic energy and secondary losses (ΔH_s [kJ/kg] related to the friction and ventilation, partial admission and humidity), and also the internal enthalpy drop H_i [kJ/kg], the internal turbine efficiency η_i [-] and the electric generator power P_g [kW]:

$$H_i = H_t - \Delta H_m - \Delta H_s \quad (5)$$

$$\eta_i = \frac{H_i}{H_t} \quad (6)$$

$$P_g = \eta_m \cdot \eta_g \cdot \dot{m} \cdot H_i \quad (7)$$

where η_m – mechanical efficiency [-];
η_g – electric generator efficiency [-];

2.3 Results and discussions

The common data of the two turbines with no connection with the rotation speed are presented in Table 1:

Table 1. Common parameters non related to the speed

Parameter	Unit	Rateau	Curtis
x_{opt}	[-]	0.442	0.215
c_1	[m/s]	734.3	743.8
u	[m/s]	324.3	160.0
α_1	[°]	11	17
L_{n1}	[m]	0.01	0.02

The other parameters depending to the rotation speed are shown in the following figures:

 https://doi.org/10.1051/e3sconf/202018001017

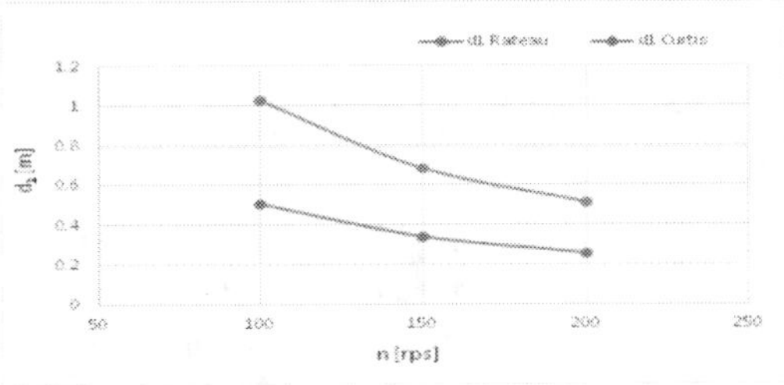

Fig. **3** Average diameter versus rotation speed

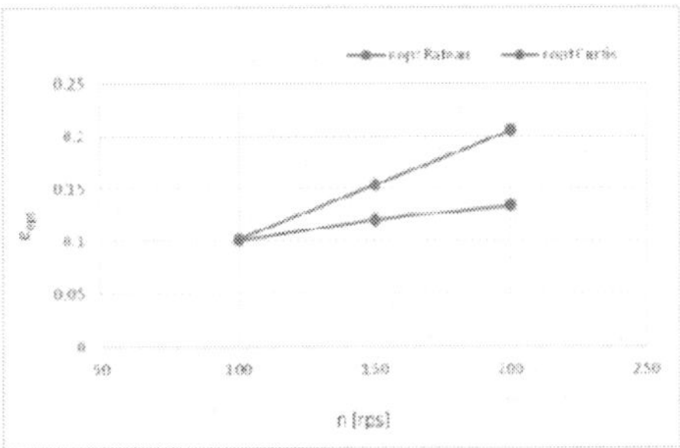

Fig. 4. Admission rate versus rotation speed

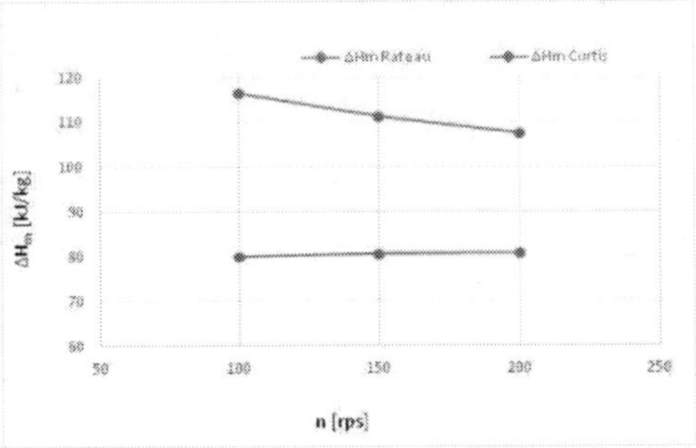

Fig. **5** Main losses versus rotation speed

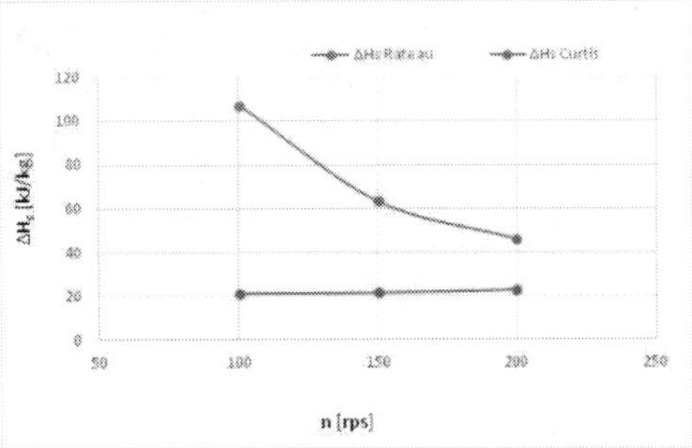

Fig. **6** Secondary losses versus rotation speed

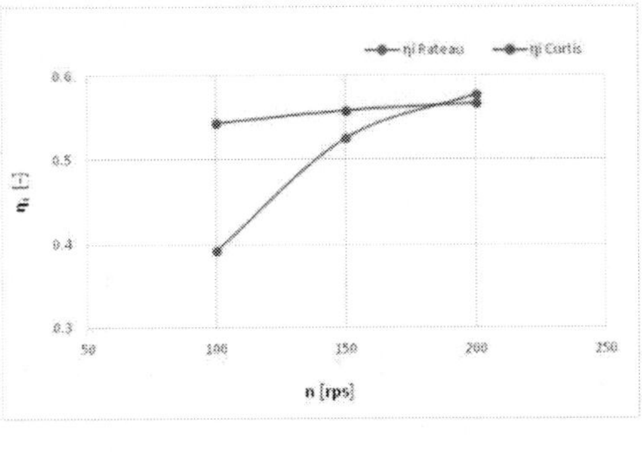

Fig. 7. Internal efficiency versus rotation speed

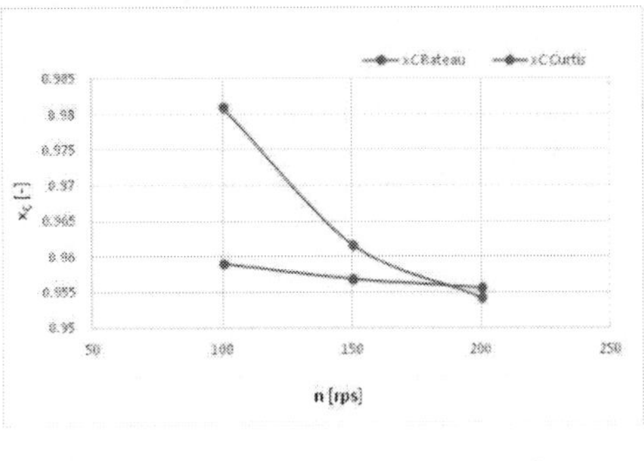

Fig. 8 Exhaust vapour content versus rotation speed

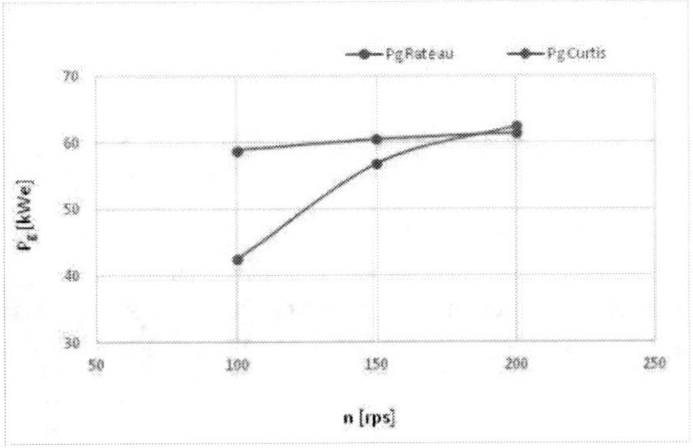

Fig. 9. Electric power versus rotation speed

In Figure 3 is displayed the average diameter of the stage versus the rotation speed. It can be seen the diameter decrease when the speed increases, and also the smaller diameter for Curtis than Rateau.

The admission rate (Figure 4) seems to be higher for Curtis than for Rateau, but only at rotation speeds smaller than 200 rps.

The main losses, presented in Figure 5, are higher for Curtis than for Rateau, due to the three rows of blades instead of only one.

But the secondary losses, shown in Figure 6 are significant more important for Rateau than for Curtiss, due to the weight of the friction and ventilation loss, depending with u^3.

Corroborating the two last sentences, the internal efficiencies are the same at 200 rps, but Curtiss beats Rateau at lower speeds. (Figure 7)

An important parameter to be checked is represented by the vapour content at the turbine's exhaust. In order to protect the blades against the water droplets erosion its value should be higher than 0.88. In the Figure 8, this condition is fully respected, the smaller value registered for both turbines is about 0.955 at 200 rps. A supplementary erosion test with Esher-Wyss and Hitachi criteria confirmed this affirmation.

The final and important parameter is represented by the power delivered by the electric generator (Figure 9). At 200 rps, the values are identical for both turbines (about 62 kW_e), but at lower speeds Curtiss is more powerful than Rateau.

The overall efficiency (electric+thermal) is the same for all configuration at the value of 0.914, that means that the biomass conversion into energy has a high efficiency in comparison with other energy conversion chains.

3 Design characteristics of radial turbines

A third solution analyzed was a radial-inflow turbine type. The centripetal effect makes possible to change as much energy as possible from the fluid into useful work. Some other significant advantages can be mentioned in comparison with the axial-flow turbine: a greater amount of work per stage, a simplest manufacturing, a high ruggedness and a lower cost [12]. The radial turbines are more efficient than the axial one for small mass flows and can perform large enthalpy drops with relatively low peripheral speeds. The turbine performance is a function of the similarity parameters expressed by specific speed n_s and specific diameter D_s, and for compressible flow Reynolds number R_e^* and Mach number M_a^*.

$$n_S = \frac{n \cdot \dot{V}_3^{1/2}}{H_t^{3/4}} \tag{8}$$

$$D_S = \frac{D \cdot H_t^{1/4}}{\dot{V}_3^{1/2}} \tag{9}$$

For the project data the isentropic enthalpy drop is $H_t = 325.6$ kJ/kg and the volumetric rate of flow relative to the rotor exit is $V_3 = 0.22$ m^3/s; the spouting velocity $c_0 = 807$ m/s. In the case of ideal radial turbine, with complete recovery of the exhaust energy the optimum velocity ratio is 0.707 and the typical range is indicated $0.55 < u_2^2/c_0 < 0.77$ [13].

If the normal speed is considered ($n = 50$ rps), than the specific speed is $n_s = 0.103$ and for Imperial units $n_s = 1{,}395$ or $\Omega_s = 0.0107$ rad. The Balje diagrams (n_sD_s turbine chart) shows that this low value of specific speed is outside of the domain for the radial-inflow turbine with good performance [12-15]. The solution with maximum efficiency is indicated to be a partial admission axial turbine.

A high speed could be a solution for the inflow-radial turbine. A desired minimum efficiency and a corresponding specific speed can be determine on the Balje $N_S - D_S$ and Rohlick diagrams, like as $n_s = 0.6$ rad. for a total to static efficiency $\eta_{ts} = 0.8$. Substituting this value, the imposed isentropic enthalpy drop and mass rate of steam flow into the n_s equation the rotating speed of about n = 3320 rps is found. The thermal computation of the stage for a reaction degree of $\rho = 0.55$ gives the fluid velocities and geometric dimensions. Mach numbers relative to the absolute velocities are slightly larger than 1, to the relative velocity $M_{w1} = 1.08$, $M_{w3} = 1.4$ but the inlet blade length has an unacceptable value of about 2 mm. In these conditions a new evaluation of stage losses can decrease substantially the efficiency and the turbine solution became inacceptable.

As a conclusion the very low steam flow and the too large enthalpy drop are incompatible with the inflow radial turbine type. Splitting the enthalpy drop between two radial turbines does not significantly improve the results.

4 Conclusions

A lot of agricultural waste are generated yearly. The farmers use it as fuel for stoves and hot water boilers. Some of them need electricity and heat for adjacent activities (bakery, milk/meat processing, wood industry, etc.)

A good opportunity is to transform the flame tubes hot water boilers into steam generators and to couple it with microturbines.

The novelty of this paper is represented by the comparative analysis of axial and radial turbines regarding the design characteristics and operating performances. The main parameter is represented by the rotation speed.

One of the conclusions was that axial turbines are more suitable for this task in comparison with radial ones. Rateau is more simple and cheaper, Curtis is more complicate and expensive, but more robust.

The second conclusion is that higher rotation speeds (200 rps) offers compactness and better efficiencies than lower rotation speeds (100 rps).

The market offers microturbines in the power range of 15-300 kW_e, in multiple constructive technologies and performances. This paper can help the investors to choose one of it.

This work was supported by a grant from the Romanian Ministry of Research and Innovation, CCCDI—UEFISCDI, project number PN-III-P1-1.2-PCCDI-2017-0404/31PCCD/2018, within PNCDI.

References

1. G. Negreanu, V. Berbece, L. Mihăescu, I. Oprea, I. Pîşă, D. Andreescu, Thermal power plant for energy willow use: design, performances, Proceedings of 3rd International Conference on Thermal Equipment, Renewable Energy and Rural Development TE-RE-RD 2014, 12-14 iunie 2014, Mamaia pp. 101-104, ISSN 2359-7941.
2. L. Mihăescu, G. Negreanu, I. Pîşă , V. Berbece, G. Lăzăroiu, New design of a flame-tube steam generator based on a hot water boiler burning agricultural waste, Proceedings of "2020 7th International Conference on Energy Efficiency and Agricultural Engineering (EE&AE) (to be published)

3. * * *, Microturbine technology - Reduce steam pressure and generate electricity (White Paper), (www.spiraxsarco.com/uk).
4. * * *, MICROSTEAM® POWER SYSTEM - Opening a New Market for Energy Recovery and On-Site Power Generation, Carrier 2008, (www.carrier.com)
5. * * *, Small size Steam Turbine now available, (www.innecs.com)
6. * * *, Steam turbine MK-3, (https://www.etgroup.nl/en/steam-turbine-mk3/)
7. * * *, Steam turbine MK-7, (https://www.etgroup.nl/en/steam-turbine-mk7/)
8. H. Kuwabara, M. Nishimura, M. Matsukuma, T. Matsui, Micro Steam Energy Generator, KOBELCO TECHNOLOGY REVIEW NO. 29 DEC. 2010, (https://www.kobelco.co.jp/english/ktr/pdf/ktr_29/016-022.pdf)
9. * * *, S2E I. 50-200 kW - Micro Steam Turbine & Generator Module, (https://technopa.eu/s2e-i-50-200-kw/)
10. A. Alford, P. Nichol, B. Frisby, The Development of a Small High Speed Steam Microturbine Generator System, 2015 IOP Conf. Ser.: Mater. Sci. Eng. 90 012062 (2015) (http://iopscience.iop.org/1757-899X/90/1/012062)
11. T. Grecu, M. Cârdu, I. Nicolau, Turbine cu abur (Steam turbines) – In Romanian), Editura tehnică (Technical Publishing House), Bucharest, 1976.
12. S. L. Dixon, „Fluid Mechanics and Thermodynamics of Turbomachinery", 4th Edition. Butterworth-Heinemann, Woburn, MA, 1998
13. S. A. Korpela, „Principles of Turbomachinery", John Wiley & Sons, Inc., Hoboken, New Jersey, 2011
14. H. E. Rohlik, „Analytical determination of radial-inflow turbine design geometry for maximum efficiency" *NASA TN D-4384*, (1968);
15. E. Kenneth, P.E. Nichols, „How to Select Turbomachinery for Your Application", Barber-Nichols Inc. Technical Papers.

Analysis of rotors' critical mode of operation to be employed in the design of a wind generation control unit

Alina Fazylova[1], *Gani* Balbayev[2*], *Diana* Ilieva[3], and *Madina* Aliyarova[4]

[1] A. Burkitbaev Institute of Industrial Engineering, Satbayev University, Almaty, Kazakhstan
[2,4] Almaty University of Power Engineering and Telecommunication, Almaty, Kazakhstan
[3] University of Telecommunications and Post, Sofia, Bulgaria

Abstract. This paper gives an overview and comparative characteristics of the main rotors available today, with properties suitable for the climatic conditions of Almaty and Almaty region, along with their critical mode of operation. A number of studies is carried out analysing the intrinsic features and capabilities of rotors, through which the problem of choosing an optimal rotor model for the subsequent design of a wind generator control unit is solved. One of these studies is a computer simulation of the main existing rotors operating in the same climatic conditions. The selected model must fully comply with the given environmental conditions of the selected area, in which the maximum wind speed and its critical value are set for all the studied rotor models. Thanks to the comparative studies on the technical characteristics of the most commonly used rotors, as well as to the modelling results, conclusions are drawn for the most effective and optimal choice to be employed in the design of a wind generator for the climatic conditions of the Almaty region.

1 Introduction

Currently, the whole world is in search of new energy sources, since the main energy sources used at this time lead to irreversible pollution of the environment and pose a serious threat to humanity and the ecology of the entire planet. In addition, the demand for energy is growing every year, so it is necessary to look for new, more environmentally friendly and inexpensive energy sources. There are many different environmental energy sources in nature, including wind energy. Since the generation of wind generators with characteristics suitable for certain natural and climatic regions is the most productive and, most importantly, safe way to use wind flow energy, it is necessary to work out in more detail the choice of rotors for the subsequent design of the wind generator control unit.

We selected the Darrieus Rotor as the main turbine under study, since the main results of the work [1] are suitable for the climatic conditions of the city of Almaty and the Almaty region. Patented by French aviation engineer Georges Jean-Marie Darrieus in 1931, the Darrieus wind turbine is often called the "egg whisk" because of its appearance. It consists of a few vertically directed blades that rotate around a Central axis. In [1], it was found that by adjusting certain parameters, it is possible to increase the efficiency and a several energy

*Corresponding author: gani_b@mail.ru

indicators. However, as soon as important technical requirements for wind turbines are taken into account (for example, strength and wear resistance, gear ratio, requirements for bearings, generator), the limit of the amount of energy that can be obtained from wind energy is reduced to 10-30% of the actual wind energy. This limit is called the "capacity factor", which is unique for each type of wind turbine. Based on 3D modelling, it is possible to calculate the power coefficient depending on the speed [2]. There are also developments on the study of a vertical-axis wind generator at Nanyang University of Technology (Singapore), where comparative characteristics of the experimental data of the wind tunnel and output parameters are obtained by means of simulation [3]. This work allows us to verify the correctness of the selected method for the study of rotors. There are also developments in aerodynamic experimental studies [4]. Their main purpose is to measure the pressure distribution in different places of the blades of the wind turbine and to measure the flow through the turbine. The main measurement errors and measurement uncertainty are also discussed. This explains the difficulty in determining the angle of attack and dynamic pressure. This paper presents as well the results of studies of existing rotors in a critical mode of operation for the subsequent development of a system of control units for optimizing the operation of a wind generator. The critical mode of operation involves the operation of a wind generator at the maximum permissible value of wind speed. The results of research on the climatic conditions of the selected area in the Almaty region (Kazakhstan) are used. Based on the available and obtained data, a specific family of rotors are selected for the subsequent construction of a complete control system.

2 Virtual experiment

During the experiment we set the same boundary conditions for all measurements, one of which the wind speed of 35 m/s - a critical value for the city of Almaty [5]. The others relate to the turbine rotating blades - all made from the same material - steel, with rotor diameter of 3m and length of a single blade 4m.

A basic modelling has been formulated to characterize both its design and operation in [6] and main design variables and constraints have been selected by referring to a practical application as a speed reducer on a small windmill [7].

With environmental studies for the city of Almaty, the vertical-axis wind generators are the optimal type of wind generators for the city of Almaty and the Almaty region. Therefore, the main vertically axial rotor models were selected for research [8]. All rotor research and modelling have been conducted using SolidWorks software.

The first model under study was the H-type Darrieus rotor [9]. A 3D model with the distribution of the wind flow over the blades is shown in Fig. 1. The flow entering the rotor is shown on the right, and the flow coming out — on the left. The colours show the distribution of the wind flow acting on the rotor. Since the study is carried out at the maximum permissible value of wind speed, a flow of 35.007 m/s has been set, which is shown in red. Such a significant speed can be reached in the presence of flat terrain [10].

In the figure, the red arrow indicates the space where the wind flow does not exert pressure on the rotor. The negative value of the wind speed (marked in blue in the figure) shows the flow repelling from the rotor opposite to the direction of the reference. The simulation results were obtained using SolidWorks software. This program allows to automatically calculate all parameters and display graphs.

Figure 2 shows the rotor dynamic characteristics in a normalized form at the start of the rotor for 80 iterations, where the Y axis normalizes from 0 to 1 such parameters as pressure acting on different parts of the wind turbine, wind speed, turbulence and power. In the figure, the graph of the pressure change acting on the blades is shown in blue, the graph of the pressure change on the main axis is red, the pressure acting on the entire installation is

 https://doi.org/10.1051/e3sconf/202018002001

green; gray color is for turbulence, the speed of the air flow is indicated in yellow (that is, its conversion into electricity); turquoise color shows the flow rate acting along the Z axis (it does not actually matter much, since the main flow is directed along the Y axis), dark green - the force acting on the blades of the wind turbine. Black shows the maximum dynamic pressure on the blades, orange - the total maximum pressure on the entire turbine, pink - the maximum speed along the Y axis, gray - the maximum speed along the Z axis, dark blue - the maximum pressure on the blades. Maximum values do not play a role in this study. Since the study is dynamic, iterations are indicated along the X axis. One iteration corresponds to 1 second. Their actual values are shown in Figure 3.

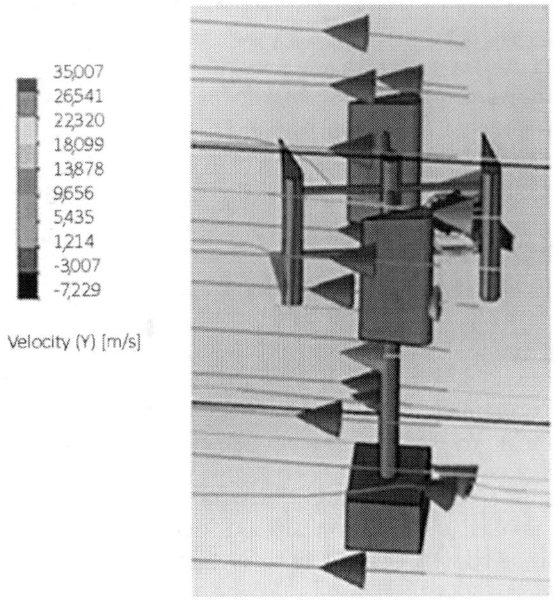

Fig. 1. 3D model with the distribution of the wind flow over the blades.

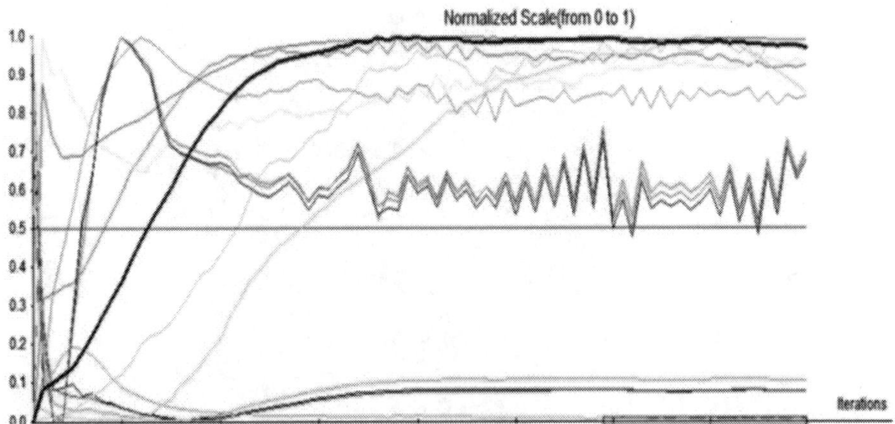

Fig. 2. Simulation plots of the wind flow.

Name	Current Value	Progress	Criterion	Averaged Value
GG Av Dynamic Pressure 1	267.971 Pa	Achieved (IT = 41)	2.97306 Pa	268.243 Pa
GG Av Heat Flux 1	0 W/m^2	Achieved (IT = 40)	0 W/m^2	0 W/m^2
GG Av Static Pressure 1	101370 Pa	Achieved (IT = 54)	3.83442 Pa	101370 Pa
GG Av Total Pressure 1	101638 Pa	Achieved (IT = 54)	4.51555 Pa	101639 Pa
GG Av Turbulence Intensity 1	7.10725 %	Achieved (IT = 53)	0.571865 %	7.07384 %
GG Av Velocity (Y) 1	20.1664 m/s	Achieved (IT = 80)	0.00255442 m/s	20.1659 m/s
GG Av Velocity (Z) 1	0.234487 m/s	Achieved (IT = 79)	0.0297928 m/s	0.262034 m/s
GG Av Velocity 1	20.3726 m/s	Achieved (IT = 40)	0.0230425 m/s	20.3754 m/s
GG Friction Force (Y) 1	9.3762 N	Achieved (IT = 42)	0.204746 N	9.38799 N
GG Max Dynamic Pressure 1	592.499 Pa	Achieved (IT = 40)	34.3555 Pa	568.328 Pa
GG Max Total Pressure 1	102188 Pa	Achieved (IT = 40)	62.7313 Pa	102194 Pa
GG Max Velocity (Y) 1	31.1091 m/s	Achieved (IT = 40)	1.01021 m/s	30.5578 m/s
GG Max Velocity (Z) 1	19.6333 m/s	Achieved (IT = 40)	1.09542 m/s	19.6183 m/s
GG Max Velocity 1	31.3689 m/s	Achieved (IT = 40)	0.943034 m/s	30.721 m/s
SG Av Total Pressure 1	101373 Pa	Achieved (IT = 53)	5.12907 Pa	101373 Pa

Fig. 3. Simulation results of the Darrieus rotor.

The graph shows that such indicators as pressure, turbulence, the force acting on the blades from different sides, are normative. Of great interest in this study is the value of speed. The flow velocity acting on the rotor was set at 35 m/s, of which 20 m/s cannot be used to convert electrical energy, that is, the graph shows 40% of the efficiency of this rotor. This is enough for the conditions of the city of Almaty and Almaty region. Nevertheless, the Darrieus rotor can withstand a wind speed of 44 m/s in the idle position. The minimum wind speed is zero, respectively, the wind generator does not move. At a 0-30 iteration interval, the wind generator is in acceleration mode. In the range of 40-60 iterations along the X axis, a wind generator is installed and various fluctuations of indicators are possible here, which is a normal process. After 60 iterations, the wind generator is set for normal operation. Also, according to the results of creating this model, it would be fair to mention that there are problems with the self-starting of this rotor.

The second investigated model of a vertical-axis wind generator is a multi-blade wind generator [10]. Similar to the Darrieus rotor, identical initial parameters have been set in accordance with the climate analysis data for the city of Almaty and a 3D model has been constructed (Fig. 4). Simulation results have also been obtained (Fig. 5 and Fig. 6).

Fig. 4. 3D model of a multi-blade wind generator.

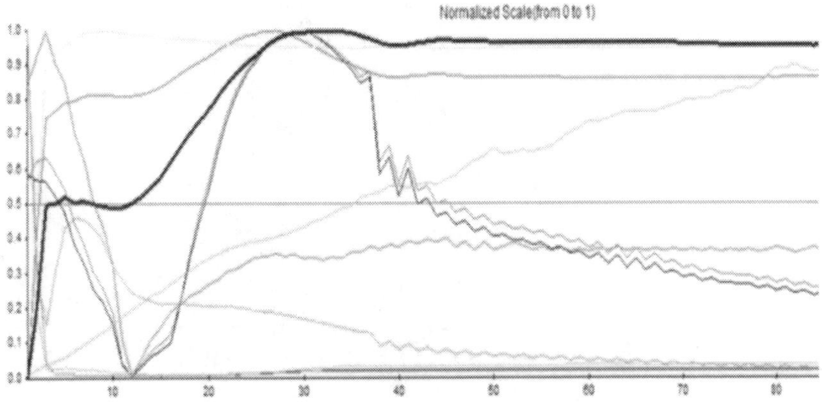

Fig. 5. Simulation plots with model in Fig.4.

Name	Current Value	Progress	Criterion	Averaged Value
GG Average Dynamic Pressure 1	308.537 Pa	Achieved (IT = 45)	2.38038 Pa	308.629 Pa
GG Average Heat Flux 1	0 W/m^2	Achieved (IT = 40)	0 W/m^2	0 W/m^2
GG Average Static Pressure 1	101435 Pa	Achieved (IT = 49)	32.8446 Pa	101436 Pa
GG Average Total Pressure 1	101744 Pa	Achieved (IT = 49)	32.5171 Pa	101745 Pa
GG Average Turbulence Intensity 1	6.12699 %	Achieved (IT = 113)	0.443678 %	5.94674 %
GG Average Velocity (Y) 1	21.214 m/s	Achieved (IT = 40)	0.0237198 m/s	21.2147 m/s
GG Average Velocity 1	21.6939 m/s	Achieved (IT = 56)	0.0638448 m/s	21.6923 m/s
GG Maximum Dynamic Pressure 1	761.472 Pa	Achieved (IT = 68)	104.332 Pa	766.457 Pa
GG Maximum Total Pressure 1	104035 Pa	Achieved (IT = 58)	440.821 Pa	104042 Pa
GG Maximum Velocity (Y) 1	35.2356 m/s	Achieved (IT = 40)	1.36812 m/s	35.2776 m/s
GG Maximum Velocity 1	35.5836 m/s	Achieved (IT = 68)	2.08425 m/s	35.6268 m/s
SG Average Total Pressure 1	101175 Pa	Achieved (IT = 53)	21.5578 Pa	101173 Pa

Fig. 6. Simulation results of a multi-blade wind generator.

Since the wind flow is set along the Y axis, the main energy is distributed on this axis, a wind speed of 0.23 m/s can be neglected as in the previous study. The graph shows that the maximum speed with which the wind generator works effectively is 21 m/s, which is 1 m/s more than for the Darrieus rotor. But this indicator is not paramount for the conditions of the city of Almaty and the Almaty region. The minimum wind speed as well as for the Darrieus rotor is zero. At the interval 0-5 iteration, the wind generator is in acceleration mode. In range of 5-30 iterations, small fluctuations are observed, but they do not greatly affect the operation of the wind generator and we can say that after the 5th iteration the wind generator works in normal mode. The efficiency of a multi-blade wind generator is 50% [12]. Also, the results of creating this model give us reasons to deduce that the wind generator can work at low speeds, as it accelerates quickly enough. That is, if the Darrieus rotor accelerates to 60 iterations, then a multi-blade wind generator takes 5 iterations. Among the shortcomings of this model, one can single out the cost of materials for constructing a rotor, since it has two rows of 12 (or more) blades each.

The third investigated model of a vertical-axis wind generator is the Magen Air Rotor System (MARS) [13]. Similarly to previous studies, identical initial parameters were set in accordance with the analysis of climatic conditions for the city of Almaty and a 3D model was constructed with the selected wind flow along the Y axis from the bottom up (Fig. 7).

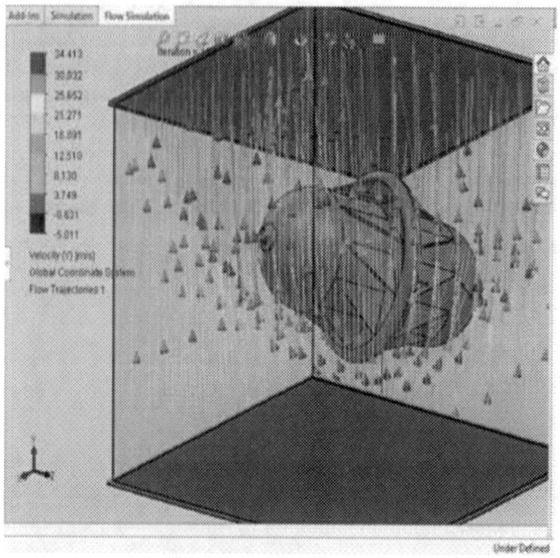

Fig. 7. 3D model with the selected wind flow.

3 Conclusion

Thus, for the study of rotors in a critical operating mode, the main representatives of different families of vertically axial wind generators were selected. The simulation results show that the Darrieus rotor is a slow running model of a wind generator, and even with the maximum allowable wind value applied to it accelerates very slowly. If the Darrieus rotor accelerates to 60 iterations, the multiblade wind generator accelerates to 5 iterations (12 times slower than the multiblade wind generator). During the simulation it has been found that the Darrieus rotor operates efficiently at 20 m/s, and the multi-blade wind generator is capable of working at 21 m/s. This means that the limit for efficient operation of a multi-blade wind generator is higher than that of the Darrieus rotor. But this indicator is of low priority as there are practically no such blustery winds in the studied region. Thus, according to the modelling results, we can conclude that in the conditions of the Almaty region a multi-blade wind generator will work more efficiently and more reliably than the other studied models of vertical-axis wind generators.

References

1. K. S. Pyankov, *Mathematical modeling of flows in wind turbines with a vertical axis /* //Fluid Dynamics, №. 2, **vol. 49,** pp. 249–258, (2014)
2. D.V. Korobatov, A.S. Martyanov, E.S. Bodrova, S.A. Chetoshnikov, *Study of the aerodynamic characteristics of a micro-wind power plant, Journal of South Ural State University* (National Research University)", ESaC. **№1(34)** (2017)
3. S. Roy, U.K. Saha, *Wind tunnel experiments of a newly developed two-bladed Savonius-style wind turbine. Appl Energy* **137**: pp.117-125, (2015)
4. J. G. Schepers, S. J. Schreck, *Aerodynamic measurements on wind turbines, wires Energy and Environment/***Volume 8,** Issue 1 (2018)
5. *A record figure reached power generation in Kazakhstan.* ww.zakon.kz/kazakhstan/4477173.html (visited on 03.01.2020)
6. G. Balbayev, M. Ceccarelli, *Design and characterization of a new planetary gear box, Mechanisms and Machine Science* **No.17**, pp. 91-98 (2014)

7. G. Balbayev, G. Carbone, *A dynamic simulation of a novel continuous variable transmission, Mechanisms and Machine Science* **No.17,** pp. 109-116 (2014)
8. *Official site of the President of The Republic of Kazakhstan*. www.akorda.kz/ru/republic_of_kazakhstan [02.01.2020]
9. Yu. V. Kozhukhov, A.A. Lebedev, A.M. Danylyshyn, E.V. Davletgareev. *Auditing the characteristics of wind generators using CFD modeling on a supercomputer*, a magazine. **No. 7** (107), pp. 81-87 (2016)
10. Terziev A., Analysis of the possibilities of using reference wind measurements in the calculation of energy production from wind farms on low complex terrains, E3S Web Conf., **Volume** 112, 2019, https://doi.org/10.1051/e3sconf/201911202004
11. J. Twidell, A. Weir. *Wind Energy // Renewable Energy* - M .: Energoatomizdat, 393 p. ISBN 5-283-02469-5 (1990)
12. B. Saravanan, S. D. Sikrid, P. Kothari. *A solution to the unit commitment problem – a review, Front. Energy*, no. **7**(2), p. 223–236 (2013)
13. R. D. Zimmerman, C. E. Murillo-S´anchez. *Matpower Optimal Scheduling Tool MOST* 1.0 User's Manual, Power Systems Engineering Research Center (PSerc) (2016)

Compressed air energy storage facility with water tank for thermal recovery

Iulian Vlăducă[1], *Claudia* Borzea[1,*], *Dan* Ionescu[1], *Alexandra* Țăranu[1], *Răzvan* Ciobanu[1], *Vicențiu* Ringheanu[1], and *Răzvan* Nedelcu[2]

[1]Romanian Research and Development Institute for Gas Turbines COMOTI, 061126 Bucharest, Romania
[2]TURBONED S.R.L., 210253 Târgu Jiu, Romania

Abstract. The paper presents the prototype of the first Romanian Compressed Air Energy Storage (CAES) installation. The relatively small scale facility consists of a twin-screw compressor, driven by a 110 kW three-phase asynchronous motor, which supplies pressurized air into a $50m^3$ reservoir, of 20 bar maximum pressure. The air from the vessel is released into a twin-screw expander, whose shaft spins a 132 kW electric generator. The demonstrative model makes use of a $5m^3$ water tank acting as heat transfer unit, for minimising losses and increasing efficiency and the electric power generated. Air compression and decompression induce energy losses, resulting in a low efficiency, mainly caused by air heating during compression, waste heat being released into the atmosphere. A similar problem is air cooling during decompression, lowering the electric power generated. Thus, using a thermal storage unit plays an essential role in the proper functioning of the facility and in generating maximum electric power. Supervisory control and data acquisition is performed from the automation cabinets. During commissioning tests, a constant stable power of around 50 kW with an 80 kW peak was recorded.

1 Introduction

Electrical energy storage [1, 2] is a process in which electrical energy from a power network is converted into a form that can be stored, and then converted back to electrical energy when needed. Such a process produces electricity at times of low demand, when the cost is low, or for the backup of intermittent energy sources such as wind or solar. The energy is stored until a high demand arises or no other energy source is available.

Compressed air energy storage (CAES) is a technique for supplying electric power to the grid for meeting peak load requirements, conventional compressed air energy storage being a practicable technology for electric load levelling. A CAES power generation facility uses electric motor-driven compressors to inject air into a reservoir, later releasing the compressed air to turn turbines and generate electricity into the grid. Even though CAES concept dates back more than four decades ago [3], environmental concerns and technological advances are encouraging their research and development interest to increase worldwide [4-7].

* Corresponding author: claudia.borzea@comoti.ro, borzeaclaudia@gmail.com

Compressing and decompressing air introduce energy losses, resulting in a reduced electric-to-electric efficiency. The low efficiency is mainly caused by air heating during compression. This waste heat, which holds a large share of the energy input, is released into the atmosphere. A similar problem is air cooling during decompression, lowering the electricity generated with the risk of water vapours freezing in the air.

The present paper proposes the relatively small-scale ROCAES system, with compressed air storage in high pressure vessels, and with thermal energy recovery by means of a water tank. In the compression and expansion processes occurring in a CAES installation, the heat loss in the compression process, respectively the useful energy loss in the expansion process, are high, in a direct dependency with compressor and expander systems. The screw compressor and screw expander used are machines that work in isothermal cycle, with oil injection for inner rotors lubrication and air temperature control during operation. Depending on the process heat recovery solution chosen, the efficiency of an energy storage station can vary between 41% and 75%, and a share of more than 65% of the heat resulted from compression can be recovered in the form of transferred energy [8].

2 Power plant design with thermal storage

The CAES system presented consists of a 100 kW twin-screw compressor that supplies compressed air into a reservoir, and a 132 kW twin-screw expander. The compressor is driven by a 110 kW asynchronous three-phase motor. The expander is driven by the compressed air released from the 50m^3 reservoir, spinning a 132 kW asynchronous three-phase generator. The system's operation has been explained in detail in our previous paper [9]. The block diagram of ROCAES station is presented in Figure 1. Two air storage tanks have been envisaged, currently the installation being commissioned with only one pressurized air tank.

At hourly intervals, when the demand for electricity increases and hence the energy price is higher, the screw expander driving the electric generator starts its operation. The air compression is accompanied by an important energy loss due to the heating of the compressed air in the compression process. The heat of the cooling oil is partially recovered by the 5 m^3 water tank, that stores the thermal energy of the oil tray, and transfers it to the oil heating the expander, since air expansion is accompanied by a severe temperature drop.

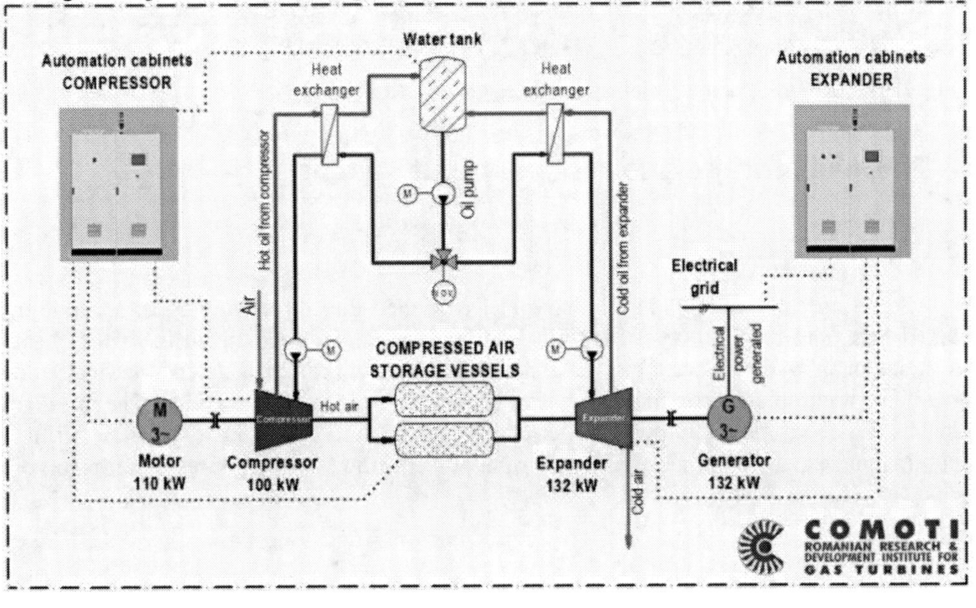

Fig. 1. ROCAES block diagram [10]

The compressed air is stored in a thermally insulated environment, in order to recover a share of the process heat. In the current configuration, with a 50 m^3 air reservoir, it is estimated that the compressor will be able to operate around 4 hours per day, and the expander about 30 minutes, in case of peak electricity demand. During operation, heated oil is necessary to be provided at expander's air inlet, for a better expansion work. The water tank is insulated as well, for reducing heat losses. At a temperature loss gradient of ~1°C/h, it is estimated that the temperature in the tank after 20 hours will be 20°C lower, considering that the thermal insulation consists of a 50 mm thick layer of basaltic mineral wool. In Figure 2 below, we present the ROCAES pieces of equipment installed at the commissioning location. The screw compressor, screw expander and the control room are placed inside the production hall, while the compressed air vessel is situated outside. The overall assembly of the ROCAES facility installed in the industrial hall of SC Popeci Heavy Equipment SA (Craiova, Romania) is presented in Figure 2 below.

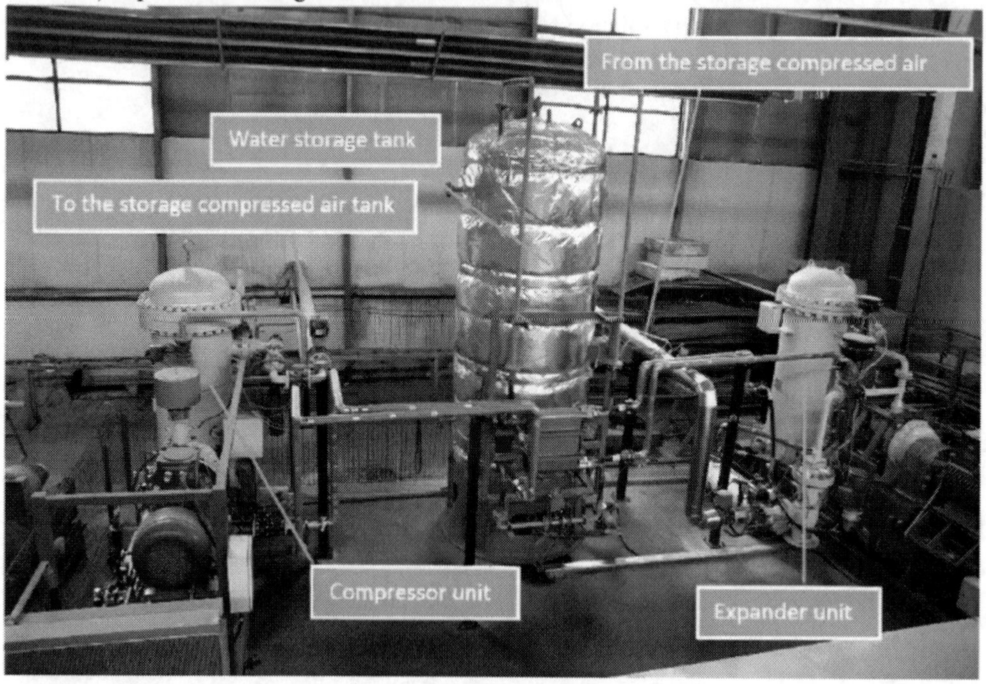

Fig. 2. The ROCAES facility installed at Popeci, Craiova (Romania)

3 Thermal storage and recovery unit

3.1 Water tank design

Stored thermal energy [11-13] is typically divided into three categories: sensible heat, latent heat, and reaction heat. Sensible heat storage can be recognized in district heating systems or in water heater tanks. Every unit of heat added makes a sensible heat storage medium (water in our case) increase with a certain temperature amount [11]. The installation makes use of a 5 m^3 (5000 litres) water tank (Figure 3 left) for oil heat recovery during the cooling and expansion process. Figure 3 right presents the 50 m^3 compressed air reservoir.

Fig. 3. Heat recovery water tank (left) and compressed air reservoir (right)

The cold and hot water in the thermal storage tank are considered to be fully mixed because its volume is relatively low, hence the temperature in the tank is considered to be uniform. This assumption simplifies the analysis because the storage and the load temperature are assumed to be equal [14]. The purpose of the heat recovery module is to store a volume of water (storage medium) resulted from the heated oil resulted from compression, and, at the required time, respectively during the functioning of the expander, to transfer the heat to the oil injected at its inlet. The three-way motor operated valve opens the circuits comprising the heat exchangers. The heat exchange takes place between the hot oil from the compressor and the storage medium represented by the water in the heat recovery module. The water pump mounted on the circuit forces water to flow through the heat exchangers. When the expander is started up, the three-way motor operated valve opens the water circuit that passes through the heat exchanger on the expander circuit. A heat exchange occurs between the storage medium and the oil injected into the expander, a significant part being recovered from the process heat. When the temperature in the separator vessel increases over 60°C, the PLC (Programmable Logic Controller), through its implemented and programmed software, automatically gives the command of opening the valve to recirculate the oil on the bypass circuit (see the block diagram in Figure 1).

3.2 Mathematical model

The computations were realised relying on the following input data: initial water temperature in the vessel 12°C (285.15K), maximum heated water temperature is 60°C (333.15K), hence a temperature difference of $\Delta t_{w,max}$ = 48°C = 48K.

The theoretical evolution of the pressure in the air tank, during compressor operating time, t_R , was estimated to increase from 6 bar to 17.6 bar in 3837 s. The water quantity rate circulated in t_R=1.07 hours is m_{vr} = 1.303 kg/s.

The amount of heat rate introduced into the water vessel by the hot oil of the compressor is calculated as:

$$Q_{u,c}^{t} = m_{vr} \cdot c_w \cdot \Delta t_{w,max} = 257.16 \text{ kW} \qquad (1)$$

where: m_{vr} [kg/s] – water quantity rate circulated in the tank volume ($m_{vr} = V_{w,tank}/t_R$ where t_R is the compressor operation time, and $V_{w,tank}$ = 5 m^3 is the volume of the water tank); c_w = 4184.4 J/(kgK) – specific water heat.

The functioning time for the pressurized air reservoir is important for the heat stored in the water heat recovery tank. Relying on the pressure in the air tank, the functioning of the expander will be analysed in five time intervals extracted from the data acquired during commissioning tests. The energy of the heated water, in the recirculation time t_R=1.07 hours, is thus calculated as:

$$W_a = Q^t_{u,c} \cdot t_R = 275.16 \text{ kWh} \quad (2)$$

The expander air flow is calculated with the formula below, as the flow was measured with an orifice plate, according to ISO 5167-2 standard.

$$\dot{M}_a = C \times E \times \varepsilon \times \frac{\pi}{4} \times d^2 \times (2\Delta P \times \rho_a)^{\frac{1}{2}} \text{ [kg/s]} \quad (3)$$

where: $\dot{M}_a$ [kg/s] – air mass flow; C – discharge coefficient (=0.65); $E = 1/(1-\beta^4)^{1/2}$=1.072; β = d/D = 0.754; d – diaphragm diameter (= 49 mm); D – pipe diameter (=65mm); ε – expansion coefficient (=0.98); ρ_a – air density on the upstream of the diaphragm; ΔP – differential pressure on the diaphragm.

The resulting measurement accuracy is about ±1%. From air thermal equation of state, we find the air density:

$$p \cdot V = R \cdot T \cdot m/M \Rightarrow p = \rho \cdot R \cdot T/M \Rightarrow \rho = p \cdot M \,/\, (R \cdot T) \quad (4)$$

where: R – the universal gas constant = 8314.32 J/(kmol·K); T – gas temperature in Kelvin (T=273+t (°C)); M – molar mass (for air ~ 29 kg/kmol). We made the assumption that air is an ideal gas. The computations were realised on five functioning periods, relying on the experimental data, extracted in Table 1.

Table 1. Medium air mass flow and volume flow within the experimental operation periods T1…T5

Period	**T1**	**T2**	**T3**	**T4**	**T5**
Time [s]	92	93	152	172	208
Iterations	155	158	259	295	342
$\overline{\dot{M}_a}$ **[kg/s]**	0.278	0.361	1.201	0.638	0.444

The heat rate introduced into the expander oil system is:

$$Q^t_{u,exp} = \dot{M}_{oil} \cdot c_{oil} \cdot \Delta t_{oil} \text{ [W]} \quad (5)$$

where: $\dot{M}_{oil}$ [kg/s] – oil flow mass circulated in the expander; c_{oil} = 2054 J/(kgK) – oil specific heat; Δt_{oil} [K] – temperature range for oil cooling in the expander. The oil mass flow was calculated using the air heat extracted during expander operation, relying on temperature differences between theoretical cycle without heating through oil injection, and the practical temperature reached during expander operation, with oil injection. The theoretic expansion temperature without air heating is given by the equation:

$$T_{ev,teor} = T_{inlet} \cdot \pi_t^{\frac{k}{k-1}} \quad (6)$$

where: π_t – expansion ratio; k – isentropic coefficient.

The heat difference is given by the subsequent equation:

$$\Delta Q_{diff} = \dot{M}_a \cdot \frac{k}{k-1} \cdot \frac{R}{M} \cdot \frac{k}{k-1} \cdot \left(T_{ev,exp} - T_{ev,teor}\right) \quad (7)$$

where: $T_{ev,exp}$ value is acquired by the temperature transducer; $T_{ev,teor}$ is given by equation (6); $\dot{M}_a$ is given by the equation (3); R – the universal gas constant; M – molecular mass of air. The oil flow mass is given by the equation:

$$\dot{M}_{oil} = \frac{\Delta Q_{diff} \cdot (k-1) \cdot M}{k \cdot R \cdot (T_{oil,exp} - T_{ev,exp})} \tag{8}$$

4 Automation system and commissioning tests

The purpose of the automated SCADA (Supervisory Control and Data Acquisition) system, represented by the independent sets of drives and automation cabinets for compressor and expander automation respectively, is to control and monitor the operation parameters of thermal, mechanical and electrical processes, to regulate and keep these ones within certain operation limits, in order to ensure maximum safety conditions. On compressor and expander skids, transducers are mounted for monitoring the important parameters of the installation. The information from the transducers is transmitted to the two separate PLCs located in the automation cabinets, for the independent operation of expander and compressor units.

After filling the compressed air reservoir with the compressor operating for about 4 hours, reaching a pressure of about 16 bar, the compressor was shut down. Then, the expander air inlet valve was opened, and the asynchronous three-phase machine was started in motor regime, with a softstarter, until reaching no load speed (~1470 rpm). Compressed air from the storage tank was released into the expander by opening the reservoir valve towards expander. The pressurized air released turns the expander, whose shaft spins the shaft of the electric machine above the synchronous speed of 1470 rpm, this one entering automatically in generator regime and injecting electric power into the grid.

The minimum pressure condition set at expander inlet is 2 bar. If not meeting this condition, the expander goes through an automatic shutdown sequence. The condition was set considering the minimum pressure in the air reservoir dropping down to 4-5 bar, considering the pressure losses until air reaching expander's inlet. In Figure 4 are presented the control panels interfacing the two PLCs, during compressor and expander operation tests.

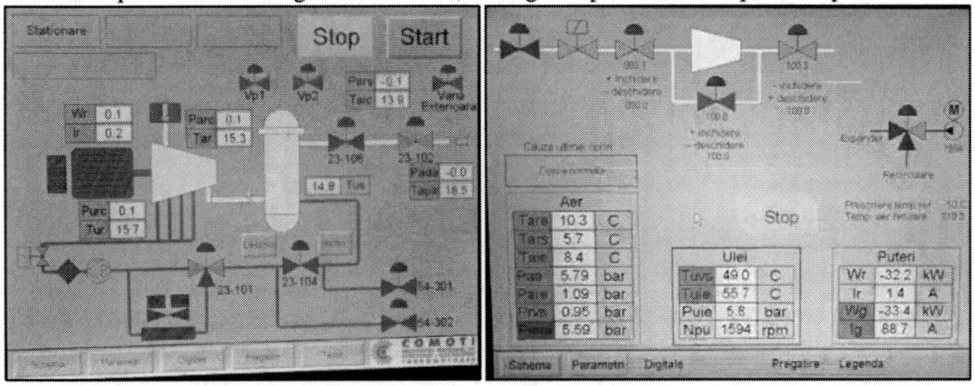

Fig. 4. Compressor (left) and expander (right) control panels showing real-time parameters

In the subsequent figure, the generated power at the first experimental start-up is presented, for commissioning period T1. The positive power shows that the asynchronous electric machine is launched as motor, absorbing power from the grid. Then, as the expander inlet valve gradually opens to suction air from the reservoir, the electric machine is spun over its synchronous speed as generator. It has a normal instantaneous spike of almost 40 kW generated power when reaching synchronous speed, and then it begins to stabilize in generator regime, injecting almost 80 kW maximum electric power.

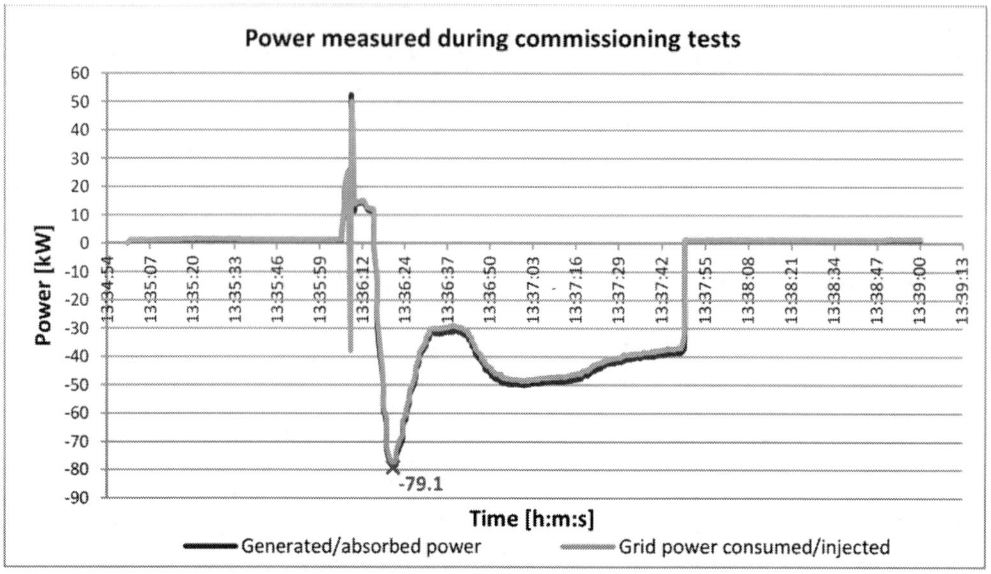

Fig. 5. Power plot relying on acquired data during commissioning test for period T1

The two power curves in Figure 5 are very close, but the injected power into the grid is slightly lower because electric consumers in the electric cabinets also need to be supplied with the power produced. The other T2…T5 periods have similar power graphs.

5 Experimental and theoretical results

Both the maximum power and the stabilization power (Figure 5 above) were significantly higher than the ones obtained during bench tests without thermal recovery tank, even with preheating oil with a heating resistance [9, 10]. In the tables below are given the parameters ranges (minimum and maximum values), measured during commissioning tests.

Table 2. The parameters ranges acquired during operation periods T1…T5

	T1=92s	T2=93s	T3=152s	T4=172s	T5=208s
Electric grid power [kW]	56.7…50.4	77.4…30.2	65…6	15.7…14.1	39.1…18.8
Expander inlet oil temperature [°C]	33.5…42.3	65.6…60	48.4…52	44.8…51.4	50.6…55.8
Expander outlet oil temperature [°C]	15.7…15.2	12.9…-20	9.6…-7.6	-2…-16.8	15.2…-18.3
Air temperature at separator outlet [°C]	14.2…11.7	13.7…9.3	9.6…8.9	16…9.7	12.3…7.8
Expander inlet air temperature [°C]	14.7…11.7	16.1…12.6	14…9.8	13.9…8.1	13.1…6.2
Expander outlet pressure [bar]	2.9…2.3	2.6…3.6	1.8…3.2	4.1…2.8	1.9…1.3
Air pressure before flowmeter [bar]	11.7…9.6	12.6…10.2	10.6…8.8	11.1…8.2	7.7…5.1
Expander inlet pressure [bar]	10.3…9	11…9.2	8.9…7.7	9.6…7.7	7.3…4.7

In Table 2 we selected the minimum and maximum values from a total of 24 values used for iterations in PTC Mathcad software, where the computations were performed. The power of the expander is given by the following relation:

$$P_{exp,teor} = \dot{M}_a \cdot \frac{k}{k-1} \cdot \frac{R}{M} \cdot T_{asp} \cdot \frac{\left[(\pi_t)^{\frac{k-1}{k}} - 1\right]}{\eta_{correction}} / 1000 \text{ kW} \tag{9}$$

where π_t is the expansion coefficient, having the expression:

$$\pi_t = \frac{p_{ev,exp}+1}{p_{inlet,exp}+1} \tag{10}$$

We introduced a correction factor, $\eta_{correction}$, for the oil injection flow. A small deviation from the air flow rate $\dot{M}_a$ induces a significant error in the difference of heat exchanged with the oil ΔQ_{diff} calculated in equation (7). The air mass flow $\dot{M}_a$ depends on the Reynolds number air flow measured with the orifice plate in the diaphragm. An adjustment in Reynolds number gives the $\eta_{correction}$ factor used to calculate more accurately the overall efficiency. In Table 3 are given the maximum values of the oil flow rate $\dot{M}_{oil}$ (l/min) at a density ρ_{oil} of 905 kg/m^3 for ISO VG 68 oil type used, the injection oil heat rate $Q_{oil\_injection}$ and the theoretical calculated power of the expander $P_{exp,teor}$, as well as the expansion coefficient π_t.

Table 3. The medium injected oil heat and expansion coefficient π_t vs. theoretical expander power

Period	**T1**	**T2**	**T3**	**T4**	**T5**
seconds	92	93	152	172	208
$\overline{\dot{M}_{oil}}$ **[kg/s]**	0.207	0.149	0.632	0.175	0.241
$\overline{Q_{oil\_injection}}$ **[kW]**	15.9	18.9	60.8	18.9	24.7
$\overline{\pi_t}$ **[ND]**	0.335	0.391	0.456	0.436	0.384
$\overline{P_{exp,teor}}$ **[kW]**	61.1	73.1	79.0	36.6	37.9

It is observed from Table 3 that at a maximum π_t and a maximum heat oil rate injected $Q_{oil\_injection}$, we obtain the maximum theoretical power. The medium heat rate necessary for expander heating during T1…T5 periods is estimated as $Q_{oil\_injection,medium} = 27.84$ kW. The efficiency was determined iteratively, for a positive and linear increasing of the ΔQ_{diff} with the oil mass flow $\dot{M}_{oil}$, as η_{medium} being the medium ratio between theoretical power P_{teor} and experimental power P_{exp} in 24 iterations, and:

$$\eta_{global} = \eta_{medium} \cdot \eta_{correction} \tag{11}$$

In Table 4 are given the efficiencies calculated for all experimental operation ranges.

Table 4. The efficiencies calculated for T1…T5 periods

Period	**T1**	**T2**	**T3**	**T4**	**T5**
seconds	92	93	152	172	208
$\eta_{correction}$	0.35	0.35	1	0.8	0.8
η_{medium}	83.2	79.1	31.0	53.7	64.9
η_{global}	29.1	27.7	31.0	43.0	51.9

The expander oil was heated from -10°C to 56°C with the water stored in the storage water tank. The oil used for recirculation was drawn from the separator vessel of the expander, where the temperature was ~12°C. If we consider that the energy of the oil from

-10°C to 12°C was lost in the process, we obtain the values in Table 5. The medium heat rate lost in the process was $Q_{loss} \cong 8.6$ kW.

Table 5. Heat lost during T1...T5 commissioning periods

Period	**T1**	**T2**	**T3**	**T4**	**T5**
seconds	92	93	152	172	208
$Q_{oil\_injection}$ **[kW] (lost)**	6.0	5.98	14.2	10.98	5.85

The water pump circulates the water in the expander heat exchanger with the same flow as the compressor heat exchanger, that is $m_{vr} = 1.303$ kg/s.

The energy that can be produced for a functioning with 1°C decrease in the water temperature, with the rate of m_{vr}, is:

$$Q_{u,c}^{t} = \dot{M}_{vr} \cdot c_w \cdot \Delta t_{w,max} = 5452.2 \text{ W} \tag{12}$$

In order to achieve the medium heat rate $Q_{oil\_injection,medium}$ of 27.84 kW needed for the expander, the water circulation rate should be:

$$\dot{M}_{vr} = m_{vr} \cdot \frac{Q_{oil\_injection,medium}}{Q_{u,c}^{t}} = 6.65 \text{ kg/s} \tag{13}$$

The amount of energy extracted from the 5000 litres of water, with a water mass flow of 6.65 kg/s (399.2 l/min), requires a total time of functioning $t_{R,total}$ of ~12.5 minutes for 1°C of water temperature decrease This time should be enough for the expander operating at the recorded parameters, for a total functioning time in generator regime, that was:

$$t_{R,exp} = \sum_{i=1}^{5} T_i = T1 + T2 + \ldots + T5 = 717 \text{ s} \tag{14}$$

The total energy required for cooling the expander, during the total operation time of about 0.2h (12 minutes), is:

$$W_{total} = (Q_{oil,injection,medium} + Q_{loss,medium}) \cdot t_{R,exp} = 7.3 \text{ kWh} \tag{15}$$

6 Conclusions

The ROCAES power facility was installed and commissioned at the industrial hall of Popeci Heavy Equipment plant in Craiova, Romania. It comprises a screw electro-compressor, driven by a 110 kW motor and a screw expander driving a 132 kW electric generator. The compressed air is stored in a 50 m^3 reservoir, of 20 bar maximum pressure. The demonstrative model makes use of a 5 m^3 water tank for minimising thermal energy losses, increasing the efficiency and generated power.

Air compression and decompression processes are accompanied by energy losses due to air heating during compression, releasing waste heat into the atmosphere, and air cooling during decompression, lowering the electric power generated. This results in a low efficiency, hence using a thermal storage tank has an important part in the operation of the installation and in generating higher electric power.

A stabilized power of around 50 kW with a maximum of 80 kW was obtained. Both the peak power and the stabilization power recorded were found to be significantly higher than the ones obtained during in-house bench tests without the water tank, even if oil preheating was ensured.

For a reliable CAES energetic process, the aimed performances are of minimum 30 minutes of continuous operation at around 80 kW, considering 80% generator efficiency. A water temperature of 60°C requires higher oil and water mass flow through the heat exchangers, thus increasing the electrical energy consumed. For more power, a higher capacity of the air storage reservoir(s) is required, able to attain a higher pressure. Introducing an additional compressor stage as in [7], of higher pressure, would also increase the overall performances.

Further researches will consider increasing the stored capacity of compressed air, using two or more reservoirs. As well, an efficient solution for increasing the temperature of the stored water to a maximum of 90°C will be sought, and in case of not having enough energy for the expansion process, a parallel oil heating device will be ensured.

The research work presented herein was funded by Competitiveness Operational Program 2014-2020, Priority Axis 1 – Research, Technological Development and Innovation in Support of Economic Competitiveness and Business Development, Action 1.2.3, Partnerships for Knowledge Transfer, contract POC 79 / 08.09.2016, carried out with the support of Romanian Ministry of European Funds.

References

1. H. Chen, X. Zhang, J. Liu, T. Chunqing, Compressed Air Energy Storage, in *Energy Storage - Technologies and Applications*, 101-112 (IntechOpen, 2013), DOI:10.5772/52221.
2. S. D. Garvey, A. Pimm, Compressed Air Energy Storage, in *Storing Energy*, 87-111 (Elsevier, 2016). DOI:10.1016/B978-0-12-803440-8.00005-1.
3. R. Allen, T. Doherty, L. Kannberg, *Summary of selected compressed air energy storage studies*, (Pacific Northwest National Laboratory (U.S.), 1985), https://www.osti.gov/servlets/purl/5872515.
4. C. Guo, Y. Xu, X. Zhang, H. Guo, X. Zhou, C. Liu, W. Qin, W. Li, B. Dou, H. Chen, *Energy* **135**, 876-888 (2017), DOI:10.1016/j.energy.2017.06.145.
5. G. Venkataramani, V. Ramalingam, K. Viswanathan, Sci. Rep., **8**(1), 9981 (2018), DOI:10.1038/s41598-018-28025-5.
6. Chen, T. Zheng, S. Mei, X. Xue, B. Liu, Q. Lu, J, Modern Power Syst. Clean Ener. **4**(4) 529-541 (2016), DOI:10.1007/s40565-016-0240-5.
7. G. Dib, P. Haberschill, R. Rullière, Q. Perroit, S. Davies, R. Revellin, *Appl. Ener.* **260**, 114248 (2020). DOI: 10.1016/j.apenergy.2019.114248.
8. M. Broniszewski, S. Werle, E3S Web of Conf. **70**, 03001 (2018), DOI:10.1051/e3sconf/20187003001.
9. C. Borzea, I. Vlăducă, D. Ionescu, V. Petrescu, F. Niculescu, C. Nechifor, G. Vătăşelu, M. Hanek, E3S Web of Conf **112**, 02010 (2019), DOI:10.1051/e3sconf/201911202010.
10. C. Borzea, A. Săvescu, C. Hărăguță, A. Găzdac, R. Nedelcu, Elect. Mach., Mater. and Drives – Pres. and Trends), **1**(1), 1-12 (2020), DOI:10.36801/apme.2019.1.2.
11. I. Dincer, M. Ezan, Thermal Energy Storage Applications, in *Heat Storage: A Unique Solution for Energy Systems*, 85-135 (Springer, 2018). DOI:10.1007/978-3-319-91893-8_4.
12. O. Burheim, Thermal Energy Storage, in *Engineering Energy Storage*, 47-61 (Elsevier, 2017). DOI:10.1016/b978-0-12-814100-7.00004-3.
13. P. Li, C. Chan, Thermal Energy Storage Materials, in *Thermal Energy Storage Analyses and Designs*, 21-63 (Elsevier, 2017). DOI:10.1016/b978-0-12-805344-7.00003-1.
14. M. Sameti, A. Kasaeian, S. Mohammadi, N. Sharifi, Sust. Ener., **2**(2), 52-56 (2014). DOI:10.12691/rse-2-2-3.

Influence of four modern Flettner rotors, used as wind energy capturing system, on container ship stability

Mihail-Vlad Vasilescu[1], *Ionut* Voicu [2]*, *Cornel* Panait[3] and *Violeta-Vali* Ciucur[4]

[1]Constanta Maritime University, Doctoral School of Mechanical Engineering, 104 Mircea cel Batran street, ZC 900663, Constanta, Romania
[2]Constanta Maritime University, Department of Engineering Sciences in Mechanics and Environment, 104 Mircea cel Batran street, ZC 900663, Constanta, Romania
[3]Constanta Maritime University, Department of General Engineering Sciences, 104 Mircea cel Batran street, ZC 900663, Constanta, Romania
[4] Constanta Maritime University, Department of Electrical Engineering Sciences, 104 Mircea cel Batran street, ZC 900663, Constanta, Romania

Abstract. This article is presenting the influence on the stability of a container ship, by connecting four modern Flettner rotors, as wind energy capturing system. Modern Flettner rotors, are energy power generators. They are one of the various equipment used to capture and harness wind energy. By reading the article, the reader can discover what are the forces which influence the modern Flettner rotors and how they influence the ship stability.This article major points are: calculation the influence of the modern Flettner rotors on the ship transverse and longitudinal stability, calculation of the ship new displacement, new draft, new metacentric height and period of rolling.

1 The benefits of using the Flettner rotors

Flettner rotors are vertical cylinders that position in the wind front work on the principle of Magnus depending on the wind speed and its direction, propelling the ship [1]. We have installed on a container ship four modern Flettner rotors (Fig.1), with the role of capturing and harnessing wind energy, in order to use it for ship propulsion. These rotors are not the main source of propulsion but contribute to reducing fuel consumption with 3% to 15 % depending on vessel size, navigation area and mode of operation. This mode of propulsion is only applicable to those ships that have a free deck, because the wind front captured by the Flettner rotors must not be disturbed. To apply this additional mode of propulsion, the characteristics of the chosen vessel must be carefully analyzed.
By using Flettner rotors, forces appear in the horizontal plane, which, combined with the Magnus effect, modify the stability of the ship, and can also be a danger of shearing the rotor support poles.

* Corresponding author: panaitescumariana1@gmail.com

The results of wind propulsion suggest that when this technology is implemented at scale, it can produce up to 20% net savings in fuel costs with a payback period of less than four years at current fuel prices, confirming that wind technologies are commercially-viable solutions that reduce fuel and carbon emissions in the industry [3]. When the speed of wind are favourable, this solution allow the main engines to be throttled back, saving fuel and reducing emissions while providing the power needed to maintain speed and voyage time. The Flettner rotors can be used with new vessels or with existing ships without off-hire costs. The existing installation on the ship can be completed with the Flettner rotor support foundation when desired and there is a stationary quay. There are many reccommended solutions by the international companies (e.g. Norsepower and Bore Ltd of Finland).
This case study is an example of the use of renewable energy resources with an impact on the environment and climate change that we analyzed during the *HORESEC* project PN-III-P1-1.2-PCCDI-2017-0404 /31PCCDI/2018, *Holistic on the Impact of Renewable Energy Sources on Environment and Climate*.

1.1 Principle characteristics of the ship and modern Flettner rotors

For the study of propulsion with Flettner rotors, a container was chosen as the study vessel, having the following technical characteristics (Fig. 1) [2]:

Fig. 1. Container ship with four modern Flettner rotors installed on deck

- Ship displacement: 39449 mt
- Ship width: 29.4 m
- Ship length:168.56 m
- Height of the rotor: 24 m
- Diameter of the rotor: 4 m
- Weight of the rotor: 40 kg.

2 Calculation of the new displacement D`

$$D`=D+4 \times m=39449+160= 39609 \ mt \quad (1)$$

where:
"m" is the mass of one modern Flettner rotor.

3 Calculation of the ship draft at the new displacement D`

The new displacement is D`=39609 mt.

By interpolation from hydrostatic tables we extracted for any draft the mass which must be loaded or discharged to change a ship's mean draft by one centimeter [1]. We will note this size with TPC.
The TPC for any draft isTPC= 45.47 . The difference between displacements is:

$$D` - D_{9,59} = 23.96\ mt \tag{2}$$

For new draft

$$T'P = 9.59 + (D` - D_{9,59}/TPC) \times 0.01 = 9.5952\ m. \tag{3}$$

4 Calculation of the ship longitudinal stability

A surface ship's stability can be divided into two parts [4]: intact stability and damaged stability. The intact stability is when the intactness of its hull is maintained. The damaged stability means that the some compartments or tanks are flooded by seawater [5].

4.1 Determination of KG

The vertical distance between the keel and the center of gravity is expressed as 'KG'[4].
We note with
- "GM" the metacentric height; it is a measurement of the initial static stability of a floating body; it represent the distance between the centre of gravity G of a ship and its metacentre M (Fig. 2)[4];
- "TKM" the transverse stability; it is the distance between the keel K and the tranverse metacenter M;
- "KM"the distance from keel to the metacentre.

The ship stability condition are preseted in the Figure 2 and the ship stability conditions with the four modern Flettner rotor are presented in the Figure 3.

$$GM = KM_T - KG \tag{4}$$

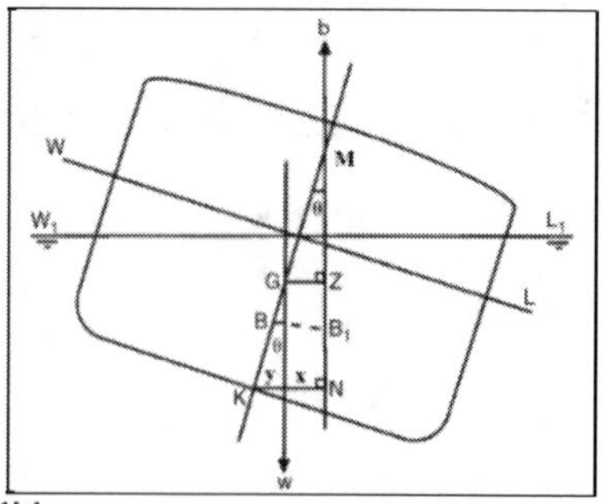

Fig. 2. Ship stability initial condition

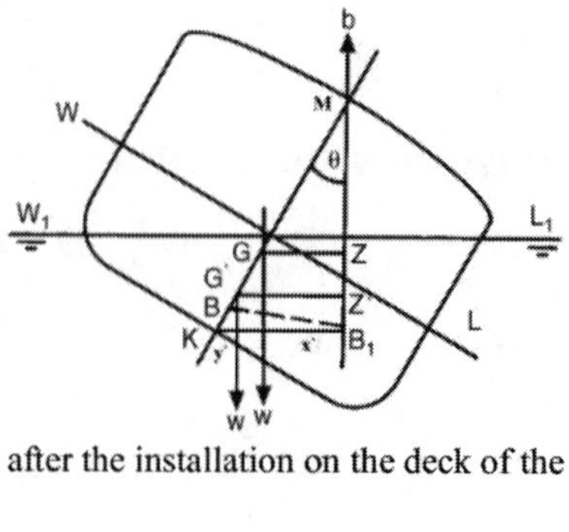

Fig. 3. Ship stability condition after the installation on the deck of the four modern Flettner rotors

So, for our case study: GM= 4.32 m; TKM=KM_T = 12.32 m . According to equation (4) this means KG = 8 m.

4.2 Determination of the transverse stability and longitudinal stability for the new draft

For transverse metacentre T'_P = 9.5952 m, by interpolation from hydrostatic tables we extracted for the new draft the transverse stabilityTKM` and longitudinal stability LKM`.
TKM` = $KM_T$` = 12.3548 m;
LKM` = $KM_L$` = 231.7072 m,
where:
Transverse metacentre- the point of intersection of the vertical through the center of buoyancy of a ship in the position of equilibrium with the vertical through the new center of buoyancy when the ship is slightly heeled.
Longitudinal metacentre - the point in a vertical line through the centre of gravity of a ship where this line is intersected by a second vertical line through the centre of buoyancy when the ship is inclined at a very small angle in a fore and aft direction.
LKM- longitudinal stability is the distance between the keel K and the longitudinal metacenter M.

4.3 Determination of new metacentric height GM`

We note with:
- "ML" the longitudinal metacentre meaning the metacentre of the ship in its longitudinal direction;
- "GML" the longitudinal metacentric height of the ship;
- "KM" the distance from keel K to the metacentre M;
- "KG" the distance from keel K to the centre of gravity G;
- "G'M" the new metacentric height;

For our case study, with KG = 8 we obtain the the value of new metacentric height:

$$G'M = KM_T - KG = 4.3548\ m \quad . \tag{5}$$

4.4 Determination of the ship initial righting lever GZ, in order to calculate the ship transverse stability

$$KN = x + y\ ; \tag{6}$$
$$x = GZ\ ; \tag{7}$$
$$y = KG\ x\ \sin\theta\ ; \tag{8}$$
$$KN = GZ + KG\ x\ \sin\theta \tag{9}$$
$$GZ = KN - KG\ x\ \sin\theta \tag{10}$$

where
θ - the angle of heel (Fig. 2; Fig. 3); GZ- the righting lever; B- the centre of buoyancy; WL- the water line; w- the weight; KG- the vertical distance between the keel and the center of gravity [5].
For our case study:
GM = 4.3200 m; D = 39449 mt. The variation of GZ for different values of the angle of heel is presented in the Table 1:

Table 1. GZ calculated for different values of the θ angle from 0^0 to 90^0:

θ	Sin θ	KG	KN	KG x sin θ	GZ
0	0.000	8	0.000	0.000	0.000
10	0.174	8	2.113	1.392	0.721
20	0.342	8	4.260	2.736	1.524
30	0.500	8	6.310	4.000	2.310
40	0.640	8	7.830	5.120	2.710
45	0.707	8	8.396	5.656	2.740
50	0.766	8	8.828	6.128	2.700
60	0.866	8	9.330	6.928	2.402
70	0.939	8	9.332	7.512	1.820
80	0.984	8	8.892	7.872	1.020
90	1.000	8	8.010	8.000	0.010

4.5 Determination of the new righting lever G`Z `, in order to determine the ship transverse stability

$$KN= x` + y` \quad (11)$$
$$x`= G`Z` \quad (12)$$
$$y`= KG` x \sin\theta \quad (13)$$
$$KN= G`Z` + KG` x \sin\theta \quad (14)$$
$$G`Z`=KN-KG` x \sin\theta \quad (15)$$

For our case study : GM`= 4.3548 m ; D`= 39609 mt ,
where
θ – the angle of heel; G`Z`- the new righting lever; B – the centre of buoyancy [5]; KG`- distance from keel K to the centre of gravity G; G`Z`- the new righting lever after the centre of gravity moved; G`M- the new metacentre height, it is a measurement of the initial static stability of a floating body. It is the distance between the centre of gravity G` of a ship and its metacentre M; WL- water line; w- weight; KM- distance from keel to the metacentre.

Table 2. G`Z` calculated for different values of the θ angle from 0 to 90^0

θ	Sin θ	KG`	KN`	KG` x sin θ	G`Z`
0	0.000	8.0348	0.000	0.000	0.000
10	0.174	8.0348	2.127	1.398	0.729
20	0.342	8.0348	4.280	2.749	1.531
30	0.500	8.0348	6.337	4.017	2.320
40	0.640	8.0348	7.853	5.142	2.711
45	0.707	8.0348	8.423	5.680	2.743
50	0.766	8.0348	8.866	6.154	2.712
60	0.866	8.0348	9.368	6.958	2.410
70	0.939	8.0348	9.365	7.544	1.821
80	0.984	8.0348	8.931	7.9060	1.025
90	1.000	8.0348	8.045	8.034	0.011

The transverse stability curve is plotted on the graph with black (Fig. 4).

https://doi.org/10.1051/e3sconf/202018002003

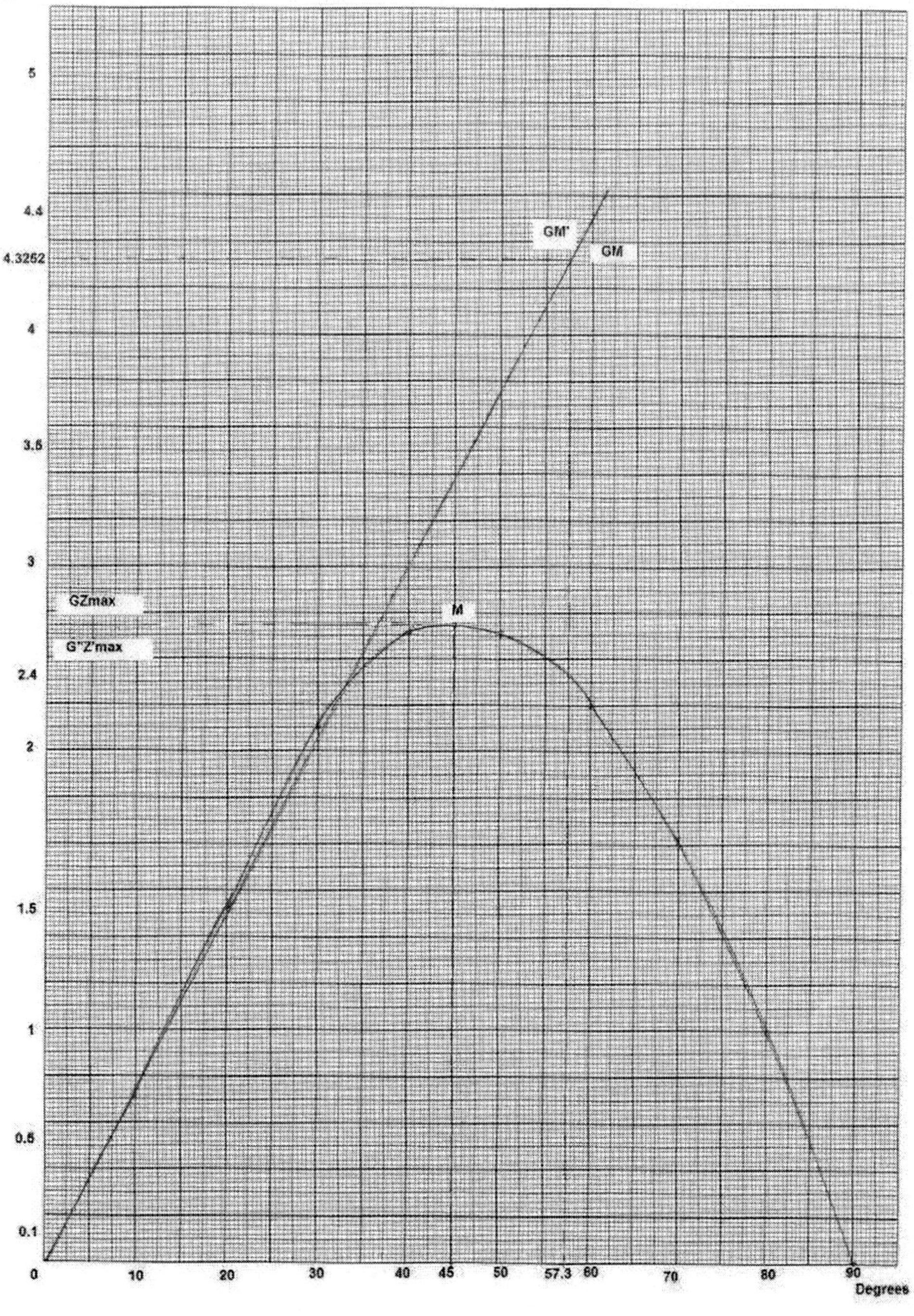

Fig. 4. Transverse stability graph

The graph represents the curve of the static stability of the ship and the variation of the stability righting lever according to the variation of the transverse tilt angle of the ship. The initial stability is presented without the energy generating equipment. The final stability is presented with the energy generating equipment installed.

The interpretation of graph is:

- Point O, represents the origin of the curve;
- Point M and M`, maximum point of the curve, critical angle of roll or maximum angle of canary; the point where the recovery arm has the maximum value;
- Point V, represents the point of sunset or decline;
- Increasing branch, is between O and M respectively O and M`;
- φ_v the maximum inclination angle, at which the vessel, left free, returns to its initial equilibrium position;
- φ_m, is the angle of inclination at which the openings in the body of the ship, superstructures which cannot be sealed, are flooded;
- The area delimited by the static stability curve and by the abscissa represents the total mechanical work of the recovery moment, meaning the mechanical work with which the ship is able to withstand the external moments applied dynamical; this area represents the dynamic stability reserve of the ship;
- Flood angle of the deck is identified on the static stability curve as the point at which the curve changes its shape from rising to decreasing on the rising branch.
- It is also known as the turning point of the curve. This point can be determined by drawing series of vertical lines on the stability curve.
- Analyzing each delimited area can determine the point at which the curve changes its tendency from ascending to descending. It is a difficult problem to approximate and always leaves room for interpretation.

For

$$G'M > GM; \qquad 4.3548 > 4.32$$

ship transverse stability has been increased, after the placement of the four modern Flettner rotors, comparing with initial stability [6].

4.6 Determination of the rolling period

The rolling period can be calculated on formula [7]

$$T=2xlxCx1/GM^{1/2} \tag{16}$$

Where:

l- ship width; l = 29.4 m; C-the coefficient describing ships transverse gyration radius; C = 0.371 [7] for T_p' = 9.5952 m; GM = 4.3200 m; G'M = 4.3548 m; T- initial rolling period; T =10.4956 s; T`- the rolling period after the placement of the four modern Flettner rotors that generates green, unconventional energy; T`=10.4536 s.

From the graph we can also observe that the ship has an initial faster rolling period comparing with the rolling period after the placement of the four modern Flettner rotors, which is slower [2].

$$T>T` \qquad 10.4956>10.4536.$$

4.7 Determination of the inclining moments of the ship

4.7.1 Stable wind

$$l_{w1}=0,0514xAxZx1/D \quad (17)$$

$$l_{w1}^{'}=0,0514xA^{'}xZ^{'}x1/D^{'} \quad (18)$$

where:
D - initial ship displacement; D\`- displacement of the ship, after the connection of the balloon with the ship; A- the wind pressure area on the hull of the ship and on the cargo on the deck above the waterline, calculated for a draft $T_p$= 9.56 m and a D = 39449 mt; A\`- the wind pressure area on the hull of the ship and on the cargo on the deck above the waterline, calculated for a draft $T_{p'}$\`= 9.5952 m and a D\`= 39609 mt; Z - vertical distance between the center of A and the center of the lateral area of the keel,at a draft of T_p= 9.56 m and a D = 39449 mt; Z\`- vertical distance between and the center of the lateral area of the keel, at a draft of $T_{p'}$\`= 9.5952 m and D\`= 39609 mt; T_p= 9.56 m; D = 39449 mt.

Calculation of A
By interpolation from wind area table (Beaufort Scale), for a draft T_p = 9.56 m, results A = 1478.9936 m^2 ~ 1479 m^2.

Calculation of Z
By interpolation from wind area table, for a draft T_p= 9.56m results Z = 10.2272 m ~ 10.23m.
l_{w1} = 0.019708; Tp''= 9.5952 m; D\`= 39609 mt.

Calculation of A\`
By interpolation from wind area table, for a draft T_P\`= 9.5952m, results A\`= 1472.9589m^2 ~ 1473m.

Calculation of Z\`
By interpolation from wind area table, for a draft T_P\`= 9.5952m, results Z\`= 10.231424 ~10.231m
$l_{w1}^{'}$ = 0.0195567172 m ~ 0.0196m.

4.7.2 In case of a gust of wind

$$l_{w2}=1.5\, l_{w1} \quad (19)$$
$$l_{w2}^{'}=1.5\, l_{w1}^{'} \quad (20)$$

T_P = 9.56 m; D = 39449 mt; l_{w2} = 0.029562 m ~ 0.296 m; T_p\`= 9.5952 m; D\`= 39609 mt; $l_{w2}^{'}$ = 0.0293350758 m ~ 0.0293 m.

5 Calculation of the ship longitudinal stability and its effect on the ship trim

The moment of change trim one centimeter is given by equation (21)

$$MTC^{l}cm=DxGMLx1/100L \quad (21)$$

where:
MTC- the moment of change trim one centimeter; D- ship displacement; L- ship length; GML- GM longitudinal.

5.1 Calculation of the MTC` for $T_{'P}$= 9,5952 m

By interpolation in hydrostatic tables at $T_{P'}$, results MTC`= 532,8804 mt x m; MTC`- final MTC after the installation of four modern Flettner rotors on the ship main deck; MTC = 531.92 x m.
We can observe that:

$$MTC<MTC` \qquad 531.9200 < 532.8804.$$

5.2 Calculation of initial GML

$$MTC^{l}cm=DxGMLx1/100L$$

L = 168.56 m; D = 39449 mt;

$$GM_L=MTCx100xLx1/D^{=}227.281\ m\ . \quad (22)$$

5.3 Calculation of G'ML

$$MTC^{l}cm=DxGM\ L\ x1/D' \quad (23)$$

D`= 39609 mt;

$$GML^{'}=MTC^{'}x100xLx1/D^{'}=226.773\ m. \quad (24)$$

Because GML >GML$^{'}$, 227.281 > 226.7725 it means that the longitudinal stability of the ship, after the positioning the four modern Flettner rotors, has reduced.

6 Conclusions

From this paper we can see that the four modern Flettner rotors, positioned on the ship are not influencing to much the ship stability.
From the stability point of view, transverse stability has improved:

$$G^{'}M > GM \qquad 4.3548 > 4.3200$$

but the value is insignificant.

Longitudinal stability of the ship, after the four modern Flettner rotors, were positioned on the ship, was reduced:

$$GML > G^{'}ML \qquad 227.281 > 226.7725$$

After positioning the four modern Flettner rotors on the container ship, we can observe that the rolling period has decreased:

$$T > T` \qquad 10.4956 > 10.4536$$

We consider that the installation of four modern Flettner rotors on a container ship, it is possible, beneficial in terms of reducing fossil fuel consumption, reducing greenhouse gas emissions, saving money for the owners and is safe for navigation.

This paper has financial support from the Constanta Maritime University under assistance project PN-III-P1-1.2-PCCDI-2017-0404 /31PCCDI/2018, *Holistic on the Impact of Renewable Energy Sources on Environment and Climate-HORESEC.*

References

[1] GlopMEEP, *Technology. Flettner rotors*, Global maritime energy efficiency partnerships, Information portal J. **4** (2018)
[2] M.V. Vasilescu, M. Panaitescu, F.V. Panaitescu, C. Panait, *Efficiency of using a hybrid marine propulsion system vs. conventional system*, SPIE Proc. Advanced Topics in Optoelectronics, Microelectronics and Nanotechnologie (to be published)
[3] J. Frith, *Wind rotors for ships,* Maritime J. **6** 12 (2015)
[4] M. Rhodes, *Ship Stability for Mates/Masters,* Seamanship International Ltd, Glasgow, 21 (2003).
[5] S. Chakraborty, *Ship Stability – Understanding Intact Stability of Ships*, Naval architecture J. **27** (2019)
[6] B. Barras, D. Derrett, *Ship Stability (Masters and Mates),* sixth edition, Elsevier, London, **44** (2006).
[7] W. Wawrzyński, P. Krata, *Method for ship's rolling period prediction with regard to non-linearity of GZ curve,* Journal of Theoretical and Applied Mechanics **54(4)** 1329–1343, (2016), DOI: https://doi.org/10.15632/jtam-pl.54.4.1329
[8] M.V. Vasilescu, M. Panaitescu, F.V. Panaitescu, C. Panait, *Choose the best electricity sources for a container ship, by using a hybrid optimization model for electric renewable* , SPIE Proc. Advanced Topics in Optoelectronics, Microelectronics and Nanotechnologie (to be published)
[9] A. Maruccia, *Optimisation model for a ship's hybrid energy system with a Flettner rotor,* Master thesys, **1** 81(2019).

Influence of Flettner balloon, used as wind energy capturing system, on container ship stability

Mihail-Vlad Vasilescu [1,*] and *Dumitru* Dinu [1]

[1] Constanta Maritime University, Doctoral School of Mechanical Engineering, str. Mircea cel Batran, No. 104, ZC 900663, Constanta, Romania

Abstract. This article presents the influence on the stability of a container ship, by connecting a Flettner balloon, as wind energy capturing system. Flettner balloon is a generator of electric power, filled with helium that rotates around a horizontal axis and sends the electricity using a cable. It rotates around a horizontal axis in response to the wind, effectively generating clean, renewable electricity at a lower cost than all competing systems. These article major points determined by the authors are: calculation of the forces which influence the balloon, calculation the influence of the balloon on the ship transverse and longitudinal stability, calculation of the ship new displacement, new draft, new GM and period of rolling. As a conclusion of the article, the reader will discover that the transverse stability of the ship will insignificant decreased with a small value of 0.01 and the longitudinal stability will be improved by 0.7532. The article demonstrates that a Flettner balloon, installed on a container ship, is a possible concept of capturing wind energy.

1 Introduction

On a container ship we have installed a Flettner balloon, which is situated above the vessel, anchored with some lines. The Flettner balloon has some wind stabilizers which help him maintain his position.
The container ship initial characteristics, without Flattner balloon connected are:

- L=168.56 m;
- l=29.4 m;
- D=39449 mt;
- GM=4.32 m;
- T_p=9.56 m;

* Corresponding author: vladmihail_2005@yahoo.com

Where:
L= Length of the ship; l= Berth of the ship; D= Displacement; GM= Metacentric height is a measurement of the initial static stability of a floating body. It is calculated as the distance between the centre of gravity of a ship G and its metacentre M. T_p=Ship draft.
Due to high altitude at which the Flettner balloon rises and because of the helium gas, the forces which appear in the balloon are:
F_a =Lifting force; G_{He}=Helium weight which is inside the balloon; G_B =Balloon weight structure; G_{cp}=Electrical cables and mooring lines weight.
By using Computational Fluid Dynamics (CFD) we can visualize the velocity and pressure distribution around the section of the Flettner balloon.

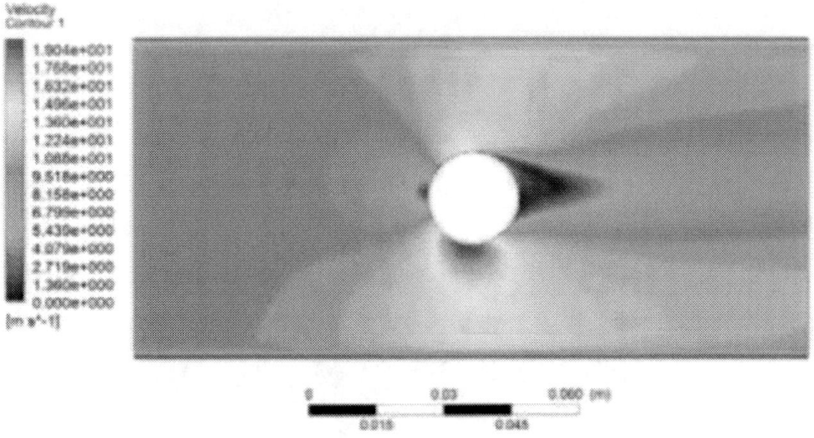

Fig.1. Velocity around the section of Flettner balloon with circulation [1]

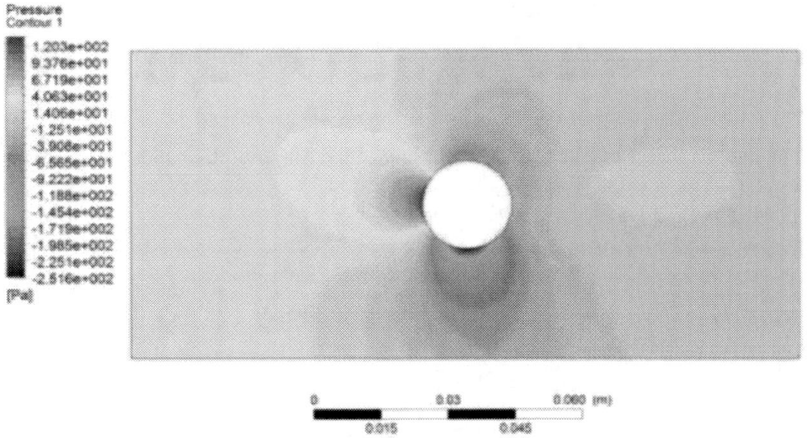

Fig.2. Pressure distribtion around the section of Flettner balloon with circulation [1]

2 Calculation of the ship draft at the new displacement D`

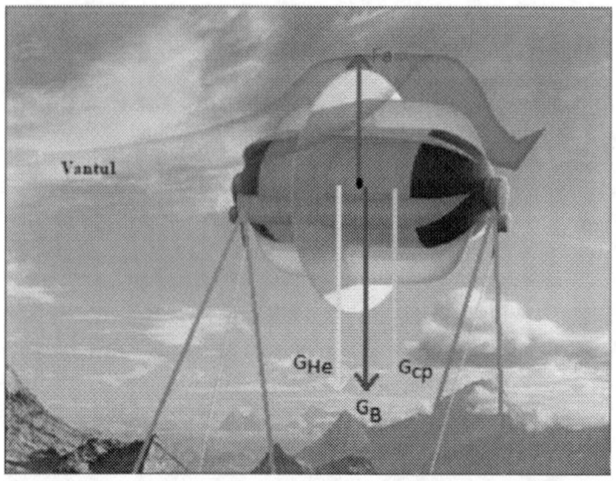

Fig. 3. The forces which appear and influence the lifting of the balloon[2]

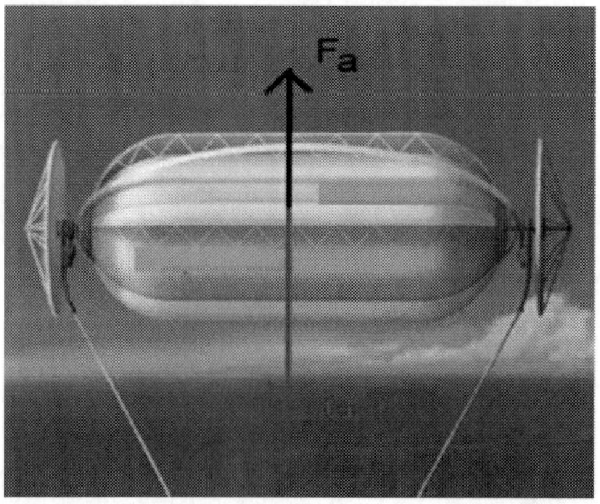

Fig. 4. The merged forces which appear and influence the lifting of the balloon[3]

$$F_a = \rho_A V_b g = 2528.4kN = 2528400N \quad (1)$$

$$G = G_{He} + G_B + G_{cp} = mg \quad (2)$$

m=12.626t

$$G = mg = 12.626x9.8 = 123.7348\ kN = 123734.8\ N \quad (3)$$

$$F_a^{``} = F_a - G \cong 245\ mt \quad (4)$$

Where:

$F_a^{``}$= Resulting lifting force, with the extracted value of the weights; m= mass; g= gravitational acceleration; $\rho_A = 1.29\ \frac{kg}{m^3}$= water density; $V_b = 200000m^3$ = balloon volume;

2.1.Calculation of the new displacement D`

$$D^{`} = D - F_a^{``} = 39204\ mt \quad (5)$$

$D^{`} = 39204\, mt$

By interpolation from hydrostatic tables we extracted the immersion value - TPC.

For any draft is the mass which must be loaded or discharged to change a ship's mean draft in water by one centimetre.

$TPC = 45.39\, mt$

$$D^{`} - D_{9.50} = 14.78\, mt \quad (6)$$

New draft $T_p^{`} = 9.50 + \frac{D^{`} - D_{9.50}}{TPC} x0.01 = 9.5032m$ (7)

3 Determination of the ship transverse stability

3.1 Calculation of KG

$$GM = KM - KG \quad (8)$$

Where:

$GM = 4.32\, m;$

$TKM = 12.32\, m \Rightarrow KG = 8\, m$

Where:

KG= The distance from keel K to G center of gravity;

3.2 Calculation of TKM for the new draft

$T_p^{`} = 9.5032\, m$

By interpolation from hydrostatic tables we extracted the TKM` and LKM` for $T_p^{`}$.

$TKM^{`} = KM_{T^{`}} = 12.31\, m$

Where:

Transverse metacentre= The point of intersection of the vertical through the centre of buoyancy of a ship in the position of equilibrium with the vertical through the new centre of buoyancy when the ship is slightly heeled.

TKM= Transverse stability. The distance between the keel K and the tranverse metacenter M;

3.3 Calculation of new GM (Metacentric height)

KG = 8 m

$$GM^{`} = KM_T - KG = 4.31\, m \quad (9)$$

3.4 Determination of the initial righting lever GZ [Fig.5]

Transverse statistic stability is a term used to describe the ability of a ship to return to the upright position when it has been forcibly heeled by an external force and is momentarily at rest when floating in still water.[4]

GZ= Righting lever. It is defined as the horizontal distance, measured in meters, between the centre of gravity G and the vertical line of action of the buoyancy force Bf, acting through the centre of buoyancy B when the ship is heeled.

The righting moment at any angle of heel represents the instantaneous value of the ship`s ability to return to the upright position, when the ship is in still water conditions and is momentarily at rest. It is expressed in tonnes-meter.

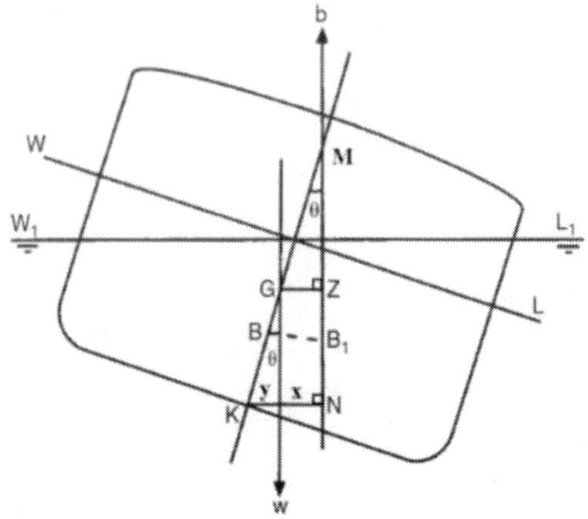

Fig. 5. Ship Stability initial condition

$$KN = x + y \quad (10)$$
$$x = GZ \quad (11)$$
$$y = KG\ x\ sin\theta \quad (12)$$
$$KN = GZ + KG\ x\ sin\theta \quad (13)$$
$$GZ = KN - KG\ x \sin\theta \quad (14)$$
$GM = 4.32\ m;\ D = 39449\ mt$

Where:

θ = Angle of heel; GZ= Right lever; B= Centre of buoyancy before trimming; B_1 = Centre of buoyancy after trimming; KG= Distance from keel K to the centre of gravity G; GM= Metacentric height, it is a measurement of the initial static stability of a floating body. It is the distance between the centre of gravity G of a ship and its metacentre M. KN=Being the righting lever measured from the keel K; WL= Water line;
w= Weight; KM= Distance from keel K to the metacenter M.

Table 1. GZ calculated for different values of the θ angle from 0 to 90.

θ	Sin θ	KG	KN	KG x sin θ	GZ
0	0.000	8	0.000	0.000	0.000
10	0.174	8	2.113	1.392	0.721
20	0.342	8	4.260	2.736	1.524
30	0.500	8	6.310	4.000	2.310
40	0.640	8	7.830	5.120	2.710
45	0.707	8	8.396	5.656	2.740
50	0.766	8	8.828	6.128	2.700
60	0.866	8	9.330	6.928	2.402
70	0.939	8	9.332	7.512	1.820
80	0.984	8	8.892	7.872	1.020
90	1.000	8	8.010	8.000	0.010

The transverse stability curve is plotted on the graph with black (Fig.7). The values from table 1 are taken from KN cross curves of ship stability booklet.[5]

3.5 Calculation of the new G`Z ` [Fig.4]

$GM^{`} = 4.31\ m$

$D^{`} = 39204\ mt$

$$KN = x^{`} + y^{`} \quad (15)$$

$$x^{`} = G^{`}Z^{`} \quad (16)$$

$$y^{`} = KG^{`} x\ sin\theta \quad (17)$$

$$KN = G^{`}Z^{`} + KG^{`}\ x\ sin\theta \quad (18)$$

$$G^{`}Z^{`} = KN - KG^{`}\ x\ sin\theta \quad (19)$$

Where:

G`Z`= Right lever; KG`= Distance from keel to the new center of mass/gravity; G`Z`=New right lever after the center of gravity moved; G`M= New metacentric height, it is a measurement of the initial static stability of a floating body. It is the distance between the centre of gravity G` of a ship and its metacentre M.

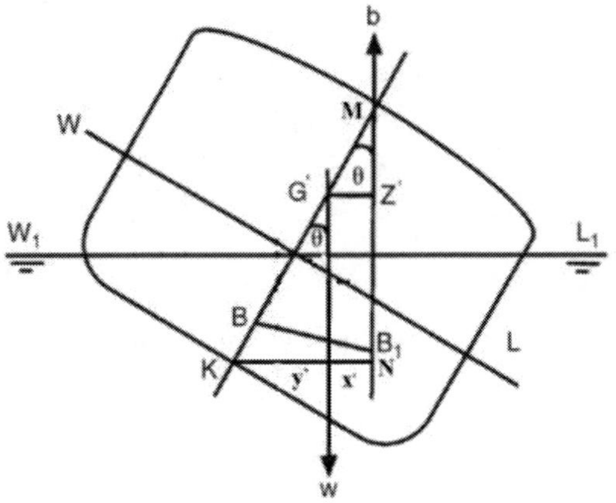

Fig. 6. Ship Stability condition after the connection of Flettner balloon

Table 2. G`Z` calculated for different values of the *θ* angle from 0 to 90.

θ	Sin θ	KG`	KN`	KG` x sin θ	G`Z`
0	0.000	8.01	0.000	0.000	0.000
10	0.174	8.01	2.113	1.393	0.720
20	0.342	8.01	4.250	2.739	1.511
30	0.500	8.01	6.300	4.005	2.295
40	0.640	8.01	7.820	5.126	2.694
45	0.707	8.01	8.383	5.663	2.720
50	0.766	8.01	8.825	6.135	2.690
60	0.866	8.01	9.330	6.936	2.394
70	0.939	8.01	9.331	7.521	1.810
80	0.984	8.01	8.899	7.881	1.018
90	1.000	8.01	8.010	8.010	0.000

The transverse stability curve is plotted on the graph with blue (Fig.7). The values from table 2 are taken from KN` cross curves of ship stability booklet.[5]

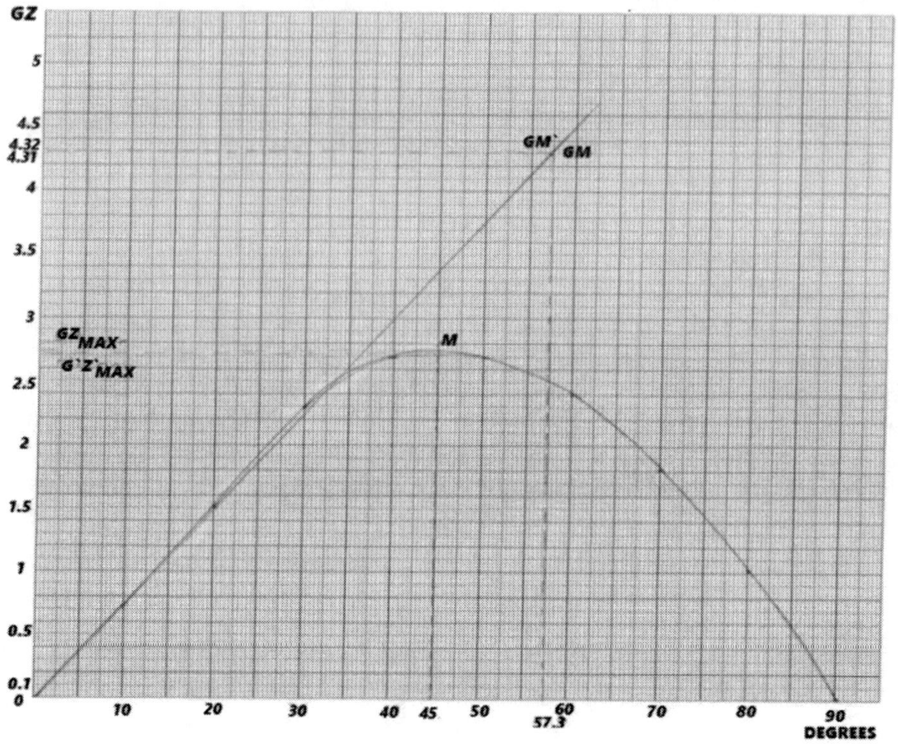

Fig. 7. Transverse stability graph

The graph represents the curve of the static stability of the ship and the variation of the stability right lever according to the variation of the transverse tilt angle of the ship. The initial stability is shown in black, without the energy generating equipment. With blue is shown the final stability, with the energy generating equipment installed.

Where:

- Point O, represents the origin of the curve;
- Point M and M`, maximum point of the curve, critical angle of roll or maximum angle of canary; the point where the recovery arm has the maximum value;
- For GM`< GM 4.3184< 4.32 ship transverse stability is lower connection of the ship with the balloon that generates green, unconventional energy.

3.6 Rolling period

$$T = \frac{2xlxC}{\sqrt{GM}} \qquad [6,7] \qquad (20)$$

Where:

$C = 0.371$ for $T_P^{`} = 9.5032\ m$; $GM = 4.32\ m$ and $GM^{`} = 4.31\ m$

T= initial rolling period; T`= rolling period after the connection of the ship with the balloon that generates green, unconventional energy; GM=Metacentric height, it is a measurement of the initial static stability of a floating body. It is the distance between the centre of gravity G of a ship and its metacentre M. GM`=New metacentric height. $T = 10.4956\ s$

and $T^{`} = 10.5078\ s$.
From the graph we can also observe that the ship has initial a slower period of rolling, comparing with the rolling period after the connection of the ship with the balloon.

$T < T^{`}$ $10.4956 < 10.5078$

3.7 Inclining moments of the ship

3.7.1 Stable wind

$$lw_1 = \frac{0.0514\ x\ A\ x\ Z}{D} \quad (21)$$

$$lw_1^{`} = \frac{0.0514\ x\ A^{`}\ x\ Z^{`}}{D^{`}} \quad (22)$$

Where:
D = initial ship displacement; D`= displacement of the ship, after the connection of the balloon with the ship; A= the wind pressure area on the hull of the ship and on the cargo on the deck above the waterline, calculated for a draft $T_p = 9.56\ m$ and a $D = 39449\ mt$; $A$ `= the wind pressure area on the hull of the ship and on the cargo on the deck above the waterline, calculated for a draft $T_p^{`} = 9.5032\ m$ and a $D^{`} = 39204\ mt$; Z= vertical distance [5] between the center of A and the center of the lateral area of the keel, at a draft of $T_p = 9.56\ m$ and a $D = 39449\ mt$;
Z`= vertical distance between and the center of the lateral area of the keel, at a draft of $T_p^{`} = 9.5032\ m$ and $D^{`} = 39204\ mt$; $T_p = 9.56\ m$ and $D = 39449\ mt$;
Calculation of A
By interpolation from wind area table [5], for a draft $T_p = 9.56\ m$ => $A = 1478.9936\ m^2$
Calculation of Z
By interpolation from wind area table[5], for a draft $T_p = 9.56\ m$ =>$Z = 10.2272\ m$; $lw_1 = 0.019708\ m$; $T_p^{`} = 9.5032\ m$ and $D^{`} = 39204\ mt$
Calculation of A`
By interpolation from wind area table [5], for a draft $T_p^{`} = 9.5032\ m$ =>
$A^{`} = 1488.7313\ m^2$
Calculation of Z`
By interpolation from wind area table[5], for a draft $T_p^{`} = 9.5032\ m$ =>$Z^{`} = 10.2203\ m$;
$lw_1^{`} = 0.019948\ m$

3.7.2 In case of a gust of wind

$$lw_2 = 1.5\ x\ lw_1 \quad (23)$$

$$lw_2^{`} = 1.5\ x lw_1^{`} \quad (24)$$

$T_p = 9.56\ m$ and $D = 39449\ mt$
$lw_2 = 0.029562\ m$
$T_p^{`} = 9.5032\ m$ and $D^{`} = 39204\ mt$
$lw_2^{`} = 0.029922\ m$

4 Ship longitudinal stability and its effect on the ship trim

$$MTC1cm = \frac{D\ x\ GM_L}{100\ x\ L} \quad (25)$$

Where:
MTC= moment of change trim one centimeter
D= ship displacement
L= ship length [8]
GM_L =GM longitudinal

4.1 Determination of LKM for the new draft

$T_p^` = 9.5032\ m$
By interpolation from hydrostatic tables we extracted the LKM` for $T_p^`$.
$LKM = KM_L = 232.9952\ m$
Where:
Longitudinal metacenter= The point in a vertical line through the centre of gravity of a ship where this line is intersected by a second vertical line through the centre of buoyancy when the ship is inclined at a very small angle in a fore and aft direction. LKM= Longitudinal stability. The distance between the keel K and the longitudinal metacenter M; [9]

4.2 Calculation of the MTC` for $T_p^` = 9.5032\ m$

By interpolation in hydrostatic tables at $T_p^`$ => $MTC^` = \mathbf{530.3664\ mt\ x\ m}$
$MTC^`$ = Final MTC after the connection of the balloon with the ship
$MTC = 531.92\ mt\ x\ m$
We can observe that:
$MTC > MTC^`$ $531.92 > 530.3664$

4.3 Calculation of initial GM_L

$$GM_L = \frac{MTC\ x\ 100\ x\ L}{D} = 227.2810 \quad (26)$$

4.4 Determination of $GM_L^`$

$$MTC1cm = \frac{D\ x\ GM_L^`}{100\ x\ L} \quad (27)$$

$$GM_L^` = \frac{MTC^`\ x\ 100\ x\ L}{D^`} = 228.0342 \quad (28)$$

Because $GM_L < GM_L^`$ 227.281< 228.0342 it means that the longitudinal stability of the ship, after the connection of the Flettner balloon has been improved. Thus we have reduced the effect of stresses affecting the longitudinal axis of the ship.

5 Conclusions

From this paper we can see that the Flettner balloon connected to the ship in not influencing to much the ship stability.
From the stability point of view, transverse stability has decreased:
$GM^{`} < GM$ 4.31 < 4.32
but the value is insignificant.
Longitudinal stability of the ship, after the connection of the balloon, was improved: $GM_L^{`} > GM_L$ 228.0342 > 227.2810
After the connection of the ship with the balloon, we can observe that the rolling period has been increased:
$T^{`} > T$ 10.5078 > 10.4956
By connecting the Flettner balloon to the ship main deck, the ship will be taken out from the water, GM will be slightly reduce by 0.01 because the center of gravity G will rise. The righting lever GZ` will be reduced.
Above study we demonstrate that the installation of a Flettner balloon on a container ship, is a possible method of reducing fossil fuel consumption and greenhouse gas emissions. The study also proves that a Flattner balloon installed on a container ship with the above characteristics is safe for navigation.

References

1. A. A. Scupi, *Magnus effect studied with analytical and numerical methods,* JMTE, **1**, Nautica, Constanta, 41 (2017), ISSN 18844-6116.
2. https://www.betterworldsolutions.eu/10-amazing-wind-turbines/magenn-mars-wind-turbine/, accessed on: 05 January 2020.
3. https://techaneesh.blogspot.com/2012/08/magenn-air-rotor-system-marsmars-is-new.html, accessed on: 05 January 2020.
4. B. Barras, D. Derrett, *Ship Stability (Masters and Mates)*, sixth edition, Elsevier, London, 84 (2006).
5. Hakodate Shipyard, *Manual of Stability*, Japan, 56 (2003).
6. M. Rhodes, *Ship Stability for Mates/Masters*, Seamanship International Ltd, Glasgow, 58 (2003).
7. W. Wawrzyński, P. Krata, *Method for ship's rolling period prediction with regard to non-linearity of GZ curve,* J.T.A.M. **54(4),** 1329–1343 (2016), DOI: https://doi.org/10.15632/jtam-pl.54.4.1329.
8. M.V. Vasilescu, M. Panaitescu, F.V. Panaitescu, C. Panait, *Choose the best electricity sources for a container ship, by using a hybrid optimization model for electric renewable* , SPIE Proc. Advanced Topics in Optoelectronics, Microelectronics and Nanotechnologie (to be published).
9. A. Maruccia, *Optimisation model for a ship's hybrid energy system with a Flettner rotor*, Master thesis, **1**, 81(2019).

Analysis of energy potential of biofuels obtained from camelina sativa

Madalina Ghilvacs[1], *Razvan* Carlanescu[2,*], *Maria* Paraschiv[1], and *Malina* Prisecaru[1]

[1]Politehnica University of Bucharest, Faculty of Mechanical and Mechatronics Engineering, Splaiul Independenței 313, Bucharest 060042, Romania
[2]Romanian Research and Development Institute for Gas Turbines COMOTI, 220 D Iuliu Maniu Bd., sector 6, Bucharest, Romania

Abstract. In a world where the energy demand is increasing daily, energy efficiency and renewable energy play a major role. Our research comes to help fighting the biggest problem that our world is facing today, global warming. This study aims to evaluate the energy potential of biofuels obtained from camelina sativa in order to reduce fossil fuel dependence, greenhouse gas emissions and increase energy efficiency. Camelina (Camelina sativa L.) is a cruciferous oilseed plant belonging to the Brassicaceae (mustard) family with the common name false flax and gold of pleasure. Since camelina can grow under different conditions without being a competitor of food land, this energetic plant can be used for production of both, first- and second-generation biofuels. In our study we determined the elemental analysis of the subproducts obtained from camelina sativa, through experimental tests, and we built a mathematical model to determine the calorific power and to simulate the burning of the biofuels into a steam generator in order to analyse the energy potential. The results show us that the biofuels obtained from camelina sativa can represent an alternative to fossil fuels and should be considered in the transition to green energy.

1 Introduction

According with the national energy strategy 2019-2030, with the prospect of the year 2050, beside energy efficiency, promoting energy from renewable sources plays an important role in order to reach the long-term global goal agreed at Paris in 2015, limiting the rise of the global average temperature to 2 ° C, compared to the pre-industrial level. The total energy production based on biomass and waste presents, in all scenarios, a consistent increase during the analysed period, 2020-2050. The trend of accelerating biomass production after 2030 is notable, by developing modern and efficient technologies on a large scale, especially in the rural area.

For the period 2020 - 2050 - electricity production based on renewable sources (biomass) will be 0.9 TWh (1.3% of the share of energy resources in electricity production)

* Corresponding author: razvan.carlanescu@comoti.ro

compared to 0.4 TWh (0.7% of the share of energy resources in the production of electricity) in 2017. By 2030 small power stations powered exclusively with biomass, bioliquids, biogas, and waste will be developed, until such plants will have a total installed capacity of 139 MW. The boilers of some of the current thermoelectric power stations will be adapted to allow the burning of a biomass addition [1].

The promoting of energy from renewable sources will be done in particular for heating/cooling and transportation sector (biofuels). For heating/cooling sector, the goal is a rate of increase in the share of renewable energy sources in the total energy demand for heating and cooling by 1.3% annually by 2030. Regarding transportation sector, the share of alternative fuels in the total energy demand for transport will increase from 5.4% in 2015 to 12.8% in 2030; thus, a 2.5-fold increase in the demand for biofuels is expected [1].

Renewed interest in camelina sativa is based on the interest of producing cost-effective biofuels [2]. Camelina sativa seeds have a high level of oil content, which is about 40% [3], can be grown as an annual summer or biannual winter crop under different climatic and soil conditions, can be grown on poorer soils and require lower inputs of fertilizers and pesticides than traditional crops which led to a substantial cost reduction [2] [4]. Many researchers have been evaluated the properties of biofuels obtained from camelina sativa and the possibility of using these biofuels in engines. The direct use of camelina sativa vegetable oil on a CI Perkins 1104C-44 engine was investigated in 2013. Results highlighted the fact that it's possible to obtain good engine performance when using camelina sativa vegetable oil, if some calibration parameters are adjusted [5]. There are several papers which investigated if the properties of bio-diesel obtained from camelina sativa oil are within the European standard for bio-Diesel limits [6] [7]. Studies on bio-kerosene production from camellia sativa oil have been also reported. Corporan E. et al. (2011) have reported that the bio-kerosen obtained from camelina sativa has better thermal oxidation stability than the classic aviation fuel, but to produce bio-kerosen from camellia sativa oil is still too expensive [8].

Through our work we would like to add a plus in promoting the using of biomass for both heating/cooling purpose and transportation sector. The first step in our work and this paper's focus is to evaluate the energy potential of camelina sativa subproducts. Our future work, which is already ongoing, will focus on obtaining biomass for second generation biofuels.

2 Description of biomass

Today's focus on engineering, genetic engineering, biology, chemistry, agricultural sciences is to find new energy sources, new methods of obtaining energy, more efficient biomass in order to replace the conventional fuels with biofuels.

Camelina, also known as lemongrass, yolk, egg yolk, fake, wild, German sesame, and Siberian sesame are a very old agricultural plant, cultivated for its oily seeds. The plant was cultivated in Europe before 1950 [9]. The main area of expansion was from Eastern Europe to Central Asia. Its cultivation area ranges from 36° north latitude to 70° north latitude. Today, camelina is grown in Romania, Slovenia, Ukraine, China, Finland, Germany, Austria and Montana in the US.

Camelina sativa was chosen for our research based on the following characteristics [10] [11] [12]:

- Camelina can be grown in short marginal periods without being a competitor of food land since it can be used as a natural fertilizer when it is combined with edible crops on good arable land;
- Camelina possesses a short-season crop, (85-100 days);
- Camelina can be grown as an annual summer or biannual winter crop under different climatic and soil conditions;
- Camelina sativa can be grown on poorer soils and require lower inputs of fertilizers and pesticides than traditional crops;
- Camelina needs little water;
- Camelina sativa may be used as a rotation crop to improve the health of the soil.

The main product of Camelina sativa is the oil produced by crushing and pressing the seeds, the oil content of Camelina seeds ranged from 30 to 40 %. Extraction of oil from oilseeds by pressing yield a by-product called Oil cakes/meal.

The biofuel obtained from Camelina was used till now for two main purposes [13]:

- Camelina oil as feedstock for Renewable Jet Fuel;
- Camelina oil as feedstock for Renewable Biodiesel.

3 Material and methods

For this work, we used Camelia seeds, first Romanian camelina (Camelina sativa, L. Crantz) cultivar. The seeds, oil and meal (cake) for our experiments were provided by NARDI Fundulea. Camelia sativa seeds contain mainly, 4,25 – 5,24% ash, 11,06 – 15,24% cellulose, 18,87-21,97% proteins, 30,1 – 49,7% fats, and 25,83 – 28,21 mg/100 g vitamin E (gama tocopherol), as well as minerals – 0,18% Calcium (Ca), 0,53 % Phosphorus (P), 0,49% Cooper (Cu), 1,39 % Manganese (Mn), 4,47 % Iron (Fe) and 2,56 % Zinc (Zn) [14].

The object of this research is to evaluate the potential of using camelina sativa subproducts (seeds, oil, meal) as replacements for fossil fuels in small power stations and to identify energy potential and pollutant emission variation when comparing with the ones of conventional fuels. A mathematical model was built in order to simulate the burning of camelina sativa subproducts in a small power station [15] [16] [17] [18] [19]. The mathematical model used for our research was built in MATLAB; the flow chart of the simulation is presented in figure 1.

We used as entry data for our simulation the elemental analysis of camelina sativa subproducts, which we determined experimentally in a previous work [14]. The elemental analysis with reference to the initial state for Camelina oil, seeds and cake is presented in table 1.

For our experimental test we used THERMOSTAHL MCL-N boiler, 55 kW. The boiler is made of steel, ideal material for thermal fluctuations, resistant to expansion and shrinkage. The monobloc construction, without elements and other welded structures is an advantage of the MCL-N boiler. This boiler is also protected against heat shock.

To determine the pollutant emissions, we used the method of discontinuous analysis - on the 55 kW boilers using the AFRIXO MAXILYZER analyser - is a method of analysis with electrochemical cells. Throughout this method we determine the values of the pollutant emissions upon evacuation and due to the discontinuous character of the analysis a single value of the emissions representing an average value will be recorded.

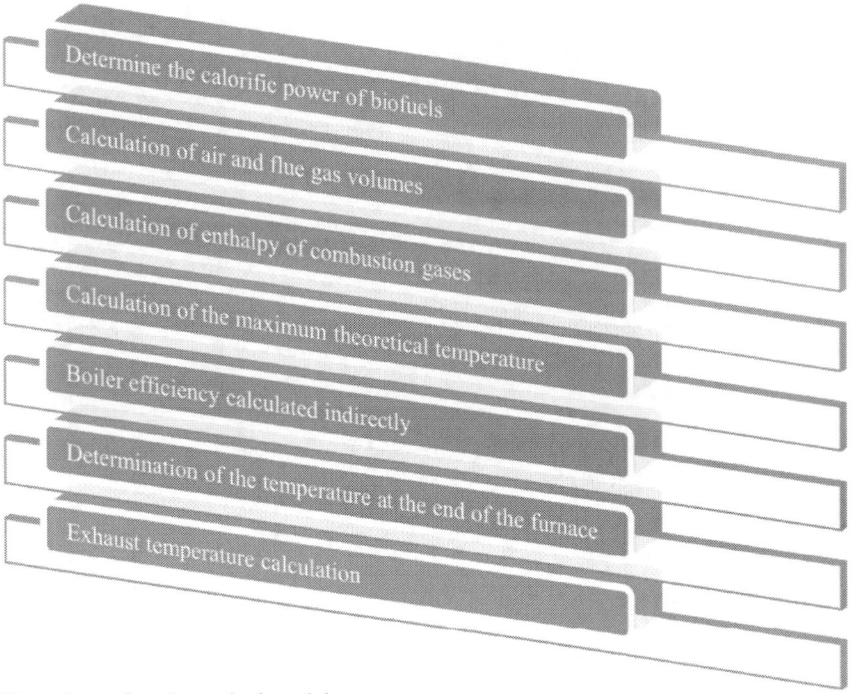

Fig. 1. Flow chart of mathematical model

Table 1. Elemental analysis of biofuels obtained from camelina sativa

Biomass subproducts	Ci [%]	Hi [%]	Ni [%]	Si [%]	Oi [%]	Ai [%]	Wt [%]
Oil	75.5	11.6	0.3	0	12.6	0	0
Meal (cake)	51	7.4	4.7	0.3	23.117	6.686	6.797
Seeds	58.9	8.6	3.7	0.3	20.045	4.193	4.262

4 Results and discussion

4.1 Calorific power

As can be seen from table 2, from calorific power point of view, camelina sativa subproducts can compete with conventional fuels. Because camelina sativa oil has a high calorific power value, it can easily be used as a superior fuel for industrial boilers, ships and even airplanes.

Table 2. Calorific power value for various fuels

Fuels	Calorific power [MJ/kg]
Wood	9 – 16
Brown coal (Rovinari)	9,80
Pit coal (Aninoasa)	20,30
Camelina meal	22,30
Methanol	22,70
Pit coal (Vulcan)	23,10
Camelina seeds	26,60
Ethanol	29,70
Camelina Oil	36,20
Naphtha	39,40
Diesel fuel	44,80

4.2 Temperature variation inside the boiler and energy potential

Our mathematical model allowed as calculating the following parameters:

- The volumes of air and combustion gases;
- The enthalpy of flue gases;
- The maximum theoretical temperature;
- The fuel flow;
- The temperature at the end of the furnace;
- The exhaust temperature.

Based on the fact that camelina cold oil has a high theoretical maximum temperature and requires a larger space for heat transfer by radiation, it is desirable to be used in big power station. In small boilers it is much better to burn meal and seeds as the exhaust temperature of the flue gases are lower, while the oil which has a higher energy potential can't transfer all the heat, and thus the temperature of the exhaust gases is bigger.

Table 3 present us the results of mathematical model simulation in terms of air and combustion gases volumes for all camelina sativa subproducts (seeds, meal and oil). The volumes of air and combustion gases are expressed in m^3_N / kg. In table 4 the values for combustion parameters resulted from our mathematical model are presented: the enthalpy of fuel gases, the maximum theoretical temperature, the fuel flow, the boiler efficiency, the temperature at the end of the furnace and the exhaust temperature.

Table 3. Results mathematical simulation - Volumes of air and combustion gases

Volumes of air and combustion gases	Unit	Camelina seeds	Camelina meal	Camelina Oil
V_{Omin}	$[m_N^3/kg]$	1,443	1,207	1,971
V_{a0}	$[m_N^3/kg]$	6,872	5,747	9,386
V_{awet0}	$[m_N^3/kg]$	6,982	5,839	9,537
V_{awet}	$[m_N^3/kg]$	8,719	7,292	11,908
V_{CO2}	$[m_N^3/kg]$	1,1	0,952	1,41
V_{SO2}	$[m_N^3/kg]$	2,1 x 10^{-3}	2,1 x 10^{-3}	0
V_{RO2}	$[m_N^3/kg]$	1,102	0,954	1,41
V_{N2}	$[m_N^3/kg]$	5,458	4,578	7,417
V_{H2O}	$[m_N^3/kg]$	1,127	1,006	1,45
V_{0g}	$[m_N^3/kg]$	7,687	6,538	10,277
V_g	$[m_N^3/kg]$	9,408	7,977	12,629

Table 4. Results mathematical simulation – Combustion parameters

Combustion parameters	Unit	Camelina seeds	Camelina meal	Camelina Oil
h_g	$[kJ/kg]$	1,442 x 10^4	1,229 x 10^4	1,923 x 10^4
t_{tm}	[°C]	1,597 x 10^3	1,576 x 10^3	1,622 x 10^3
η	[%]	78,773	78,583	78,993
B	$[kg/s]$	2,664 x 10^{-3}	3,187 x 10^{-3}	1,953 x 10^{-3}
t_{ef}	[°C]	644	642	647
$t_{exhaust}$	[°C]	100	101	111

As shown in figure 2, the highest temperature value is recorded in furnace, around 1600 °C. In the furnace, the combustion gases are formed after the combustion process; and on their way to the evacuation the flue gases go through the pipes of smoke. The smoke pipes are introduced into a water tank, where part of the heat of the flue gas is transferred to the working agent and thus the temperature drops considerably.

Since the temperature of the flue gases at the exhaust is still high enough, 110 °C, it is necessary to have two routes for the flue gases. In this case we must use a fan which means additional costs and high energy consumption. As can be seen from figure 3, the highest velocity of the gases is at the end of the furnace.

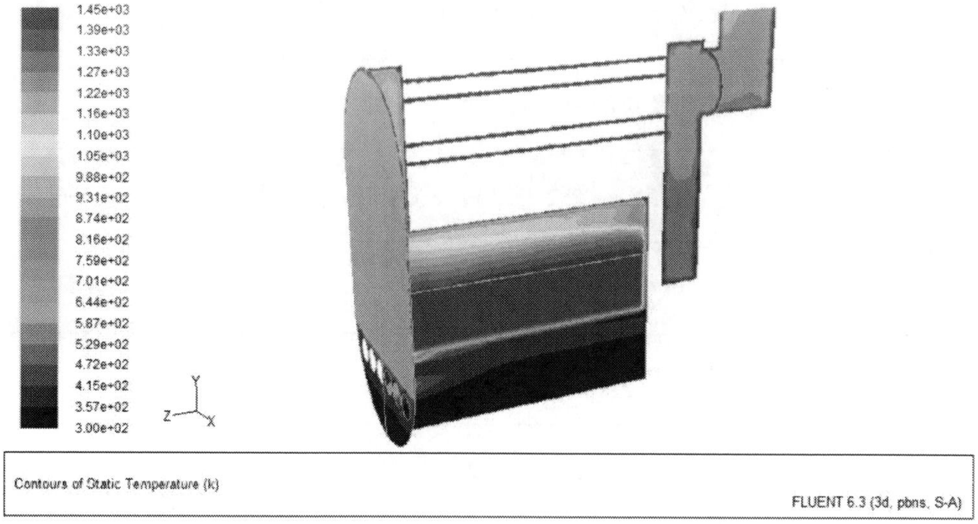

Fig. 2. Temperature variation inside the boiler

Fig. 3. Variation of gases velocity inside the boiler

4.3 Pollutant emissions

From the point of view of polluting emissions, their values are not higher than other fuels; however, the use of chemical enhancers can lead to the appearance of harmful substances in the flue gases. The comparative analysis for all pollutant emissions for willow, acacia and camelina sativa biofuels (meal and seeds) is presented in table 5. Table 5 present also the air temperature and the

Table 5. Comparative analysis between the functioning of the boiler on willow, acacia, camelina sativa meal and seeds

Pollutant emissions Temperatures	Willow	Acacia	Camelina meal	Camelina seeds
O2	16 %	15 %	8.3 %	8.9 %
CO	0.11 %	0.499 %	0.012 %	0.014 %
CO2	4.8 %	5.8 %	4.7 %	5.1 %
SO2	7 ppm	0 ppm	280 ppm	265 ppm
NO	42 ppm	117 ppm	121 ppm	140 ppm
λ	4.2	3.5	1.65	1.74
T air	31.3 °C	32.1 °C	29 °C	29.5 °C
T gaze	257 °C	329 °C	251 °C	242 °C

5 Conclusions

Based on our results, Camelina sativa can be taken into consideration for replacing the conventional fuels in order to meet the 2030 targets with regards to biomass for energy production.

The calorific power values of Camelia main commercial products classify the Camelia seeds and meal besides Pit coal and Camelia cold oil between ethanol and fuel oil (naphtha).

The oil has the highest efficiency of use due to the conditions of the combustion process, mainly through the possibility of using excess air. Camelina oil should be used in big power station while camelina seeds and cake should be used in small power stations. Boiler efficiency calculated indirectly when using biofuels obtained from camelina sativa is around 78 %.

Since the main disadvantage of first-generation biofuels is that they come in direct competition with food, our future work will focus in obtaining camelina biomass for second generation biofuels and evaluate its perspective in terms of biofuels production and remediating the toxicity of soils.

This work has been funded by the European Social Fund from the Sectoral Operational Programme Human Capital 2014-2020, through the Financial Agreement with the title "*Scholarships for entrepreneurial education among doctoral students and postdoctoral researchers (Be Antreprenor!)*", Contract no. 51680/09.07.2019 - SMIS code: 124539.

References

1. Ministerul Energiei, Strategia energetică a României 2019-2030, cu perspectiva anului 2050 (2019).
2. J.K. Crowley, EU Research Contract No. AIR3-CT **94** 2178 (1995).
3. J.T. Budni, W.M. Breene, D.H. Putman. J. Am. Oil Chem. Soc. **72 (3)**, 309–315 (1995).
4. H. Prankl, K. Krammer, M. Worgetter, ALTENER Final Report, **XVII/4.1030/Z/96-013**, 44–64 (1999).
5. S. W. Kruczynski, Energy Conversion and Management, **65**, 1–6 (2013).
6. B. R. Moser, S. F. Vaughn, Bioresource Technology, **101**, 646–653 (2010).
7. C. Ciubota-Rosie, J. R. Ruiz, M. J. Ramos, A. Perez, Fuel, **105**, 572–577 (2013).
8. E. Corporan, T. Edwards L. Shafer, M. J. Dewitt, C. Klingshirn, Energy and Fuels, **25**, 955–966 (2011).
9. A.G. Plessers, W.G. McGregor, R.B. Carson, W. Nakoneshny, Can. J. Plant Sci. **42**, 452–459 (1962).
10. H. Abramovicˇ, V. Abram, FoodTechnology and Biotechnology, **43**, 63–70 (2005).
11. J. Zubr, B. Mattha¨us, Industrial Crops and Products, **15**, 155–162 (2002).
12. A.L. Pilgeram, D.C. Sands, D. Boss, N. Dale, D. Wichman, P. Lamb, C. Lu, R. Barrows, M. Kirkpatrick, B. Thompson, D.L. Johnson, , ASHS Press 129-131 (2007).
13. H. Schindlebauer, P. Hodl, Institute for Chemistry and Technology of Petroleum Products (Ed.), **9** (1994).
14. I. Toncea, D. Necsariu, T. Prisecaru, L. N. Balint, M. Ghilvacs, M. Popa, Romanian Biotechnological Letters, **5,** 8594 – 8602 (2013).
15. I. Pîsa, C. Neaga, L. Mihaescu, T. Prisecaru, *Generatoare de abur, Vol. I, Îndrumar de laborator* (Editura U.P.B., 1994).
16. C. C. Neaga, *Tratat de generatoare de abur*, **vol I**, ISBN 973-8130-67-0 (Editura AGIR, 2001).
17. C. C. Neaga, *Cazane şi combustibili. Combustibilii energetici şi arderea* (Centrul de multiplicare IPB, 1980).
18. C. C. Neaga, *Determinarea randamentului unui cazan de abur în conformitate cu Recomandarea ISO* (Energetica, 1976).
19. D. Stefanescu, A. Leca, L. Luca, A. Badea, M. Marinescu, *Transfer de căldură şi masă* (EDP, 983).

A combined approach for assessment the functionality of photovoltaic modules in real-world operation

Ivaylo Nedelchev[1*] and *Hristo* Zhivomirov[1]

[1]Technical University of Varna, Department of Theory of Electrical Engineering and Measurements, Bulgaria

Abstract. The energy production from the solar photovoltaic power plants is highly dependent on the meteorological and physical conditions such as solar radiation, ambient and panel's surface temperature, inclination of the photovoltaic (PV) panel construction *etc.* The I-V curves are the most important for estimation the functionality and production of each PV panel, as well as finding the maximum power point (MPP) of it. Because of the continuous generation of energy, the silicon crystal of the PV modules begins to depreciate and this decreases the energy production. In real-world operating conditions, revealing the energy state of the PV panels is the main point for estimating the PV panels' efficiency. Furthermore, one complete approach, including electrical measurements and temperature distribution information over the PV panels' surface, can reveal the risky elements (subcells) and provide data for prevention damages and working interruptions. This paper presents a method for PV monitoring in which conventional electrical instrumentation devises and thermographic camera are used, in order to estimate the real physical state of the PV panels'.

1 Introduction

The most of the PV power plants are all yearly exposed of the nature's influence such as wind, rain, snow and accumulation of dust. The depreciation of the PV panels is a crucial factor for energy production losses. The decreasing of the quality of the PV panel's crystal is direct consequence of this process. It is expressed in decreasing of the efficiency and fill factor (FF), existing of higher operating temperature of some panel's cells, expansion the resistive effects over the short circuit current (I_{SC}) and open circuit voltage (V_{OC}) *etc.* Fast assessment of all these effects and influence factors over the energy generation during dynamically changing operating conditions, allows quality maintenance of the large PV power plants and can be useful for prevention damages and energy losses [1-3].

The diagnostic and monitoring of the photovoltaic panels in working conditions requires many functional parameters to be known, as well as the meteorological data and the parameters of the location [1-3]. This paper represents a simple combined approach for PV panel monitoring involving measurements of the main electrical quantities of the PV panels

* Corresponding author: iynedelchev@abv.bg

and thermographic measurements. The preliminary assessment about the location and solar radiation are also taken into account in the study.

2 The observed physical parameters of the study

2.1 PV plant location parameters

The main parameters of the PV plant location and specific parameters are shown in the Table 1 [4].

Table 1. Parameters of the PV power plant's position, inverters and local time.

Parameter	Value	Parameter	Value
Longitude, λ	41,47°	Number of inverters	15
Zenith angle, θ	48,53°	Model of the installed invertors	Woodward IDS Solo 250
Tilt angle, δ	25°		Delta Solivia EC G3 20 TL
Day number of the year, D	341	Nominal AC power of the installed inverters, kW	250 kW (10 inverters)
Altitude, z	217 ÷ 228 m		20 kW (5 inverters)
Azimuth orientation (Direct North - 0°), φ_A	180°	Number of string boxes (SSB)	35
Installed peak power, MW	2,3	Number of strings in each SSB	13 or 14
Number of power substations	5	Total number of the strings	407
Nominal power of the installed power substations	500 kVA	Number of panels in each string	24

Considering the short daytime of the season time period (in the beginning of December) and location (Southwest Bulgaria), the measurements was conducted during the day interval between 10 a.m. and 2 p.m.

2.2 Parameters of the PV panel and measurement approach

The objects of the measurement are PV panels, which location is chosen in random manner all over the PV power plant area. All tested modules belong to one kind of model. The operating period till the moment of the experiment is 5 years. Their electrical and physical parameters are given in the Table 2 [5]. The most important parameters of the tested PV panels are: peak power (measured by standard test conditions - P_m), peak efficiency (η), short circuit current (I_{sc}), open circuit voltage (V_{OC}), maximum power current (I_m), maximum power voltage (V_m), effective surface area (A), panel type, spectral sensitivity, system losses (according the panel's age, pollution, snow, shading, wiring, mismatch, connections, availability) *etc.* [1, 2, 3, 6, 7].

One of the functional parameters of each module is the surface temperature (T). It is an important indicator for the resistive losses in each cell into the PV panel. The simultaneous measurement of the surface temperature can derives crucial information about the depreciation of the PV panel's elements. It is well known fact that the tilt of the construction is a precondition for unevenly collection of dust, snow or some other pollution on the panel's surface. Especially in snowy conditions the bottom side of the construction is still shaded, while the top side is illuminated. This fact can cause unequal aging of each panel's subcells and thus to decrease of the quality parameters of the entire module. It could be expected, that this process increases the intrinsic resistance of some cells of the bottom side of the module and therefore the operating temperature of the affected cells will rise slowly during the aging. Along with the conventional electrical measurements, there were included thermographic measurements, in order the temperature of the cells to be simultaneously detected and the temperature differences (ΔT) between them to be received. In this case, the temperature distribution along the vertical direction of the panel's construction will shows the maximum value of the ΔT.

Table 2. Parameters of the tested modules.

Parameter	Value	Parameter	Value
STC power, P	235 W	effective surface area, A	1,5 m^2
peak efficiency, η	14,5 %	Array type	fixed
short circuit current, I_{SC}	8,45 A	Panel type	a-Si
open circuit voltage, V_{OC}	37,3 V	System losses, L_S	20%
maximum power current, I_M	7,98 A	Number of cells, N	60
maximum power voltage, V_M	29,5 V	effective surface area, A	1,5 m^2

The signal processing of the received thermographic images consists of the histogram and vertical thermal curve (VTC) of each image (each module respectively), as well as the biggest temperature difference (ΔT_M) on them. The thermographic signal processing was executed by the original software of the thermographic camera.

3 Measurement approach and received results

3.1 Preliminary assessment and measurement of the solar radiation

The executed monitoring of the panels consists of two stages: measurement of the physical quantities, which are used for estimation of the parameters of each tested PV panel and thermography. Firstly, V_{OC}, I_{SC} and solar radiation of the tested panel are measured. On the second stage via thermography, the temperature distribution over the panel's surface was captured. Assessing the energy production requires the preliminary data for the solar radiation. For the purpose of the study some empirical approaches is used in order to find the level of the solar radiation at the location of measurements.

The global incident solar light reached an object laying on the Earth's surface consists of three main components: direct, diffused and reflected light. It was taken three approaches in order to assess the global solar radiation I_G: empirical assessment [8 - 10], calculation from the global data bases [12 - 14] and direct measurement.

Empirically, the daily maximum of the solar radiation I_D can be obtained by the formula [9, 10]:

$$I_D = 1353 \cdot 0.7^{AM^{0.678}}, \tag{1}$$

The variable in the expression (1) is called air mass factor - *AM*. The parameter represents the path from the atmosphere that the light must pass through for reaching the Earth's surface, relative to the direct path through the zenith. The air mass factor can be found from [8, 9]:

$$AM = \frac{1}{\cos(\theta)}, \tag{2}$$

where θ is the zenith angel: between the vertical of the position and the sun beam direction.

Considering other two components (diffused and reflected solar radiation), the global incident solar radiation on the Earth's surface object I_G increases proportionally. Thus, the relation between them is [2, 3, 14]:

$$I_G = 1{,}1 \cdot I_D. \tag{3}$$

Another source of information for assessing the incident global solar radiation I_G is the global data bases [11 - 13], which represent the average value of I_G for seasonal or daily time period. For calculating the incident global solar radiation, which effect on the tilted PV panel's surface, have to be used the expression:

$$I_{GPV} = I_G \cdot \sin(\alpha + \delta), \tag{4}$$

where I_{GPV} is the global solar radiation which impacts on the PV panel, α is the elevation angle of the sun position, δ is the tilt angle of the PV construction across the Earth surface. According the number of the day *D* in the year and latitude λ of the measurement location, α is calculated to be 25,9°.

The real way to assess the solar radiation is direct measurement. The solar radiation was measured at the time period between 10 a.m. and 2 p.m. During the experiment in the real conditions, the moisture and clouds affected the measured data.

On the Table 3 are shown measurement data and the results of the both calculating approaches, reduced to PV panel's surface according (4). All the values are reduced to the tilted surface of the PV panel - I_{GPV}. The received data express daily maximum level of the I_{GPV}. For the further calculations measured value of I_{GPV} were used.

Table 3. PV panel's data.

Calculated via Empirical approach		Calculation via Global data base (CMSAF)		Measured	
I_G, W/m^2	I_{GPV}, W/m^2	I_G, W/m^2	I_{GPV}, W/m^2	I_G, W/m^2	I_{GPV}, W/m^2
931,7	715,3	1020	783,7	655	503,6

3.2 Electrical measurements and efficiency calculations

During the research, two electrical parameters were measured: short circuit current I_{SC} and open circuit voltage V_{OC} of more than hundred PV panels. In the Table 4 are shown the results of eight of them as a representative data of the whole electrical measurement. By use of the measured electrical values and catalogue data, can be obtained the next PV panel's parameters:

- Normalized V_{OC} - υ_{oc}, which is calculated as [9]:

$$\upsilon_{OC} = \frac{q}{m \cdot k_B \cdot T} \cdot V_{OC}, \tag{5}$$

In (5) $q=1,602 \cdot 10^{-19}$ C is the elementary charge, m is the ideality factor, $k_B = 1,381 \cdot 10^{-23}$ J/K is Boltzmann's constant and T is absolute temperature. According to the model for the calculation of the identity factor m and taking into account the measurement situation – direct solar radiation (I_D) and ambient temperature (T), the ideality factor was accepted to be $1,8$ [9, 15].

- Fill factor – FF - express the ratio between maximum output power and product of V_{OC}, I_{SC}. The next expression shows the empirical way for calculation the FF by υ_{oc} [9, 16, 17]:

$$FF = \frac{\upsilon_{oc} - l\ (\upsilon_{oc}+0,72)}{\upsilon_{oc}}, \tag{6}$$

- Maximum output power – P_M, obtained from the equation:

$$P_M = FF \cdot V_{OC} \cdot I_{SC}, \tag{7}$$

- Efficiency – η. It depends on all parameters obtained above – measured electrical values and calculated FF and I_{GPV} [3, 9]. The expression for efficiency is:

$$\eta = \frac{V_{OC} \cdot I_{SC} \cdot FF}{A.I_{GPV}}, \tag{8}$$

In (8), A is the parameter which shows the effective PV panel's working surface. For tested modules A is received from catalog datasheet ($A = 1,5\ m^2$) [5]. The calculated, as well as the measured data are shown on the Table 4. The value of the ambient absolute temperature T during the experiment is within the interval $280\ K \div 285\ K$.

Table 4. PV panel's data.

№	1	2	3	4	5	6	7	8
V_{OC}, V	34,9	35,3	35,2	35,5	35,3	35,1	36,2	35,7
I_{SC}, A	0,61	0,97	1,6	2,3	2,3	2,6	2,97	3,44
I_{GPV}, W/m^2	92,9	135,1	251,8	336,6	343,9	402,9	464,3	503,6
υ_{oc}	14,03	14,19	14,15	14,27	14,19	14,11	14,55	14,35
FF	0,754	0,756	0,756	0,757	0,756	0,755	0,760	0,758
P_M, W	16,1	25,9	42,6	61,8	61,4	68,9	81,8	93,1
η, %	11,5	12,8	11,3	12,3	11,9	11,4	11,7	12,3

4 Thermography

For the study there were captured more than hundred thermographic images of the PV modules. The images were received via thermal imager type TESTO 890 B. For measurements, from each formed string coupled in every string boxes (SSB) were chosen one panel. In that manner, the obtained results were collected from equally distanced areas, located all over the whole region of the PV plant. From the other hand, the measurement panels are chosen to belong to higher and lower rows of the existing metal construction. This type of distribution of the measurement objects was applied in order applicable statistical data to be obtained. By this type of string coupling, inverters distribution and specific area landscape (more than 11 m. of denivelation of the terrain), it was tested only 1,5 % of the total number of the PV panels to be obtained the statistical relevant data for the entire PV plant.

On the Fig. 1 ÷ Fig. 8, are represented the temperature distribution of the eight of them.

The shown patterns belong to distanced areas and strings and they are involved in a representative sample for the whole PV plant for further thermographic calculations and image processing. From the thermography it visible, that the temperature distribution shows higher temperature levels on the most of the bottom located cells of the PV panels. This phenomenon is connected with the tilt of the construction, which implies cumulating of the pollutions and longer time of shading during the snow season. The bottom area of some images shows more dissipated power and therefore losses, which is indicator for raised value of the intrinsic resistance of the more heated cells. This shows that the bottom cells probably have a different level of depreciation compared to the top located cells into the most of the PV panels. The bottom cells with the raised intrinsic resistance will reduce the output panel's current and thus the generated power and the PV panel's efficiency will be smaller.

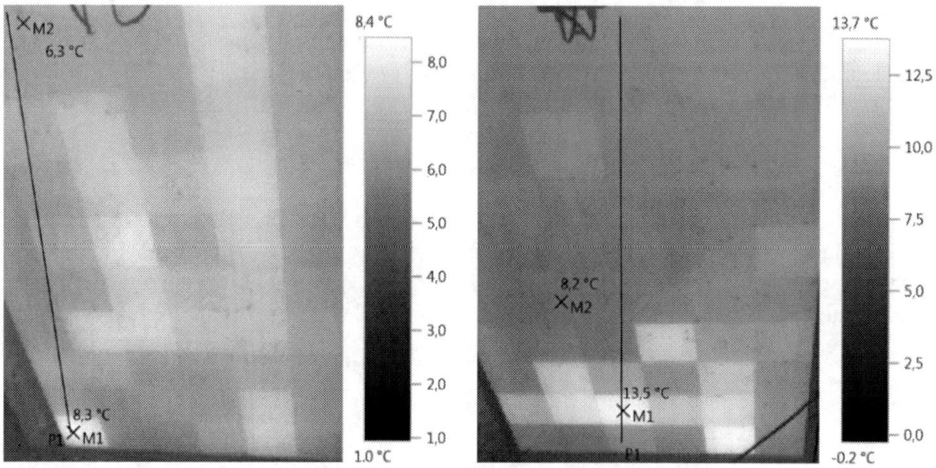

Fig. 1, Fig. 2. Temperature distribution of the tested PV panels № 1and № 2.

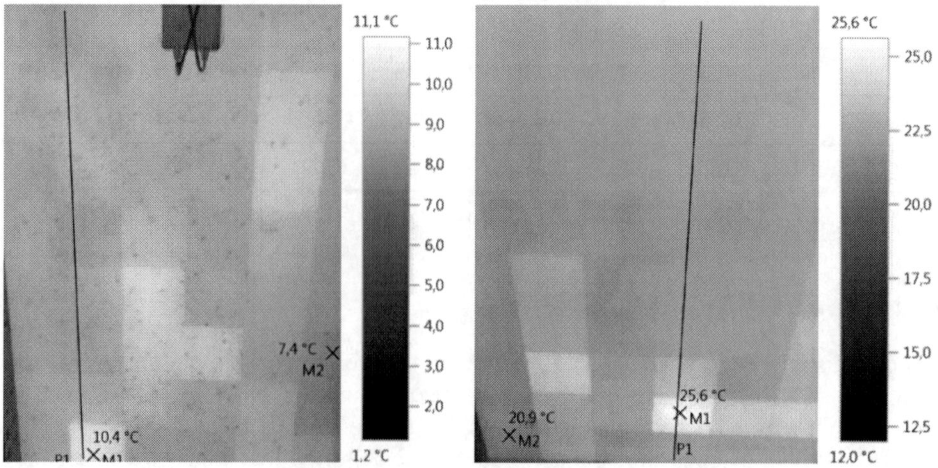

Fig. 3, Fig. 4. Temperature distribution of the tested PV panels № 3 and № 4.

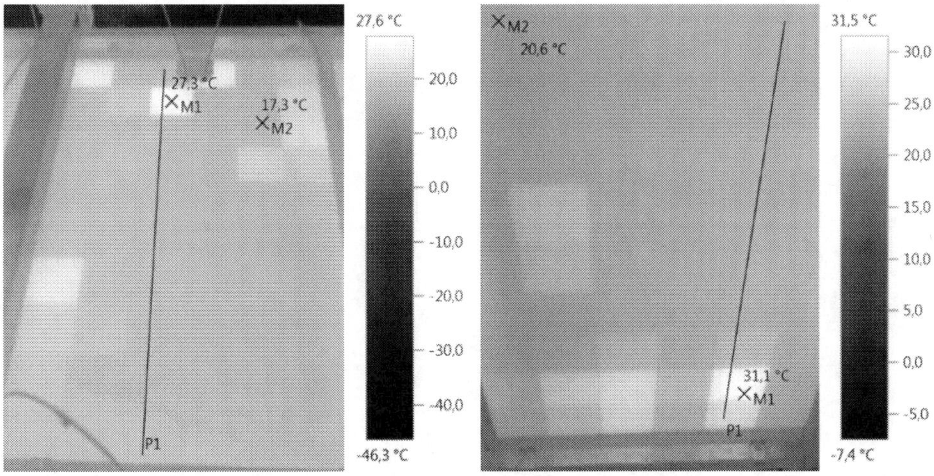

Fig. 5, Fig. 6. Temperature distribution of the tested PV panels № 5and № 6.

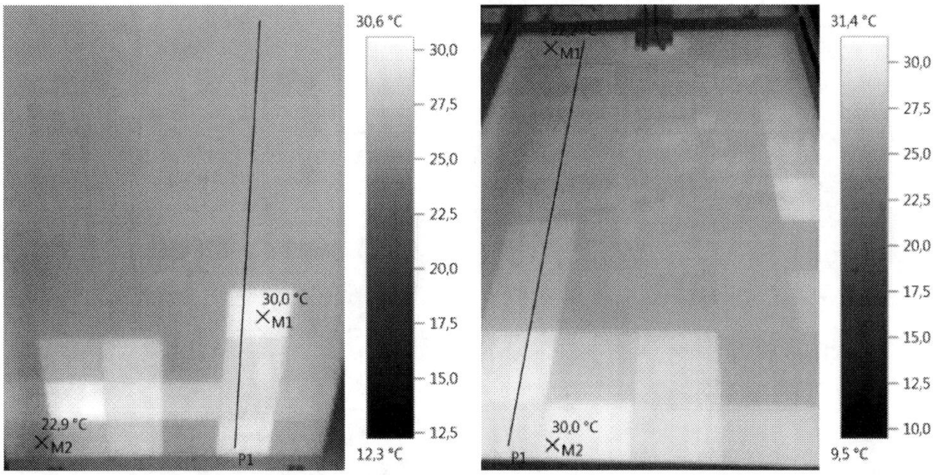

Fig. 7, Fig. 8. Temperature distribution of the tested PV panels № 7 and № 8.

5. Results from processed thermographic images

The processed data of the thermographic images include the histograms on two of the modules and their VTC's in direction from the top to the bottom. These charts are shown on the Fig. 9 – Fig. 12. The histograms (Fig. 9, Fig. 10) represent almost Laplace's distributions, with some exceptions and irregularities. These side components on the right of the maximum (on the higher temperature levels) express the overheated cells or these of them which are exposed of risk of damage. Vertical thermal curves of the tested PV panels № 2 and № 8 are shown on the Fig. 11, Fig.12. Calculated maximum temperature differences (ΔT_M) via thermography over the panel's surface are given in the Table 5.

Table 5. Maximum temperature differences on the surface of the tested PV panels.

№	1	2	3	4	5	6	7	8
ΔT_M, °	2	5,4	35,2	4,7	10,1	10,5	7,1	7,9

They vary from 2 °C to 10,5 °C, which mainly depends on the solar radiation and from the PV panels depreciation. The bigger is ΔT_M, the higher is level of the depreciation of some subcells.

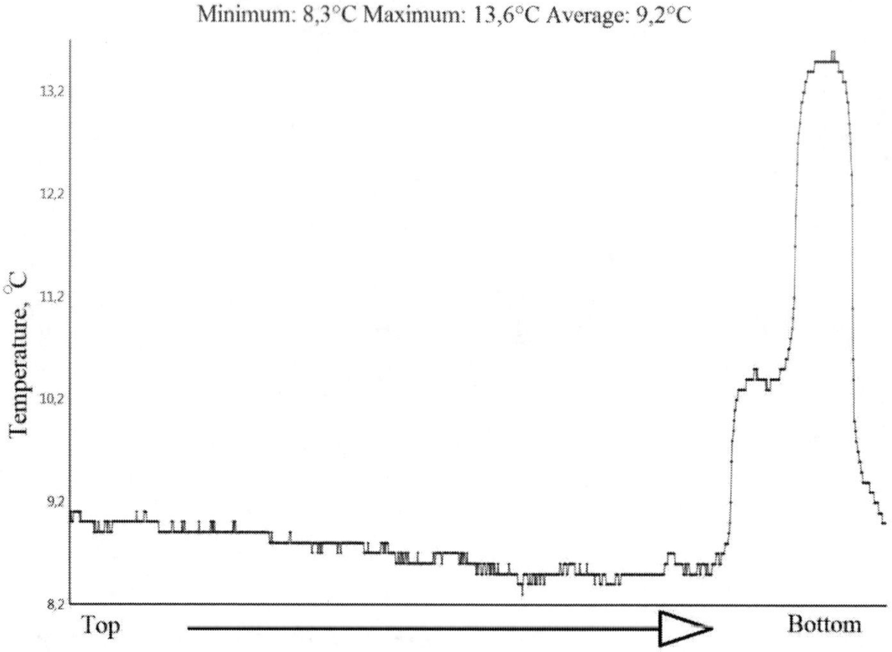

Fig. 9. Histogram of the tested PV panel № 2.

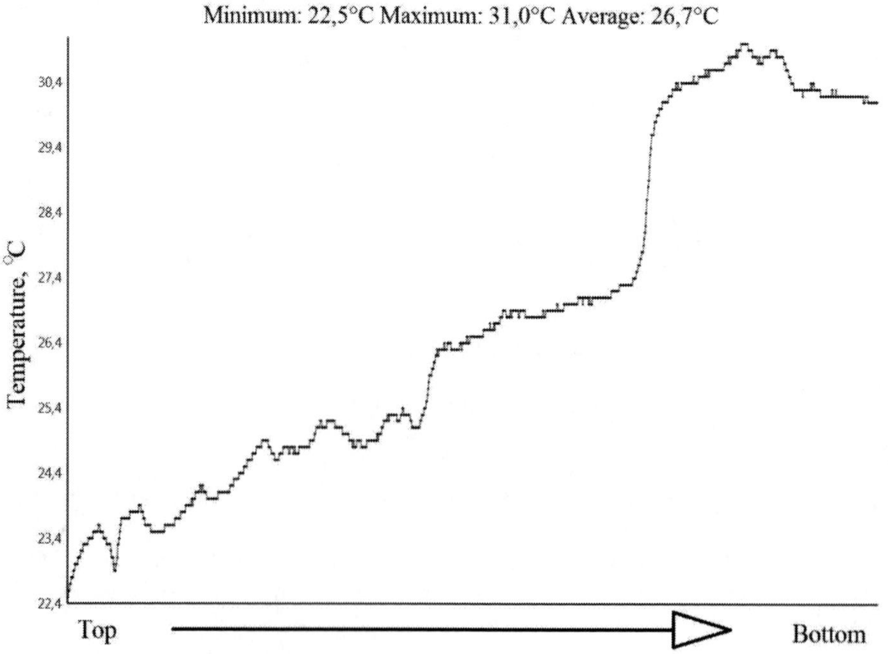

Fig. 10. Histogram of the tested PV panel № 8.

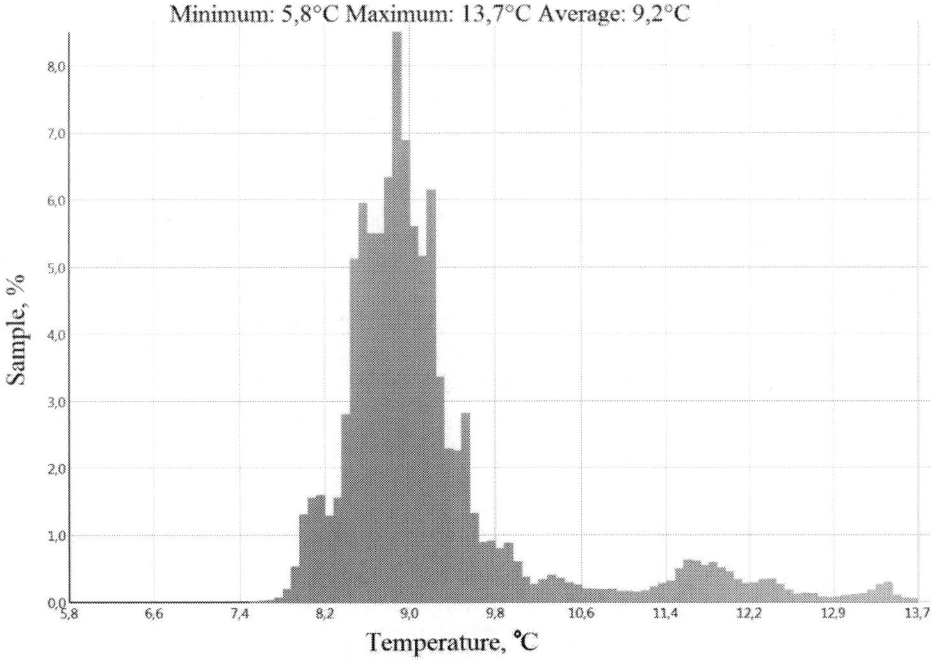

Fig. 11. Vertical thermal curve (VTC) of the tested PV panel № 2.

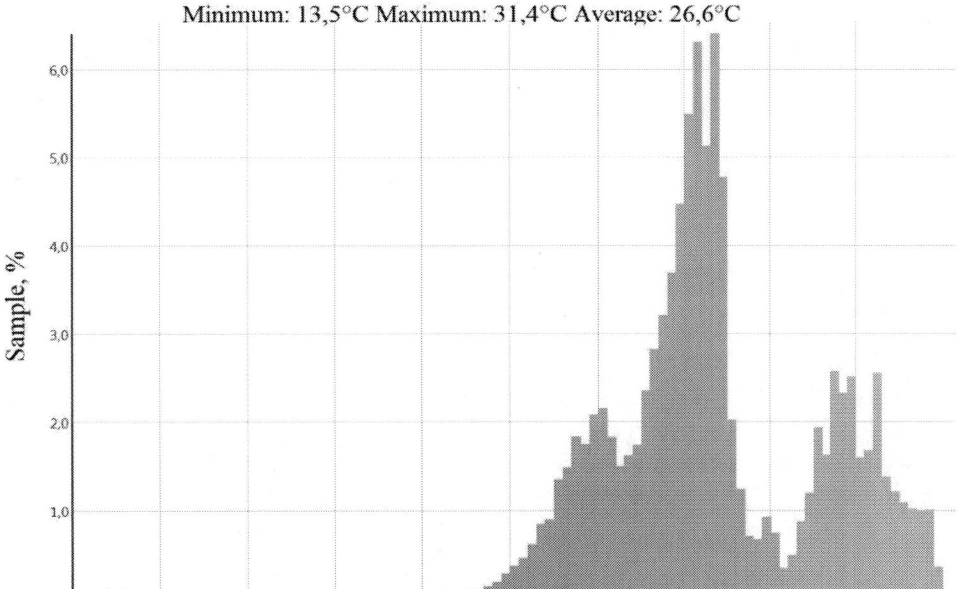

Fig. 12. Vertical thermal curve (VTC) of the tested PV panel № 8.

6 Conclusions

The combined approach used for monitoring of the PV power plant in real field conditions gave interesting results about the efficiency, operating mode and temperature distribution over the PV panels. The main conclusions derived from this study are:

- The calculated solar radiation using Global base data have a large deviation from measured levels. The changing local meteorological conditions impact the measurements and reduce the solar radiation in a big scale.
- Maximum level of the measured solar radiation on the tilted PV panels' surface is 503,6 W/m^2.
- On this time season and after a 5 year of period of exploitation, the efficiency of the PV panels is between 11,3 % and 12,8 %.
- Obtained values for the Fill factor (*FF*) for the tested PV models are within the interval 0,75 ÷ 0,76.
- The maximum temperature differences over the panels' surface received via thermography is within the interval 2°C ÷ 10,5°C.
- The most of the tested panels show the higher level of the temperature on the bottom located subcells – in the lower levels of the construction. Furthermore, some separate cells of the panels № 5, № 6, № 7 and № 8, shows indications for overheating.
- Using histogram of the thermographic images, can be located the risky area on the panel's surface. Deviations on the right side of the main distribution show area with higher risk of damage. By the histogram using further calculations the ratio between the risky and whole area of the PV panel can be obtained.
- The vertical thermal curve (VTC) could be useful tool for assessing reliability of the whole string into the tested PV panel.

The used combined approach and the received results can be used for monitoring and estimation functionality and depreciation of each module of a PV power plant at the subcell level in short time. It is clear, that for more relevant data the frequently measurements during the appropriate time periods have to be done, but this study reveals a comparatively fast method for assessment the current state of each PV plant and the received data are limited in a short-time period respectively.

References

1. T. Markvart, L. Castaner, *Practical Handbook of Photovoltaics: Fundamentals & Aplications*, Elsevier, (2003)
2. R. Messenger, J. Ventre, *Photovoltaic Systems Engineering*, CRC Press, (2004)
3. M. A. Green, *Silicon Solar Cells: Advanced Principles & Practice,* University of New South Wales, (1995)
4. https://www.enfsolar.com/pv/inverter-datasheet/5129
5. https://www.enfsolar.com/pv/panel-datasheet/crystalline/2503
6. P. Kadar, A, Varga, *Intelligent Systems and Informatics*, IEEE, 549-552, (Subotica 2012)
7. C. A. Gueymard, D. Myers, K. Emery, Solar Energy, **73**, 443-467, (2002)
8. F. Kasten, A. T.Young, Applied Optics, **28**, 4735–4738, (1989)
9. A. B. Meinel, M.P. Meinel, *Applied Solar Energy*, Addison Wesley, (1976)
10. E. G. Laue, Solar Energy, **13**, 43 - 50, IN1-IN4, 51-57, (1970)
11. https://wui.cmsaf.eu/safira/action/viewProduktDetails?eid=21839&fid=19
12. https://www.nrel.gov/grid/solar-resource/spectra-astm-e490.html
13. https://re.jrc.ec.europa.eu/pvg_tools/en/tools.html#PVP
14. R. E. Bird, C. Riordan, J. Climate and Appl. Met., **25**, (1986)
15. Z. Hameiri, K. McIntosh, *Photovoltaic Specialists Conf.*, IEEE, 1416-1412, (2013)
16. M. A. Green, Solid-State Electronics, **24**, 788 - 789, (1981)
17. M. Leilaeioun, Z. C. Holman, J. Appl. Physics **120**, 123111, (2016)

Predictive maintenance of photovoltaic plants using multirotor drones with automatic recharging system of Li-Po batteries

Vasile Pleşca[1*] and *Cristian* Purece[2]

[1]Politehnica University of Bucharest, Electrical Engineering PhD School, Bucharest, Romania
[2]INCDE ICEMENERG, Bucharest, Romania

Abstract. Creating the correct maintenance of the strings that form the electric generators, from a photovoltaic park, it requires scanning of the equipment with thermal vision cameras, in the visible and infrared field, at an optimum angle determined precisely for each location depending on the geographic position and the architecture of the power station. In general, this angle is difficult to maintain from the ground so it is necessary to mount the cameras on a support, or most efficiently on a multi-rotor drone that is very stable and easy to handle. Maintenance using multi-rotor drones and heat sensitive cameras is no longer a novelty and has multiple advantages: low costs, minimization of work accidents, and the data acquired can be analyzed in real time or can be stored for later analysis. But this process also has disadvantages: limited flight autonomy of drones, generated by the hard development of battery technology and a small number of specialized personnel in this field. To minimize these disadvantages, we have developed an automatic recharging system of batteries, without disconnecting them, which allows the automation of the scanning process and offers the possibility of taking control of the drones and from a distance.

1. Introduction

Romania has an estimated solar energy potential of 1433 thousand TEP, the equivalent to 60 x 106 GJ / year and 103.2 thousand TEP electricity, which is about 1200 GWh / year. Due to its geographical position and the diversified structure of the relief forms, Romania is well suited to all renewable energy sources, and from the point of view of solar energy, our country has a potential with over 60% sunny days, that is 210 days out of a total year, almost 20% better than Germany, which has long been a leader in the field. [4, 6, 8, 10]

According to Transelectica, 962 photovoltaic parks were built in Romania between 2013 and 2017, representing approximately 4,871.66 MW installed power. Statistically, of these, 212 have a power installed below 1 mw, and 112 have a power installed below two mw. On development regions, the South-Muntenia region has the most photovoltaic parks

* Corresponding author: adi_plesca@yahoo.com

and the highest installed power, being followed by the Central and North-West region. On the counties, the first places are owned by Brasov, Prahova and Giurgiu counties, but when it comes to the installed power, Giurgiu owns the first place, with an installed power of 79.2 MW, followed by Brasov, with an installed power of 61 MW and Calarasi 60 MW. From the point of view of the surface occupied by the photovoltaic park, the average is between 20 ÷ 50 hectares, and the largest parks are found in Sebiş - Arad, with an area of 202 ha, Recaş - Timiş with an area of 200 ha and Livada - Satu Mare with an area of 135 ha. [6, 8, 10, 11].

Fig. 1. Photovoltaic installation with installed power - 22kWp - INCIDE ICEMENERG [15]

As all these photovoltaic parks are approaching half their lifetime, depending on the quality of the installed photovoltaic panels, concrete measures are required to manage the possible defects that are assimilated with this type of plants. Thus, performing predictive maintenance can limit yield losses and prevent escalation of defects from one equipment to another or from one area to another. The previous picture shows a small installation, located on the terrace of the laboratory body, of INCDE ICEMENERG, with PIF in 2007, in which tests were performed on its equipment.

2. Complex system for the maintenance of photovoltaic power plants

Through the method of interconnecting the cells at the panels, the panels in strings and the strings in the inverter systems, during the exploitation of this equipment, a series of defects may appear, some easy to see with the naked eye but others difficult to identify, practically only when the defect has already caused major losses by propagating to other equipment or generating fires with a negative impact on the entire plant or its vicinity.

Because these installations are usually distributed on rather large surfaces or at great heights, in the case of those installed on the roof, without any special equipment, normal or preventive maintenance it is very difficult to achieve this. Many of the defects can be determined by other comparative methods but these are enough that if they are not detected and remedied in a short time from their emergence, they can generate other defects with a much greater impact.

2.1 General notions used in the maintenance process

Generally, in a photovoltaic installation, the defects have as a secondary effect in the first phase, a local overheating. Thus, a scan with thermal imaging cameras, at the optimum angle and distance, generally detects 99% of the defects. Due to the size of the photovoltaic power plants and the large number of elements to scan, the manual inspection is not very efficient so it is necessary to scan from air as well, at the same time capturing images in the IR and visible spectrum.

For aerial scanning you can use several methods or better said more aircraft: with human crew, various types of drones but also balloons properly equipped. Analyzing these variants it becomes obvious that the balloon is excluded from the beginning because the time required for a complete scan is very large but also because it would be very difficult to keep the same trajectory and optimal distance from the surface of the panels, at the next scan, a useful function for comparing the captured images and issuing a diagnosis as close to reality as possible. The variant of flying equipment with human crew is an option, but it has some disadvantages that are difficult to overlook. These are expensive and the captured images are made from a rather large distance (30 ÷ 50 m), the distance that introduces errors in the final analysis because not all the defects are highlighted. Thus, the only viable solution are drones that have the ability to scan automatically or remotely, on predetermined routes or at predefined angles, and the cost of such a service is much lower than the option of using a solution with human personnel on board. Although this solution has many advantages overall, it also has a number of disadvantages: the speed of flight is slower than the version with human personnel on board but it compensates by obtaining images with much better resolutions. However, the biggest disadvantage of the drones is their low autonomy, so the total scan time of the entire park will be quite long taking a few minutes to recharge / change such a battery.

Because such scanning equipment does not cost very much compared to a human-powered aircraft, or relative to the rest of the equipment in a facility, each park can afford itself to purchase such a system. In addition, the operation of such equipment can be done automatically by creating points in the area of the park and where the drone is desired to be controlled by the operating personnel with a minimum of training. In this case, the flight authorizations are obtained very easily, because the flight height does not exceed 30 m and in general the locations of these stations are located in the outer-urban environment.

A complex system for scanning photovoltaic parks is composed of:
- multirotor drone support;
- high resolution visible camera;
- thermal imaging camera;
- system of storage and processing acquired data, located inside the park or at a distance;
- system of automatic recharging of the batteries located in the center of the park, preferably near an inverter system that provides the necessary energy and the efferent communications with the rest of the systems;
- digital weather station;

2.2 Thermal-vision cameras used to scan photovoltaic panels

In general, the companies specialized in the production of thermal-vision cameras also have dual cameras which include an IR sensor but also a high-resolution sensor in the visible spectrum. [12, 13] Thus, the scanning process, but especially the processing of the acquired data, is a much faster and more accurate one, as the camera can eliminate a series of scanning errors automatically. If no such equipment is available, a high-resolution camera can be used, in parallel with a thermal-vision one, and by superimposing the two visible and thermal-vision spectra and using specific algorithms for anomalous detection, through the overlapping of acquired images and comparisons with control images from the database, we can identify a number of differences that may have as a cause the following defects:
- defective cells at the level of a photovoltaic panel;

- photovoltaic panels in short circuit;
- weak connections at the level of strings or blocks of connections;
- defects in the connection cables due to mechanical, accidental actions;
- pollution level by identifying deposits on the surface of photovoltaic panels, which can be: dust layer or chemical, industrial deposits from industry in the area;
- cracks of the protective surface or cells of a photovoltaic panel;
- possible problems due to total or partial shading due to the vegetation in the location or from the layers of snow, frost, ice deposited on the active surface of the panels.

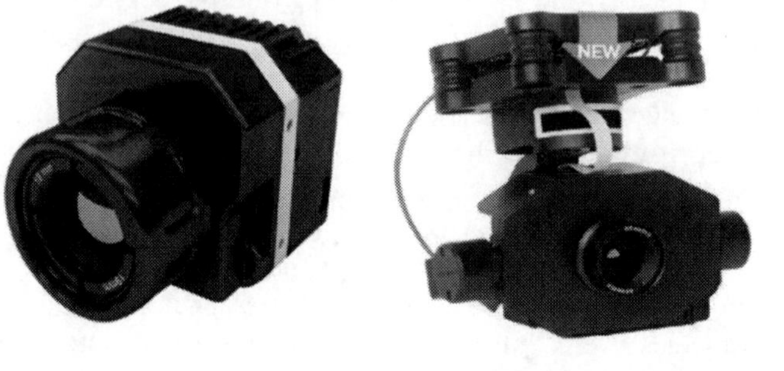

Fig. 2. Thermal imaging camera specific for scanning photovoltaic panels, in simple form or mounted on a gimbal. [12]

For an optimal inspection of a photovoltaic panel using the thermal-vision scanning system, there must be a solar radiation of 500W / m^2, preferably as high as possible, without variations greater than 100 W / m^2 during the measurements, so the cloudy days but especially the windy days should be avoided because it randomly changes the thermal gradient on the active surface of the panel. Also, in order to easily highlight the temperature differences at the panel level, a thermal sensitivity of less than 0.08K is required. During the summer, these conditions are quite easy to reach, because the days are longer, but in the case of frosty winter days, an optimal scan can be done only a few tens of minutes a day, in which case the process takes quite a long time. If these conditions can not be met, the scanning process is slow, which can not be done automatically because too many errors occur and uncertainties. Thus, the specialist who operates the scanning system must be present at the location and eliminate any errors that may occur through repeated scans at that point.

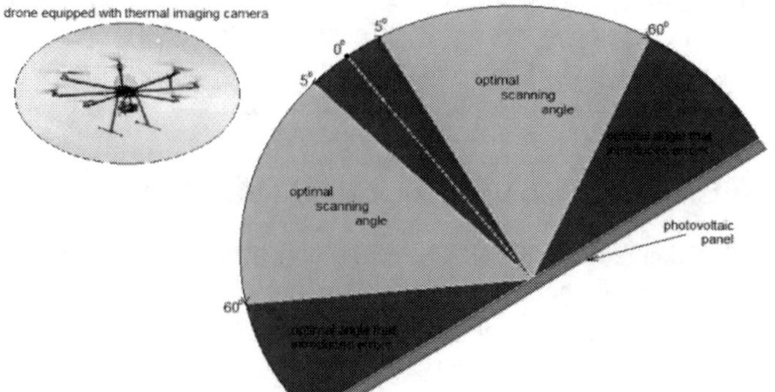

Fig. 3. Determining the optimum angle for scanning the photovoltaic panels with thermal imaging camera

In order to avoid occasional reflections and shadows that can return images with false defects, during the scanning process, it is recommended that the angle of passage not be perpendicular to the surface of the panel and be indicated within in the ranges: 5 ÷ 60 degrees or -5 ÷ - 60 degrees, compared to the perpendicular formed with the surface of the panel, avoiding the interval ± 5 degrees (fig.3).

The minimum resolution of the thermal imaging camera must be better than 640 x 512 physical pixels, and the recording frequency of the images 8.3 Hz PAL or 7.5Hz NTSC. If a better scan frequency is desired on the market there are also a variant of 25 Hz PAL and 30 Hz NTSC but for this you need a license. Obviously, since we are talking about a camera that is mounted on a drone, every gram counts so the camera along with the associated lens is advised not to exceed 150 grams, and the dimensions to be as compact as possible. Very important is its instantaneous consumption which is preferable to be below 2W and at a voltage around 5Vcc. Also, both cameras must be mounted solidly on a gimbal that must ensure mechanical stability by eliminating vibrations and that can be controlled fine from a distance. Also, from a distance, access to the entire menu of the cameras must be ensured so that optimum optical and / or digital zoom can be achieved, including setting the main parameters during the scan: color palettes, camera mode, camera optimization, polarity control, etc. Thus, the camera must be 100% compatible with the autopilot system of the drone, but also with the telemetry system installed on it. [4, 9, 12, 13]

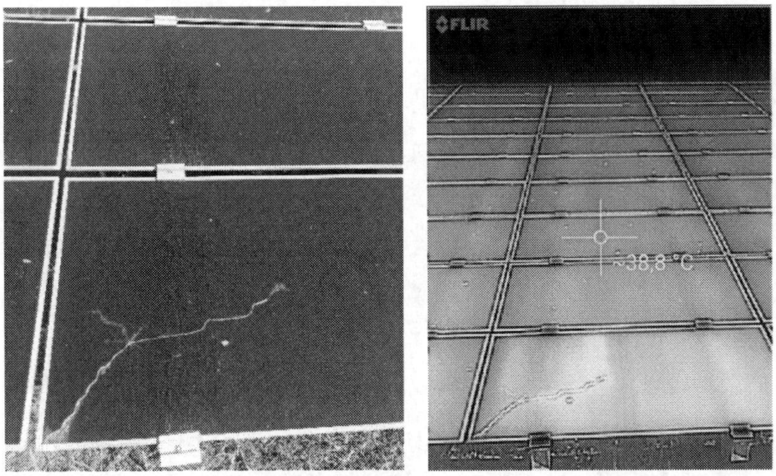

Fig. 4. Crack in the photovoltaic panel seen in the visible and infrared spectrum - Mavrodin photovoltaic plant [15]

Fig. 5. Exposure of the connections panel in the visible and infrared spectrum - Mavrodin photovoltaic plant [15]

2.3 Drone selection as scanning system support

Depending on the weight of the sensors and the size of the photovoltaic park, the support drone can be one with a quad-copter, hex-copter or octo-copter type, with four or eight arms. For this to be able to operate autonomously and to follow the geo-referential map of the photovoltaic park, it must be equipped with a high precision GPS sensor, of the order of centimeters on the horizontal but also on the vertical. Data registration will be done at the drone level but will also be transmitted remotely to a database and a processing server. In addition, the telemetry system installed on the drone will retrieve data from the autopilot and video cameras and transmit it to the ground: GPS position, flight speed, altitude, instantaneous battery capacity and camera status; an example of such a drone can be seen in the following figure. The ground processing unit estimates based on the data received from the drone, collaborated with the data from the weather system located in the park, the drone deviation from the predetermined route for scanning and the estimation of the remaining flight time. When the estimated flight time or battery capacity falls below pre-set thresholds, the system sends the drone a return command to the docking station where the battery recharging process will begin. When the recharging process is complete, the system tests all parameters, including weather data and if all is well, it sends the drone an order to restart the scanning process from where it left off, taking into account the new position of the Sun.

Fig. 6. Quad-copter type drone used during testing [15]

During the scan, the system analyzes the acquired data and if the anomalies appear on the image, they mark the place on the geo-referential map. An operator will determine the type of defect, the priority of remedy and the analysis of the cause that caused the defect. Defects are identified using specialized algorithms that take into account the references previously registered in the databases: dimension, location and spectral references. [1, 4, 9, 12, 13].

Table 3: Classifying the defects according to the spectral images at a new photovoltaic panel

Image recorded	Possible cause	Possible damage
Hot Spot	Dust deposit	Temporary coverage of the active area
	Layer deposited due to local pollution	
	Deposit of bird droppings	
	Excessive humidity	
Hot spot	Cracks of the cells	Production defects
Hot spot / cold spot	Impurities or bubbles of gas on the sandwich panels	
Areas hotter than the rest of the panel	Cracks on the surface of the panel at the level of a cell/ or several cells	Production defects, cracks of cells, short circuit and bypass diode defect
Photovoltaic panels or strings hotter the rest	Electrical contact problems or the panels are not connected	Mounting defects or electrical defects

2.4 Automatic battery recharging system

The automatic reloading system is a prototype made especially for multi-engine drones that perform missions independently. The system is a fixed one, based on a series of sensors and two microcontrollers, one located on a charging base and one at the drone level.

Fig.7. Experimental automatic charging system of the monitoring drone batteries [15]

When the drone arrives for recharging, the two microcontrollers connect to perform the steps required to perform the task:
- fixed landing with precise positioning in the horizontal plane. At the drone level and on a fixed base, special sensors are mounted for the precise positioning of the drone on the loading area;
- drone status check after landing confirmation - zero current passed on the main path (engines stopped). This eliminates any confirmations that may appear wrong and which may lead to the system crashing;

- battery polarity testing / synchronization. In the event of a landing in strong wind conditions, the position of the drone on the loading base may be different from that established, and the probes may be in the wrong position. Risk of battery explosion and drone system damage. If there is an error, it will wait until there are flight conditions and restart the drone on the basis of loading. If everything goes normally the system starts the charging process;
- start of the process of charging the batteries with the tracking of charging curve specified in the documentation;
- testing / balancing battery cell voltage throughout the charging process;
- when the system confirms the full charge of the battery, the weather conditions and the hour set for the scan are tested. If all the required conditions are met the drone takes off and resumes the scan process from the last position, taking into account the position of the Sun.

The contact between the polarized loading pads and the drone is made by a probe system, mounted on the drone landing gear, which, using the weight of the entire drone-mounted equipment, helps to make a firm contact.

2.5 Weather station

The digital weather station must be minimally equipped with anemometer, wind vane, pyranometer, thermometer and hygrometer and must return results in digital format with an accepted accuracy: wind speed and direction, solar radiation, temperature and humidity of the environment. If the photovoltaic park has a large surface area or if there are large differences in level, the presence of several sensors to eliminate errors is required.

3. Conclusions

The increasing number of new photovoltaic installations, installed in Romania, but especially the large number of applications at maturity, with an operating age of approximately 7 years, makes the predictive maintenance service still in high demand from now on. The use of multi-engine drones equipped with heat-sensitive cameras for maintenance of such installations is no longer a novelty, but it has been proven in time that it is a cheap and reliable solution that can detect early defects that cannot be identified with other monitoring equipment. Thus, an acceptable yield is maintained at the level of the entire installation, and the beneficiary has minimal losses in the operation of the respective installation, maintaining the operating costs within acceptable limits.

By automating the entire scanning process there are several advantages:
- eliminates any human errors that may occur due to the routine or the conditions in the site;
- the optimal positioning of the video equipment relative to the elements to be scanned depends on the position of the Sun and the angle of the panels with horizontal;
- automatic limitation of the scans depending on the weather conditions: wind speed, precipitation, solar radiation level, the state of the plant and the level of charge of the power flowed into the system.
- reducing maintenance costs by automatically performing predictive maintenance;
- increasing or maintaining the efficiency of the photovoltaic power plant by identifying in a timely manner the defects that have occurred without reaching major defects with interruptions in the partial or total functioning of the power plant;

The work has been funded by the Operational Programme Human Capital of the Ministry of Europe Funds through the Financial Agreement 51675/09.07.2019, SMIS code 125125.

5. References

1. Yevgen Barsukov, Jinrong Qian - *Battery Power Management for Portable Devices*, ISBN 13: 978-1-60807-491-4, © 2013 ARTECH HOUSE 685 Canton Street Norwood, MA 02062 (2013)
2. Kimon P. Valavanis, John Evans, Daniel Felix, George J. Vachtsevanos – "*Handbook of Unmanned Aerial Vehicles*", ISBN: 978-90-481-9706-4, 978-90-481-9707-1, Springer Science + Business Media Dordrecht (2015)
3. B. Clough, Metrics, schematics! How do you track a UAV's autonomy? In Proceedings of the AIAA 1st Technical Conference and Workshop on Unmanned Aerospace Vehicles, Portsmouth, (2002)
4. Ghe. Olteanu, V. Plesca and others – *Energy in the post-oil era.* ISBN 978-973-1741-23-9, editura ICEMENERG (2010)
5. *1625-2008 - IEEE Standard for Rechargeable Batteries for Multi-Cell Mobile Computing Devices* – IEEE 1625-2008 - https://ieeexplore.ieee.org/stamp/stamp.jsp?tp=&arnumber=4657368 (2020)
6. https://appsso.eurostat.ec.europa.eu/nui/show.do?dataset=nrg_ind_ren&lang=en (2020)
7. https://www.autogreen.ro/energie/nivelul-de-insolatie-si-dimensionare-panourilor/ (2019)
8. https://ec.europa.eu/eurostat/statistics-explained/index.php?title=Renewable_energy_statistics/ro (2019)
9. http://solarcenter.ro/blog/panourile-solare-mod-de-functionare-si-beneficii/ (2019)
10. http://free-energy-monitor.com/index.php/energy/fotovoltaice (2019)
11. http://www.transelectrica.ro (2020)
12. https://www.flir.com/products/vue-pro-r/ (2020)
13. https://www.dronebase.it/prodotto/dual-sensor-ir-eo-gimbal-gemini-ir-2/ (2020)
14. http://www.dummies.com/consumer-electronics/drones/battery-considerations-for-your-drone/ (2020)
15. Personal photos

Researches on the development of an optical system for improving the efficiency of photovoltaic panels

Dana Rizescu[1], *Ciprian Ion* Rizescu[1], *Cristian Gabriel* Alionte[1,*]

[1]University POLITEHNICA of Bucharest, Mechatronics and Precision Mechanics Department, Splaiul Independentei no. 313, sector 6, Bucharest, Romania.

Abstract. The paper aims to create a unit using Fresnel lenses. This module consist of an arrangement of planar-convex Fresnel lenses, made of methyl polymethacrylate, mounted in a hexagonal raster and bent to form a dome for collecting and amplifying solar radiation at several focal points and with the possibility of adjusting the focal distance. The module has three aluminum alloy legs that are vertically adjustable. This construction using Fresnel lenses mounted in a hexagonal raster will increase the energy collected by a photovoltaic panel, thus significantly reducing energy losses and long-term costs. Compared to normal lenses, these lenses are considerably lighter because the material in the center of the lens is reduced, and for manufacturing can be used methyl polymethacrylate or polycarbonate, thus greatly reducing manufacturing costs. To show the functioning of the unit the authors presents in the paper a MATLAB simulation of the Fresnel lens model and a ray propagation.

1 Introduction

The energy of solar radiation is an inexhaustible source of energy, which we can use to replace known fossil fuels or uranium ore, having an annual energy of 23,000 times greater than the actual energy consumption produced by them.

Solar power is the conversion of this energy given by the sun into electricity by using photovoltaic systems, concentrating solar power or a combination of the two. Solar panel concentrators use lens systems, mirrors, or sun tracking devices to focus a wide area of sunlight in a small radius (area).

Advantages of using a solar power concentration system:

- Increase the amount of solar energy received;
- Replacement of pollutant fossil fuels;
- Recovery of the investment in a relatively short time;
- It is environmentally friendly;
- They can be accessed by a considerable number of people for both personal and professional purposes.

* Corresponding author: cristian.alionte@upb.ro

In the literature exist some new approaches [1] which use Fresnel lenses to collect and focus the sunlight rays onto very high efficiency tandem-junction solar cells. By using a relatively inexpensive Fresnel lens to collect the sunlight and to focus it onto much smaller solar cells, the cost and weight of the cells are dramatically reduced. By using very high efficiency cells, the required array area is minimized, reducing overall system weight and launch volume. The advantages of the new Fresnel lens concentrating solar electric power supply are now being widely recognized [2].

In [1,3,4] are presented some solar energy systems with Fresnel lenses solar concentrators, used for space applications, for the satellites power feeding. More specifically, this patent relates to for space power. These systems are high-efficiency, extremely light and robust. For example, in Fig. 1 [1] is shown a Fresnel flexible lens attached to the end supports, in which the end supports extend the lens to maintain their proper position and shape on orbit in space. The main part of the system is the flexible, stretched Fresnel lens optical concentrator fastened on a curved spring sustained by a vertical support and two structural elements of compression that support the pillars and springs. Solar energy is focused on receiver / converter which is fastened on the surface of the radiator for the dissipation of waste heat in space. In this embodiment of this concentrator, before the deployment, the lenses flexible and the end brackets are folded into a flat, low volume package for launch efficient in space.

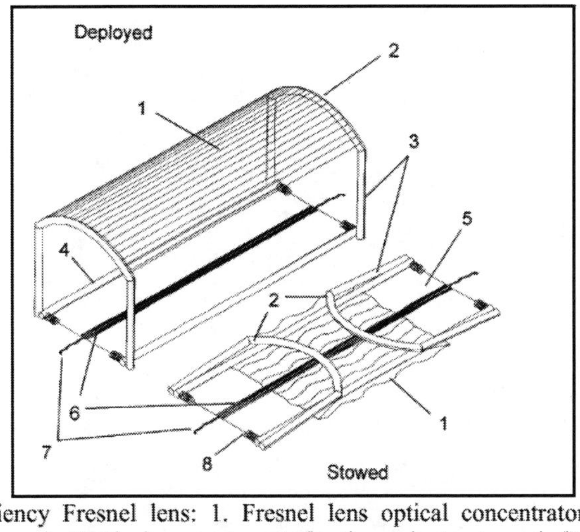

Fig. 1. High efficiency Fresnel lens: 1. Fresnel lens optical concentrator; 2. Curved spring attached to Fresnel lenses; 3. Vertical support posts for the arch structure; 4. Structural elements of compression that support the pillars and springs; 5. The surface of the radiator for the dissipation of waste heat in space; 6. The receiver / converter; 7. Wires for electricity transport generated by the receiver.

The current paper is focused on the Fresnel lenses configuration which are part on a new stationary photovoltaic system. This system does not have any mobile parts and through a complex mirror system that surround the dome, oriented so that the solar rays are normal on the surface of each lens. This system will be presented in another paper.

2 Unit construction

Following the detailed research developed by the authors regarding the solution for the concentration of the light beam, the following constructive solution was reached. An arrangement of planar-convex Fresnel lenses, made of methyl polymethacrylate, mounted in a hexagonal raster, bent so as to form a dome for collecting and amplifying solar radiation at several focal points and with the possibility of adjusting the focal length with three attached aluminum alloy legs that are vertically adjustable. The aim is to increase the energy collected by a photovoltaic panel, thus significantly reducing energy losses and long-term costs, but also the consumption of other polluting energy sources.

The series of hexagonal Fresnel lenses of the dome is inclined at an angle that allows the incident beam of solar radiation to be inclined on the surface of the solar cell, on the surface of the solar panel, Fig. 2.The reason why the solution was chosen for the lenses to be twice as large in diameter than the size of the solar panel is so that there is no loss of light and the concentration is on the entire surface of the panel. This theory of overlap can be properly implemented if the inclination of each illumination lens is in the normal direction of the vector and this can be achieved by using the inclination angle of each illumination area. The system consists of:

1. Fresnel lens arrangement which formed a dome using hexagonal lenses;
2. Adjustable foot to tilt and dome orientation;
3. Photovoltaic panel on which the solar rays are concentrate.

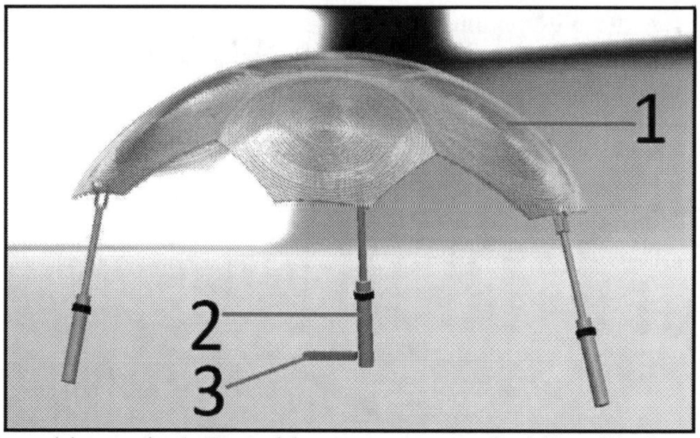

Fig. 2. Designed Fresnel lens unit: 1. Fresnel lens arrangement; 2. Adjustable foot; 3. Photovoltaic panel.

The hexagonal concentration form of the Fresnel lens resembles the beehive and it is more resistant due to the shape and allows an easy detachment or attachment of other lenses, thus increasing assembly much easier, reducing costs [6]. The designed model was developed in Solidworks and is shown in Fig. 3.

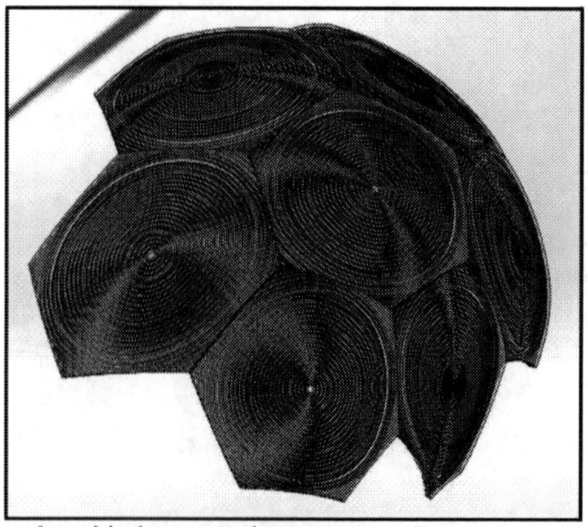

Fig. 3. The Fresnel lenses placed in hexagonal arrangement

3 Materials for Fresnel lenses

This method of concentration was chosen using Fresnel lenses, because of the advantages they offer over conventional lenses. Compared to normal lenses, these lenses are considerably lighter, the material in the center of the lens being reduced, using only its faces for the collection and concentration of light.

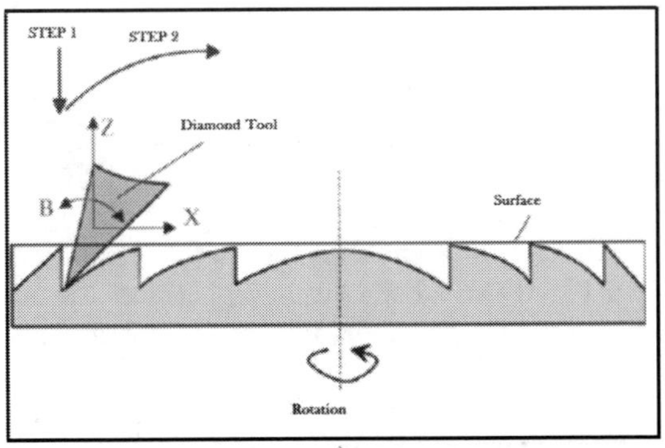

a)

b)

Fig. 4. The manufacturing of the Fresnel lens: a) the manufacturing process; b) hexagonal Fresnel lens

Fresnel lenses can also be made of glass, but most are made of acrylic or polycarbonate, thus greatly reducing manufacturing costs. At the same time, Fresnel lenses can be easier to handle compared to ordinary lenses, having greater flexibility and being more durable [5]. They are several manufacturing possibilities for Methyl polymethacrylate plate (PMMA). One solution is using a numerically controlled machine based on a CAD model. In the process can be used a diamond tool to lower and raise the corners of the future Fresnel lenses, while the PMMA plate is rotated, as can be seen in Fig. 4a. Another solution is the injection molding of PMMA, as can be seen in Fig.4b.

4 Design of Fresnel lens and ray propagation simulation

To design the experimental setup, it was developed a Fresnel model using MATLAB environment.

Through the Fresnel of the dome will pass those "useful" rays which has an angle of incidence below the limit values bellow the total reflection of the lens surface. In the Fig. 5 can be seen that these rays (represented with red color) are distancing from the normal rays (represented with blue color). These rays are influencing the dimensions of the solar cell.

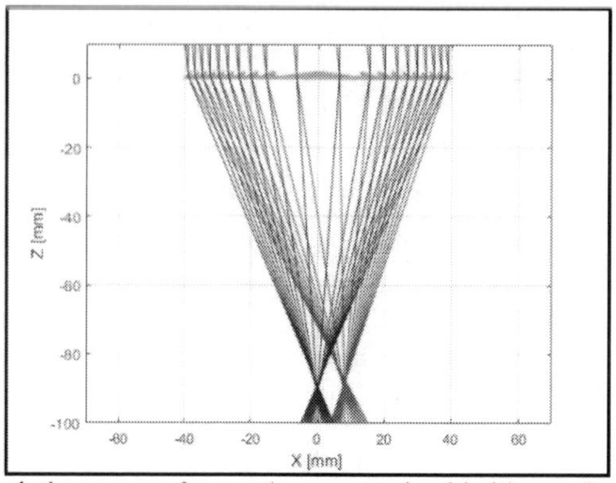

Fig. 5. MATLAB simulation: normal rays (represented with blue color) versus "useful" rays (represented with red color)

To have a minimum dimension of the solar cell, it was considered in the simulation only the normal rays. To achieve this scope, we are currently researching and building a complex mirror system around the dome and will be presented in another paper.

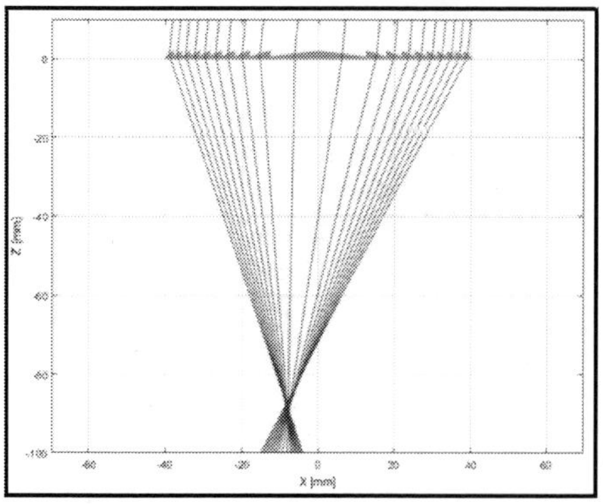

a)

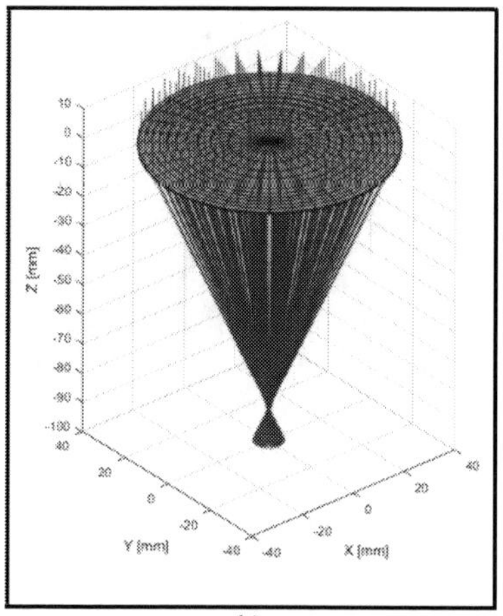

b)

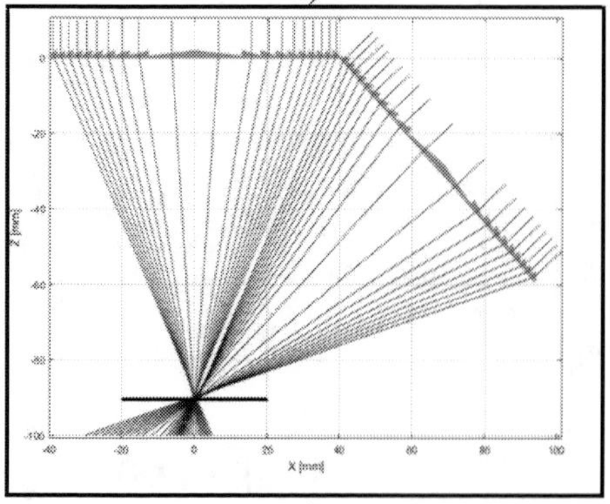

c)

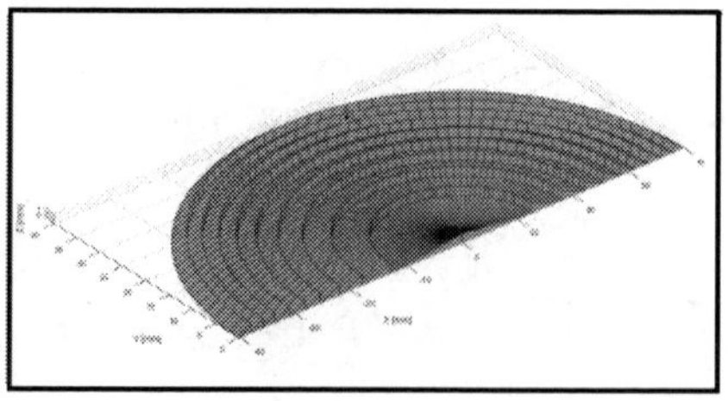

d)

Fig. 6. MATLAB simulation: a) the Fresnel lens ray propagation simulation 2D; b) The Fresnel lens ray propagation simulation 3D; c) The two Fresnel lenses ray propagation simulation 2D; d) The Fresnel lens 3D model

The simulation has the following steps:

- design the Fresnel lens with a proper geometry, as can be seen in fig. 6a;
- simulate a ray propagation simulation, as can be seen in fig. 6b;
- simulate the ray propagation for the dome using a hexagonal arrangement according to [7], as can be seen in fig. 6c.

5 Conclusions

In the paper is presented a new solution which proves that there are:

- long-term financial advantages, because is an alternative to creating electricity and replacing fossil fuels in order to help the environment by using photovoltaic panels;
- by using Fresnel lens the structure has:
  - a reduce the cost;
  - lighter weight due to the material;
  - more compact structure;
  - flexible structure;
  - easy to manufacture and to assembly
  - by using polymethylmethacrylate instead of glass for a lower refractive index, cost reduction and a half weight;
- the structure is adjustable due to the supports which has the possibility of adjusting them vertically to change the focal length.

The way of obtaining normal beams on the surfaces of the Fresnel lenses that make up the dome is a continuation of the research theme. Also, the relative orientation of the Fresnel lenses that make up the dome, the characteristics of the lenses, the geometry of the profile of the Fresnel surfaces in order to increase the number of useful rays for different orientations of the Sun are further research directions.

References

1. M. J. O'Neill, *Stretched Fresnel Lens Solar Concentrator for Space Power* (US Patent, US6075200, 2000)
2. L. L. C. Chen, *Stationary Photovoltaic Array Module Design for Solar Electric Power Generation Systems* (US Patent, US6653551 2003)
3. V. Shaoulov, J. Rolland, *Compact relay lenses using microlenslet arrays*, in International Optical Design Conference 2002, Proc. SPIE **4832** (2002)
4. N. Lindlein, *Simulation of micro-optical systems including microlens arrays*, Journal of Optics A: Pure and Applied Optics **4** (2002)
5. J. P. Rolland, V. Shaoulov, F. J. Gonzalez, *The art of back-of-the-envelope paraxial raytracing*, IEEE Trans. Educ. **44** (2001)
6. K. Mersereau, C. R. Nijander, W. P. Townsend, R. J. Crisci, A. Y. Feldblum , D. Daly *Design, fabrication, and testing of refractive microlens arrays*, IOP Short Meeting, **30** (1991)
7. A. E. Bruno, S. Barnard, M. Ehrat, R. Vőlkel, P. Nussbaum, H. P. Herzig, *Optische Detektionsvorrichtung für chemische Analysen an kleinvolumigen Proben* (European patent application no 96810341.6. 1996)

Transforming Romanian school buildings in prosumers. An opportunity for increasing the energy efficiency in electrical networks and reducing the environmental impact of the power distribution sector

Mircea Scripcariu[1*], *Cristian* Gheorghiu[2], *Ioan-Sevastian* Bitir-Istrate[3] and *Mihaela-Madalina* Bonea[4]

[1]Associate Prof., Dept. of Energy Production and Use, University POLITEHNICA of Bucharest, Romania
[2]PhD. Student, Dept. of Energy Production and Use, University POLITEHNICA of Bucharest, Romania
[3]Lecturer, Dept. of Energy Production and Use, University POLITEHNICA of Bucharest, Romania
[4]Master Degree Student, Dept. of Energy Production and Use, University POLITEHNICA of Bucharest, Romania

Abstract. In the last decades, electricity systems suffer crucial transformations from integrated networks with a unidirectional flow, to decentralized, bidirectional flow networks. The stakeholders of this large-scale business have changed or improved their role. Electricity distribution companies are becoming service providers and end-users are becoming electricity producers. With the legislation issued in 2018, a new stakeholder was born on the Romanian electricity market, the prosumer. So, the decentralization reached an extreme by giving the smallest end-user the opportunity to become an electricity producer. In the first paragraph of this paper, authors summarize provisions of most recent energy strategies issued by the Romanian Government and the regulatory framework for incentivizing this market player. Given the growing interest of local administration in energy efficiency and pollution reduction, authors consider that schools can play a major role in fulfilling ambitious environmental targets. Most school buildings date from the communist regime and have a rather similar design. In their energy management programmes, municipalities consider energy efficiency in school buildings and authors present the case study of a school in the city of Ploiesti, one of the largest in southern Romania. A brief description of the electricity end-use systems is presented. Before any measure is taken on the supply-side, school's administration needs to implement the demand-side measures. Next, these measures are investigated, and the results of the technical-economic analysis are presented. In southern Romania, photovoltaic (PV) solutions have better operating conditions and for this case study the solution for local generation with PV panels is presented. In the paper, authors are making an in-depth investigation of extending the prosumers

*Corresponding author : mirceas1960@yahoo.com

model, with local PV generation on school buildings, over most of Romanian geographical regions. The carbon footprint is calculated, and the environmental benefits of this integrated approach are highlighted.

1 Strategical options and Regulatory Framework for Prosumers

The transition towards clean energy is essential in the fight against climate change. Economic performance and global geopolitical relationship are modelled by the energy sector, thus making the monitoring and understanding the past and future determining factors of electricity prices and costs an extremely important action. Each effort to improve the environmental impact counts. Modern electricity systems are switching from a unidirectional energy flow to a bi-directional one and from electricity supply to complex services. In this framework, modern renewable electricity generation technologies have continuously improved their economical performances allowing the smallest consumer to become a "prosumer", i.e. an active player in the electricity market. This opportunity will stimulate end-users to actively participate to the reduction of world's carbon footprint.

As per the Romanian legislation [1], the prosumer is the end-user which simultaneously uses, produces, stores and can sell electricity produced from Renewable Energy Sources (RES). As [2] points out, the prosumer can optimize the power demand and respectively the power injection in the local grid, thus flattening the global power demand curve and reducing the network balancing costs.

The main objectives of [1] are to develop the RES integration capacity and mechanisms into the National Power Grid, to encourage the transition from *consumer* to *prosumer* for residential, industrial, agricultural and public entities, on par with the development of smart metering systems, thus leading to the development of a nation-wide Smart Grid. The maximum installed capacity in prosumers is set to reach 750 MW by 2030 in the European Union alone.

The European Union, by it's RES support Directive (CE 2016b), states that the prosumer transition should be encouraged by a *feed-in-tariff* scheme, offering financing and grants, and by modifying the fiscal legislation in order to ensure the bi-directional transactions economic compensations between the prosumers and the Distribution System Operators (DSO).

DSO's must also continue the refurbishment process of their grids, while considering the target to develop the Smart Grid concept and ensuring the real-time interactions with the existing and future prosumers.

In the European Union, the prosumers are divided into [3]: *residential prosumers* – which produce electricity by photovoltaic systems – residential associations, foundations, or other civil entities with no lucrative purpose which only produce electricity to cover their own needs, *commercial prosumers* – small and medium enterprises (SME), universal stores, office buildings and other similar sites which are defined by a different main activity (NACE Code) than electricity production and *public prosumers* – schools, hospitals and other public institutions which can generate electricity.

In this paper, the public prosumer sector will be in-depth analysed, as densely populated urban areas have a significant energy economy potential and represent an optimal platform for environmental impact reduction, as Rotterdam has already proven [4] by implementing the urban symbiosis concept (see Figure 1). This notion has also been implemented, by the EU Celsius Cities Project to test the smart energy planning in Köln, Genova, Göteborg and London.

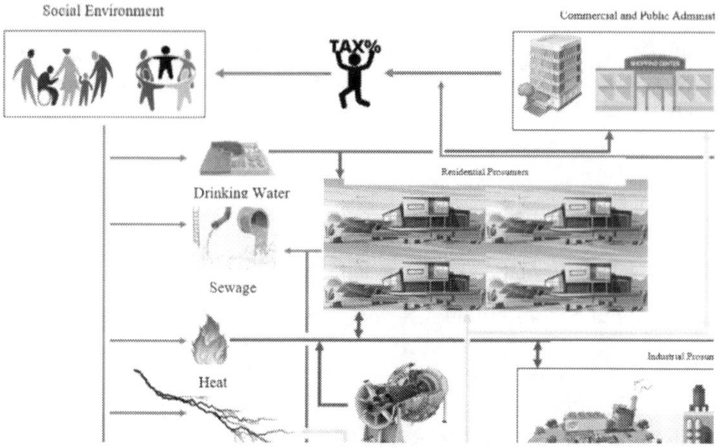

Fig. 1. Urban symbiosis concept where taxes (green), drinking water (blue), heat (red), power (yellow) and services / production (grey) sectors function as a whole, reusing and thus increasing the efficiency of energy use

The Romanian legislation [1] states that a prosumer which has an installed capacity of 27 kW or under is entitled to sell the electricity produced and injected into the power grid to the energy suppliers with which they have an ongoing energy supply contract. The legislation also states that Public Authorities which have an installed capacity of 3 MW or under, financed partially or fully from structural funds and which don't have an "income generator" status can also sell the produced electricity in the power grid by means of bilateral contracting. However, at this moment, the Romanian legislation does not provide the possibility for the prosumer to sell energy on the market, at a competitive price as it only allows the compensation of the energy injected in the network at a yearly regulated price, set as the average daily closing price of the market for the previous year.

In order to evaluate the potential of public prosumers, this paper will analyse the Romanian school building sub-sector. In Section 2, the project development potential represented by Romania's school-building sub-sector will be investigated, with a case-study of a relevant school. The technic and economic impact of the development of such public prosumers, starting from the case study presented and extrapolating for all the similar sites in Romania will be also presented. In Section 3, a brief review for possible funding sources will pe presented, while in Section 4 the main conclusions and the future work needed will be presented.

2 Technic and economic analysis for a national programme

The paper based its analysis on a case study represented by a school located in one of the largest cities in Romania, Ploiesti, in the southern part of the country. "Grigore Moisil" primary school, was part of a project in which several Romanian primary schools were selected for building gym halls – new buildings, out of the main building. Throughout Romania, 726 primary schools out of the total of 3,039 existing primary schools took part in the Gym Hall building project which took place between 2008 – 2012 and was coordinated by the Ministry of Development and was funded by the Ministry of Development and by the National Investment Company.

The first stage of the analysis consisted in an energy performance benchmarking for the energy boundary of the Grigore Moisil Primary school. As of March 2020, no school in Romania is engaged in pro-active energy efficiency by the means of drafting up a yearly energy-efficiency programme. In essence, a preliminary energy audit was conducted, by means of in-site visits, in order to determine the Significant Energy Uses (SEU) and gather the required necessary data. The SEUs in the yearly electricity consumption are presented in Table 1.

Table 1. SEUs in the case study Primary School

Subsystem	Equipment	Quantity [pcs.]	Rated Power [kW]	Installed Power [kW]	Sub-System Weight in the total Energy Consumption [%]
HVAC	Air Conditioning Units	30	2.5	75	35
Lighting	CFL 4 x 18 W fixtures	640	0.072	46.08	23
	T8 CFL tubes	120	0.036	4.32	
IT&C	PC	82	0.12	9.84	19
	Laptop	12	0.06	0.72	
	Monitors + TVs	90	0.04	3.6	
	Printers	12	1.5	18	
	Modem, Routers, Switches	-	0.519	0.519	
	Paper Shredder	4	2	8	
Appliances	Dishwashing machine	1	1.8	1.8	23
	Refrigerators	2	0.186	0.372	
	Microwave oven	4	3.5	14	
	Coffee filter	2	2.1	4.2	
	Water Dispenser	10	2.8	28	
Total				**214.45**	**100**

Before considering the transition from a consumer to a prosumer, all the available and economically viable Energy Performance Improvement Actions (EPIA) must be implemented, in order to minimize and flatten the power curve of the energy boundary, thus minimizing the environmental impact of the boundary – Environmental Impact Reduction (EIR). Several EPIAs were analysed from a technic and economic point of view and the results are presented in Table 2. A conversion factor of 0.314 tonnes of CO_2 equivalent per MWh was used.

Table 2. EPIA technic-economic analysis results

EPIA Type*	EPIA	IC** [EUR]	Annual Savings			EIR
			[MWh/year]	[toe/year]	[EUR/year]	[tCO_{2eq}/year]
O	Improving the energy efficiency conduit of personnel, energy consumption accounting etc.	0	2.05	0.176	1,024	0.64
LIC	Installing temporization systems for the IT&C and Appliances sub-systems	307	13.45	1.16	1,156	4.02
HIC	Replacing the 4x18W CFL *** fixtures with 30W LED*** panels	2,357	5.02	0.43	431	1.5
HIC	Replacing the common spaces CFL tubes with 8W LED tubes	196	1.89	0.163	163	0.566

* O = Organizational EPIAs, LIC = Low Investment Costs EPIAs, HIC = High Investment Costs EPIAs, ** IC = Investment Costs *** CFL = Compact Fluorescent Lights / LED = Light Emitting Diode

The analysis began by determining the monthly solar potential for all the regions in which the 726 standard gym halls are located, with a minimum result for Cluj-Napoca –

1.33 kWh/m^2/day in December and maximum value of Constanta – 6.57 kWh/m^2/day in July.

In order to determine the forecasted energy production for each school (1) was used:

$$W_{produced}^{monthly} = A \cdot \eta_n \cdot I_{avg} \cdot n_{days} \cdot Cp \left[\frac{kWh}{month}\right] \quad (1)$$

where A [m^2] is the total surface of the PV system, η_n [%] is the rated PV panel efficiency, I_{avg} [kWh/m^2/day] is the average daily irradiance, determined by using RetScreen software, n_{days} [-] is the number of days in the month and Cp [-] is the performance coefficient of the PV system. The performance coefficient of the PV system includes all types of power and energy losses of the system, out of which the most important are: Inverter losses – between 4% and 10%, temperature generate PV cells losses – between 5% and 20%, DC lines losses – between 1% and 3%, AC line losses – between 1% and 3%, Shading losses – between 0% and 80%, Low-load losses in inverters – between 3% and 7%, dust / snow / atmospheric pollution losses in PV cells – approximatively 2%.

The forecasted production, respectively the power demand curve for Ploiesti's Primary School – the case study – is presented in Figure 2.

As it can be observed, due to the permanent use of the gym (Physical Education classes during the school year and Basketball / Martial Arts / Volley and Football classes throughout the year), the electricity consumption is higher than expected during the summer break (July-September).

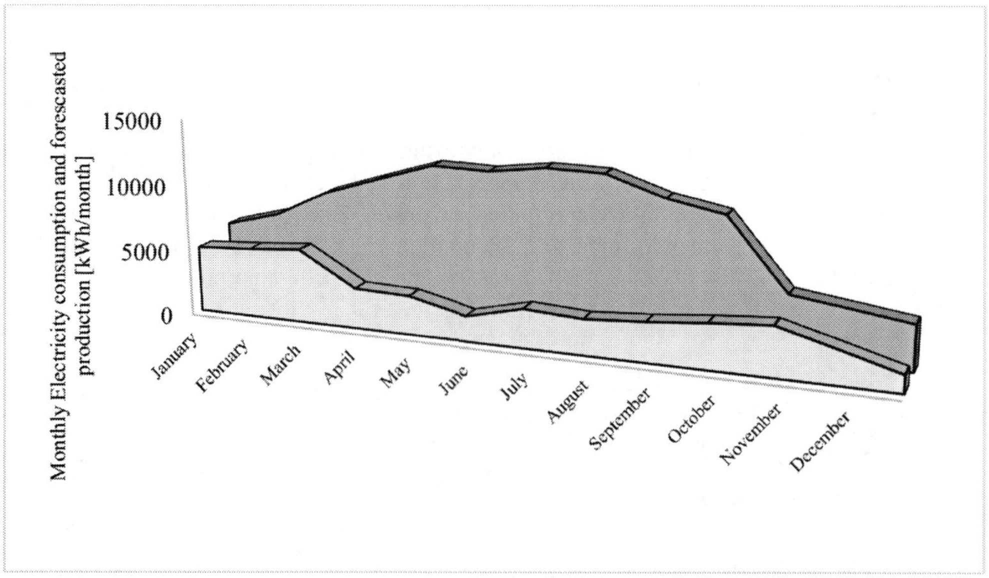

Fig. 2. Monthly electricity consumption (grey) and electricity production forecast (green) for Grigore Moisil Primary School in Ploiesti

In table 3 the forecasted energy production of Grigore Moisil's project and the extrapolated results for all the 726 schools are presented, as determined by using the RETScreen Expert software (https://www.nrcan.gc.ca/maps-tools-publications/tools/data-analysis-software-modelling/retscreen/7465).

Table 3. The monthly energy production forecast for Girgore Moisil and for all the 726 schools

Month	Ploieşti Forecasted Production [MWh]	Global Forecasted Production [MWh]
January	5.79	3.552
February	6.923	4.450
March	8.977	6.083
April	10.401	6.669
May	11.768	7.876
June	11.63	7.850
July	12.226	8.313
August	12.164	8.091
September	10.723	6.513
October	9.914	5.338
November	4.636	3.360
December	3.458	2.756
TOTAL	**105.152**	**68,103**

The main criterions used in the technic and economic analysis of the project were the Net Present Value – NPV (2), the Internal Rate of Return – IRR (3), the Simple Payback Period (4), determined by considering a variable annual net income and the Benefit – Cost Analysis – BCA (5). Levelized Cost of Electricity - LCOE (6) [5] was also determined, in order to compare the project with other relevant existing projects.

$$NPV = \sum_{t=1}^{tst} \frac{I_t - C_t}{(1+a)^t} - IC\ [EUR] \tag{2}$$

where t_{tst} is the analysis time-frame, in years, I_t is the yearly income in the tth year, in EUR/year, C_t are the yearly expenditures in the tth year, in EUR/year, a is the discount rate, in %/year and IC is the investment cost, in EUR.

$$NPV = \sum_{t=1}^{tst} \frac{I_t - C_t}{(1+IRR)^t} = 0\ [EUR] \tag{3}$$

where the Investment Cost can be included in the yearly expenditures as a depreciation cost.

$$SPP = \frac{IC}{\frac{\sum_{i=1}^{t} I_i - C_i}{t}} [years] \tag{4}$$

$$BCA = \frac{IC}{NPV} [-] \tag{5}$$

$$LCOE = \sum_{t=1}^{n} \frac{\frac{I_t + C_{O\&M,t} + F_{C,t}}{(1+dr)^t}}{\frac{E_t}{(1+dr)^t}} [EUR/MWh] \tag{6}$$

where the $C_{O\&M}$ [EUR/year] is the annual Operation and Maintenance Cost, F_C [EUR/year] are the annual Fuel Costs, E_t is the annual generated power [MWh/year], dr [%] is the degradation rate of the PV panels and n [years] is the average lifecycle of the project – 25 years in this case [6], based on the standard PV panel Degradation graph.

The investment costs are divided into three categories:

- PV Panel investment cost – starting from the average useful surface of the gym hall roof – 800 m^2, and considering a PV panel with a surface of 1.97 m^2, which includes the mounting distances, 203 PV panels (with an installed power of 97.44 kW) are considered, with an associated investment cost of approximatively 35,000 EUR;
- Inverter investment cost – based on the PV panel installed power and the rated efficiency of the inverters (approximatively 94%), in order to maximize the functioning times, fifteen 10 kW inverters were considered for installation, with an average price of 450 EUR/pc. – a total investment cost of approximatively 6,700 EUR;
- PV System Installation costs – considering the required materials (metallic profiles, electric cables, auxiliary components, distribution panels, a smart meter, insulation protection etc.) these costs range from between 5% to 10% of the total PV System cost – for this analysis a total installation cost of 4,000 EUR was considered.

The yearly Operation and Maintenance Costs were considered to be 500 EUR/year. The discount rate was considered $r = 5.6$ [%/year] [7]. The average electricity acquisition price was considered $p_{Wb} = 104$ EUR/MWh, for Ploiesti's Grigore Moisil Primary School. The average electricity selling price, as per [8], is $p_{Ws} = 47$ EUR/MWh.

The analysis was conducted for a period of 25 years, as per the technical specifications of the PV cells, respectively of the power inverter units. A degradation factor of 0.6%/year was considered (with a minimum efficiency of 85% of the rated efficiency guaranteed throughout the 25 years analysis period).

The results obtained for Grigore Moisil Primary School and for all the other 725 primary schools (summed up as a global result, for the entire project) are presented in Table 4.

Table 4. The technic – economic analysis results

Indicator	Ploiești	Nation-wide programme
Net Present Value – NPV [EUR]	44,665	27,640,207
Internal Rate of Return – IRR [%/year]	14.88	13.61
Simple Payback Period – SPP [years]	6.04	6.49
Benefit-Cost Analysis – BCA [-]	1	0.85
Levelized Cost of Electricity – LCOE [EUR/MWh]	98.12	119

For a nation-wide programme the IRR, SPP and BCA indicators have worse values than for the Grigore Moisil Primary School project. These results are justified by the different solar potential of the various regions in which the analysed primary schools are located – as it can also be seen in Figure 3.

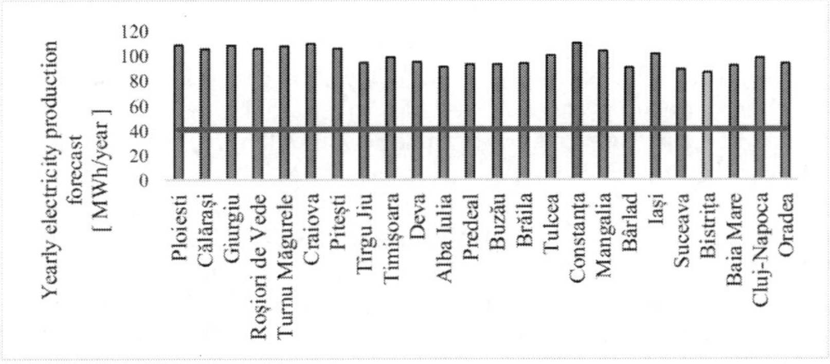

Fig. 3. Yearly electricity production forecast for the geographic areas in which the schools are situated with the average yearly electricity consumption for a standard primary school (red) and the most productive (green) respectively less productive (orange) areas.

3 Project's Environmental Impact Reduction

Based on the results presented in Section 2, the Environmental Impact (EI) of the case-study Primary School integration in a nation-wide programme of transition for Primary Schools from consumers to prosumers – Primary School Photovoltaic Initiative – PSPI, in four scenarios: do nothing / EPIA only / PV only / PSPI is presented in Table 5.

Table 5. EI analysis results

Scenario	-	EI* Case Study	EI* PSPI	M.U.
Do-Nothing	(1)	12.84	9,325	[tonnes CO_{2eq}/year]
EPIA only	(2)	7.32	5,315	[tonnes CO_{2eq}/year]
PV only	(3)	-21.10	-12,060	[tonnes CO_{2eq}/year]
PSPI	(4)	-28.59	-17,375	[tonnes CO_{2eq}/year]
Total Reduction	**(1)-(4)**	**41.43**	**26,700**	**[tonnes CO_{2eq}/year]**

*where the negative values represent the decrease of the Environmental Impact of the DSO, as a result of injecting the additional RES generated power into the power grid

PSPI's impact on the efficiency of the DSO's Power Grid was also taken into account. Based on the annual reports of the DSOs, the average Own Technological Consumption (OTC), which includes the Power and Energy losses and Commercial Losses is situated at 14% – 15% of the transited low-voltage electricity [9]. The same reports state that the Power and Energy losses are of about 4%, on average, which leads to an additional energy saving of up to 2,700 MWh/year, respectively 280,800 EUR/year and 856 tonnes of CO_2 equivalent per year. As it can be observed, the environmental impact of the project is significant. The PSPI reduces the national EI by 41.43 tonnes of CO_2 equivalent per year for the case study Primary School (from the initial EI of 12.84 tonnes of CO2 equivalent per year to a negative EI of 28.59 tonnes of CO2 equivalent per year). At a global level, the PSPI would lead to a reduction of the environmental impact of up to 30,000 tonnes of CO_2 equivalent per year – approximatively 637,500 tonnes of CO_2 equivalent over the analysis period (25 years).

4 Project development mechanisms

The PSPI development should follow a clear and specific plan, in order to maximize it's technic, economic and environmental impact. Firstly, the Primary Schools involved should, with the help of Energy Services Companies – ESCOs, determine their baselines, via an Energy Management contract. Secondly, each Organisation should engage in a Complex Energy Auditing contract, in order to determine the optimal EPIA plan and evaluate the real technic and economic viability of installing a PV plant. Finally, the PSPI should finance the Primary Schools which satisfy a set of technic and economic viability targets.

The financing can be ensured by at least three accessible means. The first and easiest way of financing the PSPI is the Energy Performance Contracting. between ESCOs and the targeted Primary Schools. The PV project can be approached with a pay-back scheme similar to BOOT (Build-Own-Operate and Transfer) developments, in which ESCOs would get a return on investment (with a reasonable, project-wide, regulated profit margin) for the contract period, then the Primary School would get the ownership of the PV installation for free – a method which is fully accepted by the Romanian National Agency for Financial Administration (NAFA) [10]. If the energy efficiency improvement actions are approached in a different EPC – for example, a shared-energy savings contract, the Romanian Legislation should be modified so that the Public Administration can end the fiscal year with a variable result (different from the initial budgeted funds) – as the energy economy may be variable – thus the required energy related spending may also vary. At present, a Public Administration Entity must close the fiscal year with an economic indicator as close as possible to zero – any deviation (a decrease in energy consumption – thus a decrease in energy bills) leading to serious alterations of the next year available capital offered by the State Budget. The second option for financing the PSPI is by setting up a cooperation between the Ministry of Education and Research and the Ministry of Environment, Waters and Forests via it's Environment Fund Administration (EFA). At present, EFA finances with up to 20,000 RON/household (approximatively 4,200 EUR/household) up to 9,000 residential prosumer projects [11].The third option for financing the programme is by contracting a lease / credit from the EBRD, which has numerous PV installation projects financed and leased. A similar financing scheme is also available in Romania, the funding being covered by the Romanian Energy Efficiency Fund (REEF https://free.org.ro/) – however, this option implies that the leased Organisation must co-finance the project with up to 20% of the total value – this might be less attractive for the Public Administration Entities. At present, in Romania, efforts are directed towards drawing an Energy Performance Contract. For over a year, the organization that represents some of the most representative ESCOs, ESCOROM (http://escorom.ro/eng/), in partnership with ANRE (Romanian National Energy Regulatory Authority) are struggling to issue a draft of an EPC. Such a model can give participants in project development confidence that the interest of all parties is protected.

5 Conclusions and further research

This paper presents the main technical and economic aspects that can describe a national programme to transform public consumers into prosumers. With the new legislation issued by the Romanian Parliament this is legally possible, and authors have investigated this opportunity and are presenting their results in this paper.

Starting from the case study, a concept that consists of implementing energy efficiency measures followed by the installation of PV panels was presented. Costs and benefits are more easily estimated, being given the fact that excess electricity is better managed over a longer period. The financial indicators for the case study are positive, with an IRR of almost 15%/year and a payback period of around 6 years. The analysis of expanding such a project at a larger scale through a national programme, show financial indicators with smaller values but still attractive. This is given by the fact that the case study building is in southern Romania, with one of the best solar irradiances. LCOE is similar to the ASEAN [13] and other EU countries' [14] values of 130 EUR/MWh, respectively 107 EUR/MWh. The proposed PSPI could lead to an environmental impact reduction of approximatively 30,000 tonnes of CO_2 equivalent per year, from the implementation in all the 726 Primary Schools.

While the value might seem irrelevant when compared to Romania's annual CO_2 equivalent emissions, which are of approximatively 78.800.000 tonnes of CO_2 equivalent per year, one must consider the fact that, in a Public / Local Administration, Primary Schools are only one of the multiple types of buildings that would be suitable for the transition from simple consumers to environmental friendly prosumers. For example, in Bucharest alone there are 419 state owned education-sector buildings (124 Kindergartens, 171 Primary Schools, 103 Secondary Schools, 2 Professional Schools, 1 Post-Secondary School, 7 children clubs, 7 sports clubs, 5 Art and Trade Schools), 86 state owned healthcare-sector buildings, 6 city halls, 6 NAFA buildings, 7 Courthouses, and around 200 other buildings for various other national or local institutions, with an additional, significantly higher, available sites in University Campuses. An important factor in the success of such a programme is the energy efficiency education of the Public Administration clerks. Ensuring an energy efficiency education for the personnel, corelated with properly applying the law [15],[16] for Public Authorities, will lead to a better understanding of the necessity and benefits of improving the EPIs and EIRs of each energy boundary owned and operated by the Romanian State.

As further research, other projects transitioning consumers to prosumers for the Public Administration should be assessed, such as waste to energy solutions, as photovoltaic systems are not applicable and viable from a technic and economic point of view to all state owned buildings, due to their geographic position, shading by other buildings, small available roof surface or because they are heritage buildings.

References

1. The Romanian Parliament, "Law no. 184/2018 for approving the Government Emergency Ordinance no. 24/2017 regarding the modification and updating of Law no. 220/2008 for determining the incentive system for producing energy from renewable energy sources and the modification of other normative acts," Official Gazette, Part I, No. 635/20.07.2018.
2. Ministry of Energy, Economy and Business Environment, "2019 – 2030 Romania's Energy Strategy with the perspective of 2050," 2019.
3. Nikolina Sajn, "Briefing – Electricity "Prosumers"," European Union's Research Service, 2016.
4. Lenhart, J., van Vliet, B. & Mol, A., "New roles for local authorities in a time of climate change: the Rotterdam Energy Approach and Planning as a case of urban symbiosis. Journal of Cleaner Production," vol. 107, pp.593- 601, 2015.
5. Aishwarya S. Mundada. Kunal K. Shah. Joshua. M. Pearce. "Levelized Cost of Electricity for Solar Photovoltaic. Battery and Cogen Hybrid Systems." Renewable and Sustainable Energy Reviews 57. 2016.
6. The Romanian Government, "Decision no. 2139/30.10.2004 for approving the normal period of use for capital goods," Oficial Gazette, Part I, No. 46/13.01.2005.
7. Public Procurement National Agency, "Order no. 2717/318/2018 for the revision of the discount rate," 2019.

8. Romanian National Energy Regulatory Authority, "Order no. 227/2018 regarding the approval of the framework contract for buying and selling electricity produced by the prosumers which own a renewable sources power generation installation of up to 27 kW per site and to modify previous regulations of the electricity sector," Official Gazette, Part 1, No. 1114/28.12.2018.
9. https://www.anre.ro/ro/energie-electrica/legislatie/documente-de-discutie-ee/metodologii-proceduri-tarifare/ [accessed in January 2020]
10. The Romanian Parliament, "Law no. 50/2019 regarding the State Budget of Romania," Official Gazette, Part I, No. 209/15.03.2020.
11. Environment Fund Administration, "Photovoltaic Green House Programme," 2019.
12. The Emission Database for Global Atmospheric Research (EDGAR), https://edgar.jrc.ec.europa.eu/ [accessed in January 2020].
13. European Technology and Innovation Platform, "PhotoVoltaic," PV Manufacturing in Europe Conference, 2017.
14. Md Abdullah AL MATIN, Shutaro TAKEDA, Yugo TANAKA, Shigeki SAKURAI and Tetsuo TEZUKA, "LCOE Analysis for Grid-Connected PV Systems of Utility Scale Across Selected ASEAN Countries," ERIA Discussion Paper Series no. 305, ERIA-DP-2019-19, 2019.
15. The Romanian Parliament, "The Energy Efficiency Law no. 121/2014," Official Gazette, Part I, No. 574 from 01.08.2014.
16. The Romanian Parliament, "Law no. 160/2016 for modifying and updating the Energy Efficiency Law no. 121/2014," Official Gazette, Part I, No. 562 from 26.07.2016.

E3S Web of Conferences **180**, 02010 (2020) https://doi.org/10.1051/e3sconf/202018002010
TE-RE-RD 2020

Impact of pollutants in the energy sector on the environment and technologies for treating liquid and solid waste from power plants

Irina Alina Chera-Anghel[1*], *Loredana* Popescu[2], *Florentina* Condrea[3]

[1] Ph.D. student, Research Assistant, Energy Research and Modernization Institute – ICEMENERG, Bucharest, Romania
[2] Senior researcher, Energy Research and Modernization Institute – ICEMENERG, Bucharest, Romania
[3] Senior researcher, Energy Research and Modernization Institute – ICEMENERG, Bucharest, Romania

Abstract. The main pollutants in the soil are composed of ash, clay, and calcium sulfur, which results from the desulfurization of flue gases. However, the infiltration of liquid fuels and lubricants, as well as transformation oil, may be neglected. Liquid residues result from several technological processes. The main source of pollution is the water treatment process that must be used in the water-steam circuit. Another important source of liquid residues is the cooling process of the thermal agent. A third major source of water pollution is the flue gas desulphurization process and the transport of slag and ash. Thus, it should not be neglected that power plants are large industrial installations, which require permanent maintenance, liquid residues resulting from cleaning operations of various equipment. To have the technology for treatment, disposal and storage of high-efficiency waste, we must take into account the following factors: the volume and type of waste, their properties, the likely evolution of the installation, the volume of waste generated by it, treatment possibilities and even limits imposed by environmental regulations.

1 Introduction

The energy sector, worldwide, through the complexity of the installations and technological processes, the characteristics of the raw materials, which are entering the process, and of the resulting waste, is one of the most important industrial processes, whose impact on the environment is major. Electricity production is a technological process efficient, fully conditions are considered the impacts on the environment. The multitude of chemical reactions, which take place in all the technological processes in the power plant, changes the characteristics of the materials, entered into the contour, thus generating residues, which must be stored, in-situ, or released into the environment, polluting, to some extent, all environmental factors.

* Corresponding author: irina.chera@icemenerg.ro; irina.alina.chera@gmail.com.

The analysis of the pollution generated by power plants should take into account the following sources of pollution: solid particulate air pollution and toxic gases (SO_2, NO_2,NO, CO, CO_2, NH_3, HCl, etc.) and evaporation, the pollution discharge areas wastewater, thermal emission pollution, soil pollution, pollution caused by hydro-technical circuits, which can become sources of air pollution, by evaporation, at the cooling towers pollution caused by slag and ash deposits, which contaminate the soil and groundwater, these are the most important sources of pollution that we will present in this article.

At the level of thermoelectric power stations, several sources of pollution can be identified, such as transport installations; preparation and storage of fuels; combustion plants; water supply installations; technological water treatment plants; wastewater treatment and neutralization systems and associated sewers; slag deposits, ash, and by-products of neutralization reactions.

The entries in the contour of the chemical balance, similar to a mass balance, are the fuel; limestone, necessary for SO_2 emission control; water, necessary for the production of steam, cooling of the thermal circuit and other inherent uses; ammonia and catalysts, necessary to control N emission; different chemicals used in water treatment, softening and neutralization processes, as well as in maintenance operations. electricity supply, it is found that this process is not efficient, to a large extent.

The chemical substances, frequently used in the technological flow of some classic power plants, are coagulants, for water pretreatment; sulfuric acid; HCl; caustic soda; ammonia; sodium phosphates, and corrosion inhibitors. In addition to these substances, various acids, bases, and organic substances are also used to perform periodic maintenance operations, including washing of heat exchangers, on the water circulation side, of air preheaters, assuming the risk of sulfate deposition. of ammonium is high, of the electrodes of deposition of electro filters, etc.

The useful results from the outline of the global balance of a thermal power plant are, of course, electricity and, finally, thermal energy, to which can be added certain quantities of ammonium salts, usable as chemical fertilizers, gypsum and various agglomerates, usable in building materials in the industry, as well as certain amounts of sulfur and sulfuric acid, if the recovery is more complex. However, there is a large amount of waste, which is a source of pollution, for all environmental factors, whether these substances are discharged into the atmosphere or emissions, or stored in situ.

So to remove these sources of pollution, this article presents some treatments that are applied to remove this waste from the energy sector, to protect the environment.

2 Sources of pollutants from power plants

From ecologically, there are important differences between the various categories of pollutants. Thus, they differ:

- pollutants by direct action on the human body, such as nitrogen oxides, sulfur oxides, carbon monoxide, and some heavy metals;
- pollutants with direct action on vegetation, such as sulfur dioxide and combinations of chlorine with hydrogen;
- pollutants underlying the formation of acids, such as SO_2, SO_3, NO, and NO_2;
- pollutants that become factors of influence of the climate, such as CO_2, NO_2, and N_2O, as well as important factors in accentuating the greenhouse effect or contributing to the destruction of stratospheric ozone.

The main process in the plant, that of the combustion of the fuel, is a major source of air pollution. Besides the pollution of the atmosphere, the combustion of the fuel, as well as all the processes that contribute to the development of the main process, are important sources of liquid and solid waste, which can pollute, severely, the water and the soil. Quantitatively, the most important solid wastes from a power plant are slag and ash, resulting from the combustion of fuel and calcium salts, resulting from the process of reducing sulfur oxide emissions [1].

Even though slag and ash storage requires considerable economic effort, soil pollution generated by fuel preparation facilities, fuel depots, fuels, lubricants, various chemicals should not be neglected. Also, the residues resulting from the electrical installations of the plants should not be neglected. The soil can be polluted by the leakage of transformer oil, the storage of ceramic materials, and some more or less oxidized metal waste.

Another important source of liquid waste is the process of treating the water used in the thermal circuit. Thus, the wastewater, resulting from this process, contains traces of the reagents used, as well as acid and basic demineralize. Another important source of polluted water is the cooling circuit, as well as the leakage from the main circuit. A third important source of water pollution is represented by the installations for the treatment of gaseous emissions, in particular for the treatment of the emission of sulfur oxides, as well as the installations for the transport of slag and ash.

Water vapors, as well as water droplets, driven by the flow of vapors, are a source of local, pollution of the atmosphere. Continuous evaporation of a flow of water causes the concentration of salts in the coolant to increase. When the solution reaches saturation, the salts can settle on the condenser tubing and other cooling circuit equipment. Sedimentation can be avoided by introducing chemical inhibitors or by periodically purging an amount of cooling water. The discharge of purged water, with a high concentration of salts, represents a second pollutant residue, resulting from the cooling circuit. Regardless of whether the cooling is performed in an open or closed circuit, the temperature of the cooling water favors the development of microorganisms, which can clog the condenser pipes.

To avoid the exaggerated development of these microorganisms, the cooling water is chlorinated, a certain amount of chlorine can be found in the water directed to the emitter or in the purge. Due to the high toxicity of chlorine, at least in the case of open-circuit power plants, it is necessary to de-chlorinate the cooling water. Relatively large amounts of contaminated wastewater result from the ash and slag transport facilities, obviously in the case of solid fuel combustion. The amount of ash and slag is strictly dependent, at the same amount of energy provided, on the concentration of the coal tailings. It can be appreciated, however, that about 20%, of the total of the solid substance, is retained, in the form of slag, at the base of the outbreak, 4%, in the form of flying ashes, at the exit of the gases from the economizer, the rest of 76% being retained by electro filter or filter bag.

The retained ash from the flue gas can be transported hydraulically, in the form of sludge, or pneumatically, in the case of new plants, the latter option is preferred. The slag, retained at the base of the outbreak, is cooled by a continuous flow of water. Periodically, the slurry, thus formed, is transported to the final depot or an intermediate storage area.

Under these conditions, the resulting water is heavily polluted with suspended solids, dissolved salts, and heavy metals, the pH of the residual water can vary within very wide limits. Some of the wastewater used in the transport of slag and ash are listed in Table 1.

Table 1. The level of pollutants in the water used to transport the slag and ash

No. crt.	Substance	Water transport slag from the bottom of the furnace		Water transport fly ash	
		minim	Maxim	minim	Maxim
1	Water flow	18 m^3/min	88 m^3/min	11.5 m^3/min	33.2 m^3/min
2	pH	4.1	7.9	3.6	6.3
3	Dissolved solid	70 mg/l	405 mg/l	142 mg/l	821 mg/l
4	Solid substance in suspension	5.2 mg/l	658 mg/l	2.2 mg/l	258 mg/l
5	Aluminum	0.5 mg/l	8.0 mg/l	3.2 mg/l	8.8 mg/l
6	Ammonia (in nitrogen equivalent)	0.02 mg/l	0.32 mg/l	0.04 mg/l	0.43 mg/l
7	Arsenic	0.003 mg/l	0.16 mg/l	0.006 mg/l	0.024 mg/l
8	Cadmium	0.002 mg/l	0.003 mg/l	0.024 mg/l	0.053 mg/l
9	Chlorine	6 mg/l	16 mg/l	6 mg/l	15 mg/l
10	Copper	0.02 mg/l	0.15 mg/l	0.17 mg/l	0.46 mg/l
11	Cyanide	0.02 mg/l	0.02 mg/l	0.02 mg/l	0.02 mg/l
12	Iron	1.8 mg/l	12 mg/l	0.34 mg/l	6.7 mg/l
13	Lead	0.01 mg/l	0.04 mg/l	0.02 mg/l	0.3 mg/l
14	Magnesium	0.4 mg/l	9.4 mg/l	9.5 mg/l	21 mg/l
15	Mercury	0.0003 mg/l	0.0027 mg/l	0.0003 mg/l	0.0007 mg/l
16	Sulfates	42 mg/l	81 mg/l	241 mg/l	442 mg/l
17	Zinc	0.02 mg/l	0.17 mg/l	1.2 mg/l	2.8 mg/l

The high concentration of the solid suspensions requires, for the majority of the power stations, in which coal is burned, the creation of artificial lakes, in which the solid substance settles, the resulting water can be reused or discharged. Sulfur oxide emission reduction plants, whether based on wet or dry technologies, are generating polluted wastewater.

The amount of wastewater is significantly higher in the case of wet technologies, but also the dry and quasi-dried technologies can be considered as generating wastewater, polluted, as long as the reaction by-products are transported to the storage site through a flow of waste. the water. Frequently, the water resulting from the desulphurization processes, especially from the wet ones, has a high content of dissolved and suspended solids and are saturated with calcium sulfate. Because coal also contains certain amounts of chlorine and fluoride, which are quickly retained during the wet desulphurization process, the wastewater will also contain chemical combinations of these substances.

Also, the content of unreacted additive, usually calcium carbonate, as well as certain heavy metal content, should not be neglected. Typical concentrations of pollutants in the water resulting from the process of wet secondary desulphurization are given in Table 2 [1].

Table 2. Typical concentrations of the pollutants in the water from the wet desulfurization systems

Substance	Concentration (mg/l)	Substance	Concentration (mg/l)	Substance	Concentration (mg/l)
Aluminum	0.02 ÷ 2.0	Lead	0.02 ÷ 0.52	Chlorine	421 ÷33000
Arsenic	0.005 ÷ 1.8	Magnesium	4.1 ÷ 750	Fluorine	0,6 ÷ 58
Calcium	181 ÷ 2600	Mercury	0.0005 ÷ 0.07	Sulphites	0.9 ÷ 3500
Chromium	0.014 ÷ 0.5	Sodium	11 ÷ 29000	Sulphates	601 ÷ 35000
Copper	0.003 ÷ 0.56	Zinc	0.02 ÷ 0.59	Total dissolved solids	2801 ÷ 92500

If, in addition to the previous observations, the ones related to the relatively large quantities of wastewater, resulting from the process of preparing the fuel, directly or indirectly, as well as those from the maintenance operations, which flow through the drainage systems, the necessity arises achieving a strict control of the wastewater, at the level of the whole plant.

3 TREATMENT OF LIQUID AND SOLID WASTE

3.1 Treatment of liquid waste

Numerous technologies can be used to treat the liquid residues of the thermoelectric power stations, to eliminate the suspended solids, separate and retaining the petroleum residues, resulting from the loss of fuel or lubricants, the dechlorination, the retention of the heavy metal residues. Some of the commonly used techniques are presented in the following sections.

3.1.1 Elimination of solid suspensions

The oldest and most widespread for the elimination of solid suspensions is that of the wastewater decanting, in the settling basins. These tanks must be so designed as to provide residence time necessary to carry out the physical phenomenon of gravitational settling.

In the current installations, the decanting process is associated with coagulation and flocculation processes, both for reducing the time spent in the reagent, in the raw water, and for reducing the size of the treatment plants. In many situations, coagulation and flocculation are not independent technological processes but are embedded in the constructive structure and technological flow of the decanter [2,3].

The fine particles create colloidal suspensions in water. The kinetics of colloidal dispersion is influenced by several parameters, such as [2]:

- the nature of the two-phase environment;

- the temperature at which the decanting, coagulation, flocculation processes take place;
- colloid concentration;
- the presence of substances that act as inhibitors;
- the intensity of agitation;
- the residence time of the mixture of water and reagent, in the treatment tanks.

The initiation and development of the decanting, coagulation and flocculation processes can be characterized by a dimensionless parameter, called the Camp number [2]

$$G = K \cdot t, \text{ where:} \tag{1}$$

- G is the velocity gradient, and t represents the residence time.

The velocity gradient is defined as the difference between the velocities of two adjacent liquid layers, is temperature dependent, through the dynamic viscosity coefficient, and is one of the most important parameters of the coagulation and flocculation processes. The velocity gradient is defined as the difference between the velocities of two adjacent liquid layers, is temperature dependent, through the dynamic viscosity coefficient, and is one of the most important parameters of the coagulation and flocculation processes. The average velocity gradient, corresponding to the turbulent regime, can be calculated with the relation [2]:

$$G=\sqrt{\frac{P}{V \cdot \mu}} \quad (s^{-1}), \text{ where:} \tag{2}$$

- P is the power dissipated for shaking a volume;

- V of water and μ is the coefficient of dynamic viscosity.

Flocculation is a slow kinetic process so that the agitation does not have to be violent, to allow the formation of micro flocs, and then their volume increase. Unlike flocculation, coagulation is an irreversible process, which requires small times for the reaction to be complete under intense agitation. Under these conditions, the adoption of a certain value for the Camp number depends on the type of reaction, the time it takes, the energy consumption, the nature of the coagulant, and the amount of coagulant. The station time and Camp number are also chosen according to the clarification technology, which is used subsequently [4,5].

3.1.2 Separation of oils

Lubricating oils and greases, for the maintenance of central mechanical equipment, pollute the drain water in general and sewage network. Other sources of pollution, of this type, are the liquid fuel storage facilities, as well as the transformers in the power outlet station in the power plant. All leaks, which may be contaminated with such residues, must be collected and treated in a separation facility. Because most oils have less specific weight than water, the separation consists, in fact, in the temporary, temporary retention of contaminated water, in some separation tanks. In Figure 1 a schematic representation is given of an oil separator, the gravity-type construction.

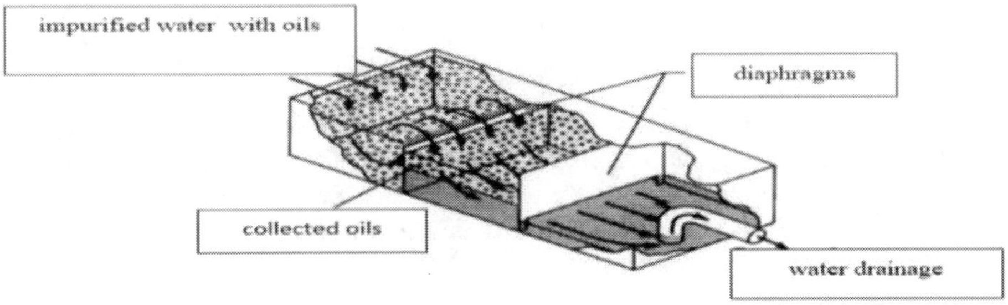

Fig. 1. The principle of gravitational separation of oils from wastewater

To prevent the oil residue from leaking, simultaneously with the water, when it is discharged from the separator, the separator is provided with diaphragms and with the possibility of partial discharge. The collected oil is pumped outside at certain intervals using portable pumps. When the residual oil is finely dispersed in water, forming emulsions, and the limits imposed on the effluent pollution level are very drastic, fusion oil separators can be used. These separators are also of gravitational type, except that some elements are used, which cause the agglomeration of the small oil particles. The agglomeration process continues until the particle sizes are large enough to achieve effective gravitational separation.

3.1.3 Dechlorination and reduce the heavy metal content

In addition to the processes described above, the wastewater from power plants can browse the dechlorination process steps, evaporation, and distillation, crystallization, and disposal of heavy metals [1,6,7]. Current standards impose limits on the quality of the water of chlorine compounds reduced, precise control is required so that the presence of these compounds in the wastewater to be discharged into the environment. In power plants, residual chlorine can exist both in the water in the cooling circuit, either closed or open, and in the wastewater resulting from the processes of treating the thermal agent and transporting the solid pollutants. Dechlorination can be achieved by the addition of sulfur dioxide, sodium sulfite, or sodium metabisulphite. Due to the high flow rates of water in the cooling circuit, it is preferable to use the cheapest additive such as SO_2. In certain situations, activated carbon can also be used [8,9].

The chemical reactions specific to the dechlorination process are the following:

$$SO_2+H_2O \rightarrow H_2SO_4 \tag{3}$$

$$H_2SO_3+HOCl \rightarrow H_2SO_4+HCl \tag{4}$$

$$NH_2Cl+ H_2SO_3 \rightarrow NH_4H_2SO_4+HCl \tag{5}$$

All of these reactions have very fast kinetics. From a stoichiometric point of view, 0.9 ppm SO_2 is required to retain 1 ppm chlorine.

The equipment used to supply the SO_2 system is similar to that used for liquid chlorine supply. Keeping the storage containers at a temperature of 22 ÷ 28°C avoids relaying the gas into the pipelines (the boiling temperature of SO_2 is - 7°C, at atmospheric pressure). If the gas consumption is high, several containers can be used at the same time, the pressure reduction in them not being large enough that the relic phenomenon occurs.

Due to the toxicity of sulfur dioxide, several measures must be taken to prevent contamination of the enclosures in which the installation is located. First of all, it is preferable to install, in separate rooms, the pressurized part and the vacuum part of the installation. The rooms must be equipped with exhaust fans, installed close to the floor because sulfur dioxide is heavier than air. It is also preferable to install SO_2 detectors, whose transducers are also installed near the floor of the room.

Metals can be removed from wastewater by various processes, such as alkaline precipitation, sulfur precipitation or ion exchange. Alkaline precipitation uses reagents, such as CaO or NaOH, leading to the formation of heavy metal hydroxides. Because in such reactions there are relatively few metals, alkaline precipitation is rarely used in the technological flow of wastewater treatment plants. It is preferable to use the iron co-precipitation process [1,10].

In the first step of the process, the pH of the wastewater is corrected, up to that value at which the reaction between the metals, which is about to be eliminated, and the iron salts have good kinetics. After the addition of ferric chloride and an oxidation process, ferrous sulfate is formed, which precipitates in the presence of flocculating agents. The dissolved and suspended metals are retained by the flammable ferrous sulfate flakes at the bottom of the reaction and decanting chamber.

3.2 Solid waste treatment

The adoption of a particular system or technology for the treatment, disposal, and storage of solid residues must take into account the following factors: volume and type of residues, their properties, possible restrictions between residual substances and those of the site, the environmental impact of the deposit, the environmental norms in force, but also factors of prognosis, such as the probable evolution of the boiler, of the volume of residues generated by it, of the possibilities of treatment and even of the limits imposed, by the environmental norms [10, 11].

The evaluation of the operational factors refers to the following aspects:

- types of solid residues generated; these may change, during the operation of the boiler, in particular, due to the adoption of new technologies for the treatment of the flue gases, but also due to the change of the combustion technique or the type of fuel;
- the weight of each type of solid residue; these weights can change over time due to the use of different qualitative fuels or mixtures of fuels and due to the modification of desulfurization techniques;
- if these changes are long-term changes, the weight of each type of residue may also differ in the short term, the effect of the dynamic character of the pregnancy;

- the location of the points of discharge and storage, temporary, each type of solid residue;
- the operating procedures of the boiler and its load curve;
- the expected life of the plant.

Regarding solid waste properties, interested in a wide range of physical and chemical properties, from wettability, and to potentially toxic gases, through spontaneous release or by any processes of gases. The adoption of technical solutions must take into account all these properties and, also, the fact that they can change over time.
The restrictions imposed on the solid waste storage site are related to the following issues:

- the actual use of the land in question;
- possibility of flooding, drainage or stagnation of water;
- the proximity to the natural habitat of the different species of animals, as well as the risk that the deposit for each species may present;
- the storage capacity of the deposit;
- the topography of the place and how favorable it is for further development, possibly other than the one for the realization of a solid residue deposit;
- the existing climatic conditions and how they can be changed, at the local level, by placing the residue depot;
- how and the extent to which groundwater can be affected;
- the costs of the transport of the residues, to the place of their storage.

Regardless of whether the landfill is surface or located in a natural depression, the landfill must be dry before storage, with a solid content of more than 85%. Drying simplifies the transport of residues, allows easier distribution in the warehouse, and more compact compaction, the density of the deposited material can reach 85 ÷ 95% of the maximum density of solid waste.

During periods when the boiler is switched off (during the hot season, if it is a district heating plant) or when an energy block is in reserve, the ash hydro transport pipes can be used to wet the surface of the depot through two or three holes. spill. The operation of the flood system must be carried out carefully, by permanently changing the position of the spillways, to avoid the formation of water mesh, which may endanger the stability of the deposit. A part of the fine ash, separated by the electrophiles, can be mixed with water, in a ratio of 1: 1, the dense fluid thus obtained, being able to be disposed on the surface of the deposit.

Thus, hardened layers are formed, whose thickness is of the order of 15 ÷ 20 cm, and which is characterized by good stability. In this final stage, three types of operations are performed: final compaction and leveling, to ensure drainage; placement of infiltration barriers; coating with a vegetal soil layer, and anti-erosion seeding [12].

4 Conclusions

Power plants face major environmental problems, the solution to which depends on obtaining an environmental permit to operate under the conditions imposed by the European Union. Romania, in the effort to join the European Union, has established the environmental protection strategy, the priority directions for energy.
As measures to prevent environmental pollution are the following:
- use of coal and low-sulfur fuel;
- increasing the efficiency of electricity production through rehabilitation and modernization works;
- improving the combustion control and regulation processes, with the completion of maintenance, rehabilitation and modernization works;
- increasing the performance of electro filters;
- introduction of burners with low N emissions;
- installation of desulphurization systems;
- improving energy-environment management;
- elimination of advanced, physical, and moral wear and tear installations.

References

[1] L.F. DRBAL, P.G. BOSTON, K.L. WESTRA, R.B. ERIKSON, *Power plant engineering*, by Black & Veatch, Chapman & Hall, International Thomson Publishing Company, (1996).
[2] O. BLĂGOI, E.L. PUŞCAŞ, *Surface water treatment - chemical methods,* Editura Dosoftei, Iaşi, (1997).
[3] G.M. MASTERS, *Introduction to Environmental Engineering and Science,* Prentice-Hall, Englewood Cliffs, New Jersey, (1991).
[4] B.T. RAY, *Environmental Engineering*, PWS Publishing Company, An International Thomson Publishing Company, Boston, (1995).
[5] C.A. WENTZ, *Hazardous Waste Management*, McGraw-Hill, New York, (1989).
[6] H.M. FREEMAN, ed., *Standard Handbook of Hazardous Waste Treatment and Disposal*, McGrawHill, New York, (1989).
[7] E.J. HALLER, *Simplified Wastewater Treatment Plant Operations*, Technocomic Publishing Company, Lancaster, Pennsylvania, (1995).
[8] H.S. PEAVY, D.R. ROWE, G. TCHOBANOGLOUS, *Environmental engineering*, McGraw-Hill, (1985).
[9] P.A. VESILING, J.J. PIERCE, R.F. WEINER, *Environmental engineering–Third Edition*, McGraw Hill, (1993).
[10] M.D. BROWN, T.D. WENCE, T.C. REILLY, *Solid Waste Transfer Fundamentals*, Ann Arbor Science, Ann Arbor, Michigan, (1980).
[11] A.P. SINCERO, G.A. SINCERO, *Environmental Engineering–A Design Approach*, Prentice-Hall Inc., (1996).
[12] D.C. WILSON, *Waste Management: Planning, Evaluation, Technologies*, Clarendon Press, Oxford, England, (1981).

Solar collectors network map: shape and time dynamics of stored thermal energy for different user loads

Camelia Stanciu[1], *Dorin* Stanciu[1,*], *Claudia* Ioniţă[1], and *Diana-Maria* Olaru[2]

[1] Department of Engineering Thermodynamics, Heat Engines, Thermal and Refrigeration Equipment, University Politehnica of Bucharest, 313 Spl Independentei, Bucharest, Romania

[2] Faculty of Engineering in Foreign Languages, Mechanical Engineering field, University Politehnica of Bucharest, 313 Spl Independentei, Bucharest, Romania

Abstract. Solar flat plate collectors are commonly used to transform solar energy into thermal one by heating a heat transfer fluid (HTF). The available thermal energy to the user is characterized by a certain thermal capacity and a certain temperature level. Comparing to the case of a single collector, when a higher thermal capacity is required in the user process, more collectors should be coupled in parallel, so that the HTF mass flow rate is increased, exiting the solar collector at the same temperature. Another case may be encountered when the user process requires the same mass flow rate of HTF, but at a higher temperature. In this last case, a series connection is considered. The aim of the present paper is to compare these cases and to build a solar panel map that could be used to choose the best arrangements for a specific user need. A study is performed regarding the map dynamics with respect to HTF consumption to the user for three storage tank capacities of 100L, 200L and 300L respectively. The results reveal a recommendation regarding collectors coupling function on user needs.

1 Introduction

Flat plate collectors (FPC) are commonly used to transform solar energy into thermal one by heating a heat transfer fluid (HTF). For domestic applications, the temperature and mass flow rates are limited to a relatively narrow range of values, but for industrial ones, the range might depend on the system operating conditions and might encounter more important fluctuations along a day. Thus, in some cases the HTF temperature level or mass flow rate might need to be adjusted depending on the required parameters or operating constraints. This is the aim of this paper, to study how such adjustment could be proceeded.

In the literature, there are numerous studies regarding the behaviour of solar collectors under different operating constraints such as input solar radiation or thermal loads delivered to the user, particularly for flat plate collectors.

A recent study regarding the flexible design of flat plate solar collectors for industrial applications is presented in [1] for two industrial processes. The paper proposes an approach

* Corresponding author: dorin.stanciu@upb.ro

for the collector network structure defined by adjusting the mass flow rate and the inlet temperature of the working fluid to achieve the smallest collector surface area, and thus minimum investment cost. The results reported regard the number of lines in parallel and the number of collectors in series in each line to answer the specific load defined for the two industrial processes under the above-mentioned optimisation criterion. Other similar works are described in [1], for industrial and agricultural applications. The great majority of them are regarding solar fields of FPC or other solar collectors types.

By "flexible design", Martínez-Rodríguez et.al. referred to "the capability of the system to deliver the required thermal duties over the year as the ambient conditions vary" [1].

In a similar manner, the target of the present paper is to study the possibility of coupling more FPC either in series or in parallel in order to ensure the required thermal energy to the consumer, both in terms of mass flow rate and temperature level. Obviously the FPC operation is always constraint by the input solar energy or ambient conditions and by the user thermal energy load. Solar panel maps are built in order for the user to appropriately choose the best arrangement needed for a certain thermal load at a certain temperature level.

2 System description

The studied system is composed by a FPC, a uniform temperature storage tank and the output /return water network to the user, as sketched in Figure 1.

A fraction of the incoming solar radiation on the FPC area, $G_{T,t}$, is absorbed by the collector tubes and used to heat the HTF, water here, from an inlet temperature $T_{f,in}$ to an outlet temperature, $T_{f,o}$. The mass flow rate of HTF exiting the collector, $\dot{m}_{FPC}$, enters the storage tank where it instantaneously mixes with the HTF mass existing there. As a result, the entire HTF mass inside the storage tank volume is at the same temperature T_{ST}.

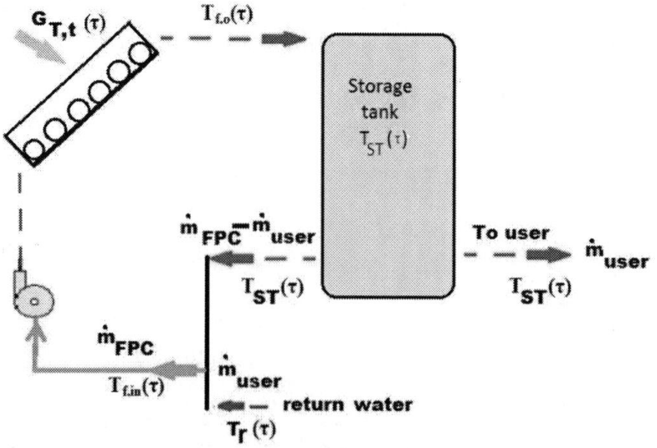

Fig. 1. Flat plate collector with uniform temperature water storage system and user delivery system.

Depending on the user demand, a mass flow rate $\dot{m}_{user}$, is delivered at the storage tank fluid temperature and returns at a certain temperature T_r towards the inflow section of FPC. At the same time, a HTF mass flow rate $\dot{m}_{FPC}$-$\dot{m}_{user}$ is extracted from storage tank. Before entering the FPC, the two streams are mixed, and the inflow HTF temperature $T_{f,in}$ is obtained. The return fluid temperature T_r can be the ambient temperature T_a or a higher value resulting from the user consuming network.

When a higher thermal energy is required to the user, coupling more collectors could be advantageous. Two arrangements are studied in this regard.

The first one is a series coupling of two or more collectors, as presented in Figure 2.

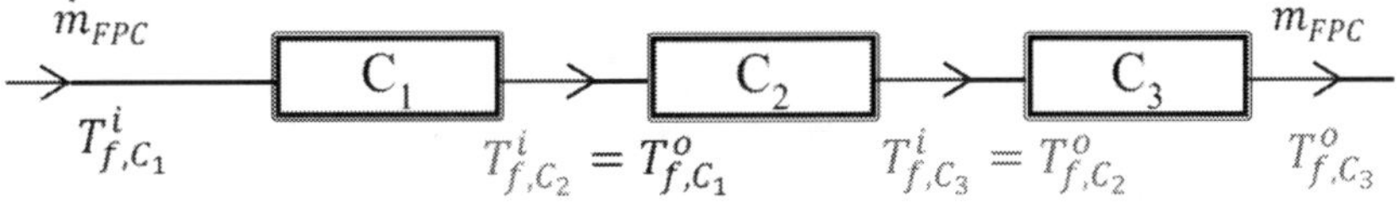

Fig. 2. Series coupling of flat plate collectors.

The same mass flow rate of HTF passes through all collectors. Due to the fact that the inlet HTF temperature to the second collector is equal to the outlet one from the first collector, the outlet temperature from the second collector will be higher. Thus, at the end of the coupling, the HTF temperature is significantly higher compared to the case of using a single collector. Consequently, the available thermal energy is higher.

The second case represents a parallel coupling of two or more collectors, as presented in Figure 3.

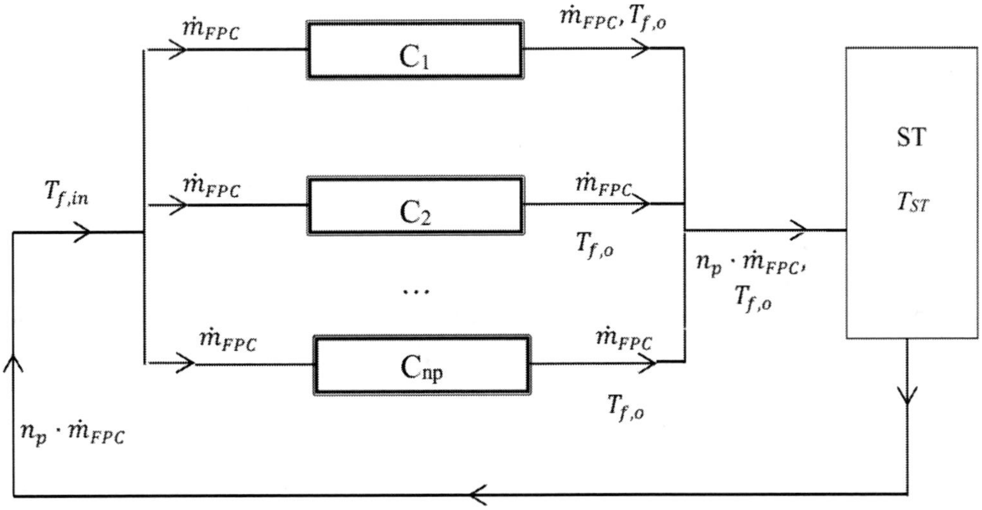

Fig. 3. Parallel coupling of flat plate collectors.

In this case, the outlet HTF temperature is the same as compared to the case of one collector, but the mass flow rate is increased proportional to the number of coupled collectors. Consequently, the available thermal energy is increased in this case, too.

3 Mathematical modelling

The total solar incoming radiation (beam and diffuse) under the chosen tilt angle, $G_{T,t}$ and ambient air temperature, T_a, are generated using METEONORM V7.1.8.29631 software [2]. These data represent 10-years measured average values and were extracted for Bucharest location (44.25°N latitude), on July 15th.

The heat transfer mechanisms inside the FPC is presented in Figure 4. The overall heat flux incident on the collector surface is $\dot{Q}_s = A_{FPC}G_{T,t}$, where A_{FPC} represents the surface area of FPC. The heat flux passing the FPC cover towards the absorber plate is expressed as $\dot{Q}_{Ab} = (\tau\alpha)\dot{Q}_s$ computed as function of the absorbance - transmittance property *(τα)* of the glass cover. Further, this heat flux is divided into two parts. The useful heat flux, $\dot{Q}_u$, is taken over by HTF, while the heat flux loss, $\dot{Q}_{loss,p-c} = \dot{Q}_{loss,c-a}$, is transferred to the environment.

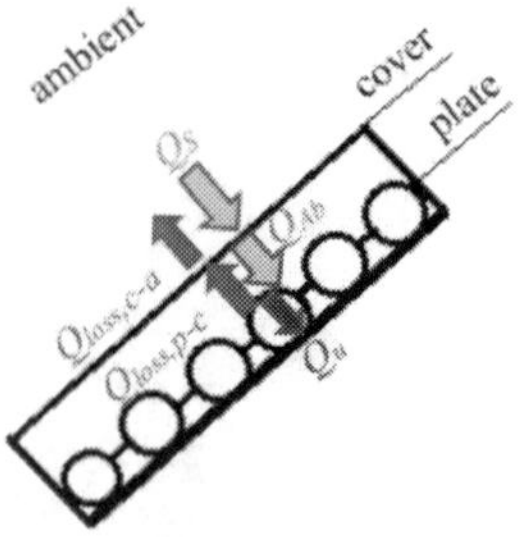

Fig. 4. Losses in the considered flat plate collector.

The losses between collector's plate and cover $\dot{Q}_{loss,p-c}$ occur by both free convection and radiation mechanisms. The free convection losses are computed by using the Hollands correlation [4] , which links the Nusselt number, $Nu_{cv,p-c}$ to the Rayleigh number $Ra_{L,p-c}$ and the collector tilt angle β. Accordingly, the convection heat transfer coefficient is expressed by:

$$h_{cv,p-c} = Nu_{cv,p-c} \cdot k/L_{p-c} \quad (1)$$

where L_{p-c} represents the distance between the plate and the cover and k is the air thermal conductivity. The radiation heat transfer coefficient between the collector plate and cover is computes as:

$$h_{rad,p-c} = \sigma\left(T_p + T_c\right)\left(T_p^2 + T_p^2\right)\left(\frac{1}{\varepsilon_p} + \frac{1}{\varepsilon_c} - 1\right)^{-1} \quad (2)$$

These losses are further transmitted from the cover towards ambient by the same heat transfer mechanisms (convection and radiation). The corresponding heat transfer coefficients are:

$$h_{cv,c-a} = 5.7 + 3.8w \quad (3)$$

$$h_{rad,c-a} = \varepsilon_c\sigma(T_c + T_a)(T_c^2 + T_a^2) \quad (4)$$

At each time moment the two heat flux losses are equal. Assuming that the plate and cover areas are the same, the corresponding equation looks like:

$$\dot{q}_{loss} = \left(h_{cv,p-c} + h_{rad,p-c}\right)\left(T_p - T_c\right) = (h_{cv,c-a} + h_{rad,c-a})(T_c - T_a) \quad (5)$$

The useful heat rate transformed into thermal energy carried by the HTF can be expressed in two ways. One way is to apply the balance energy equation to the plate surface, and another is to consider the forced convection heat transfer between the tubes wall and the HTF. Neglecting the conduction through tubes wall, one obtains the following equation:

$$\dot{Q}_u = \dot{Q}_{Ab} - A_{FPC}\dot{q}_{loss} = N_{tube}(\pi D_{tube}L_{tube})h_{HTF}(T_p - \bar{T}_f) \quad (6)$$

where N_{tube} represents the number of tubes inside FPC through which the HTF flows, D_{tube} and L_{tube} are the tube inner diameter and length respectively, while $\bar{T}_f = (T_{f,o} + T_{f,in})/2$ is the mean temperature of the HTF between inlet and outlet of FPC tubes. Assuming that the convection heat transfer occurs under constant heat flux boundary condition and that the flow through tubes is laminar and fully developed, the value of Nusselt number considered to compute h_f is Nu_f=4.36. Further, the outflow HTF temperature is computed as:

$$T_{f,o} = T_{f,in} + \frac{\dot{Q}_u}{(\dot{m}_{FPC} c_{p,FPC})} \tag{7}$$

The mean plate temperature T_p, cover temperature T_c, and HTF temperature at the collector's outlet $T_{f,o}$ are unknown and determined by solving the system of equations at each time step. The inlet fluid temperature to the collector $T_{f,in}$ is computed after each time iteration, resulting from the mixing of two HTF streams, as presented in Figure 1:

$$T_{f,in} = T_{ST} - \frac{\dot{m}_{user}}{\dot{m}_{FPC}}(T_{ST} - T_r) \tag{8}$$

The storage tank temperature T_{ST} is computed by integrating the expression resulting from the First Law of Thermodynamics applied in transient conditions:

$$T_{ST} = T_{ST}^{(\tau-1)} + \frac{\dot{m}_{FPC} c_{p,FPC} (T_{f,o} - T_{ST}^{(\tau-1)}) - (UA)_{ST} \left(T_{ST}^{(\tau-1)} - T_a\right)}{m_{ST} c_{p,ST}} \Delta\tau \tag{9}$$

where m_{ST} represents the mass of fluid stored in the storage tank. Capacities of 100L for each squared meter of FPC are recommended [3].

The available thermal energy in the storage tank at each time step is:

$$Q_{ST} = m_{ST} c_{p,ST} (T_{ST} - T_a) \tag{10}$$

assuming that the ambient temperature is used as reference.

In all above equations, the material properties and geometric parameters are known: cover absorbance-transmittance property $(\tau\alpha) = 0.8$, plate emittance $\varepsilon_p = 0.05$, cover emittance $\varepsilon_c = 0.89$, plate-cover distance $L_{p\text{-}c} = 0.04$m, collector tilt angle $\beta = 30°$, FPC receiving area $A_{FPC} = 2\text{m}^2$, tube diameter and length $D_{tube} = 0.00881$ m and $L_{tube} = 2$m respectively, number of tubes $N_{tube} = 12$. Fluid properties (air conduction coefficient k, HTF specific heat $c_{p,FPC}$, etc) are determined as temperature dependent, using EES software [5]. A HTF mass flow rate $\dot{m}_{FPC} = 0.035$kg/s was considered inside the tubes. For the storage tank, constant overall losses were considered as $(UA)_{ST} = 11$W/K. A wind velocity of $w = 1$ m/s was also considered for computing the forced cover-ambient convection.

In order to obtain the necessary thermal load and HTF temperature to the consumer, different coupling schemes of FPC are studied. The useful heat rate to the user $\dot{Q}_{user}$ is directly proportional to the mass flow rate $\dot{m}_{user}$ and the temperature difference between delivery and return streams:

$$\dot{Q}_{user} = \dot{m}_{user} c_{p,ST} (T_{ST} - T_r) \tag{11}$$

When user needs a higher thermal energy rate at a certain temperature level T_{ST}, the only way to increase it is by increasing the mass flow rate. Thus, a parallel coupling of more collectors could be useful. In case of higher temperature level needs, a series coupling could be sought.

3.1 Series collectors coupling

When two or more collectors are coupled in series, as presented in Figure 2, the mass flow rate passing through each FPC is the same, but the inlet temperature in a collector is equal to the outlet temperature from the previous one. Thus, the HTF temperature is gradually increased along the FPC series and checked at each time for safety operating conditions.

In this case, the computing algorithm is the same as the one presented for the single FPC case, the difference consisting in successively repeating it for each collector. Consequently the same mass flow rate of HTF $\dot{m}_{FPC}$ is entering the storage tank, but at a higher temperature $T_{f,o}$ compared to the single FPC case.

3.2 Parallel collectors coupling

The parallel connection is presented in Figure 3. When a certain number n_p of collectors are coupled, the storage tank is fed by n_p streams of HTF existing each collector at the same final temperature $T_{f,o}$. Consequently, the thermal energy rate available in the storage tank for consumer use is:

$$\dot{Q}_{ST,FPCp} = n_p \dot{m}_{user} c_{p,ST} (T_{ST} - T_r) = n_p \dot{Q}_{ST} \tag{12}$$

Results are further presented.

4 Results and discussions

The numerical simulation was firstly performed for a single FPC connected to a storage tank of 100L capacity. In this case, a constant volume flow rate of 5L/h is delivered to the user at the storage tank temperature. Water was chosen as heat transfer fluid. The return temperature to the FPC was considered the ambient one. Results in terms of inlet and outlet fluid temperatures, storage tank one and available thermal energy in the storage tank were obtained.

For this particular case, it was noticed that the temperature in the storage tank, initially equal to the ambient one, began increasing at sunrise. It reached the value of 60°C at 11:10 AM and continued increasing until 15:40 when the maximum value of 84.53°C was reached. After this time, it began decreasing, as the solar radiation decreased as well. It reached again 60°C at 21:00. When performing this calculation for different storage tank capacities (200L and 300L) and for the same constant user consumption of 5L/h from the storage tank, Figure 5 was obtained. One may notice some important aspects regarding the whole system behaviour:

- Firstly, the maximum storage tank temperature decreases as the storage tank capacity increases. Obviously, heating a more important quantity of HTF with the same available solar energy conducts to a lower temperature.
- Another observation is the fact that the same fluid temperature in the storage tank is reached later in the case of higher capacities. For example, when a 100L storage tank was used, a temperature of 60°C was reached at 11:20 AM, while in the case of 200L this value was reached at 13:40. In the last case of 300L, the maximum temperature reached by the storage tank fluid was 58.83°C at 17:30. In conclusion, the storage tank capacity should be correctly linked to the size of the system (FPC area and mass flow rates) in order to reach a desired value for the HTF temperature.
- The third notice regards the capacity of the storage tank to ensure a longer use of the stored thermal energy. One may observe that the decrease in thermal energy Q_{ST} in larger storage tanks is slower and more damped after the maximum value was reached.

When considering two or three collectors connected in series, the results are revealed by Figure 6. The same behaviour is met, but higher storage tank fluid temperatures are reached. One may notice that 120°C could be achieved when connecting two FPC in series to a 100L storage tank. The results require the following observation: the HTF circuit in the FPC and ST is pressurized at 3bar in order to avoid steam production and system damage, while keeping the ability to produce hotter fluid. These results show the ability of the network to cover different thermal energy values at different fluid temperatures. It's importance is revealed when choosing the correct size of the network for a certain application.

In Figure 7 the results are plotted for the 200L storage tank, but for all considered collector networks. Parallel couplings are mentioned by "*FPCp*", while series ones by "*FPCs*". This chart reveals that when connecting two or three collectors in parallel, the same temperature level is reached in the storage tank in the same delay from start-up, but the thermal energy stored is more important due to a larger mass flow rate of fluid feeding the storage tank.

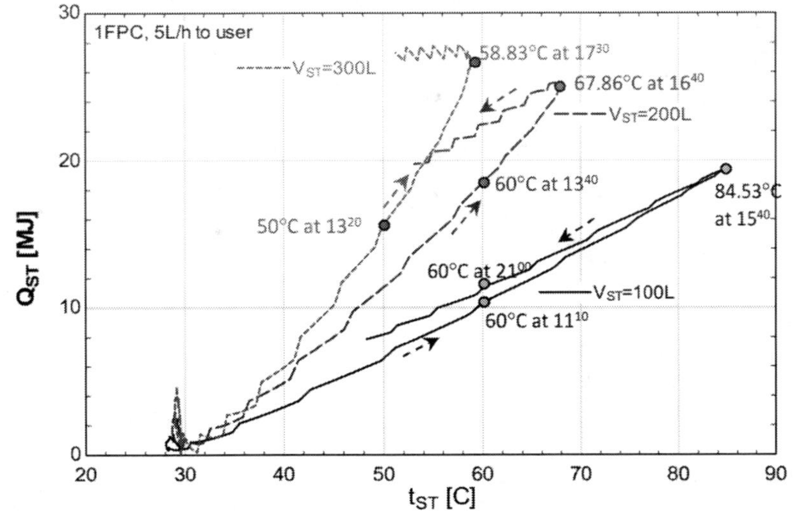

Fig. 5. Available thermal energy stored in the storage tank (5l/h volume flow rate for user delivery); sensitivity study with respect to storage tank capacity.

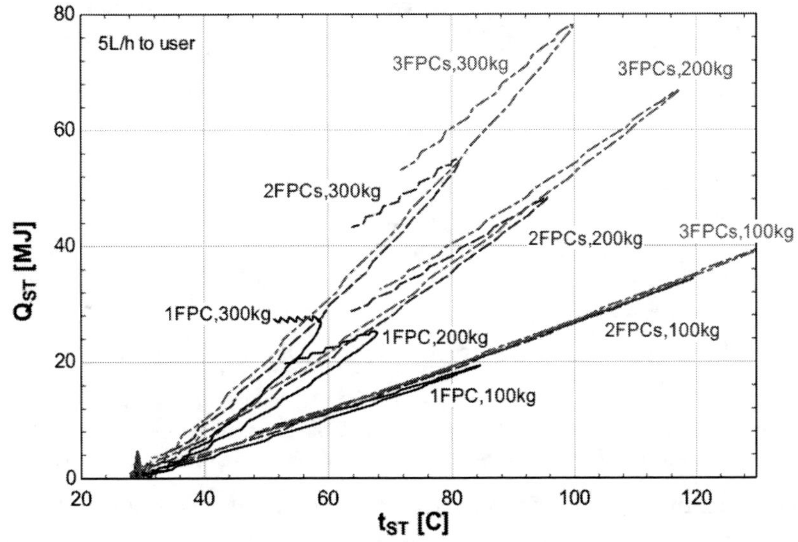

Fig. 6. Available thermal energy stored in the storage tank (5l/h volume flow rate for user delivery); sensitivity study with respect to storage capacity and number of collectors in series.

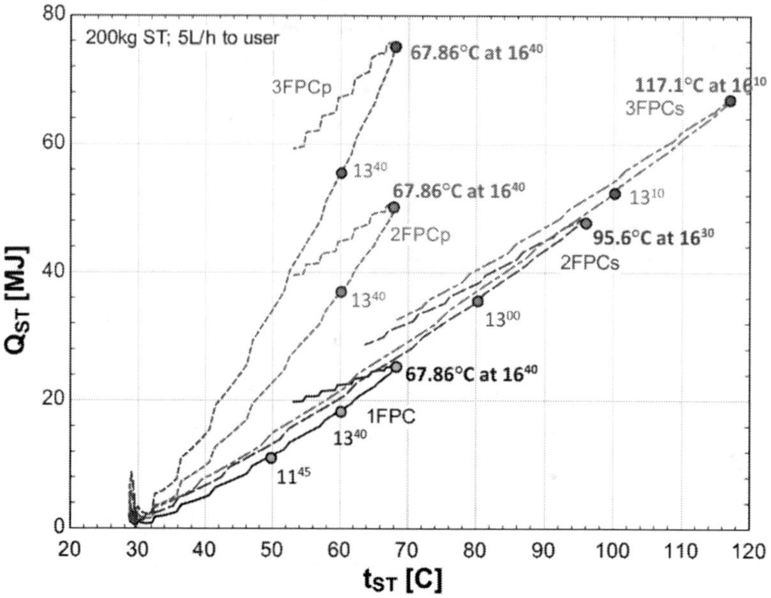

Fig. 7. Solar collector network optimization chart: available thermal energy stored in the 200kg storage tank (5l/h volume flow rate for user delivery) - comparison of different collector couplings.

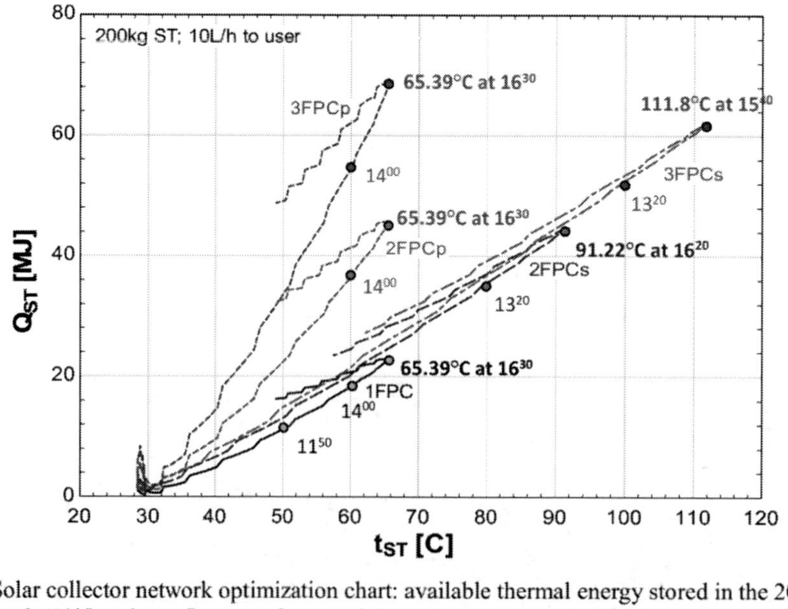

Fig. 8. Solar collector network optimization chart: available thermal energy stored in the 200kg storage tank (10l/h volume flow rate for user delivery) - comparison of different collector couplings.

When increasing the consumption to the user to 10L/h of HTF from the storage tank, the results are similar, but lower temperature values are reached, as revealed by Figure 8. Also, the value of the thermal energy available in the storage tank is lower, due to an increased flow rate constantly delivered to the user.

One may conclude that the solar collector maps built serve as selection charts for the user depending on the necessary thermal energy level and also on the temperature at which this energy is delivered.

5 Conclusions

The paper presents the results of a numerical simulation performed in the case of water solar heating using a flat plate collector provided with uniform storage tank. The results obtained are compared in terms of available thermal energy to the user, characterized by a certain thermal capacity and a certain temperature level. Few solar collector networks maps are built when comparing the use of a single collector to the cases in which more collectors are coupled in parallel or in series. These maps might be of interest when choosing the best arrangements for the system components. This choice depends on the thermal capacity and HTF temperature level required by the user. The results revealed the dynamic behaviour of these networks with respect to HTF consumption.

References

1. G. Martínez-Rodríguez, A.L. Fuentes-Silva, J.R. Lizárraga-Morazán, M. Picón-Núñez, Energies **12**, 570 (2019)
2. METEONORM V7.1.8.29631 software
3. J.A. Duffie, W.A. Beckman, *Solar Engineering of Thermal Processes* (Hoboken: John Wiley & Sons, 2006)
4. K.G.T. Hollands, T.E. Unny, G.D. Raithby, L. Konicek, ASME J. Heat Transfer **98**,189-193 (1976)
5. EES, Engineering Equation Solver, academic licence, V10.055-3D

IoT solution for monitoring indoor climate parameters in open space offices

Ioana Udrea[1,*], *Viorel Ionut* Gheorghe[1], *Laurentiu Adrian* Cartal[1], *Despina* Duminica[1], *Silviu* Petrache[1], *Tudor Catalin* Apostolescu[2], and *Valeru Friedemann* Kraus[3]

[1] Mechatronics and Precision Mechanics Department, Faculty of Mechanical Engineering and Mechatronics, University "POLITEHNICA" of Bucharest, RO

[2] Faculty of Informatics, Titu Maiorescu University, Bucharest, RO

[3] ARCHIBUS Solution Center - Romania, Bucharest, RO

Abstract. Although there are numerous high performance BMS (Building Management System), which monitor the indoor climate parameters, data access, sensor positioning, and other aspects may not be under control. On the other hand, IoT (Internet of Things) is experiencing exponential growth, as more and more devices and sensors are connected to the cloud. Thus, a sensor monitoring solution for indoor climate parameters was developed. The proposed solution is not expensive, and it is based on a Raspberry Pi board endowed with temperature, humidity, and pressure sensors. The developed application reads the values detected by the sensors, processes the date, and afterwards transmit the information to the IoT ThingSpeak platform. The large area is characteristic of open space offices, so the influence of radiant walls is small, and the operative temperature can be approximate with air temperature. This type of building is conditioned by the air conditioning system, so the air speed in this indoor environment is usually low and could be approximated by the design. So, with the data read by the developed solution the thermal comfort parameters can be approximated. If inadequate values are found, teams that carry out complex and precise measurements could be sent to the site. To achieve this goal a PMV calculator software is developed. Its validity is tested in accordance with the European standard ISO 7730. After that, the PMV computer is used with data read from sensors. Both the data read from the sensors and the newly calculated PMV are sent to the ThingSpeak IoT platform.

1 Introduction

The Internet of Things (IoT) is growing exponentially and it is considered by many as the most important technology initiative of today [1]. The main objective of IoT based environment monitoring system is to provide environmental parameters at remote location

* Corresponding author: ioana.udrea@upb.ro, udreaioana@yahoo.com

using internet. The IoT does not function without cloud-based applications to interpret and transmit the data coming from multiple sensors. The IoT solution provides the possibility of logging measurements from locations all over the world and of visualizing and analyzing the gathered data from any device connected to the Internet [2].

Among the most popular publicly available IoT platforms, which collect data from sensors used in countless fields including heating, ventilation, and air conditioning (HVAC), indoor air quality (IAQ) and thermal comfort (TC), are the following: Thingwork, SensorCloud, RealTime.io, OnePlatform [3].

Another aspect to be considered in sensors monitoring is represented by the way in which the sensors communicate the data. Among the technologies available in IoT, RFID and sensor networks are considered the fastest growing information and communication technologies [4]. A comparison between four different communication technologies, NFC, RFID, Bluetooth and Wi-Fi is done in [2], the authors concluded that low-power Wi-Fi promises multiple years of battery lifetime while providing easy integration to existing infrastructure with built-in IP-network compatibility.

In modern buildings management systems (BMS) are used. The most intelligent systems integrate various sensors such as temperature, humidity, lighting level, CO2 and others. Currently most of the data are analyzed directly by the BMS software or at most related to the building information model (BIM) [5].

An important human thermal comfort index, that take into account thermal indoor parameters that could be read by the sensors, is predicted mean vote (PMV). The PMV model is developed by professor Fanger [6] and its work is considered as classical model of thermal comfort. It considers that human comfort depends on the quantitative, combined influence of six parameters (air temperature, mean radiant temperature, water vapor pressure, air velocity, clothing level, and metabolic rate).

Measurement of thermal comfort parameters inside a building could be done using performant devices, like ComfortSense [7]. The device can use high-performance probes for measuring the operating temperature, relative humidity, air temperature, and velocity [8]. ComfortSense consists of a main frame with a built-in A/D (analog/digital) converter and USB 2.0 interface. The measurement probes, connected with the main frame, are positioned on the system stand, a tripod. The measurement is set up using a PC and application software. If you want real time data from multiple locations, ConfortSense cannot be used. Once due to the fact that the application it has does not have an API with which to transmit data on an IoT platform (probably this was not wanted because such systems are not intended to work without stopping). The second reason would be the high costs of the device, because to transmit simultaneously from several rooms would require a system for each of them.

For this reason, this paper is proposed to develop a less expensive solution that is suitable for IoT. It involves reading the thermal comfort parameters using a sensor system connected on a Raspberry Pi device and transmitting the read values in real time to an IoT platform. At the same time, based on these parameters, the PMV thermal comfort index will be calculated and published in real time on the chosen IoT platform.

Some of the parameters that are part of the PMV index will be approximated. For a better approximation it is recommended that the system be used in the central area of an open space office, where the influence of the radiant temperature of the outer elements of the building envelope is very small.

Due to the approximations it introduces, this solution is not thought of as a precise solution for calculating PMV, it is thought of as an indicator of places where comfort problems could occur. In these places, teams could be sent later to make precise measurements of the comfort parameters. The main advantage of the solution proposed in this paper is that the IoT solution can centralize data from many sensor systems and identify in real time, the zones with potential problems in a wide area.

2 Method and procedure

2.1 Sensor system presentation and ThingSpeak platform

The sensor system used in this study is composed of a Raspberry Pi board, a multiplexing connection board and a breadboard with two sensors module *DHT22* (composed of a temperature and a humidity sensor) [9] and *BMP180* [10] (compose of a temperature and a humidity sensor). In order to facilitate the Raspberry Pi connection with the two sensors and the fan, a multiplexing board is used, see fig. 1.

The Raspberry Pi used is a Model 4, B [11]. It is a credit-card-sized single-board computer. The CPU on the board is an ARM v8 processor with 1.5 GHz speed. It has a variety of interfacing peripherals, including 2 USB 3.0 ports, 2 USB 2.0 ports, 2 micro-HDMI ports, 4 GB RAM, Micro-SD Card storage and 40 pin GPIO header. For GPIO port expansion a multiplexing connection board is used.

a)

b)

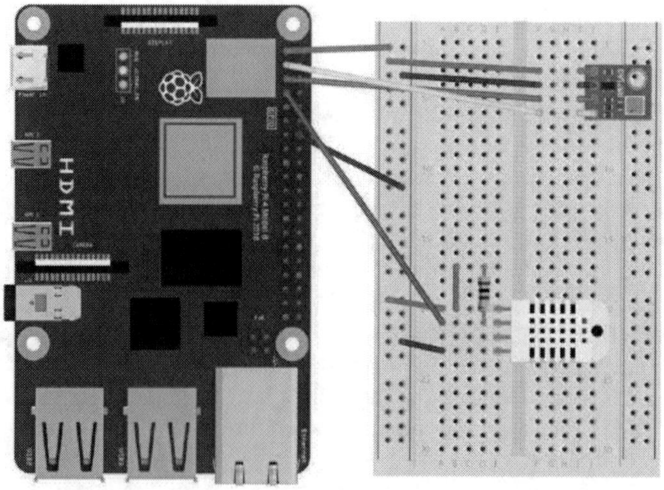

c)

Fig. 1. Sensor system: a) photo - all components: Raspberry Pi, multiplexing board, breadboard with sensors; b) photo - detail of sensor's area; c) schematic representation of the system.

Two monitors, keyboard, and mouse can be connected to Raspberry Pi through micro-HDMI and USB connectors and it can be used like a desktop computer. It supports a number of operating systems including a Debian-based Linux distro, Raspbian, which is used in our design. Raspberry Pi can be connected to a local area network through Ethernet cable or USB Wi-Fi adapter, and then it can be accessed through SSH remote login.
Additionally, the Raspberry Pi is equipped with a housing and a 5V fan.
In figure 1.c) schematic representation of the system is presented, one can see the connections that are made between the Raspberry Pi board with the DHT22 sensor and with the BMP180 sensor. The diagrams were made using Fritzing software.
ThingSpeak [12] is an IoT analytics platform service that allows in an easy way to aggregate, visualize, and analyse live data streams in the cloud. One can send data to ThingSpeak from his devices, create instant visualization of live data, and send alerts.

2.2 PMV equation in the classical approach of thermal comfort

The equations for PMV index, established by Fanger in his doctoral thesis [6], are used in the European standard ISO-7330 [13], see equation (1) to (4) of this standard. A clear explanation of the terms involved in the expression PMV can be found in equation (2.1) of Webb [14], the formula taken also in equation (1) of this paper.

$$
\begin{aligned}
PMV = &\underbrace{\left[0.303 \cdot e^{(-0.036 \cdot M)} + 0.028\right]}_{thermal\ sensation\ coefficient} \cdot \\
&\cdot \{ \underbrace{(M - W)}_{internal\ heat\ production} \\
&\underbrace{-3.05 \cdot 10^{-3} \cdot [5733 - 6.99 \cdot (M - W) - p_a]}_{heat\ loss\ trough\ skin} \\
&\underbrace{-0.42 \cdot [(M - W) - 58.15]}_{heat\ loss\ by\ sweating} \\
&\underbrace{-1.7 \cdot 10^{-5} \cdot M \cdot (5867 - p_a)}_{latent\ respiration\ heat\ loss} \\
&\underbrace{-0.0014 \cdot M \cdot (34 - t_a)}_{dry\ respiration\ heat\ loss} \\
&\underbrace{-3.96 \cdot 10^{-8} \cdot f_{cl} \cdot [(t_{cl} + 273)^4 - (\bar{t_r} + 273)^4)]}_{heat\ loss\ by\ radiation} \\
&\underbrace{+f_{cl} \cdot h_c \cdot (t_{cl} - t_a)}_{term\ description\ below}\}
\end{aligned}
\tag{1}
$$

$$
\begin{aligned}
t_{cl} = 35.7 - 0.028 \cdot (M - W) - I_{cl} \cdot \{3.96 \cdot 10^{-8} \cdot f_{cl} \cdot [(t_{cl} + 273)^4 - \\
(\bar{t_r} + 273)^4] + f_{cl} \cdot h_c \cdot (t_{cl} - t_a)\}
\end{aligned}
\tag{2}
$$

$$
h_c = \begin{cases} 2.38 \cdot |t_{cl} - t_a|^{0.25} & for\ 2.38 \cdot |t_{cl} - t_a|^{0.25} > 12.1 \cdot \sqrt{v_{ar}} \\ 12.1 \cdot \sqrt{v_{ar}} & for\ 2.38 \cdot |t_{cl} - t_a|^{0.25} < 12.1 \cdot \sqrt{v_{ar}} \end{cases}
\tag{3}
$$

$$f_{cl} = \begin{cases} 1.00 + 1.290 \cdot I_{cl} & for\ I_{cl} \leq 0.078\ m^2K/W \\ 1.05 + 0.645 \cdot I_{cl} & for\ I_{cl} > 0.078\ m^2K/W \end{cases} \quad (4)$$

where,
M – metabolic rate [W/m^2];
W – effective mechanical power [W/m^2];
I_{cl} – clothing insulation [m^{2}K/W];
f_{cl} – clothing surface area factor;
t_a – air temperature [°C];
$\overline{t_r}$ – mean radiant temperature [°C];
v_{ar} – relative air velocity [m/s];
p_a – water vapour partial pressure [Pa];
h_c – convective heat transfer coefficient [W/(m^{2}K];
t_{cl} – clothing surface temperature [°C].

In ISO 7730 [13], Annex D, is presented a basic computer program for calculating PMV and PPD. In this program, the four-degree equation of t_{cl} is solved by iterations.
The ISO 7330 standard declares that the basic computer software for calculating PMV and PPD given it's just a recommendation, other programming languages may also be used. The output should be verified with the example values given in Table D.1 from this standard.
Regarding the PMV equation, this paper aims two things. The first to perform in Python a function that can be called with all the parameters included in the PMV calculation, in this sense the steps in Annex D of the standard will be followed, this is presented in chap. 3.2. After testing, the function will be used by software that reads data from sensors, chap. 3.1. Here some parameters can still be set and others (temperature, humidity) are read from the sensors. Chapter 4. shows how to use the software and specifies all the values we used as input data for calculate PMV.

3. Software development

3.1 Sensor system integration in ThingSpeak

Raspberry Pi4 operating system (Raspbian, from Unix/Linux OS family [15]) - Raspbian comes pre-installed with plenty of software for education, programming, and general use. It has Python, Scratch, Sonic Pi, Java and more. A special developed Python script runs on this Linux system.
First a ThingSpeak account was settled and afterwards a new Channel on ThingSpeak was defined and configured. A Channel is used to collect data from a device, from another channel, or from the web. Each channel includes eight fields that can hold any type of data, plus three fields for location data and one for status data.
In fig. 2 one can see the details of new defined Channel, named *DHT22&BMP180_RaspberryPi*, that is used for showing the data received from sensors and some computed data (PMV value, the calculation of which will be presented in the next subchapter). For each to be shown, a field was defined, thus appearing the fields named: *Temperature DHT22 [°C], Humidity DHT22 [%], Temperature BMP180 [°C], Pressure BMP180 [mBar], PMV.*

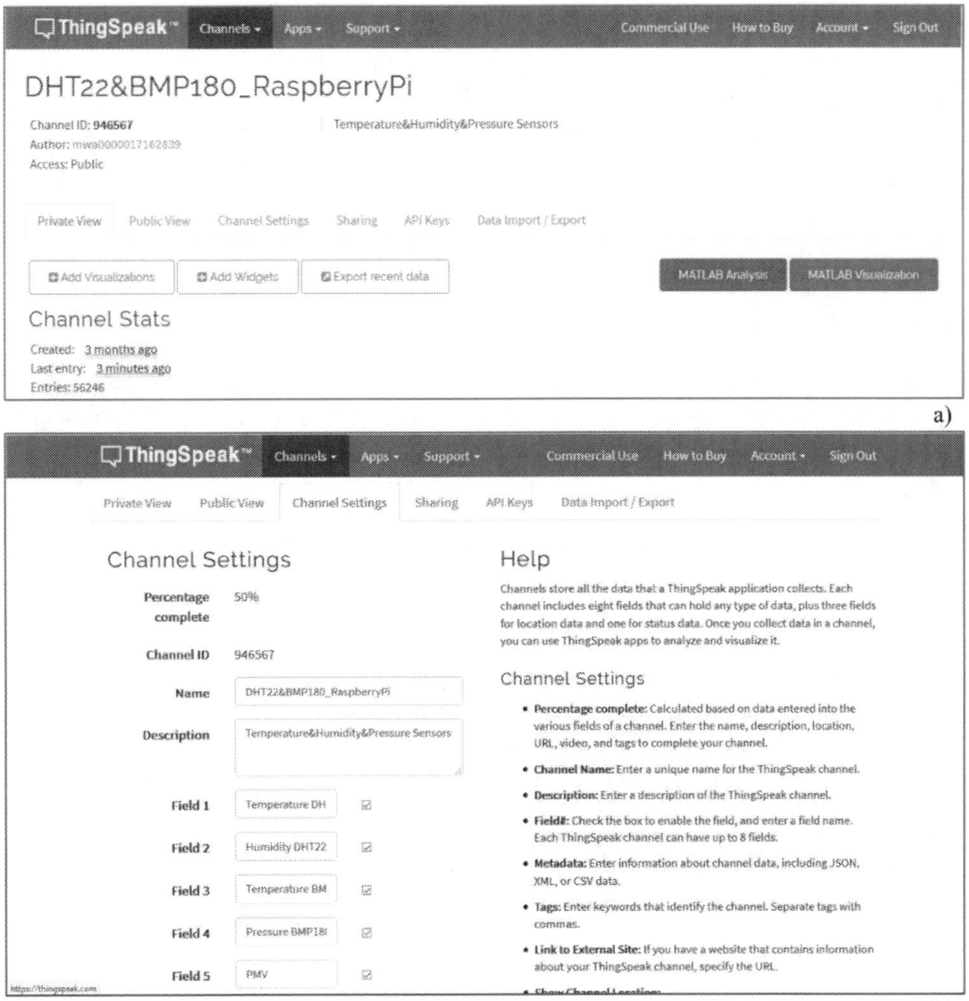

b)

Fig. 2. ThingSpeak configuration: a) Channel details; b) Fields details

A software was developed for the sensor system presented in subchap. 2.1. It is programmed in Python that comes built-in with Raspbian. For this, libraries developed by sensor manufacturers were used and were integrated with the technology made available by ThingSpeak. ThingSpeak not only provides communication through the "GET" and "POST" methods, available in just about any programming language, but in the case of Python it provides a library with methods that can be accessed directly in the code. The result is the publication in the ThingSpeak Channel of the values read by the sensors in the form of graphs corresponding to each Field of the Channel, see fig. 2.

Two libraries, named bmp180 (that is used for BMP180 sensor) and Adafruit_DHT (that is used for DHT22 sensor) are imported. For those libraries, some Raspberry Pi pins number that correspond to sensors pins could be specificity in code. For example, we specified pin GPIO4 on Raspberry Pi board that correspond to data pin of DHT22. For others sensor's pins the default values proposed by libraries are used (and the scheme is constructed according to those default values).

The function that calculates PMV, presented in subchapter 3.2 is called with the parameters read from the sensors. The resulted PMV index is published on another Field of the sensor's Channel.

3.2 PMV calculator development

Using equations (1) to (4) of this paper, the basic computer program given by ISO 7730 [13], and CISBE Guide [16] (Appendix 1.A1: Determination of predicted mean vote (PMV)), a function that calculates the PMV index was developed. It is written in Python software and is added to the software presented in subchapter 3.1.

The signification of the *input* data in the function are:
clo = Clothing, clo (1 clo = 0.155 W/m^2K);
met = Metabolic rate, met (1met=58.15W/m^2), range(58 to 232 W/m^2);
wme = External work, met;
ta = Air temperature, °C;
tr = Mean radiant temperature, °C;
vel = Relative air velocity, m/s;
rh = Relative humidity, %.

The *output* is PMV.

The partial water vapor pressure, pa, whose formula is not specified in the basic computer program given by ISO 7730 [13], is calculated using equation (5).

$$pa = rh \cdot e^{\frac{16.6536-4030.183}{ta+235}} \quad (5)$$

The result must be validated with ISO7730-Table D.1 as requested by the standard. For this validation, a lot of combination of input data imposed by the standard must be used and the result must be compared with the standard values.
The function that calculate PMV index is called with the values imposed by the standard and the results are shown in fig. 3.

Python Console

clo	met	wme	ta	tr	vel	rh	PMV
0.5	1.2	0.0	22	22	0.1	60	-0.75
0.5	1.2	0.0	27	27	0.1	60	0.76
0.5	1.2	0.0	27	27	0.3	60	0.43
0.5	1.2	0.0	23.5	25.5	0.1	60	-0.01
0.5	1.2	0.0	23.5	25.5	0.3	60	-0.55
1	1.2	0.0	19	19	0.1	40	-0.6
1	1.2	0.0	23.5	23.5	0.1	40	0.36
1	1.2	0.0	23.5	23.5	0.3	40	0.12
1	1.2	0.0	23	21	0.1	40	0.05
1	1.2	0.0	23	21	0.3	40	-0.17
0.5	1.6	0.0	22	22	0.1	60	0.05
0.5	1.6	0.0	27	27	0.1	60	1.17
0.5	1.6	0.0	27	27	0.3	60	0.95

Fig. 3. Result of testing PMV calculator

A comparison of our results with values given standard ISO7730 is presented in table 1.
In addition, it is desired to compare with other existing PMV calculators. A study about software applications that compute PMV index is done in [17]. The authors classify them

according to the environment / operating system (OS) where they are implemented. Thus, there are web applications, Windows or for mobile devices OS. Eight applications have been identified, which is not much around the world. Main input and output parameters for the investigated software are specified.
Besides ISO7730-Table D.1, the output of the new developed software is compared with results obtained from two other PMV calculators: JAVA applet for ISO 7730 [18] and CBE Thermal Comfort Tool [19].
A good correspondence with the ISO7730-Table D.1 values and an almost perfect match with the CBE Thermal Comfort Tool values are remarked. The ISO7730 requirements are met so it can be stated that a PMV calculator has been made in accordance with ISO7730.

Table 1. Validation of the new developed PMV calculator software and comparison with other PMV calculators

PMV value			
ISO7730-Table D.1	JAVA applet for ISO 7730	CBE Thermal Comfort Tool	New developed software
-0.75	-0.79	-0.75	-0.75
0.77	0.74	0.77	0.76
0.44	0.42	0.43	0.43
-0.01	-0.04	-0.01	-0.01
-0.55	-0.57	-0.55	-0.55
-0.6	-0.63	-0.6	-0.6
0.5	0.34	0.36	0.36
0.12	0.11	0.12	0.12
0.05	0.02	0.05	0.05
-0.16	-0.19	-0.17	-0.17
0.05	0.02	0.05	0.05
1.17	1.16	1.17	1.17
0.95	0.94	0.95	0.95

4. Results and discussion

As input for PMV calculator some parameters are estimated, some are read from sensors and some are approximated. Usually the level of clothing and metabolism should be approximated in the calculation of PMV.
For the graphics in the fig. 4 we used a value of 0.5 clo for the level of clothing, that means that the clothes are light. Metabolic rate used is 1.2 met, it is specific to a standing person with low activity. The air velocity used is approximated with 0.1 m/s, a low value, suited to open space offices. Any other values could be used as input data.
The air temperature and relative humidity are read by sensors. The mean radiant temperature is considered equal with air temperature, especially in the center of an open plan office, where the influence of outdoor radiant construction elements is very small.
Thus, the input data in the Python program is the following. (The meaning of the terms is presented in subchapter 3.2.)

```
clo = 0.5;
```

 https://doi.org/10.1051/e3sconf/202018002012

met = 1.2;
wme = 0.0;
ta = value given by sensor DTH22;
tr = ta;
vel = 0.1;
rh = value given by sensor DTH22.

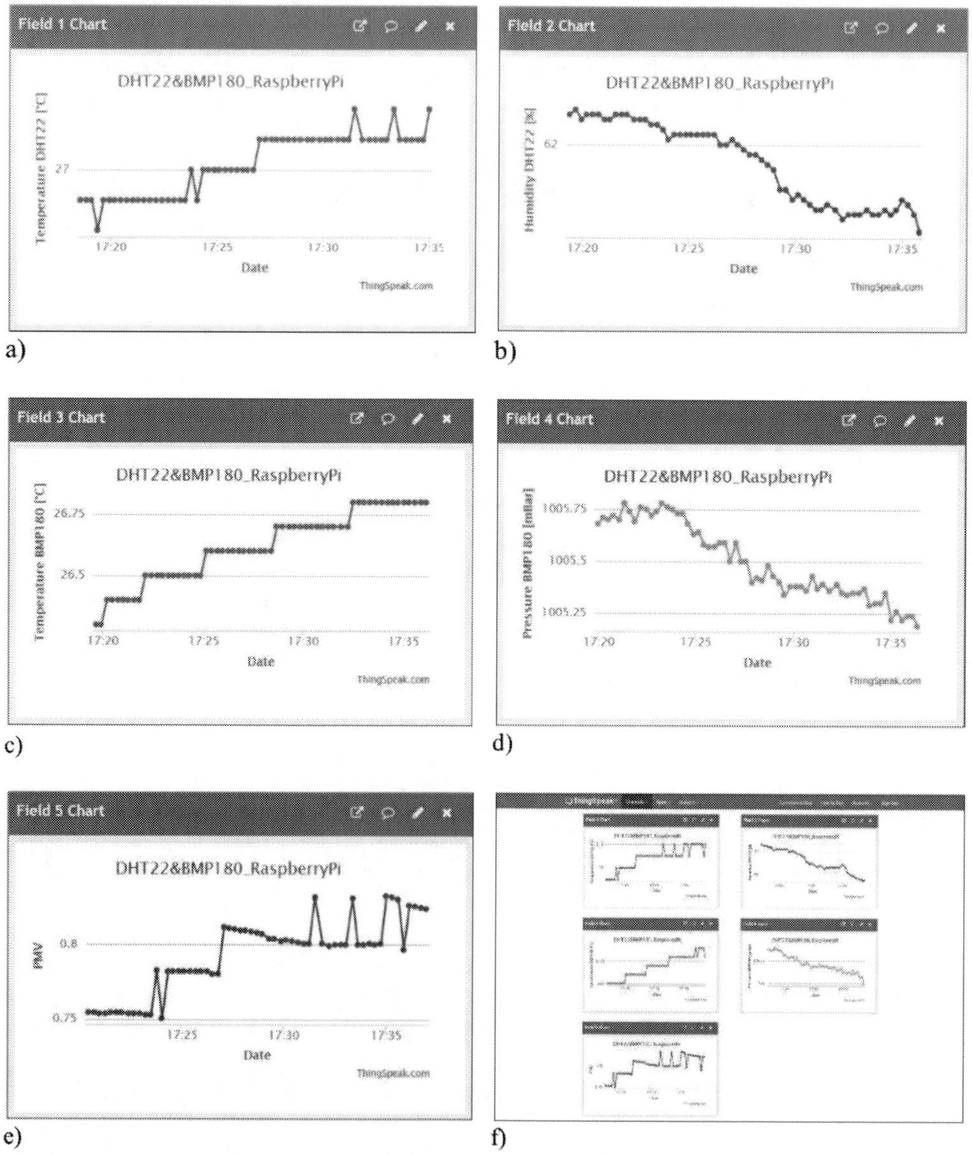

Fig. 4. Charts on ThingSpeak platform: a) Temperature given by DHT22; b) Humidity given by DHT22; c) Temperature given by BMP180; d) Pressure given by BMP180; e) computed PMV; f) "Channels" web page with all charts on it

The graphs in figure 4 represent values measured in a laboratory in an interval of 15 minutes. The data transmission interval to the ThingSpeak platform is 15 seconds. From the graphs in figure 4 we can observe the increasing tendency of the temperature and the decreasing tendency of the temperature. It is observed that PMV also has an increasing tendency, it follows the temperature trend but with small variations because the relative humidity also enters in the PMV calculation.

5. Conclusions

The PMV calculator software is developed according to ISO 7730 [13] and is validated using the standard validation method. After validation it is used to calculate the PMV having as input data the values read from the sensors and certain estimated values.
The novelty of this paper is the real-time publication, on an IoT platform, of the PMV thermal comfort index. The use of less expensive sensors is considered an advantage, it allows mounting the sensors in countless spaces and the transmission of data in real time. Unlike the performance measuring instruments that must be moved from one area / space of a building to another as the measurements are performed. It should not be forgotten that this system is not regarded as a final measure of the thermal comfort index, it is only an indicator on the places where problems might arise, in those areas teams will be sent to perform high performance measurements.
In addition to the advantage of real-time data transmission, IoT platforms also offer the advantage of centralizing information from multiple locations, a simultaneous tracking of all the spaces in a building, why not the entire portfolio of buildings.
We propose that in further studies we install several such systems in the open space offices of a company or in the halls of a university and to continuously record data over a longer period, of a few weeks, maybe even months. During this period, we also aim to compare the results with measurements made with the ConfortSense device [7] to see what are actually the errors introduced by this solution.

References

1. J. H. Nord, A. Koohang, J. Paliszkiewicz, The Internet of Things: Review and theoretical framework, Expert Systems With Applications, 133 (2019) 97–108, doi: https://doi.org/10.1016/j.eswa.2019.05.014
2. S. R. Shinde, A. H. Karode and S. R. Suralkar, Review on-IOT Based Environment Monitoring System, International Journal of Electronics and Communication Engineering and Technology, 8(2), 2017, pp. 103–108. http://www.iaeme.com/IJECET/issues.asp?JType=IJECET&VType=8&IType=2
3. A. Al-Fuqaha, M. Guizani, M. Mohammadi, M. Aledhari and M. Ayyash, "Internet of Things: A Survey on Enabling Technologies, Protocols, and Applications," in *IEEE Communications Surveys & Tutorials*, vol. 17, no. 4, pp. 2347-2376, Fourthquarter 2015, doi: 10.1109/COMST.2015.2444095.
4. J. Gubbi, R. Buyya, S. Marusic, and M. Palaniswami, "Internet of Things (IoT): A vision, architectural elements, and future directions," Future Generation Computer Systems, vol. 29, no. 7, pp. 1645-1660, 2013/09/01/ 2013, doi: https://doi.org/10.1016/j.future.2013.01.010.
5. S. Tang, D. R. Shelden, C. M. Eastman, P. Pishdad-Bozorgi, and X. Gao, "A review of building information modeling (BIM) and the internet of things (IoT) devices integration: Present status and future trends," Automation in Construction, vol. 101, pp. 127-139, 2019/05/01/ 2019, doi: https://doi.org/10.1016/j.autcon.2019.01.020.
6. Fanger P.O (1970), Thermal Comfort. Analysis and Applications in Environmental Engineering. Published in the United States by McGraw-Hill Book Company 1972. Originally presented as the author's thesis, Danmarks Teknishe Højskole 1970

7. ComfortSense, measurement system, developed by Dantec Dynamics, https://www.dantecdynamics.com/solutions-applications/solutions/thermal-comfort/comfortsense/, (accessed Mai 2020)
8. I. Udrea, V. Badescu, Usage of solar shading devices to improve the thermal comfort in summer in a Romanian PassivHaus, Simulation: Transactions of the Society for Modeling and Simulation International 1–16, (2019), DOI: 10.1177/0037549719887790
9. DHT 22, temperature-humidity sensor, https://www.adafruit.com/product/385, (accessed March 2020)
10. BMP180, Barometric Pressure/Temperature/Altitude Sensor, https://www.adafruit.com/product/1603, (accessed March 2020)
11. Raspberry Pi4, desktop computer, https://www.raspberrypi.org/products/raspberry-pi-4-model-b/, (accessed March 2020)ThingSpeak (2020), IoT Analytics Platform, https://thingspeak.com/, (accessed March 2020)
12. ThingSpeak (2020), IoT Analytics Platform, https://thingspeak.com/, (accessed March 2020)
13. ISO-7330, (2005), International Standard ISO 7330: Ergonomic of the thermal environment - Analytical determination and interpretation of thermal comfort using calculation of the PMV and PPD indices and local thermal comfort criteria.
14. Webb, A. L. (2012), Mapping Comfort: An Analysis Method for Understanding Diversity in the Thermal Environment, Master in Science in Architecture Studies at the Massachusetts Institute of Technology
15. Raspbian (2020), official supported operating system for Raspberry Pi, https://www.raspberrypi.org/downloads/raspbian/, (accessed March 2020)
16. Environmental design, CIBSE Guide A, The Chartered Institution of Building Services Engineers London, The Yale Press Ltd, Norwood, London SE25 5LY, ISBN 0 900953 96 9, (1999)- Carte, unde in cap Appendix 1.A1: Determination of predicted mean vote (PMV) am gasit programul recomandat de ISO 7730 reformulat si cu greseli corectate.
17. F.R.A. Alfano, B. I. Palella, G. Riccio, Notes on the Calculation of the PMV index by means of Apps, 71st Conference of the Italian Thermal Machines Engineering Association, ATI2016, 14-16 September 2016, Turin, Italy, published in Energy Procedia 101 (2016) 249 – 256 (2016)
18. JAVA applet for ISO 7730, ISO 7730 Ergonomics of the thermal environment. Analytical determination and interpretation of thermal comfort using calculation of the PMV and PPD indices and local thermal comfort, PMV 2008 ver. 1.0, Ingvar Holmer, http://www.eat.lth.se/fileadmin/eat/Termisk_miljoe/PMV-PPD.html, (accessed March 2020)
19. T. Hoyt, S. Schiavon, F. Tartarini, T. Cheung, K. Steinfeld, A. Piccioli, and D. Moon, 2019, CBE Thermal Comfort Tool. Center for the Built Environment, University of California Berkeley, https://comfort.cbe.berkeley.edu/EN, (accessed March 2020)

E3S Web of Conferences **180**, 02013 (2020) https://doi.org/10.1051/e3sconf/202018002013
TE-RE-RD 2020

Solution for controlling a hydraulic motor using cloud data

Ioana Udrea[1,*], *Viorel Ionut* Gheorghe[1], *Mihai* Avram[1], *Silviu* Petrache[1], *Alina* Popescu-Cuta[1], *Romeo Traian* Popa[1]

[1] Mechatronics and Precision Mechanics Department, Faculty of Mechanical Engineering and Mechatronics, University "POLITEHNICA" of Bucharest, RO

Abstract. In the last period, Internet of Things (IoT) is experiencing an increasing development, and the tendency of hosting applications and especially data, in cloud computing systems is growing. Currently, most of the Building Management System (BMS) analyse directly data or transfer it to the cloud.

In this context, the proposed solution aims to read indoor thermal climate data from cloud computing platform and based on them operate a hydraulic motor. The input data could be generated by an in-house sensor monitoring solution or by a foreign BMS system that transmits open source data to cloud.

In this study some conditions are imposed to the input data and as a result, control current is generated and transmitted in order to command proportional electro-hydraulic valve that drive a hydraulic motor. The electro-hydraulic valve is controlled with the help of an amplifier, by a data acquisition board (DAQ) that runs LabView application. With the help of the developed software, the motor speed and the direction of rotation can be controlled.

In last years, the number of requests for architectures projects with moveable elements has increased because they multiply the possible use of the buildings. With the information generated by the sensors that measure the indoor climate parameters, the hydraulic motor could be used in such a construction. Examples of atypical buildings with special functionalities that the architects would propose and that needs high torque actuation are presented in this paper.

1 Introduction

Internet of Things (IoT), considered by many to be one of the most important technologies of our day, has grown exponentially in recent years [1]. RFID and sensor networks are considered the fastest growing information and communication technologies among the technologies available in IoT [2]. In recent works is emphasized the development and

* Corresponding author: ioana.udrea@upb.ro, udreaioana@yahoo.com

performance of wireless sensors that measure indoor climate parameters [3]. Environmental monitoring is an important IoT application which involves monitoring the surrounding environment and reporting this data for effective short term measures such as remotely controlling the heating or cooling devices and long term data analyses and measures [4]. For the past decades, as a research highlight, networked control systems (NCSs) not only have attracted the attention of many scholars but also have generated huge profits in the industry. Compared with the traditional control systems, NCSs have obvious advantages: low cost, flexibility, easy re-configurability, natural reliability, robustness to failure, and adaptation capability [5]. On the other hand, hydraulic machines are widely used in construction. Computer control can improve motion performance of a hydraulic device in this respect [6].
In this context, the present paper aims to realize a solution for controlling a hydraulic motor using cloud data.
The data will be read from an IoT platform, named ThingSpeak, and according to one of the read parameters, temperature, the control will be done. For this, a computer and a data acquisition board (DAQ) will be used. LabVIEW software will be used to read data from cloud and transform the read value into digital signal for DAQ.
It is intended that DAQ to support automatic control, according to temperature and manual control from LabVIEW [7] software interface (Front Panel). The DAQ will transform digital signal into analog one and will give it to the command unit of a hydraulic motor, see fig. 1.

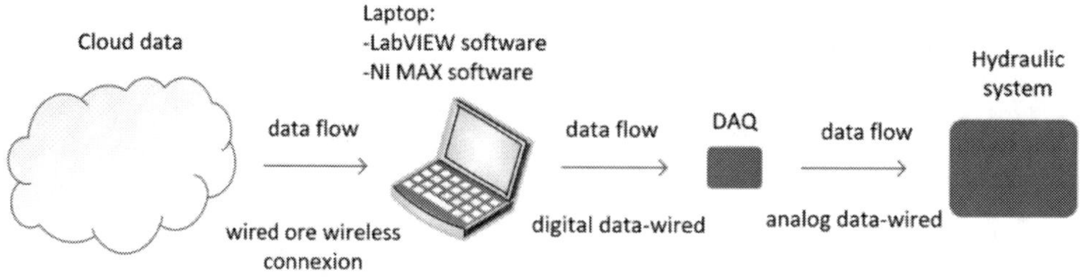

Fig. 1. Data flow in the process developed in this work

2 Method and procedure

This chapter presents the physical elements of the experiment: hydraulic system and data acquisition board and the virtual elements: ThingSpeak platform and data LabVIEW software.

2.1 Hydraulic system description

The entire hydraulic system is composed of the hydraulic part and a proportional amplifier. In fig. 2 a) is presented the hydraulic diagram, composed of: 1 - Pump unit; 2 - 4/3-way proportional valve; 3 - Rotary hydraulic motor (*Sauer Danfoss OMM 8 – Orbital Motors, 8.2 cm³, 1950 min⁻¹*); 4 - Flow meter.
In fig. 2 b) is presented the proportional amplifier and its related elements diagram: 1 - Control block made with NI6009; 2 - Power supply 24Vdc; 3 - Two channel proportional amplifier; 4 - Proportional electromagnetic actuators of the 4/3-way proportional valve (*FESTO, P_{max}: 12 MPa , ISO/DIN 4401 size 02*). In fig. 2 c) a detail photo of pump unit can be seen (*FESTO Hydraulikaggregat TN 152962, P_{max}=12mPa, V=8cm³*).

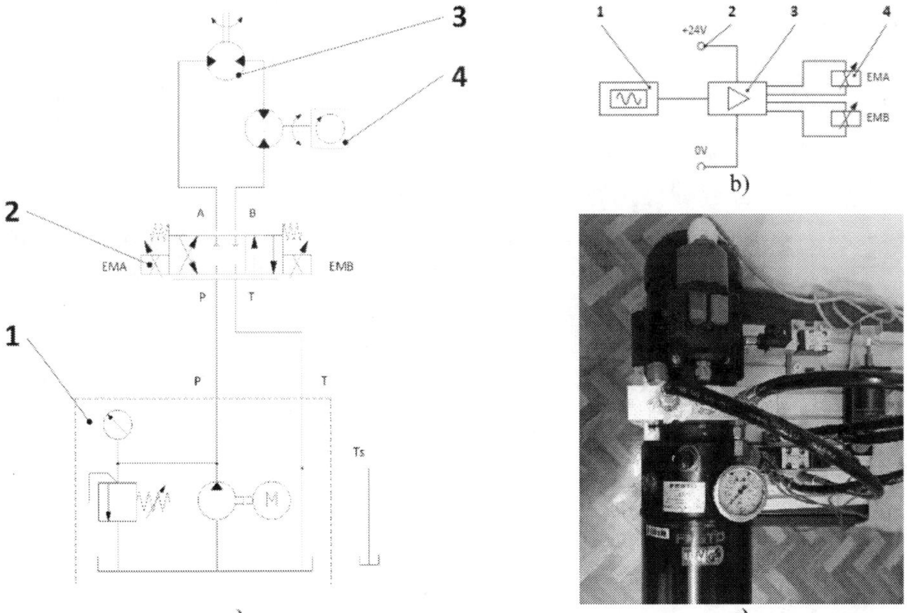

a) b) c)

Fig. 2. Elements of hydraulic system

The amplifier allows control of proportional valves. It is designed so that it can either be driven by two independent solenoids (1 channel) or a valve with two solenoids (2 channel) such as a 4/3-way proportional valve. The proportional amplifier thus works either as a 1-channel amplifier or a 2-channel amplifier. The inputs are short-circuit protected or surge-proof up to 24 V.

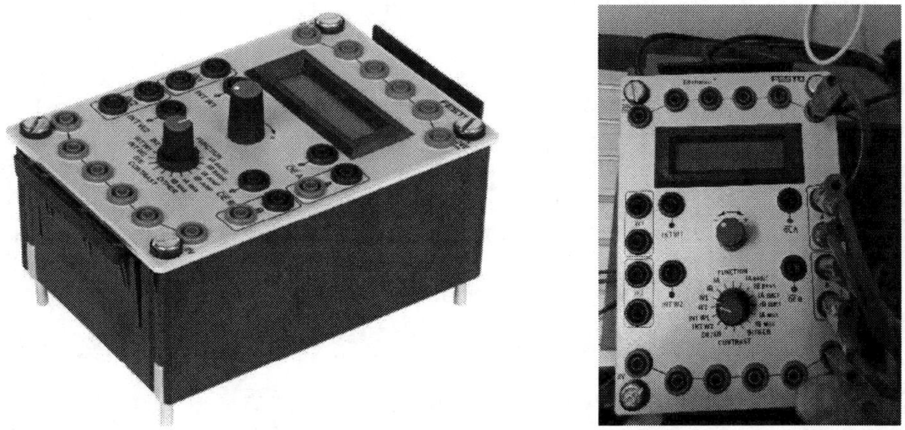

a) b)

Fig. 3. Proportional amplifier 162255 produced by Festo a) image from Festo web site [8]; b) photo of the amplifier used in this experiment

2.2 ThingSpeak platform

ThingSpeak [9] is IoT analytics platform service that allows someone to aggregate, visualize, and analyse live data streams in the cloud. One can send data to ThingSpeak from his devices, create instant visualization of live data, and send alerts. In a recent review paper [10] ThingSpeak is described as an open source platform to recover and store data over the internet.

From the ThingsSpeak platform, channel *DHT22 & BMP180_RaspberryPi* will be used, see fig. 4. Here more fields are created. They show the data transmitted by sensors of indoor thermal climate parameters: temperature, humidity and pressure. A field displays the calculated value of the PMV (Predicted Mean Vote) index, based on the indoor climate parameters read by the sensors and other approximate parameters.

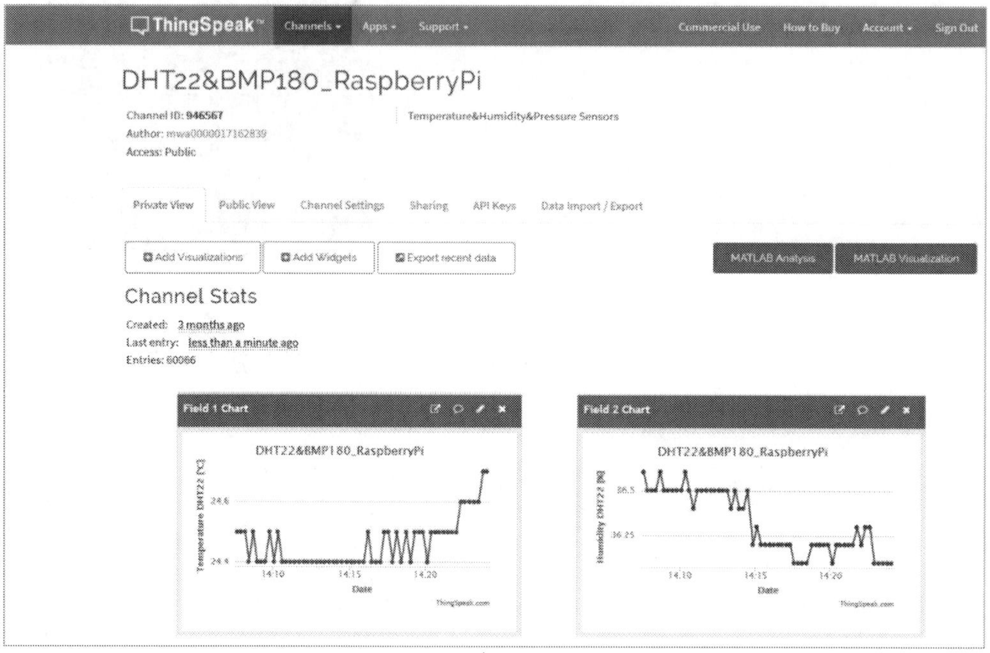

Fig. 4. The channel *DHT22&BMP180_RaspberryPi*, created on ThingSpeak

2.3 LabView software

LabVIEW is systems engineering software for applications that require test, measurement, and control with rapid access to hardware and data insights [7]. In this work was used LabVIEW 2013 release [11]. LabVIEW offers a graphical programming approach structured in two windows: front panel which looks like a control panel in which the IN and OUT data are displayed in a user-friendly way and diagram block where one can see the actual code of the program, realized in a combination of elements and logical structures and graphical links in the form of wires between them, see fig. 8.

2.4 Data acquisition board (DAQ) presentation

The National Instruments (NI) USB-6009 DAQ devices provide eight single-ended analog input (AI) channels, two analog output (AO) channels, 12 DIO channels, and a 32-bit counter with a full-speed USB interface [12]. For this DAQ National Instruments provides a software, named Measurement & Automation Explorer (NI MAX) [13], used for managing devices and interfaces, manage installed NI software, manage virtual channels or

tasks for devices and others. NI MAX version must be compatible with DAQ driver. For studied DAQ the installed version is NI-DAQmx 16.0.
In fig. 5 is presented an image with NI USB-6009 device, used in this work. In fig. 5 b) one can see the DAQ connected to proportional amplifier used for hydraulic system control. The DAQ pins AO 0, AO 1 and GND are connected to this device.

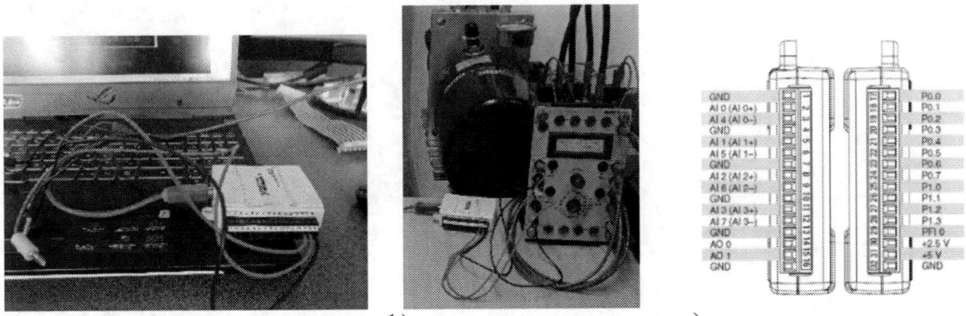

a) b) c)

Fig. 5. Data acquisition board NI USB-6009 used in this work: a) pin connectors used; b) connected with proportional amplifier; c) pinout

In fix. 6 is presented DAQ configuration, realized in NI MAX software. In fig. 6 a) a new channel, for NI USB-6009 is defined. In fig. 6 b) the DAQ output AO 0 is configured. The proportional amplifier accepts input signals named w1 and w2 with values between 0 and 5 V. For this reason, two analog outputs have been configured for DAQ, both having values in the range 0-5 V.

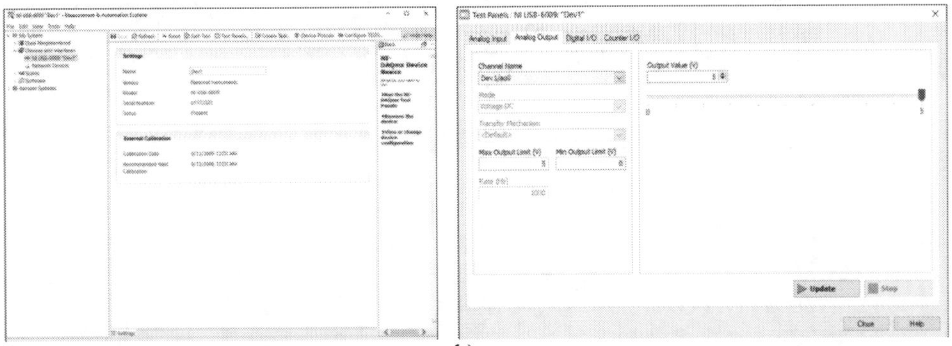

a) b)

Fig. 6. DAQ configuration in NI MAX: a) Channel developed for NI USB-6009; b) Analog Output

3 Results and discussion

The ThingSpeak channel, DHT22 & BMP180_RaspberryPi, presented in subchap. 2.2 provides the data in json format, also.
From the data provided by json file four fields are used in the application developed in LabVIEW software. With this data, in LabVIEW are created the following graphs: Temperature [°C], Relative Humidity [%], Pressure [mBar] and PMV, see fig. 7 a).
The code used for this can be seen in fig. 8. Here a subprogram, named *Read from JSON(SubV).vi* is used. The subprogram has eight outputs, corresponding to instant value

of: temperature, relative humidity, pressure, PMV and graph that shows history data for the instant values.

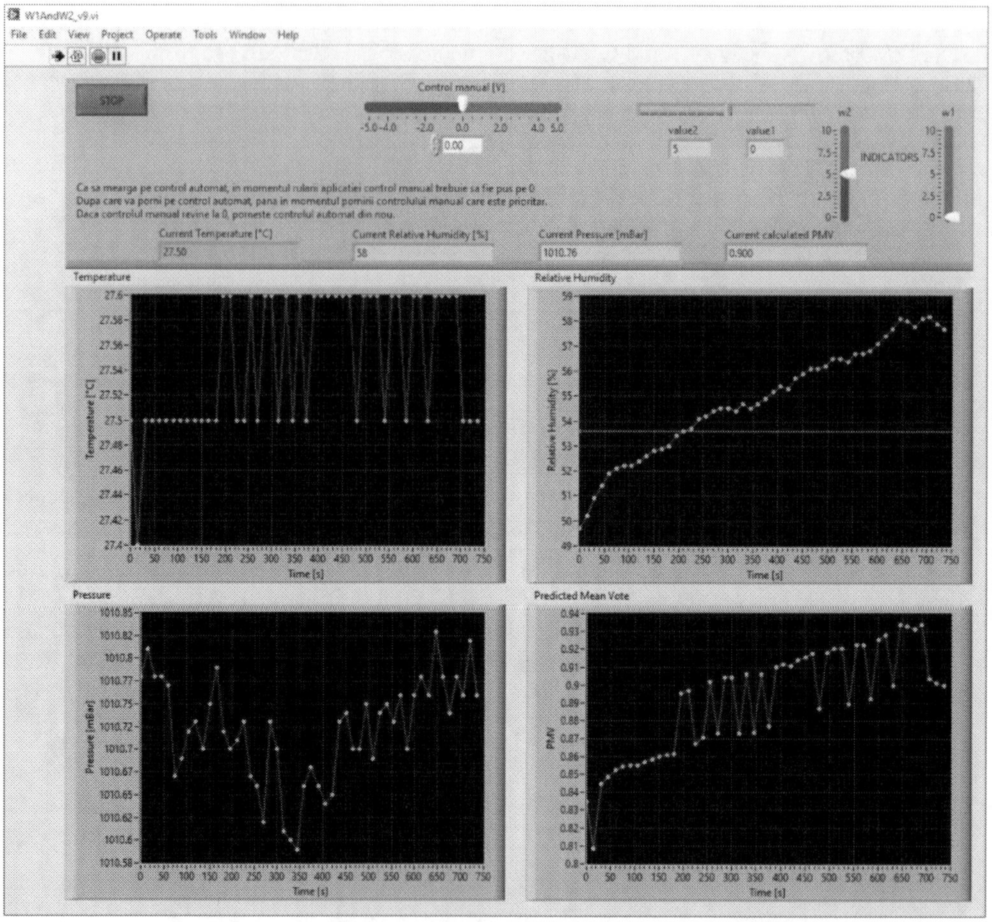

a)

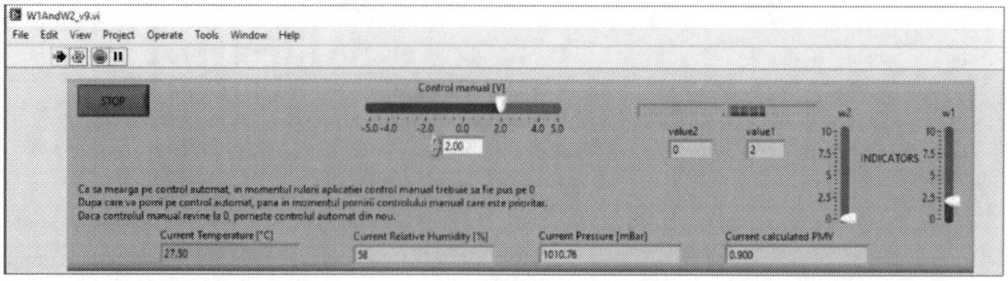

b)

Fig. 7 Application developed in LabVIEW-Front Panel a) automated control; b) manual control

The LabView main application (that use the subprogram) displays the data provided in four data fields and four graphs. It also has a decision block (Formula Node) that determines whether DAQ will receive decisions made based on temperature values or manual controls. The manual control can be in the range -5.0 - 5.0 V, see fig. 7. The data are processed and w1 and w2 take values in range 0 – 5.0V each. An algorithm for temperature is also used.

Manual control	
manual input w ∈ (0, 5)	w1=w; w2=0
manual input w ∈ (-5, 0)	w1=0; w2=-w
Control based on temperature values (when w=0)	
temp < 19	w1=5; w2=0
temp ∈ [19, 21)	w1=(temp-19)*5/2; w2=0;
temp ∈ [21, 25.5]	w1=0; w2=0
temp ∈ (25.5, 27]	w1=0; w2=abs((temp-27)*5/2);
temp > 27	w1=0; w2=5

Tab. 1 Algorithm that transform input data (manual control or automated control based on temperature) in appropriate values for DAQ

As can be seen from the tab. 1 engine speed is variable depending on the temperature value. The temperature intervals in tab. 1 are chosen according to EN15251 category limits [14].

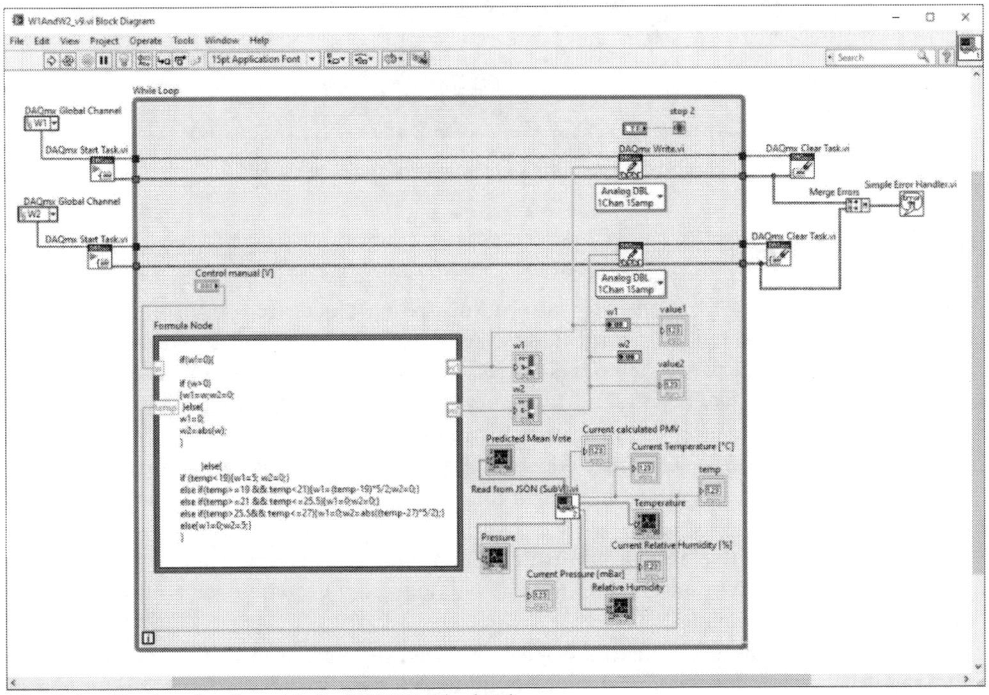

Fig. 8 Application developed in LabVIEW-Block Diagram

In fig. 7, is presented the running application. In fig. 7 a) one can see that instant temperature value is above 20°C threshold, so, according to tab. 1 the values for w1 and w2 should be w1=0; w2=5. The values were calculated correctly, this is shown by the indicators w1 and w2 located in the top right of the image. In fig. 7 b) it is observed that manual control is set to a value of 2 V. In this situation, according to tab. 1 will result w1=2; w2=0. These values are also found for the two indicators w1 and w2.
Manual control is a priority for automatic control. Automatic control only works when there is no manual control (when its corresponding variable is 0).
The elements in the photograph in fig. 9 can be found in the diagram in fig. 1, excepting "cloud", of course.

Fig. 9. Image with all elements of the experiment

4 Conclusions

In this paper real data from an IoT platform located on cloud are used. In this application cloud data is represented by a multi-field channel of the ThingSpeak platform. This data consists of sensor read value: temperature, relative humidity, pressure and a comfort index, PMV. Data is taken from the cloud by interpreting a json file using LabVIEW software [11]. Here is an application that processes the read data and with the help of NI-MAX [13] commands are send to a DAQ that converts the digital signal into an analog one. This is required for a digital amplifier that gives the signal to a distributor that control a rotary hydraulic motor.

Similar studies have not been found in the literature, such as applications presenting the command of a hydraulic motor based on the indoor climate parameters read by sensors, parameters located on cloud.

It is known that hydraulic actuators perform high torque actuation so they can be used for moving heavy elements of an atypical building in correlation with the indoor thermal climate parameters read by the sensors. Such as opening an atrium when exceeding certain values of temperature or thermal comfort indexes, lowering some lifts, moving some surfaces by orienting them towards certain cardinal points and other applications.

An example of an atypical construction that has a retractable roof and uses hydraulic actuation to move it, is the Amsterdam Arena [15]. The control of the hydraulic motor that determines the opening of such a roof could be made based on the interior temperature. In the case of a very large construction, such as a stadium, the cloud solution is useful, it allows the systematization of a large number of sensor systems that would be positioned in different locations of the stadium. Of course, the temperature should be correlated with the outdoor temperature so that the effect is to increase it when it is too small and decrease it when it is too high. Decisions regarding the control of the hydraulic motor can be easily taken depending on the PMV control parameter, which is provided by the studied sensor system. At a lower level, an example of a construction (village hut) whose hydraulically

operated walls close and open depending on the temperature is in Xiahuayuan Village, China [16].
Recently, the number of requests for architectures projects with moveable elements has increased. They multiply the possible use of the buildings, in this sense, a presentation of various types of architectures with moveable elements is made by Christoph Paech [17].
Although less commonly used in buildings, hydraulic fans can be used here for ventilation [18] or as firefighting equipment [19]. In this case it is useful to control them according to the parameters of the indoor climate and even increase or decrease the speed of the fan drive motor according to indoor temperature (which is the input parameter in the case of the acting presented in this paper).

References

1. J. H. Nord, A. Koohang, J. Paliszkiewicz, The Internet of Things: Review and theoretical framework, Expert Systems With Applications, 133 (2019) 97–108, doi: https://doi.org/10.1016/j.eswa.2019.05.014
2. J. Gubbi, R. Buyya, S. Marusic, and M. Palaniswami, "Internet of Things (IoT): A vision, architectural elements, and future directions," Future Generation Computer Systems, vol. 29, no. 7, pp. 1645-1660, 2013/09/01/ 2013, doi: https://doi.org/10.1016/j.future.2013.01.010.
3. A. Mylonas, O. B. Kazanci, R. K. Andersen, B. J. Olesen, Capabilities and limitations of wireless CO2, temperature and relative humidity sensors, Building and Environment, (2019), doi: https://doi.org/10.1016/j.buildenv.2019.03.012
4. S. Zafar, G. Miraj, R. Baloch, D. Murtaza, K. Arshad, An IoT Based Real-Time Environmental Monitoring System Using Arduino and Cloud Service, Engineering, Technology & Applied Science Research, Vol. **8**, No. 4, (2018), 3238-3242
5. Y. Q. Xia, Y. L. Gao, L. P. Yan, M. Y. Fu, Recent Progress in Networked Control Systems – A Survey, International Journal of Automation and Computing, DOI: 10.1007/s11633-015-0894-x, (2015)
6. J. Činkelj, R. Kamnik, P. Čepon, M. Mihelj, M. Munih, Closed-loop control of hydraulic telescopic handler, Automation in Construction, **19** (2010) 954–963
7. National Instrument, LabVIEW software release 2019, https://www.ni.com/ro-ro/shop/labview.html, (accessed March 2020)
8. Festo, Proportional amplifier 162255 (https://www.festo-didactic.com/int-en/learning-systems/equipment-sets/hydraulics/components/proportional-amplifier.htm?fbid=aW50LmVuLjU1Ny4xNy) (accessed Feb. 2020)
9. ThingSpeak, IoT Analytics Platform, https://thingspeak.com/, (2020), (accessed March 2020)
10. R. K. Dwivedi, S. Singh, R. Kumar, Integration of Wireless Sensor Networks with Cloud: A Review, 2019 9th International Conference on Cloud Computing, Data Science & Engineering (Confluence), publisher IEEE, DOI: 10.1109/CONFLUENCE.2019.8776968
11. NI LabVIEW 2013 Release Details, http://www.ni.com/labview/release-archive/2013/, (accessed March 2020)
12. NI USB-6009, Bus-Powered Multifunction DAQ, User Guide, National Instruments, 2015, https://www.google.com/url?sa=t&rct=j&q=&esrc=s&source=web&cd=1&ved=2ahUKEwjGgsbXnsToAhVJxosKHbZqDsAQFjAAegQIARAB&url=http%3A%2F%2Fwww.ni.com%2Fpdf%2Fmanuals%2F371303n.pdf&usg=AOvVaw3oYX2843098LnQYRFcHhZb, (accessed March 2020)

13. NI MAX, National Instruments, Measurement&Automation Explorer, version 16.0.0f0, 2016
14. EN 15251, Indoor environmental input parameters for design and assessment of energy performance of buildings addressing indoor air quality, thermal environment, lighting and acoustics, EN, 2007.
15. Retroth, A Bosh Company-References, Amsterdam Arena, https://www.boschrexroth.com/en/xc/industries/machinery-applications-and-engineering/architecture-in-motion/references/references-50, (accessed Mai 2020)
16. J. Astbury, Operable Interactive Village Hut opens and closes depending on the temperature, https://www.dezeen.com/2019/06/10/operable-interactive-village-hut-weiguo-xu-china/, (accessed Mai 2020)
17. C. Paech, Moveable membranes - smart solutions in the field of architecture, TensiNet Symposium 2019 "Softening the Habitats: Sustainable Innovations in Minimal Mass Structures and Lightweight Architectures", 3-5 June 2019, Politecnico di Milano, Milan, Italy, DOI: 10.30448/ts2019.3245.53
18. T. A. Dill, Hydraulic fan control, US Patent - US4441651A, 1982, https://patents.google.com/patent/US4441651A/en#patentCitations, (accessed Mai 2020)
19. Leader - Hydraulic Fans, https://www.leader-group.company/en/firefighting-equipment/fire-ventilators/hydraulic-fans, (accessed Mai 2020)

Obtaining thermal energy from renewable sources in rural areas using a combined energy system

Liliana Dumitrescu[1,*], *Corneliu* Cristescu[1], *Radu* Rădoi[1], *Ioan* Pavel[1], *Florian* Dragomir[2] and *Oana Diana* Cristea[3]

[1]National R&D Institute for Optoelectronics, Subsidiary Hydraulics and Pneumatics Research Institute INOE 2000-IHP, 040558, Cutitul de Argint 14, Bucharest, Romania
[2]SC PROMECO SD SRL, 062382, Drumul Bacriului 16, Bucharest, Romania

[3]INMA Bucharest, No 6 Bd Ion Ionescu de la Brad Bucharest, Romania

Abstract. The article presents an energy system solution in which the heat energy required to get domestic hot water needed for a family home is obtained by the cumulative conversion of energy from 2 renewable sources: solar energy and biomass energy. Compared to the classic variants of thermal systems, which exploit a single source and provide only partially the energy needed, the combined use consists of energy supply from one source when the other is unavailable or insufficient. In the present case, when the domestic hot water requirement is no longer covered by the solar thermal panels (consumption during the night, days without sun or temporary consumption higher than the usual one), energy from biomass is used in the same system. It is converted to thermal energy using a generator where the gasification module uses the TLUD process. In addition to a higher yield of biomass, this process allows the use of a wide range of biomass types, existing in rural areas. The article presents the results obtained from the experimentation of the system that uses 4 solar panels and a thermal energy generator with a power of approx. 10 kW.

1 Introduction

Renewable energy covered in 2018 a percentage of 18.1% of the total final energy consumed. On the other hand, the renewable energy sector provides jobs for approx. 11 million employees worldwide.

Despite progress in implementing renewable energy systems, increasing energy efficiency and access to conversion systems, the world is not on track to meet the goals of the Paris Agreement or the Sustainable Development Goal 7. Global carbon emissions (CO2) related to energy production increased by about 1.7% in 2018 due to increased consumption of fossil fuels. Global subsidies for fossil fuel consumption have increased by 11% since 2017, and companies selling fossil fuels have continued to spend hundreds of

* Corresponding author: lilianad.ihp@fluidas.ro

millions of dollars lobbying for delaying, controlling or blocking climate change policies and advertising to influence public opinion [1].

It is therefore clear that the solution cannot come only from national or global policies, and that in addition to these, at the micro level there is a large room for maneuver for the population to use renewable energy to help reduce the negative impact of fossil fuel use, on the one hand, and on the other hand to reduce the final energy bill [2]. The use of one or more types of renewable energy, as far as possible in a unitary system, is a solution to this problem. The article presents some aspects related to the development of an energy system that exploits 2 types of renewable energy, in order to produce heat for a small final consumer, such as a home for a family of 3 ... 4 persons. The system covers the need for domestic hot water (DHW) consumption, in conditions of medium comfort.

2 Presentation of the principle underlying the proposed system

The proposed system is based on modern, high-efficiency solar thermal panels, to which is added, for completion, an energy module in which biomass in the form of pellets or briquettes is gasified and burned, when the energy requirement exceeds the amount provided by the panels. The main component is represented by thermal solar panels because solar energy in Romania is in large quantities, superior to other countries, due to its geographical location. The combination of solar energy and biomass energy is also recommended by the large amount of biomass of various types, available at national level (agricultural, forestry, fruit growing, wood processing, etc.) [3, 4].

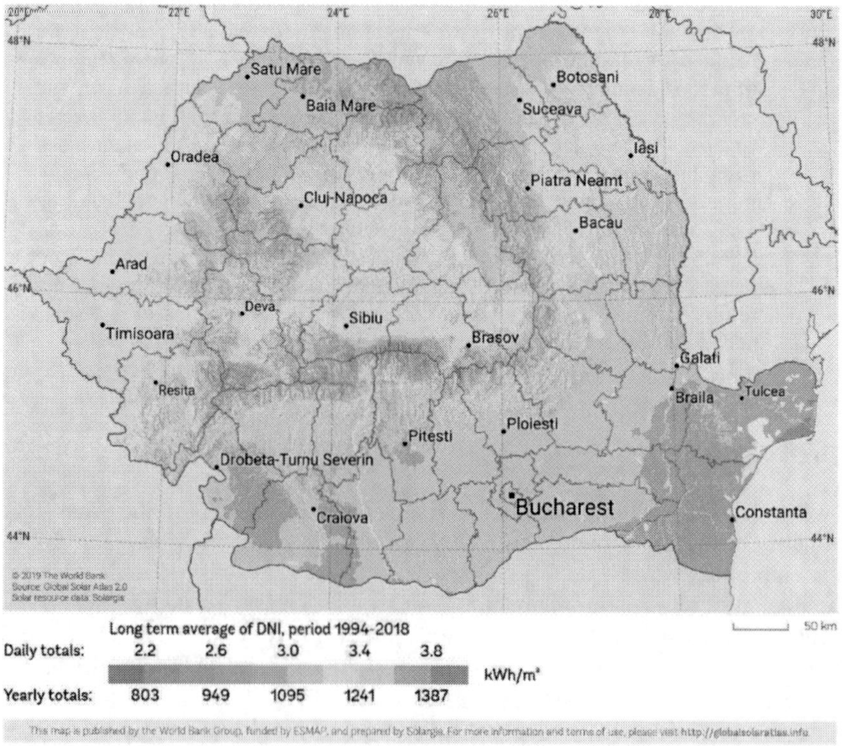

Fig. 1. Map of normal solar radiation in Romania - © 2019 The World Bank, Source: Global Solar Atlas 2.0, Solar resource data: Solargis.

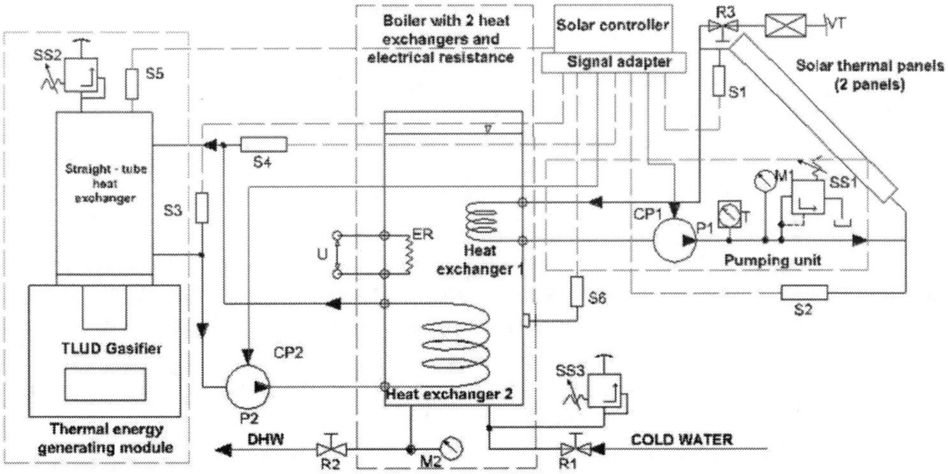

Fig. 2. Schematic diagram of the combined system.

In the figure above one can see the entire structure of the combined system. In addition to the main components, others appear in the diagram: • SS 1,2,3 – Relief valves; • S 1,2,3,4,5,6 – Temperature sensors; • ER – Electrical resistor; • CP 1,2 – Circulation pumps; • VT – Air-vent valve; • M 1,2 – Manometers; • T – Termometer; • DHW – Domestic hot water. Besides the solar thermal panels, the system comprises an energy module that converts energy from biomass into thermal energy. This energy, together with that produced by the solar panels, is stored in a bivalent tank (boiler with 2 coils). The system is managed by a solar regulator, which controls the operation of the circulation pumps that transfer energy from the production areas (solar panels and the energy module, respectively) to the storage area.

3 System description and sizing of main components

The pressurized thermal solar panels (2 pieces) will each have 12 vacuum tubes. Current solar panels with vacuum tubes are "heat pipe"- type solar panels. The vacuum tubes consist of two concentric glass tubes between which a vacuum is formed. The inner tube is covered by an absorbent surface and inside of it there is a copper tube through which a thermal agent circulates. The vacuum between the two tubes minimizes heat loss through convection and conduction, allowing superior performance (higher efficiency and temperatures). Such a solar thermal panel produces from 569 to 780 kWh/m^2/year, if the solar radiation is 1000 kWh/m^2/year [5]. Considering that in Romania there are higher values, of 1000...1450 kWh/m^2/ year, it was considered reasonable to choose a calculation value of 1200 kWh/m^2/year, a value that covers most of the regions of the country where the system can be installed.

If we intend the purchasing of medium performance solar panels, a conversion value of at least 600 kWh/m^2/year can be taken into account.

Given that a 12-tube panel has a minimum surface area of 2 m^2, the amount of energy that the panel can produce annually under the above conditions results as calculated:

$$E_p = 2m^2 \cdot 600 \cdot \frac{kWh}{m^2 \cdot year} \cdot 1\, year = 1200\ kWh \tag{1}$$

For 2 panels we will have:

$$E_{2p} = 2 \cdot 1200 kWh = 2400 kWh \tag{2}$$

If we consider an overall heat transfer factor of 0.9, it results that the useful energy transferred to the water to be heated will be:

$$E_u = 0.9 \cdot\ 2400 kWh = 2160 kWh \tag{3}$$

On the other hand, the need for hot water to be produced was calculated at the level of an average family with 3 members, each using approx. 35 liters of DHW per day, according to current standards used in this field. Therefore, the combined system will have to produce a minimum of 105 liters / day, at a temperature of max. 60°C.

To convert cold water to DHW, heating with 45°C is required (e.g. from 15 to 60°C); therefore, the energy required for annual consumption will be:

$$Q_a = 365 \cdot m_d \cdot C \cdot \Delta T \tag{4}$$

where m_d = 105 kg, C = 4.181 J / kg • K and ΔT = 45°C, hence it results:

$$Q_a = 365 \cdot 105 \cdot 4181 \frac{J}{kg \cdot K} \cdot 45\, K = 365 \cdot 105 \cdot 4.181 \cdot 45 = 7.21 \cdot 10^9 J \tag{5}$$

or Qa = 2003 kWh, taking into account the relation:

$$1\, kWh = 3.6\, MJ \tag{6}$$

If we compare this value with the value of 2160 kWh, we find that the energy produced by the solar panels and transferred is higher than the one required only for use as DHW.

For systems where another energy source is used, whether renewable or conventional, the coverage of the primary source demand is calculated at 80% [6].

The DHW production capacity of the 2 panels was correlated with that of the bivalent boiler, which will have a capacity of at least 120 l. This capacity is considered sufficient to cover the DHW requirement of a family with 3 members, given that the system has 2 sources of thermal energy; thus, during periods when solar radiation, which is the main source of energy, is absent or insufficient, the energy module is put into operation, which also transfers heat energy to the boiler, raising the water temperature. The boiler will also have an electrical resistance.

The energy module is the second source of thermal energy in the combined system. It was designed to produce energy from biomass that is found in various forms, in nature or processed [7].

The energy module consists of 2 main parts, which can be identified in the Figure 3 below:

1. Gasifier made using the TLUD (Top-Lit Up-Draft) principle
2. Heat exchanger

The main characteristics of the gas generator, as it results from the calculation algorithm for a gas generator on the TLUD principle of 10 kWth and operating time of 3.5 h, are the following:

If the specific power of the reactor is approx. 300 kWth / m², then for the power of 10 kWth, proposed in the project, results the reactor section of 10/300 = 0.033 m². This section led to a reactor diameter of 0.2 m.

The input data for the calculation algorithm for the 10 kWth gas generator will be:

Reactor diameter: $D_r = 0.2$ m; biomass layer height (loading height): $H_{rbm} = 0.5$ m.

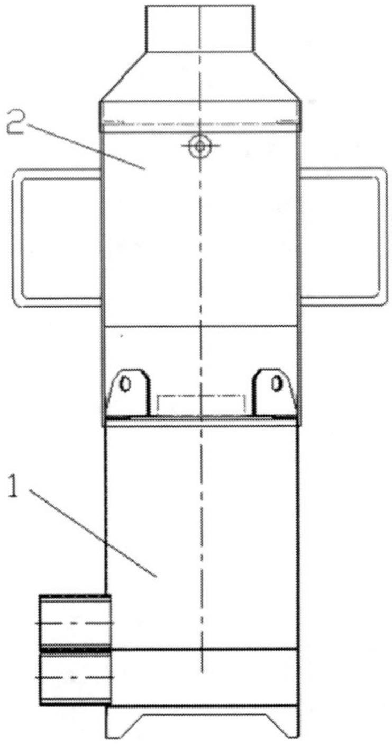

Fig. 3. Thermal energy generating module - part of the combined energy system.

Knowing the diameter of the reactor we obtain the area of its section:

$$S_r = \frac{\pi \cdot D_r^2}{4} = \frac{\pi \cdot 0.2^2}{4} = \frac{0.19625}{4} = 0.0314\, m^2 \quad (7)$$

Volume of biomass in the reactor:

$$V_{rbm} = H_{rbm} \cdot S_r = 0.5 \cdot 0.0314 = 0.0157\, m^3 \quad (8)$$

Since in this phase we intend to use pellets with density $\rho_{pel} = 600$ kg / m³, the initial mass in the reactor will be:

$$M_{bm0} = \rho_{pel} \cdot V_{rbm} = 600 \frac{kg}{m^3} \cdot 0.0157\, m^3 = 600 \cdot 0.0157 = 9.42\ kg \quad (9)$$

For the designed gas generator with the main geometric parameters above, the amount of biomass consumed during a cycle was calculated:

The specific hourly consumption of gasified biomass is $C_{hs} = 85 \frac{kg}{m^2} \cdot h$; therefore, for the surface of the reactor we will have an hourly consumption of:

$$C_{hgbm} = C_{hs} \cdot S_r = \frac{85\,kg}{m^2 \cdot h} \cdot 0.0314\,m^2 = 2.669\,\frac{\text{kg}}{\text{h}} \tag{10}$$

The functioning time will be:

$$T_f = \frac{M_{bm0}}{C_{hgbm}} = \frac{9.42\,kg}{2.669\,\frac{\text{kg}}{\text{h}}} = 3.53\,h \tag{11}$$

Energy from gasified biomass:

$$E_{gbm} = M_{bm0} \cdot P_{Cibm} = 9.42 \cdot 17 = 160.14 \text{ MJ} \tag{12}$$

where the pellet biomass was considered to have a calorific value of 17 MJ / kg [8].

Thermal power of hot gases, expressed in kWth:

$$P_g = \frac{E_{gbm}}{T_f \cdot 3.6} \cdot \eta_{gTLUD} = \frac{160.14 \cdot 0.93}{3.53 \cdot 3.6} = 11.719\ kWth \tag{13}$$

The thermal power at the burner, which takes into account the combustion gas combustion efficiency (η_{burn} = 0.95) and the insulation efficiency (external losses η_{insul}= 0.96), will be:

$$P_{burn} = P_g \cdot \eta_{\text{burn}} \cdot \eta_{insul} = 11.719 \cdot 0.95 \cdot 0.96 = 10.68\ kWth \tag{14}$$

The heat exchanger takes the heat resulting from the combustion of the combustible gas at the top of the gas generator and transfers it to the fluid circulating between the exchanger and the bivalent boiler. It is a tubular construction, with 5 tubes with a diameter of 60 mm, which is mounted above the gas generator, so that the flame produced in the upper part of the gas generator heats the fluid in the exchanger.

4 Conclusions

The combined system, in the proposed structure, covers the domestic hot water needs of a family of 3 ... 4 persons, even in periods when solar energy, which is normally the most important source of energy, is diminished or absent. The system allows easy scaling to increase the amount of DHW. Depending on the area where the system is located and the availability of sources, the share of the 2 primary energy sources (and therefore the size of the 2 subsystems) may vary. At the same time, the system can be used to reduce the consumption of the home heating system, performing a preheating of the heating medium.

This paper has been funded by the Romanian Ministry of Research and Innovation under Programme I-Development of national R&D system, Subprogramme 1.2–Institutional performance–Projects financing excellence in R&D&I, Financial agreement no. 19PFE/17.10.2018, Phase 4, while the scientific results have been obtained in the framework of a project co-financed by the European Union through the European Regional Development Fund, under Competitiveness Operational Programme 2014-2020, Priority Axis 1: Research, technological development and innovation (RD&I) to support economic competitiveness and business development, Action 1.1.4 – Attracting high-level personnel from abroad in order to enhance the R&D capacity, project title: *Establishing a high level proficiency nucleus in the field of increasing renewable energy conversion efficiency and energy independence by using combined resources*, project acronym: CONVENER, Financial agreement no. 37/02.09.2016.

References

1. ***, *Renewables 2018. Global Status Report*, https://www.ren21.net/wp-content/uploads/2019/05/GSR2018_Full-Report_English.pdf (2018)
2. I. Bostan, V. Dulgheru, I. Sobor, V. Bostan, A. Sochireanu, *Renewable energy conversion systems. Wind, Solar, Hydraulic Energy* (Tehnica-Info, Chişinău, 2007)
3. L. Dumitrescu, C. Cristescu, V. Barbu, M. Mateescu, *Proceedings of International Symposium ISB-INMA TEH' 2018*, pp. 965-970 (2018)
4. E. Maican, *Renewable energy systems* (Printech, Bucharest, 2015)
5. ***, http://www.westech-solar.ro, accessed on 12.03.2020
6. ***, https://ec.europa.eu/energy/intelligent/projects/en/projects/east-gsr, Training Manual Romania, accessed on 21.02.2020
7. H.S. Mukunda, S. Dasappa, P.J. Paul, N.K.S. Rajan, M. Yagnaraman, D.R. Kumar, M. Deogaonkar, Curr. Sci. **98**, 5, pp. 627–638 (2010)
8. T.B. Grîu, *Appreciation and increasing the wooden biomass calorific power* (PhD. Thesis, Braşov, Romania: Transilvania University of Braşov, 2014)

Simulation analysis of thermal performance of the solar air/water collector by using computational fluid dynamics

Jafer Shandal[1], *Qahtan* A Abed[2,*], and *Dhafer M.* Al-Shamkhee[3]

[1,3]Engineering technical college/ Najaf, Al - Furat Al- Awsat Technical University, 31001, Iraq
[2]Technical Institute/ Al-Rumaitha, Al - Furat Al- Awsat Technical University, 31001, Iraq

Abstract. Utilizing solar energy has received a giant activity by energy applications into distinctive states within the world. The sun is regarded strong supply providing uninterrupted coherent energy. In this research, simulation study has been done to estimate the performance of dual purpose solar collector (DPSC). Flat plate dual purpose solar collector with dimensions (194 cm length, 95 cm width and 14 cm thickness) has been used for heating the air and water for different using. Rectangular fins are used in air duct to increase heat transfer coefficient in air heater part. COMSOL Multiphysics 5.4 computer program has been used to calculate the theoretical results. Many parameters such as a different flow rate and temperatures for the inlet working fluids are investigated.

1 Introduction

One of the most serious matters in the world is energy. As we know, there are many sources to get the energy. The most important of which is the most common fossil fuel, but, because of the many problem caused by fossil fuel, the attention have been given to searchers for another sores. Solar energy is considered more renewable energy using thermal collectors. These collectors are most common type of solar energy applications. Assari et al [1] Studied an experimental and theoretical investigation performance of two working fluid solar collector to product hot air and hot water. To improve the performance of the hybrid air and water collector, the authors used various air forms of air ducts. They showed that the rectangular fin types have better performance. Nematollahi et al [2] Published an experimental study to investigate the two different types of flat solar collector's effectiveness. Using dual purpose solar collector (DPSC) decreases costs and required space. Venkatesh and Christraj [3] Carry out an experimental study on same collector which design by M.R. Assari et al. But in this study, the authors studied dray a mixture of three vegetable types; dill the first one, parsley second and the last one was coriander at constant air and water flow rates. The obtained results were evaluated with other results had been obtained from the solar drier collector have an electrical heater. The results appear that the dryer solar collector with an electrical heater has 20 °C -30 °C

* Corresponding author: qahtan.abed@atu.edu.iq

temperature more than the dryer solar collector without a heater. Ref.[4] Conducted an experimental work to increase solar collector's efficiency by investigating a combined solar collector. The testing of this system included as solar water and air heater in two cases under the load conditions. The work results show that the solar water heater with 0.015 kg/s water mass flowrate has thermal efficiency equal to 67.69%, while solar air heater with 0.0104 kg/s has an 85% thermal efficiency. However, the rate of thermal efficiency for the hybrid solar collector was 70%. Ref.[5] Design and study experimentally and theoretically of the dual solar collector. Deferent tests are done to investigate dual solar collector efficiency and outlet temperature or both fluids. The authors were found perfect agreement between the theoretical and experimental results. They showed a minimum fluid flow rate it cases to enhance the temperature and efficiency. The solar air with the triangular duct can be providing enhance in the collector efficiency and fluid temperature near to 10% more of the other collectors. The dual solar collector has many interesting benefits. This collector is modifying from the classic geometry to the solar water heater with L-shaped fins fixed in the absorber plate [6]. It's working to raise the amount of heat transfer in air heater side. The authors offer that the DPSC can increase the water (100L) temperature around 30°C during the day. In water heating, daily efficiency reached 50%. While the daily efficiency of air heating reached 52%. Ref.[7] Conducted a numerical simulation work on the DPSC integrated with a porous media. The simulation is carry out by ANSYS 13 software program. The suggestion system is compared with dual solar collector designed by M.R. Assari, et al. The compared show that the increase in the temperature rise is gained for the working fluids (air and water) under different operating conditions. Ref.[8] Carried out an experimental and theoretical study to investigate the DPSC incorporated with the walls. This study conducted to controlled and non-controlled the indoor temperature. The numerical and experimental results were validated. The experimental shows that the temperature of the air inside the room can be increased by about 3.43 °C during the winter season, while the power consumption of the DPSC was 3.5 KWh, when setting the air temperature inside the room to 18 °C. The numerical results show that means daily thermal efficiency of DPSC was 44.3%. Ref.[9] Studied flat plate solar collector experimentally and theoretically. The three modes had been presented, air heating, water heating and air-water compound heating. The mathematical model is analysis the effect of working fluid mass flow rate on the behaviour of air and water heating modes. The thermal collector efficiency is 51.3% and 51.4% for the first and second mode. In the third mode, the rate of the collector efficiency equal to 73.4%. Also, the work results present that the fluid flow rate has affected significantly on fluid outlet temperature and collector efficiency. In the first mode and a second mode, the authors had been recommended the airflow rate reached 0.02 kg/s - 0.25 kg/s and the water flow rate reached 0.06 kg/s - 0.08 kg/s respectively. the solar hybrid collector has been studied experimentally with dehumidification system by Ref.[10]. The system has tested with different value of the airflow rate (50.4 kg/h - 64.8 kg/h) and water flow rate (60 kg/h to 120 kg/h).The authors concluded that the maximum productivity of the distillate during the day of 15.23 kg/m^2 and 14.14 kg/m^2 was obtained from the concave system and convex system respectively. The total thermal efficiency of the system ranges about 68% for the collector integrated with turbulators. The purpose of this study is to prove the enhancement of a DPSC by computational fluid dynamics. A three-dimensional CFD analysis is validated with the previous studies for the DPSC. This collector combined the employment; therefore it can reduce the cost and working area. The collector geometry and weather conditions were studied then

2 Description of the system

The performance of any solar thermal collector may be developed by modified conventional design to air-water solar collector simultaneously. The suggested collector used as boundary conditions detailed by Assari et al [1]. The modified design of air-water solar collector is decrease cost and required space. In this study, the parts of solar air-water collector used are shown in figure 1. The main parts of the suggested dual purpose solar air-water collector are metal pipes and the absorber plate. The absorber plate (94×194×1 mm) consists of triangular fins which have angle equal to °60.

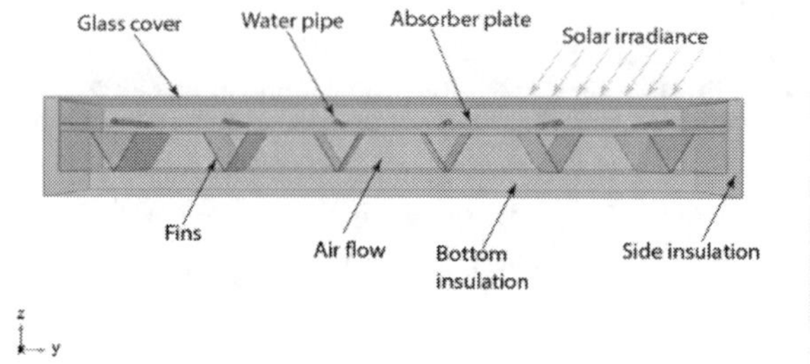

Fig.1. Scheme sectional view of the air-water heater collector model.

The collector is consisting of flat plate solar collector with two parts; one of them for water heating has water pipes, inlet water side and outlet water side. These pipes bonded on the absorber plate. While the other part is air heating with triangular fins to increase the heat transfer coefficient. Table 1 presents the details of geometrical parameters and materials used for the suggested collector.

Table 1. Dual purpose solar air water collector geometric specification

Collector component	Material	Quantity	Value
Absorber plate	Aluminium	High	0.94m
		Width	1.94m
		Thick	0.001m
Water pipes	Zinc	Diameter	7mm
Fins	Aluminium	Triangle with the angle of	60°
Glazing	Float glass	Thick	4mm
Bottom insulation	Glass wool	Thick	5cm
Sid insulation	Silicon rubber	Thick	2cm
Main frame	Stainless steel	Thick	1 mm

3. Modeling of DPSC

The geometry of the dual purpose solar air-water collector is present in Fig. 1. They operate and performance of the solar air-water collector were modelled by using the COMSOL Multiphysics software. It is consists of an absorber plate, water pipe, air channel and thermal insulation layers. The solar irradiance is passing through the glass cover. Therefore is incident on the absorber plate and transformed into internal thermal energy.

The ambient air at temperature T_{amb} is entering through the air channel, it is streaming down the absorber plate and the water pipe, and it gets the heat by two sources; the first one from the absorber plate and the second from triangular fins. At the same time, the water is flowing at inlet temperature $T_{w,in}$ through the pipes; it's heating by the absorber plate and pipe surface. The first modelling point consists of present components of the DPSC. They are shown in Fig. 2. These components correspond to four domains: the water pipes (A), absorber plate (B), triangular fins (C) and air channel (D).

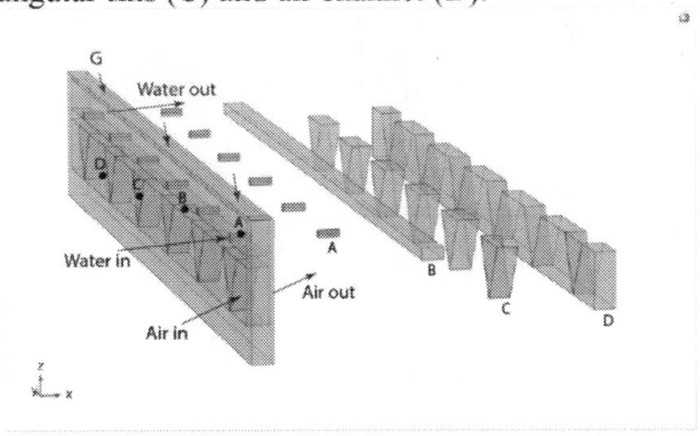

Fig. 2. Shows the main components of the DPSC. (A) water pipes, (B) absorber plate, (C) triangular fins, (D) air channel

The water flow inside the circular pipes was laminar flow type. It is depending on the Reynolds number (2000 to 4000). This is defined as shows:

$$R_e = \frac{\rho D V}{\mu} \tag{1}$$

Where: ρ and V is the density and velocity of working fluid (water), respectively, D circular pipe diameter and water dynamic viscosity.

3.1 Geometry and grid generation

The system modeling and simulation was carried out by COMSOL Multiphysics 5.4. It is a good environment to solve set of equation systems of unsteady-state or stationary in space partial differential equations in two or three dimensions. The primary stage of the built DPSC model is the type of mesh. Therefor many types of CFD mesh have been created (see figure 1). The type and number of mesh are larger in this model where fluid velocity gradients, temperatures are presumed to be significant. Figure 3. shows the mesh close to the water pipes, the triangular fins and the absorber plate is small sizes to the accuracy of the solution.

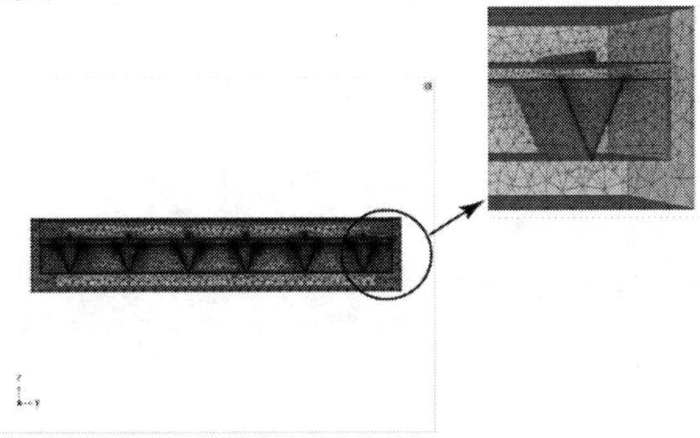

Fig. 3. The mesh type for the DPSC model.

The complete mesh consists of 1186525 domain elements, 186108 boundary elements, and 20144 edge elements. The assumptions used in simulation analyses of air-water solar collector are listed below:

- The solar air-water collector is 3D flow.
- The mean temperature value for all metal parts of the collector ($T_{p,m}$).
- The flow of the working fluid is laminar
- Neglecting the heat losses by convection and radiation from the metal parts.

3.2 Governing equations

The equations using in the fluid flow is incompressible Navier-Stokes equations. The governing equations of modelled:

- Continuity equation

$$\rho\left(\frac{\partial u}{\partial x}+\frac{\partial v}{\partial y}+\frac{\partial w}{\partial z}\right)=0 \tag{2}$$

Where ρ is fluid density (kg/m^3), u, v, w are velocity vectors (m/s).

- Momentum conservation equation

x- direction momentum equation

$$\rho\left(\frac{\partial u}{\partial t}+u\frac{\partial u}{\partial x}+v\frac{\partial u}{\partial y}+w\frac{\partial u}{\partial z}\right)=-\frac{\partial p}{\partial x}+\mu\left(\frac{\partial^2 u}{\partial x^2}+\frac{\partial^2 u}{\partial y^2}+\frac{\partial^2 u}{\partial z^2}\right) \tag{3}$$

y- direction momentum equation

$$\rho\left(\frac{\partial v}{\partial t}+u\frac{\partial v}{\partial x}+v\frac{\partial v}{\partial y}+w\frac{\partial v}{\partial z}\right)=-\frac{\partial p}{\partial y}+\mu\left(\frac{\partial^2 v}{\partial x^2}+\frac{\partial^2 v}{\partial y^2}+\frac{\partial^2 v}{\partial z^2}\right) \tag{4}$$

z- direction momentum equation

$$\rho\left(\frac{\partial w}{\partial t}+u\frac{\partial w}{\partial x}+v\frac{\partial w}{\partial y}+w\frac{\partial w}{\partial z}\right)=-\frac{\partial p}{\partial z}+\mu\left(\frac{\partial^2 w}{\partial x^2}+\frac{\partial^2 w}{\partial y^2}+\frac{\partial^2 w}{\partial z^2}\right) \tag{5}$$

- Energy conservation equation

$$\rho.C_p\left(\frac{\partial T}{\partial t}+u\frac{\partial T}{\partial x}+v\frac{\partial T}{\partial y}+w\frac{\partial T}{\partial z}\right)=k\left(\frac{\partial^2 T}{\partial x^2}+\frac{\partial^2 T}{\partial y^2}+\frac{\partial^2 T}{\partial z^2}\right) \tag{6}$$

Where C_p, T, μ and k are specific heat of fluid (J/(kg.K)), fluid temperature (K), dynamic viscosity of fluid (kg/(m.s)), and thermal conductivity of fluid (W/(m.K)) respectively.

3.3 Initial conditions

The above equations have been solved by the following initial conditions.

At first the initial temperature of the initial air, water inlet and absorber plate equal the ambient air temperature at the starting time (which t=0).

$$T_{air,w,p}(x,y,z,t=0)=T_{amb}(t=0) \tag{7}$$

While, the initial air and water velocity field (in 3D) are equal to zero:

$$u_{air,w}(x,y,z,t=0)=v_{air,w}(x,y,z,t=0)=w_{air,w}(x,y,z,t=0)=0 \tag{8}$$

3.4 Boundary conditions

The constant inlet water temperature, however the inlet air temperature is equal to ambient temperature, while the inlet pressure equals the measured atmospheric pressure:

$$T_w(x=0,y,z,t)=20\ ^oC \tag{9}$$

$$T_{air}(x=0,y,z,t)=T_{amb}(t) \tag{10}$$

$$p(x=0,y,z,t)=p_{atm}(t) \tag{11}$$

The air channel walls which do not in contact with the absorber plate or the has good thermal insulation, also there are No-slip condition at the air chanal walls, therefore

$$-k_{air}\nabla T=0 \tag{12}$$

$$u_{air}(x,y,z,t)=v_{air}(x,y,z,t)=w_{air}(x,y,z,t)=0 \tag{13}$$

4 Model validation

In order to identify differences between the experimental results of the previous work and the simulation results of this study, prepared the chart described below which shows the absorber plate temperature of the two studies and change with time. We note good convergence between the two results. (See fig.4)

Two indicators have been used to show the model accuracy: the first is Mean Bias Error (MBE) as follows:

$$MBE=\frac{1}{n}\sum_{i=1}^{n}(T_{i,comp}-T_{i,exp}) \tag{14}$$

where: $T_{i,\exp}$ and $T_{i,comp}$ are the experimented and computed absorber plate temperature, respectively, while n refer to the total number of values.

While the second indicators is Root Mean Square Error (RMSE) is defined as:

$$RMSE=\sqrt{\frac{1}{n}\sum_{i=1}^{n}(T_{i,comp}-T_{i,exp})^2} \tag{15}$$

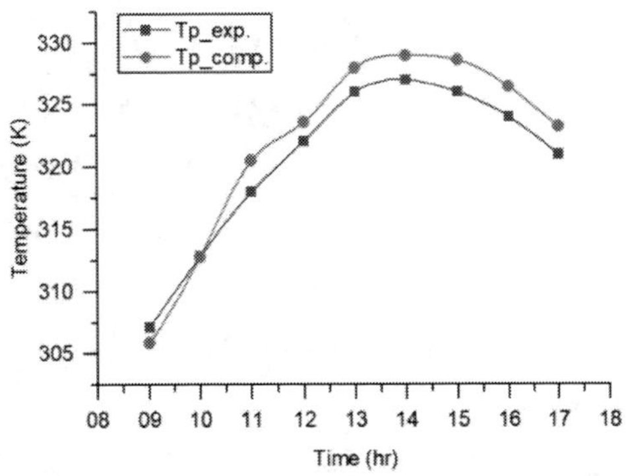

Fig. 4. Computed and experimental results acquired during 21 May 2018 for a DPSC.

We note a good convergence between the experimental results and the simulation results; this is due to the adoption of the same conditions in simulation.

5 Result and discussion

The solar air-water collector under test was directed southward in the city of Najaf, Iraq (latitude 32°02', longitude 44°20') and the tilt angle is 32° .Three days of the year, summer, winter and spring data were used to compare results. The simulation results for the modified the DPSC is presented in this section. The meteorological (weather condition) and operation parameters (outlet temperature) during three days on January, May and July are shown in figures 5-7.

Figure 5 (A) shows that during an experimental winter (5 January) day the solar irradiance is lower on the dual purpose solar collector surface than the other two days.

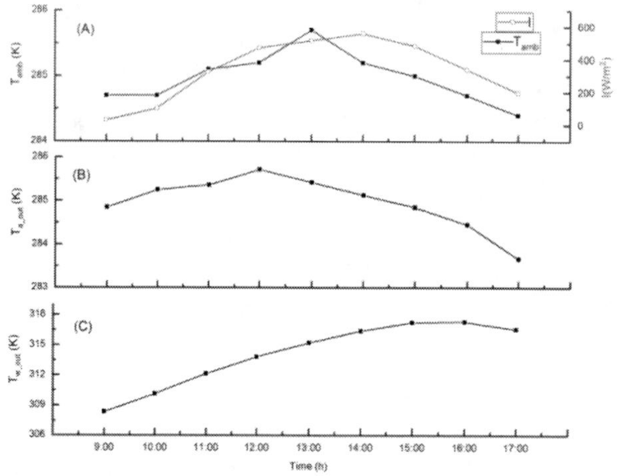

Fig. 5. (A) Solar irradiance and ambient temperature on cloudy day 5 Jan. 2018, (B) outlet air temperature and (C) outlet water temperature

The simulation outlet water and air temperatures increases have a time variation that is analogous to the direct solar irradiance curve (like Fig. 5B and Fig. 5C). Next figure 6, the DPSC operation during clear sky day on May is presented. The outlet water and air temperatures are smaller in the morning for the DPSC than the end of the day. This meaning lower solar irradiance on the morning and accumulated the heat in the collector at the end of the day.

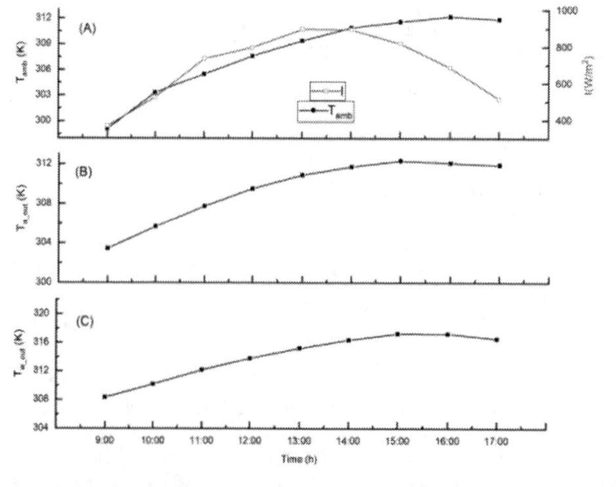

Fig. 6. (A) Solar irradiance and ambient temperature on nearly clear sky day 1 May 2018, (B) outlet air temperature and (C) outlet water temperature

The contours draw of the DPSC temperatures obtained from the 3D CFD analysis are described in figures 7, 8 and 9. They are presenting the temperature gradient for the two working fluids (water and air) for three days. The contour DPSC of magnitudes temperatures assist to understand the thermal heat phenomenon of the solar collector during the day.

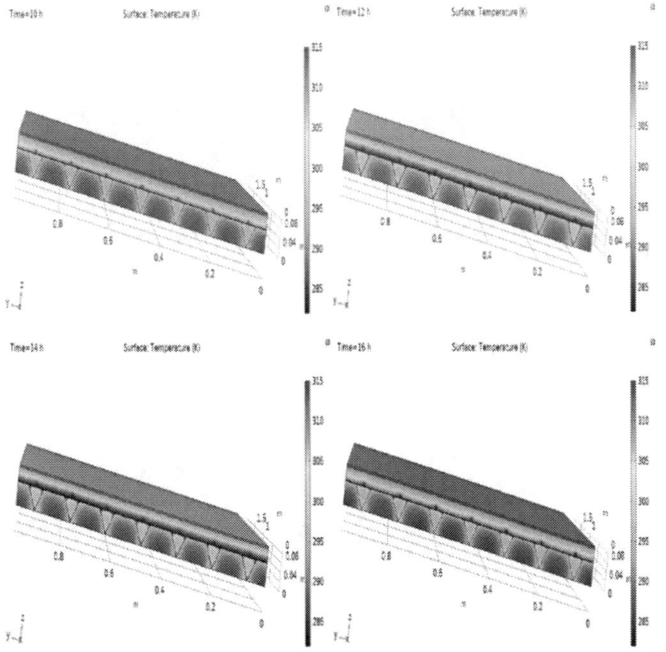

Fig. 7. Temperature distribution for the fluids stream along the flow at summer season

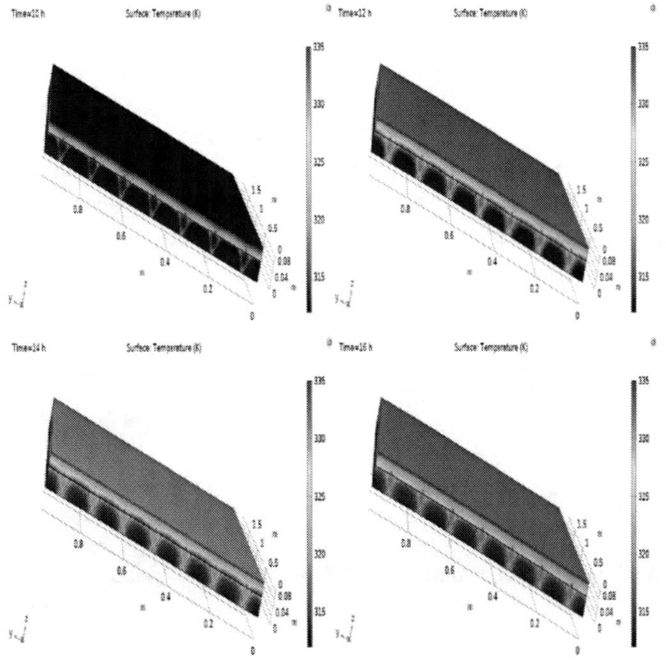

Fig. 8. Temperature distribution for the fluids stream along the flow at winter season

The fluid temperature distribution in a DPSC with various times during the day has been shown in Fig. 9. At the beginning of the day, the absorber plate and water pipe temperature are higher than in other parts. The absorber plate sends heat energy to the water pipes and air channel that transmit this energy to water and air. They observed that higher temperature values are expected at 2 PM for three days.

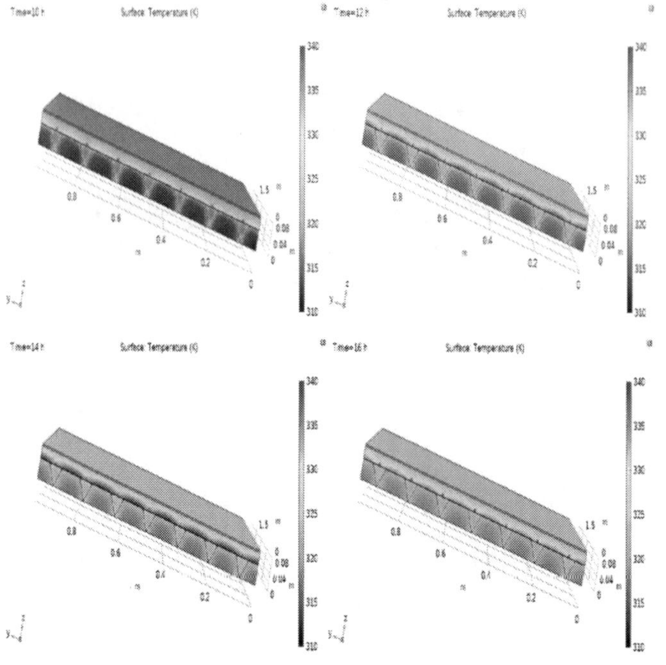

Fig. 9. Temperature distribution for the fluids stream along the flow at spring season

The above figures show that the temperature distribution in the DPSC three days for three seasons. The maximum temperature about a (341k, 319K and 335K) in the absorber plate in Summer, winter and Spring respectively. It starts with the gradual escalation with the solar radiation increase. The distribution of air velocity is showing in figure 10. Air velocity varies between 0.02 m/s - 0.18 m/s. The velocity of air near the fins wall is closed to zero; it then gradually increases until it reaches the greatest value in the middle of the fins. Air velocity effects on the air outlet temperature, when the velocity increases, the air outlet temperature decreases.

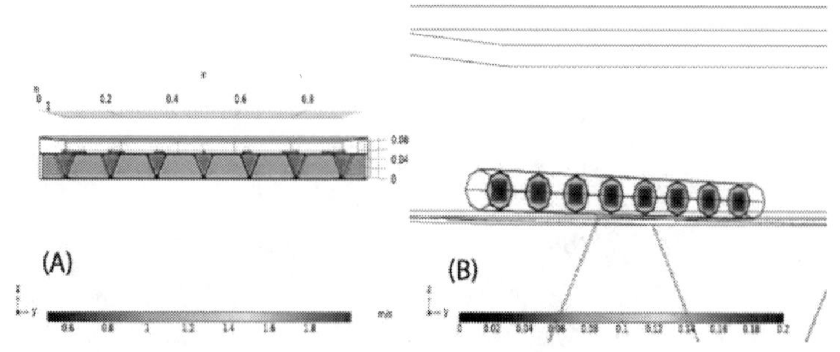

Fig. 10. The fluid velocity distribution inside DPSC: (A) air, (B) water.

Figure 10 shows the flow of water in the pipes. Because the velocity is equal in all pipes according to the simulation assumptions, we took one pipe to show the velocity change.

The velocity in the pipe is fully developed where the greatest velocity is concentrated in the middle.

6 Conclusion

A simulation analysis has been study to investigate the thermal performance of the solar air\water collector by using computational fluid dynamics was consider in this paper. The system modelling and simulation studies have been performing by using software COMSOL Multiphysics 5.4. Temperature and velocity of fluid streams are present under time dependent situation. The reduction of airflow by fins increases the outlet air temperature. Parametric analyses on the system are fluid flow rates, solar radiation and water inlet temperature. When increases water inlet temperature, efficiency decreases, other ways the efficiency increases with increases the mass flow rate. The differences between actual and simulation results for the dual-purpose solar collector is 2.23%. The higher efficiency was 69.42% in summer.

References

1. M. R. Assari, H. Basirat Tabrizi, and I. Jafari, 'Experimental and theoretical investigation of dual purpose solar collector', Sol. Energy, **85**, (2011).
2. O. Nematollahi, P. Alamdari, and M. R. Assari, 'Experimental investigation of a dual purpose solar heating system', Energy Convers. Manag., **78,** (2014).
3. A. Mohajer, O. Nematollahi, M. M. Joybari, S. A. Hashemi, and M. R. Assari, 'Experimental investigation of a Hybrid Solar Drier and Water Heater System', Energy Convers. Manag., **76,** (2013).
4. R. Venkatesh and W. Christraj, 'Experimental Investigation of Multipurpose Solar Heating System', J. Energy Eng., **141,** 3, (2013).
5. H. K. and M. M. M.R. Assari, H. Basirat Tabrizi, 'Design and Performance of Dual-Purpose Solar Collector', no. August, (2007).
6. J. Ma, W. Sun, J. Ji, Y. Zhang, A. Zhang, and W. Fan, 'Experimental and theoretical study of the efficiency of a dual-function solar collector', Appl. Therm. Eng., **31,** 10, (2011).
7. A. Venu and P. Arun, 'Simulation Studies on Porous Medium Integrated Dual Purpose Solar Collector', **3,** 1, (2013).
8. J. Ma et al., 'The thermal behavior of a dual-function solar collector integrated with building: An experimental and numerical study on the air heating mode', Energies, **11**, 9, (2018).
9. D. Zhang, J. Li, Z. Gao, L. Wang, and J. Nan, 'Thermal performance investigation of modified flat plate solar collector with dual-function', Appl. Therm. Eng. **5,** 108, (2016).
10. T. Rajaseenivasan and K. Srithar, 'Potential of a dual purpose solar collector on humidification dehumidification desalination system', Desalination, **28**, 404. (2017).

Energy and exergy analysis of a flat plate collector using collector test equations

Adina-Teodora Gheorghian[1,*], *Tudor* Prisecaru[1], and *Mihaela Cristina* Ciobanu[1]

[1] "Politehnica" University of Bucharest, Department of Engineering Thermodynamics, 313 Splaiul Independentei, 060042, Bucharest, Romania

– **Abstract.** This paper presents the energetic and exergetic analysis of a widely used flat plate collector based on the collector testing equations. The evaluation of the performances of the collector is done under the climatic conditions corresponding to a typical meteorological year in Constanta, Romania. Using a dimensionless-geometric optimization criterion, the climatic conditions for which the energy efficiency and exergetic efficiency have the maximum values are determined. The optimal weather conditions for the collector's operation are: the solar irradiation of 916 W/m^2 and the air temperature of 15.4 °C; under these conditions, the energy and exergy efficiencies of the collector are 57% and 4.8%, respectively.

1 Introduction

Solar collectors are usually investigated based on energy analysis (the first law of thermodynamics). Unfortunately, this analysis does not take into account the degradation of the energy quality that takes place in the process of heating the working fluid [1]. For this, an exergetic analysis (the second law of thermodynamics) should be performed. Exergy analysis is useful for identification of thermodynamic imperfections and for the design improvement of solar systems [2].

In recent years, several papers have been published regarding the thermo-economic modelling and optimization of flat plate collectors (FPC) using energy and exergy analyses.

A procedure for designing and optimizing a FPC based on exergetic analysis is described in the paper [3]. The obtained results show that the energy efficiency increases with the increase of the operating parameters. Exergetic efficiency increases with increasing solar irradiation and decreases rapidly as air temperature and wind speed increase. The exergetic efficiency increases with the inlet fluid temperature until the temperature reaches approximately 315 K, after which it rapidly decreases.

The paper [4] studies the effect of the design parameters on the performance of a FPC based on a theoretical energy and exergy efficiency model. The results of the study showed that the overall performance of the manifold improves when the fluid inlet temperature is 40 degrees higher than the ambient temperature and the mass flow rate is low.

Optimization through maximizing energy efficiency and minimizing the annual cost of a solar heater using flat plate collectors is presented in [5]. To solve the multi objective

* Corresponding author: adina_gheorghian@yahoo.com

optimization problem, the particle swarm optimization (PSO) algorithm is used. The obtained results showed that the highest efficiency and the lowest annual cost are obtained for a lower heat transfer rate.

The paper [6] performs a comprehensive analysis of the energetic, exergetic and financial performances of a flat plate collector and heat pipes located in Fez, Morocco. The mathematical model of the solar collector has been implemented in Matlab based on the coupled heat transfer equations. Using hourly and monthly performance indices, the authors study the effect of the weather, operating and design parameters on the performance collector's. It was found that the best combinations of these parameters are: the temperature of 12.18 °C, the wind speed of 1.5 m/s, the solar irradiation of 986.93 W/m^2, the fluid's inlet temperature of 45 °C, the mass flow of 0.036 kg/s and 13 heat pipes. For this combination, the performance of the collector is optimum (the thermal efficiency is 54.01%, the exergy efficiency is 6.70% and the overall cost is 771.5 $).

The energy efficiency analysis of two planar solar collectors based on the collector testing equations and on a theoretical model is presented in paper [7]. The authors point out that the thermal efficiency increases with the mass flow rate, especially at high average temperature values of the working fluid.

The paper [8] presents the exergetic and economic analysis of a FPC that uses riser tubes with two different heat transfer enhancers (a rod and a tube). It has been found experimentally that the rod heat transfer enhancer provides a relatively higher exergy efficiency than the tube heat transfer enhancer. The maximum exergetic efficiency is 11.3% for the rod and 10.9% for the tube, compared to only 8.3% for the rising tube (plane, without enhancer).

This study presents the energetic and exergetic analysis of a FPC based on the collector testing equations. The evaluation of the performances of the collector is carried out under the climatic conditions corresponding to a typical meteorological year in Constanta, Romania. Using a dimensionless-geometric optimization criterion, the climatic conditions for which the energy efficiency and exergetic efficiency have maximum values are determined.

2 Methods

2.1 Solar collector

The studied solar collector is a FPC type, which has good performances and is mostly often used for domestic hot water preparation.

The technical characteristics of a good quality FPC, frequently met on the local market, are provided in Table 1.

Table 1. Main characteristics of the studied flat plate solar collector

Characteristic	UM	Value
Dimensions (Length x Width x Height)	mm	1970 x 970 x 86
Gross area, A_G	m^2	1.91
Aperture area, A_A	m^2	1.73
Absorber	-	Cooper with blue selective coating
Cover	-	Tempered, low iron prismatic glass 4 mm
Insulation (rock wool)	-	back : 40 mm, edge : 20 mm glass
Weight	kg	41

The performance parameters, determined by the steady state method, mentioned in the test certificate of the collector, are shown in Table 2.

Table 2. Performance parameters (related to gross area) of the studied FPC

Parameter	UM	Value
Peak collector efficiency, $\eta_{0,b}$	-	0.672
Heat loss coefficient, a_1	$W/(m^2 \times K)$	0.310
Temperature dependence of heat loss coefficient, a_2	$W/(m^2 \times K^2)$	0.014
Incidence angle modifier for beam irradiance, $K_b(\theta)$ for θ = [10 20 30 40 50 60 70 80] degree	-	[1.00 0.99 0.97 0.93 0.88 0.78 0.58 0.00]
Incidence angle modifier for diffuse irradiance, K_d	-	0.82
Flow rate for testing	$kg/(s \times m^2)$	0.02
Heat transfer fluid	-	water-glycol

These performance parameters were used as input data for the mathematical model described below.

2.2 Location and weather data

The location chosen for this study was the city of Constanta (44.177 °N, 28.633 °E, III 9 climate region), on the Black Sea coast.

A set of hourly meteorological data corresponding to a Typical Meteorological Year (TMY) for this location was retrieved using Photovoltaic Geographical Information System (PVGIS) TMY tool [9]. The annual range used to define TMY was 2007-2016.

The meteorological parameters used as input data in the mathematical model were: time, t [h], ambient air temperature, T_a [°C], solar irradiance on a horizontal surface, G_{horiz} [W/m^2] and solar irradiance on a surface always normal to the Sun, G_{bn} [W/m^2].

2.3 Mathematical model

The useful power extracted from the solar collector according to ISO 9806, valid for stead-state testing, is:

$$\dot{Q}_u = A_G\left[\eta_{0,b} K_b(\theta) G_b + \eta_{0,b} K_d G_d - a_1(T_m - T_a) - a_2(T_m - T_a)^2\right], \quad (1)$$

where A_G [m^2] is the collector's gross area, $\eta_{0,b}$ – peak collector's efficiency, $K_b(\theta)$ – the incidence angle modifier for beam solar irradiance, G_b [W/m^2] – beam solar irradiance, K_d – the incidence angle modifier for diffuse solar irradiance, G_d [W/m^2] – diffuse solar irradiance, a_1 [W/m^2/K] – the heat loss coefficient, a_2 [W/m^2/K^2] – temperature dependence on the heat loss coefficient, T_m [°C] – mean temperature of heat transfer fluid, T_a [°C] – ambient air temperature. Details of computing beam and diffuse solar irradiance can be found in reference [10]. The incidence angle modifier is computed by linear interpolation using the input data.

The annual useful energy per square meter of the solar collector is obtained by summing the heat produced every hour of a year:

$$\frac{Q}{A_G} = \sum_{t=0}^{t=8760} \frac{\dot{Q}_u}{A_G} \cdot t \quad (2)$$

The thermal efficiency of the collector, which is defined as the ratio of the useful gain over one hour to the incident solar energy over the same time period is:

$$\eta = \frac{\dot{Q}_{u}}{\dot{Q}_{S}} = \frac{\dot{m}C_{p}(T_{out} - T_{in})}{A_{G} \cdot G_{T}} , \tag{3}$$

where T_{out} [°C] is the working fluid's outlet temperature, T_{in} [°C] is the working fluid's inlet temperature, G_T [W/m^2] – global solar irradiation on the tilted surface, [kg/s/m^2] – the mass $\dot{m}$ flow rate of the working fluid per m^2 of collector, C_p [J/kg/K] – the specific heat of the fluid, [W] – the power $\dot{Q}_S$ received from the Sun.

The exergy efficiency of the collector is the ratio of the useful exergy $\dot{Ex}_u$ over the absorbed exergy $\dot{Ex}_S$ [2]:

$$\eta_{ex} = \frac{\dot{Ex}_{u}}{\dot{Ex}_{S}} = \frac{\dot{m}C_{p}\left[(T_{out} - T_{in}) - T_{a} ln\left(\frac{T_{out}}{T_{in}}\right)\right]}{A_{G} \cdot G_{T}\left[1 - \frac{4}{3}\left(\frac{T_{a}}{T_{S}}\right) + \frac{1}{3}\left(\frac{T_{a}}{T_{S}}\right)^{4}\right]} , \tag{4}$$

where T_S=6273 [°C] is the Sun's temperature.

The geometric criterion used for multi-objective optimization evaluates the distance between each objective function point and the ideal point using the expression [11]:

$$C = \sqrt{\left(\frac{\eta_{ex,max} - \eta_{ex}}{\eta_{ex,max} - \eta_{ex,min}}\right)^{2} + \left(\frac{\eta_{max} - \eta}{\eta_{max} - \eta_{min}}\right)^{2}} , \tag{5}$$

where η_{max}, η_{min} are the thermal efficiency's maximum and minimum values during a year and $\eta_{ex,max}$, $\eta_{ex,min}$ – the exergy efficiency's maximum and minimum values.

3 Results and discussions

3.1 Meteorological data

In order to characterize the weather conditions in Constanta, air temperature and global irradiation on a horizontal surface and on a tilted surface are plotted, hourly, for the recommended average day of each month of the year [12].

The air temperature values 2 m above the ground in Constanta during the average day of each month of the year are presented in Fig. 1.

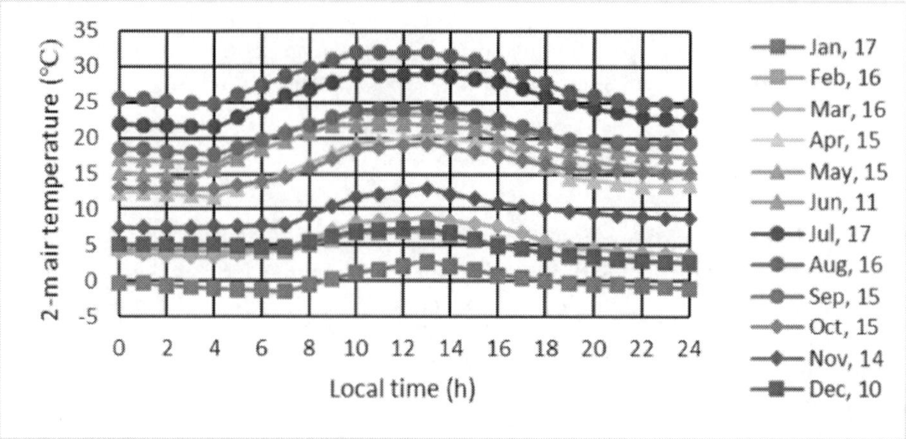

Fig. 1. Air temperature as a function of the hour of the day.

As one may see in Figure 1, the temperature difference between day and night is on average 6 °C. The maximum air temperature is recorded in August 16 and it is about 32 °C, while the minimum temperature is -2 °C in January 17. The air temperature reaches the maximum at noon, between 12-14 hours, while the minimum value is recorded at dawn, between 04-07 hours.

Figs. 2 and 3 show the hourly variation of global solar irradiation on a horizontal surface and on a 30° tilted surface, respectively on recommended average days. Solar irradiation has the highest values during summer: in May, June, and July. Maximum solar irradiation occurs at noon on June 11 (940 W/m^2 on a horizontal surface) and on May 15 (990 W/m^2 on a 30° tilted surface).

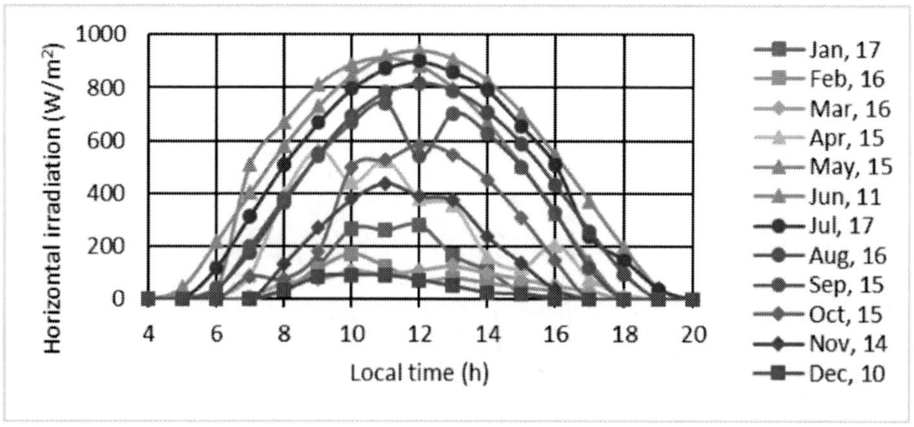

Fig. 2. Overall solar irradiation on a horizontal surface as a function of the hour of the day.

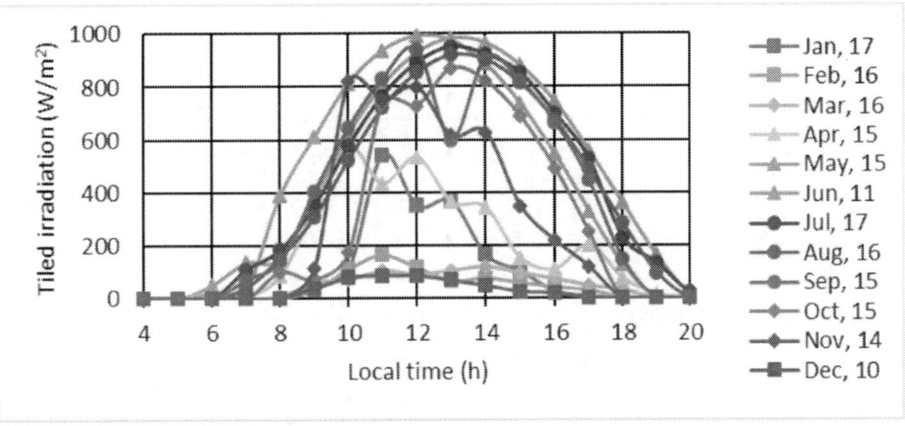

Fig. 3. Global solar irradiation on a 30°tilted surface as a function of the hour of the day.

The difference between the irradiation on a horizontal and a tilted surface increases considerably during the autumn months. On 14th of November at 11 o'clock, the maximum irradiation on a horizontal surface is 440 W/m^2 and at 12 o'clock on the same day, the maximum irradiation on a 30° tilted surface is 800 W/m^2. In the winter months, solar irradiation is low, the maximum value does not exceed 200 W/m^2.

The variation of the water temperature at the exit of the collector in the weather conditions described above is graphically represented in Fig. 4.

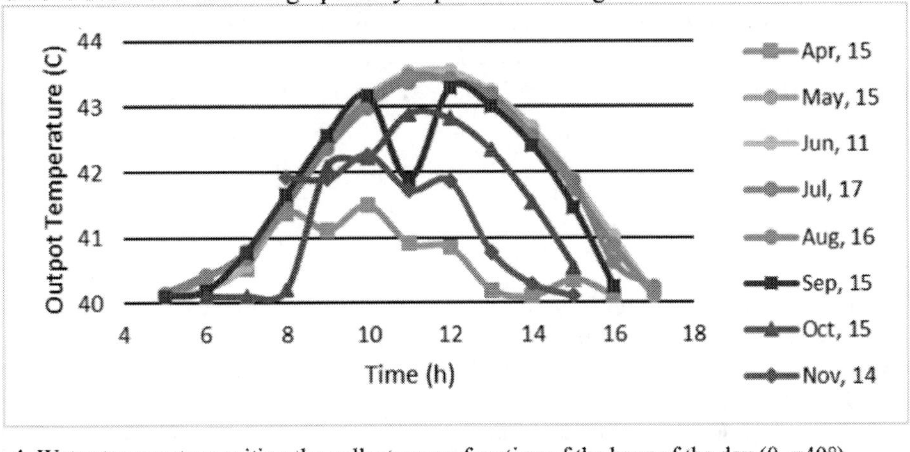

Fig. 4. Water temperature exiting the collector as a function of the hour of the day (θ_m=40°).

The calculations were performed for a mass flow of 0.02 kg/s/m^2 and an average water temperature (between the inlet and the outlet of the collector) of 40 °C. It can be noticed that the collector can produce hot water with a temperature above 42 °C from May to October. For the rest of the months of the year it is necessary to use a gas or electric boiler to heat the water up to this level.

The influence of air temperature and solar irradiation on the energy and exergy efficiencies of the FPC is investigated below. During this investigation, the mass flow and the average water temperature were kept constant, at the values mentioned above.

3.2 Air temperature's effect

In order to analyze the influence of the air temperature on the efficiency of the FPC, an annual simulation of the collector's operation was performed. The air temperature was corresponding to the typical meteorological year (TMY) previously mentioned and the solar irradiation was set to a constant value of 500 W/m^2.

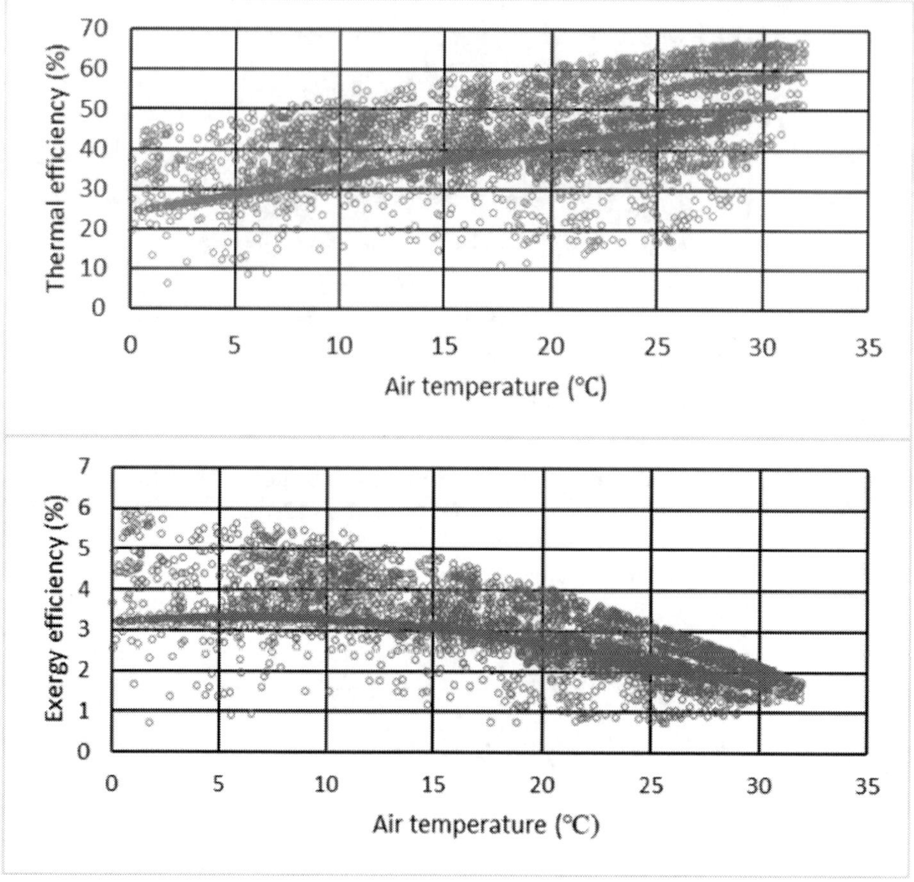

Fig. 5. Air temperature's effect on thermal and exergy efficiencies (solar irradiance 500 W/m^2).

As it can be seen from Fig. 5, an increase in the air temperature leads to an almost linear increase in the energy efficiency. For an increase of 10 °C in air temperature, the efficiency increases by about 10%. This dependence is due to the decrease of the thermal losses of the collector as the air temperature increases.

The air temperature increase determines the decrease of the exergetic efficiency of the collector, initially slower and after reaching the value of 15°C, more pronounced. This is because the exergy destruction in solar collectors increases with the increase of the air temperature [5].

3.3 Effect of solar irradiation

This analysis was done by simulating the operation of the FPC for a whole year, at a constant air temperature of 15 °C and a solar irradiation corresponding to the typical meteorological year.

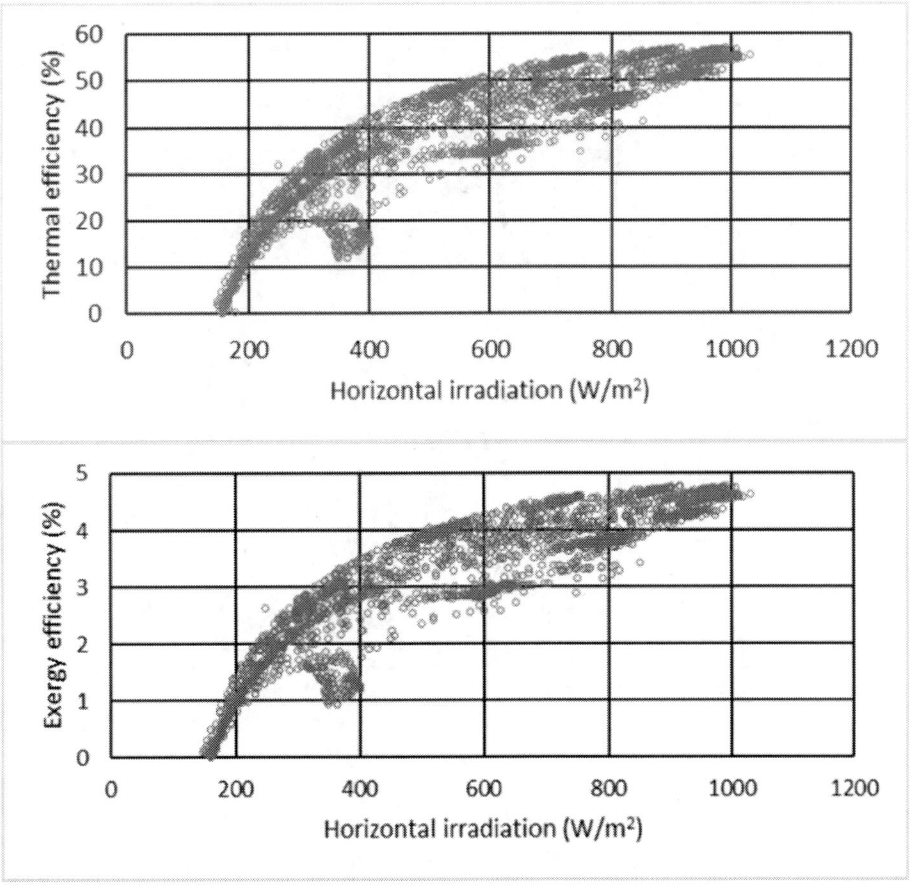

Fig. 6. Solar irradiation's effect on thermal and exergy efficiencies (air temperature 20 °C).

The efficiency of the collector for different values of solar irradiation recorded hourly during TMY is graphically represented in Figure 6. Both efficiencies, energetic and exergetic, increase with the increase of solar irradiation, following a similar pattern. The slope of the efficiency decreases continuously as the solar irradiation increases, at the beginning faster and after reaching the value of 350 W/m^2 slower. The maximum limits of the FPC's efficiency are reached at a solar irradiation of 1000 W, when the energy efficiency is 55% and the exergetic one is 4.5%.

The monthly energy provided by the FPC at air temperatures and solar irradiations corresponding to the typical meteorological year mentioned are studied below. The simulations were performed for a mass flow rate of 0.02 kg/s/m^2 and an average water temperature of 40 °C.

3.4 Energy and exergy produced by the FPC

The values of the useful energy and exergy produced by the FPC were calculated for each month of the year by numerical integration. The results are graphically presented in Fig. 7.

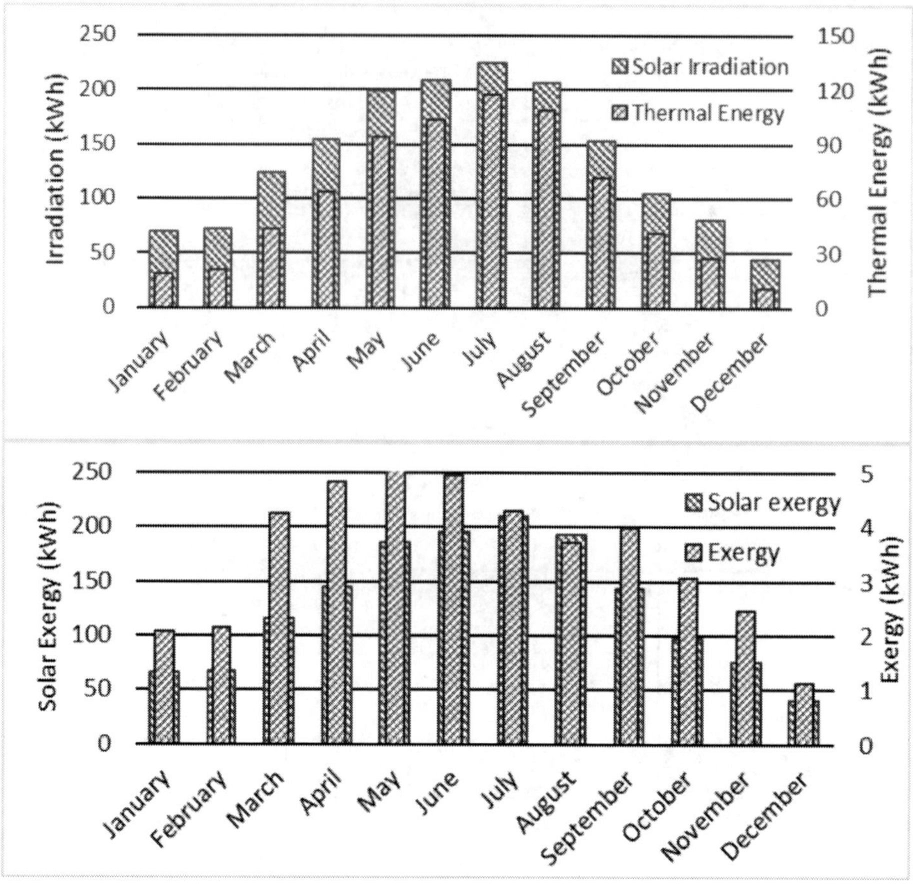

Fig. 7. Energy and exergy values produced by the FPC as a function of the month of the year.

The sunniest month of the year is July, when the solar irradiation reaches 230 kWh and the useful energy exceeds 120 kWh. Solar exergy is maximum in July, having a value of 210 kWh, but the output exergy is maximum in May (5 kWh), when the air temperature is lower than it is in July.

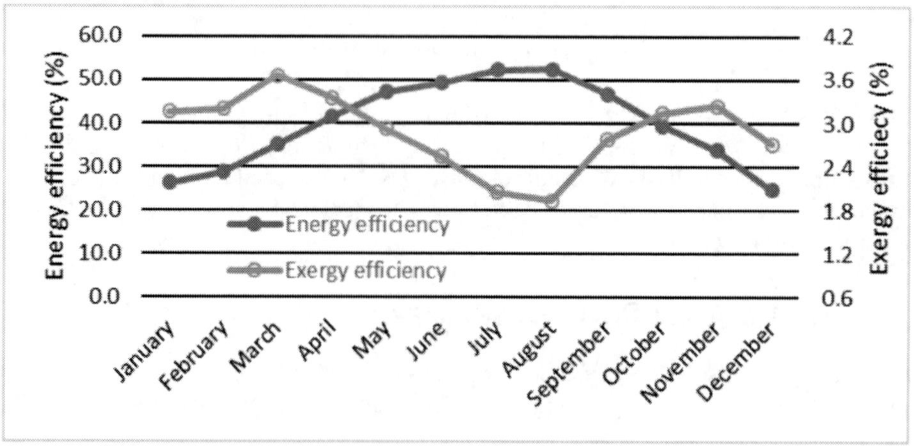

Fig. 8. Energy and exergy efficiencies of the FPC as a function of the month of the year.

The monthly energy and energy efficiency of the collector were computed based on the data presented in Fig. 7 (see Fig. 8). It was found that the energy efficiency has a maximum value during the summer months (over 50%), while the exergetic efficiency reaches maximum values in the spring-autumn months (3.6%), when the solar exergy is still significant (over 100 kWh) and the air temperature is not very high.

During winter months (December-February) the energy efficiency of the collector is around 25% and the energy efficiency is 3%.

The influence of the average temperature of the water on the efficiency of the collector is studied below.

3.5 Effect of mean temperature

The results of the simulation of the FPC operating under the meteorological conditions defined by TMY, for a constant mass flow of 0.02 kg/s m^2 and three average water temperature values are summarized in Table 3.

Table 3. Efficiency at three different mean temperatures of water.

Month	Mean temperature, °C					
	30		40		50	
	η	η_{ex}	η	η_{ex}	η	η_{ex}
January	32.8	2.9	26.3	3.2	20.3	3.0
February	35.1	2.8	28.8	3.2	23.4	3.3
March	41.6	3.0	35.0	3.7	28.9	3.9
April	48.2	2.3	41.4	3.4	35.3	3.9
May	53.4	1.6	47.1	2.9	41.4	3.9
June	55.6	1.0	49.4	2.5	43.4	3.6
July	58.2	0.3	52.3	2.1	46.4	3.3
August	58.7	0.2	52.6	1.9	46.7	3.2
September	53.4	1.4	46.8	2.8	40.6	3.7
October	46.4	2.2	39.5	3.1	33.0	3.6
November	40.8	2.6	33.9	3.2	27.5	3.5
December	31.0	2.4	24.7	2.7	18.9	2.6

One may see that the increase of the working fluid's average temperature leads to the decrease of the energy efficiency. That is because, with the increase of the average temperature, the temperature difference at which the heat transfer takes place is reduced. Thus, the useful heat gained decreases and, consequently, the thermal efficiency decreases.

Also, it is found that the increase of the average temperature leads to the increase of the exergetic efficiency. This is due to the decrease of the exergy losses to the environment with the increase of the average fluid's temperature.

3.6 Optimal weather conditions

For finding the appropriate meteorological conditions for which the energetic and exergetic efficiencies are maximum, the geometric criterion (5) was used.

Fig. 9 shows the variation of the exergetic efficiency according to the energy efficiency calculated hourly, throughout the typical meteorological year.

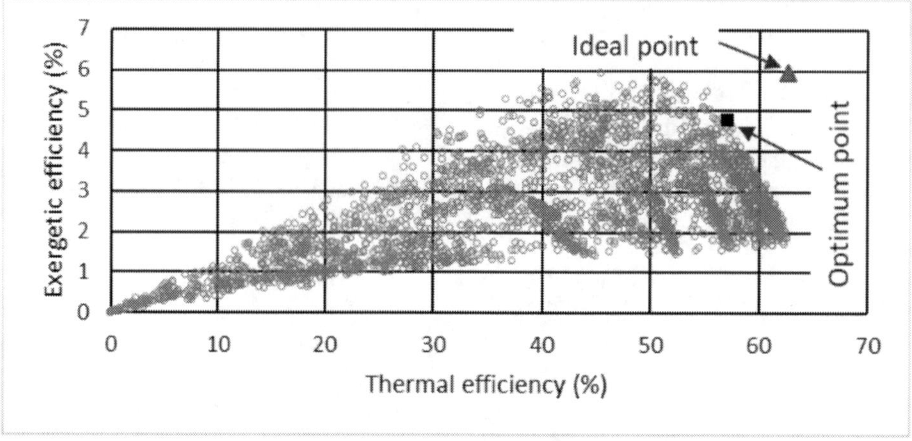

Fig. 9. Exergetic efficiency of the FPC as a function of its thermal efficiency.

The ideal point of coordinates (η_{max} = 62%, $\eta_{ex,max}$ = 6%) is represented on the graph by a triangle. The position of the closest point to the ideal point, the so-called optimal point with the coordinates (η_{opt} = 57%, $\eta_{ex,opt}$ = 4.8%) represented by a square, was found using the geometric criterion. The optimal meteorological conditions corresponding to the optimum point are: the solar irradiation of 916 W/m^2 and the air temperature of 15.4 °C.

4 Conclusions

The use of the test equations allows the achievement of the energetic and exergetic analyses of a FPC mounted in a location for which the weather data for a typical meteorological year are known.

Based on the analyses carried out for a good quality FPC, located on the Black Sea coast, in the city of Constanta, the following conclusions can be formulated:

- the increase in the air temperature leads to an almost linear increase of the energy efficiency; at an increase in air temperature of 10 °C, the efficiency increases by about 10%;
- increasing the air temperature causes the collector's exergetic efficiency to decrease, initially slower and after reaching 15 °C, more pronounced;
- energy and exergy efficiencies increase with increasing solar irradiation, initially faster and after the irradiation reaches 350 W/m^2, slower;
- the sunniest month of the year is July, when the solar irradiation reaches 230 kWh and the useful energy exceeds 120 kWh; the solar exergy is maximum in July, when it has a value of 210 kWh, while the output exergy is maximum in May, having a value of 5 kWh;
- energy efficiency is maximum during the summer months (over 50%), while the exergetic efficiency reaches maximum values in the spring-autumn months (3.6%);
- the increase in the working fluid's average temperature leads to the decrease of the energy efficiency and the increase of the exergetic efficiency;
- optimal weather conditions for the collector's operation are: the solar irradiation of 916 W/m^2 and the air temperature of 15.4 °C; under these conditions, the energy and exergy efficiencies of the collector are 57% and 4.8%, respectively.

For a more detailed analysis of the performance of a FPC, it would be necessary to use the complete mathematical model, implemented in a high programming language coupled with an hourly weather data set corresponding to a TMY.

Acknowledgments

This paper is a result of the research conducted within the project ME15-18-01, within the national program GNaC 2018 ARUT.

References

1. S. A. Kalogirou, S. Karellas, K. Braimakis, C. Stanciu, V. Badescu, *Exergy analysis of solar thermal collectors and processes*, Progress in Energy and Combustion Science, Volume **56**, Pages 106-137, (2016)
2. H. Gunerhan, A. Hepbasli, *Exergetic modeling and performance evaluation of solar water heating systems for building applications,* Energy Build, **39**, 509–516, (2007)
3. S. Farahat, F. Sarhaddi, H. Ajam, *Exergetic optimization of flat plate solar collectors*, Renewable Energy **34,** 1169–1174, (2009)
4. F. Jafarkazemi, E. Ahmadifard, *Energetic and exergetic evaluation of flat plate solar collectors,* Renewable Energy, Volume **56**, Pages 55-63, (2013)
5. Z. Hajabdollahi, H. Hajabdollahi, *Thermo-economic modeling and multi-objective optimization of solar water heater using flat plate collectors*, Solar Energy **155,** 191–202, (2017)
6. A. Allouhi, M. B. Amine, *Effect analysis on energetic, exergetic and financial performance of a flat plate collector with heat pipes*, Energy Conversion and Management **195,** 274–289, (2019)
7. Z. Chen, S. Furbo, B. Perers, J. Fan, E. Andersen, *Efficiencies of flat plate solar collectors at different flow rates*, Energy Procedia **30,** 65 – 72, (2012)
8. K. Balaji, S. Iniyan, M. V. Swami, *Exergy, economic and environmental analysis of forced circulation flat plate solar collector using heat transfer enhancer in riser tube*, Journal of Cleaner Production **171,** 1118-11 27, (2018)
9. PVGIS Interactive tools TMY, retrieved from https://re.jrc.ec.europa.eu/pvg_tools/en/tools.html#TMY, February 10, (2020)
10. J.E. Hay, J.A. *Davies, Calculations of the solar radiation incident on an inclined surface.* In: Hay, J.E., Won, T.K. (Eds.), Proc. of First Canadian Solar Radiation Data Workshop, **59**. Ministry of Supply and Services, Canada, (1980)
11. Bellos E., Tzivanidis C., Multi-objectiv optimization of a solar driven trigeneration system. Energy 2018, 149:47-62, (2018)
12. J.A. Duffie, W.A. Beckman, *Solar Engineering of Thermal Processes*, **4**th ed., John Wiley & Sons, Hoboken, (2013)

Gas yield variation in wood biomass co-current air gasification process – continuous operation

Gabriela Ionescu*, *Raluca Nicoleta* Tirtea , *Cora* Gheorghe-Bulmau, and *Cosmin* Marculescu

University Politehnica of Bucharest, Department of Energy Production and Use, Splaiul Independenței 313, Romania

Abstract. The purpose of this work was to study the gas yield variation resulted from the cherry wood gasification with air using a lab-scale rotary kiln gasifier. The feedstock was continuously fed into the preheated reactor at 600°C, in co-current configuration, using atmospheric air as a gasifying agent. The results indicate the importance of oxidation reaction control, through the feeding flow rates of biomass and air and the reactants mixing rate. From the experiment, the hydrogen yields were about 2-4%, while the carbon monoxide varied between 8-21%. Additionally, the paper provides process observations based on the continuous monitoring of gas composition. The specific flow rates of substances and installation operating conditions were linked to process run through syngas composition.

1 Introduction

Recent reports and former scientific studies, including the authors, have outlined the advantages and challenges of bioenergy usage: i) biomass is an important energy source but its structure could cause a chain of issues mainly associate at their phase change behavior affecting the overall process energy efficiency [1]; ii) globally, the bionergy usage could decrease the greenhouse gases with 65% by 2050 [2], iii) in 2017, bioenergy accounted for roughly 70% of the renewable energy consumption, worldwide [3]. Among the existing biomass thermochemical conversion technologies, gasification provides gas fuel (syngas) that can be used in furnaces, power machines and electrochemical fuel cells, for both heat and power generation [4]. The studies conducted on biomass waste gasification have highlighted the following: i) the syngas quality and overall process efficiency is influenced mainly by biomass composition and its particle size, gasifying agents, operating conditions, type of gasifier and their link with each other[5]; ii) its usage in a direct electrical heating system could offer low emissions [6]; iii) systems with capacities lower than 200 kWe are more preferable due to local direct usage, while higher than 2 MW are more favorably due to investment efficiency, although the feedstock collection implies high costs [7].

Previous studies on biomass gasification have used agricultural straws, crop residues, woody forestry and waste resulted from both mention activities or municipal solid stream

* Corresponding author: gabriela_ionescu@ymail.com

[8,9]. Most of the experimental studies have been carried out on biomass air/oxygen gasification, at high temperatures typically ranging between of 800–1000°C [10], more specifically agricultural waste 750–850 °C and woody biomass 850–950 °C [11].

According to Ren et al., 2019 review [5], the syngas heating values resulted from air biomass gasification ranges between 4 and 7 MJ/Nm^3. Before its usage in internal combustion engines, a cleaning-cooling system is necessary, due to the moisture, tar, dust and temperature of the syngas. Hydrocarbons compounds with a molecular weight above benzene (tars) could appear at low temperatures, typically between 200°C-500°C[10], reaching an average level of 1 and 50 g/Nm^3 [12]. Higher gasification operating temperatures (≥800°C) is recommended for tar reduction, although, this can favor the formation of slag from ash agglomeration. [13]. The exit gas temperature from the gasifier can vary between 200-700°C [10], while the ideal gas temperature for its usage in internal combustion engines should be around room temperature [14]. When operating on syngas as a substitute for petroleum fuels, the internal combustion engines are derated from about 15% to 40% (15% to 20% dual fuel diesel, 30% to 40% spark ignition)[15].

Studies have shown that perennial crops, such as pruned fruits and woody fruits, including cherry [14] could be explored as primary feedstock in the thermochemical conversion schemes. Same remark has been done for the Prunus avium tree family, commonly called sweet cherry, wild cherry and gean, due to its woody based structure [15] including also their stones [16] and their seeds [17, 18]. Apart from their role in the food supply chain, the Prunus avium trees could reach to their end life, becoming an agroforestry residue. Another cherry wood residue source is also the furniture manufacturing sector. In the current study the effects of cherry wood components on air gasification was investigated using a lab-scale rotary kiln gasifier at 600°C. The study findings are related to process set-up and installation operation influence on gasification reactions and products formation.

2 Material and Methods

The cherry wood waste originated from Prahova county, Romania. The schematic diagram of the experimental set-up is presented in the following figure. First, the feedstock was reduced in size by using a Fritsch cutting mill (Pulverisette 15). This step facilitated the feedstock waste characterization procedures, gasifier feeding system and degradation reactions. The primary analysis was carried out conforming to ASTM standard (E870 – 82). The high heating value (HHV) of the feedstock was determined using IKA C 200 calorimeter.

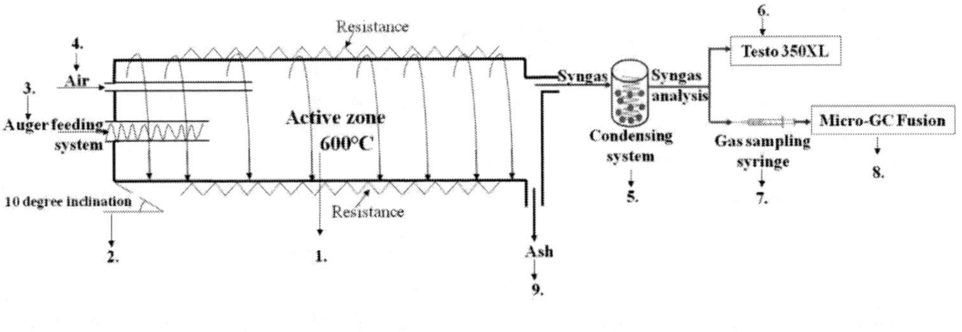

Fig. 1. Air gasification of cherry wood chips experimental set-up (adapted after [21]: 1. Rotary kiln reactor; 2. Reactor inclination degree; 3. Auger feeding system; 4. Air supply; 5. Syngas condensing system; 6. Syngas analysis (Testo 350XL). 7. Syngas sampling syringe; 8. Syngas analysis (Micro-GC Fusion gas analyser). 9. Ash.

The air gasification of cherry wood chips experiments were conducted in a rotary kiln gasifier lab-scale pilot plant. The rotary kiln gasifier (1) main characteristics have been presented previously by the authors [21]. In sum, the lab-scale pilot reactor allows the variation of the feedstock feeding rate and reactor rotation speed, its inclination degree and temperature. In this work, the cherry wood waste chips were continuously fed into the reactor (2), in co-current configuration, using atmospheric air as gasifying agent (3). The waste feeding rate was determined based on preliminary trials made on the hopper screw conveyor, considering also the reactor rotational speed and its 10°inclination (4). To assure the partial oxidation of the feedstock the Equivalent Ratio (ER) was calculated:

$$ER=\frac{Air\ introduced\ in\ the\ process}{Stoichiometric\ air} \quad (1)\ [22]$$

The calculations were done based on the elemental composition of the feedstock used for the air required for theoretically complete combustion (stoichiometric) determination. The cherry wood residues elemental composition expressed in percentage [%], used in the aforementioned determination was : C-49.52, H-5.81, O-42.97, N-0.31,S-0.02, Cl-0.02, Ash-1,35, Moisture-0 [23]. Cheery wood ultimate analysis is similar to other types of woody biomass, according to a study based on 95 different literature sources, regarding the ultimate analysis of woody biomass. According to this study woody biomass contains 41.78 – 57.00% Carbon, 4.85 – 8.11% Hydrogen, 35.85 – 52.26% Oxygen, 0.04 – 1.43% Nitrogen, 33ppm – 0.2% Sulphur, and 0.0004 – 0.08% Chlorine [24]. According to the formula presented above, the amount of air required for a complete combustion (stoichiometric condition) of cherry wood residues needs to be determined. To determined it, all combustible components of the material are considered. The gasification process temperature was set at 600°C. The syngas was passed through an ice water cooling system (5), aiming to retain the condensable fraction that could be formed during the process. The composition of the non-condensable fraction has been analyzed using two modalities presented in Figure 1., marked with 7-9. The Testo-350XL gas analyzer allowed real-time measurements, offering immediate information on CO, CO_2, O_2, H_2, SO_2, NO and NO_2 concentrations, hence, gasification process state and stability. Additionally, at fixed time intervals, syngas samples were extracted after gas cooling using a syringe (7) and injected to micro-gas chromatograph (Micro-GC Fusion, Inficon (8) immediately. Due to its features, the instrument is suitable for syngas fast and accurate composition analysis. Hence, the instrument allowed the detection of H_2, O_2, N_2, CH_4, CO, CO_2, and light hydrocarbons (C_1 to C_{12}).

Six gasification experiments were conducted at different reactor mechanic parameters and ERs in order to determine the optimal reactor settings for air cherry wood residues gasification. The process parameters modified were the reactor rotation speed, the biomass and air feeding rate (Table 1). Process duration may be controlled in this type of installation by reactor rotation rate and its inclination Since the most important gasification reactions are also the slowest, there must be ensured enough process time for all char oxidation reaction to take place [25]. As it is stated above, the gas composition was determined using two gas analyzers. Testo 350XL gas analyzer has the advantage of real time measurements, while the Micro-GC has the advantage of accurate measurements. The use of Testo 350XL provides real time information allowing users to observe and quantify the experiment results during process run. Gas-chromatography provides more accurate results and can determine far more gaseous species.

Table 1. Rotary kiln reactor operating parameters for cherry wood residues air gasification

No. experiment	Gasifying agent (air) mass flow	Feedstock mass flow	ER
	[kg/s]	[kg/s]	[-]
1	0.00156	0.00036	0.74
2	0.00104	0.00033	0.54
3	0.00104	0.00033	0.54
4	0.00031	0.00021	0.25
5	0.00104	0.00031	0.57
6	0.00021	0.00020	0.18

In the first experiment conducted, only the Testo gas analyzer was used, while for the experiments 2 to 6, both Testo and Micro-GC were used to determine the gas composition. During the second experiment, Testo measurements were interrupted to collect the gas samples for gas-chromatography analysis. For experiments 3 to 6 gas analysis by Testo and Micro-GC was conducted concomitant. To characterize the gasification experiments, oxygen, carbon dioxide, carbon monoxide and hydrogen concentrations in the produced gas were analyzed. Sulphur and nitrogen oxides registered very small concentration. Moreover, these gas species concentrations are very low and don't modify the process energy balance.

Since Testo350XL is a gas analyzer designed for combustion processes its results may not be perfectly accurate for gasification processes. Its use in gasification processes helps to characterize in real time the process run, and to determine if the gasification reactions occur.

3 Results and discussion

The physicochemical properties of cherry wood residues are presented in Table 2. The volatile matter (VM), fixed carbon (FC), ash content and HHV as received (a.r) values are closely with the ones reported by Telmo et.al., 2010 [26] on 84.9%, 15%, 0.1% and 17500 kJ/kg.

Table 2. Physicochemical properties of cherry wood residues average values

Property	Moisture [%]	$VM_{a.r.}$ [%]	$FC^*_{a.r.}$ [%]	$Ash_{a.r.}$ [%]	$HHV_{a.r.}$ [kJ/kg]	$HHV_{d.b.}$ [kJ/kg]
Average	0.32	81.95	15.74	1.98	17323	18498

a.r.=as received
d.b.=dry basis
*=calculated by difference

Depending on process parameters the syngas flow rate varied between 0.00038 and 0.001887 kg/s. That corresponds to solid-gas conversion rate of about 93.31% - 98.23% [27]. Continuous operated systems provide process observation advanced over batched operated systems. The permanent monitoring of gas composition offers information about process status throughout the entire experiment. So, both specific flow rates of substances and installation operating conditions will be reflected in syngas composition. In the graphs below is presented the variation of principal gas species concentration in time for the experiments conducted.

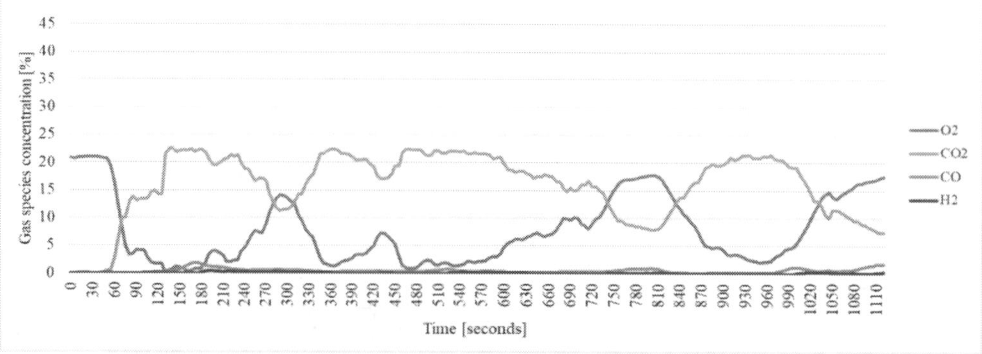

Fig. 2. Gas species concentration in time – experiment no. 1

According to first experiment results presented in Fig. 2, char gasification stage didn't occur, leading to an incomplete gasification process. High levels of CO_2 and the absence of H_2 and CO in the produced gas can be the result of short process duration. Both biomass feeding and the reactor rotation speed were set at high values, therefore the feedstock was discharged from the reactor at a faster rate than the char gasification minimum reaction time. Also, high O_2 concentration indicates an insufficient reactants residence time inside the reactor.

At lower gases flow rate (increase reaction time) both CO_2 and CO are formed (second experiment, Fig. 3). According to Testo measurements CO_2 maximum concentration increased with almost 8 percentage points, while the CO maximum concentration increased with more than 10 percentage points, when the air flow rate decreased. At lower air flow rates there is also a lower Nitrogen quota in the syngas. That will increase all other gas species yield in the syngas, including CO_2 and CO. Small H_2 concentration is also registered. Even though the H_2 concentration is low, comparing to usual biomass air gasification processes (6 – 22% according to Heidenreich et al. [28]), it is higher compared to the first experiment conducted.

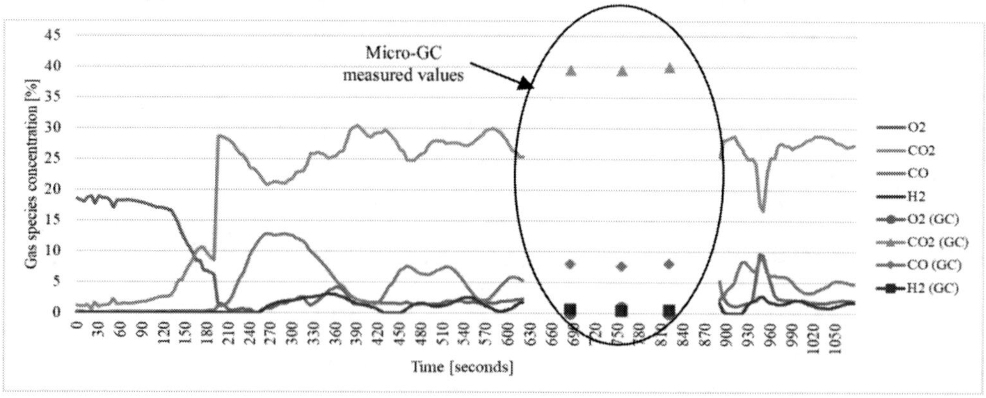

Fig. 3. Gas species concentration in time – experiment no. 2

In Fig. 4 is presented the composition of syngas obtained from the third gasification experiment. According to the Micro-GC analysis, longer residence time of the reactants increases CO and H_2 concentrations, while CO_2 yield decreases.

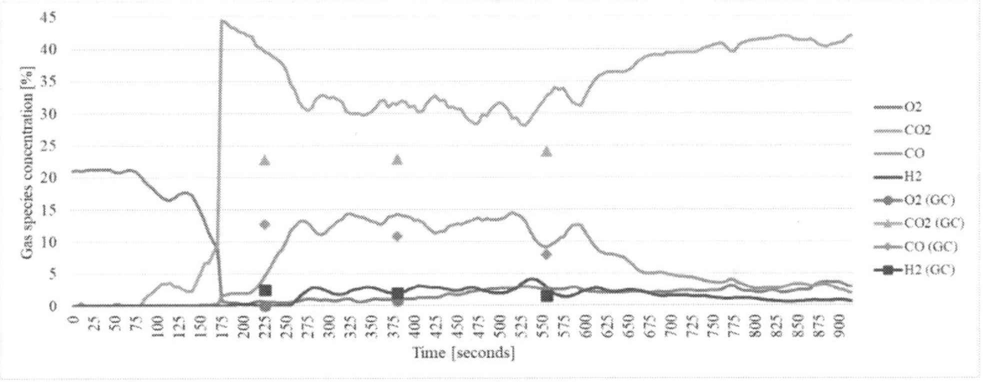

Fig. 4. Gas species concentration in time – experiment no. 3

The decrease of feed-in substances flow rates along with the increase of reaction time (experiment no. 4), enhance the CO formation (14% - 15%), while the CO_2 concentration decreases to 28% (Fig.5). Hydrogen yield reaches 4%. During experiment no. 4 higher CO concentrations were obtained (about 20%) with an important decrease in CO_2 yield down to 17% while the Hydrogen quota maintained constant as in experiment no. 3.

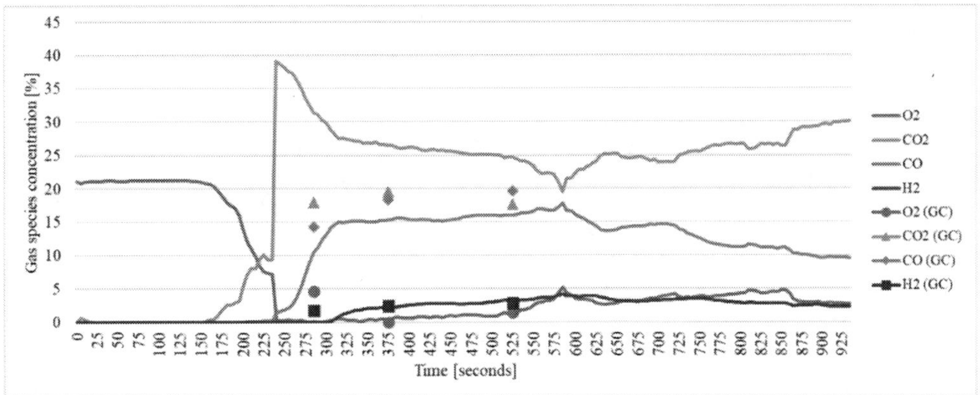

Fig. 5. Gas species concentration in time – experiment no. 4

The control of oxidation reactions, through the feeding flow rates of biomass and air and the reactants mixing rate is very important. This can be best observed by comparing especially the results obtained from the experiments no. 3 and 5. So, by comparing these experiments can observe that for shorter residence time CO_2 tends to increase (to 30%), while CO decreases fast (to about 2%), because of limited $C+CO_2 \rightarrow 2CO$ reaction. Although it is well known that at small process temperature, less than 700-800°C, CO_2 formation is favored to the detriment of the CO. However, the longer process duration can increase the CO formation. In the combustion stage of the gasification process CO_2 is produced. In the char gasification stage CO_2 reacts with the char to produce CO. Char-gas reaction rates are very slow, therefore in order to ensure a complete char conversion, process duration must be correlated to these reaction rates [28, 29]. The current results show a positive influence of a low reactor rotation speed on CO formation.

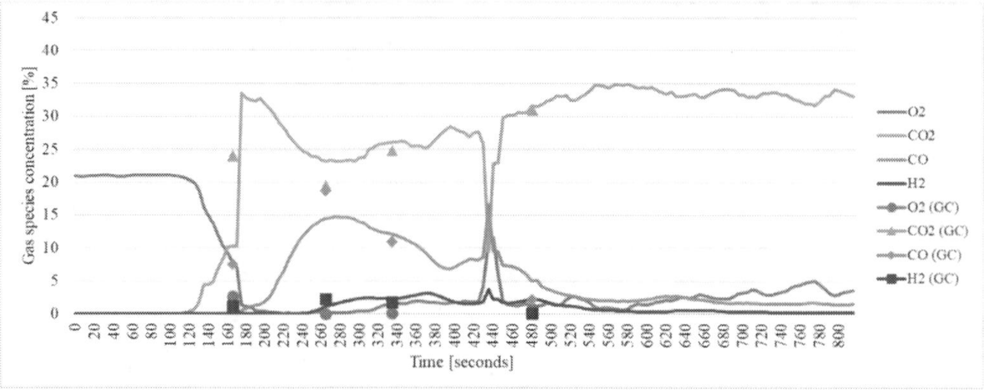

Fig. 6. Gas species concentration in time – experiment no. 5

Best gasification results were obtained for experiment no. 6 conducted at higher syngas retention time. Hydrogen concentrations determined by both gas analyzers records very small values, less than 5% with CO yield (20%) overtaken the CO_2 that decreased at about 15%.

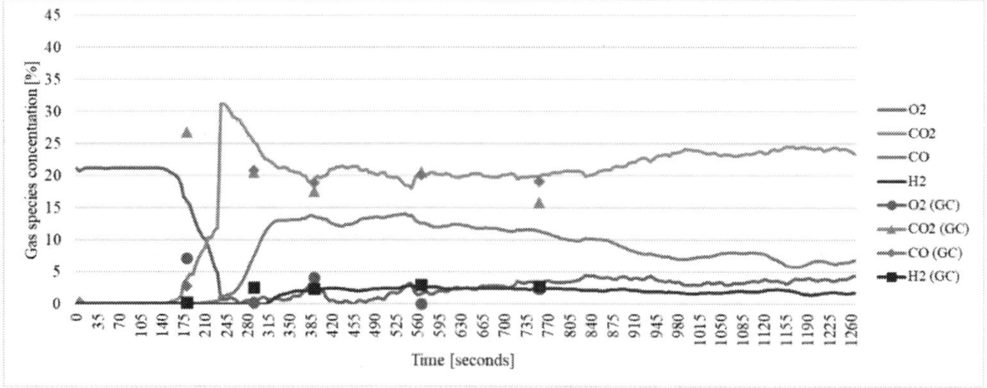

Fig. 7. Gas species concentration in time – experiment no. 6

Even if the Hydrogen yields across all experiments are not typical for biomass air gasification processes [28, 30-32], their values are within expected range related to process temperature. Although H_2 production registered was so low, gas energy content was up to 3.50 MJ/Nm^3 mainly because of CO relative high concentration [27]. As previously stated, the presence of Nitrogen from the air reactant also limits the syngas specific energy content.

Process energy efficiency was computed for all 6 gasification processes, according to process mass balance and reaction products properties, resulting a maximum process energy efficiency of 74.08% including the syngas sensitive heat calculated at 375°C (exhausted temperature) [27]. Therefore, we can assume that at higher process temperature, carbon monoxide and hydrogen production will increase, leading to a higher calorific value of the gas, and a higher process energy efficiency.

4 Conclusions

The core findings of the study are related to process set-up and installation operation influence on gasification reactions and products formation. The specific operating characteristics of the utilized installation were synthesized as common control parameters available in any gasification facility (feedstock residence time, gas flow rate, solid – gas mixing rate etc). The results indicate the importance of oxidation reactions control, through the feeding flow rates of biomass and air and the reactants mixing rate. Constant Hydrogen yield of about 2-4% was obtained for all experiments while the CO concentration varied from 8 - 21%. The lower heating value of the produced gas reached 3.5 MJ/Nm^3 despite the possible false air infiltrations (because of installation construction characteristics) and high Nitrogen quota from the air used as reactant. As expected, the permanent monitoring of gas composition offers information about process status throughout the entire continuous operated system experiment providing process observation. Consequently, the specific flow rates of substances and installation operating conditions were linked to process run through syngas composition.

Funding provided through Romania's "Competitiveness Operational Programme 2014-2020" Priority Axis 1: Research, Technological Development and Innovation (RD&I) to Support Economic Competitiveness and Business Development Action 1.1.4. Attracting high-level personnel from abroad in order to enhance the RD capacity ID/Cod My SMIS: P_37_768/103651; No. Contract: 39/02.09.2016

References

1. C. Mărculescu, V. Cenuşă, F. Alexe, Waste Manage., **47**, 133-140 (2016).
2. J Bogner, R Pipatti, S Hashimoto., C. Diaz, K. Mareckova, L. Diaz, P. Kjeldsen, S.Monni A. Faaij,Q. Gao, T. Zhang, M. Abdelrafie Ahmed, R.T.M. Sutamihardja, R. Gregory, Waste Manage& Res, **26**(1), 11-32, (2008)
3. Global Bioenergy Statistics 2019,World Bioenergy Association, https://worldbioenergy.org/uploads/191129%20WBA%20GBS%202019_LQ.pdf
4. A. Kozlov, O. Marchenko, S. Solomin, Energy Procedia, **158**, 1004-1008 (2019).
5. J. Ren, J.P. Cao, X.Y. Zhao, F.L. Yang, X.Y. Wei. Renew Sust Energ Rev, **116**, 109426 (2019)
6. S. K.Sansaniwal , K. Pal, M. A. Rosen, S.K. Tyagi, Renew Sust Energ Rev, **72**, 363-384, (2017).
7. Y.A. Situmorang, Z. Zhao, A. Yoshida, A. Abudula, G. Guan. Renew Sust Energ Rev, **117**, 109486 (2020).
8. M.L.V. Rios, A.M. González, E.E.S. Lora, O.A.A. Olmo. Biomass Bioenerg, **108**, 345-370. (2018).
9. D. Baruah, D. C. Baruah, Renew Sust Energ Rev, **39**, 806-815,(2014).
10. T.A. Milne, R. J. Evans, N. Abatzaglou. *Biomass gasifier"Tars": their nature, formation, and conversion* (National Renewable Energy Laboratory, Golden, CO (US), (1998).
11. P. McKendry Bioresource Technol, **83(1)**, 55-63 (2002)
12. P. Basu, *Biomass gasification, pyrolysis and torrefaction: practical design and theory,3rd Edition* (Academic press,2018)
13. J. M. Bermudez, B. Fidalgo, *Production of bio-syngas and bio-hydrogen via gasification. In Handbook of Biofuels Production* ,(Woodhead Publishing, 431-494, 2016).

14. A. Pradhan, P. Baredar, A. Kumar, J Pure Appl Sci Tech, **5**(2), 51-66, (2015)
15. R. P. Bates, K. Dölle. Adv Res, **10**(1), 1-8, (2017).
16. N. Bilandzija, Span J Agric Res, **2**, 292-298 (2012)
17. P. Wang, B. H. Howard, Energies, **11(1)**, 25, (2018)
18. C.J. Duran-Valle, M. Gomez-Corzo, V. Gomez-Serrano, J. Pastor-Villegas, M.L. Rojas- Cervantes, Appl. Surf. Sci. 252, 5957–5960, (2006).
19. G. Duman, C. Okutucu, S. Ucar, R. Stahl, J. Yanik Bioresource Technol, **102(2)**, 1869-1878. (2011).
20. J.F. Gonzalez, J.M. Encinar, L.C. Jose, E. Sabio, M. Chacon, J. Anal. Appl. Pyrolysis **67**, 165–190 (2003)
21. D. Boldor, C. Marculescu, R. State, M. Patrascu. Proceedings of AIChE Annual Meeting. AIChE, 2018, October)
22. R. N. Tîrtea, C. Mărculescu, A. Badea (2017). Aspects on residual biomass to gas fuel conversion using air as oxidizer. UPB Sci. Bull., Series C, 79(1), 315-322.
23. G. Ionescu,C. Bulmău. *Estimation of Energy Potential for Solid Pyrolysis By-Products Using Analytical Methods.* (In Analytical Pyrolysis. IntechOpen, 2019)
24. R. N. Tîrtea, *Study of optimal thermal conversion processes of biomass by maximizing global energy efficiency*. (PhD Thesis, University Politehnica of Bucharest, 2019)
25. N. P. Cheremisinoff, J. Rezaiyan. *Gasification technologies: a primer for engineers and scientists.* (CRC press, 2005).
26. C. Telmo, J. Lousada, N. Moreira, Bioresource Technol, **101**(11), 3808-3815, 2010.
27. C. Bulmău, R.N. Tîrtea, G. Ionescu, C. Mărculescu, Process characterization and energy balance of air wood residues gasification using continuous operated pilot scale reactor, 9th International Conference on Thermal Equipment, Renewable Energy and Rural Development, TE-RE-RD 2020, eISSN: 2267-1242, under review.
28. Heidenreich, Steffen, Michael Müller, and Pier Ugo Foscolo. Advanced biomass gasification: New concepts for efficiency increase and product flexibility. Academic Press, 2016.
29. P. Basu. *Biomass gasification, pyrolysis and torrefaction: practical design and theory.2nd Edition* (Academic press, 2013).
30. M. Balat, M. Balat, E. Kırtay, H. Balat, Energ Convers Manage **50(**12), 3158-3168, (2009).
31. M. Puig-Arnavat, J. C. Bruno, A. Coronas, Renew Sust Energ Rev, **14**(9), 2841-2851, (2010).
32. E. Bocci, M. Sisinni, M. Moneti, L. Vecchione, A. Di Carlo, M. Villarini, Enrgy Proced, **45**, 247-256, (2014).

Process characterization and energy balance of air wood residues gasification using continuous operated pilot scale reactor

Cora Gheorghe-Bulmau*, *Raluca* Nicoleta Tirtea, *Gabriela* Ionescu, *Cosmin* Marculescu
University Politehnica of Bucharest, Department of Energy Production and Use,
Splaiul Independenţei 313, Romania

Abstract. This work aimed to study the effect of gasification process operating conditions on syngas composition and properties, and process efficiency. A rotary kiln gasifier lab-scale pilot plant with capacity ≅ 30 kg/h and a power of 30 kWe was used for gasification tests applied to cherry wood at different loads, for a temperature of about 600°C, while the air was used as gasification agent for all tests. The syngas composition was measured and analyzed. The results have shown that conversion of wood cherry through gasification lead to a lean fuel gas of 3.5 MJ/Nm^3 and installation characteristics have a major influence both on process and syngas properties. This is happened because the rotary kiln gasifier allows some air infiltrations, and consequently a high N_2 content in the syngas composition. The energy balance of the cherry biomass gasification processes was calculated. It was found also that gas density varies slightly from 1.26 to 1.43 kg/m^3, while the specific heat of the gas varies from 1.04 to 1.34 kJ/kgK.

1 Introduction

Wood biomass presently dominates the renewable energy sources for different uses in heat and power generation, transport fuels production [1, 2], wood and wood products accounting for 6.0 % of the total energy consumed within the EU in 2016 [3]. Referring on biomass in the wood processing sector, it can be said that it continues to be an unexploited source of the wood biomass potential. There are significant amounts of wood residues generated by wood-processing industry, produced by manufacturing of veneer, furniture sawn and timber. A considerable part of these residues is generally used on site for energy in factory driven heating installations or sold directly to energy producers [4]. Since one of the late focuses of policy makers is development and deployment of enabling technologies to facilitate the integration of variable renewable energy, residues from wood processing industry may be a viable source for energy production. The furniture industry is the most complex activity of the wood industry and the most performing [5] and depends on the regional differences in wood by-products flow due to diversification of forestry resources, wood –processing industry and pulp-industry around Europe (Figure 1).

* Corresponding author: cora4cora@gmail.com

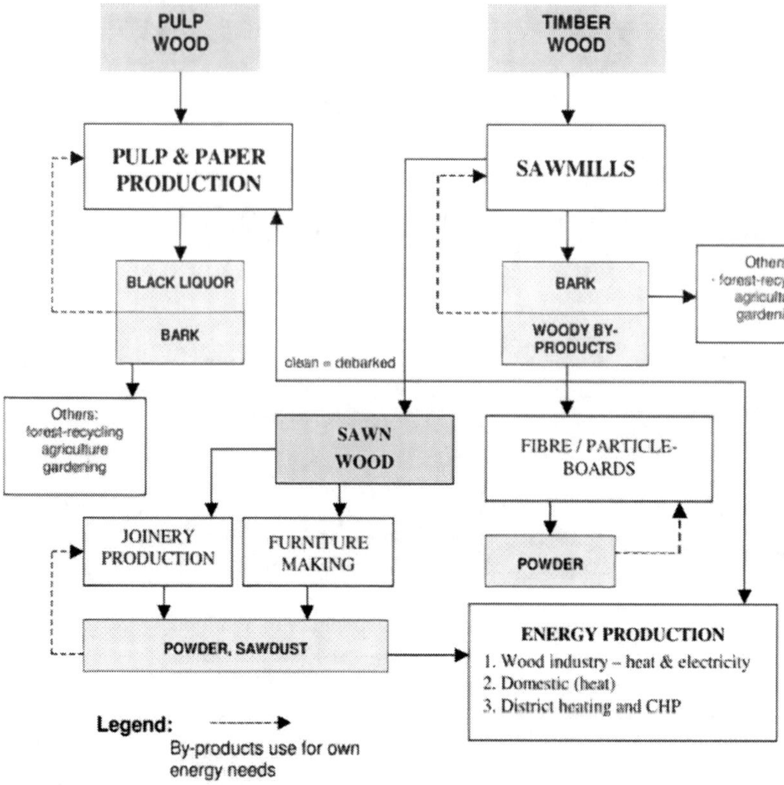

Fig. 1. EU market demands and flows of wood-industry by-products [Eurostat, 2018] [3]

In Romania the furniture manufacturing activity is well represented among the EU states, occupying the third position in 2018 [6]. The main wood species used for furniture industry in Romania are: solid oak, but also beech, cherry, maple, chestnut, fir-tree etc. As furniture industry may very well choices cherry wood, the resulted cherry wood residues may be a considerable resource and so its potential, unexploited, can be valuable with different direction of use.

Gasification is an advanced thermal conversion process that produces a combustible gas obtained from different solid fuels, this being the key of its importance [7]. There are various ways for use of gasification process, and this can be integrated into many systems. The more often application of gasification process is for energy production and electricity in cogeneration stations. Biomass gasification technologies met a rapid development due to constantly diminishing of the fossil fuel resources use along with a constant increase in the demand for electricity [8]. Over incineration, gasification has viable advantages because of the flexibility in the way in which the energy is utilized. Moreover, there are many possibilities for syngas use: it can either be combusted directly, used as a fuel in gas engines/turbines, stored, or processed through catalytic processes (for example, Fischer-Tropsch) to produce liquid fuels or chemicals. The quality and quantity of syngas depends on gasifying agent, that can also affect the composition of the gas, tar content, and heating value [9 - 11]. A widely gasifying agent is air because it is cheaper than other gasifying agents and it is easy to provide to the systems. But use of air as gasification agent lead to a syngas, which comprises large amounts of nitrogen and therefore reduces the heat of the gas combustion [8]. Recently some studies have been published in order to validate the design and to optimize the operation conditions of biomass gasification processes [12 – 14].

From these a conclusion raises: the unit design represents one of the most important factors in determining the syngas quality and heating value [15]. Updraft, downdraft, and CFB gasifiers provide a maximum efficiency of about 75 %, while the maximum energy output is 2.7 MW, 1.1 MW, and 11.1 MW, respectively [16]. Both fixed-bed and downdraft gasifiers are appropriate for producing low heating value gas [17 – 19], first in case of medium size applications [18, 19], the second for generating electricity of small-scale systems in the range of 10 kW up to 1 MW [18 - 21].

This research evaluates the feasibility of using cherry wood biomass in a developed gasifier at semi-industrial scale and how process operational conditions affect the properties of the syngas or the global efficiency of the process. The results generated from the gasification tests indicated differences in the composition, density and calorific value of the syngas produced from the wood biomass and revealed also the impact of equivalent ratio on process energy efficiency.

2 Material and Methods

Data implying the presentation, analysis and characterization of the feedstock used in the current research have been already summarized in a previously manuscript [22]. The gasifier specifications and experimental set-up applied during this research were also described in detail. Here follows a short presentation of these. Tests of cherry wood air gasification were operated in a rotary kiln reactor lab-scale pilot plant. The reactor has a capacity of 30 kg/h, a power of 30 kWel and it allows the variation of the feedstock feeding rate, rotation speed of the reactors, inclination degree, and temperature. In the current tests, air gasification (co-current configuration mode) was applied on cherry wood, at 600°C, the reactor being set at 10° inclination. The modified process parameters were the reactor rotation speed, the biomass and air feeding rate. The syngas was passed through an ice water cooling system, while the composition of the non-condensable fraction has been analyzed using Testo-350XL (real-time measurements) and micro -gas chromatograph-Micro-GC Fusion, Inficon (time-sequential measurements). Testo 350XL gas analyzer cannot determine simultaneously the Hydrogen and hydrocarbons concentrations, therefore only the Hydrogen concentration was determined. In the cases where the micro gas-chromatographer (Micro-GC) was used, the presence of Methane, Ethane and Propane were detected. Based on the measurements, the average syngas compositions were determined.

To establish the mass balance of the process, ash and tar were collected and weighted for each gasification process. Ash was collected in a water vessel in order to be cooled immediately and not to react in the presence of air. Afterwards, the ash was dried, weight and analyzed. The tar fraction was trapped in the condensing system placed between the gasification reactor and gas extraction and sampling [22]. The tar was also weight and analyzed.

When exiting the gasification reactor, syngas has important sensitive energy content. Regenerative applications use this sensitive energy of the syngas. For this type of applications, hot-gas efficiency (HGE) is computed according to equation (1). HGE takes into account both chemical and sensitive energy content of the gas [23 - 26]:

$$\mathrm{HGE} = \frac{\mathrm{LHV_{gas}} \cdot \mathrm{M_{gas}} + \mathrm{M_{gas}} \cdot \mathrm{c_{gas}} \cdot (\mathrm{T_{gas}} - \mathrm{T_{ref}})}{\mathrm{LHV_{fuel}} \cdot \mathrm{M_{fuel}}} \quad (1)$$

where: LHVgas/LHVfuel is the low heating value of gas/fuel [kJ/kg]; Mgas [kg] is the mass of gas produced from gasifying a specific amount of fuel Mfuel [kg]; c_{gas} [kJ/kgK] is the specific heat of gas; Tgas is the gas temperature exiting the reactor [°C], and Tref is the reference temperature [°C].

3 Results and discussion

In gasification processes the reactor type has major influence on syngas properties. Rotary kilns are not utilized in gasification processes because of air and fuel poor mixing level. They are currently used in pyrolysis processes or cement industry. Nevertheless, the process evolution trend can be observed and modified in real time because of continuous operation of the pilot reactor used in our study. The differences between our results and other studies are because of reactor type, but the influence of process parameters on reaction products are following other research studies [27, 28].

3.1. Syngas composition

For the first experiment conducted, gas composition was determined using a Testo 350XL gas analyzer, while for experiment no. 2 to no. 6, the gas composition was determined through gas-chromatography. During the first experiment, according to the gas composition, a preponderant combustion process occurred in the reactor. High amount of CO_2 was produced, and unreacted air was found in the gas (O_2 concentration exceed 7 %). CO concentration did not reach 2 %, while the H_2 concentration was below 1 %. All of these are a consequence of a highly amount of air introduced in the process. In the 2nd experiment the equivalence ratio (ER) was reduced and the syngas composition changed considerably (Figure 2). CO_2 concentration has almost doubled, CO concentration increased more than 4 times, while the O_2 concentration was almost zero. H_2 concentration slightly increased from 0.40 % to 0.63 %, but the presence of light hydrocarbons may be observed, of which the most notable is the ethane with a concentration of 1.92 %.

When the ER decreases, and the time residence increases – compared to the first case – (Experiment no. 3 – Figure 3), an improvement in gas quality can be observed. The CO concentration exceeds 10 %, while the H_2 concentration reaches a medium value of 2 %. Carbon dioxide average concentration grows slightly comparing to Experiment no. 1.

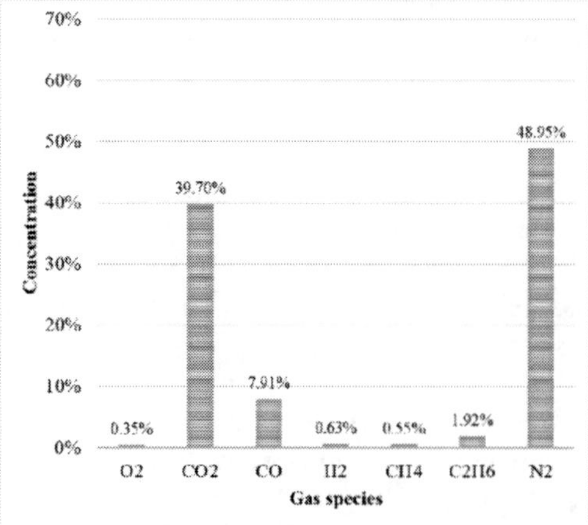

Fig. 2. Syngas composition – Experiment no. 2

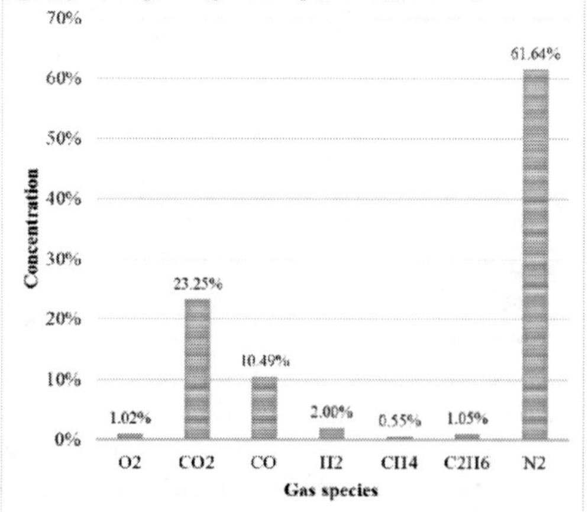

Fig. 3. Syngas composition – Experiment no. 3

In Figure 4 is presented the medium composition of the syngas produced during the 4th gasification experiment. In this case, the ER was further reduced, while the residence time was maintained high. Since less air was present in the reactor and the gases where maintained in the reactor longer, the CO_2 resulted from the combustion stage reacted with the char and more CO was produced. The concentrations of carbon oxides are approximately equal, 18.35 % vs. 17.39 %. Both H_2 and CH_4 concentrations exceed 2 %.

In the 5th experiment (Figure 5) the residence time was slightly decreased, and the ER was almost doubled compared to experiment no. 4. In consequence, CO_2 concentration increased (~ 25 %) and CO concentration decreased (less than 10 %). Hydrogen and methane concentrations also decreased to 1.35 %, respectively 0.67 %. However, the ethane concentration exceeds 1 %.

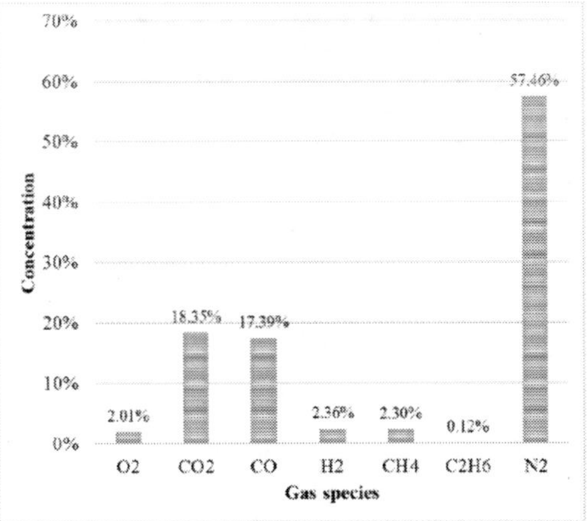

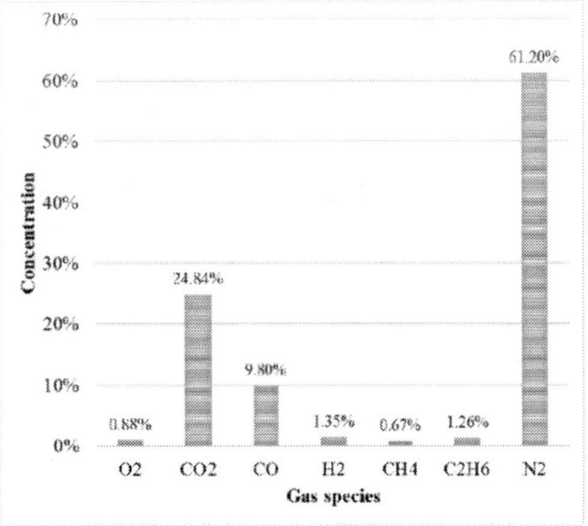

Fig. 4. Syngas composition – Experiment no. 4

Fig. 5. Syngas composition – Experiment no. 5

In Figure 6 is presented the average syngas composition resulted from the 6th gasification process. Compared to the previous cases the residence time was the highest, and the ER was lowest. CO_2 and CO medium concentration slightly exceeds 20 %, respectively 16 %.

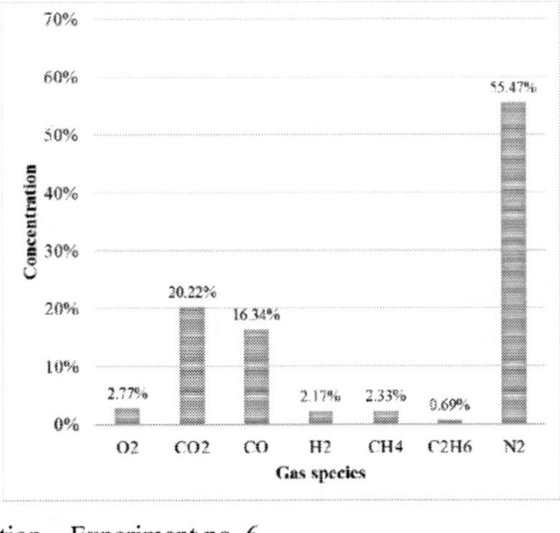

Fig. 6. Syngas composition – Experiment no. 6

According to the carbon oxides concentrations, better results were obtained in the 4th experiment. Hydrogen concentration is also smaller compared to Experiment no. 4 (2.17 %), while the light hydrocarbons concentrations (methane and ethane) are slightly higher (2.33 % vs. 2.30 % and, respectively 0.69 % vs. 0.12 %).

3.2. Syngas properties

Gas properties were determined considering the average composition of the gas presented in the previous section. The low heating value (LHV), density and specific heat of the gas were determined and presented in Figure 7 and Figure 8.

Figure 7 reveals how gas composition influences the heating value of the syngas. Gas heating value increases with combustible gas species concentrations. In the first experiment, the combustion process was preponderant; therefore, the gas heating value has not reached 1 MJ/m^3. The syngas produced from gasification experiments no. 4 and no. 6 registered a heating value that exceeds 3 MJ/m^3, while the heating value of the gas resulted from experiments no. 2, no. 3 and no. 5 is approximately 2.4 MJ/m^3. The highest LHV obtained is of 3.5 MJ/m^3.

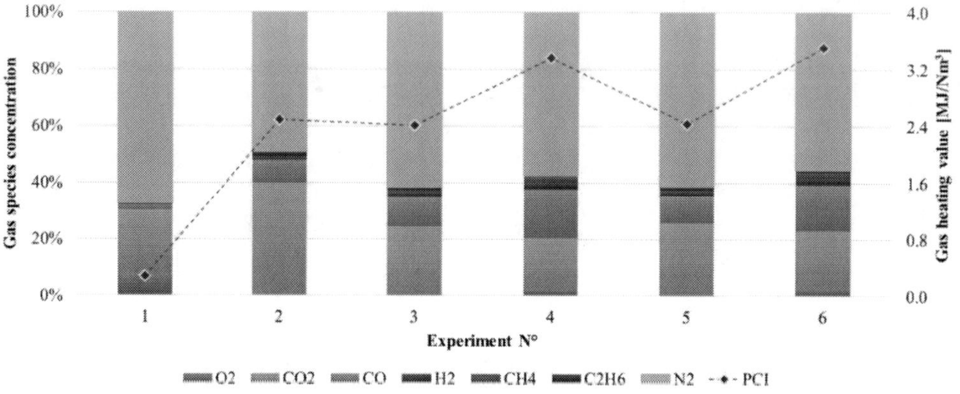

Fig. 7. Low heating value of the syngas as a function of gas species concentration

It is known that use of air as gasification agent lead to a syngas with a high content of Nitrogen, which reduces the LHV of the gas to a value of 3-6.5 MJ/Nm^3 [29]. Although the value obtained from our experiments is low, it follows values already reported by other researchers [27, 28] for syngas resulted from wood and correlated with the gasifying agent (air).

In Figure 8, the gas density and specific heat are presented for the 6 conducted gasification processes. Gas density varies slightly from 1.26 to 1.43 kg/m^3, while the specific heat of the gas varies from 1.04 to 1.34 kJ/kgK. Both density and specific heat are a function of gas composition at normal conditions of temperature and pressure.

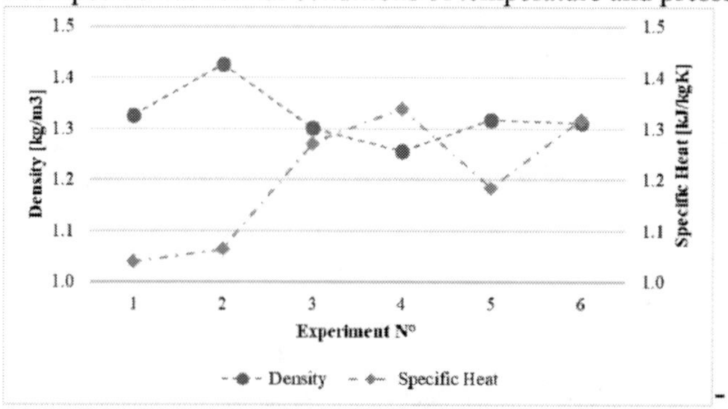

Fig. 8. Density and specific heat of the syngas

Natural gas density at standard temperature and pressure varies between 0.7 and 0.9 kg/m^3 [30], while its specific heat is 2.34 kJ/kgK [31]. Compared to the types of syngas presented in this paper, natural gas has lower density, but higher specific heat.

3.3. Mass and energy balance

For each gasification experiment, mass and energy balance were determined. According to mass balance (Figure 9) gas production ranges between 93.31 % and 98.23 %, while the tar production averages 1.18 % and the ash production averages 2.81 %.

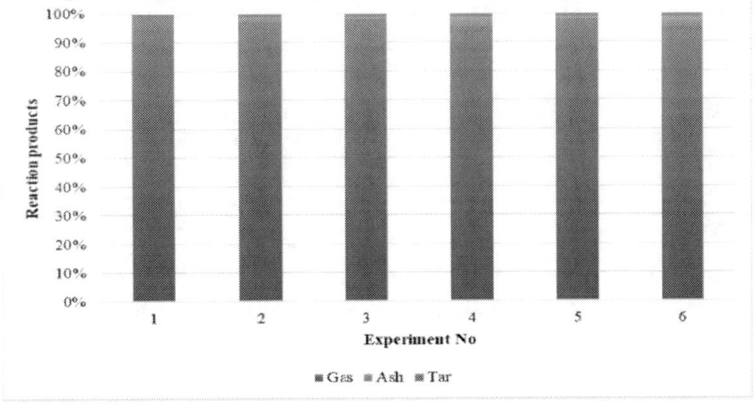

Fig. 9. Reaction products distribution of the gasification experiments

Highest gas production corresponds to the first experiment conducted which is characterized by the highest ER. Lowest gas production 93.31 % and 93.89 % correspond to experiment no. 4 and no. 6, when the ER value set was at its lowest. Even though the gas production is lower in these two cases, the LHV of the gas was considerably higher (Figure 7). Tar production varies slightly between 1.15 % and 1.20 %, while the ash had a quota of 0.59 – 5.53 % (experiment no.1 and, respectively no. 4) of all reaction products.

In Figure 10, the input and output energy flows of the gasification reactor are presented. Experiments no. 2, no. 3 and no. 5 are similar regarding the input/output energy flows values, but their process energy efficiency varies between 69.04 % and 74.08 %. As it is mentioned above, the syngas resulted from these three experiments have a medium LHV, compared to the other experiments conducted, but these processes registered the highest energy efficiency.

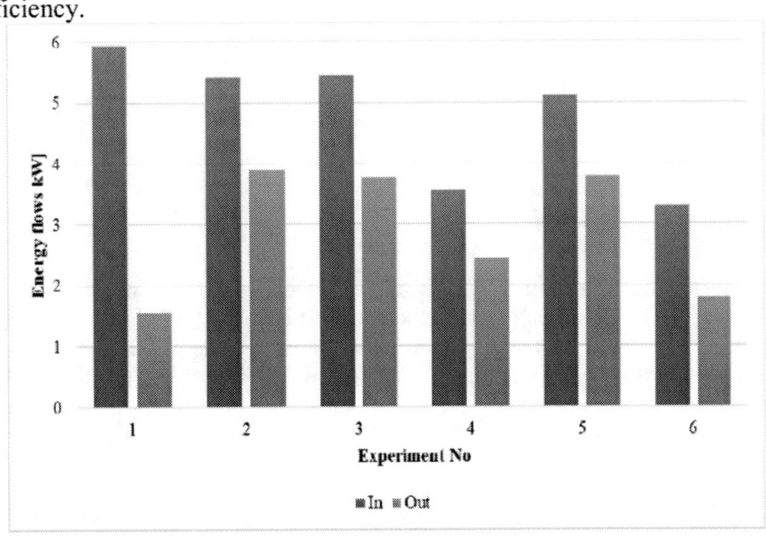

Fig. 10. Energy balance of the gasification experiments

Colum 1 in Figure 10 corresponds to combustion process, being out of range of current gasification energy efficiency. The process energy efficiency of experiment no. 4 and no. 6, which corresponds to the highest gas LHV, was 68.28 % and, respectively 54.39 %.

Wood gasification in rotary kiln reactors it is not preferred due to lack of proper agent – feedstock mixing. Improper mixing of input flows, and between the gas phases, leads to low carbon conversion efficiencies. In the gasification experiments conducted, the ash Carbon content varied from 10 to 27 %. Therefore, process energy efficiency determined resulted quite low.

The low energy efficiency can be also explained through the absence of air preheating before injection into reactor leading to a cooling effect and oxidation reactions delay.

4 Conclusions

Continuous operated reactors offer undeniable advantages in process observation over batch or fed-batch reactor. The real time gas sampling and process parameters monitoring give accurate information on process run and reaction products formation. While the collected are closer to industrial ones, the reactor / installation type strongly influences the process and restrains the validity of the results to the specific experimental set-up. The rotary kiln reactor used in the study introduced a series of perturbations through the false air infiltration and low fuel-air mixing level but enabled the precise control and observation of the process.

The results show, as expected, the importance of Equivalent Ratio on process energy efficiency as well as on carbon conversion ratio. Low values of ER led to a longer residence time in the rotary kiln reactor and so a longer reaction between the CO_2 resulted from the combustion stage reacted and char leading to high CO production.

Mass balance indicated a gas production in the range of 93.31 % and 98.23 %, while the average production for tar registered a value of 1.18 % and of 2.81 % for ash, respectively.

The calculated value of gas density recorded a slow variation from 1.26 to 1.43 kg/m^3, while a reduced difference was obtained for the gas specific heat of the gas, between 1.04 and 1.34 kJ/kgK. The results show that process energy efficiency is not sufficient for gasification process, as it is for combustion, and it must be correlated with the syngas specific energy content. The process energy efficiency maximized at 74 %, while de syngas LHV was maximum when the process energy efficiency was about 55 %.

Relevant information on air-gasification process was achieved with respect to operating parameters, syngas quality and global energy efficiency.

Acknowledgment

Funding provided through Romania's "Competitiveness Operational Programme 2014-2020" Priority Axis 1: Research, Technological Development and Innovation (RD&I) to Support Economic Competitiveness and Business Development Action 1.1.4. Attracting high-level personnel from abroad in order to enhance the RD capacity ID/Cod My SMIS: P_37_768/103651; No. Contract: 39/02.09.2016

References

1. P. Lauri, P. Havlík, G. Kindermann, N. Forsell, H. Böttcher, & M. Obersteiner, Energy Policy, **66**, 19-31, (2014).

2. WBA, World Bioenergy Association, Sweden. (2016)
3. Eurostat, Report 2018, Wood as a source of energy, https://ec.europa.eu/eurostat/statistics-explained/index.php?title=Wood_as_a_source_of_energy&oldid=427588
4. EU Report, 2017, Sustainable and optimal use of biomass for energy in the EU beyond 2020, Annexes. file:///C:/D_inexistent/A_POC_E_2016/Conferinte/TERERD%202020/Materiale/biosustain_annexes_final.pdf
5. V. Burja, R. Mărginean, Economica Journal, **65,** 4 (2013).
6. EOS, Report 2018-2019, EOS (European Organisation of Sawmill Industry) Annual Report of the European Sawmill Industry, (2019). https://www.eos-oes.eu/downloads/EOS-annual-report-2018-2019.pdf
7. C. J. Roos, Northwest CHP Application Center (2010).
8. S. Poskrobko, D. Krol, A. Borsukiewicz-Gozdur, Drewno: prace naukowe, doniesienia, komunikaty **59** (2016).
9. F. Meng, Q. Ma, H. Wang, Y. Liu, & D. Wang, Fuel **249** (2019).
10. T. M.Ismail, A. Ramos, E. Monteiro, M.A. El-Salam & A. Rouboa, Renew. Energ. **147** (2020).
11. R. N. Tirtea, C. Bulmau, G. Ionescu & C. Marculescu, In E3S Web of Conferences **112**, EDP Sciences (2019).
12. A. A. Ahmad, N. A.Zawawi, F. H. Kasim, A. Inayat & A. Khasri, Renew. Sustain. Energ. Rev. **53** (2016).
13. S. Heidenreich & P. U. Foscolo, Prog. Energ. Combust. **46** (2015).
14. B. Steubing, R. Zah & C. Ludwig, Biomass and bioenergy **35**, 7 (2011).
15. W. Zhang, Fuel Proc. Technol. **91**, 8 (2010).
16. M. A. F. Hamad, A. M., Radwan, & A. Amin, IntechOpen (2017).
17. A. Jain, S. Chopra, J. Agric. Eng, **IX,** 5 (2007).
18. F.M. Guangul, S.A. Sulaiman, A. Ramli, Bioresour. Technol. **126** (2012).
19. C. Marculescu, V. Cenuşă, F. Alexe, Waste Manag. **47** (2016).
20. M. Asadullah, Renew. Sustain. Energ. Rev. **29** (2014).
21. A.P. Faaij, Energy Policy **34** **(**2006).
22. G. Ionescu, R. Tirtea, C. Bulmau, C. Marculescu, 9th International Conference on Thermal Equipment, Renewable Energy and Rural Development (TE-RE-RD), 2020 – Under Review.
23. P. Basu, Academic Press (2013).
24. C. Stevens, Robert C. Brown. John Wiley & SonS (2011).
25. M. de Souza-Santos. CRC Press (2010).
26. R. N. Tirtea, PhD Thesis, University Politehnica of Bucharest (2019).
27. N. Harihastuti, In E3S Web of Conference **73**, EDP Sciences (2018).
28. E. Oveisi, S. Sokhansanj, A. Lau, J. Lim, X. Bi, F. Preto & C. Mui, Environments **5** 7 (2018).
29. S. Heidenreich, M. Müller, and P.U. Foscolo, Academic Press, 2016.
30. The Engineering ToolBox, https://www.engineeringtoolbox.com/gas-density-d_158.html
31. The Engineering ToolBox, https://www.engineeringtoolbox.com/specific-heat-capacity-gases-d_159.html

The generation of a mathematical model of the biogas production process from organic municipal solid waste

Marzhan Temirbekova[1], *Madina* Aliyarova[1], *Iliya* Iliev[2*], *Aliya* Yelemanova[1], and *Saule* Sagintayeva[1]

[1]AUPET named after G. Daukeev, 050000 Almaty, Kazakhstan

[2]University of Ruse, Department of Thermotechnics, Hydraulics and Environmental Engineering 7017 Ruse, 8 Studentska Str., Bulgaria

Abstract. This paper justifies the efficiency of the biogas collection and utilization at the MSW (municipal solid waste) landfill in Almaty with the installation of several modern biogas plants. The optimal mode of processes occurring in a biogas plant is determined by computer generated simulations. Mathematical model parameters were identified to describe biochemical processes occurring in a biogas plant. Two approaches are used to resolve the mathematical model: the finite-difference method for solving the system of differential equations and simulation modeling by using the Any Logic package. A program is written in the algorithmic language C ++. Numerous calculations were carried out, the results of which are presented in curves and their qualitative picture is consistent with the ongoing processes. The created computer program allows to make a preliminary forecast of anaerobic fermentation occurring in the bioreactor depending on volume of the substrate, methane microorganisms and temperature conditions.

1 Introduction

The Joint Stock Company «Tartip», is engaged in the collection, recycling and disposal of municipal solid waste in the city of Almaty, a large metropolis of the Republic of Kazakhstan. The organization of household waste removing and its scientific-based recycling is a topical issue. Environmental conservation is one of the major challenges of our time. The JSC «Tartip» has 200 units of special-purpose machines, located at seven production capacities in six urban districts. The motor traffic is controlled by using the GPS monitoring system to ensure the timely removal, improve the speed and to optimize the logistics of removal routes. The Company carries out work on the municipal solid waste (MSW) and other waste removing as carefully as possible, without harming the

*Corresponding author: iiliev@enconservices.com

environment. More than 2,000 tons of garbage are dumping daily in Almaty, which must be collected and removed out of the city. The JSC «Tartip» constantly improves the processes being carried out, gives priority to preventive measures to prevent or minimize pollution on the environment and energy usage reducing, and also the Company conducts open and transparent activities [1].

In accordance with practice all over the world, several options can be proposed for the biogas collection and disposal at the municipal solid waste (MSW) landfill in Almaty, which is the largest waste disposal site serving the city and villages in the surrounding area. About 13.5 million tons of waste has been accumulated within this landfill, from which 1740 m^3 / h of biogas can be produced.

In connection with this, the most attractive projects are the biogas production. For this purpose, it is sufficient to install several biogas plants. In this work, based on the well-known mathematical models and Any Logic simulation modelling, computer programs were created for preliminary forecast of the anaerobic fermentation process in biogas units with overpressure protection and a combined frame-blade structure of an anchor type agitator with fully automatic electric drive [2,8].

2 Mathematical model

The morphological composition of waste was obtained as a result of time - studying for Almaty. The morphological composition studies of waste in Almaty, shows that food waste is about 30% [3]. Table 1 presents the Morphological composition of waste in Almaty.

Table 1. Morphological composition of waste in Almaty

Component name	**Almaty, 2008 (%)**	**Almaty, 2012 (%)**	**Arithmetic mean value (%)**
Food waste	32.50	23.40	27.95
Paper and cardboard paper	27.60	27.80	27.70
Park and garden waste	0.00	0.00	0.00
Wood waste	1.20	2.40	1.80
Rubber, leather, bones	0.70	0.04	0.37
Textile material	2.40	1.30	1.85
Other organic waste*	4.30	5.53	4.92
Plastic material	8.30	14.30	11.30
Metal	2.90	1.60	2.25
Glass	4.30	8.60	6.45
Construction debris	10.80	9.40	10.10
Stone screening dust	0.70	0.00	0.35
Inorganic waste*	4.30	5.53	4.92
Total	100	100	100
The sum of an organic components	68.7	60.5	64.6
The sum of an inorganic components	31.3	39.5	35.4

As the table shows, the significant specific gravity of food waste once again confirms the effectiveness of the biogas plants.

Mathematical and computer simulations using mathematical models are the most efficient ways of conducting simulation experiments, as they are less expensive compared to full-scale experiments. In the papers of L.I. Ruzhinskaya and A.A. Fomenkov, the analysis of known models of organic matter anaerobic fermentation with the biogas production is executed. The models that are often used in the references describing the kinetics of methane fermentation and the parameters of the kinetic process are demonstrated. Different models describe the growth and development of micro population, the components degeneration, and the preparation of products in an anaerobic bioreactor [4,10,11]. There are papers related to simulated exponential curves, the modified Gompertz curves of the biogas production process from MSW in anaerobic reactors [5]. In addition, models are generated for the biogas production process optimization, comprehensive models including the microbial flora growth kinetics, which take into account various charging mechanisms and time frame, temperature of process and inhibitory effects [6].

In the papers of A. G. Topage, V.A. Wigant, L.A. Khvorov [7] the mathematical model is shown as an ordinary differential system with initial conditions. For simulation, the Any Logic simulation process package is used.

Consider the following task [7,9] for modeling the process of biofuel production from organic municipal waste:

$$
\begin{aligned}
&\frac{\partial W_i}{\partial t}=-k_i\cdot W_i\cdot f_H(S),\ i=1,\dots,M,\ \ W(0)=W_0,\\
&\frac{\partial S}{\partial t}=\gamma\sum_{i=1}^{M}k_i\cdot W_i\cdot f_H(S)-\rho_M\cdot f_M(S)\cdot\frac{S\cdot B}{K_S+S},\ S(0)=0,\\
&\frac{\partial P}{\partial t}=Y\cdot(1-\theta)\cdot\rho_M\cdot f_M(S)\cdot\frac{S\cdot B}{K_S+S},\ P(0)=0,\\
&\frac{\partial B}{\partial t}=\theta\cdot\rho_M\cdot f_M(S)\cdot\frac{S\cdot B}{K_S+S}-K_D\cdot B,\ B(0)=B_0,\\
&f_H(S)=\left(1+\left(\frac{S}{A_H}\right)^{N_H}\right)^{-1},\\
&f_H(S)=\left(1+\left(\frac{S}{A_M}\right)^{N_M}\right)^{-1}.
\end{aligned}
\qquad (1)
$$

where

W– is the initial substrate concentration(g/l);

S – is the hydrolysis product concentration (g/l);

W_i– is the vector components of the various types raw material concentrations;

P– is the total biogas yield, (g/l);

k – is the hydrolysis kinetic constant;

k_i– *is* the hydrolysis kinetic constants specific to W_i types;

M– is the number of considered raw material types (in this case 3);

B– is the concentration of the methane microorganisms in biomass;

γ – is the conversion coefficient of the substrate into fatty acids (stoichiometric coefficient);

ρ_M – is the methanogenesis maximum specific rate in the terms of the volatile fatty acid biomass utilization;

K_s– is the half saturation constant in the Monod equation for the methanogenesis intensity;

$(1-\Theta)$- is the substrate fraction used for biogas formation;

Y– is the conversion coefficient of the fatty acids utilization flow to final product yield units (biogas);

K_D- is the decay factor;
$f_H(S)$ and $f_M(S)$– are describe the hydrolysis reactions inhibition and the microbial fermentation process by fatty acids (acidification).

The values of the constant coefficients used in the model (1) are given in the form of a system and dynamic table in the paper [7]. Using model (1), we generate a model in Any Logic. The simulation model and curves of the key process indicators are given in Figure 1. In Fig. 1, the concentrations of various types of feedstock in the form of the W_i vector components are indicated through W1, W2, W3.

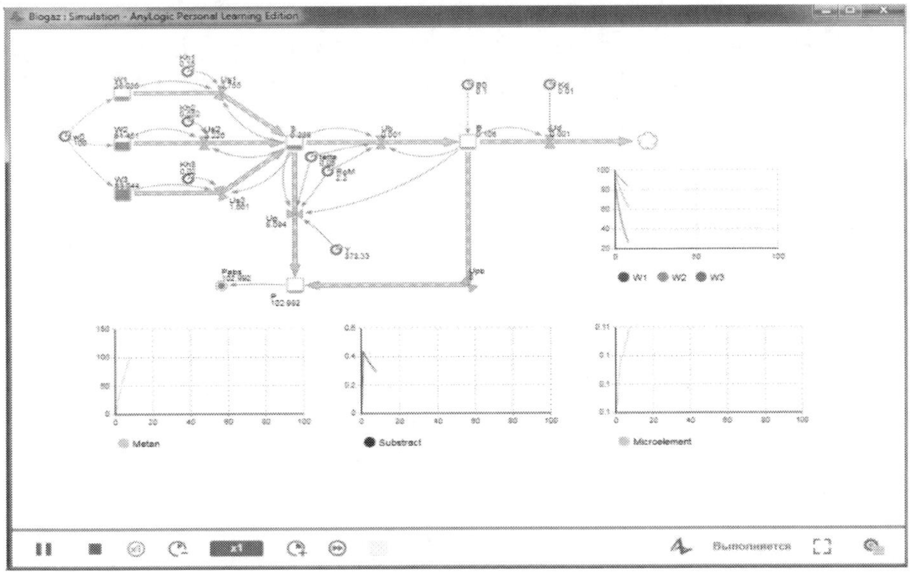

Fig.1. The biogas model generated in Any Logic

For the model generation, the «days» unit of time was selected, since all the coefficients are given in the 1 / day format. For the differential equation system (1) a corresponding semi-implicit difference scheme of the following form has been built:

$$
\begin{aligned}
&W_i^{n+1}=W_i^{n+1}-\tau\cdot k_i\cdot W_i^n\cdot f_H(S^n), \quad i=1,2,3,\\
&S^{n+1}=S^n+\tau\cdot\left(\gamma\sum_{i=1}^{3}k_i\cdot W_i^n\cdot f_H(S^n)-\rho_M\cdot f_M(S^n)\cdot\frac{S^n\cdot B^n}{K_S+S^n}\right),\\
&P^{n+1}=P^n+\tau\cdot\left(Y\cdot(1-\theta)\cdot\rho_M\cdot f_M(S^{n+1})\cdot\frac{S^{n+1}\cdot B^n}{K_S+S^{n+1}}\right), \qquad (2)\\
&B^{n+1}=B^n+\tau\cdot\left(\theta\cdot\rho_M\cdot f_M(S^{n+1})\cdot\frac{S^{n+1}\cdot B^n}{K_S+S^{n+1}}-K_DB^n\right),\\
&n=1,2,3,\dots,N.
\end{aligned}
$$

For the numerical solution of difference equations, a program has been developed using the algorithmic language C++. Unknown values of the concentration vector of the source feedstock of various types are specified in the form of a one-dimensional array, the concentration of hydrolysis products, i.e. the total biogas yield, the biomass concentration of methanogenic microorganisms, are considered in the program as variables depending on the discrete time. Based on the given values of initial conditions, the desired values are calculated according to recursive formulas (2) depending on time. Then the calculated results are imported into a graphical editor to build graphs.

The identification of the coefficients included in the mathematical model was carried out through a computing experiment. Numerical computations on the mathematical model (1) may be performed in a wide range of τ time step and equation coefficients meeting the stability conditions of the difference scheme (2). Therefore, it is necessary to determine the value of the time step corresponding to real time.

By using the initial conditions $W_0 = 100$ g / l, $B_0 = 0.1$ g / l, we obtain the *P, S, B* dynamic pattern which are written as the following curves (Figure 2). In this point, at the grid step of $\tau = 0.01$ and N = 50,000, gives a result corresponding to 500 days.

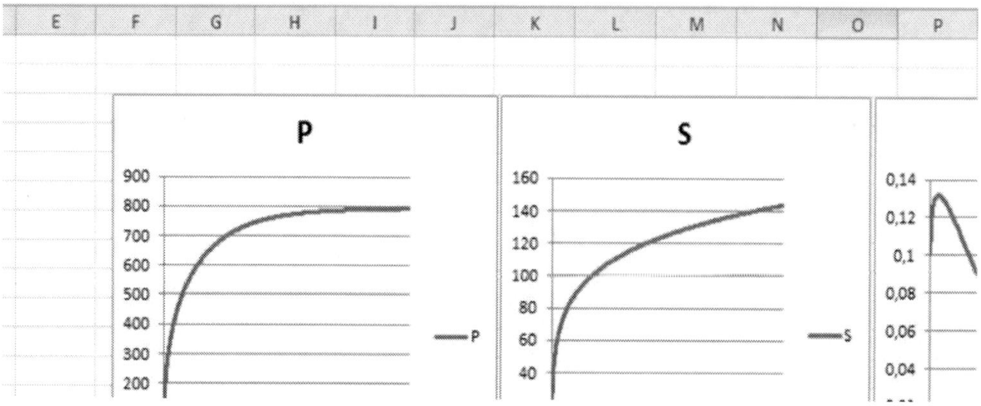

Fig. 2. Change of *P* – is the total biogas yield (g / l), *S* is the hydrolysis product concentration (g / l) *B* – is the concentration of the methane microorganisms in biomass

Set the value γ = 0.32, and Y = 0.85, and obtain the values of the total biogas yield, the hydrolysis product concentration, and the methane microorganisms biomass concentration in 500 days.

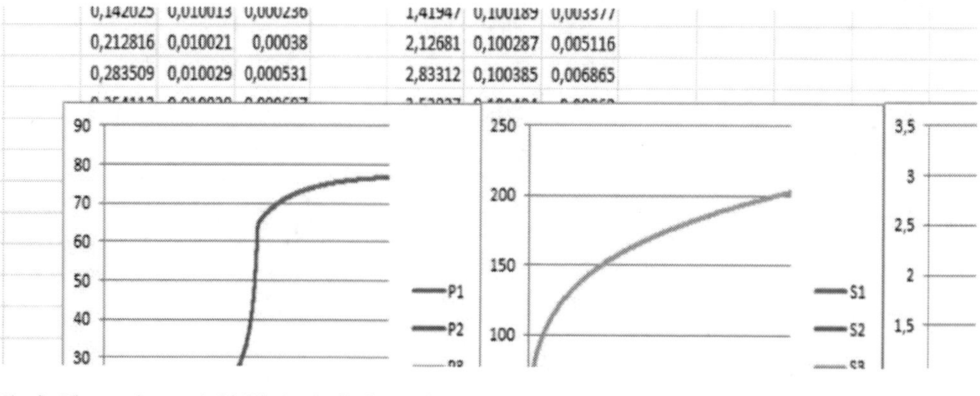

Fig.3. The methane yield (*P*), hydrolysis product concentration (*S*) and (*B*) -concentration of methanogenic bacterium changes.

The blue lines in the figure show the curves corresponding to the W_0 = 300g / l, B_0 = 0.3 g / l initial values. Under these conditions, the reactor is operating normally.

If too few methane microorganisms are introduced (red line, W_0 = 300g / l, B_0 = 0.005 g / l, the cork effect is observed in this case. If too much organic raw material is introduced, W_0 = 1000 g / l, B_0 = 0.2 g / l, then the curve will be such as shown by the green line (Figure 3).

3 Conclusions

Thus, developed software product in the C++ algorithmic language and simulation model using Any Logic created on the basis of a mathematical model in the form of a system of ordinary differential equations allow to carry out numerous computing experiments to simulate various scenarios of processes in biogas units. The software products developed by the authors of the article have been tested on previously known test examples. The correspondence of the non-dimensional time step to real time has been established and computing experiments have been carried out to identify other coefficients of the equations. The created set of application programs allows to make a preliminary and predictive computer modeling of the process of biogas production from various morphological composition of organic waste.

References

1. Web site: www.tartyp.kz/o-nas/
2. S.A. Keshuov, V.I. Barkov, A.B. Tokmoldayev, *The biogas plants application in livestock business*, Almaty, (2012)
3. Zh.M. Zhapparova, EcoInform, Materials of the 5th International Conference "Cooperation for Solving the Waste Challenge", 195–197, (2008)
4. L. I. Ruzhinskaya, A. A. Fomenkova, Scientific Journal «Science rise» **№4/2(4)**, 63-69, (2014)
5. H.M. Lo, T.A. Kurniawan, M.E.T. Sillanpää, T.Y. Pai, C.F. Chiang, K.P. Chao, M.H. Liu, S.H. Chuang, C.J. Banks, S.C. Wang, K.C. Lin, C.Y. Lin, W.F. Liu, P.H. Cheng, C.K. Chen, H.Y. Chiu, H.Y. Wu, Bioresource Technology, Volume **101**, Issue 16, 6329-6335, (2010)
6. Mandy Gerber, Roland Span, *An Analysis of Available Mathematical Models for Anaerobic Digestion of Organic Substances for Production of Biogas*, IGRC, (2008)
7. A. G. Topage, V. A. Vigant, L. A. Khvorova, Chemistry of vegetable raw material, No. **1**, 171-184, (2018)
8. I. K. Iliev, V. I. Kamburova, A. K. Terziev, *Analyses of landfill gas potential in Bulgaria*. IN: EEA & AE 2013- International Scientific Conference 17-18.05.2013, Ruse, Bulgaria, Printing Centre at Ruse University, 230-235, (2013)
9. V. I. Kamburova, I. K. Iliev, A. K. Terziev. *Schemes for project financing of combined heat and power production through biomass gasification*. IN: 3rd International Conference of Thermal Equipment, Renewable Energy and Rural Development, Mamaia-Romania 12-14 June 2014 (the best paper), Politehnica Press, 65-71 (2014)
10. Dimitrov R., Zlateva P. *Possible Uses of Biogas for Power Purposes*, IN: BulEF-2019-11th Electrical Engineering Faculty Conference, 1-4, (2019)
11. Dimitrov R., Ivanov Z., Zlateva P., Mihaylov V. *Optimization of biogas composition in experimental studies*. 8th International Conference on TE-RE-RD 2019, E3S Web of Conferences, vol. **112**, (2019)

https://doi.org/10.1051/e3sconf/202018002019

Operation on battery packs of damaged installations on a tugboat

Violeta-Vali Ciucur[1*] and *Andreea* Jenaru[2]

[1] Constanta Maritime University, Faculty of Naval Electromechanics, 104 Mircea cel Batran Street, 900663 Constanta, Romania
[2] Constanta Maritime University, Faculty of Navigation and Naval Transport, 104 Mircea cel Batran Street, 900663 Constanta, Romania

Abstract. The supply of own services of direct current is made from at least two sources of DC power, the independent and capable sources of ensuring, each of them, the entire consumption. As a normal power source, are used AC/DC rectifiers and DC/DC converters, as the case. The provision of safety power supplies its able to take over, for a limited time, normalized, the whole consumption in the DC of this objective, is mandatory for all objectives. As a safe power supply are used battery packs. Each battery pack supplies a single network; it is not supplied from the same battery, two or more distribution networks, with different nominal voltages through intermediate sockets. For voltage supply, in case of dismissal of the basic source, they can be used as local sources, which provide electricity directly, the battery, built to ensure the required voltage and capacity in battery packs. The ships present the damaged installation which can operate supplied by battery packs. Electrical control circuits, signalization circuits, protection circuits and automation circuits are supplied with DC voltage. In this article we will calculate the battery capacity of a ship.

1 Overview

Safety in the power supply depends on the reliability of the electricity production, transport and distribution installations.

As the electricity obtained from the alternators cannot be stored, it means that when the voltage disappears to the consumer it can only be supplied from another local source which implies the existence of a generator, an automatic starter installation, etc.

This solution is used to ensure the supply of vital and relatively high power consumers, but which do not have a widespread.

Known, however, that the continuous electricity can be "accumulated", it is practiced the solution of continuous voltage supply of the importing consumers, but with powers not too great, having the advantage that, for supply, in case of dismissal of the basic source, they can be used as local sources, which provides electricity directly, battery, built to ensure the voltage and capacity required in battery packs. [1]

* Corresponding author: violetaciucur_umc@yahoo.com

Usually, electrical control circuits, signalization, protection and automation are supplied with DC voltage.
The battery packs ensure the direct current supply of the electric receivers the supply of which must be ensured during the disappearance of the voltage of its own AC services.
For those electrical receivers whose power must not be interrupted at any time and high-power electrical receivers and short-lived, which does not support the voltage variations that occur if their supply is made exclusively by rectifiers or converters.
Usually, each battery pack supplies a single network, with a single nominal voltage and they cannot be powered from the same battery two or more distribution networks with different nominal voltages, through intermediate sockets.
Networks with different voltages are powered by different batteries. The networks with different nominal voltages will be galvanic separated.
For electrical stations with voltage ≥110 V with large number of cells (≥5) in the case of a single working battery, this system can be divided into two sections.
The battery is connected to both bar sections and the charging sources are either basic or backup, each one is connected to one of the sections (Fig. 1.).
Usually in the case of a single working battery, the two sections are permanently connected.
The connection between the sections can be opened during operation for short periods of time for the galvanic separation of the receivers, for earthquake detection, or during the revision work maintenance or repair at one of the bar sections.
For each working battery there is a simple system of main collector bars divided into two or three sections by one or two longitudinal couplings.
In the case of a single working battery, the principle of bar connection of the power supplies is applied. [2]

2 Fire extinguishing system in the machine compartment with battery packs

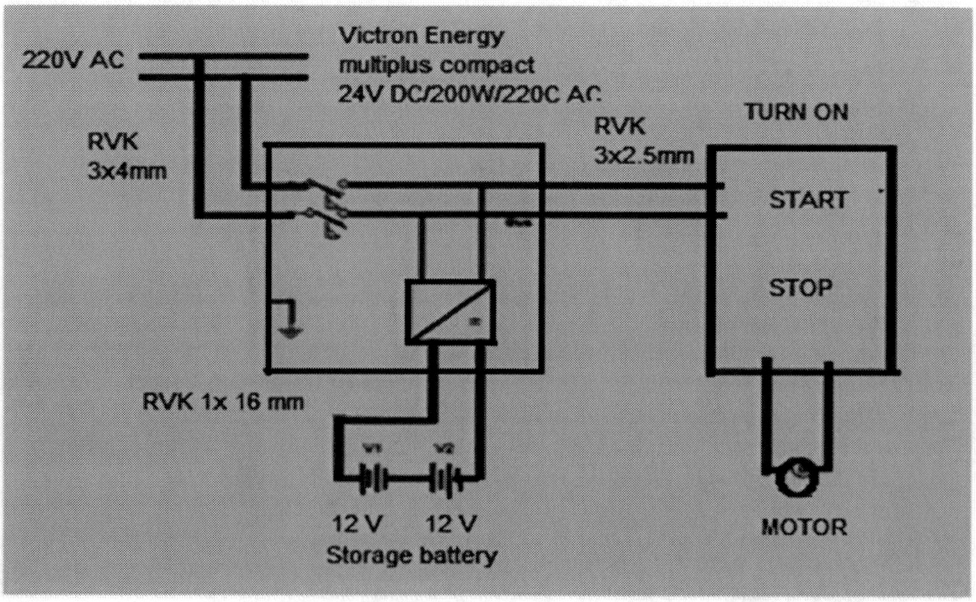

Fig. 1. Fire installation for extinguishing the fire in the machine compartment with battery packs

The power supply is 220V AC and the two batteries are charged by a Victron energy multiplier compact 24 V DC / 2200 W / 220 V.
The batteries are sized 12 V every 180 Ah.
The conversion from 24 V to 220 V AC is performed.
In this way the operation of the pump motor in case of failure is ensured.
The pump motor is 1.1 kW, 2800 rpm operating at a voltage of 220 V.
The driven pump draws water from a freshwater tank supplied from the dock and pushing it on the system with sprinklers, with a pressure of 6 bars.
The sprinkler system consists of a manually operated sprinkler head which starts only when the system start button is pressed. They are located above the main engines and the DG of damage.
The inverter regulates the AC output frequency for the systems with coupled photovoltaic inverters (Fig. 2.)

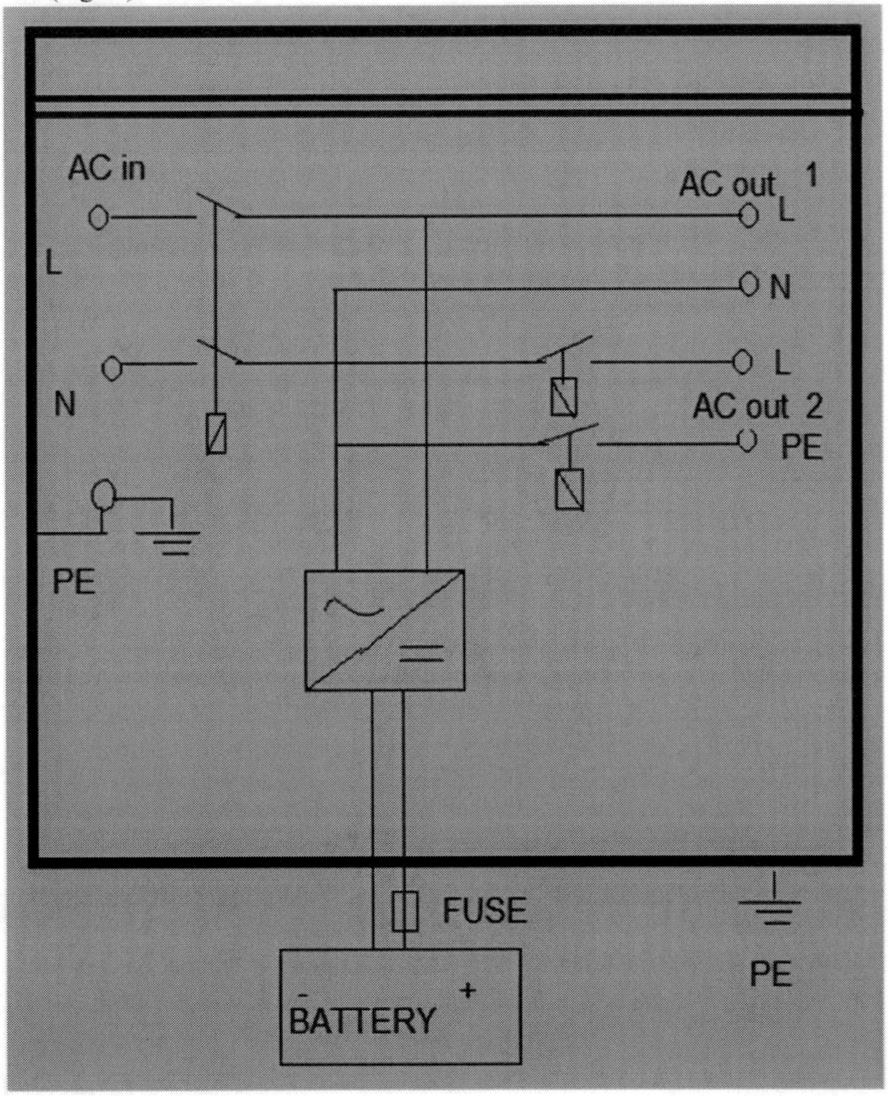

Fig. 2. Inverter block diagram

MultiInverter can be used in the solar converter connected to the network that is, a photovoltaic inverter, to make an island system. [4]

Thus the charge will flow to the battery.
This can cause the batteries to charge.
To prevent this, the solar converter must be powered or stop when the battery voltage becomes too much.
Most brands and types of PV inverters can be power limited by adjusting the output frequency of the MultiInverter, that is, by changing the frequency, the loading process can be adjusted.
The transfer switch can affect the frequency range.
When the system is in inverter mode the output power of the inverter connecting to the electrical network is controlled by changing the frequency of alternating current output of the system. A mechanism commonly referred to as "frequency change".

The conditions that must be met to change the frequency:
1. The battery voltage decreases below the recharged voltage threshold: the voltage that causes the system to restart the charging cycle.
For lithium batteries, this threshold is defined as the floating voltage - 0.8V.
When configured at 54.8 V or more, then the battery voltage should fall below 54V.
For lead batteries, the recharged voltage is defined as 5.2V below the configured floating voltage.
Unless it is set to 56.8 V or greater: then the battery voltage must decrease below 51.6 V.
2. The system is connected to the mains supply or to the generator
As soon as you apply a voltage to the AC input, it will synchronize the AC frequency to 50Hz and then connect.
List of probably consumers:

- general alarm
- navigation lights
- fog signaling facility
- CO_2 launch signaling facility
- navigation equipment
- interphone
- lighting damaged
- various signals

The *alarm bell installation* is used to set the alarm on board the ship or to transmit different signals on the ship.
The alarm is transmitted from the wheelhouse.
The installation of the alarm bells is supplied at 24V DC from the battery charger.
The alarm goes off on the ship by actuating the switch in the navigation desk where in parallel with the alarm switch the lamp is connected which signals the functioning of the bells.
At least every 10 days but every time before the ship at sea the proper functioning of the alarm bell system must be checked.
The *super typhoon siren installation* is used for signaling provided for by the "London 1974 Convention for the Protection of Human Life at Sea" under low visibility conditions, the ship being equipped with super typhoon siren installation.
The super typhoon siren installation is composed of:

- air siren
- automatic control block
- fog lamp
- button

The system is powered at 220 V, 50 Hz.
Radio goniometer installation OG 132 is used to determine the position of the ship based on the bearings, the ship being equipped with a goniometer radio installation.
The installation consists of:

- receiver
- radiometer
- connection box
- food block
- frame antenna

The radio goniometer installation is powered at 220 V, 50 Hz and 24 V.
To ensure the proper functioning of the installation, before leaving for the drive it is done position determinations at 1-2 radios, determining the correct functioning of the device and maintaining the viability of the broadcasting curve.
Periodically, the status of the frame antenna, the compensation antenna, the receiver and their connections is examined.

Radiotelephony installation UUSR 6401

To ensure the short distance connection between ships at sea in the case of convoy navigation in port with tugboats and piles, in the wheel and in the port basin with the port dispatcher, the ship is equipped with a UUS radio station.
Radiotelephony facility in UUS for ships type R6401 t is supplied at 220 V, 50 Hz.
The connection to the ship's on-board network is made through the power cord of the transmission-reception unit.

Radiotelephone RR3906 – 2

To ensure radio links on the ship, a RR3906 - 2. is installed. The installation consists of:

- radio phone UI
- power supply and charging batteries
- battery batteries
- antenna stick

The system is supplied at 220 V, 50 Hz.

3 Calculations to determine the battery capacity of radio batteries

- VHF / DSC radiotelephone cl. A - emission
- reception

3h x 5.5 A = 16.5 Ah (3.1.)
3h x1.0 A = 3.0 Ah (3.2.)

- VHF / DSC radiotelephone cl. D - emission
- reception

1 h x 5.5 A = 5.5 Ah (3.3.)
1h x 1.0 A = 1.0 Ah (3.4.)

- MF / HF radio handset - broadcast
- reception

3h x 8.0 A = 24.0 Ah (3.5.)
3h x 1.2 A = 3.6 Ah (3.6.)

- NAVTEXT

6 h x 0.1 A = 0.6 Ah (3.7.)

- GPS

6 h x 0.1 A = 0.6 Ah (3.8.)
Total: 54.8 x 1.5 = 82.2 Ah (3.9.)

Battery 2x12 V Capacity 90 Ah

The battery forming charge must ensure the nominal capacity of the battery.
Charging is also done at a lower current than the maximum charging current, but in this case the charging time is extended proportionally.
It does not charge at a load current of less than 40% of the maximum load current.
Generally for battery formation, a step-by-step charge is used.
The load current is a current of 7 [A] for each element.
The total current flowed from the load source is obtained by multiplying by 7 the number of elements placed in parallel to the load.
If the battery is discharged with a high current, the discharge will stop when the voltage reaches 1.8 [V] per element.
If the unloading is done in one hour, the voltage on the element should not fall below 1.75 [V]. After the battery is discharged, the battery is charged.
Charging starts after at least 20 minutes and at most two hours when the download is stopped. It is most favorable charge is the one with constant current value and equal to 1/3 of the maximum load current.

4 Conclusions

On board ships, the electricity sources are represented by the electric generators.
These are either DC or AC.
They are driven by diesel engines or turbines, as well as battery packs.
If the ship runs out of power it switches automatically/manually on battery packs.[3]
They provide the minimum electricity for moving the ship.

References

1. Harrould-Kolieb, E., „Shipping impacts on climate: a source with solutions", Sydney, 2019, OCEANA
2. Eykerman, A., „Fuel flexible solution efficient shipping", Wärtsilä Corporation, Helsinki, 2012
3. Hensher, D. A. & Button, K. J, „Handbook of transport and the environment", Amsterdam, Pergamon, 2013
4. Specific documentation on board ship.

Intensity of soil erosion in Romania, predictions in time and space, European and indigenous perspectives

In memoriam Vasile Herea

Petru Cardei[1,*], *Raluca* Sfiru[1] , and *Vergil* Muraru[1]

[1]INMA Bucharest, Informatics and Engineering, No 6 Bd Ion Ionescu de la Brad Bucharest, Romania

Abstract. The subject of this article is the estimation of the water erosion given by different sources during the history of over one hundred years of observations. The differences between the estimates made at the near times, or at the appreciably different times, are viewed for the approximation, of the random behaviour of the factors involved in the water erosion process, but also for the changes (apparently in time) of the intensity of the factors involved. The so-called climatic changes, characterized mainly by apparently (at the scale of human life) non-periodic changes of meteorological factors, produce effects including on the factors involved in water erosion, which are not in the category of meteorological parameters, such as soil erodibility, but also the geometric parameters of the slopes. By default, there are effects in vegetal cover and management parameters. From this point of view, the influencing factors of mathematical models for predicting water erosion should be recalculated or periodically reviewed.

1 Introduction

The idea of writing this article started from the differences of appreciation of the intensity of the water erosion in time and space, which means, from the perspective of the specialists from different geographical areas. It is a definite fact that estimating and forecasting the intensity of soil erosion is a very difficult test for any specialist. Over time, the estimation and the prediction of soil erosion have become more precise, benefiting from a multidisciplinary contribution of unquestionable. However, when we compare the values of the estimates at large intervals, must we consider that today's soil is not the same as fifty or one hundred years ago, and the soil over fifty years will not be the same as today. In this context, in which we are aware the irreversibility of the transformation of the eroded subject (soil) but also of the main erosive factor (climatic conditions) and secondary (man and his habits, however a component of nature) is can be made a comparison between different estimates over time and by different authors. Is trying a possible explanation of the differences in the estimates in time and space (regarding authors from different areas) and,

* Corresponding author: petru_cardei@yahoo.com

by this, some modifications are suggested in the formulation of the mathematical models of erosion, which will, possibly, give them a dynamic character.

Research on soil erosion in Romania has a history of over half a century, [1], [12], and [15]. Results of the researches on the water erosion evolution on the Romanian territory also appear in books containing collections of advanced research in the field, [13-14], [16-18].

2 Estimates and measurements of the intensity of water erosion in Romania

Estimating the intensity of erosion has been a permanent concern in the world of science since the beginning of the 20th century. A beautiful fresco of the history of the development of the scientific field of soil erosion is presented in [11]. Now, after more than a hundred years from the beginning, it is tempting to compare the estimates made in the same geographical area, at different times and with different tools, which are becoming more and more refined. Such tests are possible for many geographical areas. I will limit myself to the one they belong to, for now, Romania.

3 Results

To benefit from several sources of erosion estimation, we restricted the analysis to the territory of Romania, where I participated in 2005 - 2012 in some research projects that had as a subject the water erosion of the soil. For simplicity, we will present the results found in the temporal order.
In [1], page 92, table 8, it is shown that at the level of the years 1950 - 1960, the annual loss of soil by erosion in the Valea Calugareasca wine-growing area was 172.90 t/ha, for lands with a 25% slope, without anti-erosion land management. It is remarkable that, at that time, the author separates the losses by seasons (December-March, 0 t/ha, April-May, 96.0 t/ha, June-September, 76.0 t/ha and October-November, 0.900 t/ha).
An estimate of water erosion since 1983, [2], indicates for the above area an average value of 30-45 t/ha per year.

In a research project dedicated to the study of water erosion and landslides, we estimated the risk of erosion and represented as a map this parameter for the entire territory of the Research Institute for Vine and Viticulture, ICDVV Valea Calugareasca from Prahova county. We have presented the results in [3] as a risk map of water erosion resulting from the research. The values of the risk of water erosion indicated by the map in [3] are between 0 and 30 t/ha per year. However, 48,205% of the mapped area has soil losses through water erosion below 3 t/ha per year and less than 22% have soil losses through water erosion of over 7 t/ha per year. These statistical conclusions can be seen in fig. 1. For the calculation the USLE mathematical model was used and applied to each vineyard plots, including for the plots which at the time were not cultivated, having only uncontrolled plant cover. For the calculation, each plot was considered as a whole slope surface with constant inclination. The slope and inclination were determined on the images from the Google Earth archive, so the approximation is coarse.

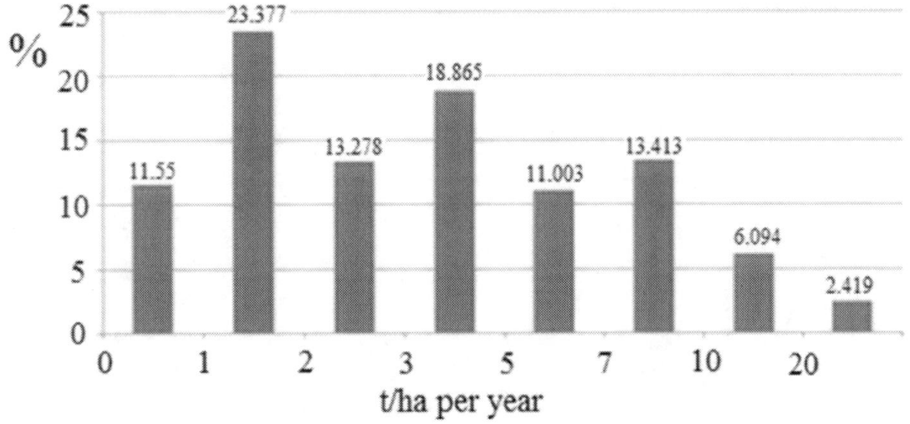

Fig. 1 Distribution of vineyards plots by risk intervals of water erosion, in the territory of the ICDVV Valea Calugareasca.

Using the ESDAC databases, the average assessment of water erosion at the county level in Romania can be extracted (data valid for micro-regions and in the EU). From the Excel public data files extracted from the ESDAC† address, [4-6], the synthetic representation of erosion in Romania can be obtained, table 1. For Prahova county, we have the soil value lost through water erosion of 2.45 t/ha per year.

Table 1. Counties of Romania in order of average value of soil loss through water erosion, in t / ha per year.

Vaslui	7.75	Neamt	2.75	Arges	2.09
Iasi	5.97	Hunedoara	2.73	Tulcea	1.95
Botosani	5.92	Harghita	2.72	Olt	1.79
Cluj	5.26	Suceava	2.59	Covasna	1.66
Salaj	5.00	Valcea	2.55	Timis	1.63
Mures	4.72	Prahova	2.45	Dambovita	1.51
Bistrita-Nasaud	4.30	Satu Mare	2.42	Arad	1.51
Bacau	4.20	Buzau	2.41	Teleorman	1.29
Galati	4.09	Caras-Severin	2.37	Bucuresti	1.13
Alba	3.96	Constanta	2.32	Ilfov	1.07
Sibiu	3.57	Brasov	2.31	Giurgiu	1.06
Mehedinti	3.12	Dolj	2.22	Calarasi	1.02
Maramures	2.95	Gorj	2.22	Ialomita	0.93
Bihor	2.79	Vrancea	2.20	Braila	0.84

The problem of the relatively large differences between the presented estimates is not so much one of an order of magnitude, but rather one that is related to the dynamics of the erosion process, the dynamics of the components involved in the process: the climatic factor, but can, especially, the soil evolution (structure, texture, composition). The differences can be seen in fig. 2. The assessments given in the presented documents are in agreement with [7], the SIDASS project.
An obvious observation, which results from the ranking given in fig. 3 or table 1, refers to the fact that compared to the hierarchies of the 1950s, 1980s, where the counties of Vrancea and Buzau were the first, in the ESDAC map, the 2015 edition, they are halfway ranking.

† The resolution of public maps based on ESDAC is insufficient to estimate the intensity of erosion on the slopes of a normal farm. They are for guidance only.

https://doi.org/10.1051/e3sconf/202018003001

For example, compared to the 1983 estimate, [2], the values of the ESDAC map are approximately 7 times lower, relative to the maximum value.

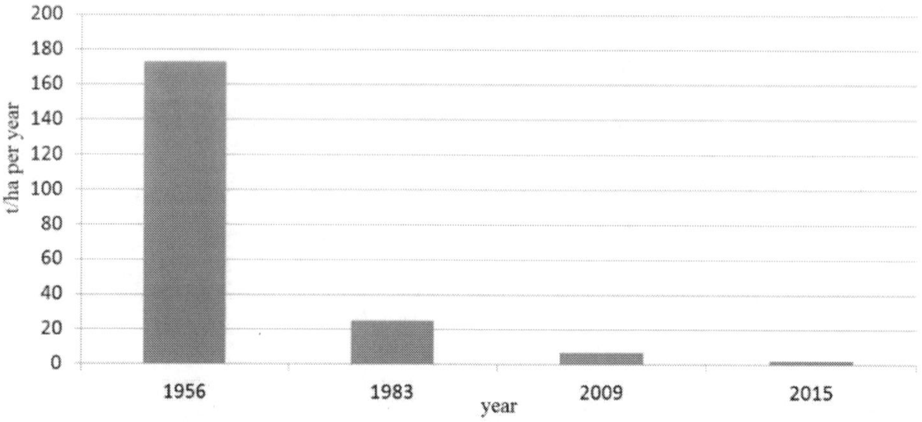

Fig. 2 A dynamic image of soil loss estimates by water erosion in the Valea Calugareasca wine-growing area, Prahova county, Romania.

The problem of the relatively large differences between the presented estimates is not so much one of an order of magnitude, but rather one that relates to the dynamics of the erosion process, the dynamics of the components involved in the process: the climatic factor, but may, in particular, the soil evolution (structure, texture, composition). The differences can be seen in fig. 2. The data appraisals in the documents presented are in agreement with [7], the SIDASS project.
The problem of estimating the intensity of erosion extends to the whole country, Romania. Using the data from table 1, the graphical representation of fig. 3 is obtained, which facilitates easier observation of the ranking of counties after soil loss through erosion.
In 2009, as part of a research project in the field of soil erosion and landslides, we tested for the Valea Calugareasca wine area, a somewhat new mathematical model, [8]. The mathematical model [8], of empirical type, is a linear model of the sum of effects type (unlike USLE, which is a factorial model):

$$A = 41.73 + 0.181 \cdot p + 0.046 \cdot I - 0.387 \cdot clay - 8.125 \cdot aveg \qquad (1)$$

where:
- p is the slope in percentages;
- I is the amount of annual precipitation, in cm;
- $clay$ is the percentage of clay contained in the soil;
- $aveg$ is a coefficient of vegetal cover established by the authors [1], as follows: 1, for soil not covered and ploughed, 2 for partially covered soil in a proportion of maximum 25%, 3, for a cover between 25 and 50% , 4, for the vegetation of forest type between 50 and 75% and 5, for very dense vegetation;
- A is the annual loss of soil in tonnes per hectare.

For the Valea Calugareasca area, the clay content in the soil, according to [9], varies between 9 and 43%. Also from [9] is obtain the multiannual average precipitation for the years 2000 - 2012 is 660.8 mm, respectively 604.7 mm, the annual average. The slopes of agricultural land (wine-growing area) in Valea Calugareasca are between 8% and 30%, according to [10].

It can be seen that the simple formula proposed by the authors [8], leads to results that fit well with the erosion intensity estimates in the Valea Calugareasca area without a recalculation of the coefficients of formula (1) with the help of experimental results.

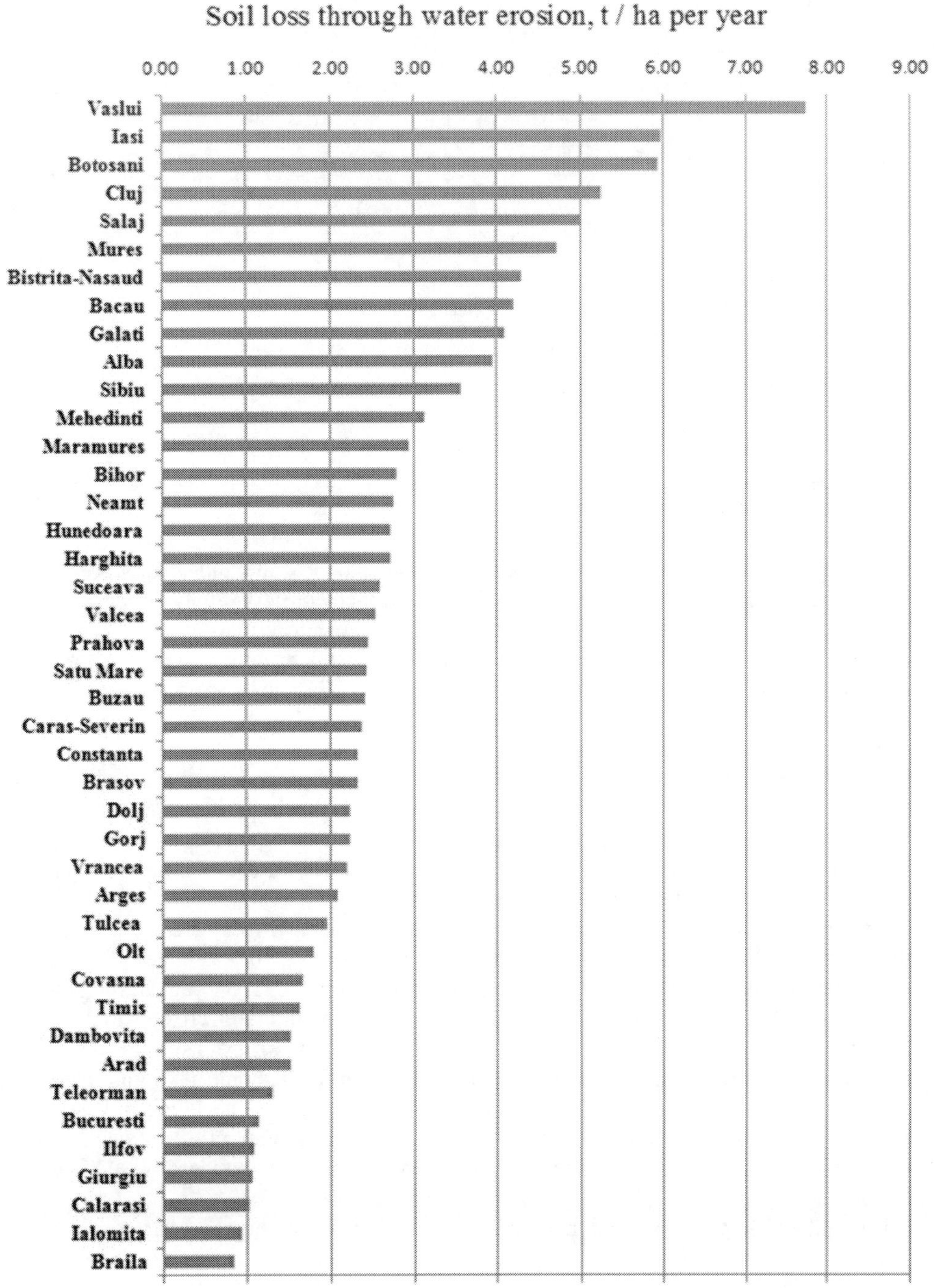

Fig. 3 Graphic representation for the hierarchy of water erosion at the county level in Romania, [7].

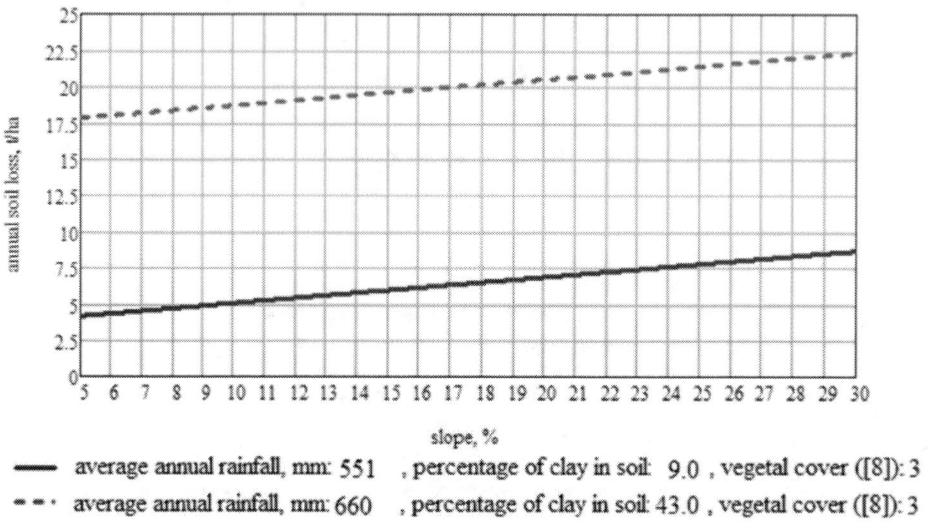

Fig. 4 The annual loss of soil by water erosion, in extreme cases, in the Valea Calugareasca wine area, calculated using formula (1).

4 Conclusions

One of the possible explanations for the differences between the estimates of soil quantities lost annually through erosion, over time, is the application of the anti-erosion measures on increasingly larger surfaces (cultivation of slope surfaces on level lines, terraces, complex vegetal coverings, landscaping walls, slope support or anchoring).
Another explanation is the modification in time of the soil structure: texture, composition, natural regeneration capacity, etc. The compacting of the soil increases the resistance to erosion, and these phenomena have increased naturally, but also due to the excessive mechanized intensive exploitation.
Certainly, we cannot ask the problem of the inaccuracy of procedure or method, as long as we do not take into account the natural changes produced in the components of the analyzed process. I refer here to a series of natural processes that take place in the soil, listed above, and which should be taken into account in the calculation and measurements made to estimate the intensity of soil erosion.
For the above reasons, when re-evaluating the erosion intensity in an area, it will start from the updated values of the factors that influence the process. For example, after each soil work, its erodibility should be recalculated, seasonal change of vegetation should be considered, etc. Given these evolutions of influencing factors, soil erosion loss estimates should be revised from season to season or even more often. An extension of the tools and methods of calculating soil erosion losses to a dynamic form should take into account the dynamic characteristics of the factors involved in the process. The long-term monitoring process would be a solution for developing a dynamic model of the erosion process, capable of giving long-term predictions. An initial attempt, in this regard, was made by the authors.
It is possible that, to study the dynamics of soil erosion, it may be necessary to switch to an intrinsic independent variable of the process, which replaces time, because it is not the time the factor that produces the erosion, but the factors involved in the process: precipitation, wind, also may be temperature, etc.

A possible candidate for the intrinsic parameter as the independent variable of the erosion process would be, the total variation of the soil layer thickness (this is an increasing monotone function, but not linear in general), the total variation of a function, for example. The total variation of the amount of rain and / or of the wind flow could be considered as an intrinsic parameter of the erosion process. Due to the collaboration of these physical factors in the erosion process, only one characterization parameter may not be sufficient for this process, two or more such parameters may be needed. It is also important to study the possible correlations between the influential parameters (factors) of the erosion process, considered in its various mathematical models, to avoid the use as independent variables of some dependent factors.

References

1. M. Motoc, Fl. Trasculescu, *Eroziunea solului pe terenurile agricole si combaterea ei*, Ministerul Agriculturii si Silviculturii, Editura Agro-Silvica de Stat, (1959)
2. I. Ionita, *Material Prezentare Conferinta* , Universitatea Al. I. Cuza Iasi, http://www.unserboden.at/files/dlt2011_13.pdf, (2011)
3. R. Sfiru, P. Cardei, V. Muraru, V. Herea, *Methods and techniques of drawing up risk maps for surface rain erosion phenomenon*, INMATEH - Agricultural Engineering, pp. 35 - 42, **2**/ (2010)
4. https://esdac.jrc.ec.europa.eu/themes/indicators-soil-erosion
5. P. Panagos, P. Borrelli, J. Poesen, C. Ballabio, E. Lugato, K. Meusburger, L. Montanarella, C. Alewell, *The new assessment of soil loss by water erosion in Europe, Environmental Science & Policy*, **54**: 438-447. DOI: 10.1016/j.envsci.2015.08.01, (2015)
6. P. Panagos, A. Imeson, K. Meusburger, P. Borrelli, J. Poesen, C. Alewell. *Soil Conservation in Europe: Wish or Reality? , Land Degradation and Development*, **27** (6): 1547-1551, 2(2016)
7. https://esdac.jrc.ec.europa.eu/Library/Data/EIONET/Meeting032009/country/Romania.pdf
8. U. C. Sharma, V. Sharma, *Mathematical model for predicting soil erosion by flowing water in ungauged watersheds, Erosion Prediction in Unguaged Basins:Integrating Methods and Thechniques*, Proceedings of symposium HSO held during IUGG2003, Sapporo,July 2003, IAHS Publ. No. 279, (2003)
9. ICDVV, Elaborarea tehnologiilor vitivinicole adaptate pentru diminuarea efectului perturbator al schimbarilor climatice, proiect sectorial ADER 2020, (2011) http://www.madr.ro/attachments/article/117/ICDVV-Valea%20Calugaresca-ADER-116.pdf
10. https://www.onvpv.ro/sites/default/files/pdfs/caiet_de_sarcini_doc_dealu_mare_271ro.pdf
11. Y. Li, X. Bai, Y. Tian, G. Luo, *Review and Future Research Directions about Major Monitoring Method of Soil Erosion*, 2017 International Conference on Environmental and Energy Engineering (IC3E 2017), IOP Conf. Series: Earth and Environmental Science **63** (2017)
12. A. Lup, L. Miron, I. Alim, *Soil Erosion in Romanian Agriculture*, Lucrari Stiintifice Universitatea din Iasi, **60**(2) (2017)
13. D. Gadone, Stanchi S., *Soil erosion Issues in Agriculture*, IntechOpen, (2011)

14. D. Bucur, J. Gerard, A. Costica, *ASoil Erosion Control on Arable Landsfrom North-East Romania*, Soil Erosion in Agriculture, IntechOpen, (2011)
15. R. Blanchard, L. Farrell, *Guide to Sources for Agricultural and Biological Research*, University of California Press, (1981)
16. C.A. Eftene, S. Dumitru, A. Manea, D. Raducu, *A Review of the Impacts of soil sealing on Soil Properties in Romania*, SGEM2016 Conference Proceedings, **2**, (2016)
17. M. Sevastel, N. Petrescu, M. Musat, Sarbu N., *Soil erosion and conservation in Romania - some figures, facts and its impact on environment*, Annals: Food Science and Technology, **11**, (2010)
18. https://ec.europa.eu/eurostat/statistics-explained/index.php/Agri-environmental_indicator_-_soil_erosion

E3S Web of Conferences **180**, 03002 (2020) https://doi.org/10.1051/e3sconf/202018003002
TE-RE-RD 2020

Soil moisture influence in the soil tillage operations

Petru Cardei[1,*], *Raluca* Sfiru[1], *Sebastian* Muraru[1], *Paula* Condruz[1]

[1]INMA Bucharest, Informatics and engineering, No 6 Bd Ion Ionescu de la Brad Bucharest, Romania

Abstract. This article presents results obtained by the author in an attempt to introduce new parameters that influence the working process of agricultural soil working machines, in the classical formulas that mathematically model the tillage draft force. This article naturally introduces soil moisture. To introduce soil moisture into the calculation formulas of the tillage draft force, other physical parameters affected by soil moisture were used: density, cohesion, adhesion, etc. In this article, is explicitly describes, only the variation of soil density with its moisture. Considering some phenomena such as soil *swelling* or *shrinkage*, required the development of an adequate mathematical model of the expression of the density of soil-water mixture, concerning the concentration of water and the densities of dry soil and water. The final formulas obtained are tested to experimental results from the literature. Finally, the author sets out his opinion regarding the use of the results in design calculations and regarding the elevation to the rank of physical law of a mathematical model represented by a formula with a large number of parameters, which in turn are still dependent on many others parameters.

1 Introduction

The process of soil processing of agricultural land surfaces is fundamental in agriculture. This process has been studied and modelled in many variants. All these variants have certain converging ideas, such as the dependence of the draft force on the second power of the working speed, the depth and the working width, as well as on other parameters which, in one form or another, appear in different models. In [1] I tried to sketch a lot of the parameters that influence the process of interaction between the working organs of the tillage machines and the soil. The summary of the influential parameters is incomplete, especially since some of the parameters are not mechanical, being pedological (texture, structure, compaction and soil composition, for example, organic matter content). The mechanical parameters of the soil depend implicitly on the soil moisture and its pedological parameters. Specifically, mechanical parameters are functions of humidity, but not only of this.
Generally, the model parameters of the main mathematical models of the draft forces generated by the soil tillage machines are themselves functions, which depend on the parameters that characterize the soil. The parameters that characterize the soil are many; in

* Corresponding author: author@email.org

addition, not all fully quantified (texture). Without simplifying, we would have as model parameters, functions of 20-30 variables (maybe, not all independent). Such entities are very difficult to study, but especially to be known, as an expression, as a form. To determine an approximate dependence, many hypotheses, many calculations, but also many expensive and complex experiences are needed. On the other hand, many of the parameters characteristic of the soil are not, generally, independent. Pedological research has shown some dependencies between these parameters and between these and other groups of parameters: meteorological (as can be considered humidity), agricultural technology used culture, etc. Therefore soil moisture is a very important parameter, which does not appear explicitly in the most popular models of draft force, [12], [11], [29], and [32]. On the other hand, the literature dedicated to soil cultivation abounds with observations on optimal humidity, or optimum humidity ranges for soil tillage, [2], [3], [35-38], and so on. These were reasons for the approach presented in this article, see [39]. The list of names, notations and units of measurement and of the parameters used in this article is given in table 1.

Table 1. The list of names, notations and units of the parameters used.

Name	Notation	Unit
The volumetric content of water	θ	-
Volume of water	V_w	m^3
The total volume of wet material	V_{wet}	m^3
Volume of soil and vegetation particles	V_s	m^3
The volume of air	V_a	m^3
Volume of dry soil in the oven	V_{dry}	m^3
Gravimetric water content	u, c_w	-
Water mass	m_w	kg
Total mass of wet material	m_{wet}	kg
Mass of dry sample	m_{dry}	kg
Gravimetric water content	u'	-
Gravimetric moisture content	u''	-
Specific density of a material	d	-
Specific density of wet material (sample)	d_{wet}	-
The density of any material	ρ	kg/m^3
Water density	ρ_w	kg/m^3
Expansion-shrinkage coefficient of soil due to humidity	δ	-
Model parameter for the coefficient of expansion and contraction of the soil	δ_0	-
Mass concentration of dry soil in moist soil	c_{dry}	-
Mass concentration of water in moist soil	c_w	-
Draft force	F	N
Soil cohesion	c	Pa
Soil-working body adhesion	c_a	Pa
Horizontal blade angle (rake angle)	α	rad
Model parameters	x, y, z, t, u_0, B	-
Coefficient analogous to the friction coefficient	f	-
Plough weight	G	N
Coefficient that characterizes specific soil deformation resistance	k	Pa
Working body width	b	m
Working depth	a	m
Coefficient which depends on the shape of the active surface of the body and the soil properties	ε	kg/m^3
Dimensionless coefficient	ε'	-

2 Soil moisture - a profoundly involved parameter in the process of interaction between the working organs of the machines for soil tillage and the soil

According to [5], the moisture or the water content of the soil is the amount of water that is physically bound to the earth, when harvesting and evaporating at 105°C.
There are many defined parameters to quantify the water concentration in the soil. Thus, for example, according to [4], the volumetric water content is defined by the formula (1):

$$\theta = \frac{V_w}{V_{wet}} \tag{1}$$

where:

$$V_{wet} = V_s + V_w + V_a \tag{2}$$

The gravimetric content of water, after [6], is given by the formula:

$$u = \frac{m_w}{m} \tag{3}$$

For materials that exhibit volume changes when they contain water, such as coal, the *gravimetric water content* has the modified formula:

$$u' = \frac{m_w}{m_{wet}} \tag{4}$$

In the field of wood processing, geotechnical and soil science, according to [4], the gravimetric humidity formula is used:

$$u'' = \frac{m_w}{m_{dry}} \tag{5}$$

If the actual specific density of a material is introduced by the formula:

$$d = \frac{\rho}{\rho_w} \tag{6}$$

Is obtaining the relation (7):

$$\theta = u \cdot d_{wet}. \tag{7}$$

The gravimetric water content is calculated taking into account the formula:

$$m_w = m_{wet} - m_{dry}. \tag{8}$$

Table 2. The optimum moisture range for ploughing, by soil types, after [2].

Type of the soil	Optimal moisture for ploughing, %
Sandy	8.0 – 30.0
Sandy - clay	10.0 – 28.0
Clay-sandy	13.0 – 26.0
Clay	15.0 – 25.0
Clay-loamy	17.0 – 23.0
Loamy	18.0 – 20.0

In [2] it is shown that performing the ploughing work outside the optimum humidity range leads to an increase of 30 - 50% of the ploughing resistance, see also [39]. The same author, in [3], shows that in the case of heavy, clay soils, the optimum humidity range for the soil works is 19 - 21%, and for light, sandy soils, between 7 and 21%, even up to 28%. In [3], is interesting the following indication of optimality: the optimum moisture at the plough is achieved when the specific resistance at the plough is between 35000 Pa and 45000 Pa for light soils and 61000 - 75000 Pa for heavy soils. This minimum resistance increases by 10% on easyily compacted soils and by 30% on compacted soils.
The emergence of an optimum point in the soil cultivation process is the result of considering an optimization problem with several objective functions, as we have already

shown: the tensile strength (must be minimized), the quality of the soil processing (must be brought to a value as high as possible, closer to the value quantified by specialists), tractor wheels slip (it must be minimized), the working capacity (it must be maximized), the energy consumption (it must be minimized), the soil compaction (it must be minimized), the wear of the cars (it must be minimized), etc. To take into account several objectives, a multi-objective functional is built, whose minimization guarantees the achievement of a compromise point in the space of the optimization parameters, taking into account the possible restrictions imposed, [25-28].

The optimization of the working regime of the soil tillage machines, having as an objective function only the draft soil tillage force, so far, is not confirmed experimentally. There are works that show that certain factors involved in the mathematical modelling of the working process, present extreme values in usual humidity intervals, [7-8].

One way to obtain the behaviour of some mechanical parameters of the soil in relation to humidity, as described above ([7-8]), is to introduce a phenomenon of variation of the volume of the soil through the absorption or loss of water. According to [9-10] or [34], the property of soils to increase their volume by soaking with water has been called swelling, and the contraction of the soil is the property of the soil to decrease its volume by losing water (drying), an inverse phenomenon of the swelling phenomenon.

The volume variation of the soil is introduced through the variation of the density of the wet soil. The next densities are defined:

$$\rho_{dry} = \frac{m_{dry}}{V_{dry}}, \rho_w = \frac{m_w}{V_w}. \tag{9}$$

For the calculation of wet soil density, is proposed the following formula:

$$\rho_{wet} = \frac{m_{wet}}{V_{wet}} = \frac{m_{dry}+m_w}{\delta \cdot V_{dry}+V_w}. \tag{10}$$

In many technical fields, the density calculation of a mixture is calculated with formulas similar to formula (10), considering $\delta = 1$. In the hypothesis introduced in this article, if $\delta < 1$, then volume contraction occurs, and if $\delta > 1$, then the volume expansion of the soil occurs. The mass concentrations of soil and water from the wet soil are defined by the reports:

$$c_{dry} = \frac{m_{dry}}{m_{wet}}, c_w = \frac{m_w}{m_{wet}}. \tag{11}$$

Using (11) and amplifying the last fraction of (10) with $m_{wet}{}^{-1}$, we obtain the expression of the wet soil density depending on the densities of the dry soil, and of the water, the mass content of water and the coefficient of expansion and contraction of the volume of soil at humidification:

$$\rho_{wet} = \frac{\rho_{dry}\rho_w}{\delta \cdot (1-c_w)\rho_w + c_w\rho_{dry}}. \tag{12}$$

It is noted that the mixture containing only water ($c_w = 1$), has the density of water. For the mixture that is reduced to dry soil ($c_w = 0$), to have the density of the dry soil, the dilation-contraction coefficient is required to have the property: $\delta(0) = 1$.

As a result, the coefficient of expansion - volume shrinkage of the soil upon humidification must be a function of humidity, in this case: $\delta = \delta(c_w)$. The coefficient of expansion-shrinkage of the soil when humidifying, also depends on other parameters, such as soil texture, its composition and structure.

For some graphical representations that try to give a more understandable image of the model (12) for calculating the density of the wet soil, a coefficient of expansion-contraction coefficient of form is considered:

$$\delta(c_w) = 1 + c_w \cdot \frac{\delta_0 - 1}{u_s(u_s - 2u_{vf})} \cdot (c_w - 2u_{vf}) \tag{13}$$

3 Ways to introduce soil moisture in the formulas of the draft force of the soil tillage machines

For the mathematical models of the tensile strength in which the soil density explicitly appears, it is easy to introduce moisture by replacing the density in the formula with the form (12) of the density, taking into account also (13). An example is a formula proposed in [11]:

$$F = x\rho b^2 v^2 \left(\frac{c+c_a}{\rho v^2}\right)^y \left(\frac{a}{b}\right)^z (\sin\alpha)^t. \tag{14}$$

where x, y, z, t are model parameters (they are established for the experimental data corresponding to a particular soil type). An elementary variant containing moisture for formula (14) would be:

$$F = x b^2 v^{2-2y}\left(c(c_w) + c_a(c_w)\right)^y \left(\frac{a}{b}\right)^z (\sin\alpha)^t \cdot \left(\frac{\rho_{dry}\rho_w}{1 + c_w \cdot \frac{\delta_0 - 1}{u_s(u_s - 2u_{vf})} \cdot (c_w - 2u_{vf}) \cdot (1 - c_w)\rho_w + c_w\rho_{dry}}\right)^{1-y} \tag{15}$$

In the expression (15) of the draft soil tillage force, the parameters x, y, z, t (which can be deduced by the method of the smallest squares, for example, using an experimental data set) may also depend on the soil moisture, texture, structure and composition of the soil. In the same way, due to the density of the wet soil, soil moisture dependence can also be introduced in Goriacikin's formula. Taking into account the statements in [12], Goriacikin's mathematical model for tensile strength can be written as:

$$F = fG + kab + \varepsilon abv^2. \tag{16}$$

where also after [12],

$$\varepsilon = \varepsilon'\rho. \tag{17}$$

From (16) and (17) is obtained:

$$F = fG + kab + \varepsilon'\rho abv^2. \tag{18}$$

Considering that in (18), the density is that of the wet soil, given by (12):

$$\rho = \rho_{wet}. \tag{19}$$

the equation that models the tensile strength after Goriacikin becomes:

$$F = f(c_w)G + k(c_w)ab + \varepsilon'\frac{\rho_{dry}\rho_w}{1 + c_w \cdot \frac{\delta_0 - 1}{u_s(u_s - 2u_{vf})} \cdot (c_w - 2u_{vf}) \cdot (1 - c_w)\rho_w + c_w\rho_{dry}} abv^2. \tag{20}$$

About the parameters f and k, it can be assumed that they also depend on the texture of the soil, and k is also variable with the working depth, a. After (20), the tensile strength is a function with 10 - 13 arguments, becoming more complicated than the original version, (16). The density of the soil, as well as the tension at vertical penetration, decrease with the organic matter content of the soil, [17] which should be taken into account in a possible extension of the models presented in this article.

4 Results

The primary results are mathematical formulas or models (12), (13), (15) and (20).

For soils with variable volume depending on humidity, the variation of the density and the coefficient of expansion-contraction with a mass content of water are given in fig. 1-2. For the value of parameter δ_0 smaller than 1, the variation of soil density is of convex type, and that of the expansion-contraction coefficient is of concave type. If the parameter δ_0 takes values greater than 1, the situation is reversed. This model of variation of soil density with moisture therefore allows convex or concave variations.

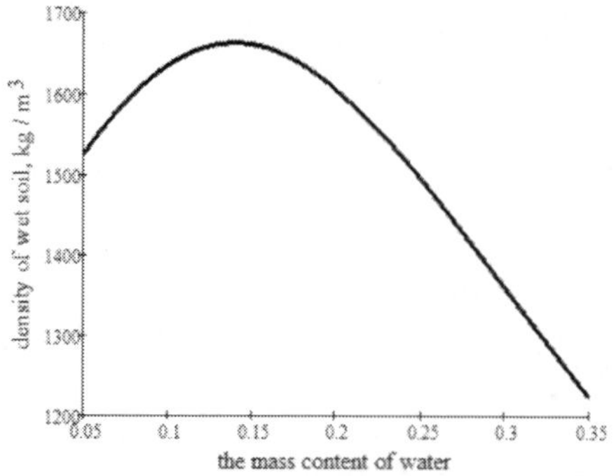

Fig. 1 The dependence of wet soil density on the mass content of water.

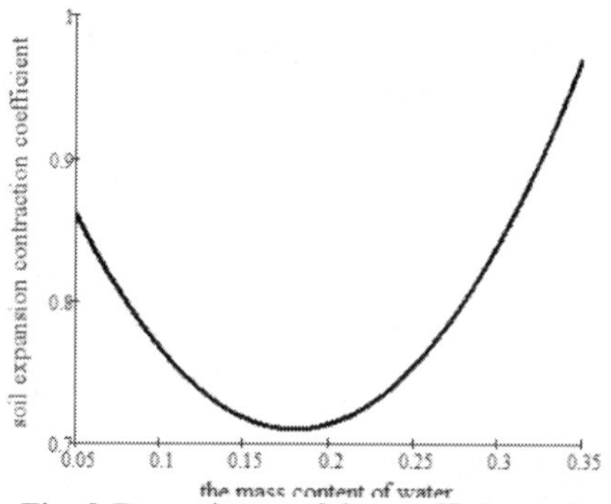

Fig. 2 Dependence of the coefficient of expansion-contraction on the water content.

$\rho_{dry} = 1350$ kg/m^3, water density, $\rho_w = 1000$ kg/m^3, $\delta_0 = 0.85$, $u_s = 0.055$,$u_{vf} = 0.18$

The model of the tensile strength (15) is considered, having as its origin the model proposed by [11]. We used the model proposed in [11] with the parameters x, y, z, t determined in the precision and order tests from [14], for the experimental data from [13]. These values may theoretically depend on soil moisture. Because we did not study their dependence on humidity, we take the average values characteristic for the humidity interval in which we worked, $x = 0.99882, y = 0.88324, z = 0.448778, t = 0.001703$. Also, for the cohesion and adhesion of the soil to the surfaces of contact with the soil, the average values in the range of humidity in which it was worked are considered: $c = 41000$ Pa, $c_a = 32500$ Pa. The angle of the working body with the horizontal is kept constant, $\alpha = 30°$. In these conditions, the dependence of the draft force, (15), with the soil moisture, for three values of working depth, is given in the fig. 5. The dependence of the same model of the draft force with the working speed is given graphically, in the fig. 6.

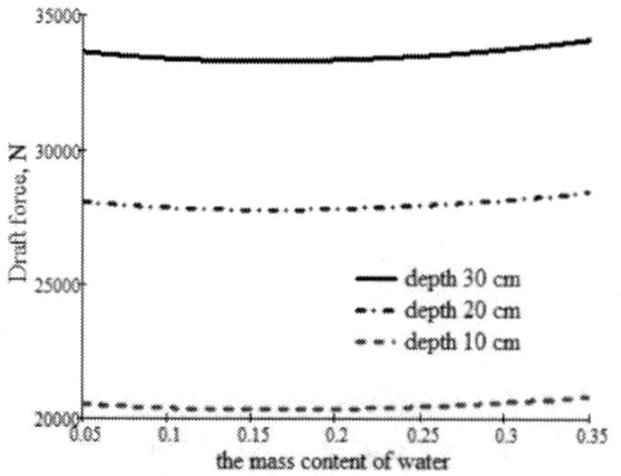

Fig. 3 Variation of the draft force with the soil moisture (three working depth).

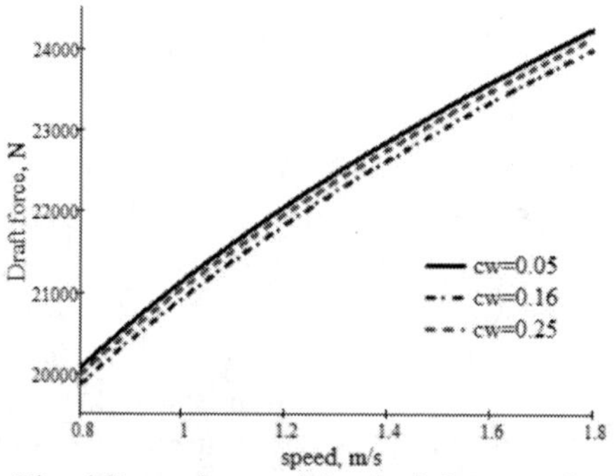

Fig. 4 Dependence of the draft force on the working speed, for three soil moisture.

$\rho_{dry} = 1350$ kg/m^3, water density, $\rho_w = 1000$ kg/m^3, $\delta_0 = 1.15$, $u_s = 0.055$, $u_{vf} = 0.18$

It can be observed that the tensile strength (15) has, during the used humidity range, a convex variation, which allows the optimum working humidity to be highlighted, if only the tensile strength is considered as an objective function (fig. 3). The optimum humidity is in the vicinity of the horizontal coordinate of the top of the parabola (13), which models the variation of the density of the wet soil to the mass content of water. The model (15) shows the increase of the tensile strength with the working speed, highlighting the existence of an optimal humidity relative to the objective function of the draft force (fig. 4).

If the parameter δ_0 is smaller than 1, the influence of humidity is transmitted to the draft soil tillage force, which varies concave to the soil moisture, having a maximum point in the working range, [39]. The result is transferred also to the variation of the draft force to the working speed, [39]. Similar variations of the tensile strength (both horizontal and vertical), to soil density and moisture, have been reported in [18]. We remind that the density of the moist soil and its moisture are not independent parameters.

Generally, in theoretical and experimental works, we are accustomed to variations of the draft force in report to the working speed, which have the convex form (that is, the second derivative concerning the slightly negative speed), [15-16], [12], [20], [23], [24], and [32]. Some variations from this rule (some small portions on which the curve is concave) appear, for example in [19], and some experimental, laboratory data even show curves that appear to be slightly concave, at higher soil moisture (20%), [31]. Moreover, in [32] (p. 74), after [33], it can be observed that the convex form is highlighted by inertial forces, and the concave component manifests itself due to static or other nature components. This behaviour is very interesting and can be a subject of research with many useful results. Results of the change in soil density at humidification similar to those given graphically in [32], page 110, can be obtained using the model proposed in this paper, using instead of a parabolic function, an exponential function of Gaussian type, for (13) :

$$\delta(c_w) = 1 + B \cdot \left(e^{-\left(\frac{u_0 - c_w}{c}\right)^2} - e^{-\left(\frac{u_0}{c}\right)^2}\right) \tag{21}$$

When following the approximation of an experimental curve, it is recommended to calculate the parameters of the functions (13) and (23), using the least squares method.

For the variation of soil cohesion, [15] suggests by experimental results a law similar to density, and after cohesion through simple relationships, the dependence of soil adhesion on the metal body of the car can be calculated concerning the water content (moisture). We leave it to those interested to make this extension for the formula (model) (14) of the draft force. Also for a comfortable extension, it is possible to determine a dependence of the soil density on the working depth, which is easy to suspect due to the variation of the penetration resistance with the depth, noticed in numerous experimental works, for example [21] or [22].

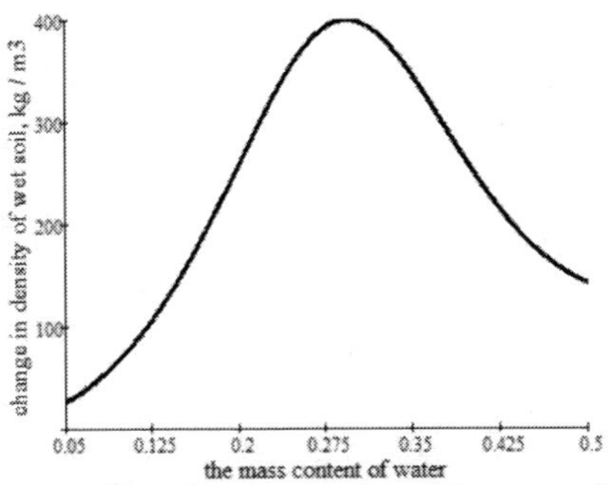

Fig. 5 Changing the density of the wet soil according to the water mass content.

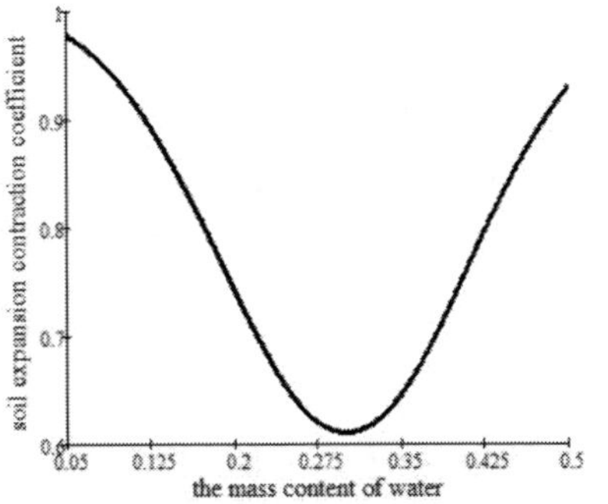

Fig. 6 Changing the density of the wet soil according to the water mass content.

$\rho_{dry} = 750$ kg/m^3, $\rho_w = 1000$ kg/m^3, $u_0 = 0.30$, $B = -0.4$, $c = 0.158$.

For the mathematical model of the draft force given in [29] and named in [30], "ASABE Universal Draft Equation", the situation is a little more complex since all the coefficients listed (four in numbers) are, hypothetically, dependent on humidity, texture, composition and soil structure. Further details about this case are given in [39].

5 Conclusions

This article describes a natural way of introducing soil moisture into formulas (mathematical models) of the draft tillage force. To this end, for better modelling of the phenomena of swelling or shrinkage of the volume of moistened soils, a mathematical model of the change of the density of the humidified soil in relation to the density of the same soil in the dry state was introduced. The proposed model has a number of important consequences in soil behaviour and implicitly in the behaviour of draft tillage force models. Using the model of the variation of soil density with its humidity, we succeeded in satisfying mathematical modelling of some experimental facts, especially the variation of the draft tillage force with the humidity. By this, it was shown that there is the possibility to approach the problem of optimum humidity even only from the narrow point of view of minimizing the draft tillage force (objective function).

It has been shown that the introduction of humidity into the alternative models (formulas) of the draft force can be done naturally, that is, through the known variation of some physical parameters of the soil with the moisture: soil density, cohesion, adhesion. In formulas in which not all these parameters appear explicitly, the model parameters (specific strengths, friction coefficients) can be expressed as a function of density, cohesion and adhesion and, consequently, the dependence of soil moisture appears.

In the same way that soil moisture was introduced in the tensile strength models, the texture, composition and structure of the soil can be introduced.

The phenomenon of the interaction between the soil and the working organs of the tillage machines is influenced by several tens of parameters. If each of them were introduced in the manner in which, in this article, soil moisture was introduced, very long calculation formulas would be obtained. The large length of the formulas favours the occurrence of writing errors (text and software) and increases the computation time. Therefore, it is recommended to keep a series of short, easy-to-write and calculating relationships, accompanied by an explanatory logic scheme, to use these increasingly refined models of the phenomenon as efficiently as possible.

In other words, such a multitude of relationships that can even reach the order of tens (also considering alternative formulas) loses the chances of constituting a physical law, at most it can "break" into several laws, supplemented by relations connection.

This work was supported by a grant of the Romanian Research and Innovation Ministry (RRIM), through Programme 1 – Development of the national research-development system, subprogram 1.2 – Institutional performance – Projects for financing excellence in RDI, contract no. 16PFE, and was done by "NUCLEU" Programme, developed with the support of the RRIM, project PN 19 10 01 02.

References

1. P. Cardei, Tests for physical laws of the draft force generated in the tillage operations, preprint, DOI:10.13140/RG.2.2.16582.01607,(2019)
2. V. Popescu, Perioada optimă pentru lucrările solului, Lumea Satului, (2015)
3. V. Popescu, Lucrările solului ca la carte, Lumea Satului, (2014)
4. https://en.wikipedia.org/wiki/Water_content
5. https://graduo.ro/referate/ecologie/determinare-umiditate-in-sol-156286
6. T.W. Lambe,R.V. Whitman, Soil Mechanics, John Wiley & Sons, Inc., p. 553, (1969)
7. C. M. P. Vaz, J.M. Manieri, I. Clerici De Maria, M.Th.Van Genuchten, Scaling the Dependency of Soil Penetration Resistance on Water Content and Bulk Density of Different Soils, Soil Science Society of America Journal, (2013)

8. S. Manuwa, O., C. Ademosun, Draught and Soil Disturbance of Model Tillage Tines Under Varying Soil Parameters, Agricultural Engineering International: the CIGR Ejournal. Manuscript PM 06 016. Vol. IX. March, (2007)
9. S. Stadnic, Pedologie Curs de prelegeri, Universitatea de stat din Bălţi, (2010)
10. T. Demeter, Pedologie generala, Universitatea din Bucuresti, (2009)
11. A. Moeenifar, S.R. Mousavi-Seyedi, D. Kalantari, Influence of tillage depth, penetration angle and forward speed on the soil/thin-blade interaction force, Agric Eng Int: CIGR Journal, vol. **16**, no.1, p. 69-74, (2014)
12. M.N. Letosnev, Masini Agricole, Editura Agro-Silvica de Stat, Bucuresti (1959)
13. P. Cardei, M. Matache M., C. Nutescu, Optimum working conditions for variable width ploughs, Preprint RG, (2017)
14. Cardei P., Draft force on soil tillage, model tests and relative ordering, DOI: 10.13140/RG.2.2.11951.89769, (2019)
15. G. Gebresenbet, Analysis of Forces Acting on Moulboard Ploghs and Seed Drill Coulters in Relation to Speed, Depth and Soil Conditions, Static and Dynamic Studies, Swedish University of Agricultural Sciences, Department of Agricultural Engineering, Uppsala, SLU Info/Repro, (1991)
16. P. Okoko, E.A. Ajav, W.A. Olosunde, Draft and power requirement for some tillage implements operating in clay loam soil, Agricultural Engineering International: CIGR Journal, **20**(1): 95–102, (2018)
17. Lungu S., Consideratii privind cresterea durabilitatii organelor active de la masinile agricole pentru prelucrat solul si semanat, Teza de doctorat, Universitatea Tehnica Gheorghe Asachi din Iasi, (2014)
18. E.L. Bravo, H.M. Suarez, O.G. Cueto, C.I. Coronel, H. Ramon, Effect of Moisture and Soil Compaction on Tillage Operations, Revista Ciencias Técnicas Agropecuarias, Vol. **25**, No. 1, pp. 32-37, (2016)
19. S. Ranjbarian, M. Askari, J. Jannatkhan, Performance of tractor and tillage implements inclay soil, Journal of the Saudi Society of Agricultural Science **16**, 154-162, (2017)
20. G.T. Owen, Subsoiling forces and tool speed in compact soils, Can. Agric. Eng., **31**, 15-20, (1988)
21. M.A. Ghazavi, Evaluation of a New Tillage Tool; Considering Soil Physical Property, Energy Requirement and Potato Yield, Pakistan Journal of Biological Sciences, Volume **10** (22): 4050-4056, (2007)
22. H.Th. Tahir, N.H. Jeejo, H. Tariq, T.H. Karim, A Comparative Study of Conventional and Modified Tine Types of Subsoiler and Their Effect on Some Performance Characteristics, Journal of Agricultural Machinery Science, **14** (1), 47-55, 2018
23. L. Naderloo, R. Alimadani, A. Akram, P. Javadikia, H. Z. Khanghah, Tillage depth and forward speed effects on draft of three primary tillage implements in clay loam soil, Journal of Food, Agriculture & Environment Vol.**7** (3&4) : 382-385. (2009)
24. A.A.N. Ormenisan, Theoretical and experimental research concerning the influence of automatic control systems of the tractor linkage mechanisms on the dynamics and energetics of ploughing units, Thesis, University Transilvania, Brasov, (2014)
25. P. Cardei, A. Alexiou, M. Badescu, V. Vladut, N. Constantin, E. Marin, Nonlinear Friction and Resistance, Generating Sources of Optimal Points in the Energy Field of Agricultural Aggregates Working Process, Proceedings of the International Conference on Energy, Environment, Ecosystems and Development - *Recent advances in energy, environment, ecosystems and development*, (2013)

26. A. Alexiou, P. Cardei, M. Badescu, V. Vladut, K.C. Selvi, I. Kiss, N. Constantin, E. Marin, New Viewpoints in Optimizing the Energetics of Working Processes of the Agricultural Aggregates, Annals of Faculty Engineering Hunedoara – International Journal of Engineering, Tome **XI**, fasc. **4**, 2013
27. P. Cardei, A. Meca, G. Konstandinov, Working regimes of the agricultural machines designed to soil tillage: From optimization to fundamentals (1), INMATEH-Agricultural Engineering, vol. **37**, no. 2, 13-20, (2012)
28. P. Cardei, G. Konstandinov, Working regimes of the agricultural machines designed to soil tillage: From optimization to fundamentals (2), INMATEH-Agricultural Engineering, vol. **37**, no. 2, 21-28, (2012)
29. American Society of Agricultural Engineers [ASAE]. 2003a. D497.4 Agricultural machinery management data. p. 373-380. In: ASAE. ASAE standards (2003)
30. V.K. Tewari, Tractor, implement and soil force consideration for tillage implement design, Lecture – 03, Farm Machinery, https://nptel.ac.in/content/storage2/nptel_data3/html/mhrd/ict/text/126105009/lec3.pdf
31. S.R.A. Zadeh, Modelling of energy requirements by a narrow tillage tool, Thesis, (2006)
32. E. McKyes, Soil Cutting and Tillage, Elsevier, (1985)
33. R.D. Wismer, H.J. Luth, Performance of plane soil cutting blades in a clay, Transactions A.S.A.E., Vol. **15**, No. 2: pp. 211-216, (1972)
34. M. Mokhtari, M. Dehghani, Swell-Shrink Behavior of Expansive Soils, Damage and Control, Electronic Journal of Geotechnical Engineering, vol. **17**, (2012)
35. Dexter A., R., Bird N., R., A., Methods for predicting the optimum and the range of soil water contents for tillage based on the water retention curve, Soil Tillage Research, **57**(4), 203-212, (2001)
36. https://en.wikipedia.org/wiki/Optimum_water_content_for_tillage
37. P.B. Obour, J.L. Jensen, M. Lamande, C.W. Watts, L.J. Munkholm, Soil organic matter widens the range of water contents for tillage, Soil & Tillage Research, **182**:57-65, (2018)
38. W. Hoogmoed, Tillage for soil and water conservation in the semi-arid tropics, Doctoral Thesis, Wageningen University, (1999)
39. P. Cardei, Soil moisture influence in the soil tillage operations, preprint, DOI: 10.13140/RG.2.2.28028.69768, (2019)

Finite element 3D numerical simulation study of car braking systems and brake disc/drum – pad/shoe friction couple materials

George Ipate [1,*] *Filip* Ilie[2] and *Andreea Catalina* Cristescu [1]

[1] University POLITEHNICA of Bucharest, Department of Biotechnical Systems, 313 Spl. Independentei, Romania

[2] University POLITEHNICA of Bucharest, Department of Machine Elements and Tribology, 313 Spl. Independentei, Romania

Abstract. Our study uses the finite element method of modeling and analyzing the functioning of a braking system for a modern vehicle, in terms of stress ditributions, structural deformation, wear and thermal gradient of the brake disc and drum. The 3D geometric model of system brake is designed using Solidworks, and the coupled thermal and structural analysis is performed with the ANSYS Workbench R16 program. The brake was applied when the car was 85.7 km / h (ω = 125 rad · s^{-1}), the duration of braking until the car stopped was t = 5 s. For the given example, the coefficient of average friction during braking, considering the pressure on the pad p = 7.5 MPa, is μ = 0.35. It was discovered that the relative deformations of the plate and disc are larger in the area of the outer diameter than that of the inner diameter. This is also outlined by the fact that the pressure is higher on the outer sides than on the inner sides of the plate, the highest value being in the central axis zone of the outer side. Knowing the thermophysical characteristics of the disc and the plate and the working conditions, it was possible to determine the temperature variation during braking. The results of the numerical research revealed that an increase of the contact pressure and / or the relative speed between the contact surfaces implies an increase of the amplitude of the stick-slip phenomenon.

1 Introduction

The braking system is used to decrease the velocity of the car to a desired value or even to stop it, immobilizing the car in a stationary, on a horizontal road or on a slope, maintaining the constant speed of the car in the event of descending long slopes. The efficiency of the brakes is evaluated by the distance on which a car stops at a certain speed. The braking system consists of the brakes themselves and the brake actuation mechanism. Depending on the shape of the rotating part, the brakes themselves can be drum (radial), disc (axial) and combined. According to the form of the parts that produce the brakes, they are distinguished: brakes with shoes, brakes with tape and brakes with discs [1,8,9].

The main role of a rubbing material is to generate a high coefficient of friction and, at the same time, it should have a very high wear resistance. Wear of the rubbing material is one of the important characteristics of the brake pads. A good functioning is closely associated with the quality of the material, as well as with the safety in operation, and of course it is directly related to the life of the brakes [2].

Söderberg and Andersson [3] investigated how wear of the pad-to-rotor interface can be predicted using general purpose finite element analysis software. They develop a three-dimensional finite element model of the brake pad and the rotor to compute the pressure maps in the pad-disc contact, and used a comprehensive form of Archard's wear law and explicit Euler integration to reproduce the wear of the brake pad.

Hwang and Wu [4] and Belhocine [5] in their research investigate the temperature and thermal stress in the ventilated disc-pad brake during single brake. The temperature field leads to variation of the contact pressure distribution by the effects of the thermal expansion. Due to their braking stability, controllability and ability to provide a wide-ranging brake torque, braking system whit ventilated disc with straight vanes and drum brake is most popular and straightforward to make.

Zengh and Zangh [6] present a numerical model based on ANSYS APDL used for the design of drum brake. They study the static deformation and equivalent stress of drum. Zagrodzki et. all [7] present a transient finite element simulation with frictional heat generation, for the two-dimensional thermoelastic contact problem of a stationary layer between two sliding layers. By using an Abaqus commercial finite element package, they concluded that if the slip velocity is high enough, there will be uneven heating and low frequency vibrations known as "hot judder".

* Corresponding author: puiu.ipate@gmail.com

The main objective of this work is to comprehend a numerical model with 3D finite elements of a combined braking system (axial and radial) of the cars to analyze the method of friction and wear of the braking systems using the Ansys Workbench program. During this paper, a thermal analysis and a structural analysis coupled to a given braking system are performed and its stability and rigidity are investigated. Theoretical modeling of the method of using the brake pad as a viscoelastic material and of the brake disc or drum, also as determining the effect of temperature on this process were also important objectives.

2 Material and methods

2.1 Geometric and finite element modeling

A detailed three dimensional model of a Dacia Logan automobile solid system brake assembly has been studied in this research. The 3D geometric model of system brake is designed using Solidworks. The mechanical characteristics and dimension of the disc/drum brakes (Gray cast iron) and pad/shoe (lignin) materials selected in the analysis are presented in Table 1.

Table 1. Material characteristics and dimension of brake disc and pad/drum and shoe.

Dimension / Property	Disc	Pad	Drum	Shoe
Diameter inner (mm)	61	100	62	210
Diameter outer (mm)	258.0	140.2	272	235
Thickness (mm)	22.0	18	44	25
Density (kg/m^3)	7200	2030	7200	2030
Mass (kg)	4.71	0.192	8.25	0.412
Young Modulus [MPa]	110000	3180	110000	3180
Poissom's ratio	0.28	0.35	0.28	0.35
Thermal expansion coefficient [mm/°C]	$1.1\cdot10^{-5}$	$3.1\cdot10^{-6}$	$1.1\cdot10^{-5}$	$3.1\cdot10^{-6}$
Thermal conductivity (W/K.m)	37	0.275	37	0.275

A 3D tetrahedral method will be used for finite element mesh generation. Overall, 85856 nodes and 44867 elements are used for the disc and pads, respectively, 63309 nodes and 33677 elements are used for the drum and shoes (Fig.1).

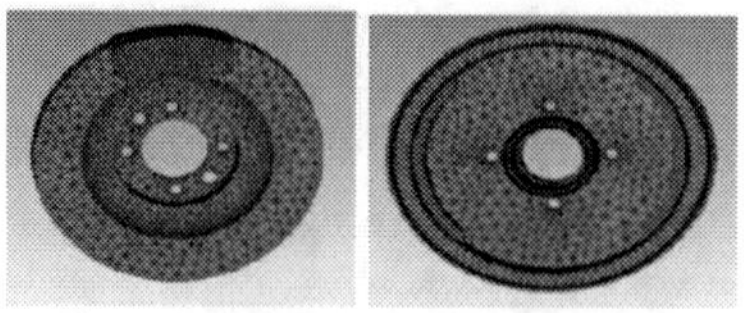

Fig. 1. Image capture of mesh of the ventilated disk (left) and drum brake (right) models.

Frictional wear can be expressed by applying a wear model between contact elements. To study tribological behaviour a frictional contact pair was established between disc-pad and drum-shoe interfaces. The used element types were Conta175 to represent contact and sliding between two surfaces and Target 170 for the associated contact elements.

2.2 Boundary condition

The investigation of the dynamic behaviour of physical systems is conducted with mathematical examples consisting of ordinary differential equations. In the present simulation, friction coefficient of contact pair is 0.35 and initial temperature is 22°C. The gray cast iron is material of disc and drum. The geometrical dimensions and material characteristics of the brake disc, pad, drum and shoe are presented in Table 1. The wheel is decelerated at an starting velocity of 87.5 km/h (125 rad/s), to the final velocity 0 km/h in 5 seconds and mechanical energy is transformed into the thermal energy. The brake pressure 7.5 MPa is imposed on the pad and shoe to generate the brake force. Figure 2 illustrates the boundary conditions applied to both geometry simulation model.

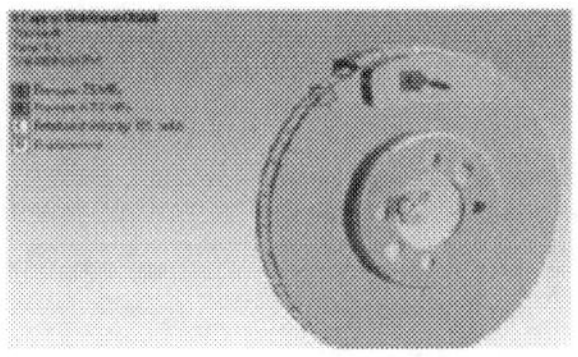

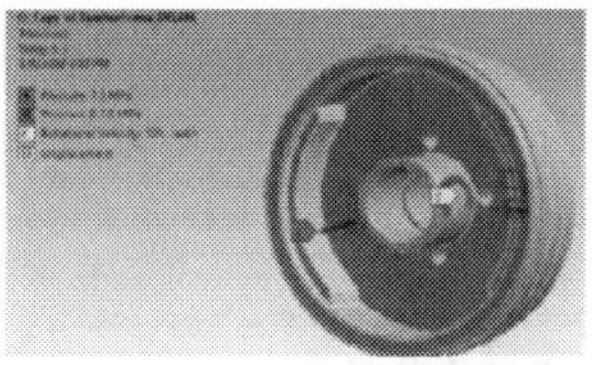

Fig. 2. Boundary condition in simulation program: disc-pad brake (left) and drum-shoe brake (right); (A) and (B) is the pressure = 7.5 MPa, (C) is rotational velocity = 125 rad/s and (D) is displacement

The transient dynamic structural analysis performed with the Ansys program determines the structure's response to tasks that vary arbitrarily (randomly) over time. In a transient dynamic analysis equation of motion is:

$$[M]\{\ddot{u}\} + [C]\{\dot{u}\} + [K]\{u\} = \{F(t)\} \tag{1}$$

where, [M] is mass matrix, [C] is damping matrix, [K] is stiffness matrix, $\{\ddot{u}\}$ is nodal acceleration vector, $\{\dot{u}\}$ is nodal velocity vector, {u} is nodal displacement vector and {F(t)} is load vector. At any given time t, these equation can be tought of a set of "static" equilibrium equations that also take into account inertia and damping forces. To solve this equation Ansys program uses the Newmark time integration algorithm, whith integration time step. Thermal solution is obtained by solving first order differential equation:

$$[C^t]\{\dot{T}\} + [K^t]\{T\} = \{Q(t)\} \tag{2}$$

where, $[C^t]$ is specific heat matrix, $[K^t]$ is thermal conductivity matrix, {T} is nodal temperature vector and {Q(t)} is sum of the element heat generation load and element convection surface heat flow vectors. If we apply to the equations of motion and conservation of the heat flux the variational principle, coupled with the thermoelastic constituent equations, will produce the following finite element matrix equation:

$$\begin{bmatrix}[M] & [0]\\ [0] & [0]\end{bmatrix}\begin{Bmatrix}\{\ddot{u}\}\\ \{\ddot{T}\}\end{Bmatrix} + \begin{bmatrix}[C] & [0]\\ [C^{tu}] & [C^t]\end{bmatrix}\begin{Bmatrix}\{\dot{u}\}\\ \{\dot{T}\}\end{Bmatrix} + \begin{bmatrix}[K] & [K^{ut}]\\ [0] & [K^t]\end{bmatrix}\begin{Bmatrix}\{u\}\\ \{T\}\end{Bmatrix} = \begin{Bmatrix}F(t)\\ Q(t)\end{Bmatrix} \tag{3}$$

where, $[C^{tu}]$ is thermoelastic damping matrix and $[K^{ut}]$ is thermoelastic stiffness matrix.

2.3 Numerical model of wear

Assuming the wear operation is a dynamic process and the estimation of the process as an initial value problem. In these situations a pattern of wear proposed by Archard is given by [3]:

$$\frac{dw(t)}{dt} = f(load, velocity, temperature, material) \tag{4}$$

The Archard wear model calculate rate of volume loss (dw/dt) due to abrasion wear friction. Introducing the dimensional wear coefficient k (specific wear rate) an expression to calculate wear local, if the contact pressure and sliding velocity is known, is obtained:

$$\frac{dw(t)}{dt} = k * p(t) * v_s(t) \tag{5}$$

where, k=K/H , K is wear coefficient addicted to the contact-pair [mm^3/m], H is material hardness [MPa], p(t) is the contact pressure [MPa] and $v_s(t)$ represents the relative sliding velocity [mm/s]. The wear is calculated by numerical integration of Eq. (5) corresponding to the explicit Euler integration scheme during the step Δt, if the contact pressure in every node is known. The wear depth after the t simulation step is then :

$$w_{t+1} = w_t + \Delta t * f(t, w_t) \tag{6}$$

The contact pressure in the wear process can be also obtained, corresponding to the law of motion, by the next equations:

$$p(t) = b_1 * t + b_2 \tag{7}$$

where b_1 = 1.4143, b_2 =4.811 are parameters that depend on the particle size and material properties. The relative sliding velocity can be obtained, by the following first-order equations:

$$v_s(t) = a_1 * t + a_2 \tag{8}$$

where a_1=0.0001 and a_2= 0.0021 are parameters computed by regression of simulated data.

3 Results and discussions

Due to their simplicity, the disc and plated brakes, respectively with the drum and the inner shoes are widespread in cars. At the action of the pads on the disc, respectively, of the sabotages on the drum, between them appear frictional forces that can give rise to a moment of braking, which is opposed to the movement of the car. Under the action of the braking moment, in the contact area of the wheel with the road, the reaction directed in the opposite direction of the movement is born. During braking, due to the friction that occurs between the drum and the friction pads of the shoes, the kinetic energy of the car turns into heat [5,10,11].

To investigate the maximum temperature, rise after the braking operation a transient thermal analysis is performed on the ventilated disc and drum brake. Boundary conditions applied for the transient thermal analysis are an heat flux transfer of 1.5395 W/mm^2 applied over the swept zone of both the pads and shoe while the direction of heat flow is towards the disc and an thermal convection applied on all the surfaces with the air coefficient of 5e-6W/mm^{2}K. The analysis starting at initial temperature of 22 ^{o}C.

Fig.3 describe the temperature maps on the friction contacts on the transient (at t=0.55 s, t=2.22 s, t=3.88 s, t=5 s) results. As expected, the tendency is for the temperature results to converge to the steady state. It can be easily observed for the disc-plate assembly, a maximum temperature of 463 ^{o}C is obtained for the initial conditions set as design variables. The temperature and total heat flux conturs for the drum -shoes assembly is shown in Fig.4. Similarly, the maximum temperature of 279.2 ^{o}C is reached for the drum on the friction surface with the brake shoe.

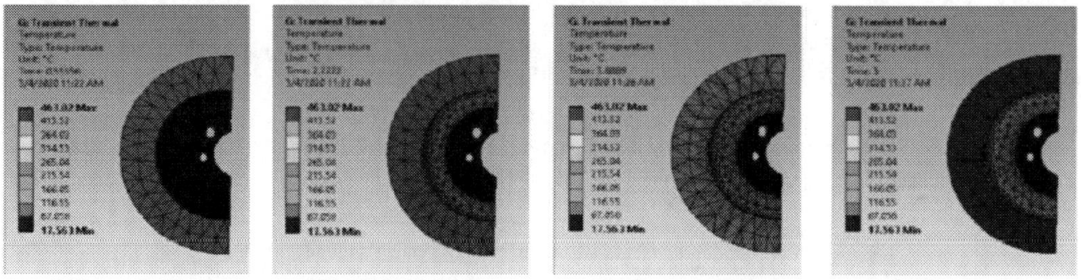

a) Time t=0.55 s b) Time t=2.22 s c) Time t=3.88 s d) Time t=5 s

Fig. 3. Temperature contours of gray cast iron disk

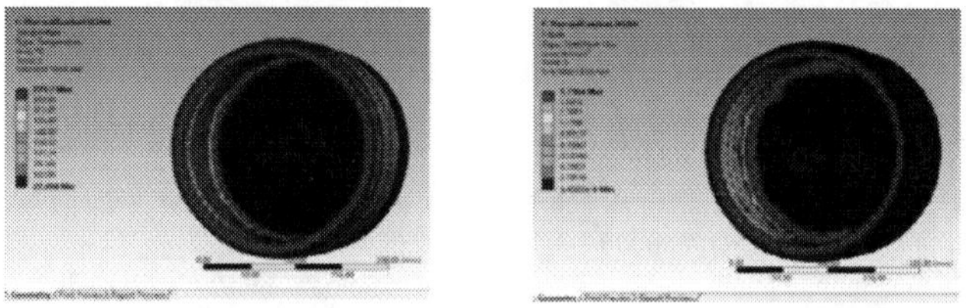

Fig. 4. Temperature (left) and total heat flux (right) contours of gray cast iron drum

Deformation analysis of the disk and drum. In accordance with the Fig. 5, it can be observed that the total displacement of pad brake is 0.3123 mm. It indicates the fact that the deformation of the pad is relatively the same in comparing whit disk, 0.417 mm. By the Fig. 6, it can be observed that the maximum total deformation of the drum is 0.2458 mm, which means that the repartition along X axis is small (in the opposite direction to gravitational acceleration). Consequently, the next drum optimization should take into account the deformation of the drum edge.

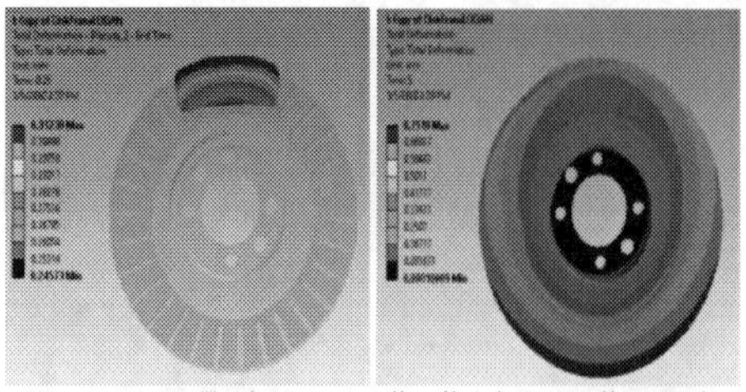

Fig. 5. Total displacement distribution on disc-pad.

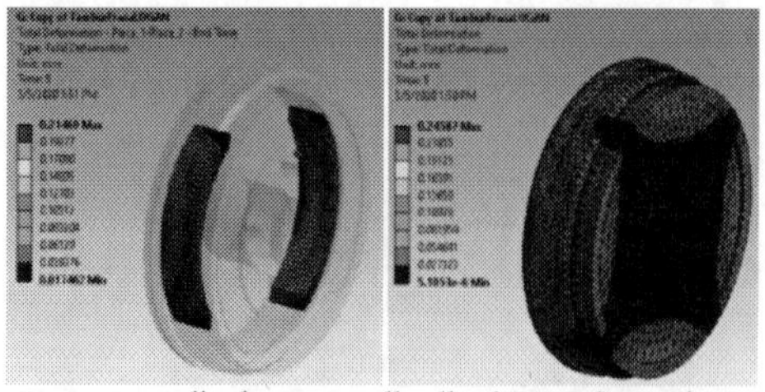

Fig. 6. Total displacement distribution on drum-shoe.

Stress analysis of the disk and drum. Fig.7-8. presents the maps of the equivalent Von Mises stress at the end of simulation, the range of results varying from 1.335 to 344.65 MPa in case of disc-pad assembly, respectively from 0 to 77.6 MPa in case of drum-shoes assembly. The maximum value, 344.46 MPa recorded during the simulations at the plate-disc torque is significantly different compared to the one obtained for the drum-shoe couple in the mechanical investigation performed under the identical conditions. It is observed the appearance of this value at the level of the disk bowl, because the disk is anchored to the wheel hub by means of screws, thus put a stop to its movement.

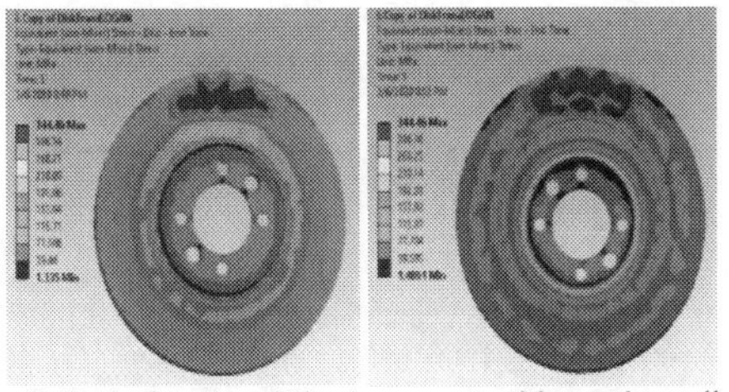

Fig. 7. Equivalent (von-Mises) stress repartition on inner disc (right) and outer disc (left)

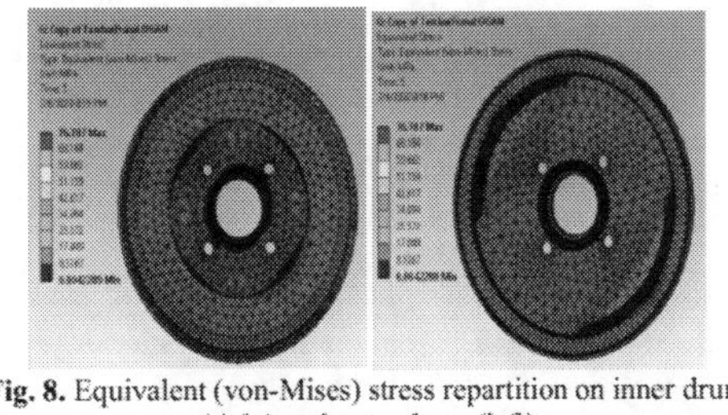

Fig. 8. Equivalent (von-Mises) stress repartition on inner drum (right) and outer drum (left)

Analyzing the stresses resulting from the simulation of the braking operation, at the contact among the shoe and the drum, the maximum tension is 77.6 MPa and appears instead of the drum clamping in the lower part of the drum. The stress contour is evenly distributed on the surface of the drum, because the drum has a symmetrical structure.

In braking systems, the speed of sliding between the disk and the plates or between the drum and the shoe is variable (Fig. 9). In the field of low and very low speeds, due to the elasticity of the elements of the braking system, the phenomenon of stick-slip appears. The amplitude of this phenomenon is influenced by the rigidity characteristics of the system, the working speed and the frictional behavior of the couple of brake disc-plate materials.

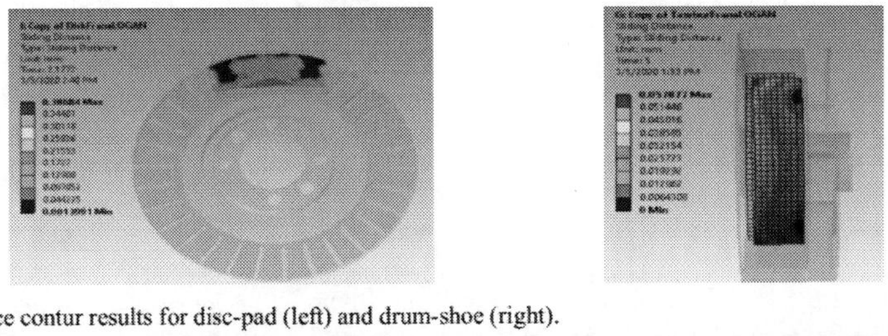

Fig. 9. Sliding distance contur results for disc-pad (left) and drum-shoe (right).

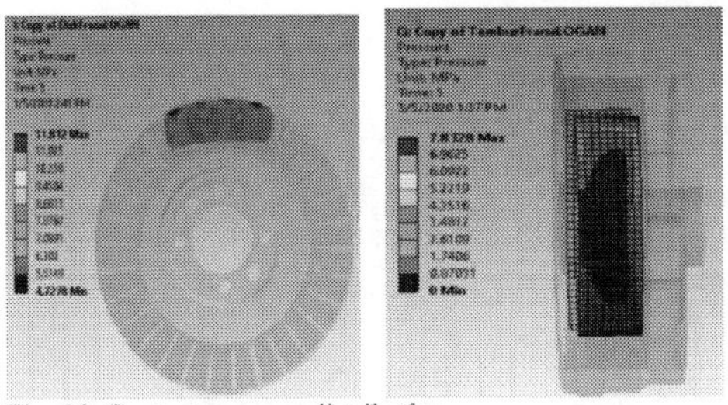

Fig. 10. Contact pressure distribution

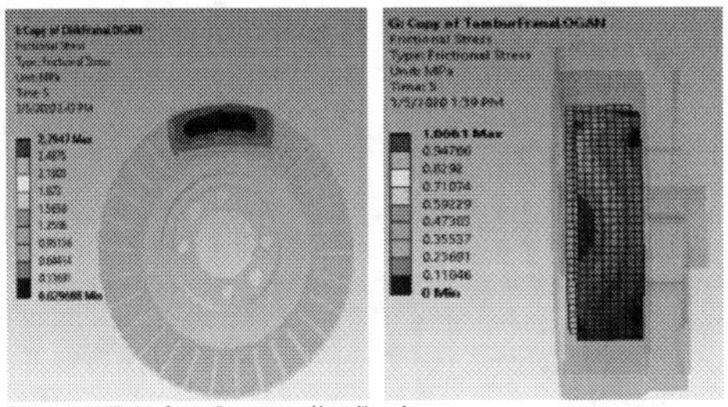

Fig. 11. Frictional stress distribution

Fig. 10 depict the contact pressure allocation in the friction interface of the inner pad taken at different times of virtual experiment. For this distribution, the range varies from 0 to 11.812 MPa and grown to a maximum value at the moment t=5 s, which matched to the car stop. The distribution of the contact pressure at the surface of the disc, shows an uneven load field at the surface contact with the brake pad. Superior pressure values are discovered in the central area of contact of the plate with the disc, that can lead to regional wear and plastic deformation results. Lower pressure results are discovered along the outer edge of the plate contact with the disc. Fig. 11 shows the frictional stress distribution for both geometrical models. Given that the model takes into account the slip phenomenon, the resulting frictional stress variation is spurious.

 https://doi.org/10.1051/e3sconf/202018003003

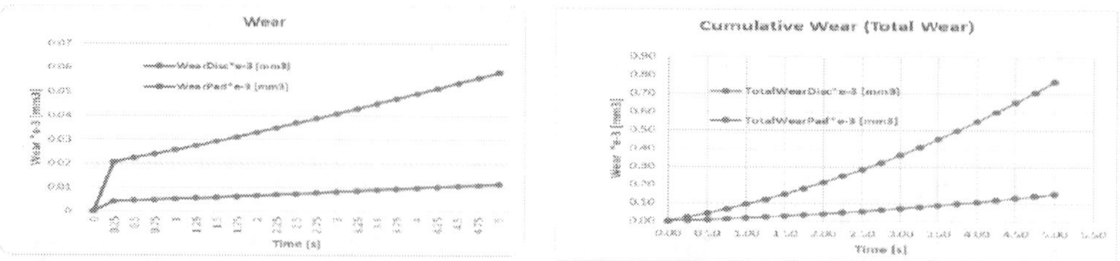

Fig. 12. The evolution in time of volumetric wear (left) and total volumetric wear (rigth) obtained during brake for disk (red line) and pad (blue line).

The simulation routine proposed is agreed to simulate the wear at the friction couple materials interface and implemented in VBExcel. Results from the virtual experiments, in the form of volumetric wear evolutions in time and the profile of cumulative wear (total volumetric wear), are shown in Fig. 12. The input data for solving the differential equation (Eq.5) starting from the initial condition $w_f(0) = 0$, were: K_disc = 1066 [mm^3 / m], H_disc = 720 [MPa], K_pad = 906 [mm^3 / m], H_pad = 120.7 [MPa], and an integration step $\Delta t = 0.25$s. Evidently, mathematical model propose is a simplification of a process that was discussed to be a function of many factors. Clearly, Eq. (5) can be made more complex by adding effects of other phenomena such as presence of oxidation, for instance.

4 Conclusions

To study structural characteristic and the tribological behavior of the braking system of a car for this study, a simulation program was created in the Ansys Workbench environment. In drum braking systems, due to the frictional forces between the front shoe and the drum, the driving force of the rear shoe is higher compared to the driving force of the previous shoe. In this way, the braking moment increases substantially. In order to equalize the uses for the two shoes, constructive solutions are used, such as: smaller pressing forces or larger friction pads on the primary shoe compared to the secondary shoe. The results obtained by the method of finite element analysis provide reference values for optimum choice of materials used in the fabrication of a braking system, making the design of the disc or drum structure more efficient. At the same time, significant improvement of work efficiency and reduction of design time by using software such as ANSYS Workbench are observed. Also, it can be noted that the wear of the brake pad is about five times greater than that of the brake disc.

References

1. G. R. K. R. Ramachandru,V. Kumars, Open j. technol. eng. discip. , *Dynamic analysis of mechanical braking system*, **2**, No. 4, 178-185, (2016).
2. R. B. Mathur, P. Thiyagarajan, T. L. Dhami, Carbon Science, *Controlling the Hardness and Tribological Behaviour of Non-asbestos Brake Lining Materials for Automobiles*, **5**, No. 1, 6-11, (2004).
3. A. Söderberg, S. Andersson, Wear, *Simulation of wear and contact pressure distribution at the pad-to-rotor interface in a disc brake using general purpose finite element analysis software*, Vol. 267, 2243–2251, (2009).
4. P. Hwang, X. Wu, J Mech Sci Technol, Investigation of temperature and thermal stress in ventilated disc brake based on 3D thermo-mechanical coupling model, **24**, 81-84, (2010).
5. A. Belhocine, M. Bouchetara, U.P.B. Sci. Bull., *Thermomechanical analysis of vehicle Braking*, Series D, Vol. 76, Iss. 1, (2014).
6. B. Zheng, J.D. Zhang, IOP Conf. Series: Materials Science and Engineering, *Finite Element Analysis of the Brake Drum Based on ANSYS APDL*, 472 012081, 1-7, (2019).
7. P. Zagrodzki, Raytech, K. B. Lam, E. Al Bahkali, J. R. Barber, J Tribol-T ASME, *Nonlinear Transient Behavior of a Sliding System With Frictionally Excited Thermoelastic Instability*, **123**, 699 – 708, (2001).
8. G. Babukanth, M. Vimal Teja, IJMIE, *Transient Analysis of Disk Brake By using Ansys Software*, **2**, Issue-1, 21-25, (2000).
9. A. Durgude, A. Vipradas, S. Kishore, S. Nimse, MAE 598-2016-11, *Design optimization of brake disc geometry*, Final Report, (2016).
10. N. A. Stoica, UPB, *Contributions to the study of the tribological processes in the disc braking system, applied to automobiles*, Doctoral thesis, București, (2018)
11. J.G.P. da Silva, C.H.S. Pereira, A. Casaril, P.E.D. Varante, H.A. Al-qureshi, SAE International, *Modeling of the Elastic Modulus of Brake Pad Materials with High Viscoelastic Content*, (2012).

Sorghum bicolor (L.) Moench. - Plant for the future in Moldavia agriculture (Romania) and for the human nutrition

Simona – Florina Isticioaia[1*], *Alexandra* Leonte[1], *Lorena - Diana* Popa[1], *Elena* Trotuş[1], *Livia* Apostol[2], *Iulian* Voicea[3] and *Gheorghe* Matei[4]

[1]Agricultural Research and Development Station Secuieni, 371 Principală Street, Neamţ, Romania
[2]National Institute of Research and Development for Food Bioresources, 5 Ancuta Baneasa Street, 2nd District, Romania
[3]National Institute of Research – Development for Machines and Installations Designed to Agriculture and Food Industry, 6 B-dul Ion Ionescu de la Brad, 1st District, Romania
[4]University of Craiova, 13 A. I. Cuza Street, nr. 13, Craiova, România (coordinator)

Abstract. In the climatic conditions of the Center of Moldavia (Romania), there are eight sorghum hybrids cultivated for grains (Arsky, Foehn, Albanus, Shamal, Kalatur şi Armorik, Elan and Alimentar) in order to establish their adaptability to the conditions of the area, by taking into account the current problems of agriculture caused by climate change and the need to find a solutions to fight hunger and improve human health. This study focused on two directions, namely: identifying genotypes that achieve the highest productions and establishing food value and the benefits of sorghum grains on human health by analysing the physico - chemical composition of the grains, by determining the mineral content of the grain and by determining the composition in amino acids. The grain production had ranged from 8623 kg/ha to 11181 kg/ha. This production had characterized by a protein content between 8.84 % and 12.80 %, in lipid content between 3.39 % and 4.15 %, in raw fibre content between 2.15 % and 3.95 % and the starch content was between 66.70 % and 74.66 %. The analysed mineral content of the sorghum samples has the following values (mg/100g a.s.): phosphorus - 280 - 330; potassium - 520 - 610; calcium - 2.4 - 3.9; magnesium - 260 - 290; iron - 11.6 - 18.9; zinc - 1.94 - 2.42; copper - 0.23 - 0.36; manganese - 1.60 - 1.97. Analysing the amino acid composition of sorghum samples it can be said that the samples contain all the essential amino acids, in concentrations that, some of them, exceed the recommended daily dose (FAO/WHO/UNU): leucine, isoleucine, histidine, valine. Also the analysed sorghum seeds add an important content of methionine, threonine and phenylalanine.

Key words: climatic change, genotypes, mineral composition, grain sorghum, yield

* Corresponding author : simonapochi@yahoo.com; matei.gheorghe@gmail.com

1 Introduction

Sorghum (Sorghum bicolor (L.) Moench.) has long been considered a "miracle culture", solving many problems that the planet has, which is increasingly populated and hungry due to global warming and in continuous search for solutions for obtaining bioenergy.

The genus Sorghum belongs to the family Poaceae (Gramineae), the subfamily Panicoideae, the tribe Andropogoneae, the subtribe Sorghinae [1]. The genus Sorghum contains a very diverse group of plants, which made it very difficult to classify sorghum varieties in domesticated and wild [2]. It originates from North - East Africa and was first cultivated 4000 years ago [3].

Sorghum is a cereal of high importance, being in fifth place in the world after the production of grains (57.6 million tonnes) after corn, wheat, rice and barley [4]. It has a great development due to its use in human nutrition, especially in the semi-arid areas of the world, where pedoclimatic conditions offer limited conditions for agriculture. Such situations are mainly in Africa, Asia and Latin America, areas that are frequently affected by drought [5], here sorghum is one of the most important crops representing the main source of energy and nutrition for people [6]. In western countries, such as the United States, Mexico and Australia, sorghum is mainly grown for animal feed, but there is a growing interest in its use for biofuel production, as well as for use in human nutrition, having given its chemical composition, which is beneficial for the development of healthy and functional foods [7].

The agronomic importance of sorghum culture is given by the resistance to drought and to the high temperatures, which gives the adaptability of this plant in the tropical and subtropical areas, in the context of the climatic changes of desertification of some areas of the globe. In addition to its high tolerance to drought and high temperatures, sorghum also has the advantage of its cultivation at high altitudes and in saline - alkaline and inert soils. These advantages are due to the fact that sorghum has a well-developed root system and the leaves are protected by wax [3]. The species can grow under the conditions of precipitation amounts below 100 mm precipitation / growth cycle, the critical drought period is about 20 days shorter than in maize (27 days in sorghum and 46 days in maize), and pollen can withstand high temperatures of 45 -50 °C, while corn pollen resists up to 35 - 40 °C. Another advantage of sorghum culture is its resistance to disease and pests. Also, sorghum has low requirements for nutrient requirements, having a well-developed root system, it allows it to extract all the nutrients it needs from the soil.

In addition to its agronomic benefits, sorghum is very nutritionally valuable. Sorghum beans can be an excellent source of starch, protein, sugar, fibre, being much cheaper compared to other crops. Grains are gluten-free, rich in starch and in nutrients and, most importantly, contain a wide range of bioactive phenolic compounds [8]. The unique phenolic profile gives sorghum a number of benefits to human health, such as reducing oxidative stress and preventing cancer [9].

Due to its superior agronomic properties and its potential to be used in human health, sorghum has attracted considerable attention in recent decades from researchers and the food and drug industries. With the ever-increasing change in consumer demand for healthy, plant-based foods, sorghum has enormous potential for exploitation and development in healthy and functional foods and food additives [10, 11]. The nutritional composition of sorghum varies depending on the cultivated hybrid. In general, carbohydrates (starch and non-starch polysaccharides), proteins and lipids are the main compounds of the grain [12]. On average, 100 g of sorghum contains about 72.1 g of carbohydrates, 12.4 g of water, 10.6 g of protein, 6.7 g of fibre and 3.5 g of fat and provides approximately 1.377 kJ of energy [13]. Starch is the dominant carbohydrate and varies depending on the hybrid, from 32.1 g to 72.5 g to 100 g of grains [14]. Sorghum is a rich source of fibres, because non-starch carbohydrate in sorghum is composed of insoluble fibres (75 % to 90 %) and soluble fibres (10 % - 25 %), which are found in the walls of the pericarp and endosperm cells - 15 g to 100 g of berries [15].

Sorghum protein can be broadly divided into prolaminic proteins (kafirins) and non-prolaminic proteins (globulins, glutelin and albumin). Kafirins are the main proteins, accounting for 70 % of the total protein, while albumin, glutelin and globulin represent the rest [16]. Sorghum beans are rich in glutamic acid, proline and leucine [17]. Sorghum proteins have low digestibility. Kafirins have high degree of polymerization and extensive disulphide bridges that are resistant to enzymatic digestion in the digestive tract and strong interaction with tannins and starch also impedes protein digestion [18]. These characteristics make sorghum a promising source of food for people suffering from obesity and diabetes [19].

Lipids from sorghum are mainly unsaturated fatty acids, being represented in particular by oleic, linoleic, palmitic, linolenic and stearic acids [13].

Sorghum is also an important source of vitamins and minerals. Vitamin B complex (pyridoxine, riboflavin and thiamine) and some fat-soluble vitamins (A, D, E and K) are the main vitamins in sorghum, and potassium, phosphorus, magnesium and zinc are the main minerals [13, 15].

Considering the current problems of agriculture caused by climate change and the need to find a solutions to fight hunger and improve human health, in the climatic conditions of the Center of Moldavia (Romania), we have conducted studies on eight sorghum hybrids cultivated for grains (Arsky, Foehn, Albanus, Shamal, Kalatur și Armorik, Elan and Alimentar). This study focused on two directions, namely: identifying genotypes that achieve the highest productions and establishing food value and the benefits of sorghum grains on human health by analysing the physico - chemical composition of the grains.

2 Methodology

In the pedoclimatic conditions from the Agricultural Research and Development Station Secuieni (Center of Moldova), researches were carried out regarding the adaptability of some grain sorghum hybrids to the conditions of the area and their influence on its production and quality. The experiments were performed on a typical Cambodian chernozem soil type, characterized as being normally supplied with mobile phosphorus (P_2O_5 - 39 ppm), and mobile potassium (K_2O - 161 ppm), moderately supplied with nitrogen (nitrogen index of soil: 2.1), weak acid (pH: 6.29) and weakly fertile (humus content 2.3 %) and in extremely adverse climatic conditions to agricultural crops. Both years of experimentation (2018 and 2019) presented during the vegetation period climatic conditions characterized as very dry and warm.

The experience was monofactorial, the factor being the sorghum hybrid for grains, and its placement in the field was carried out following the randomized blocks method, in three repetitions. As far as the experimental data are concerned, they were processed by statistical methods specific to the monofactorial experiences, and their interpretation was made by the variance analysis. The biological material came from the companies Euralis Seeds (Arsky, Foehn, Albanus, Shamal, Kalatur and Armorik hybrids) and from the Porumbeni Institute of Phytotechnics (Elan and Alimentar hybrids). The technology performed in the experimental field was the one specific for the pedoclimatic conditions in the Center of Moldova. The sowing distribution was 300000 Gg/ha and the precursor plant was the sunflower.

The sorghum berries have been analysed in 2018 - 2019 period from a physical, chemical and nutritional point of view, using modern methods of analysis, apparatus and equipment for analysis, measurement and processing specific to raw materials and products (https://erris.gov.ro/Food-Chemistry -Laboratory; https://erris.gov.ro/Microbiology-ELISA-Laboratory; https://erris.gov.ro/Food-Packaging-Laboratory), which is endowed with the National Institute for Research and Development for Food Bioresources Bucharest.

3 Results and discussions

The production results obtained under the conditions of A.R.D.S. Secuieni showed that sorghum, although it suffered during the vegetation period from unfavourable climatic conditions, showed adaptability and the obtained productions were substantial.

The average production (2018 - 2019) obtained at the sorghum showed great variations depending on the hybrid and was between 7841 kg/ha, as recorded in the variant sown with the Elan hybrid, and 9718 kg/ha, production recorded in the variant sown with Shamal hybrid. For comparison, we used the average of the experience as a witness, and the differences obtained in relation to this were statistically ensured in three of the variants, namely: in the variant sown with the Elan hybrid the difference was negative and interpreted as significant, and in the variants sown with the Foehn hybrids and Shamal, the differences were positive and interpreted as significant and distinctly significant (fig. 1).

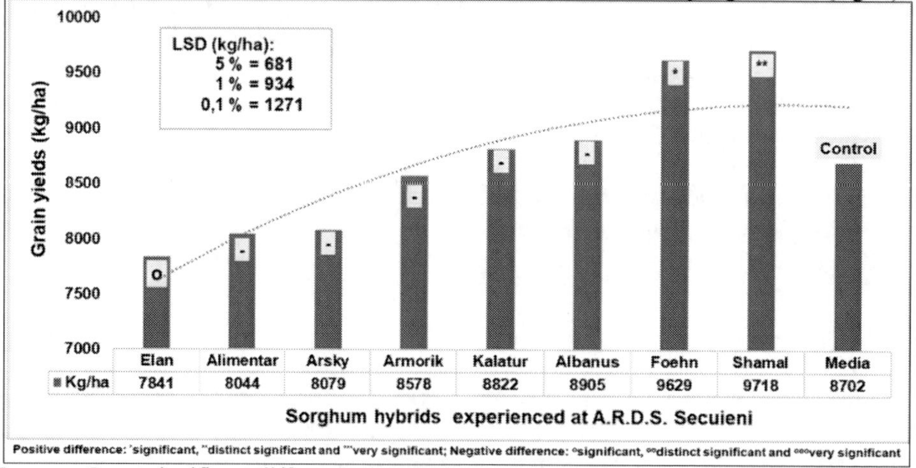

LSD means Least significant difference (For positiv values - Differences larger than LSD 5% level are significant and are indicated with *, differences larger than LSD 1% level are distinctly significant and are indicated with ** and differences larger than LSD 0,1% level are verry significant and are indicated with *** and for negativ values the *, **, *** is replaced with o, oo, ooo.

Fig. 1. Grain yields obtained from sorghum under pedoclimatic conditions of the A.R.D.S. Secuieni, average 2018 - 2019

In order to identify hybrids with high dietary value, the samples collected from the experimental field were analyzed and the results obtained indicated large variations of the physical and chemical indices depending on the genotype.

The physical indices of the seed are generally influenced by the characteristics of the hybrid and by the climatic conditions of the grain formation and filling period. Under the conditions at Secuieni, they varied according to the hybrid experienced, the values of the mass of one thousand grains being between 27.8 g (Armorik hybrid) and 33.0 g (Alimentar hybrid), and those of the hectolitric mass between 73.1 kg/hl and 76.6 kg/hl (Armorik and Foehn hybrids) (fig. 2).

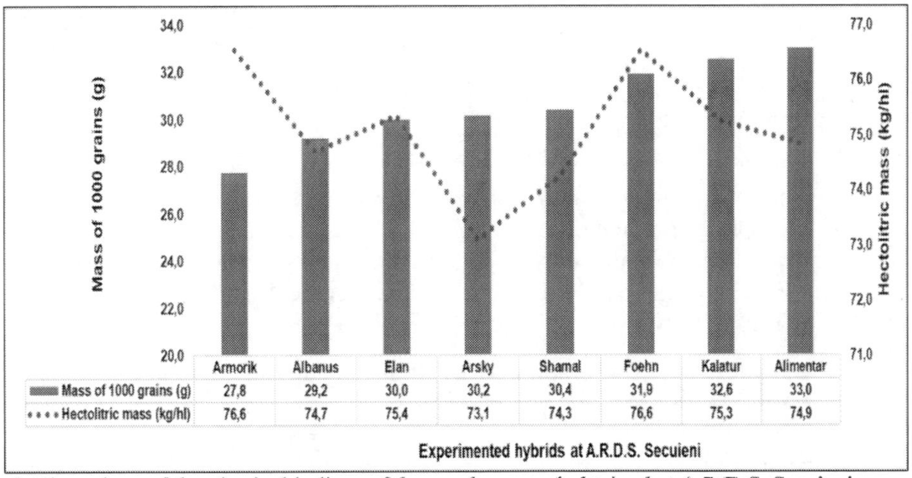

Fig. 2. The values of the physical indices of the sorghum seed obtained at A.R.D.S. Secuieni, average 2018 – 2019

Sorghum is the cereal that contains the most nutrients and, according to the latest researches, it can reduce severe asthma attacks and fights migraine. The results obtained from the analysis of the physico-chemical composition of the sorghum samples collected from the experimental field showed that the studied hybrids had a high content in starch, between 66.70 % (Foehn hybrid) and 74.66 % (Armorik hybrid) and protein total, from 8.84 % (Shamal hybrid) to 12.80 % (Elan hybrid), which makes them advantageous for use in bakery (Table 1).

Considering that the percentage of raw fiber, between 2.15 % (Arsky hybrid) and 3.95 % (Albanus hybrid), is double that of wheat flour, they take on importance in human health (Table 1).

Table 1. Physico - chemical composition of the experienced sorghum hybrids

Hybrid	Physical - chemical composition of sorghum grains (% su)				
	Proteins	Lipids	Raw fibre	Starch	Ash
Arsky	11.11	3.44	2.15	70.42	1.67
Foehn	9.64	3.90	2.69	66.70	1.60
Albanus	10.56	3.39	3.95	68.02	1.66
Shamal	8.84	3.94	2.30	67.37	1.49
Kalatur	10.57	3.82	2.64	66.94	1.61
Armorik	10.30	4.15	3.02	74.66	1.71
Elan	12.80	3.74	2.37	67.20	1.89
Alimentar	12.32	3.80	2.55	67.66	1.80

The mineral content of the sorghum samples analysed (mg/100g a.s.) had phosphorus values between 280 - 330, potassium between 520 and 610, calcium between 2.0 and 3.9, magnesium between 260 and 290, iron between 11.6 and 18.9, zinc between 1.94 and 2.42, copper between 0.23 and 0.36 and manganese between 1.60 and 1.97 (Table 2).

Table 2. Mineral content determined in sorghum hybrids experimented

Hybrid	Minerals content (mg/100 g a.s.)							
	P	K	Ca	Mg	Fe	Zn	Cu	Mn
Arsky	280	610	2.6	270	14.3	2.02	0.32	1.60
Foehn	330	530	3.9	280	18.9	2.42	0.28	1.93
Albanus	320	520	2.7	290	16.3	2.20	0.33	1.97

Shamal	280	600	3.8	260	13.6	1.94	0.23	1.70
Kalatur	290	520	2.4	270	11.6	2.07	0.35	1.82
Armorik	320	600	2.9	290	15.4	1.97	0.28	1.70
Elan	320	540	2.4	280	14.0	2.18	0.29	1.96
Alimentar	330	570	3.4	290	13.7	2.14	0.36	1.92

Comparing the results obtained (Table 2) with the daily mineral reference dose (RDI) to ensure the proper functioning of human metabolic processes (Table 3), we observe that 100 g of sorghum flour can provide more than half of the daily iron dose, the Foehn hybrid as a whole and over half the daily dose of magnesium, and the potassium content is quite satisfactory. In view of these aspects, the term "iron source" and "magnesium source" may be emitted [20, 21].

Table 3. Daily reference consumption (RDI) of macronutrients and micronutrients [22].

Mineral (maximum for sex and age groups)	Acceptable daily intake (ADI)
Calcium	1000 mg
Iron	18 mg
Magnesium	400 mg
Zinc	15 mg
Copper	2000 µg
Sodium	2400 mg
Potassium	4700 mg

The results of identifying essential amino acids in sorghum samples revealed that sorghum contains all the essential amino acids, some of which such as leucine, isoleucine, histidine and valine exceeding the recommended daily dose (FAO/ WHO/UNU). Also the sorghum seeds analyzed had an important content of methionine, threonine and phenylalanine (Table 4).

Table 4. Amino acid composition of sorghum samples collected from experimental field

	Amino Acid Composition (%)							
Amino acid	Arsky	Foehn	Albanus	Shamal	Kalatur	Armorik	Elan	Alimentar
Aspartic acid	0.60	0.55	0.56	0.48	0.59	0.57	0.60	0.50
Glutamic acid	2.12	1.95	1.84	1.63	2.01	1.93	2.08	1.80
Seryn	0.44	0.39	0.38	0.32	0.41	0.39	0.44	0.37
Glycin	0.31	0.32	0.29	0.26	0.30	0.29	0.34	0.28
Histidine	0.20	0.20	0.18	0.17	0.20	0.19	0.22	0.17
Arginine	0.36	0.36	0.35	0.30	0.36	0.34	0.38	0.33
Threonine	0.31	0.29	0.26	0.24	0.28	0.27	0.30	0.25
Alanine	0.86	0.78	0.77	0.68	0.82	0.79	0.87	0.76
Proline	0.89	0.82	0.76	0.66	0.82	0.77	0.89	0.76
Tyrosine	0.32	0.30	0.27	0.19	0.24	0.25	0.30	0.27
Valine	0.50	0.47	0.45	0.39	0.47	0.44	0.49	0.43
Methionine	0.16	0.16	0.16	0.14	0.14	0.14	0.19	0.17
Cysteine	0.11	0.10	0.11	0.12	0.12	0.14	0.13	0.12
Isoleucine	0.37	0.34	0.34	0.29	0.36	0.34	0.39	0.33
Leucine	1.30	1.16	1.16	0.97	1.22	1.18	1.34	1.16
Phenylalanine	0.51	0.44	0.42	0.32	0.42	0.43	0.50	0.42
Lysine	0.19	0.19	0.20	0.09	0.09	0.17	0.20	0.19

4 Conclusions

Sorghum is a cereal species of world importance, considering its adaptability to unfavorable environmental conditions, which is materialized by obtaining high yields that have very high nutritional qualities and improves human health.

Although the experiments at Secuieni were carried out in extremely unfavourable climatic conditions, characterized as very dry and hot, the sorghum produced large yields, ranging from 7841 kg/ha to 9718 kg/ha. Two of the eight experimented hybrids, respectively Foehn and Shamal, were noted for producing more than 9500 kg/ha.

The physical indices of the seed were influenced by the characteristics of the hybrid, and together with the high content of starch, protein and raw fibre, they represent an advantage for the use of sorghum in the baking industry. The lipid content is twice as high as wheat flour and has beneficial effects on bakery products. All eight analysed hybrids can be labelled as excellent "iron source" and "magnesium source".

The samples analysed contain all the essential amino acids, in concentrations that, some of them, exceed the recommended daily dose: leucine, isoleucine, histidine and valine. In addition to these, sorghum has a satisfactory content of methionine, threonine and phenylalanine.

This work was supported by a grant of the Romanian Ministery of Research and Innovation CCDI - UEFISCDI, "Complex system of integral capitalization of agricultural species with energy and food potential", project number PN-III-P1-1.2-PCCDI-2017-0566, Contract no. 9PCCDI/2018, within PNCDI III.

References

1. W. D. Clayton, S. A. Renvoize, Genera graminum. Grasses of the World, Kew Bull. Add. Series, 13: 389, ref. 87 (1986)
2. J. Wiersema, J. Dahlberg, The nomenclature of Sorghum bicolor (L.) Moench (Gramineae), Taxon, 56 (3): 941 - 946 (2007)
3. L. W. Rooney, R. D. Waniska, Sorghum food and industrial utilization, In Sorghum: Origin, history, technology and production, 2: 689 - 729, Hoboken, NJ: John Wiley & Sons (2000)
4. FAO, Food and Agriculture Organization of the United Nation (2017)
5. V. Taleon, L. Dykes, W. L. Rooney, L. W. Rooney, Effect of genotype and environment on flavonoid concentration and profile of black sorghum grains, J. of Cereal Sc., 56 (2): 470 – 475 (2012)
6. Y. Xiong, P. Zhang, J. Luo, S. Johnson, Z. Fang, Effect of processing on the phenolic contents, antioxidant activity and volatile compounds of sorghum grain tea, J. of Cereal Sc., 85: 6 - 14 (2019)
7. W. L. Rooney, L. W. Rooney, J. Awika, L. Dykes, Registration of Tx3362 sorghum germplasm, J. of Plant Registrations, 7 (1): 104 – 107 (2013)
8. S. Shen, R. Huang, C. Li, W. Wu, H. Chen, J. Shi, S. Chen, X. Ye, Phenolic compositions and antioxidant activities differ significantly among sorghum grains with different applications, Molecules, 23 (5): 1203, E1203 (2018)
9. F. M. Gonz´alez - Montilla, R. A. Ch´avez - Santoscoy, J. A. Guti´errez - Uribe, S. O. Serna-Saldivar, Isolation and identification of phase II enzyme inductors obtained from black Shawaya sorghum [Sorghum bicolor (L.) Moench] bran. J. of Cereal Sc., 55 (2): 126 – 131 (2012)

10. A. R. Cabral, C. Waters, H. L. Laird, L. C. Cavitt, R. K. Miller, W. L. Rooney, C. Z. Alvarado, J. M. Awika, C. R. Kerth, Sorghum bran as an antioxidant in pork and poultry products, Meat and Muscle Biology, 2 (2): 83 (2019)
11. G. Wu, Y. Shen, Y. Qi, H. Zhang, L. Wang, H. Qian, X. Qi, Y. Li, S. K. Johnson, Improvement of in vitro and cellular antioxidant properties of Chinese steamed bread through sorghum addition. LWT - Food Sci. and Techn., 91: 77 - 83 (2018)
12. H. Hill, L. S. Lee, R. J. Henry, Variation in sorghum starch synthesis genes associated with differences in starch phenotype. Food Chemistry, 131 (1): 175 - 183 (2012)
13. USDA, National nutrient database for standard reference legacy release: Full report (all nutrients) 20067, sorghum grain (2019)
14. I. S. Udachan, A. K. Sahu, F. M. Hend, Extraction and characterization of sorghum (Sorghum bicolor L. Moench) starch. Int. Food Res. J., 19 (1): 315 - 319 (2012)
15. H. S. D. Martino, P. A. Tomaz, É. A. Moraes, L. L. da Conceição, D. da. S. Oliveira, V. A. V. Queiroz, J. A. Santos Rodrigues, M. Ribeiro Pirozi, H. M. Pinheiro -Sant'Ana, S. M. R. Ribeiro, Chemical characterization and size distribution of sorghum genotypes for human consumption, Revista do Instituto Adolfo Lutz (Impresso), 71(2): 337 - 344 (2012)
16. P. S. Belton, I. Delgadillo, N. G. Halford, P. R. Shewry, Kafirin structure and functionality. J. of Cereal Sc., 44 (3): 272 - 286 (2006)
17. G. Galili, R. Amir, Fortifying plants with the essential amino acids lysine and methionine to improve nutritional quality. Plant Biotech. J., 11 (2): 211 - 222 (2013)
18. J. R. N. Taylor, M. N. Emmambux, Developments in our understanding of sorghum polysaccharides and their health benefits, Cereal Chemistry, 87 (4): 263 – 271 (2010)
19. H. Raza, I. Pasha, M. Shoaib, F. Zaaboul, S. Niazi, W. Aboshora, Review on functional and rheological attributes of kafirin for utilization in gluten free baking industry, Am. J. of Food Sc. and Nutrition Res., 4 (5): 150 – 157 (2017)
20. Regulation (EC) no. 1924/2006 of the European Parliament and of the Council of 20 December 2006 on nutrition and health claims made on foods (2006)
21. Council Directive 90/496/EEC of 24 September 1990 on nutrition labelling for foodstuffs (1990)
22. Food and drug administration, https://www.usa.gov/federal-agencies/food-and-drug-administration (2011)

E3S Web of Conferences **180**, 03005 (2020) https://doi.org/10.1051/e3sconf/202018003005
TE-RE-RD 2020

Theoretical considerations on the factors influencing the processing of hemp stalks

Paul Găgeanu, *Mihai* Olan, *Iuliana* Găgeanu*, and *Alexandru* Zaica

INMA Bucharest, 6, Ion Ionescu de la Brad Boulevard, Romania

Abstract. The present work is achieved based on the processing of the data accumulated over time, within the activity of research, technological engineering, and testing of machines for processing agricultural products in the last years, as well as studying the specialized literature with the latest achievements of some companies that make machinery for processing agricultural products. The compression of the hemp stalks for processing is analyzed, the main characteristics of the working rolls, the specific energy consumed are determined and an analysis of the obtained fiber length is performed. In order to develop a technical equipment for processing hemp stalks that ensures high performance, the following aspects were studied: the constructive and functional factors of a technical equipment that influence the working processes in order to choose a performance solution for processing hemp stalks. In the last part of the paper, the main technical and functional characteristics of a high-performance technical equipment for processing the hemp stalks harvested in green are established.

1 Introduction

Hemp is part of the group of textile plants of great value for human and industrial use.

Hemp (Figure 1) is one of the oldest cultivated plants (over 2000 years), being used mainly to obtain fibers used in clothing making. Hemp strains from indigenous (local) varieties and wild hemp contain 10-12% fiber, while improved varieties have 26-32% [1], [2].

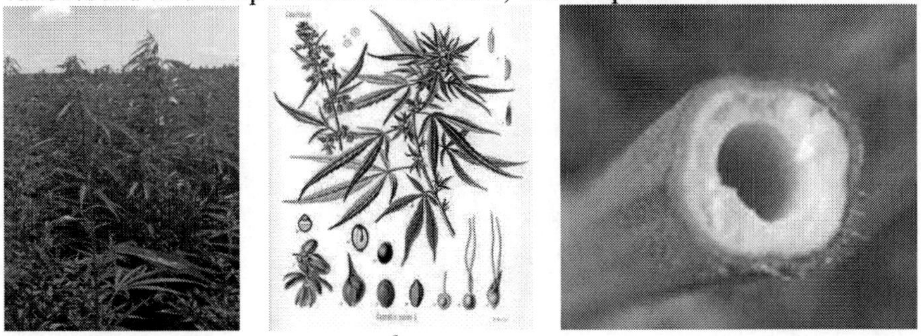

a b c

Fig. 1. Hemp plants
a) aerial parts; b) COmponents: root, stalks and fruits; c)green stalk section, [20], [21]

* Corresponding author: iuliadml@yahoo.com

The fibers are very strong and that is why the stalk is difficult to cut, tangling on the blades of the combine - meaning that powerful cutting machines must be used, otherwise harvesting will be cumbersome [3]-[5].

The stalks should be cut as long as possible - long fibers are better and more useful than short ones. The areas cultivated with hemp in the world have been continuously reduced, reaching from over one million hectares in 1938, to about 425,000 ha in 1988 now under 400,000 ha are cultivated with hemp. The main cultivating countries are: Russia, India, Pakistan, Turkey, North Korea, Poland, Hungary, Bulgaria, Chile [6].

The hemp stalk from which the fibers are extracted is straight and has ramifications at the bottom. Its length is between 1 and 3 m, and the thickness varies from 4 to 8 mm. The anatomical tissues of the epidermis are more complex compared to that of flax [7].

Hemp stalks for textile fibers are assessed by: predominant color, which is usually yellow - green or yellow, length at a certain percentage of plants; thickness and moisture.

Mechanical breakage of the stems is known as decortication and results in the separation of free fibers from the less valuable wood core (waste) [8], [11].

The fiber content in the stalks is influenced by the variety, the technological and pedoclimatic conditions. Fibers have a number of properties that are particularly valuable in strength (traction, torsion, friction, rotting), extensibility (elastic and plastic), spinning capacity, length greater than the sisal, jute, manila or cotton fibers, making them usable in a great number of fields: in the textile industry, in the manufacturing industry, in the automotive industry, etc., [12], [13]

Basically, hemp processing for fiber must follow the procedures of harvesting, cutting, baling, pre-processing, peeling and cleaning. Leaving it longer in the field to dry further helps decortication. But recent research has shown that non-twisted "green decortication" is also acceptable for some industrial applications [14], [15]. In this case, green hemp decortication requires more power and energy than dry hemp and is less efficient [16].

2 Material and method

In order to obtain fibers from hemp stalks, they must go through a combined process of compressing the stems and crushing the woody part, detaching and separating them from the fibers. For this, two pairs of rollers are proposed to perform the above operations.

a) **Compression**: for the first phase (compression of stalks), the stalks from their initial thickness *d* are pressed to thickness *s* after passing through the first pair of rollers (Figure 2).

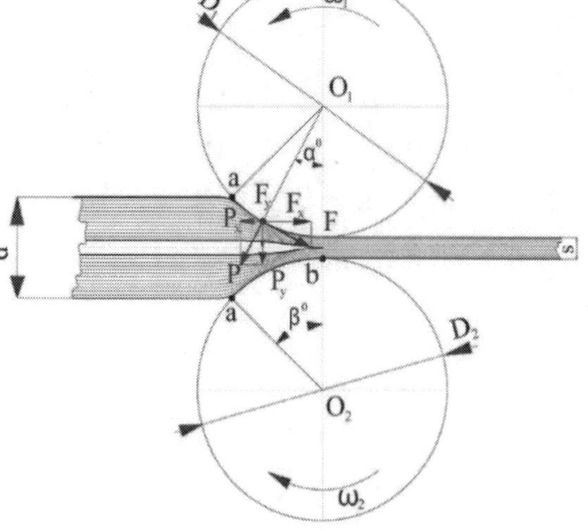

Fig. 2. Schematic diagram of compressing hemp stalks

The forces with which the rollers act on the stalk layer for each zone are directed perpendicular to the contact surface and can be reduced to a resultant force *P*. By decomposing this force in two directions perpendicular to the material compression and respectively its displacement, we obtain the components *Py* which compresses the material and *Px* which pushes the material from the passage. At the same time, the frictional forces acting on the surface of the rollers act tangentially, which can be presented by the resultant *F*. By decomposing the resultant force *F*, we obtain the components *Fx*, which drive the material between the rollers and *Fy* which contributes to the pressing of the material.

Analyzing the process of compression (pressing) it is found that, when the rollers approach, simultaneous with decreasing the space *s*, the contact angle α^o increases. As it turns out, the condition of the material between the rollers is:

$$F_x \geq P_x \quad (1)$$

or

$$F\ cos_\alpha \geq P\ sin_\alpha \quad (2)$$

Friction force, is $F=P_{tg\varphi}$

Replacing, we obtain

$$P\ tg_\varphi\ cos_\alpha \geq P\ sin_\alpha \quad (3)$$

namely,

$$tg_\varphi > tg_\alpha \quad (4)$$

It is found that the thickness *s* and the degree of compression (Δs) of the stems, for pressing between two rollers, depend on the diameter of the rollers D_1 and D_2, and the coefficient of friction ($F = tg\varphi$) of the surface material of the roller. Therefore, the increase in diameter *D* and the coefficient of friction, tg_φ, ensures the compression of the larger diameter stems which improves the conditions of the stems' drive and increases the productivity of the compression (pressing) process *Q* and the degree of crushing of the woody part of the stems.

b) The working width of the roller

It is estimated to work with hemp stalks where their overlap is involuntary. The working width of the rollers is given by relation:

$$B = \lambda * n * d + 2e \quad (5)$$

where: *B* - roller width, [m];

N – number of plants simultaneously processed, [-];

d– diameter of plants simultaneously processed, [m];

e – the distance at which the loading of the roller should be made in relation to the casing, [m]. (e=20-30 mm);

λ - coefficient of flattening of the stalk after pressing (λ=1,55 ÷1,59).

c) The working capacity of the rollers

The working capacity of a pair of rollers is given by the relation:

$$G=3600*B*s*V_p*\eta*\gamma \quad (6)$$

where: *s* – distance between rollers, [m];

V_p – peripheral speed of rollers, [m/s];

η – roller loading coefficient, η=0,7-0,9;

γ - volumetric weight of hemp stalks, [tf/m^3].

d) The peripheral speed of the rollers is given by the relation:

$$V_p = \pi D_n/60 \quad (7)$$

where: *D* - rollers diameter, [m];

n – rollers revolution frequency, [rot/min].

e) Specific energy consumed

The total specific energy could be simplified into two parts: the specific fiber processing energy and the specific core processing energy. Thus:

$$E_k=\mu_m E_m+\mu_f E_f \tag{8}$$

where: E_k – total energy, [J / g];
E_f – fiber specific energy, [J / g];
E_m - core specific energy, [J / g];
μ_m - the mass ratio of the basic fraction; [%];
μ_f - the mass ratio of the fiber fraction; [%].

The core ratio μ_m and the fibers ratio μ_f are constant.

Researchers have presented different models to simulate the energy requirement for stalk processing. Theories regarding the specific energy of size reduction at the initial and final dimensions of the material have been developed. Among these theories, Kick, Rittinger and Bond laws on size reduction are the most widely known and widely accepted. Their theories are presented in the following equations [17] – [19]:

$$E_k=C_k*ln(\frac{L_1}{L_2}) \tag{9}$$

$$E_R=C_R*ln\left(\frac{1}{L_2}-\frac{1}{L_1}\right) \tag{10}$$

$$E_B=C_B*ln\left(\frac{10}{\sqrt{L_2}}-\frac{10}{\sqrt{L_1}}\right) \tag{11}$$

where: E_K, E_R and E_B - values for the specific energy of each law, [W];
L_1 and L_2 - initial dimension, respectively the final dimension of the material, [m];
C_K, C_R and C_B – constant factors, [-].

The following equation was used to evaluate the total decortication energy input for 5 seconds.

$$E_{total}=y_2+y_{12}\times\Delta_t+y_3+y_{22}\times\Delta_t+\cdots+y_n+y_{n-12}\times\Delta_t\ (J) \tag{12}$$

where: y_1 = the value of the power of the first data point, [W];
y_2 = the value of the power of the second data point, [W];
y_n = the value of the power of the nth data point, [W];
Δt = time interval (periods), [s].

Specific energy is considered to be net energy divided by the supply mass.

$$E_s=\frac{E_{net}}{M} \tag{13}$$

where: E_S – specific energy, [J / g];
E_{net} – net energy, [J];
M – feeding mass, [g].

3 Results

Fiber length analysis

The ratio between the final lengths of the core and the fibers is considered to be constant. The proportions of core and fiber content, being considered uniform, lead to the hypotheses:

1) the initial length of the core L_{1m} is equal to the initial fiber length L_{1f}.
2) the relationship between the final length of the core and the fiber is:

$$L_{2m}=r\times L_{2f} \tag{14}$$

where: L_{2m} – final length of the core, [m];
L_{2f} – final length of the fiber, [m];
r – constant, [-].

By taking the fibers and the fractional core separately, Kick's original law (9) can be expressed as follows:

$$E_m=C_{Km}*ln\frac{L_{1m}}{L_{2m}} \tag{15}$$

$$E_f=C_{Kf}*ln\frac{L_{1f}}{L_{2f}} \tag{16}$$

Replacing L_{1c} and L_{2c} with the terms L_{1f} and L_{2f}, relation (8) becomes:

$$E_k=\mu_m*C_{km}*ln\frac{L_{1f}}{rL_{2f}}+\mu_f*C_{kf}*ln\frac{L_{1c}}{L_{2f}}$$

$$E_k=\left(\mu_m*C_{km}+\mu_f*C_{kf}\right)ln\frac{L_{1f}}{L_{2f}}+\mu_c*C_{km}*ln\frac{1}{r} \tag{17}$$

The final fiber length can vary over a wide range. From less than 20 mm to more than 1500 mm. To take into account the distribution of the final fiber length, E_k can be calculated as:

$$E_k=\left(\mu_m*C_{km}+\mu_f*C_{kf}\right)*ln\left(\sum_{i=1}^{n}\delta_i\frac{L_{1f}}{L_{2fi}}\right)+\mu_m*C_{kc}ln\frac{1}{r} \tag{18}$$

where: δ_i – mass proportion of the i length category, [%];
n – number of length categories, [-];
L_{2fi} – average exit fiber length for i length category, [m].

Equation (17) can be rewritten in the following general form:

$$E_K=C_K*X_K+b_K \tag{19}$$

where: C_K – constant factor determined through experimental data, [-];
b_K - constant term according to hypothesis, [J/k];
X_K - is $\ln\left(\sum_{i=1}^{n}\delta_i\frac{L_{1f}}{L_{2fi}}\right)$ and was an independent variable in Kick's modified law.

Equation (19) is Kick's modified law for determining the specific energy for hemp decortication. This modified law is related to the total energy correlated with the initial and final length of the fiber fraction.

Similarly, the modified laws of Rittinger and Bond are as follows:

$$E_R=\left(\frac{\mu_m*C_{Rm}}{r}+\mu_f C_{Rf}\right)*\sum_{i=1}^{n}\delta_i\left(\frac{1}{L_{2fi}}-\frac{1}{L_{1f}}\right)+(\frac{\mu_m*C_{Rc}}{L_{1f}}) \tag{20}$$

$$E_R=C_R*X_R+b_R \tag{21}$$

$$E_B=\left(\frac{\mu_m C_{Bm}}{\sqrt{r}}+\mu_f*C_{Bf}\right)*\sum_{i=1}^{n}\delta_i\left(\frac{10}{\sqrt{L_{2fi}}}-\frac{10}{\sqrt{L_{1f}}}\right)+(\frac{10*\mu_m*C_{Bm}}{\sqrt{L_{1fi}}*\sqrt{r}}-\frac{10*\mu_m C_{Bm}}{\sqrt{L_{1f}}} \tag{22}$$

$$E_B=C_B*X_B+b_B \tag{23}$$

B_R and b_B (J/g) variable are coefficients dependent on the initial length of fibers. In calculus, the initial length of fibers is considered unchanged.

Based on these theoretical results, the optimal technology and working diagram of a hemp stalks processing equipment were established (Figure 3). The proposed technical equipment consists of the following components: a feeding table, (1), with a hat under which the plants pass, (3), a pair of drums with discs serving as feed guides, (4), a pair of toothed rollers for the primary pressing and crushing of the stems (6), a pair of pallet rollers that provide the splitting of the wood part (7), a pair of crushing rollers, (8), a roller with displacement pallets (9), a third pair of grooved rollers to ensure a deeper crushing of the woody part of the stem, (10), another dislocation roller (11), a last pair of rollers with grooved surface which ensures the advanced crushing of the woody part (12) and a rotary dislodge with the latticework (13).

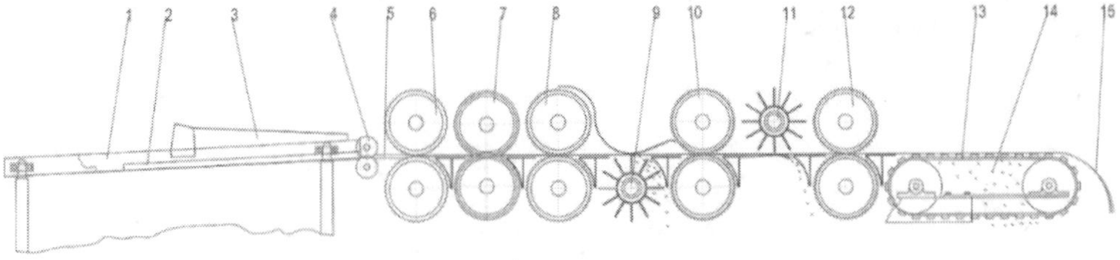

Fig. 3. Hemp stalks processing equipment

1. feeding table; 2. guides; 3. hood; 4. rollers with guide rolls; 5. hemp stalk; 6. flat rollers; 7. splitting rollers; 8. crush rollers i; 9. displacement roller i; 10. crushing rollers ii; 11. displacement roller ii; 12. crushing rollers iii; 13. dislodge iii; 14. hurd; 15. fiber

Main technical and functional characteristics

- capacity processed stalks	kg/h	1000-2000;
- feeding	-	manual
- overall dimensions:		
- length	mm	5100;
- width	mm	1600;
-height	mm	2400;
- feeding table size L x l	mm	1500x800
- number of operation stages	-	9
- installed power	kW	22
-stalks diameter	mm	6-20

In the front of the frame and adjacent to the inside end of the feeding table is mounted a pair of rollers as feed guides which consist of a pair of stationary shafts spaced apart from one another at an adjustable distance from one another depending on the thickness of plant stems.

On these stationary shafts are provided a plurality of discs mounted at a distance given by rubber sheaths mounted on the two shafts serving as separators through which the plants pass.

These shafts are positioned so that the discs on one shaft are fastened so that they are in the middle of the distance between two discs on the other shaft ensuring contact with the fed stems and pushing them to the first pair of rollers that do the processing.

The distance between the two shafts is adjustable depending on the diameter of the stalks of the material to be fed. The plant, under the pressure of the disc with which it is in contact, is drawn and, at the same time, the leaves on the stem are detached in contact with the other two discs and by free fall they are evacuated.

From here, the plant passes between the flattening rollers where it is flattened, after which it is inserted between the splitting rollers where it is longitudinally split, and the obtained strips are passed among the first crushing rollers where the woody part of the stem begins to be crushed. Part of the crushed wood is dislodged by the first pallet roller and is evacuated.

What is left is passed between the two pairs of crushing rollers. After passing through them, a large part of the wood is crushed and after the passage of the second pallet roller it is dislodged from the fibers and is evacuated.

Next, the fiber with the wood part that did not break off is passed between the rollers of the third pair of crushing rollers. The crushed wood portion is detached from the fiber mass when passing over the third dislodging roller which has different construction from the other two. It is fixed on an independent shaft on which there are two bearings with two flanges on which are mounted blades (lattices). These blades are mounted evenly on the rubber or leather straps.

In order to ensure an even better dislodging, at the extension roller head is provided with a vibrating system that during operation accelerates the process of dislodging the hurd.

The finished fiber product from the last pair of rollers is transported and removed from the machine. At the same time, a share of the wood parts is eliminated by detaching them from the fiber mass when they come in contact with the latticework.

The decortication process is completed in around 5 seconds. For each power-time curve, the 5-second width of the peak-containing curve is chosen to calculate the energy. The results are presented in Figure 4.

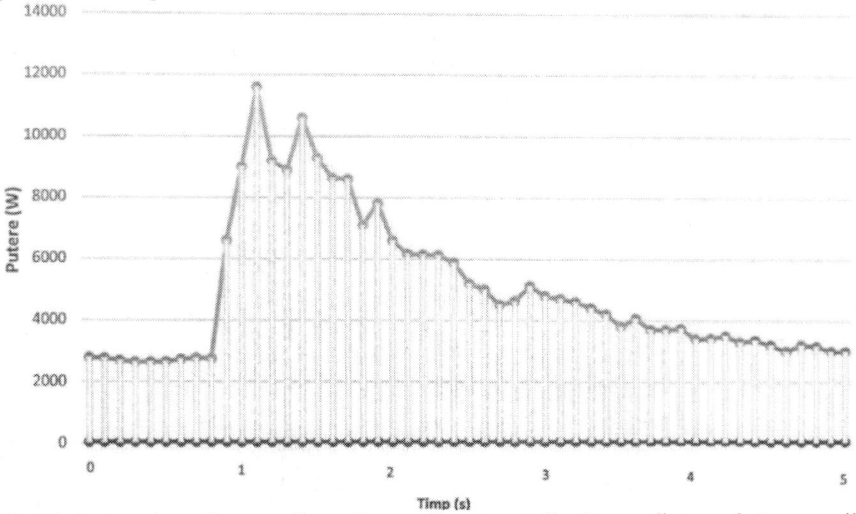

Fig. 4. Power-time diagram (for a 8 mm diameter stalk, 3 mm distance between rollers and 240 g/stalk feeding mass)

The horizontal line before the bell shape curve shown in Figure 4 represents the power consumption in idle phase. In order to determine the net energy consumption, the energy in the phase idle is deducted from the total energy obtained from the above equation.

The energy consumption in load is about 4.8 times greater compared to the idle consumption. Fiber length differs from hybrid to hybrid and depends on the length and thickness of the stalk and the peripheral speeds of the rollers.

4 Conclusions

Hemp crops are constantly growing in size, due to their multiple benefits for various industries. Therefore, the development of a technical equipment that capitalizes these crops and is accessible for farmers is very important.

By developing this equipment, we obtain a top machine for processing hemp stalks, a modern machine allowing the separation of a large amount of fibers from the mass of the stalks.

The equipment can be used with remarkable results when processing hemp stalks from small or medium-sized farms. The equipment can be transported on specific means of transport to the places where the processing of hemp stalks is required.

Acknowledgement

This work was supported by a grant of the Romanian Education and Research Ministry, on NUCLEU Program, Project "Foundation of the technology for harvesting and primary processing of hemp stalks in green", code PN 19 10 01 03, ctr. 5N/2019, and a grant on Programme 1 – Development of the national research-development system, subprogramme 1.2 – Institutional performance – Projects for financing excellence in RDI, contract no. 16PFE.

References

1. Devi B. - All about the hemp plant: 10,000 years of history, Hemp, Inc. – Stock Trading Symbol OTC HEMP, 2015;
2. Idler C., Pecenka R., Fürll C., Gusovius H.J. - Wet Processing of Hemp: An Overview, Journal of Natural Fibers, Volume 8, 2011 - Issue 2, pp. 59-80;
3. Jinke X. - Analysis and Design of Hemp Fibre Decorticators, A Thesis submitted to the Faculty of Graduate Studies of The University of Manitoba in partial fulfilment of the requirements of the degree of Master of Science, 2010;
4. Clarke R.C. - Traditional Fiber Hemp (*Cannabis*) Production, "Processing, Yarn Making, and Weaving Strategies—Functional Constraints and Regional Responses. Part 1", *Journal of Natural Fibers*, Volume 7- Issue 2, 2010, pp. 118-153;
5. Clarke R.C. - Traditional Fiber Hemp (*Cannabis*) Production, "Processing, Yarn Making, and Weaving Strategies—Functional Constraints and Regional Responses. Part 2", *Journal of Natural Fibers*, Volume 7- Issue 3, 2010, pp. 229-250;
6. Small agriculture encyclopaedia, Encyclopaedical Publishing House, Bucharest, 1988;
7. Sankari H. - Towards best fibre production in Finland: stem and fibre yields and mechanical fibre properties of selected fibre hemp and linseed genotypes; Academic Dissertation, Faculty of Agriculture and Forestry of the University of Helsinki, 24 March, 2000;
8. International Publication Number WO 2006/094359 A1 Title: Apparatus and method for processing of plant material;
9. Leduc P.J., Hill L.G., Kelly D.H., Stratton M.A. - Method for decorticating plant material, United States Patent, Patent Number: 5,720,083/Feb. 24, 1998;
10. Chen L.H., Verdes R.P. - Decorticating machine with variable speed feed and beater rollers, United States Patent nr.5465464/16. Nov.1995;
11. Patterson D.E - Machine for decorticating fiber-bearing stalks and leaves, United States Patent nr. 2480602/30 august 1948;
12. Scheleceen G. W. - Means for treating fiber-bearing plans, United States Patent nr. 1,308,376/1, July 1919;
13. Formation Ag, Decortication equipment, Updated January 8th, 2019;
14. Munder F., Furll C., Effective Processing of Bast Fiber Plants and Mechanical Properties of the Fibers, An ASAE/CSAE Meeting Presentation, Canada, 2004;
15. Riddlestone S., Stott E., Blackburn K., Brighton J. - A Technical and Economic Feasibility Study of Green Decortication of Hemp Fibre for Textile Uses, Journal of Industrial Hemp 11(2), 2006, pp. 25-55;
16. Baker M.I., Chen Y., Lague C., Landry C., Peng. Q., Zhong W., Wang J. - Hemp fibre decortications using a planetaryball mill, Canadian Biosystems Engineering. Volume 52, 2010, pp. 7-15;
17. Kick, F.M., The law of proportional resistors and its application felix. Leipzig, Germany. 1885;
18. Bond, F.C. The third theory of comminution. Trans. AIME, vol. 193, 1952. pp. 484–494;
19. von Rittinger P.R. – Lehrbuch der aufbereitungskunde in ihrer neunten entwicklung und ausbildung systematisch dargestellt, Ernst und Korn Publishing House, 1867;
20. https://www.futura-sciences.com/maison/dossiers/isolation-isolation-naturelle-solution-plein-essor-906/page/10/;
21. https://ro.wikipedia.org/wiki/C%C3%A2nep%C4%83_indian%C4%83.

Research on the use of cylindrical cutting drums with blades in cascade for Miscanthus harvesting

Anişoara Păun[1,*], *Gheorghe* Stroescu[1], *Radu* Popa[1], *Vlad* Popa, and *Constantin* Bogdanof[1]
[1]National Institute for Research - Development of Machines and Installations designed for Agriculture and Food Industry - INMA, Bucharest 013813, Romania

Abstract. In the current context of climate change, a special place is occupied by the field of energy plants. One of the renewable sources providing the biomass necessary for the generation of bioenergy, widely used in the last decades in the EU countries is the Miscanthus culture, which by capitalisation provides a series of economic benefits, but especially in terms of environment. This can be used for the production of electricity and/or thermal energy both in large power plants (30 MW +), which use thousands of tons of biomass annually, and in small domestic systems that use several tons, during the winter months. By using the Miscanthus in the combustion process, the carbon dioxide emissions are reduced and the methane emissions from the deposits are eliminated. In the case of applying the technology for Miscanthus culture direct harvesting (culture of perennial plants, over 2 m high, and high density), the technological process consists of cutting, chopping and loading in the means of transport of the harvested material. The paper presents the technological process of direct harvesting and analysis of the results obtained, achieved with the help of the trailed forage harvester CTF.

1 Introduction

In order to reduce the global warming due to the greenhouse gas emissions concentrated in the terrestrial atmosphere, a lower use of energy, especially the one derived from fossil fuels, is intended. The rapid reduction of energy resources derived from fossil fuels (coal, oil, natural gas) in parallel with the continuous increase of energy needs, requires increasing research efforts for the development of new sources of renewable energy [1].

Renewable energy is a natural renewable source, which can be produced, grown, regenerated at a comparable or faster rate than the rate of human consumption, derived from a wide range of resources such as: wind energy, solar energy, geothermal energy or biomass. Due to its variety, biomass is considered an energy source with a high degree of availability, and due to the cyclicality of the production and conversion processes it is considered a renewable source and with a positive impact on the environment. Biomass can also be produced by the special cultivation of plants for the generation of electricity, plants which are also called “energy plants”. Energy plants can be grown on agricultural land that

* Corresponding author: ani_paun@yahoo.com

is not used in agriculture (usually land that is removed from the circuit for various reasons or land considered unfit for growing food plants). Compared to traditional agricultural plants, energy plants require less care and less mineral fertilizers and pesticides. One of the renewable sources providing the biomass necessary for the generation of bioenergy, widely used in the last decades in the EU countries is the Miscanthus culture [2], which by capitalization provides a series of economic benefits, but especially in terms of environment. For capitalising Miscanthus plants, they must be harvested. Miscanthus plants are harvested starting with the third year of vegetation, in February when the stalks reached an optimum humidity of 11 ... 12%.

Depending on the requirements of using the harvested material, there are two technological variants of harvesting [3, 4, 5], namely:

1. The technology of harvesting directly from the field with the forage harvester (trailed or self-propelled), which in the working process realizes the chopping and loading of the chopped mass in the means of transport, if the transport to the place of processing takes little time. The method, similar to that used for corn harvesting, is simpler and faster.

2. The divided harvesting technology, using the rotary mower for cutting and laying stalks in continuous furrow on the ground and baling them with balers in large-size parallelepipedic bales if long-term storage until processing or transport over long distances are intended. The method has the disadvantage that it is relatively slow and requires lower expenses.

In the case of applying the technology for Miscanthus culture (culture of perennial plants with a height of over 2 m, and high density) direct harvesting, the technological process consists of cutting, chopping and loading in the means of transport of the harvested material. The technological process is realized efficiently if the forage harvester is equipped with equipment for harvesting tall plants, that can realize the complete cutting of the plants (across the entire working width). Because the Miscanthus stalks, harvested with the combine, are used as fuel in the burners of the thermal power plants, without being subjected to further processing, it is recommended that the chopped matter size be 10 ... 15 mm.

Within the technological process of the combine, the most important working part is the chopping drum, which cuts the fodder according to dimensions required and throws them into the evacuation system in the combine due to peripheral speed and air current created by it.

The main objective of the experimental research was to verify the technological efficiency of the chopping drums with straight knives placed in a cascade in the process of harvesting the Miscanthus stems.

2 Theoretical research

The technological process of direct harvesting can also be carried out with the help of the trailed Miscanthus stalk harvester - CRM1.6, figure 1, assimilated at SC MECANICA CEAHLĂU SA Piatra Neamt for which INMA has designed and made the specialized equipment for harvesting Miscanthus stalks, called EPI.

Fig. 1. Trailed Miscanthus stalk harvester - CRM1.6

The harvester works in aggregate with 65-100 HP tractors. The actuation of the working parts is done from the tractor power take-off (PTO) through a cardanic transmission.

The trailed Miscanthus stalk harvester -CRM1.6- performs, during the working process, the cutting of Miscanthus plants in the field and directs them towards the harvester working parts which perform the chopping and loading of the chopped mass in a means of transport.

It works in aggregate with 65-125 HP wheeled tractors equipped with hydraulic lifters and has two chopping sections. The trailed Miscanthus stalk harvester - CRM1.6 - (fig. 1), is composed of a rolling frame on which is mounted Miscanthus harvesting equipment, cylindrical chopping drum with small straight blades cascading, outlet pipe, transmission for operating the working parts, a device for coupling the equipment to the harvester, hydraulic installation.

The equipment for harvesting Miscanthus consists of the following subassemblies: frame; transmission sections; assembled blade; right assembled rotor; left assembled rotor; left, right, plant deflector; left, right field separator; central splitter; main transmission. The equipment is provided with a system for coupling to the harvester, which is common with that of the other pieces of equipment and the controls for driving the working parts of the harvester with the working equipment (coupling and decoupling the movement of the working parts) are operated by the tractor driver, from the tractor cab, by coupling or decoupling the PTO.

Within the technological process of the harvester, the most important working part is the chopping drum, figure 2, [6, 7] which performs the fragmentation of the fodder to the required dimensions and throws it to the system for discharging from the harvester, due to the peripheral speed and the air flow created by it.

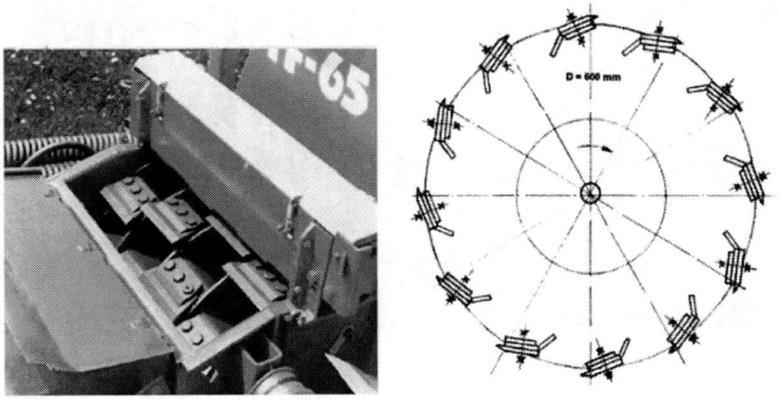

Fig. 2. Cylindrical chopping drum with small straight blades cascading

The power transmitted through the PTO, P_p, is calculated with the relation:

$$P_p = M_p \cdot \omega_{p} \cdot 10^{-3} \text{ [kW]} \qquad (1)$$

where:

- M_p is the torque measured from PTO shaft, in N·m;
- $\omega_p = \frac{\pi \cdot n_p}{30}$ - PTO shaft angular velocity rad/s;
- n_p – power take-off speed, in rpm.

The power required to tow the harvester on the horizontal field, P_t, is calculated with the relation:

$$P_t = F_t \cdot V_l \cdot 10^{-3} \quad [kW] \qquad (2)$$

where:

- F_t is the traction force measured at the coupling bar in N.
- V_l - the working (travel) speed of the aggregate in m/s

The total power P_{tot}, for towing and operating the working parts of the harvester is given by the relation:

$$P_{tot} = P_p + P_t = \left(M_p \cdot \frac{\pi \cdot n}{30} + F_t \cdot V_l \right) \cdot 10^{-3} \quad [kW] \qquad (3)$$

3 Results and discussion

The trailed Miscanthus stalk harvester was tested in two locations: INMA Bucharest experimental lot cultivated with Miscanthus where the Harvester worked in aggregate with the NEW HOLLAND TD80D tractor (figure 3a) and in Copsa Mica (figure 3b) with 120 HP tractor.

a

b

Fig. 3. Aggregate tractor - trailed Miscanthus stalk harvester, during operation

The main technical characteristics of trailed combine for harvesting strains of Miscanthus - CRM1,6:

Trailed combination type; coupling on the lateral straps of the coupling mechanism	
Power drive with , rpm	540
Type of cylindrical chopping drum with cascade knives	
Chopping drum speed, rpm	890
Chopping length, mm	10… 25 (6 steps)
Diameter of chopping drum, mm	600
Number of knives on the drum, pcs	4x10
Material discharge height, m	3.5
Tractor for drive,	CP 65 ... 125
Miscanthus harvesting equipment	
Constructive working width, m	1.6
Number of tooth rotors	2
Type of cutting machine,	with rotary disc
Effective working capacity at harvest, t / h	10
Productivity ha / hour	0.8

The working conditions and the indices realized in the test of the TRAILED MISCANTHUS STALK HARVESTER - CRM1.6 are presented in tables 1, 2 and 3.

Table 1. Working conditions

Den. No.	Specification	U M	Values
1	Harvested culture	-	Dry Miscanthus stalks
2	Plant mass production	t/ha	7.85
3	Vegetation year	-	two
4	Vegetation stage	-	Dry stalks
5	Field aspect	-	uniform without lying plants
6	Field height (average)	mm	1600 – 2800 (2000)
7	Average thickness of the plants at the cutting height	mm	8.80
8	Number of plants per m^2.	pc.	25
9	Humidity of plants when harvested	%	10.7

Table 2. Qualitative working indices

Den. No.	Specification	U M	Values			
			Test I	Test II	Test III	Average
1	Finger rotor speed	rpm	46	46.5	45.4	46
2	Blades speed	rpm	288.4	288.6	288.5	288.5
3	Working width	m	1.6	1.6	1.6	1.6
4	Plant cutting height	mm	120	130	145	120 – 145
5	Loss of material	%	≤ 5.2	≤ 5.5	≤ 5.3	≤ 5.3
6	The range of length Mscanthus	mm				
	0-15	%	22,7	24,9	23,5	23,7
	15,1-30	%	45,3	46,2	44,4	45,3
	30,1-50	%	12,3	8,6	12,2	11,0
	50,1-70	%	10,3	10,9	11,1	10,8
	70,1 -90	%	7,6	6,9	7,1	7,2
	90,1-100	%	1,2	1,9	1,1	1,4
	>100	%	0,6	0,6	0,6	0,6
	Total	%	100	100	100	100

Table 3. Operating indices

Den. No.	Size determined	U.M.	Values			
			Test I	Test II	Test III	Average
1	Working speed	km/h	5.8	6.1	6.3	6.1
2	Total power to drive the harvester in operation (without trailer coupled) (p_{tot})	kW	39.60	39.70	39.9	39.70
3	Effective working capacity (w_{ef}) (without towing the trailer)	t/h	7.6	7.9	8.1	7.9
4	Effective hourly working capacity	ha/h	0.75	0.77	0.81	0.8
5	Fuel consumption	l/ha	19.1	18.8	19.1	19

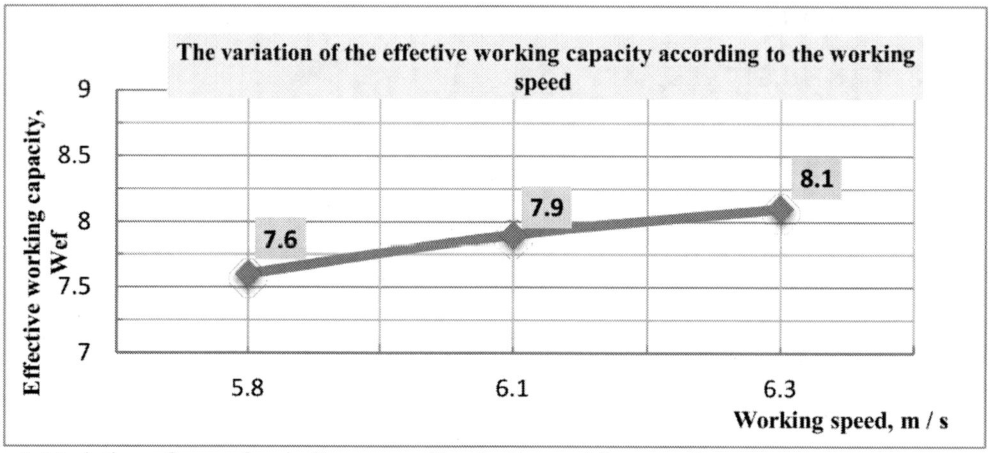

Fig. 4. Variation of operating indices according to the working speed

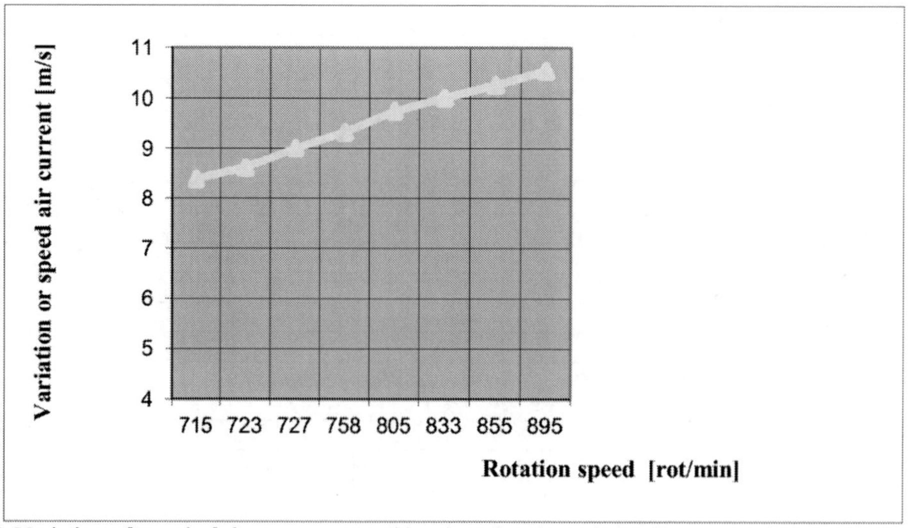

Fig. 5. Variation of speed of air current created by chopping drum depending on rotation speed

Analyzing the data in table no. 2, the following results were obtained:

-The effective working width was 1.6 m representing approx. 100% of the constructive working width;

-The cutting height was 120 - 145 mm (according to the adjustment made) and it was uniform across the entire working width.

-The chopping degree was determined for the chopped mass collected in the trailer.

The density of the chopped material, found in the collecting trailer was 78.65 kg/m3, the chopping degree of the chopped plants had values between 12 - 100 mm, and the particles with the length up to 50 mm represented approx. 80%, with a weight of 45% of the particles with the lengths between 15 - 30 mm.

Regarding the degree of chopping of plants, if we analyze the work [6] where the working process of the cylindrical chopping drum with small straight knives placed in cascade for harvesting corn for silo is studied: the chopping drum with cascade knives has the lowest power consumption, the chopping degree is comparable to the one for the Miscanthus culture with the difference that the effective working capacity is lower in the case of the Miscanthus.The effective working capacity is lower in the case of Miscanthus due to the different characteristics of the cultures approached. Also the distance of horizontal throwing of the chopped drum material with cascading knives (fig.2) is greater by approx. 1.5 m from the other drums analyzed in the paper [6]. During measurements, the Miscanthus harvester moved at a speed of 6.3 km/h.

The following conclusions regarding the operating indices were obtained from the analysis of the measurements:

- The actual torque at the tractor PTO required to operate the unloaded working bodies was determined for a tractor engine speed of 2200 rpm and a power take-off speed of 540 rpm resulting in a value of 7.85 daN m;
- During operation, the actual torque at the tractor PTO was determined under the same conditions having a value of 42.4 daN m;
- The traction force at the coupling bar of the tractor (F_t) was 210 daN if the trailer in which the harvested material was loaded moved parallel to the aggregate and 650 daN if the trailer was coupled in the back of the harvester and loaded with 600 kg;
- The pressure force of tow hitch on the tractor (F_{ap}) had a value of 310 daN in idling state and 315 daN loaded, resulting that the way of coupling the harvester to the tractor is appropriate and that the small variations that occur are due to the state of the land;
- The effective power at the PTO (P_p) was 5.98 kW idling and 28.4 kW loaded;
- The effective power for towing the harvester (P_p) was 3.9 kW if the trailer in which the harvested material was loaded moved parallel to the aggregate and 11.2 kW if the trailer was coupled in the back of the harvester;
- The total power for driving the operating harvester was determined without trailer coupled to the harvester (p_{tot}) having a value of 39.70 kW and achieving an effective working capacity (w_{ef}) of 7.9 t/h with a specific energy consumption for carrying out technological process (p_{tot}/w_{ef}) of 18 kW s/kg.

4 Conclusions

After the researches performed and results obtained, the following conclusions can be formulated:

- The cutting height was uniform and did not show large variations depending on the working speed of the tractor - trailed harvester aggregate;

- The values of the working capacity in exchange vary according to the working speed of the tractor aggregate. Thus for the average speed of 5.8 km/h the average hourly working capacity has an average value of 7.6 t/h, and for the average speed of 6.3 km/h it has an average value of 7.9 t/h.

- The optimum working speed of the aggregate is 5.8 km/h and is imposed by the real field conditions. If the terrain is well levelled the working speed can reach 6.3 km/h.
- The trailed Miscanthus stalk harvester worked properly with the working parts clogging rarely, and the transport of the plants to the feeding and chopping system of the harvester was carried out continuously, without clogging.

Acknowledgements

This paper was financed by grant of the Romanian Education and Research Ministry, through Programme 1 – Development of the national research-development system, sub-programme 1.2 – Institutional performance – Projects for financing excellence in RDI, contract no. 16PFE.

- A grant of the Romanian Research and Innovation Ministry, PN 19 10 01 03-Substantion of the technology for harvesting and primary processing green hemp stalks, contract no. 5N/07.02.2019.

References

1. A Păun, V. Vladuţ, E. Marin, D. Manea, G. Lazar, I. Gageanu, *Research people and actual tasks on multidisciplinary sciences*, Lozenec, Bulgaria (2015)
2. Clifton-Brown, J.C., S.P. Long, and U. Jørgensen. 2001. Miscanthus productivity. p. 46–67. In M.B. Jones and M. Walsh (ed.) Miscanthus—for energy and fibre. James and James (Science Publishers), London (in press).
3. I. Dănilă , V.Neculăiasa , *Harvesting agricultural machines*, IP "Traian Vuia" Timişoara, (1987)
4. Ioan Hermeneanu, Vasile Mocanu – *Technologies, machines and installations for hay harvesting and preservation of fodder on grasslands*, University of Brasov Publishing House (2008)
5. Rodney Horrocks, John Valentine, *Harvested Forages*, Academic Press, Brigham Young University, Provo, Utah, U.S.A. (1999)
6. A. Păun, A. Nedelcu, V. Neagoe, *Symposium - INMATEH Agricultural Engineering*, **38**, 33-38 (2012)
7. Voicu Emil–*The dynamics and energy of the tractor-trailed combine aggregate for fodder harvesting,* TERRA NOSTRA Publishing House, Iasi (2009)

E3S Web of Conferences **180**, 03007 (2020) https://doi.org/10.1051/e3sconf/202018003007
TE-RE-RD 2020

Desorption of the Exotic Fruit Jujube (*Ziziphus Jujube Mill*) of Bulgarian Origin

Albena Durakova[1,*], *Adelina* Bogoeva[2], *Velichka* Yanakieva[3], *Tzvetana* Gogova[4], and *Kornelia* Choroleeva[5]

[1*]Department of Process Engineering, University of Food Technologies – Plovdiv, 26 Maritza Blvd. BG-4002 Plovdiv, Bulgaria
[2]Department of Mechanical and Instrument Engineering, Technical University - Sofia, Plovdiv Branch, 25 Tsanko Duistabanov Str., Plovdiv 4000, Bulgaria
[3]Department of Microbiology, University of Food Technologies – Plovdiv, 26 Maritza Blvd. BG-4002 Plovdiv, Bulgaria
[4]Department of Technology of Grain, Fodder, Bread and Confectionery Products, University of Food Technologies – Plovdiv, 26 Maritza Blvd. BG-4002 Plovdiv, Bulgaria
[5]Kornelia Choroleeva - Language Training Centre, University of Food Technologies – Plovdiv, 26 Maritza Blvd. BG-4002 Plovdiv, Bulgaria

Abstract. The present study determined the physico-chemical parameters of the exotic fruit jujube (*Ziziphus jujube mill*) of Bulgarian origin – pH of 4.62, refractometric dry matter of 31.6 %, dry matter (in weight) of 36.78 %, TTA (total titratable acidity) – citric acid – of 0.67 %, 0.70 % malic acid, 65.95 % moisture, 0.78 % fibre, 1.41 % protein, 23.65 % carbohydrate, 0.24 % fat and energy value of 428.44 kJ/102.4 kcal. Equilibrium moisture content and monolayer moisture content were established for desorption concerning eight water activities in the 0.1 - 0.9 range and the temperatures of 10 °C, 25 °C and 40 °C. The coefficients of the modified Oswin, Chung-Pfost, Halsey and Henderson models were obtained statistically. The modified Halsey model is recommended as a description of the desorption isotherms of jujube (*Ziziphus jujube mill*) since it corresponds to the criteria for model assessment and suitability. Via BET model linearization, the values of the monolayer moisture content for desorption at the temperatures of 10 °C - 5.41 % d.b., 25 °C – 4.65 % d.b. and 40 °C - 5.08 % d.b were determined.

Key words: jujube (*Ziziphus jujube mill*), physico-chemical parameters, sorption characteristics

* Corresponding author: aldurakova@abv.bg

1. Introduction

Contemporary medicinal practices apply various methods and techniques whose efficiency was recognized and tested in China and other Asian countries thousands of years ago. The world has long acknowledged the extraordinary refreshing effects of a great number of fruits, vegetables, and herbs, including grapes, apricots, carrots, goji berry, ginkgo biloba, ginseng, pumpkin, thyme, St. John's wort, and many others [1-4].

According to Jinous Asgarpanah and Elaheh Haghighat, among the first five healthiest medicinal plants in the world we find the jujube (Chinese date) which is less common in Bulgaria although the Chinese started growing and using the plant more than 6000 years ago. There are more than 50 varieties of jujube but *Ziziphus jujube Mill* is the most popular and useful one [5].

Jujube *(Zizyphus jujuba Mill)* is among the four staple agricultural species in China where it is grown on two-million-decare areas. In these areas, there is extremely low incidence of diseases and a specific resistance to the development of respiratory tract illnesses. This may be the reason why the local population refers to the fruit of the jujube plant as "chest fruit". Selective breeding has led to the growth of bigger jujube plants whose sugar content can be three times as high as that of the ordinary jujube.

It is an interesting fact that for medicinal and prophylactic effects almost all parts of the jujube plant can be used: roots, stem, bark, leaves and fruit. It is not surprising that the fruit of the jujube is most popular because it is very tasty. It has a sweet or a slightly sweet-sour taste of the light green pulp whose consistency resembles that of bread. The jujube fruit is extremely high in flavonoids, saponins, sugars, vitamin A, vitamins C and B2, minerals (calcium, iron, and phosphorus) [6]. The vitamin C content of the jujube plant is between 8 - 18 times as high as that of tomatoes and citrus fruit and about as 200 times as high as that of apples, peaches and pears. The amount of vitamin C is extremely high – between 10 - 20 times (reaching 1700 mg/100 g) as high as that of the fruits boasting with their vitamin C content, i.e. fuzzy kiwifruit, lemons, oranges and blackcurrant. The jujube contains great amounts of P-active compounds (305-1230 mg/100 g), proteins, malic acid, pectin, tannins and rutin [7]. It is rich in carbohydrate whose composition, depending on jujube variety, ranges between 20% - 40% (over 60% in dried fruit). Such vitamin-active substances are also represented by a group of bioflavonoids (rutin, citrin, catechins, etc.) which have a positive multi-level effect on the human organism: resistance to free radicals (because of high antioxidant activity), improvement of blood cell elasticity, anti-inflammatory effect, blood pressure regulation, maintenance of the active form of vitamin C, etc. [8, 9]. Therefore, the jujube plant is a valuable source of phytochemicals, bioactive substances, vitamins, etc., which makes it useful in the treatment and prophylaxis of various diseases [10].

The literature review did not show the existence of data on the physico-chemical parameters and desorption characteristics of Bulgaria-grown jujube *(Zizyphus jujuba Mill)*, which helped identify the following objectives of the present study: analysis of the physico-chemical parameters and desorption characteristics of the exotic plant jujube *(Ziziphus jujube mill)* of Bulgarian origin.

2. Materials and Methods

2.1. Materials

The object of the present study is jujube *(Zizyphus jujuba Mill)* of Bulgarian origin supplied by an experimental facility in the town of Assenovgrad, Bulgaria.

2.2. Methods

pH – Bulgarian State Standard (BSS) 11688:1993 [11];

Dry matter, refractometric (Re) - BSS EN 12143:2000 [12],

Dry matter, in weight - BSS EN 12145:2000 [13];

TTA (total titratable acidity) BSS 6996:93 [14];

- Citric acid
- Malic acid

Fibre - BSS 11374:1986 [15];

Fat content, (%) – Soxhlet method via solvent extraction with petroleum ether. BSS 6997:1984 [16];

Carbohydrate - BSS 7169:1989 [17];

Protein – Direct Kjeldahl method analysis (determination of nitrogen content/ nitrogen determination method) – Regulation (EC) №152/2009 [18];

Energy value - Regulation 1169:2011, Appendices IVX ; XV [19].

Moisture, (%) of the products studied - standard method via drying of 5g of flour at 105°C to constant weight, according to AOAC, 1990 [20].

Determination of microbiological parameters:

Determination of *Escherichia coli*, according to BSS EN ISO 16649-2:2014 [21];

Determination of *Salmonella* sp., according to BSS EN ISO 6579-1:2017 [22];

Determination of coagulase-positive staphylococci, according to BSS EN ISO 6888-1:2005 [23];

Determination of total number of mesophilic aerobic and facultative anaerobic bacteria, according to BSS EN ISO 4833-1:2014 [24];

Determination of mould and yeast, according to BSS EN ISO 21527-2:2011 [25].

In order to analyse sorption isotherms, we used the static gravimetric method recommended for food products by Project COST 90. Due to the high initial moisture content of jujube, the samples were not moistened in advance and, after deseeding, they were weighted as per 1 ± 0.02 g in weighing bottles. Weighing bottles are placed in hygrostats above saturated solutions of 8 salts (LiCl, $MgCl_2$, K_2CO_3, NaBr, CH_3COOK, NaCl, KCl, $MgNO_3$) maintaining media with constant water activities, from 0.1 to 0.9 [26]. For high water activities, ($a_w > 0.70$) in hygrostats thymol crystals are introduced to prevent against microbiological deterioration. Hygrostats are tempered in thermostats at temperatures of 10, 25 and 40 ± 0.2 °C. When equilibrium is reached (within 20 days), the moisture content of each sample is determined via a drying method (105 °C for 24 h) as a mean value of three iterations [20].

All analyses were conducted in triplicate runs.

The experimentally obtained sorption isotherms were described with the help of the modified three-parameter models recommended for products of plant origin in Standard 245D of the American Society of Agricultural Engineers (ASAE) [26]:

Modified Chung-Pfost
$$a_w = \exp\left[\frac{-A}{t+B}\exp(-CM)\right] \quad (1)$$

Modified Halsey
$$a_w = \exp\left[\frac{-\exp(A+Bt)}{M^C}\right] \quad (2)$$

Modified Oswin
$$M = (A+Bt)\left(\frac{a_w}{1-a_w}\right)^C \quad (3)$$

Modified Henderson $$1-a_w = \exp[(-A(t+B)M^C]$$ (4)

To determine model coefficients, we applied the least squares method using a non-linear regression program. Model suitability criteria were the values of mean relative error P [%] and standard error (SEM) [27, 28]:

$$P = \frac{100}{N}\sum\left|\frac{M_i - \hat{M}_i}{M_i}\right| \quad (5)$$

$$SEM = \sqrt{\frac{\sum(M_i - \hat{M}_i)^2}{df}} \quad (6)$$

$$e_i = M_i - \hat{M}_i \quad (7)$$

where M_i and $\hat{M}_i$ are the experimentally obtained equilibrium moisture content value as calculated by the model, N being the number of test points and d*f* being the degree of freedom (number of test points minus number of constant values in the model).

To calculate monolayer moisture content, we used the linear transform on the basis of the BET (Brunauer-Emmett-Teller) model [29].

BET

$$M = \frac{M_m C a_w}{(1-a_w)(1-a_w+Ca_w)} \quad (8)$$

where M_m is monolayer moisture content, % d.b.; C - constant value.

3. Results and Discussion

The experimentally obtained results concerning the physico-chemical parameters of the exotic plant jujube *(Ziziphus jujube mill)* are presented in Table 1. The analysis shows that the plant is rich in dry matter, which accounts for the bready consistency of the pulp [6]. The results also demonstrate the presence of citric acid and malic acid. The differences in the data found in scientific literature are due to the different origin of jujube having to do with different territorial, climatic and soil characteristics. Carbohydrate content, 23.56 %, is within the range of the values reported by scientific literature [7].

Table 1. Physico-chemical parameters of Jujube *(Ziziphus jujube mill)*

№	Physico-chemical parameter	Value for 100 g
1.	pH	4.62
2.	Dry matter, refractometric (Re)	31.6 %
3.	Dry matter, in weight	36.78 %
4.	TTA (total titratable acidity) *- as citric acid* *- as malic acid*	 0.67 % 0.70 %
5.	Fibre	0.78 %
6.	Fat	0.24 %
7.	Carbohydrate	23.65 %
8.	Protein	1.41 %
9.	Moisture	65.98 %
10.	Energy value	428.44 /102.4 kJ /kcal

On receiving the raw material from the experimental facility, we analysed its microbiological profile. We determined the following microbiological parameters:

total number of mesophilic aerobic and facultative anaerobic bacteria - 1.7 x 10^4, fungi - 1.5 x 10^3, lack of *Salmonella* sp., presence of *Escherichia coli*, coagulase-positive staphylococci and yeast - below the acceptable level. The results from the analysis demonstrate that the product is not microbiologically contaminated, which makes us believe that it is suitable for direct consumption.

The experimentally obtained values of the equilibrium moisture content of jujube *(Ziziphus jujube mill)* for desorption at the temperatures of 10 °C, 25 °C and 40 °C as well as the eight values of the relative humidity of air in the 0.1 - 0.9 range are presented in Table 2.

Table 2. Equilibrium moisture content *M, %* d.b. for different temperatures *t (°C)* and water activities a_w in desorption

Sel	10 °C			25 °C			40 °C		
	a_w	M*	sd**	a_w	M*	sd**	a_w	M*	sd**
LiCl	0.113	8.54	0.05	0.113	8.23	0.11	0.112	5.59	0.06
CH_3COOK	0.234	8.81	0.13	0.225	8.36	0.16	0.201	7.45	0.09
$MgCl_2$	0.335	8.87	0.28	0.328	8.84	0.14	0.316	7.96	0.12
K_2CO_3	0.431	10.73	0.08	0.432	9.34	0.09	0.432	9.08	0.09
$MgNO_3$	0.574	12.85	0.14	0.529	10.83	0.15	0.484	9.99	0.08
NaBr	0.622	13.77	0.13	0.576	11.05	0.14	0.532	10.12	0.15
NaCl	0.757	23.46	0.24	0.753	19.06	0.18	0.747	15.75	0.08
KCl	0.868	48.88	0.15	0.843	24.75	0.23	0.823	21.13	0.21

* Mean of three replications ** Standard deviation based on three replications

The results obtained prove a correlation typical of a number of food products studied, i.e. the increase in temperature at a constant water activity is coupled with a decrease in equilibrium moisture content [30, 31].

Fig. 1 presents graphically the desorption isotherm of jujube at 10 °C which makes it clear that, according to Brunauer's classification, it belongs to type III [32].

The statistical processing with the help of a non-linear regression program using the least squares method calculated the coefficients of the modified three-parameter models applied in the analysis (Table 3). We also calculated their values for mean relative error P, %, and standard error SEM.

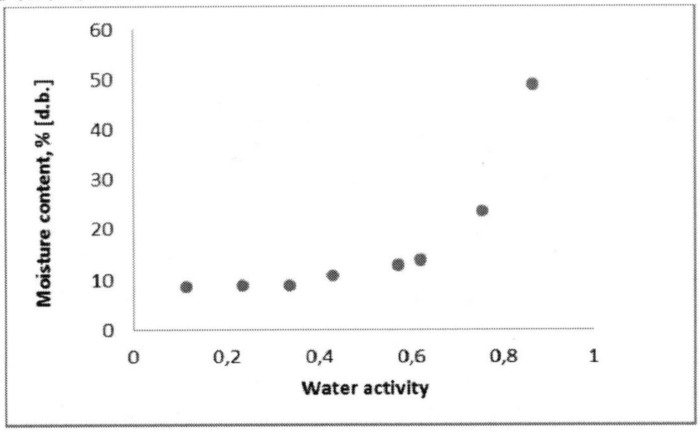

Fig. 1. Sorption isotherm of jujube at 10 °C

Table 3. Model Coefficients (*A, B, C*), mean relative error (P, %), and standard error (SEM) for desorption.

Model	*A*	*B*	*C*	*P*	*SEM*
Oswin	15,4365	-0.1726	0.5819	25.38	5.65
Halsey	**5.01127**	**-0.0123**	**2.0995**	**10.95**	**3.59**
Henderson	0.00015	7.4311	2.0327	22.85	6.18
Chung-Pfost	578.063	0.1820	76.949	16.58	7.00

The results obtained demonstrate that the lowest values of mean relative error (P, %) and standard error (SEM) were calculated for Modified Halsey. Taking into account model assessment and suitability criteria, this result allows us to recommend the three-parameter Halsey model as desorption isotherm description of jujube. The other three modified models have considerably higher values which makes them unsuitable for the desorption isotherm description of the product under study.

To calculate the monolayer moisture content (MMC) of Bulgarian jujube fruit for desorption, the BET model was linearly transformed with test data for $a_w < 0.5$, as shown in Fig. 2. On the basis of linear equation coefficients, we calculated MMC values, as presented in Table 4.

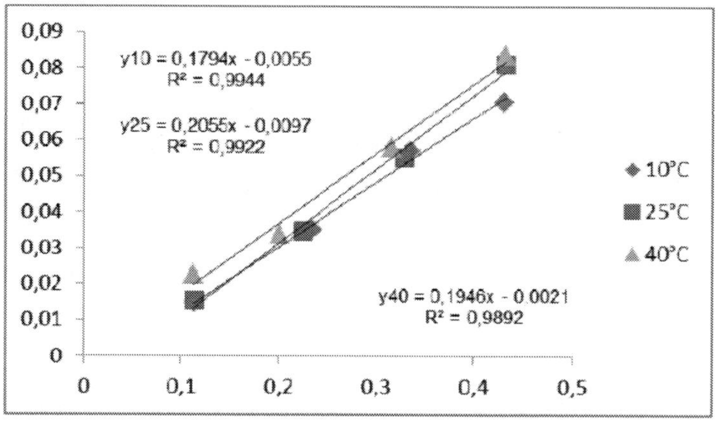

Fig. 2 BET model linearization for different temperatures in desorption

Table 4. BET monolayer moisture content, % d.b., at different temperatures t.

t (°C)	Desorption
10	5.41
25	4.65
40	5.08

On the basis of the obtained results, we can conclude that there is no direct correlation between the values of monolayer moisture content and temperature for the desorption of Bulgarian jujube *(Zizyphus jujuba Mill)*.

4. Conclusion

As a result of the present study, we obtained the equilibrium moisture content for desorption at 10 °C, 25 °C and 40 °C as well as the relative humidity in the 0.1 - 0.9 range concerning jujube (Zizyphus jujuba Mill) of Bulgarian origin. We established that temperature affects the sorption capacity of jujube (Zizyphus jujuba Mill). With the increase of the latter, at a constant water activity, the values of equilibrium moisture content are lowered. This study is the first to identify the equilibrium isotherm for desorption of the fruit of exotic Bulgarian jujube (Zizyphus jujuba Mill).

The research also determined the numerical values of the coefficients of the modified three-parameter models used, as a result of which it recommends Halsey's model as regards the description of the equilibrium isotherms of the product under study. Thus, the model can be used in engineering practice for the determination of equilibrium moisture content.

Via the linearization of the Brunauer-Emmett-Teller model, the present study also calculated the monolayer moisture content (MMC) values of the product. The moisture value corresponding to MMC is a sorption characteristic which has a positive effect on the stability of the product. The MMC values obtained are a good foundation for subsequent detailed studies on the storage regimes and shelf life of Bulgarian jujube (Zizyphus jujuba Mill).

This study was part of the Fund Scientific Research "FS N11/20-H", director project: Assoc. Prof. PhD Albena Georgieva Durakova, University of Food Technologies – Plovdiv, Bulgaria.

5. References

1. I. Dimov, ICTTE (Yambol, Bulgaria), **I**, ISSN 1314-9474, 479-485, (2013)
2. I. Dimov, A. Krasteva. Scien. W. of UFT, **LX**, ISSN 1314-7102, 151-155, (2013)
3. Z. Ghouila, S. Laurent, S. Boutry, L.Vander Elst, F. Nateche, R. N. Muller, A. Baaliouamer, Fundament Appl Sci., **9** (1), 392-410, (2017).
4. Y. Choi, J. Lee, Food Chem. **114** (4), 1386-1390, (2008).
5. J.Asgarpanah, E. Haghighat. Afr J Pharm Pharmacol, **6** (31), 2332-2339, (2012).
6. A. M. Pawlowska, F. Camangi, A. Bader, A. Braca, Food Chem., **112** 858–862, (2009)
7. J. Li, L. Shana, Y. Liua, L. Fana, L. Aib. Int. J. Biol. Macromol., **49**, 255– 259 (2011)
8. S. M. Al-Reza, J. I. Yoon, H. J. Kim, J-S. Kim, S. C. Kang, Food Chem Toxicol, **48**, 639–643, (2010)
9. H. Zhang, L. Jiang, S. Ye, Y. Ye, F. Ren, Food Chem Toxicol, **48**, 1461–1465, (2010
10. M. A. Beg, U.V.S. Teotia, S, Farooq, J. Med. Plants, **4** (5), 230-233, (2016)
11. B.S. Standard, pH, in BSS 11688:1993
12. B.S. Standard, Dry matter, refractometric, in (BSS) EN ISO 12143:2000
13. B.S. Standard, Dry matter, in weight - in (BSS) EN ISO 12145:2000
14. B.S. Standard, TTA (total titratable acidity), in BSS 6996:93
15. B.S. Standard, Fibre, in (BSS) EN ISO 11374:1986
16. B.S. Standard, Fat content, in (BSS) EN ISO 6997:1984
17. B.S. Standard, Carbohydrates in (BSS) EN ISO 7169:1989
18. B.S. Standard, Proteins, in Reglement (EO) №152/2009
19. Standard in Reglement 1169:2011. App. **IVX, XV**, (2011)
20. AOAC, 960.39 (Washington , DC), **15** ed. AOAC. (1990)
21. B.S. Standard, *Escherichia coli*, in (BSS) EN ISO 16649-2:2014
22. B.S. Standard, *Salmonella* spp., in (BSS) EN ISO 6579-1:2017
23. B.S. Standard, Coagulase-positive staphylococci, in (BSS) EN ISO 6888-1:2005,
24. B.S. Standard, Mesophilic aerobic and facultative anaerobic bacteria, in (BSS) EN ISO 4833-2:2014,
25. B.S. Standard, Yeasts and fungi, (BSS) EN ISO 21527-2:2011
26. ASAE. Standart 245D.5. St. Joseph, Michigan.
27. C.-C. Chen, R.V. Morey. Transactions of the ASAE, **32** (3), 983-0990, (1989)
28. C.-C. Chen, R.V. Morey. Transactions of the ASAE, **32**, 999-1006, (1989)
29. L. Bell, T. Labuza, AACC, 33-36, (2000)

30. A. Durakova, A. Bogoeva, R. Krasteva, T. Gogova, A. Krasteva and N. Georgieva, IOP Conf Ser Mater Sci Eng., **595**, 012058, (2019)
31. A. L. Bogoeva, A. G. Durakova. JAFR, **2**, 100026, (2020)
32. S. Brunauer, L. S. Deming, W. E. Diming, E. Troller, JACS, **62**, 1723-1732, (1940)

Sorption Characteristics of Bulgarian Penny Buns (*Boletus Edulis*)

Albena Durakova[1,*]

[1*] Department of Process Engineering, University of Food Technologies – Plovdiv, 26 Maritza Blvd. BG-4002 Plovdiv, Bulgaria

Abstract. The present study determined the sorption characteristics of Bulgarian penny buns (*Boletus*) for eight water activities in the 10 % - 90 % range and at three temperatures, of 10 °C, 25 °C and 40 °C. The coefficients of the modified Oswin, Chung-Pfost, Halsey and Henderson models were obtained statistically. The modified Halsey model is recommended as a description of adsorption and desorption isotherms since it corresponds to the criteria for model assessment and suitability. Via BET model linearization, the values of the monolayer moisture content for water activity of aw > 0.5 and temperatures of 10 °C, 25 °C and 40 °C were established. The approximate physico-chemical composition of Bulgarian penny buns (Boletus) – moisture content – 10.80 %, protein – 32.23 %, carbohydrate - 0.99 %, fat – 6.62 %, ashes – 6.20 % - was also determined.

Key words: penny buns (*Boletus*), physico-chemical parameters, sorption characteristics

[*] Corresponding author: aldurakova@abv.bg

1. Introduction

The wild mushroom penny bun (*Boletus Edulis*) grows in deciduous and coniferous woods and plains. It is among the most popular collected mushrooms in Bulgaria due to its excellent taste and high nutritive value. Penny buns are rich in carbohydrates, fats, minerals, vitamins, and fibre. They contain up to 35 % of essential proteins including glutamine, arginine, leucine and tyrosine. Penny buns are also high in PP vitamins, nicotinic acid, various macro-elements and trace elements. The consumption of penny buns supplies the body with B, C, E and D vitamins.

Penny buns are among the mushrooms with the greatest amount of amino acids, which makes them extremely suitable for vegetarian diets, as a substitute for meat. Vitamin-rich penny buns are an integral part of healthy diets. The interior of these mushrooms is easily assimilated by the gastrointestinal tract, absorbs toxins and expels them from the organism.

Traditionally, European cuisine regards penny buns as being among the most delicious wild-grown mushrooms. Due to its ability to accumulate selenium, the penny bun is a natural source of organic selenium, which makes it useful for the enrichment of various food products with micronutrients. The deficiency of selenium in the human organism can lead to many health problems [1-3]. Selenium is crucial for testosterone production and the formation and normal development of sperm cells (spermatozoa) [4]. In contrast to other edible plants, both wild-grown and cultivated mushrooms can assimilate and accumulate not only selenium but also other toxic components, such as mercury, arsenic, lead, cadmium, chrome, copper, iron, nickel, zinc, etc. [5]. Therefore, the dosage of the mushroom as an ingredient has to be very accurate since overdosages may be poisonous for human beings [6].

The literature review did not reveal data on the sorption characteristics of wild Bulgarian penny buns (*Boletus Edulis*), which motivated our research.

2. Materials and Methods

2.1. Materials

The present study focuses on wild Bulgarian penny buns (*Boletus Edulis*) from the Rhodope region of Byala Cherkva which were dried at atmospheric conditions and ground into powder.

The packaging – coextruded barrier film with copolymer covering for heat sealing, to be used in the food and flavour industry – was provided by the «ET – Ilko Tyanev Ita Plast» company.

2.2. Methods

Moisture, (%) of the products studied - standard method via drying of 5g of flour at 105 °C to constant weight, according to AOAC, 1990 [7];

Protein – Direct Kjeldahl method analysis (determination of nitrogen content/ nitrogen determination method) – Regulation (EC) №152/2009 [8];

Fat content, (%) – Soxhlet method via solvent extraction with petroleum ether. BSS 6997:1984 [9];

Carbohydrate - BSS 7169:1989 [10];

Ash, (%) – The method used for the determination of ash content in flour is according to ICC Standard №104/1 "Determination of Ash in Cereals and Cereal Products" [11];

Energy value - Regulation 1169:2011, Appendices IVX ; XV [12].

Determination of microbiological parameters:

Determination of *Escherichia coli*, according to BSS EN ISO 16649-2:2014 [13];

Determination of *Salmonella* sp., according to BSS EN ISO 6579-1:2017 [14];

Determination of coagulase-positive staphylococci, according to BSS EN ISO 6888-1:2005 [15];

Determination of total number of mesophilic aerobic and facultative anaerobic bacteria, according to BSS EN ISO 4833-1:2014 [16];

Determination of mould and yeast, according to BSS EN ISO 21527-2:2011 [17].

Grinding of experimental raw material – laboratory grinder Tefal GT110838;

Flour particle size – The determination of the size of flour particles via sieve analysis was conducted by means of "ProMel LP – 200" sieve shaker, according to BSS 754 – 80 "Milled Products. Regulations for Sampling and Testing Methods". On the basis of the preliminary analysis, the set of sieves was determined, as well as their size. A sample of 100 g was sieved by the apparatus for ten minutes.

The sorption isotherms were studied with the help of the static gravimetric method recommended for food products by Project COST 90. In order to assess the adsorption and desorption capacity of the product, the latter was subjected to preliminary forced drying and moistening. These processes took place in desiccators - above distilled water to achieve a maximally saturated moist medium and above P_2O_5 to achieve a maximally dry environment, respectively. The samples remained in the desiccators for 10 days. After the determination of the moisture content obtained via the standard drying method, they were weighted as per 1±0.02 g in weighing bottles. The weighing bottles were placed in hygrostats above saturated solutions of 8 salts (LiCl, $MgCl_2$, K_2CO_3, NaBr, CH_3COOK, NaCl, KCl, $MgNO_3$) maintaining constant water activities above the surface, in the 10% - 90 % range [18]. For high water activities, ($a_w > 0.70$) in hygrostats thymol crystals were introduced to prevent against microbiological deterioration. The hygrostats were tempered in thermostats at temperatures of 10 °C, 25 °C and 40 °C ± 0.2 °C. When equilibrium was reached within 20 days, the moisture content of each sample was determined via a drying method (105 °C for 24 h) as a mean value of three iterations [7].

All analyses were conducted in triplicate runs.

The experimentally obtained sorption isotherms were described with the help of the modified models recommended for products of plant origin in Standard 245D of the American Society of Agricultural Engineers (ASAE) [18]:

Modified Chung-Pfost
$$a_w = \exp\left[\frac{-A}{t+B}\exp(-CM)\right] \tag{1}$$

Modified Halsey
$$a_w = \exp\left[\frac{-\exp(A+Bt)}{M^C}\right] \tag{2}$$

Modified Oswin
$$M = (A+Bt)\left(\frac{a_w}{1-a_w}\right)^C \tag{3}$$

Modified Henderson
$$1-a_w = \exp[(-A(t+B)M^C] \tag{4}$$

Model suitability criteria were the values of mean relative error P [%] and standard error (SEM) [19, 20]:

$$P = \frac{100}{N}\sum\left|\frac{M_i - \hat{M}_i}{M_i}\right| \tag{5}$$

$$SEM = \sqrt{\frac{\sum (M_i - \hat{M}_i)^2}{df}} \quad (6)$$

$$e_i = M_i - \hat{M}_i \quad (7)$$

where M_i and $\widehat{M}_i$ are the experimentally obtained equilibrium moisture content value as calculated by the model, N being the number of test points and *df* being the degree of freedom (number of test points minus number of constant values in the model).

To calculate monolayer moisture content, we used the linear transform on the basis of the BET (Brunauer-Emmett-Teller) model [21].

BET

$$M = \frac{M_m C a_w}{(1 - a_w)(1 - a_w + C a_w)} \quad (8)$$

where M_m is monolayer moisture content, % d.b.; C - constant value.

3. Results and Discussion

Prior to subjecting the mushrooms to laboratory analyses, they were dried at atmospheric conditions and vacuum-packaged in packaging from coextruded barrier film with copolymer covering for heat sealing used in the food and flavour industry.

Following the acceptance of the raw material for analysis, the latter was ground by a laboratory grinder Tefal GT110838 to achieve powdery texture and particle size determined by sieve analysis. The sieve analysis determining flour particle size was conducted by means of "ProMel LP - 200" sieve shaker. On the basis of the preliminary sieve analysis of the initial sample, the set of sieves was established as well as the tolerance in mesh size. This allowed for the formation of fractions with specific particle size. Particle size is an important characteristic and a major factor in choosing storage conditions for powdery products [22, 23]. The granulometric composition of the products gives information concerning the distribution of particles in percentages with respect to total mass. The results obtained from the analysis are shown in Table 1

Table 1. Granulometric composition of flour from Bulgarian Penny Buns (*Boletus Edulis*)

Particle Size	Amount of Particles, %
1000/600	3.4
600/400	5.6
400/200	27.9
200/	63.1

On the basis of the analyses and the results obtained, we can point out that, for the most part, the fractions had particle sizes between 400 µm and 200 µm, as well as below 200 µm. The percentage of the fractions below 200 µm was most significant – 63.1 %. The data show that the product is extremely fine, which makes it suitable as an enriching ingredient of various food products.

The physico-chemical characteristics and the microbial contamination are important parameters for every food product because they are informative for both the manufacturer and the consumer. The physico-chemical analysis of the flour from Bulgarian penny buns (*Boletus Edulis*) is presented in Table 2.

Table 2. Physico-chemical parameters of flour from Bulgarian Penny Buns (*Boletus Edulis*).

Moisture content,%	Protein,%	Fat,%	Carbohydrate,%	Ash,%	Energy Value, Kcal/ KJ
10.8	32.23	6.62	0.99	6.20	192.46 / 805.25

The results show a great amount of protein and fat, typical of the analysed product, which undoubtedly makes it extremely useful for vegan diets.

The analyses of the microbiological parameters provide the following data: total number of mesophilic aerobic and facultative anaerobic bacteria - 2.0 x 10^4; fungi - 6.0 x 10^2; lack of *Salmonella* sp., presence of *Escherichia coli* - <10; coagulase-positive staphylococci - <100 and yeast - <10, i.e. all parameters studied are within acceptable levels. The results from the analysis proved that the product was not microbially contaminated, which allows us to believe that it was suitable for consumption.

The sorption isotherms and equilibrium moisture content of the flour from Bulgarian penny buns (*Boletus Edulis*) for adsorption and desorption are presented in Fig. 1 and Fig. 2. The experiment was conducted at three temperatures – 10 °C, 25 °C and 40 °C and for eight water activities in the 0.10 - 0.90 range.

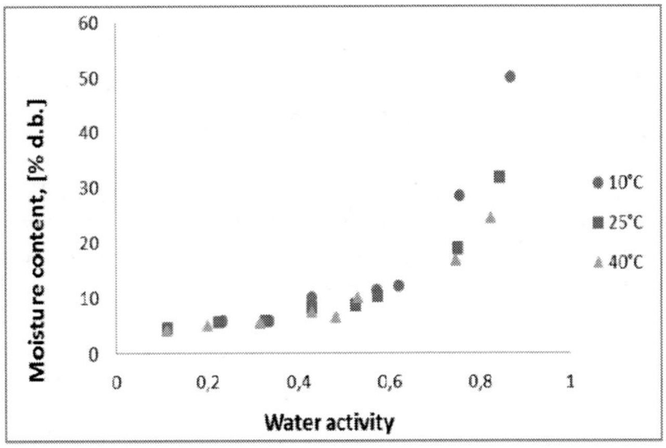

Fig. 1. Sorption isotherm for adsorption

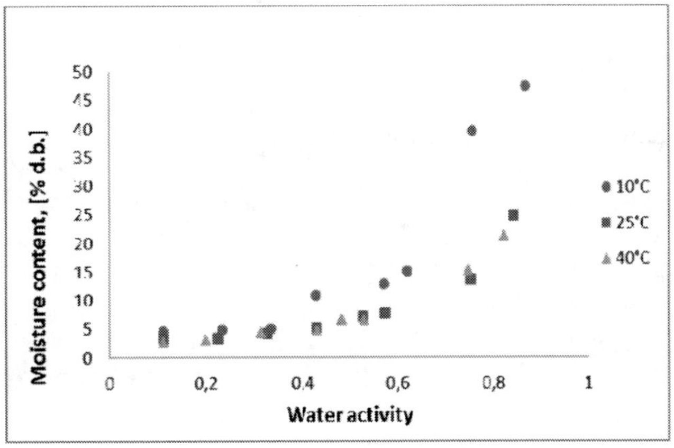

Fig. 2. Sorption isotherm for desorption

 https://doi.org/10.1051/e3sconf/202018003008

The results show that the increase in temperature at a constant water activity is coupled with a decrease in equilibrium moisture content, a dependency pointed out by many authors for various food products [24 - 30]. The isotherms of the flour from Bulgarian penny buns are typical of Type III, using Brunauer's classification [31].

The statistical processing with the help of a non-linear regression program using the least squares method calculated the coefficients of the modified models applied in the analysis. The coefficients and values of the model suitability criteria – mean relative error P, %, and standard error SEM - for adsorption and desorption are presented in Tables 3 and 4.

Table 3. Model coefficients (*A, B, C*), mean relative error (P, %), and standard error (SEM) for adsorption.

Model	*A*	*B*	*C*	*P*	*SEM*
Oswin	12.042	-0.1238	0.7732	24.94	2.28
Halsey	**2.6805**	**-0.0064**	**1.2884**	**10.63**	**2.24**
Henderson	0.00047	81.9008	1.1234	23.85	5.48
Chung-Pfost	605.4333	0.1339	196.478	27.69	5.70

Table 4. Model coefficients (*A, B, C*), mean relative error (P, %), and standard error (SEM) for desorption.

Model	*A*	*B*	*C*	*P*	*SEM*
Oswin	13.7821	-0.2066	0.7474	28.18	3.98
Halsey	**2.1665**	**-0.0165**	**1.0389**	**10.50**	**2.29**
Henderson	0.00162	30.2908	0.9553	20.33	4.74
Chung-Pfost	154.5974	0.1267	47.1888	23.09	5.06

Taking into account the analysis of the results, we can recommend the modified Halsey for adsorption and desorption isotherm description of the flour from Bulgarian penny buns (*Boletus Edulis*). Only for this model are the mean relative error values below 10 %, the standard error possessing the lowest values.

The BET model was linearly transformed to calculate monolayer moisture content for the temperatures of 10 °C, 25 °C and 40 °C and water activity within the aw < 0.5 range. The results obtained are shown in Table 5 for both processes, adsorption and desorption.

Table 5. BET monolayer moisture content, % d.b., at different temperatures *t*.

t (°C)	Adsorption	Desorption
10	3.18	4.37
25	3.58	3.54
40	3.99	3.15

Analyzing the results in Table 5, we can conclude that with respect to the product studied – flour from Bulgarian penny buns (*Boletus Edulis*) – the temperature affects the values of monolayer moisture content. At adsorption, the increase in temperature is accompanied by an increase in the values, within the 0.4 - 0.8 % range. The exactly opposite process takes place for desorption – with the increase in temperature, monolayer moisture content values are lowered from 0.4 to 1.20 %.

4. Conclusion

The data concerning flour particle size of Bulgarian penny buns, obtained via sieve analysis, show that the percentage of fractions with a particle size below 200μm is greatest – 63.1 %.

We also obtained data on the physico-chemical parameters of the flour from Bulgarian penny buns: Moisture content - 10.8 %; Protein - 32.23 %; Fat - 6.62 %; Carbohydrate - 0.99 %; Ash - 6.20 % and Energy Value - 192.46 / 805.25 Kcal/KJ.

We determined the microbiological parameters of the flour from Bulgarian penny buns: total number of mesophilic aerobic and facultative anaerobic bacteria, fungi, *Salmonella* sp. *Escherichia* coli, coagulase-positive staphylococci and yeast, all of them being within acceptable limits.

We obtained the equilibrium isotherms of the flour from Bulgarian penny buns for the temperatures of 10 °C, 25 °C and 40 °C and water activities in the 0.1 - 0.9 range.

We established that the isotherms belong to Type III, according to Brunauer's classification.

We also discovered that temperature affects the sorption capacity of the flour from Bulgarian penny buns. The increase in temperature for a constant water activity is coupled with the decrease in equilibrium moisture content.

In accordance with our criteria, the statistical processing makes us recommend the modified Halsey model for sorption isotherm description of the flour from Bulgarian penny buns.

The obtained monolayer moisture content values of the flour from Bulgarian penny buns are within the ranges of 3.18 % d.b. - 3.99 % d.b. for adsorption and 3.15 % d.b. - 4.37 % d.b. for desorption.

5. References

1. J. Rotruck, A. L. Pope, H. E. Ganther, A.B. Swason, D.G. Hafeman, W.G. Hoestra, Science, **179**, 588, (1973).
2. M. Montes-Bay´on, M. Jos´e D´ıaz Molet, E. Blanco Gonz´alez, A. Sanz-Medel, Talanta, **68**, 1287–1293, (2006).
3. S.B. Goldhaber, RTP, **38** (2), 232-242, (2003).
4. A. Saleha, Y. Yaminia, M. Faraji, S. Shariatib, M. Rezaee, J. Chromatogr. B, **877**, 1758–1764, (2009)
5. M. Navarro-Alarcon, C. Cabrera-Vique, Sci. Total Environ., **400** (1-3), 115-141, (2008).
6. L. Letavayová, V. Vlčkováb, J. Brozmanováa, Toxicology, **227** (1-2), 1-14, (2006).
7. AOAC, 960.39 (Washington , DC), 15 ed. AOAC. (1990)
8. B.S. Standard, Proteins, in Reglement (EO) №152/2009
9. B.S. Standard, Fat content, in BSS 6997:1984
10. B.S. Standard, Carbohydrates in BSS 7169:1989
11. No, ICC Standard, 104/1, (1-4), (1990)
12. Standard in Reglement 1169:2011. App. IVX, XV, (2011)
13. B.S. Standard, Escherichia coli, in (BSS) EN ISO 16649-2:2014
14. B.S. Standard, Salmonella spp., in (BSS) EN ISO 6579-1:2017
15. B.S. Standard, Coagulase-positive staphylococci, in (BSS) EN ISO 6888-1:2005,

16. B.S. Standard, Mesophilic aerobic and facultative anaerobic bacteria, in (BSS) EN ISO 4833-2:2014,
17. B.S. Standard, Yeasts and fungi, (BSS) EN ISO 21527-2:2011
18. ASAE. Standart 245D.5. St. Joseph, Michigan.
19. C.-C. Chen, R.V. Morey. Transactions of the ASAE, **32** (3), 983-0990, (1989)
20. C.-C. Chen, R.V. Morey. Transactions of the ASAE, **32**, 999-1006, (1989)
21. L. Bell, T. Labuza, AACC, 33-36, (2000)
22. A. Bogoeva, FPI Magazine, ISSN 1311-0179, 2, 27-30, (2018).
23. B. V. Bozadzhiev, A. P. Krasteva, Tsv. I. Gogova, Одеска Националъна Академия Харчових Технологий, Науковъ працъ, **30** (2), 34-39.
24. I. Dimov, ICTTE (Yambol, Bulgaria), **I**, ISSN 1314-9474, 479-485, (2013)
25. K. M. Majd, S. H. Karparvarfard, A. Farahnaky, K. Jafarpour, Food Biophys., **8** (1), 1-11, (2013).
26. M. Dupas-Langlet, J. Dupas, S. Samain, M. –I. Giardiello, V. Meunier, L. Forny, J. Food Eng., **184**, 53-62, (2016).
27. A. G. Durakova, N. D. Menkov, J. Food Eng., **68** (4), 535-539, (2005).
28. P. Staudt, C. Kechinski, I. Tessaro, L. Marczak, R. d. P. Soares, N. Cardozo, J. Food Eng., **114** (1), 139-145, (2013).
29. Z. Yan, M. J. Sousa-Gallagher, F. A. Oliveira, J. Food Eng., **86** (3), 342-348, (2008).
30. A. Bogoeva, JCEA (to be published), (2020)
31. S. Brunauer, L. S. Deming, W. E. Diming, E. Troller, JACS, **62**, 1723-1732, (1940).

Structural, kinematic and kinetostatic analysis of a manure spreader equipped by a mobile platform with articulated elements

Lucretia Popa[1,*], *Stefan-Iulian* Moise[2], *Victor* Moise[3], *Vasilica* Stefan[1] *and Ioan* Ganea-Christu[1],

[1]INMA Bucharest, 6 Ion Ionescu de la Brad Blvd., Bucharest, Romania
[2] Etudes et Productions Schlumberger, Paris, France
[3] University POLITEHNICA of Bucharest, Romania

Abstract. Organic fertilization involves working with machines by a special construction, suitable for the material to be spread, machines that have, in most cases, chain conveyors with scrapers attached to the chains. In some cases, the conveyor chain may break, due to the jams, and the need for manual unloading of the bunker (bucket) appears, which requires a great physical effort, the unloading being made by the operator, using a shovel. The construction of the platform using mobile, articulated elements would greatly facilitate the unloading at the bottom of the bunker and would eliminate the physical effort of the operator. This article presents the structural, kinematic and kinetostatic analysis of a mechanism designed for actuating a mobile platform, which is part of a manure spreader, composed of articulated elements driven by hydraulic cylinders.

For the kinematic analysis of the mobile element platform's mechanism, calculation procedures written in the Matlab syntax are used.

After the kinematic analysis of the mechanism, the kinetostatic analysis is performed, i.e. the determination of the reactions from the kinematic couplings, as well as the hydraulic force from the hydraulic cylinder. The results can be used for dimensional verification of the mechanism components, thus contributing to the optimization of the mechanism.

1 Introduction

The technological trailers that carry out the transport and distribution of bulk fertilizers (chemical fertilizers, compost etc.) have a special construction.

Over time, there have been many researches that aimed to improve the construction in order to improve the quality of the work, or for the ease of carrying out the work or the machine maintenance. [1-6].

The technological trailers, have usually a monocoque construction, and can perform only the rear discharge. Most of the time, the material is distributed with the help of a spreading device with different construction, but usually, the material is transported inside

* Corresponding author: Lucreţia Popa, lucretia_popa@yahoo.com

the trailer, to the spreading device, with the help of a conveyor with scrapers arranged on the chain.

The drag chain is on the bottom of the bin to pull the load of manure to the rear where it is shredded by a pair of beaters and spread on the ground. But there are many situations when the chain is broken, because of the manure which contains different objects like wires, stones and in this case, it is difficult to discharge the machine using un unloading equipment like a front loader mounted on the tractor or manually, using shovels.

To offer to the operator the possibility to discharge the machine when the chains are broken, this article studies the possibility to design a bin with many platforms, jointed and driven by a motor group (motor dyad), by means of parallelogram mechanisms.

This idea was transposed into a patent application submitted to the Romanian State Office for Inventions and Trademarks-OSIM. [7].

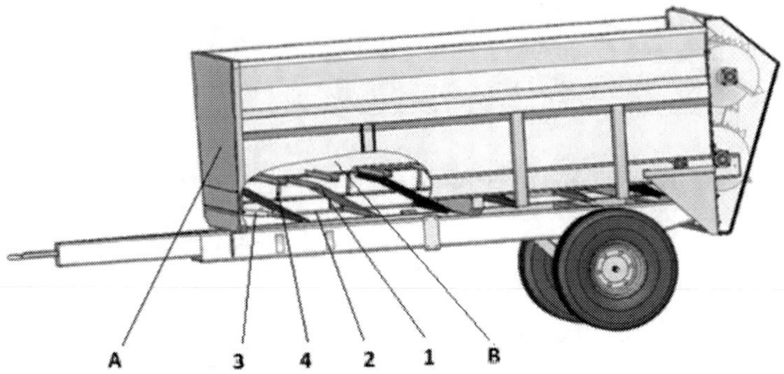

a)

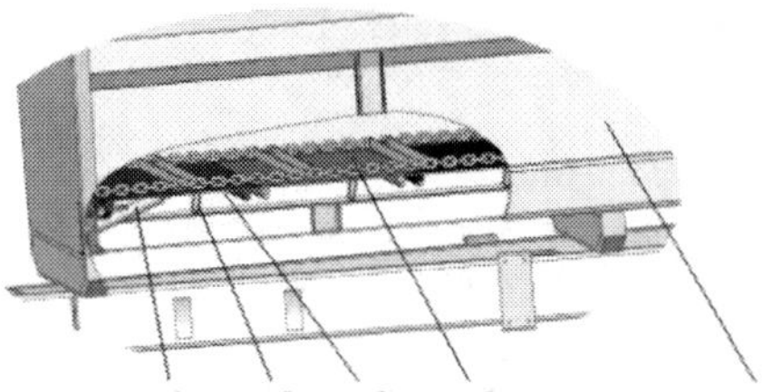

b)

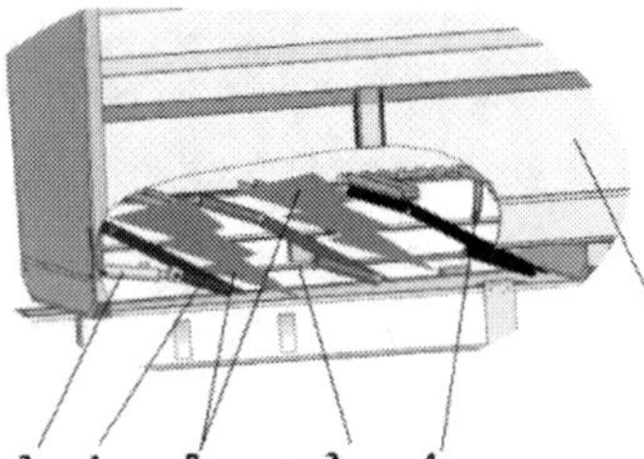

c)

Fig. 1. Manure spreading machine with joint elements [7]
a) left side perspective view; b) detail in the horizontal position; c) open platform

The construction relates to a platform with articulated movable elements, which is part of the trailer bin A, in order to improve the process of unloading the material in case of damage.

The platform consists of a number of movable elements 1, joint, the system 2 of bars that connects the movable elements and which ensures their inclination, the hydraulic cylinders 3, the lateral guides 4 which ensure the steering of the movable elements and the limitation of the maximum angle of inclination of the elements movable during the process of unloading material B remaining in the trailer.

In this article we study the mechanism characteristics, to obtain information for development of this idea and to apply it in the real design and manufacturing. In this paper, structural, kinematic and kinetostatic analysis of the mechanism acting the mobile platforms is performed. The determination of the reactions of the kinematic couplings is necessary for the dimensioning calculation. For the actuation of the entire mechanism, the size of the hydraulic cylinder motor force must be determined, which is made using the kinetostatic method and the virtual powers method.

2 Material and method

Figure 2 shows the kinematic diagram of the mechanism designed for acting the jointed movable elements, which are part of the platform.

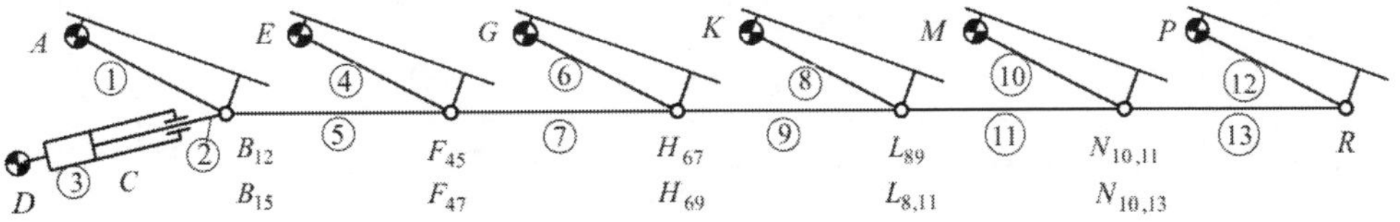

Fig. 2. Kinematic scheme of the mobile platforms driving mechanism

In order to determine the kinematic parameters of the mechanism's elements, it is necessary to know the structural groups in its composition, which means, the structural analysis of the lifting and folding mechanism of the platform jointed elements must be performed.

For the structural, kinematic and kinetostatic analysis of the mechanism, the kinematic diagram of the mechanism is known, also the dimensions of the kinematic elements and the positions of the couplings adjacent to the base, the initial position of the mechanism, the kinematic parameters in the C-motor coupling, the masses and the mechanical moments of inertia of the elements, as well as the technological forces on each individual platform.

3 Results

3.1 Structural analysis of the mechanism

The structural analysis of the mechanism consists in establishing kinematic couplings, determining the mobile elements, establishing the degree of mobility, drawing up the structural scheme, establishing the structural structures of the mechanism, constructing the multipolar scheme and establishing the structural relationship [10-15].

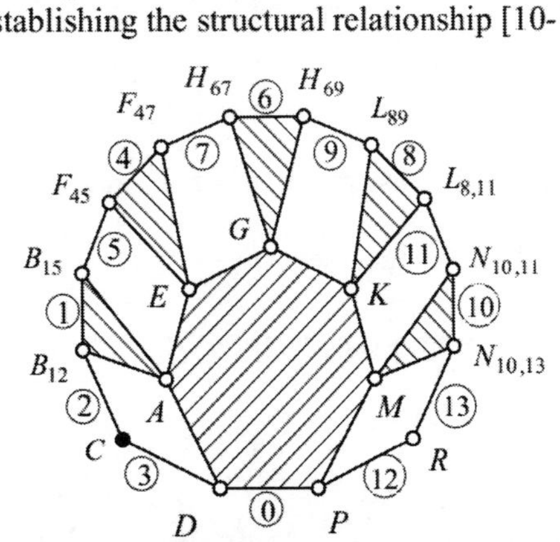

Fig. 3. Structural scheme of the mechanism

Observing the relative movements between the elements of the mechanism, (Figure 3), it is found that it has the following inferior couplings: $A(0R1)$, $B_{12}(1R2)$, $C(2T3)$, $D(3R0)$,

$B_{15}(1R5)$, $E(4R0)$, $F(4R5)$, $F(4R7)$, $G(6R0)$, $H(6R7)$, $H(6R9)$, $K(8R0)$, $L(8R9)$, $L(8R11)$, $M(10R0)$, $N(10R11)$, $N(10R13)$, $P(12R0)$ and $P(12R13)$. The number of superior couplings is zero. The mobile elements of the mechanism are: $1(A, B_{12}, B_{14})$, $2(B_{12}, C)$, $3(C, D)$, $4(E, F_{45})$, $5(B_{15}, F_{45})$, $6(G, H_{67})$, $7(F_{47}, H_{67})$, $8(K, L_{89})$, $9(H_{69}, L_{89})$, $10(M, N_{10,11})$, $11(L_{8,9}, N_{10,11})$, $12(P,R)$ and $13(B_{10,13}, R)$.

Taking into account the number of moving elements and the number of kinematic couplings, the degree of mobility of the mechanism results [10-15], namely M = 1, meaning that only one independent parameter is required for the elements of the mechanism to have unique determined movements. Figure 3 shows the structural diagram of the mechanism.

Figure 4 presents the multipolar scheme and its structural relationship.

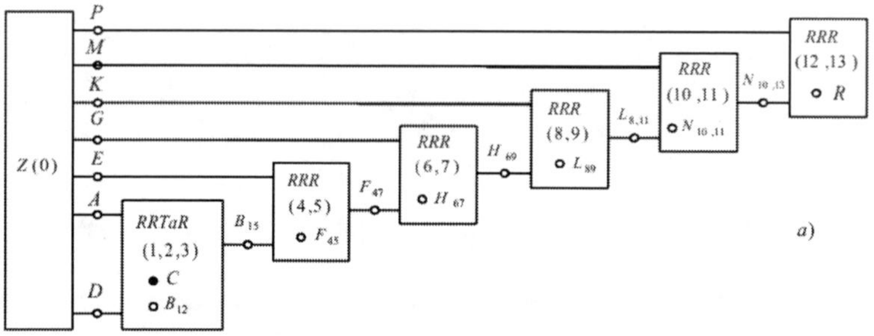

Z(0) + RRTaR (1,2,3) + RRR (4,5) + RRR (4,5) + RRR (6,7) + + RRR (8,9) + RRR (10,11) + RRR (12,13) b)

Fig. 4. Multipolar scheme of the mechanism and structural relationship

3.2 Kinematic analysis of the mechanism

Figures 2, 3 and 4 show that the mechanism consists of the base $Z(0)$ and the motor dyad $RRTaR(1,2,3)$ and five passive dyads of aspect 1 (*RRR* dyads).

Figure 5 shows the kinematic diagram of the mechanism, highlighting the kinematic parameters of position.

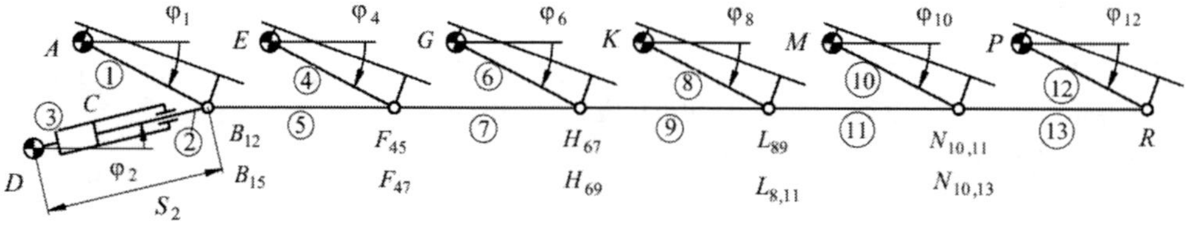

Fig. 5. The kinematic scheme of the mechanism, highlighting the position of parameters

For the kinematic analysis of the mechanism, calculation procedures written by the authors in the Matlab syntax are used. Figures 6 and 7 show diagrams of variation of angular velocities and accelerations of the motor dyad elements.

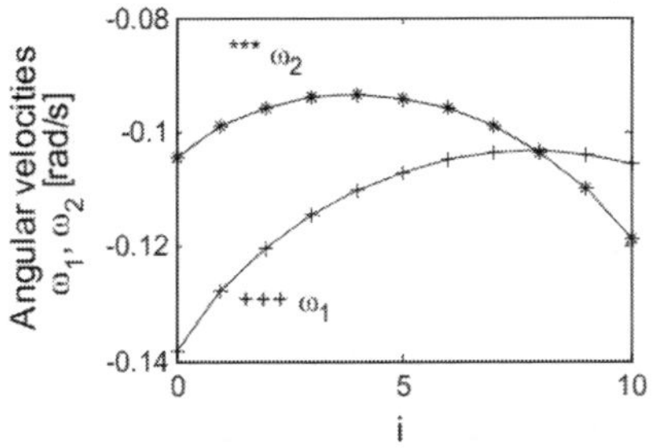

Fig. 6. Angular velocity diagrams ω_1 and ω_2

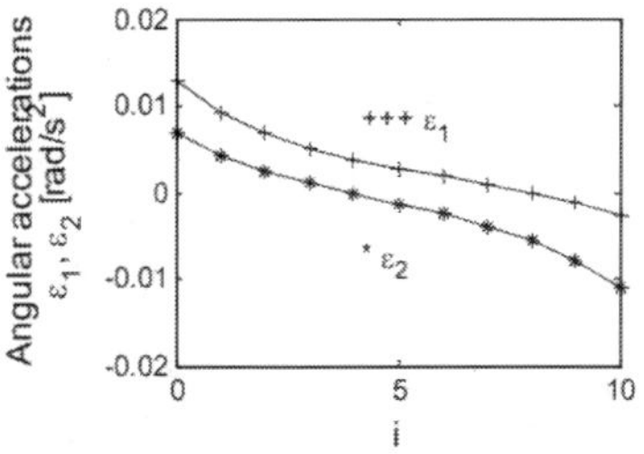

Fig. 7. Angular acceleration diagrams ε_1 and ε_2

Note: *i represents the calculation position at a given time. The distance between two calculation positions (on the x-axis) is an increment. This is equal to the stroke of the piston / (n-1), where n is the number of calculation positions (the starting position is also considered).*

3.3 Kinetostatic analysis of the mechanism

Kinetostatic analysis of the mechanism is done in the opposite direction of the kinematic analysis, that is, starting with the *RRR* (12,13) dyad and ending with the *RRTaR* (1,2,3,) motor group.

For the kinetostatic analysis, the technological forces, the forces of inertia and the forces of weight of the elements are taken into account.

Figures 8-13 show the diagrams of the forces and moments that act on the component elements of the dyads and the kinetostatic diagrams.

Table 1

Highlighting the forces and moments acting on the dyads' elements	Dyads kinetostatic diagrams
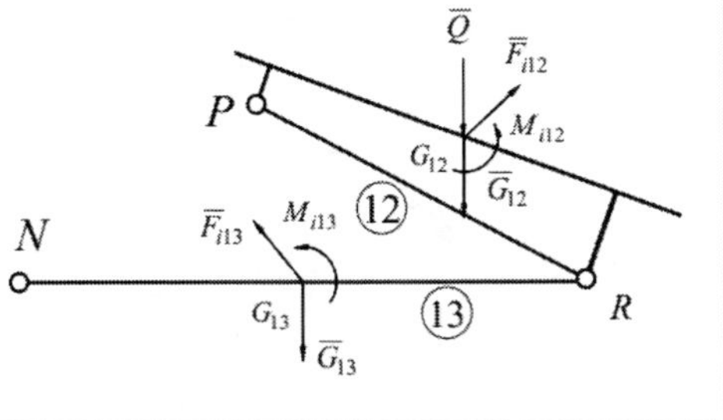	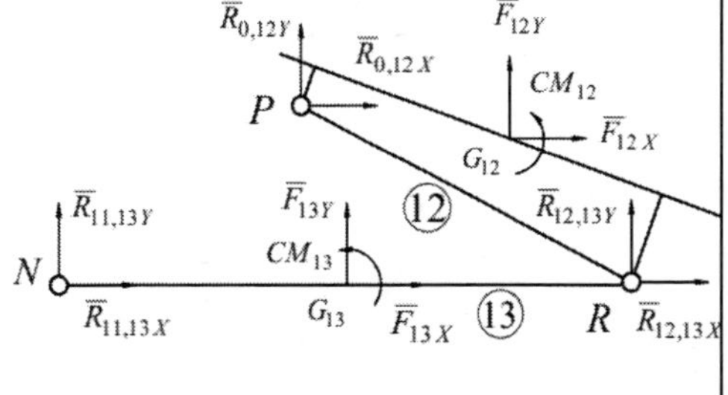
a)	*b)*
Fig. 8. *RRR* (12, 13) dyad	
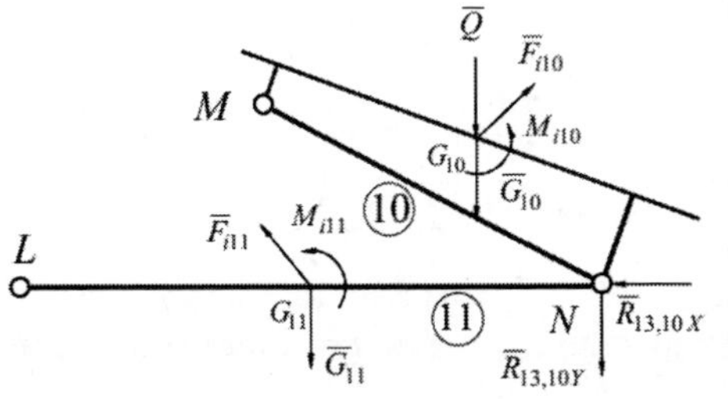	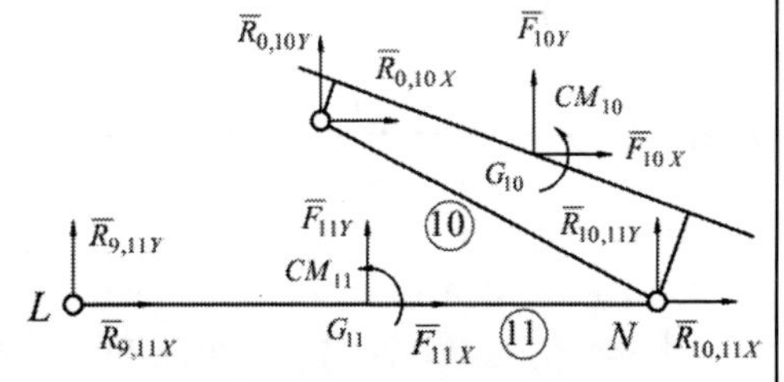
a)	*b)*
Fig. 9. *RRR* (10, 11) dyad	

Highlighting the forces and moments acting on the dyads' elements	**Dyads kinetostatic diagrams**

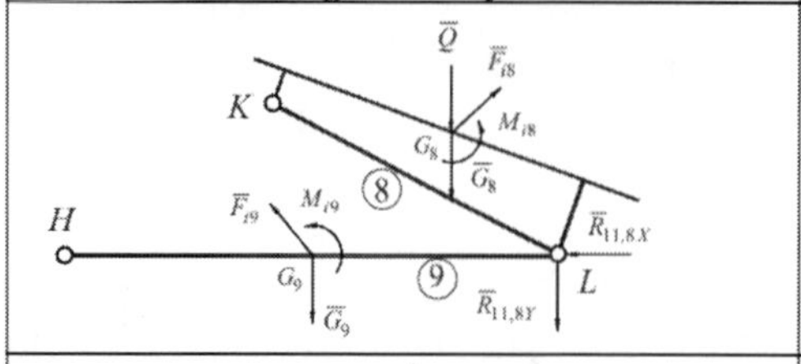

a)

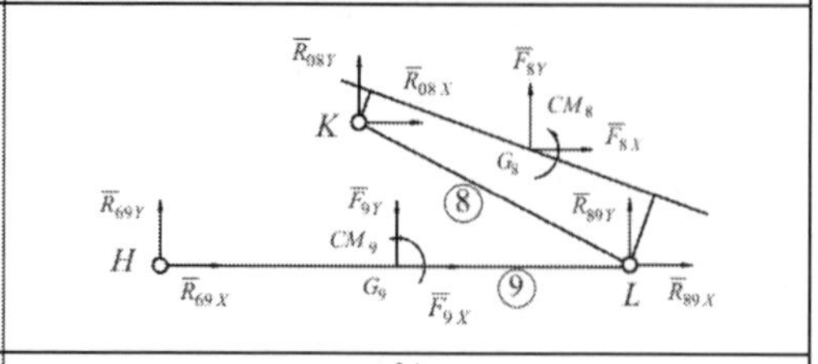

b)

Fig. 10. *RRR* (8, 9) dyad

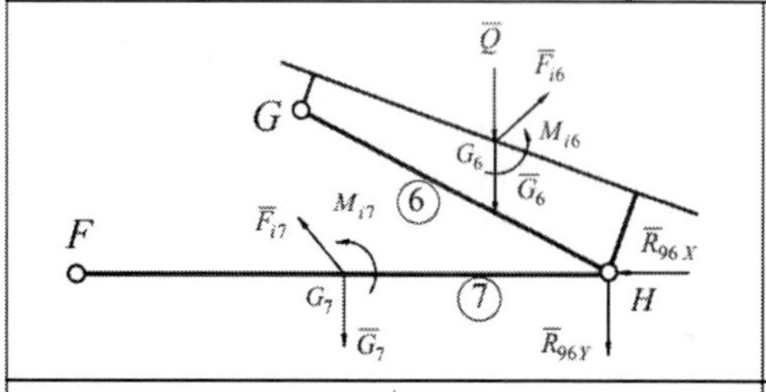

a)

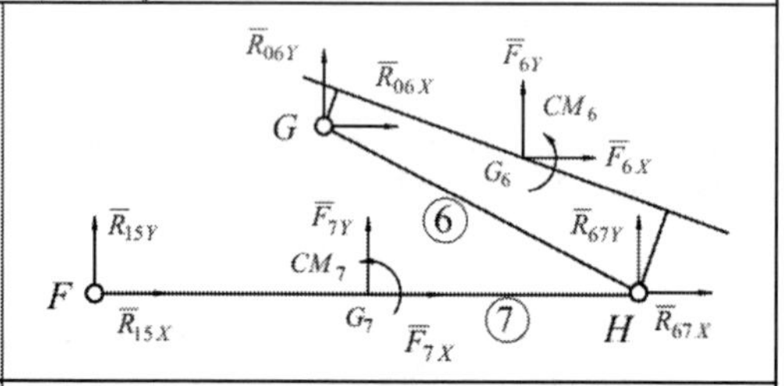

b)

Fig. 11. *RRR* (6, 7) dyad

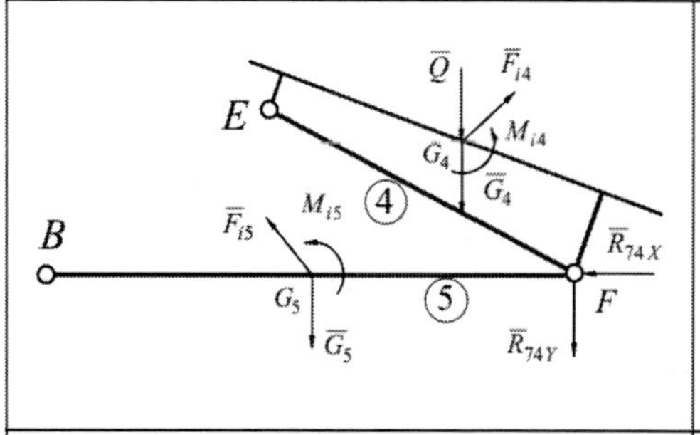

a)

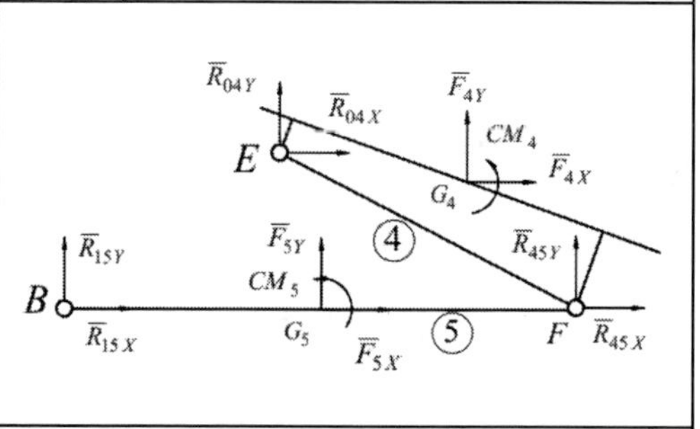

b)

Fig. 12. *RRR* (4, 5) dyad

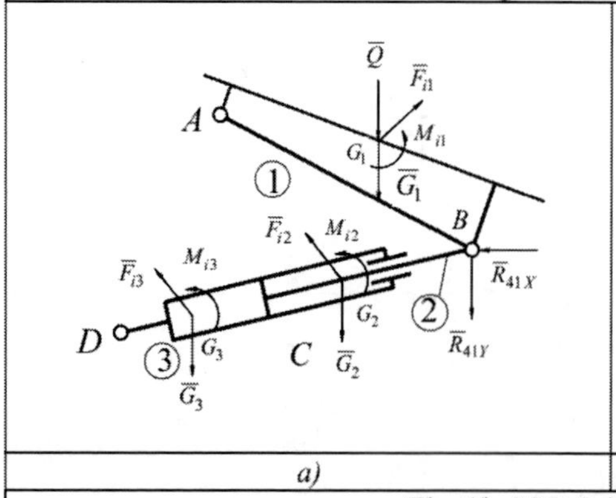

a)

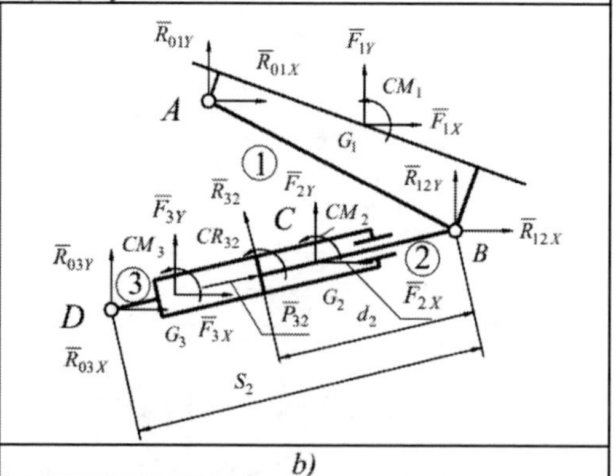

b)

Fig. 13. *RRTaR* (1,2,3) motor dyad

Figures 14, 15, 16 and 17 show the hodographs of the reaction forces of the kinematic couplings *A, B, D* and *E*, and in figure 18 the diagram of the motor force variation in the hydraulic cylinder is presented.

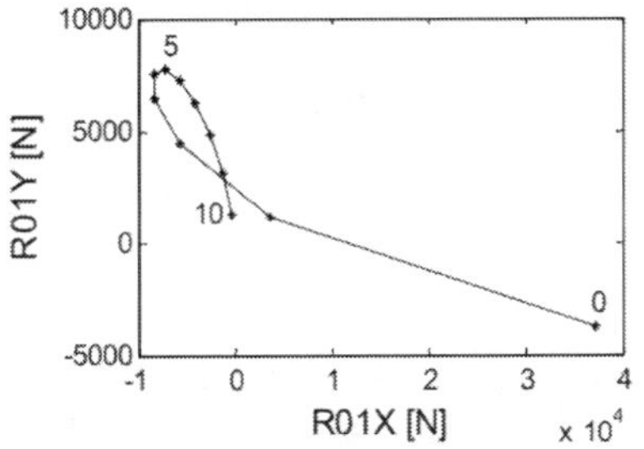

Fig. 14. Hodograph of the reactions of the coupling *A*

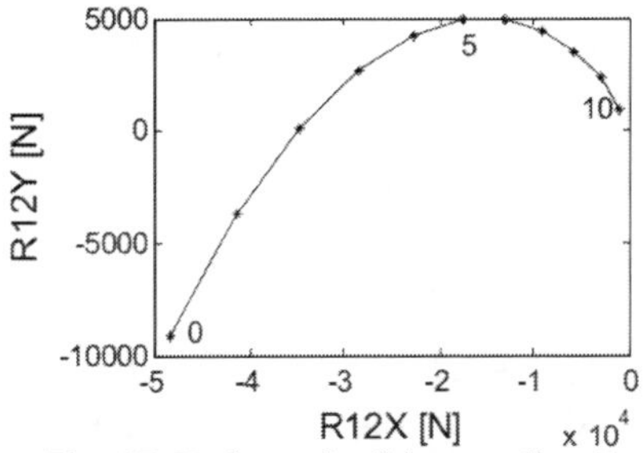

Fig. 15. Hodograph of the coupling *B*

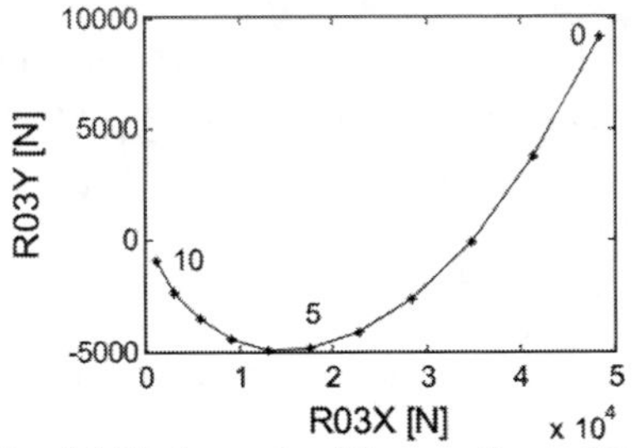

Fig. 16. Hodograph of the reactions of the coupling *D*

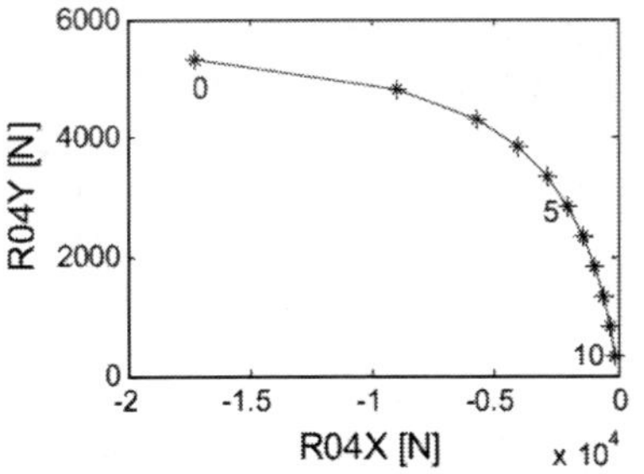

Fig. 17. Hodograph of the coupling E

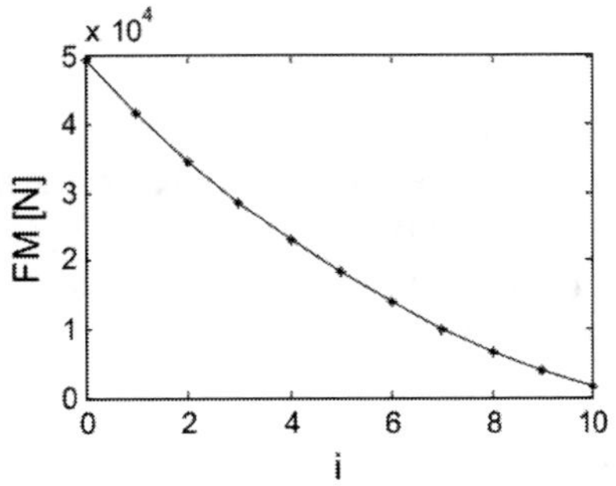

Fig. 18. Chart driving force in the coupling *C*

Table 2 presents the values of the driving force in the hydraulic cylinder, for a cycle of operation, values calculated by the kinetostatic method (FE_CINET) as well as by the virtual powers method (FE_PVIR).

Table 2

pos	FE_CINET	FE_PVIR
0	49328.67	49328.67
1	41576.80	41576.94
2	34739.41	34739.41
3	28648.28	28648.28
4	23199.73	23199.73
5	18325.45	18325.45
6	13978.50	13978.50
7	10126.10	10126.10
8	6745.94	6745.94
9	3824.86	3824.86
10	1359.63	1359.63

Analysing the data in table 2, it is observed that the same values were obtained by two methods of calculation, which means that all the calculations were correct and the data obtained can be used to dimension the mechanism that acts the platform with articulated movable elements.

4. Conclusions

The structural and kinematic analysis of the mechanism provided useful information about the functionality of the system, showing its correct functioning, if it has the degrees of freedom required for a movement without blockages and dysfunctions, the conclusion being that the mechanism was well designed in terms of structure.

The kinetostatic analysis made by the simulation enables in the future an optimal dimensioning of the elements that form the mechanism of the platform with multiple articulated mobile elements. An over dimensioning of the system can increase the price of the machine or, a subdimension causes damages of the system, and unwanted unloading of the material transported. The kinematic analysis realized in this paper is necessary for the dynamic analysis, which will be performed in the following research.

This work was supported by a grant of the Romanian Ministry of Education and Research, through Programme 1 – Development of the National Research-Development System, subprogramme 1.2 – Institutional performance – Projects for financing excellence in RDI, contract no. 16PFE.

References

1. P. Cârdei, V. Stefan, L. Popa, R. Ciuperca, *Statistical models for the working process carried out by the organic fertilizer spreading machine*, INMATEH Agricultural Engineering, **58/2**, 121-128, (2018)
2. L. Popa, L. Dudici, A. Nedelcu, A.C. Cristescu, D.M. Dumitru, *Kinetostatic analysis of ratchet mechanism that acts on conveyor of manure spreading machine,* Engineering for Rural Development-ERDev2018, 57-63, DOI: 10.22616/ERDev2018.17, (2018)
3. K. Sengottaian, N. Mohanrajhu, S. Palani, C. Jayabalan, *Design of Manure Spreader Machine for agricultural form field,* International Journal of Innovative Technology and Exploring Engineering-IJITEE, **9/1**, 4280-4284, DOI: 10.35940/ijitee.A4877.119119, (2019)
4. V. Ştefan, P. Cârdei, L. Popa, L. David, R. Ciupercă, *Influence of the manure spreading machines' working parameters on the qualitative performances of the fertilization process,* INMATEH Agricultural Engineering, **58/2**, 115-120, (2018)
5. V. Stefan, P. Cârdei, V.N. Vlăduţ, L. Popa, R. Ciuperca, N. Ungureanu, *Mathematical model for particle motion applied on a manure spreading apparatus used in environmentally friendly technology,* EEMJ, **17/1**, 217-227, (2018)
6. V. Ştefan, R. Sfîru, L. Popa, *Experimental results on the solid organic fertilizer machine MG 5,* TE-RE-RD, (2019)
7. L. Popa, A. Nedelcu, V. Ştefan, A.C. Cristescu, *Platform with mobile elements for trailers* Patent request no. A/00824/22.10.2018, OSIM, (2018)
8. I.I. Artobolevski, *Théorie des Mécanismes et des Machines (Theory of Mechanisms and Machines).* 453 p., Mir Publishing House, (1977)
9. B. Demidovich, A.I. Maron, *Elements of numerical calculation (Éléments de calcul numérique)*, 2nd edition, 717 p., Mir Publishing House, (1987)
10. C. Duca, Fl. Buium et al. *Mechanisms*, Gh. Asachi Publishing House, (2003)

11. V.Moise, E. Maican, Şt.I. Moise, *Numerical methods. MATLAB Applications (Metode numerice. Aplicaţii în MATLAB)*, 292 p., Bren Publishing House, (2016)
12. V. Moise, I. Simionescu, M. Ene, A. Rotaru, *Analysis of plane mechanisms with hitch bars. MATLAB Applications (Analiza mecanismelor plane cu bare articulate. Aplicaţii în MATLAB),* 324 p, Printech Publishing House, (2015)
13. Chr. Pelecudi. *The precision of the mechanisms*, Publishing House of the Academy of RSR, 398 p. (1975)
14. Chr. Pelecudi, D. Maroş, V. Merticaru, N. Pandre, I. Simionescu, *Mechanisms (Mecanisme*), 394 p., Didactic and Pedagogical Publishing House, (1985)
15. I. Simionescu, V. Moise, *Mechanisms*, Technical Publishing House, (1999)

Experimental Water Treatment Plant from Agrozootechnical Farm

Iulian Florin Voicea[1,*], *Ioana Corina* Moga[2], *Eugen* Marin[1] , *Dragoş* Dumitru[1], *Cătălin* Persu[1] and *Dan* Cujbescu[1]

[1]National Research - Development Institute for Machines and Installations Designed to Agriculture and Food Industry, 6 Ion Ionescu de la Brad Blvd., 013813 Bucharest, Sector 1, Romania
[2] DFR Systems S.R.L., Drumul Taberei 46, 061392 Bucharest, Romania

Abstract. According to the existing statistical data at the country level, over 65% of the wastewater (rural from agricultural livestock farms) is discharged without purification, 61% are insufficiently purified and only 10-15% are properly purified. The pollution effect of wastewater not purified or insufficiently purified on surface waters, is mainly manifested by the content of suspended matter, of organic matter, in nutritional salts, ammonium and in pathogenic microorganisms. It is well-known that the nutritional salts of nitrogen and phosphorus cause the eutrophication of surface waters, with the effect of consuming the dissolved oxygen needed to sustain the aquatic life. Ammonia is particularly toxic to aquatic life. Unsaturated or insufficiently purified wastewater pollutes groundwater, among others with nitrates, ammonium and bacteriologically. From the presented results, the purification of waste water is an essential requirement of the development of human civilization. Being a necessity with special social and ecological implications, the unitary regulation and the general provision of the necessary infrastructure is a priority, in this sense an article will be presented in an experimental treatment station that performs the collection of waste water from an agrozootechnical farm. The purification of water within this experimental model of treatment plant will be done autonomously from an energy point of view (electricity supply is carried out through a hybrid system based on photovoltaic panels and wind turbine), and the recovery. The use of purified water is done in a greenhouse, using a mixed irrigation system (dripping / spray irrigation).

1 Introduction

In the current stage of economic and social development, characterized by unprecedented dynamics, on the one hand, and the limited nature of natural resources, on the other hand, water, one of the most important natural resources, must be well managed, for to prevent the "water crisis", a term that appears more and more frequently in the forecasts of international organizations. The warning of scientists about a possible water crisis at the planetary level, has determined the United Nations to establish March 22 as

* Corresponding author: voicea_iulian@yahoo.com

"World Water Day", the goal being an action plan at the individual, national, regional and international level for use water rational, [1,2]. The concern for rational water management is due to the characteristics of the water, as a natural resource:

- water resources are limited; the water resources of a river basin (or of a region, country) are periodically restored, but they cannot grow over time;
- the water resources have an unequal distribution in time and space; they are differentiated from the distribution in time and space of water requirements;
- the water resources have limited possibilities (from a technical-economic point of view) to be transferred between different regions or hydrographic basins;
- surface waters are strongly influenced by man, both from a quantitative point of view (captures) and qualitatively (the discharge of more or less purified waste water);
- water is a reusable resource; the same quantity of water can be used - if qualitatively it corresponds - by several users, through successive catches and refunds;
- water is an important environmental factor; his good management is an obligation that derives from the requirements arising from the environmental protection legislation.

Water purification - a complex process of retention and neutralization of dissolved harmful substances, in colloidal or suspended state, present in industrial and urban wastewater, which are not accepted in the aquatic environment in which the treated water is discharged and which allow the restoration of physical properties - water chemicals before use, [3,4].

Wastewater treatment compromises two broad categories of successive operations, namely:

- the retention or neutralization of harmful or recoverable substances present in wastewater;
- processing of the material resulting from the first operation.

Thus, the treatment has as final results the waste water, in different degrees, poured into the emissary or that can be used in irrigation or other purposes and sludge, which are processed, stored, decomposed or used. The main wastewater treatment methods differ depending on the pollutants present. They can be classified, first of all, according to the mechanism that leads to the reduction of the pollutant by "conventional" physico-mechanical, physico-chemical, biochemical or biological methods. The combination of these methods allows for advanced purification, and the effluent can be re-introduced into the economic circuit, [5,6].

The agriculture currently practiced is largely polluting, and the phenomenon of pollution is largely known by specialists in the field of environmental protection. The main aspects of the environmental pollution caused by the agricultural activities can be summarized in:

- the evacuation of several million cubic meters of wastewater, not purified or incompletely purified, coming from the breeding complexes in the industrial system of pigs, birds and cattle, in surface areas and in the drying network. To these are added the infiltration of wastewater into the depths, during the storage period in ponds, battles and basins, affecting the quality of groundwater used as a source of drinking water supply in many rural localities.
- use on agricultural land, for dual purposes, of fertilization and irrigation, of sludge and waste water from livestock farmers, containing harmful salts and pathogens contaminating soil, plants, animals and humans.
- the administration on agricultural lands bordering the zootechnical complexes of exaggerated norms of manure (over 100 t / ha), at intervals of 2-3 years, which far exceed the needs of the plants and determine the accumulation of nitrates in fodder, as well as the leaching of nitrates in the groundwater.

- the use of chemical fertilizers (especially nitrogen) in doses too high and at times not correlated with the consumption in different phases of development of the cultivated plants.
- application of chemical products (pesticides), with the purpose of combating diseases, insects, rodents, nematodes and weeds in agricultural crops and apple orchards, by poorly trained people.
- the uncontrolled storage of manure and the lack of manure collection basins from the animals belonging to numerous private households have as negative effects their leakage into the running water, as well as the nitrate infestation of the groundwater.

The establishment of animal farms can be done only by observing very strict rules, especially regarding the neighbourhood, but also the conditions for the storage of waste, [7]. At EU level, but also in Romania, there are certain regulations that prohibit the placing of livestock farms at short distances from the inhabited areas. In these conditions, it becomes necessary to make some inventions in order to ensure the water and sewage needs. In order to ensure the sewerage in some situations, the agrozootechnical farm can be connected to the sewerage system of the locality, but in order to achieve this, the water discharged from the farm must comply with the norm of NTPA 002. As the evacuation of a water that is included in this norm is most often impossible, the farmer has to create, on the territory of his farm, a water treatment plant that will bring water at least within the parameters specified by the NTPA 002. In most cases, however, the agrozootechnical farm is located at very far from the centralized sewerage system the farmer being forced to find alternative solutions. Thus, a solution is the relocation of wastewater treatment technologies and the reuse of this water in agriculture.

The chemical characteristics of wastewater are basic elements in the characterization of wastewater. The chemical composition of wastewater is influenced by the specific consumption of water per capita. The higher the consumption of water per capita, the more diluted the waste water, because, in general, the amount of waste water discharged is relatively constant, [8, 9]. Important for the design of the treatment plants is the value and the state of the materials contained in the water, respectively the quantity of the solid organic matter that is suspended and separable by decanting, solid organic materials and dissolved in the water, as well as other materials specific to the waste water, such as: nitrogen under all forms of chemical combinations, chlorides, sulphides, hydrogen sulphide, etc.

2 Material and method

In order to carry out the researches presented in this article, an experimental wastewater treatment plant was used, the water treatment within this energy-independent wastewater treatment plant (powered by electricity produced by a hybrid system based on photovoltaic panels and wind turbine), and the use of purified water within an agricultural land, through the use of an irrigation system. Water quality was monitored to determine the efficiency of the system. Figure 1 presents the final technical drawing, which faithfully represents the real situation, from the field.

The treatment plant is supplied by a pump mounted in the technical room. The unit is supplied with constant flow, while the blower supplies almost continuous air to the aerobic reactor. The feed pump discharges waste water transported to the first compartment (bioreactor) of the unit. The biological treatment stage has the following sequence of compartments:

- a bioreactor with SAM technology with a slow mixer for denitrification and reduction of organic material;

- the second bioreactor with intensive aeration with SAM technology for nitrification and reduction of organic matter;
- a decanting pond with lamellar decanter;
- a sludge recirculation system.

The technical characteristics of the agro-zootechnical wastewater treatment plant are the following:
- diameter of the upper module of the water treatment plant: 2500 mm;
- length of the above ground module of the water treatment plant: 5580 mm;
- volume of the purified water storage tank: 1 m^3;
- electricity supply: renewable sources of electricity, solar panels and fully autonomous wind turbine, 230 V.c.a. and a nominal frequency of the 50 Hz network, and the power supply for control, trigger, alarm and signalling installations is 24 VAC.

The irrigation system it is composed of a system of pumping the water treatment from the retention tank that uses a surface solar electric pump and a distribution network for drip irrigation of agricultural land. The surface solar electric pump works with two 12 V, 240 W. photovoltaic panels.

Fig. 1. Installation for the recovery of waste water from small agrozootechnical farms

The irrigation method chosen was that of drip which is made localized with small flows. Drip irrigation and / or fertigation ensures the needs of the present without compromising future resources. This type of irrigation has a conservative character, offered by reducing the amount of fertilizers and chemicals needed, reducing the water consumption due to the increased capacity of the roots to retain and store water, reducing the risk of propagating diseases and pests, reducing or avoiding them. if some chemical treatments, reducing pollution, reduced leakage of chemicals in the freshwater reserve, eliminating soil erosion, applying nutrients can be controlled at the precise time when they are needed and in the desired amount, reduced costs. The irrigation was carried out inside a greenhouse and the culture that was thus irrigated was one of vegetables.

3 Results and discussions

Water quality indices. During the 3 months of the experiment were collected 10 samples of waste water and purified water, which were analyzed. Analysis of pH, suspended matter, biochemical oxygen consumption, CCOCr, nitrates, nitrites and total phosphorus were performed. Next, we have represented in graphical form by using the program Excel, the linear variation of the variation of the analyzed quantities. The biochemical oxygen consumption (CBO) is an indicator that is defined as the amount of oxygen consumed for the biochemical decomposition in aerobic conditions of the total organic solid materials in compliance with the conditions in the standards. The result of the determination is influenced by the water temperature and the duration of the determination. The most common determination is that made at 20 ^{0}C for 5 days and it is noted, CBO_5. Another indicator of appreciation of the state of the water is the so-called chemical consumption or the immediate consumption of oxygen, which represents the amount of oxygen that is consumed in the combinations of reduction of organic substances, but without the intervention of microorganisms. Usually, this determination lasts one hour. There is a high consumption of chemical oxygen in the wastewater that has a high content of hydrogen sulphide, as well as in the sludge on the bottom of the standing water, slightly agitated. Biochemical oxygen consumption indirectly appreciates the amount of organic matter that can be decomposed and directly the oxygen consumption required by the microorganisms that produce the decomposition. In urban wastewater the biochemical oxygen consumption indicator, CBO_5 varies between 100 and 400mg / dm^3, while for industrial wastewater the CBO_5 indicator does not exceed 50 mg / dm^3, except for industrial wastewater from food companies, where the values can even exceed 50 times the normal values.

Nitrates (RNO_2) and nitrates (RNO_3), wherein R represents: K, Na, etc., are contained in fresh waste water in extremely small quantities. Nitrites are unstable and can either be reduced to ammonia or oxidized to nitrates. Therefore, the presence of nitrites indicates fresh waste water being transformed. Nitrites can also occur naturally in rainwater or in snow melt. The maximum quantities of nitrites in wastewater do not exceed 0.1 mg / l. Nitrates are the most stable form of organic nitrogenous matter and generally indicate a stable water from a transformation point of view. The presence of nitrates in rivers, between certain limits is desirable, because they represent an oxygen source, in that they stimulate the growth of algae and green plants containing chlorophyll and through photosynthesis enrich the water in oxygen. In the case of treatment plants, respectively in the active sludge basins, the presence of nitrates indicates a large amount of sludge and a long time of the water treatment process.

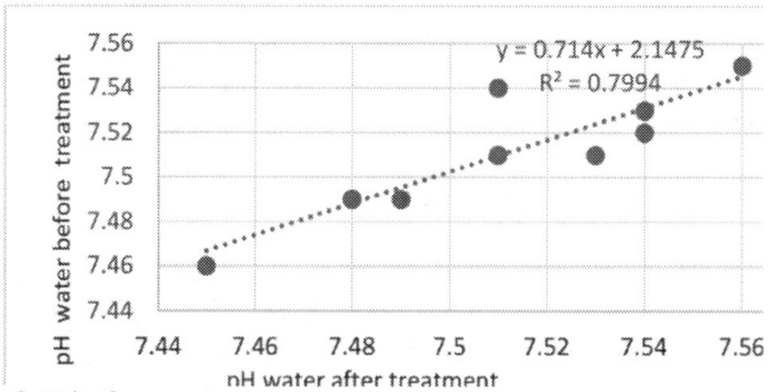

Fig.2. The evolution of pH in the water treatment

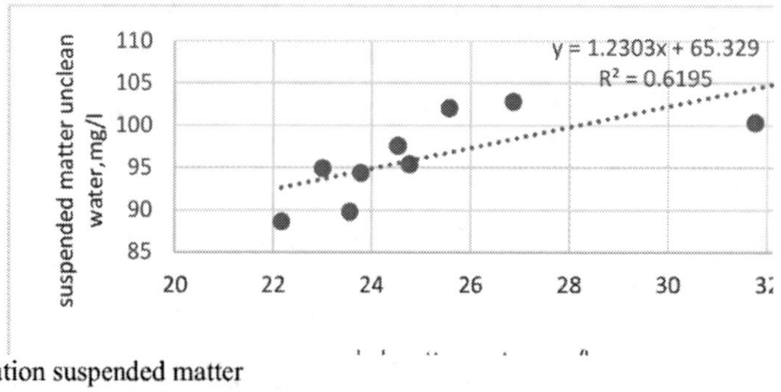

Fig.3. The evolution suspended matter

The biological filters of a well-functioning sewage treatment plant are characterized by a flow of nitrate-containing streams and nitrites in very small quantities or places. In raw wastewater, nitrate amounts range from 0.1 to 0.4 mg / l. The acidity or alkalinity of the wastewater represents their ability to neutralize acidic bases respectively. Generally, domestic wastewater is weakly alkaline, while industrial wastewater has a pronounced acidic or basic character. For the optimal conditions of the wastewater treatment process it is desirable that they be weakly alkaline, especially for the proper development of the biological treatment.

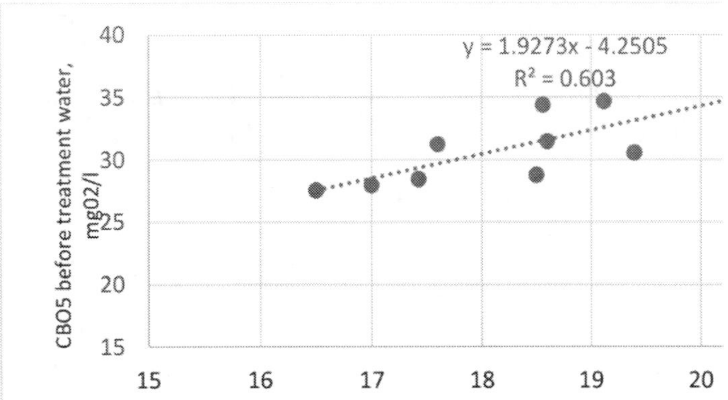

Fig.4. The evolution CBO_5

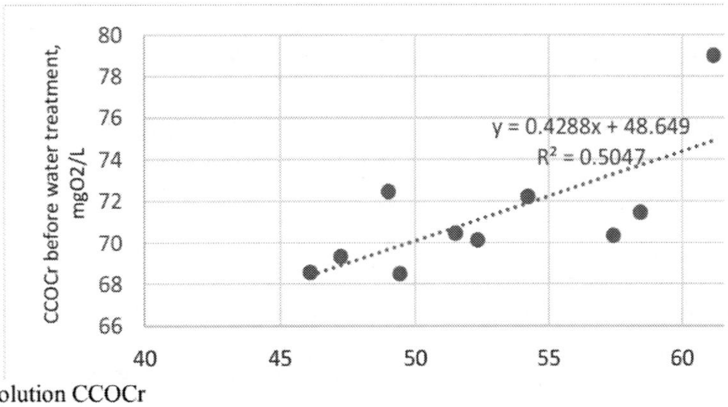

Fig.5. The evolution CCOCr

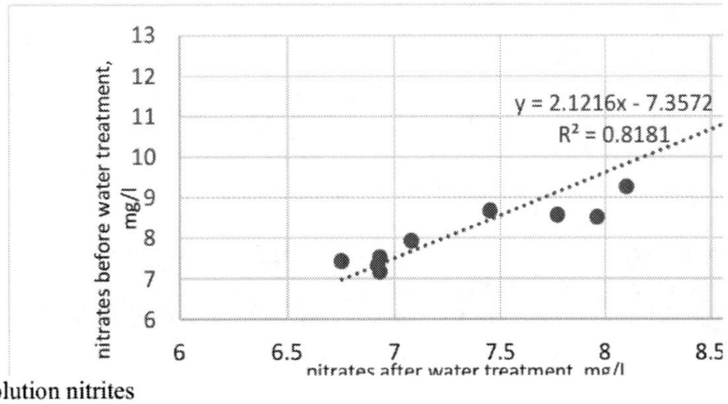

Fig.6. The evolution nitrites

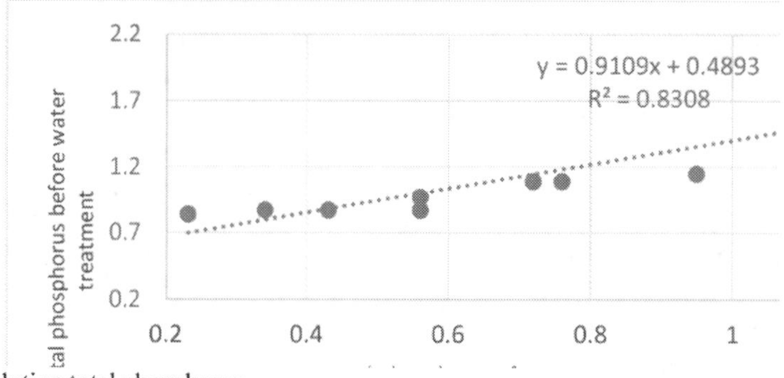

Fig.7. The evolution total phosphorus

Also, a rule of irrigation was established for the greenhouse vegetable cultivation, the results obtained being presented in the following table.

Table 1. Results for the irrigation system

Nr. variant	Moisture before watering, %	Moisture after watering, %	Theoretical norm, (n_c) m³/ha	Determined norm (n_v) m³/ha	Deviation $c = (n_c - n_v) / n_p$
I	4.1	27.5	830.34	536.76	0.55
II	8.2	34.5	778.68	447.30	0.74
III	11.1	32.8	742.14	468.72	0.58
IV	11.4	30.8	738.36	493.92	0.49
V	10.3	37.3	752.22	412.02	0.83
VI	7.4	36.2	788.76	425.88	0.85

4. Conclusions

Following the comparative analysis of the water input data with those from the exit of the experimental station model of **agro-zootechnical wastewater treatment with the maximum allowed values of the quality indicators provided in the NTPA 001 norm can be drawn the following conclusions:**

- the values of pH obtained at the exit from the station fall within the reference range 6.5-8.5, the average value obtained being 7.5.
- the suspended materials at the exit of the water from the station are below the reference value of 35 mg / l, the average value being 25.84 mg / l.
- the dissolved oxygen consumption of the water in the station was exceeded only for an analysis of the 10 performed, the average value obtained is 18.3 mg O2 / l, the indicator value of NTPA 001 being 20 mg O2 / l.
- the CCOCr values obtained at the exit from the station are located only the reference value from the normative of 70 mg O2 / l, the average value for all 10 samples analyzed being 52.68 mg O2 / l.
- nitrates at the outlet of water from the station are below the reference value of 25 mg / l, the average value being 7.45 mg / l.
- nitrites at the exit of the water from the station are below the reference value of 1 mg / l, the average value being 0.77 mg / l.
- the total phosphorus content of the water in the station was exceeded only for 2 analyses out of the 10 performed, the average value obtained being 0.67 mg / l, the indicator value in NTPA 001 being below 1 mg / l.

Regarding the deviation of the determination of the irrigation norm, it is found that it is within the limits of the range of humidity available for plants according to the agrotechnical requirements.

This work was supported by Project co-financed for the European Regional Development Fund through the Operational Program Competitiveness 2014-2020 in the financing contracts no. 80 / 08.09.2016, Axis 1, Action 1.2.3./ SMIS code 105551 / Subsidiary type contract D no. 918 / 22.08.2017 and Romanian Education and Research Ministry, through Programme 1 – Development of the national research-development system, subprogramme 1.2 – Institutional performance – Projects for financing excellence in RDI, contract no. 16PFE.

References

1. American Water Works Association, Water Quality and Treatment, *A Handbook of Community Water Suplies*, New York, USA, (2000)
2. F. Pricop, R. Scarlat, C. Ghiţuleasa, A. Popescu, M. Iordănescu, I.C. Moga, Integrated systems for control and monitoring of wastewater quality (in Romanian), *Textile Industry Magazine*, pp. 40 – 45, ISSN 1222-5347, (2013)
3. P. Popa, I. Moga, M. Rissdorfer, M. Ilis, Petrescu G., C. Crăciun, M. Matache, C. Covaliu, C.Stoian, Duckweed utilization for fresh water conservation (management) in recirculated aquaculture systems, *International Journal of Conservation Science*, vol. **(8)**, issue **4**, ISSN:2067-8223, pp. 715 – 722, (2017)
4. R. Sarbu, C. Badulescu, L. Toth, Wastewater treatment by flotation, *Acta Montanistica Slovaca*, **(1)**, pp.97-99,(2005)
5. M. Ghouti, M. Al-Kaabi, M. Ashfaq, D. Da'na, Produced water characteristics, treatment and reuse: A review, *Journal of Water Process Engineering*, Vol **(28)**, pp. 222-239, (2019)
6. N. Pichel, M. Vivar, M. Fuentes, The problem of drinking water access: A review of disinfection technologies with an emphasis on solar treatment methods, *Chemosphere*, Vol. **(218)**, pp. 1014-1030, (2019)
7. M. I. Hussain, A. Muscolo, M. Farooq, W. Ahmad, Sustainable use and management of non-conventional water resources for rehabilitation of marginal lands in arid and semiarid environments, *Agricultural Water Management*, Vol. **(221)**, pp. 462-476, (2019).
8. E. Dopp, H. Pannekens, F. Itzel, J. Tuerk, Effect-based methods in combination with state-of-the-art chemical analysis for assessment of water quality as integrated approach, *International Journal of Hygiene and Environmental Health*, vol. **(222)**, pp. 607-614, (2019)
9. S. Ricart, A. M. Rico, Assessing technical and social driving factors of water reuse in agriculture: A review on risks, regulation and the yuck factor, *Agricultural Water Management*, vol. **(217)**, pp. 426-439, (2017)
10. I. Voicea, I.C. Moga, E. Marin, I. Găgeanu, G. Petrescu, Installation for the use of waste water from small agrozootechnical farms, *International Symposium ISB-INMA-TEH*, pp. 613-618, (2019)

Determining the qualitative working indices for various forms of the active body of a rut soil compartmentation equipment

Marius Remus Oprescu [1*], *Iuliana Gageanu*[1] and *Mario* Cristea[1]

[1]National Research - Development Institute for Machines and Installations Designed to Agriculture and Food Industry, 6 Ion Ionescu de la Brad Blv., 013813 Bucharest, Sector 1, Romania

Abstract. The subdivision of ruts represents the process of periodic formation between their ridges of small soil dams, small pools of water accumulation created in the processed soil with the help of equipment to open watering channels in order to reduce the phenomenon of erosion and to make more efficient use of the water from precipitation or by using sprinkler systems. This is a method of soil and water conservation that is used on inclined lands with a slope of less than 6° or on disturbed land. The practice began to be widely adopted following the use of new herbicide technologies necessary for weed control but also for improving the mechanical equipment for dam construction. In this paper are presented the results obtained from the optimization of the dimensions and the shape of the working body of the PCVM2,2 + EMBC2 equipment used in the compartmentation of the ruts in the vine-apple orchards with calculation of the qualitative indices of obtained work.

Key words: compartmentalized rut, erosion, water, active organ optimization, qualitative indices of work

1 Introduction

The soil represents the layer from the surface of the earth's crust and is composed of mineral particles, water, air, organic matter and living organisms. It is a non-renewable resource located at the intersection of land, water and air and fulfils a number of vital functions: the production of food or biomass, source of biodiversity, habitats, species and genes, storage and transformation of many substances, source of raw materials, heritage geological and archaeological, carboniferous basin, serves as a platform or physical environment for humans and human activities. The capacity of the land is limited and the demand for natural resources from society is increasing [1,6]. Water has been identified as one of the scarce resources, which can severely restrict and even compromise agricultural production and productivity, unless is carefully managed and conserved. [13]

In the case of sloping ground, the rainwater flows to the soil surface, thus not being used by plants and at the same time causing soil degradation through erosion, an extremely serious process encountered on one third of the agricultural surface of the country and at the bottom accumulating excess water which causes flooding of plants [1,6,15].

* Corresponding author: oprescu.remusmarius@gmail.com

Recent climate change has the potential to affect agricultural production due to changes in temperature and precipitations. Changes in the distribution and quantities of precipitation are of the most critical factors that influence the global climate. Under these conditions, the need arises to develop new technologies that increase the efficiency of the use of rainwater and to support the quality of the soil and the environment in order to obtain the most profitable agricultural products. Current scenarios predict that climate change will increase the water shortage in Southern Romania [1].

In the southern part of Romania, characterized by a temperate continental climate, there is a need to conserve "in-situ" soil moisture due to insufficient rainfall for agriculture [9]. Rainwater collection has the potential to reduce soil erosion and improve the productivity of these areas. Rainwater harvesting is a general term used to describe the collection and concentration of surface runoffs for various uses, including for agricultural and household use [2].

Sloped agricultural areas can lose much of the water from precipitation by surface runoff and large amounts of soil by erosion. The process of soil loss due to erosion has been increasingly studied in recent times [5].

The interrupted watering furrows are necessary on lands with slopes or slopes that cause the drainage and accumulation of water in micro-depressions, lands that have been arranged for irrigation by sprinkling, using fixed or mobile installations [8].

The "in situ" systems, Figure 1, are the simplest and the cheapest approaches for rainwater harvesting and can be practiced in many agricultural systems. Also called water conservation works systems, they involve the use of methods to increase the amount of water stored in the soil profile by capturing or maintaining water from precipitation [3,8,10]. Higher infiltration also means higher soil moisture content. Increased soil water content allows optimal plant development [5]. Studies establishing empirical equations to estimate the conservation efficiency of compartmentalized furrows have been conducted [11,12].

The use of compartmentalised furrows in semi-arid areas has significantly influenced agricultural production [4].

Fig. 1. Comparison of plot without micro-dams (left) and plot with micro-dams (right) at the end of the season [15].

In the case of irrigated crops, developing optimal water management for irrigation is important for water productivity and improving food security. Therefore, the introduction of knowledge on the need for water for crop irrigation is essential [5,9].

In the paper are presented the results obtained from the optimization of the dimensions and the shape of the working body of the PCVM2,2 + EMBC2 equipment used in the compartmentation of ruts in the vine-apple orchards with calculation of the qualitative indices of obtained work.

2 Material and method

The equipment for moulding the soil in compartmentalised furrows in vine-orchard plantations., PCVM2,2+EMBC2-0 (Figure 2) performs compartmentalized furrows at vine-tree plantations, in order to accumulate water from rainfall in the soil on which the drops fall, thus avoiding the flow of water outside the cultivated perimeter or accumulation in depressive areas, on lands with slopes up to 5 %, on soils with light, medium or heavy texture, at a depth of at least 250 mm, at a moisture close to the minimum threshold.

Fig. 2. Device testing at the Murfatlar Vineyard Research and Development Station [14]

The equipment for moulding the soil in compartmentalized furrows at vine-orchard plantations, PCVM2,2 + EMBC2-0 has the following main subassemblies: a clearance lister, a device for realizing compartmentalized furrows provided with a control mechanism and optionally two arrow blades if hoeing is desired.

Constructive description of the device for making compartmentalized furrows (Figure 3) consists in the following main parts: control mechanism, rotor support, pallet rotor and grounding mechanism of the racquet pallet. Adjusting the furrow compartmenting mechanism will allow for soil dams along the fence at distances of 1.5; 3 or 6.

Fig.3. Device for making compartment furrows

The control mechanism consists of: wheel with spur, a transmission and drive mechanism.

Wheel with spurs is metallic and provided with steel spurs on the hoop, with the role of increasing the adhesion of the wheel to the ground, thus avoiding its skidding. The wheel with spurts will be mounted articulated at the rear central part of the frame, having the possibility of vertical oscillation around the axis that drives the cams, in order to copy the land in the working process. For the transport position, the spinning wheel will be fixed upright.

Transmission it is of the chain type and has the role of transmitting the movement from the spinning wheel to the camshaft axle. The transmission consists mainly of: support, chain wheels, 10 A chain and mudguard.

The drive mechanism (Figure 4) has the role of unlocking the pallet rotor to form the ground plug on the furrow. It is composed of the following main parts: - camshaft holder, in 3 pieces (2 cams and one piece for the wheel with spurs); - cam wheel; - lever / cable and locking bolt.

Fig. 4. Drive mechanism.

In Figure 5 is presented the kinematic diagram of the fencing compartment equipment.

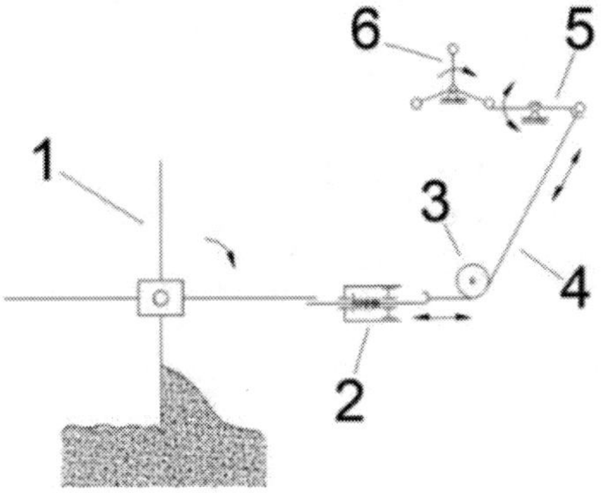

Fig. 5. Kinematic schematic of split compartment equipment.
1-pallet rotor, 2-locking element, 3-roller, 4-cable, 5-lever, 6- cam

The camshaft supports are mounted on the rear bar of the frame, behind the body and the camshaft and the lever will be mounted on the support. The steering wheel bracket is mounted on the bar behind the frame in the central position, in the direction of the steering wheel.

The cam wheel is made from a disc parallel to the bearing disc and the cam wheel is mounted on the cams (1,2 or 3), depending on the distance chosen to make the bushing compartment stops. The lever is mounted articulated on the support, in the direction of the cams and has at one end a roller and at the other end a bolt for fixing the steel cable. The cable has the role of transmitting the movement from the lever driven by the cam, to the locking pin of the rotor with pallets. The locking ratchet is made from an axis, with a welded plate at the end. The shaft slides into two translation couplings represented by two steel bushes fixed on the rotor support. An arc is mounted on the shaft which compresses when the lock mechanism is actuated and helps to lock the pallet when the control mechanism does not operate. The trigger mechanism operates safely and without locking.

The rotor support it is mounted on the sidebar of the frame, behind the body. The support consists of a vertical support, a fork and a support bar for the spring that presses the pallet rotor on the ground. The fork is hinged to the upright support, can swing freely vertically and supports the pallet rotor, the lower end of the spring of the squeegee pallet in the ground and the axle guide bushings of the rotor lock ratchet.

The rotor consists of 4 pallets of pentagonal shape arranged on an axis, the angle between 2 adjacent pallets has the value 90°. The pallets have the vertical outer side, a position that facilitates the approach of the working section to the plant row without the pallets damaging the plants.

The spring pressing the scraper pallet on the soil is mounted by means of a steel rod between the rotor support fork and the bar provided on the vertical support.

Working conditions

The experiments under operating conditions for determining the qualitative indices, were performed on the experimental group belonging to INMA-Bucharest. The research was carried out in a soil following a corn crop, the Universal 445-L tractor was used; The working speed was between 2.1; 2.5; 3.1 km / h; the working width was 1400 mm; the soil was of the chernozem type; main preliminary works consisted of ploughing and discussing; the dimensions of the initial rotor blade were: 260; 120; 260; and after optimization were: 400; 200; 260.

Determining the optimum section of the watering channel

In order to accumulate as much water as possible, it is necessary to have a surface as large as possible under the required conditions. In the case of watering furrows, the section has the form of a trapeze and in order to have minimal expenses, the surface of the trapeze must be maximum at a given perimeter. The perimeter used consists of the small base and the two oblique sides, Figure 6.

In order to express the function [11] in relation to the parameters of the problem, the following data are used: relation 1 is the area of the trapezoid section; relation 2 is the perimeter used, α, the angle between the large base B and the side l.

$$S = \frac{(B+b)h}{2} \tag{1}$$

$$P = 2l + b \tag{2}$$

The dimensions of the side *l* and the angle *α* are required so that the section of the trapezoid is maximum.

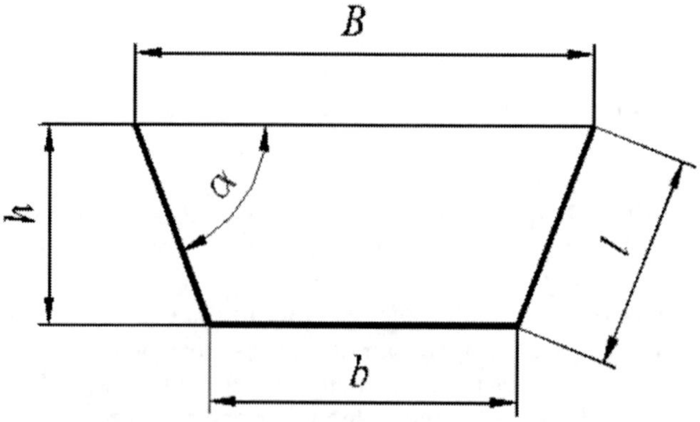

Fig. 6. The shape of the section of a channel

For a given perimeter *P*, the two variables are:

$$l = 0.333333 \cdot P \quad [m] \tag{3}$$

$$\alpha = 1.047197 \quad [\text{rad}] \tag{4}$$

and the maximum section is:

$$S = 0.144338 \cdot P^2 \; [m^2] \tag{5}$$

In this case we have h = 0.17 m, therefore:

$$l = \frac{h}{sin\alpha} = \frac{0,17}{0,866} = 0,1963 \quad [m] \tag{6}$$

Results:

$$b = l \cong 0,2 \quad [m] \tag{7}$$

$$B = 2,1 = 0,4 \quad [m] \tag{8}$$

The optimum width of the watering channel in the case of sowing crops sown at a distance between rows of 0.6 m so that the roots of the plants are not affected is 0.4 m.

Qualitative work indices for the dimensions of the rut made with the pallet rotor before and after optimization

During the tests under operating conditions, the procedures specific to the testing of the soil modelling devices were used and the following metrologically verified measuring and control devices and equipment were used: digital penetrometer with Fieldscout SC900 cone, portable HH2 moisture meter with Theta sensor ML2X samples, mechanical tape measure, mechanical chronometer, furrow-meter, rulers, square rulers, ranging rods. The soil moisture and compaction were determined at the time of testing and qualitative indications of work for the dimensions of the gullet made with the rotor with blades (Figure 7).

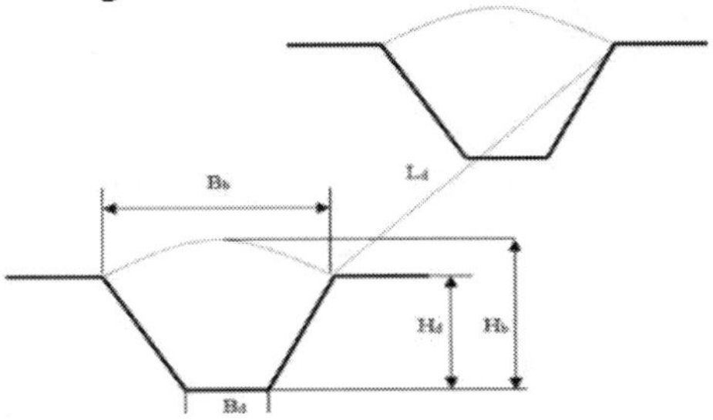

Fig. 7. The dimensions of the rut made with the rotor with blades.

(*Hb* - the depth of the watering channels, *Bd* - the width of the bottom of the furrow, *Bb* - the upper width of the furrow, *Hd* - the height of the dam, *Ld* - the distance between the dams.

Aspects from determining the qualitative indices of work are shown in Figures 8 and 9.

Fig. 8. a Determining the degree of ground clearance.

Fig. 8. b Determining the compactness of the terrain of the land.

Fig. 9. Determining qualitative indices.

The following measuring and control devices and equipment were used to determine the degree of ground clearance: sieve with 80 mm, 50 mm, 20 mm holes; 160 kg mechanical scale; metric frame; row.

The procedure for determining the degree of ground clearance is carried out as follows: at a distance of at least 20 m from the end of the plot, the metric frame is thrown and the surface of 1 m^2 is marked. Clogs larger than 100 mm are chosen and weighed and the rest of the soil, until the bottom of the furrow is taken with a shovel and passed in stages through the 80 mm, 50 mm and 20 mm sieves. All the fractions of the soil are summed and the percentage for each fraction is individually calculated: > 100 mm; 80-50 mm; 50-20 mm and < 20 mm. The procedure is repeated three times to obtain the most accurate results.

In Figure 10 are presented the four versions of blades with which the rotors were equipped and for which qualitative indices were determined for the equipment to open and compartmentalize the watering furrows.

a) straight blades with which the equipment was equipped until optimization

b) straight blade with optimized dimensions

c) inclined blade with optimized dimensions

d) curved blade with dimensions after optimization

Fig. 10. Types of blades.

3 Results and discussions

Tables 1 and 2 show the results of moisture and penetration resistance of the soil used for experiments.

Table 1. Soil moisture.

Volumetric moisture, % / depth, cm	**Value / measurement points**					
	1	**2**	**3**	**4**	**5**	**Average**
At surface	7.6	9.8	8.4	7.9	7.2	**8.18**
10	15.4	16.7	18.1	16.1	16.8	**16.62**
20	19.5	18.4	21.8	18.6	19.1	19.48

From table 1 it is shown that the moisture of the soil allows the experiments to be conducted under optimal conditions. A high soil moisture creates conditions for its sticking to pallets and therefore the impossibility of testing the equipment.

Table 2. Soil resistance to penetration.

Depth, cm	Penetration resistance, kN / Sample no.		
	1	2	3
0	16.5	17.5	17.3
2.3	17.5	18.1	17.6
5.0	22.8	21.6	22.5
7.5	38.1	28	31.5
10.0	58.3	50.1	54.1
12.5	66.7	52.6	59.3
15.0	158.2	94.7	80.2
17.5	177.4	138.8	147.5
20.0	140.4	170	155.2

From the analysis of the measured values regarding the soil resistance to penetration in conjunction with the classes of resistance of the soil to penetration (according to Methodology I.C.P.A., 1987) it resulted that in a small percentage the soil is extremely compacted.

Following the procedure for determining the degree of soil grinding on the plot where the tests were performed, the following results were obtained:

- fractions with a size greater than 100 mm, a percentage of 7 %;
- fractions with the size in the range of 50 - 80 mm, a percentage of 12 %;
- fractions with the size between 20 and 50 mm, a percentage of 38 %;
- fractions less than 20 mm in size, 43 %.

Because the fractions smaller than 50 mm are in a percentage of more than 80%, is indicated that the soil is sufficiently well-prepared for conducting the measurement in good conditions.

Figure 11 presents the qualitative indices obtained from the tests performed with the four versions of working bodies.

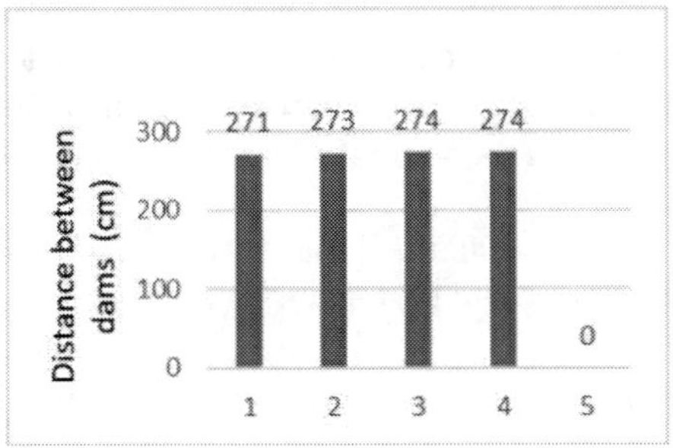

a

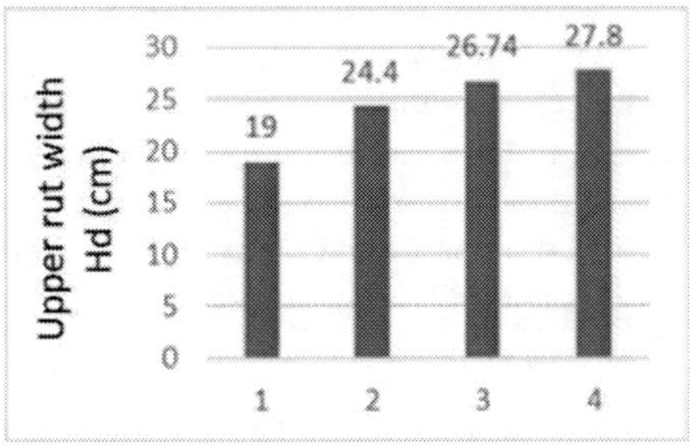

b

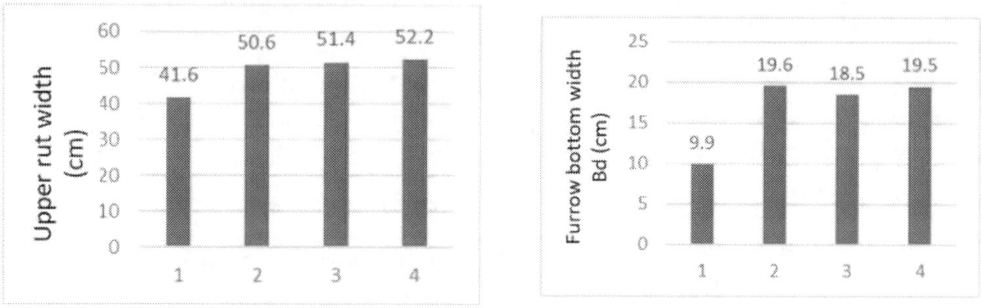

c d

Fig. 11. Qualitative indices obtained from the tests.

From Figure 11 a) it is shown that the only dimension that does not change is the distance between the dams. In Figure 11 b), the height of the dam is shown, where a significant increase from 19 cm is observed. for the rotor with the non-optimized straight blade up to 27.8.cm. In the case of the equipment equipped with rotor with curved blades to the dimensions obtained after optimization, a significant increase of about 50% due to the amount of material accumulated in front of the working body.

The upper width of the dam (Figure 11 c) has values higher than the size of the working body values obtained with the help of the opening angle of the rut, but in this case, we have differences of about 20% obtained as a result of changing the dimensions of the blades.

The width of the bottom of the furrow (Figure 11 d) varies from simple in the case of the initial equipping of the equipment's rotor to almost double in the case of using blades with the dimensions obtained after optimization (from 9.9 cm to 19.6 cm).

4 Conclusions

By measuring the qualitative indices, a significant difference is observed between the dimensions obtained with the initial version and those obtained with the optimized version, highlighting the influence that this dimension has on the volume of water to be stored by the compartmented furrow.

From the presented data, the following conclusions result:

- an increase of the section of the watering channel of more than 200 % following the modification of the dimensions of the working body and taking into account that the distance between the mini dams does not change significantly;

- a volume of water that can be stored between mini dams at least three times larger;

- comparing the height of the mini dam to the three constructive versions of blades made to the dimensions obtained from the optimization calculation results in a higher height in the case of the use of curved blades, therefore an amount of material accumulated in front of the larger working member in their case;

- After testing the operating conditions of the four versions of blades, it was shown that the version in which the curved blade was used obtained the best quality indices.

Following the experiments conducted according to the standards in force, qualitative work indices were obtained that fall within the agrotechnical requirements of the soil modelling work.

This work was supported by Romanian Education and Research Ministry, through *Programme 1 – Development of the national research-development system, subprogramme 1.2 – Institutional performance – Projects for financing excellence in RDI, contract no. 16PFE* and European Social Fund from the Sectoral Operational Programme Human Capital 2014-2020, through the Financial

Agreement with the title *"Scholarships for entrepreneurial education among doctoral students and postdoctoral researchers (Be Antreprenor!)"*, Contract no. 51680/09.07.2019 - SMIS code: 124539.

References

1. A.C. Marica, A. Busuioc, The potential of climate change on the main components of water balance relating to maize crop, Romanian J. Meteor. 6 (1-2), pp. 50-57, (2004)
2. FAO, Soil tillage in Africa: needs and challenges, FAO Soils Bulletin 69. Rome, (1993).
3. G. Brhane, C.S.Wortmann, M. Marno, H. Gebrekidan, A. Belay, Micro basin tillage for grain sorghum production in semiarid areas of northern Ethiopia, Agron. J. 98, pp. 124-128, (2006)
4. Haytham M. Salema,b,, Constantino Valero , Miguel Ángel Muñoz ,4 María Gil-Rodríguez-Effect of integrated reservoir tillage for *in-situ* rainwater harvesting and other tillage practices on soil physical Q1 properties februarie 2015
5. Hintsa Libsekal Gebremariam, Kidane Welde, Kiflom Degef Kahsay Sustainable Water Resources Management, December 2018, Volume 4, pp. 1043–1049
6. 6. I. Biolan, I. Serbu, G.C. Tusa, F. Mardare, Irrigation of agricultural crops-Technologies, AGIR Publishing House, Bucharest, (2016)
7. I. Rellinia , C. Scopesia , S. Olivarib , M. Firpoa and M. Maerker Assessment of soil erosion risk in a typical Mediterranean environment using a high resolution RUSLE approach JOURNAL OF MAPS 2019, VOL. 15, NO. 2, 356–362
8. M. R. Oprescu, D. N. Dumitru, I. F. Voicea, S. S. Biriş, Considerations regarding the use of the equipment for opening and compartmentalizing the watering furrows, International Symposium ISB-INMA TEH Bucharest 31st October – 1st November, pp. 381-386, (2019)
9. .R. M. Oprescu, I. F. Voicea, I. Dumitru, I. Dragan, C.Biolan, N. Ungureanu, E. A. Neagu, N. Bara , D. Jaloba, V. N. Arsenoaia, Rain Use Efficiency in Southern Romania after Using the Device for Soil Modelling in Interrupted Furrows for Weeding Crops - DMBC-5, E3S Web of Conferences 112, 03005 (2019), https://www.e3s-conferences.org/articles/e3sconf/pdf/2019/38/e3sconf_te-re-rd18_03005.pdf
10. S. ANDRIEŞ, V.FILIPCIUC (2016) Research in the field of soil erosion
11. Y. Liu, Yan Xin, Y Xie , W. Wang, Effects of slope and rainfall intensity on runoff and soil erosion from furrow diking under simulated rainfall, CATENA,Volume 177, June 2019, Pages 92-100, https://doi.org/10.1016/j.catena.2019.02.004
12. V. Moise, E. Maican, Şt. I. Moses, Numerical Methods. Applications in MATLAB. Printech Publishing House, Bucharest, (2016).
13. T. Gebremedhin, , Ethiopia. Effect of Drip and Surface Irrigation Methods on Yield and Water Use Efficiency of Onion (Allium Cepa L.) under Semi-Arid Condition of Northern Ethiopia Journal of Biology, Agriculture and Healthcare www.iiste.org ISSN 2224-3208 (Paper) ISSN 2225-093X (Online) Vol.5, No.14, 2015
14. https://www.agrofood.ro/Article-Murfatlar-to-device-test-to-shape-new-in-furrow soil-divided/3564/2019
15. 1https://orbi.uliege.be/bitstream/2268/163965/1/Olivier.FIWAP.2012.pdf

Evaluation of carbon dioxide released by bread dough during proving stage

Adriana Istudor [1,*], *Gheorghe* Voicu[1], *Gheorghe* Muscalu[2] *and Paula* Tudor[3]

[1] Politehnica University of Bucharest, Faculty of Biotechnical Systems Engineering, 313 Splaiul Independentei, 060042, Bucharest, Romania
[2] Bio Tehnologi Creativ, Sos. Dudesti-Pantelimon nr 42, Bucuresti, Romania
[3] Politehnica University of Bucharest, Faculty of Entrepreneurship, Business Engineering and Management, 313 Splaiul Independentei, 060042, Bucharest, Romania

Abstract. This study presents the analysis of multiple bread dough proving processes with the purpose of establishing a correlation between the concentration of released carbon dioxide during fermentation and the working parameters (time and temperature). The testing was performed using a standard recipe for white bread dough, a small capacity prover with air conditioning unit for temperature and relative humidity regulation, CO_2 and temperature sensors with data acquisition plate. In the first part are presented the results for 11 measurements of CO_2 concentration for one dough piece fermentation process at varying proving temperatures. In the second part of the paper are presented the results of 4 measurements of CO_2 for 9 dough pieces proving at temperatures between 30 and 39 °C, at 3°C intervals. The obtained measurements of CO_2 were correlated with the volume and dimensions of the finished products. The obtained results are considered relevant for this study and for the possibility of fermentation level evaluation using the quantity of CO_2 released during proving. The presented study is part of an extensive research performed for the identification of a method for automated control of working regime in industrial bread dough provers using the measurements of released CO_2.

1 Introduction

Bread making has a very important place in the food manufacturing sector and is in full process of expansion and automation. The proving process of bread dough is the largest stage of the technological process. It starts from kneading and continues during all the other operations and the first part of baking, but the final proving determines the structure of the bread, [1, 2] and its overall quality. The proving process takes place in enclosed spaces called provers, [3], which are usually air-conditioned. The purpose of proofing is to allow

*Corresponding author: istudor.adriana@mgbiotech.ro

the expansion of the gas cells present in dough, by accumulation of carbon dioxide produced by yeast. In dough, carbon dioxide can be found as gas trapped inside the cells and dissolved in the aqueous phase, [4, 5]. Acting as a sensitive foam structure carbon dioxide trapped in dough is difficult to determine. Evaluation of gas quantity produced during fermentation is frequently performed using volumetric methods, such as pressure variation [6-10]. In this study, was evaluated the influence of different temperatures on the carbon dioxide production during final fermentation of dough. This experimental research wishes to determine the possibility of correlating the quantity of carbon dioxide released during dough proving with the expansion level of dough in a given time and controlled conditions.

2 Equipment and method

During the experiments, a convection oven with proving cabinet was used. The oven is Miwe Aeromat, with 10 trays of 400 x 600 mm capacity and has an integrated command panel for temperature and humidity control during baking and proving. The measurements for CO_2 (ppm) were performed using a measuring and data acquisition device (Trotec CO_2 BZ25) which offers the possibility of downloading the data on a computer using a dedicated software. The temperature and working humidity were further monitored using a metrological verified thermos hygrometer. The equipment used is presented in figure 1.

The standard bread recipe used in the experiments had in its composition the following ingredients and percentages reported to flour: white flour, FA, 0.65 extraction, water (53%), salt (1.8%) and compact yeast (3%). The physio-chemical characteristics of the used flour are presented in table 1.

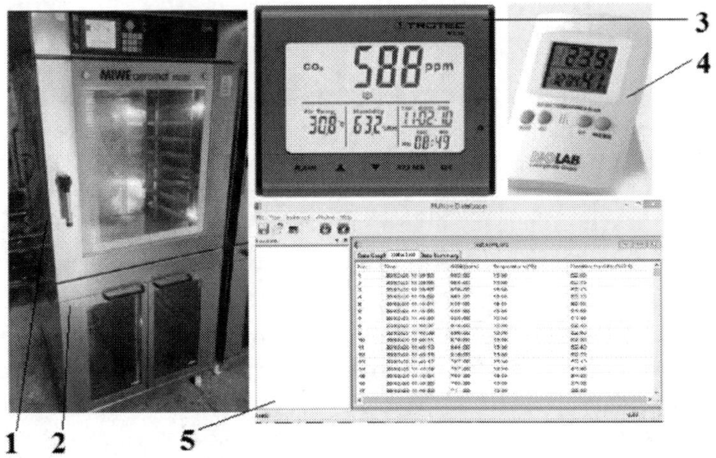

Fig. 1. The equipment used during the experiments: 1. convection oven; 2. proving cabinet; 3. CO_2 measuring device; 4. Thermos hygrometer; 5. CO_2 data software

Dough pieces with 29 °C were divided in 360 g pieces each, round shaped and after 10 minutes were shaped in long cylindrical form, at 24 cm, 4.5 cm width and 4 cm height. Each proving experiment was performed for 3600 seconds at controlled temperature, in which time the CO_2 device registered the concentration of released carbon dioxide in the proving chamber.

At first, the concentration of carbon dioxide in the environment was measured (without load). At each experiment, the measurement of carbon dioxide started at 586 ppm point value.

Table 1. The physio-chemical characteristics of the used flour

Flour	Humidity (%)	Protein (%)	Ash (%)	Falling Number (s)	Wet gluten (%)	Acidity (°Turner)
FA	13,5	10,2	0,65	308	27	1.8

In the first part of the paper, measurements of released carbon dioxide were taken for one dough piece fermented at temperatures between 30 °C and 40 °C, at 1°C interval. Based on the results obtained in the first part, the working temperatures for the second part of the paper were established. The concentration of released carbon dioxide was measured for 9 dough pieces of 360 g each at proving temperatures between 30 °C and 39 °C, at 3 °C intervals. The number of dough pieces was established taking into consideration the tray dimensions in which 3 dough pieces were placed and the proving volume necessary for optimal dough development considering the total proving volume of 0.36 m^3.

In order to establish the fermentation level in correlation with the registered CO_2 quantity and the working temperature, dough dimensions were measured (length, width, height) after each experiment. The optimal fermentation level was determined using the standard dimensions presented in STAS 91/1983 for white bread of 300g weight whose optimal dimensions are: length 28-30 cm, width 10-11 cm, height 7-8 cm.

3 Results and discussion

The results obtained in the first part of the study, where 11 measurements of CO_2 concentration were performed for one dough piece in 3600 s fermentation process at temperatures between 30 °C and 40°C are presented in figure 2. The purpose of these determinations was to evaluate if 1°C in the proving chamber has a significant influence on the fermentation level of dough and also to establish the optimal temperature interval at which the CO_2 release register modifications.

Analysing the graphics in figure 2, it can be observed that for the 30 -33 °C interval the CO_2 development curves are linear and the CO_2 quantity difference at 1 °C interval does not surpass 6 ppm at the end of fermentation time. The same phenomenon takes place at other temperature intervals. From 34 °C to 35 °C, the CO_2 difference is 7 ppm. For the next three temperature values, the CO_2 release curves become linear in the second part of the fermentation process. At 36 °C, the CO_2 value registered at the end of fermentation was 292 ppm, at 37 °C, was 307 ppm and at 38 °C the CO_2 value was 315 ppm. It can also be observed the difference between the CO_2 values registered at 33 °C and 36 °C which is 63 ppm, thus obtaining a clear difference between these two fermentation processes. At 39 °C, the final value of CO_2 was 351 ppm, with 36 ppm more than the final value of CO_2 registered at 38 °C which means that after 37 °C de development curve does not have a linear pattern.

https://doi.org/10.1051/e3sconf/202018003012

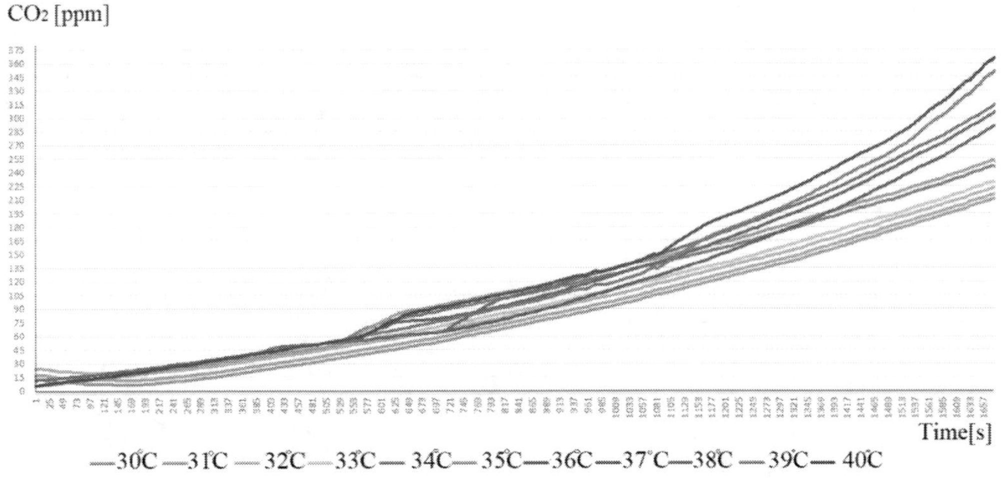

Fig. 2. CO_2 values registered during one dough piece fermentation at 11 different temperatures

In figure 3 are presented the differences between the final CO_2 values registered during these 11 measurements.

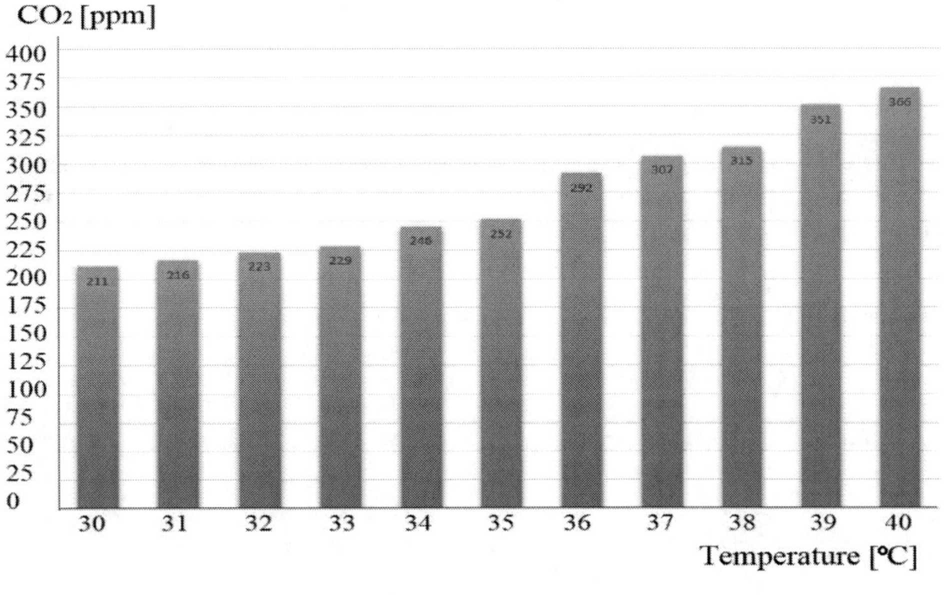

Fig. 3. The CO_2 values registered at the end of fermentation time for individual dough pieces at temperatures between 30 °C and 40 °C

Taking into consideration the results obtained in the first part of the paper, it was established that the relevant dough fermentation temperatures are at 30 °C, 33 °C, 36 °C, and 39 °C. In figure 4 are presented the comparative values between the CO_2 quantity released during the fermentation process at these temperatures and the obtained volume of bread pieces.

Taking into consideration the percentage difference between the second and the first values it can be observed that there is a correspondence between the CO_2 values and the obtained volume. Therefore, at 33 °C, in comparison with 30 °C, the concentration of released CO_2 and the volume of bread increase with 7.9 % and 7.5 % respectively. At 36°C, the obtained results show an increase of 17.3 % for the volume of bread and 21.6 % for the concentration of CO_2. From 36°C to 39°C, the increase in volume reduces to 14 % and the CO_2 release to 16.8 %.

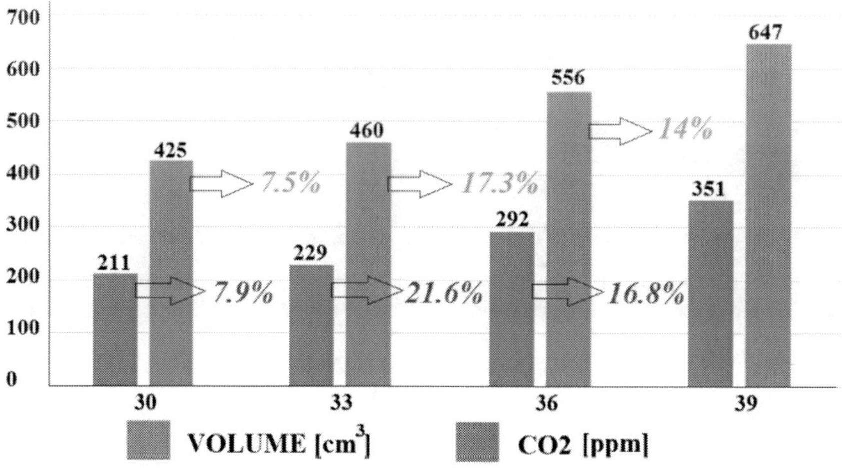

Fig. 4. The comparative percentage values between the volume of bread and the CO_2 quantity registered at 4 different temperatures

In the second part of the paper, measurements were performed for the quantity of CO_2 released during the proving process of 9 dough pieces of 360 g each. The obtained results are presented in figure 5.

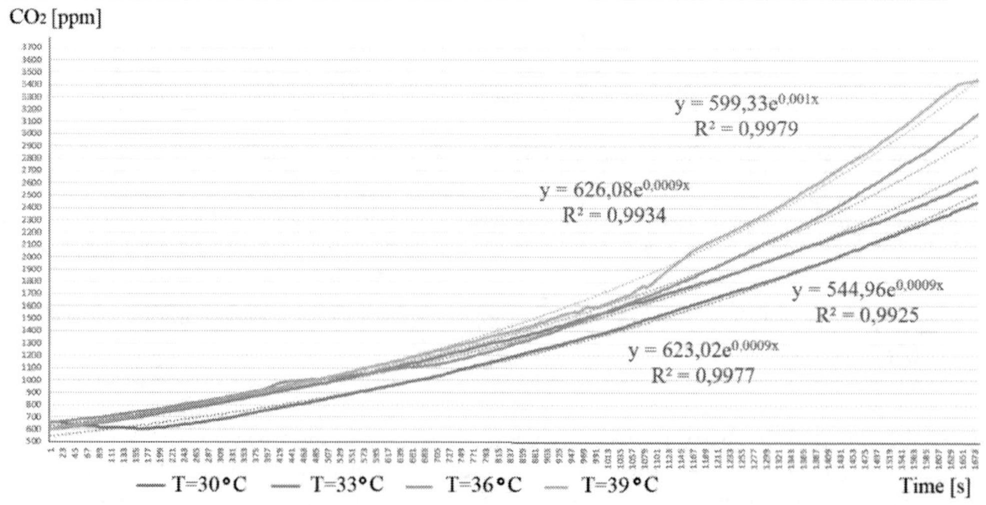

Fig. 5. CO_2 release curves for 9 dough pieces and the correlation coefficient R^2

From the analysis of CO_2 curves obtained during the process fermentation at temperatures between 30 °C and 39 °C at 3 °C intervals, it can be observed that the first 2 CO_2 curves register a constant development until the end of fermentation and the other 2 curves, at 36 °C and 39 °C register a rapid accumulation of CO_2 in the second part of the fermentation process. Thus, at the end of the fermentation time, the values obtained for the CO_2 quantity were: 1900 ppm at 30 °C, 2057 ppm at 33 °C, 2624 ppm at 36 °C and 2892 ppm at 39 °C.

The CO_2 release curves during the fermentation process of wheat dough are defined by the exponential relation: $y=b\cdot e^{a\cdot x}$. The correlation coefficients R^2 obtained for all the four experiments had over 0.99. This high values satisfy the exigence level required for these kind of measurements which have as main purpose the determination of the level of fermentation of dough in correspondence with the proving temperature and released CO_2 quantity during a specific time.

After each experiment, the dimensions of dough pieces were measured. The obtained mean values are presented in table 2.

Table 2. Dough pieces dimensions after final fermentation

Dimension	CO_2/temperature			
	1900ppm/ 30°C	2057ppm/ 33°C	2624ppm/ 36 °C	2892ppm/ 39 °C
L (cm)	29	29	29	30
l (cm)	9.5	10	11	12.4
H (cm)	6	6.5	7.5	6

3 Conclusions

In this paper it was analysed the influence of proving temperature on the final fermentation level of wheat dough based on the concentration of released carbon dioxide in the proving cabinet.

In the first part of the paper, 11 fermentation test were performed at temperatures ranging between 30 °C and 40 °C, at 1 °C interval in order to determine the influence of 1 °C on the fermentation intensity. Based on the CO_2 release curves it was observed that temperature intervals are created, at 3 °C approximately, in which the fermentation intensity level registers little differences in the CO_2 values (not more than 6 ppm). The CO_2 value difference between 33 °C and 36 °C was 63 ppm and after 37 °C, the CO_2 concentration increase up to 40 ppm at 1 °C intervals. From the analysis of the characteristic curves presented in figure 2, it can be observed that the quantity of CO_2 released during the fermentation process at temperatures between 37-39 °C increases in a more accelerated rate in the second part of the fermentation process due to the temperature increase in the dough piece.

The increase in CO_2 concentration related to higher temperatures is very much influenced by the physio-chemical characteristics of the flour used in preparing the dough, the recipe, more specifically the consistency of dough and the quantity of yeast. Of much importance is also the temperature of dough at the fermentation starting point. Therefore, using a softer dough with a high quantity of yeast will result in a higher concentration of CO_2 during the fermentation process.

Although the highest volume of bread was obtained after dough was fermented at 39 °C, it can be observed in table 2 that the dimensions of the dough pieces do not correspond with the desired quality level. Considering the measured dough dimensions, the best development of dough and of CO_2 release was registered around the 36 °C temperature point.

From the analysis of the results obtained in the second part of the paper, the fermentation process of dough is exponential. The accuracy of the performed measurements was validated using the correlation coefficients (R^2), whose values registered for all 4 experiments were higher than 0.99.

The quantity of released CO_2 during the fermentation process was correlated with the measured dimensions of dough pieces after each experiment.

Taking into consideration the value intervals registered in STAS regarding the quality level of white bread of 300g, it was established that the best product with the most correct dimensions regarding the report L/W/h was the one fermented around the 36 °C temperature, with a CO_2 release value in an hour of 2624 ppm, in a closed proving cabinet of 0.36 m^3.

Further research regard the analysis of CO_2 quantity in an industrial prover with continuous functioning during the fermentation process and the possibility of working regime control using the CO_2 released quantity.

This work has been funded by the European Social Fund from the Sectoral Operational Programme Human Capital 2014-2020, through the Financial Agreement with the title "Scholarships for entrepreneurial education among doctoral students and postdoctoral researchers (Be Antreprenor!)", Contract no. 51680/09.07.2019 - SMIS code: 124539.

References

1. A. Istudor et al, Final bread dough fermentation – requirements, conditions, equipment. A short review, INMA-TEH, (2017), pp. 519-526;
2. P. Shah, G. Campbell, S. McKee, & C. Rielly, Proving of bread dough: modelling the growth of individual bubbles. Transactions of the Institution of Chemical Engineers, Part C, 76, (1998), pp. 73–79;
3. G. Voicu, *Procese şi utilaje pentru panificaţie*, E. Bren, Bucureşti, (1999);
4. R.C. Hoseney, *Principles of cereal science and technology*, 2nd ed., Minnesota, USA: AACC Inc., (1998);
5. M.G Scanlon, M.C. Zghal, Bread properties and crumb structure, FRI, **34**, (2001), pp. 841-864;
6. A. Istudor, G. Voicu et. al, Evaluation of wheat dough fermentation process using a reactor and mathematical modelling, SGEM, **18**, (2018), pp. 161-168;
7. I. Svec, M. Hruskova, Wheat flour fermentation study, CJFS, **22**, 2004, pp. 17-23;
8. A. Ktenioudaki, F. Butler, E. Gallagher, Dough characteristics of Irish wheat varieties II. Aeration profile and baking quality. LWT-FST, **44**, 2011, pp. 602-610;
9. J. Šavel, P. Košin, A. Brož, Calculation of pressure growth during fermentation in a closed container, Kvasny prum **9**, 60, (2014), pp 233-238;
10. G. E. Rattin, J. M. Faubion, C.E. Walker, A.L. Mense, Measuring yeast CO2 production with the Risograph, CFW, **54**, 6, (2009), pp: 261-265;
11. *** STAS 91/1983, available at http://www.rompan.ro/analize-de-laborator/, accessed in 22.03.2020.

E3S Web of Conferences **180**, 03013 (2020) https://doi.org/10.1051/e3sconf/202018003013
TE-RE-RD 2020

Evaluation of hand transmitted chainsaw vibrations during wood cutting

Mihai Gabriel Matache[1*], *Mariana* Munteanu[1], *Dragos Nicolae* Dumitru[1] and *Mariana* Epure[1]

[1]INMA Bucharest, Testing Department, 013813 Bucharest, Romania

Abstract. Forestry workers are exposed to vibrations from chainsaws which usually are not taken into consideration from the negative impact point of view on their health. However, in time, these vibrations could cause trepidation disease which will affect the worker neuro-muscular and locomotory apparatus. One mean to mitigate the vibrations impact is to measure and determine the worker's exposure level to vibrations during worktime and to dispose the proper measures to reduce their negative impact. Within this paper will be presented experimental results obtained for vibration data recorded on a chainsaw during wood cutting in real life conditions. The equivalent vibration was measured at both front and end handles level and the biggest value was used for calculation of the human exposure level.

1 Introduction

In forestry there are a lot of works which include the usage of a chainsaw. A chainsaw is a portable engine driven mechanical saw which uses a set of teeth attached to a rotating chain that runs along a guide bar to cut through different materials, usually wood. It is used to do works such as tree cutting, pruning, harvesting of firewood and many other activities. The main source of vibration for a forestry worker is its chainsaw, which usually works at very high rpm and is handheld so the vibrations are transmitted to the hand-arm system through palms and fingers. Workers exposed daily to excessive vibrations transmitted to the hand-arm system could suffer, on long term, blood flow disorders in fingers and disorders of neurological functions and movements of hand and arm [1].

Hand-arm vibration occur when the handles of a power tool or the surface of a work item oscillates rapidly, and this movement is transmitted into the operator's hand-arm system. Hand-arm vibration (HAV) is usually felt by operators who regularly use handheld equipment like chainsaws, drills, rototillers and trimmers [2]. The factors which influence the effect of vibrations on human body are the oscillations intensity, the distance between the vibration source and the individual, the body position, the period of time and the exposure level, the vibration frequency spectrum and their direction. The most adverse effects are produced by vibration of frequency close to the first order vibration of the human body (6-9 Hz).

* Corresponding author: gabimatache@yahoo.com

The European Directive 2002/44/CE on the minimum health and safety requirements regarding the exposure of workers to the risks arising from vibrations impose specific requirements for the safety of workers exposed to mechanical vibrations. Through this directive two exposure indicators for the hand-arm system transmitted vibrations have been defined: the action value and the exposure limit value. The overcoming of these two values lead to a set of obligations. The daily exposure action value on the hand-arm system, standardised to an eight-hour reference period, A (8) shall be 2,5 m/s^2 RMS, while the daily exposure limit value, standardised to an eight-hour reference period, shall be 5 m/s^2 RMS [3].

The method to assess the exposure level to vibrations is standardised by EN ISO 5349 Mechanical vibration- Measurement and evaluation of human exposure to hand-transmitted vibration [4]. Based on this, various researchers studied the hand-arm transmitted vibration level by various handheld tools, usually by direct measurements during work, using three axes accelerometers mounted on the handles [5, 6, 7]. The obtained results differed in function of the testing conditions and of the used power tools. Also, there were conducted theoretical and experimental researches in order to forecast the exposure risk for workers subjected to HAV [8].

Within this paper are presented the experimental results obtained for the human exposure level during wood cutting with a chainsaw. The weighted vibrations were measured using an orthogonal reference system in full load operation mode and were used to assess the exposure level during a typical day of work.

2 Material and method

Tests were performed on the PTR 565 type chain-saw. The chainsaw was driven by a 54 cm^3, 2 strokes gas engine. The weighted vibration testing method was based on ISO 22867 - Forest Machines, Vibration Testing Code for hand-held portable machines which provides general specifications for vibration measurement of hand-held machines [9].

Vibrations were measured as accelerations in m/s^2. The weighted accelerations on the three perpendicular directions of the orthogonal measuring system: a_{hwx}, a_{hwy}, a_{hwz} were measured using two 3-axis accelerometers for both front and rear handles of the chainsaw. The total vibration value, a_{hwv} was computed for each handle.

$$a_{hwv} = \left[\left(a_{hwx}{}^2 + a_{hwy}{}^2 + a_{hwz}{}^2\right)\right]^{\frac{1}{2}} \quad (1)$$

The daily exposure level A(8) for a reference period T_0 of 8 hours was obtained from total vibration values a_{hwv} and the daily exposure time T.

$$A(8) = a_{hwv}\sqrt{\frac{T}{T_0}} \quad (2)$$

The three-axes accelerometers model 356B10 from PCB Piezotronics were mounted on the front and rear handle of the chainsaw using wax. For data recording was used a 16 channels SIRIUS data acquisition system produced by DEWESoft. The accelerometers calibration was performed using a Bruel&Kjaer calibration exciter 4294 type with a 10 m/s^2 RMS vibration level at 159.2 Hz. The digital weighing filter to the vibration data was applied directly during measurements and formula (1) was used to calculate the total vibration level. The experiments were performed on trees with diameters of approximately 0.4 m, which were cut down due to their weak condition. The chainsaw was accelerated from idle to full throttle and kept there until the work was done. The data acquisition periods were from the beginning to the finishing of the cut. For measuring the weighted vibration level were recorded 10 replicas of the cutting experiments and the mean value of the vibrations was reported. In order to estimate de daily exposure time to vibrations, the real exposure time was timed using a chronometer during a typical cutting tree cycle and then was multiplied by the number of cut trees during an 8-hour work day.

Fig. 1 Test setup for vibration measurement.

3 Results

In figure 2 are presented the unweighted time series vibration measured on the front handle of the chainsaw a_{hx}, a_{hy}, a_{hz} during one tree cutting experiment. The OY axis corresponds to the longitudinal median plane of the saw, OX axis is perpendicular on the median plane and OZ axis is perpendicular on the surface of the soil with the machine in normal position. During work, these positions were turned by approximately 60-70 degrees, in a rolling motion.

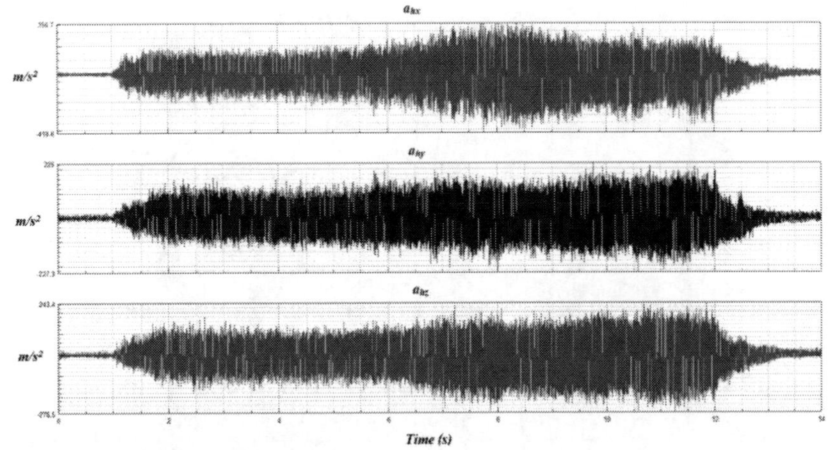

Fig. 2 Front handle chain-saw vibrations a_{hx}, a_{hy}, a_{hz}.

In figure 3 are graphically displayed the same vibrations, this time zoomed in order to better display their real oscillations. The sampling frequency for data acquisition was set at 10 kHz, so that all relevant information to be recorded.

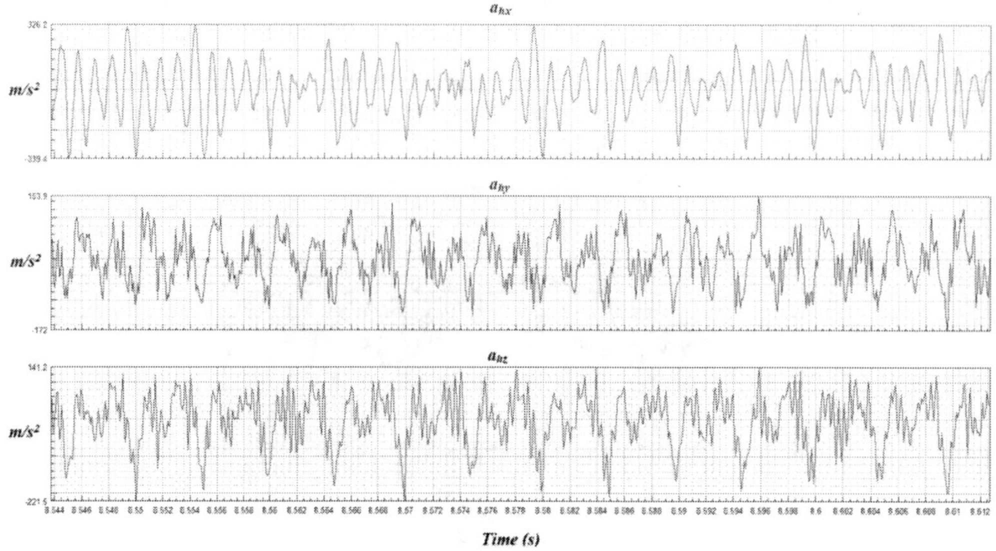

Fig. 3 Front handle chain-saw vibrations a_{hx}, a_{hy}, a_{hz} – zoom in.

In figure 4 is presented the frequency spectrum histogram, computed for the same set of data presented in figure 1. We can observe the first order frequency around 200 Hz, which is the same as that of the chainsaw full throttle 12000 rpm. The other frequencies correspond to harmonics of the first one. There frequency range presented was limited to 1250 Hz, because of the applied hand-arm weighting filter, which also works up to the same limit.

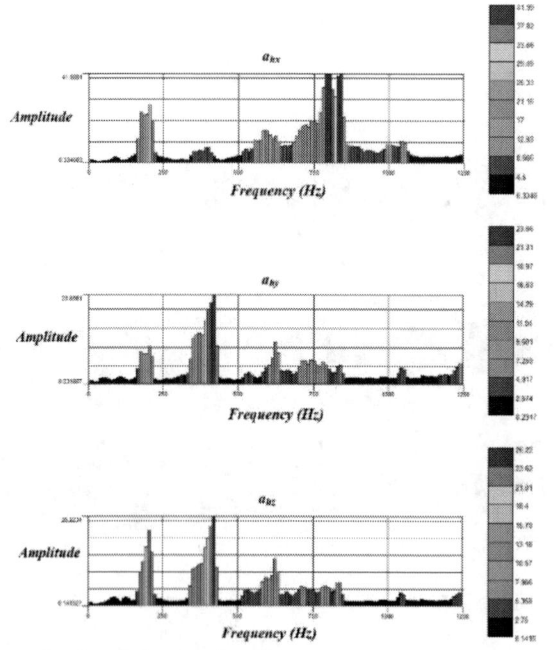

Fig. 4 Frequency spectrum for the chainsaw vibrations.

By applying the frequency weighing filter, the amplitude of vibrations dropped significantly. In figure 5 is presented graphically the applied weighing filter W_h according to EN 5349 and in figure 6 are presented the same vibrations as in figure 1, after weighting.

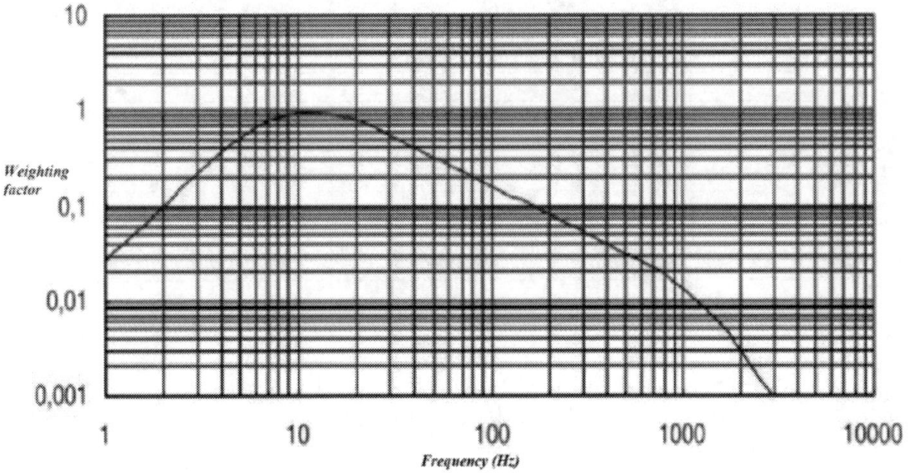

Fig. 5 Weighting filter W_h.

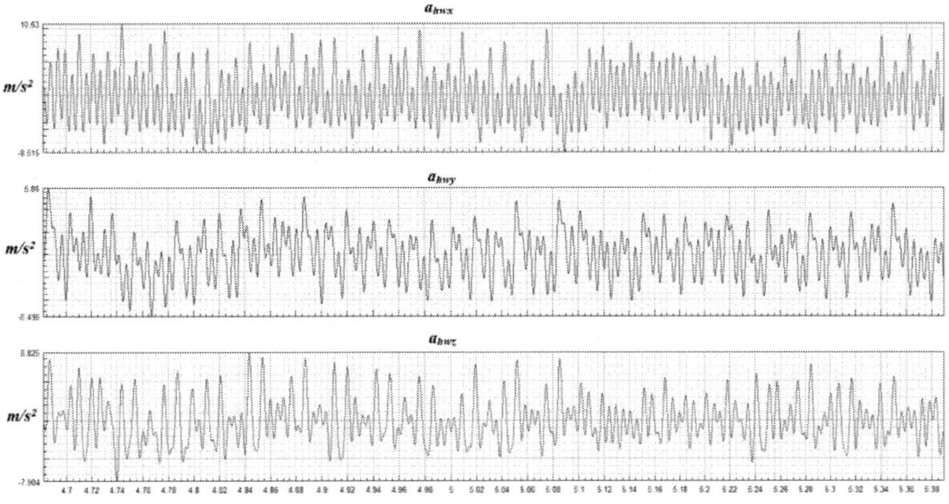

Fig. 6 Front handle chain-saw weighted vibrations a_{hwx}, a_{hwy}, a_{hwz} – zoom in.

In table 1 are presented the values obtained for the total vibration values a_{hwv} obtained for all the 10 replicas of the tree cutting experiment, both for front and rear handle and their mean value after computing the average.

Table1. Total weighted vibration values for front and rear handles of the chainsaw

Crt. no	Front handle - a_{hwv} (m/s^2)	Rear handle - a_{hwv} (m/s^2)
1.	4.84	4.24
2.	4.32	3.88
3.	4.98	4.33
4.	5.02	4.41
5.	5.21	4.48
6.	4.61	4.03
7.	4.58	4.01
8.	4.14	3.66
9.	4.35	3.78
10.	4.78	4.03
Mean value	**4.68**	**4.09**
Standard deviation	**0.34**	**0.27**

Because of the different widths of the tree trunks the duration of the experiments was also different. However, the total vibration results scatter between the experiments was of 7.26% for the front handle vibration and 6.6 % for the rear one. This scatter under 10% assures us that the results are representative for the chosen work and could be used for the assessment of the human exposure level to vibrations during chainsaw handling.

Further on, the tree felling cycles periods for each experiment were timed. The felling cycle was considered as being formed of 4 effective cutting periods: the superior wedge cutting period, the inferior wedge cutting period, rest of the trunk cutting period and trimming of the felled tree. The idle engine chainsaw handling was omitted. In table 2 are presented the results obtained after time measuring of the cutting periods.

Table2. Tree cutting felling cycle periods

Crt. no	Superior wedge cutting period (s)	Inferior wedge cutting period (s)	Trunk cutting period (s)	Trimming (s)	Total time (s)
1.	6	5	15	624	650
2.	6	6	17	748	777
3.	5	5	14	657	681
4.	7	5	16	504	532
5.	6	6	15	687	714
6.	8	6	18	724	756
7.	7	7	18	726	758
8.	6	4	14	806	830
9.	5	4	14	705	728
10.	6	6	15	741	768
Mean value	**6.20**	**5.40**	**15.60**	**692.20**	**719.40**
Standard deviation	**0.92**	**0.97**	**1.58**	**83.07**	**83.17**

The mean time consumed for one tree felling was considered to be around 20 minutes, fact which means that during a work day could be cut down around 24 trees. This means that the daily exposure time T for cutting the tree was of 17265.6 seconds, or **4.796** hours.

Applying formula (2) for calculation of daily exposure level in which we considered the mean values reported in tables 1 for total vibration, we obtained **3.60** m/s^2 for the front handle and **3.15** m/s^2 for the rear handle, values which are in the intermediate situations between the level of action and the exposure limit values (2,5 m/s^2 < A(8) < 5,0 m/s^2).

4 Conclusions

After the vibration measuring experiments performed for the wood cutting work with a chainsaw were observed high amplitude vibrations. The maximum weighted acceleration on the handles of the chainsaw showed values which could damage on long term the operator's state of health, fact that suggest the need of intervention and health monitoring.

The results for the daily exposure level of the chainsaw operator to vibrations for both handles of the equipment were located between the daily exposure action value on the hand-arm system, standardised to an eight hour reference period, A(8) = 2,5 m/s^2 RMS, and the daily exposure limit value, standardised to an eight-hour reference period, of 5 m/s^2 RMS, imposed by European Directive 2002/44/CE.

This work has been funded by the European Social Fund from the Sectoral Operational Programme Human Capital 2014-2020, through the Financial Agreement with the title "Scholarships for entrepreneurial education among doctoral students and postdoctoral researchers (Be entrepreneur!)", Contract no. 51680/09.07.2019 - SMIS code: 124539.>>

References

1. Forouharmajd F, Yadegari M, Ahmadvand M, Forouharmajd F, Pourabdian S. Hand-arm Vibration Effects on Performance, Tactile Acuity, and Temperature of Hand. J Med Signals Sens. 2017;7(4):252-260.
2. Vladimir Šinik, Željko Despotović, Slobodan Janković, "Exposure to hand-arm vibration, tools for calculating daily exposures and measurment", V International Conference „Ecology of Urban Areas“ 2016, Zrenjanin, 30 september, 2016
3. Directive 2002/44/EC of the European Parliament and of the Council of 25 June 2002 on the minimum health and safety requirements regarding the exposure of workers to the risks arising from physical agents (vibration) (sixteenth individual Directive within the meaning of Article 16 (1) of Directive 89/391/EEC)
4. EN ISO 5349 Mechanical vibration- Measurement and evaluation of human exposure to hand-transmitted vibration.
5. Monarca D., Biondi P., Cecchini M., Santi M., Guerrieri M., Colantoni A., Colopardi F., Transmission of vibrations from portable agricultural machinery to the Hand-Arm System (HAV): risk assessment and definition of exposure time for daily action and exposure limits, International Conference: September 15-17, 2008 Ragusa – Italy, "Innovation Technology to Empower Safety, Health and Welfare in Agriculture and Agro-food Systems"
6. Sorica E., Vladut V., Cardei P., Sorica C., Bracacescu C., 2017. Comparative analysis of the noise and vibration transmitted to the operator by a brush cutter, Acoustics and Vibration of Mechanical Structures—AVMS-2017: Proceedings of the 14th AVMS Conference, pp. 165-172.
7. Búrdalo Salcedo G., Ortiz E., et. al. (2014). Hand-Arm Vibration (HAV) produced by pneumatics screwdrivers used in a wind power company, International Symposium on Occupational Safety and Hygiene – SHO 2014
8. Vasilescu G. D., Ghicioi E., Simion S., Pasculescu V. (2013). Model for Forecasting the Exposure Risk of Workers to Hand-Arm Occupational Vibrations. Applied Mechanics and Materials, 430, 276–280.
9. EN ISO 22867 – Forestry and gardening machinery. Vibration test code for portable hand-held machines with internal combustion engines. Vibration on the handles.

The influence of environmental factors and heavy metals in the soil on plants' growth and development

Nicoleta Alexandra Vanghele[1], *Mirabela Augustina* Pruteanu[1,*], *Ancuța Alexandra* Petre[1], *Andreea* Matache[1] , *Dumitru Bogdan* Mihalache[1] and *Mariana Mădălina* Stanciu[1]

[1)] The National Institute of Research - Development for Machines and Installations designed to Agriculture and Food Industry - INMA Bucharest

Abstract: Plants need certain conditions that represent their living environment. When the living environment provides the conditions required by the plant, it will grow and develop properly. The growth and development of plants involve environmental factors, which represent those constituent elements of the natural environment, which actively intervene in plants' life.

The present work shows the characteristics of an agricultural soil, contaminated with heavy metals (Cu, Pb and Zn) in different concentrations, which has been divided into pots, in which were thereafter planted vegetable seedlings (tomatoes, cucumbers, parsley, spinach, carrots, radishes). During the plants' growing time, the temperature and humidity of the air inside the greenhouse, as well as the humidity and pH of the soil, were monitored. The growth and the development of the plants under certain conditions were also tracked, until the end of the growing period. The results of monitoring the plants' growth and development are important in assessing the impact of the contamination over the soil and the plants.

Key words: environmental factors, heavy metals, soil, plants

Introduction

The main environmental factors that influence plants' growth and development are: light, temperature, water, air (carbon dioxide and oxygen), nutrients, PH and toxic substances. In some situations, environmental factors may become restrictive for growth and development processes, as they become harmful if they exceed certain limits and values or they cannot meet the necessary plants' requirements. All environmental factors are relatively equal in the process of growth and development, the insufficiency of one having negative repercussions on plants.

Systemic toxic pollutants, such as heavy metals, are particularly dangerous due to their long-lasting remanence in the soil, as well as to their intake by the plants and, implicitly, by humans. The main routes of the collection of metals by plants are the radicular takeover and the foliar takeover.

[*] Corresponding author: pruteanu_augustina@yahoo.com

The **cucumber** (Cucumis sativus) is a vegetable plant, its optimum vegetation temperature is 25-28°C, and the average temperature should not fall below 15°C. The plants do not bear temperatures below 12°C or above 35°C, values outside of this range stopping the growth.

They are light-loving plants, needing 12 hours of light per day, prefer aerated soil, rich in humus, permeable, with neutral or low acid pH, with a humidity maintained at 70-80%.

The grown tomatoes come from the wild species of *Lycopersicon esculentum var. cerasiforme*, originated from the highlands of the Andes. They are annual plants that do not require special environmental conditions, in need of light and of a temperature that must not exceed 40°C nor be less than 10°C. Tomatoes prefer medium-sized sandy-clay, clay-sandy soils, rich in humus, fertile, well-structured, with a good, deep drainage, with deep groundwater and a PH level between 5.5 and 7.0, and humidity between 40 and 70%.

The **spinach** (*Spinacea oleracea*) is native in Central Asia; in Romania is cultivated in all areas. The spinach is a plant resistant to low temperatures, with the optimal growth temperature of 15-17°C. In terms of light, spinach is a less demanding species, which grows well in spring, autumn and winter, when the light radiation is low. Water plays an important role in obtaining a rich foil, being necessary that the soil moisture is constant. The spinach prefers loose, wet soils, rich in humus, with a pH of 6.5-7.5 [1].

The **parsley** (*Petroselinum crispum L*) is native in the Mediterranean Sea area since the beginning of our era and can be cultivated both for the root and the leaves. The minimum germination temperature is 2-4 °C, and the optimum vegetation temperature is 20°C. Plants can withstand temperatures of up to –30°C. The water requirements are moderate, being resistant to drought but it is a light-loving plant. In terms of soil requirements, it is a demanding plant, preferring loose, light, deep and well-fattened lands, the ideal being the clay-sandy ones.

The **carrot** (*Dancus carota L*) is not exigent to heat; its preferred temperature is between 15 and 24°C. Temperatures above 30°C are harder to tolerate, while the minimum temperature they endure is 3°C. It prefers as much light as possible: being a long day plant, the more light it receives, the more it will develop harmoniously; same thing happens in the case of humidity. For the carrot, the ideal soil should be clay-sandy, sandy, alluvial, with a pH between 5.8-7 [2].

The **radishes** (*Raphanus sativus L*) are cold-resistant plants, adapting to temperatures between 2 and 17°C. They are plants that need a lot of light and heat, and their favorite soils are those with light or medium texture, rich in humus, loosened, with a neutral pH and a good water retention capacity.

The plants capture and accumulate heavy metals, and this is due to certain factors such as: the climate, the concentrations of heavy metals within the soil, the nature of the soil in which they are grown, the atmosphere and the plants' maturity degree at their harvest [3,4,5].

The heavy metals can be grouped into essential and non-essential metals. The essential heavy metals are Co, Cr, Cu, Fe, Mn, Ni and Zn and are considered to be essential micronutrients, but they become poisonous when taken in excessive quantities. The non-essential heavy metals include Pb, Cd and Hg and are highly toxic to living organisms [6,7].

The long-term consumption of foods containing unsafe amounts of heavy metals can have repercussions on internal organs such as kidneys, liver, heart, and lead to various diseases over time such as chronical, kidney, cardiovascular, bone and nervous system diseases [8,9].

The main way of transferring the metals to the plant is by absorption through its roots.

The main routes of the collection of metals by plants are the radicular takeover and the foliar takeover. The environmental factors that make up the plants' living framework are: light, temperature, water, air and culture substrate. The temperature is one of the most important climatic factors that influence the growth and development rate of the plants. It conditions the unfolding of fundamental processes that take place in plants (growth, photosynthesis, respiration, transpiration, absorption of water and of nutritional salts, etc.). The vital processes of plants can only take place at certain temperatures [4,10]. The growth and development of plants depends on the temperature in the atmosphere. The main source of heat of the atmosphere is the solar energy, from which plants absorb only a fraction of it. Each plant species has specific thermal requirements which causes the temperature to have a limiting action on their growth and development [10].

The paper presents the monitoring of soil parameters, atmosphere and plants (vegetables) along a vegetation cycle.

Material and method

The period of planting the vegetable seedlings was done between March and April 2018. Heavy metal infestation experiments carried out between March and September 2018 were done for a number of 7 plants (vegetables), specified in Table 1.

Table 1 - Classification of the vegetable plants studied [11]

Group	Common name	Scientific name of the plant
Cucurbitaceous vegetables	Cucumber	*Cucumis sativus* L
Solano-fruit vegetables	Tomatoes	*Lycopersicon esculentum*
	Spinach	*Spinacia oleraceea* L
	Parsley	*Petroselinum crispum* L
Root vegetables	Carrot	*Dancus carota* L
	Parsley	*Petroselinum crispum* L
	Radishes	*Raphanus sativus* L

The figure #1 shows aspects of the experimental research for the vegetable plants studied.

a) Carrot b) Radishes c) Parsley-root
d) Spinach e) Cucumber f) Tomatoes

Fig. 1. - The stage of vegetable development.

All types of plants were planted in the soil and infested with the following four heavy metal concentrations: 1.5%, 3.0%, 4.5%, 6.0% and the heavy metals used were: copper, zinc, lead. In parallel, experiments were performed on plants grown in uninfested soil (0%).

The concentrations specified in the study for the study of metals were chosen taking into account the limits allowed for them, an experiment performed on salads grown on contaminated soils with different concentrations to see the behavior of plants, but also the literature in the field.

Of the metals chosen in the work, lead is the most dangerous metal, so care must be taken with its content in both soil and plant, as it can seriously disrupt human health. In the paper, it is observed that at concentrations of 3%, 4.5% and 6.0% Pb exceeds the permissible imitation in the soil which is 50 mg/kg.

In the experiments with vegetables the loading with heavy metal was carried out initially, when planting the seedlings, without supplementing it until the harvest.

The physico-chemical properties of the agricultural soil used in the experimental research were: pH 5.0-7.0, total nitrogen 1.9%, total phosphorus 0.5%, total potassium 0.9%, electrical conductivity 1.2, humidity 14.7.

The metals content of soils and plants was measured using a flame atomic absorption spectrometry

The granulometric analysis of the agricultural soil used in the experimental research was also carried out, the aspects during the work being shown in Figure 2 and the results obtained are shown in Figure 3.

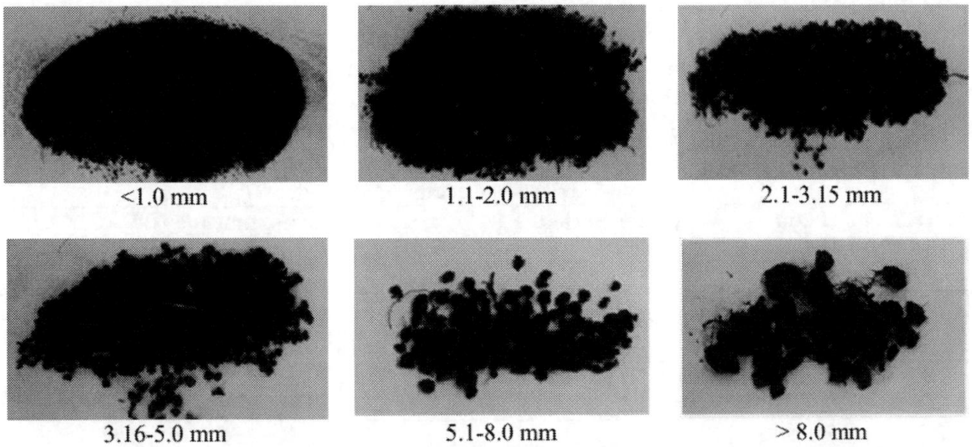

Fig. 2. Aspects during the screening of the agricultural soil used for the experimental tests

In the table 2 is presented the normal reference, alert threshold and intervention threshold values for soils of sensitive use listed in the Order no. 756/1997 of The Ministry of Waters, Forests and Environment Protection [12] and the maximum admitted values in leafy vegetables by FAO/OMS-Codex Alimentarius Commission, 2001 [13].

Table 2 - Heavy metal content (Cu, Zn, Pb) in the soil/plant, $mg \cdot kg^{-1}$

Soil Types /Plant types	Cu	Zn	Pb
Normal values [12]	20	100	20
Alert threshold [12]	100	300	50
Intervention threshold [12]	200	600	100
Maximum admissible values for leafy vegetables [13]	73.3	99.4	0.3

Results and discussions

Figure 3 showed the results obtained from the granulometric analysis of the agricultural soil used in the planting of vegetables.

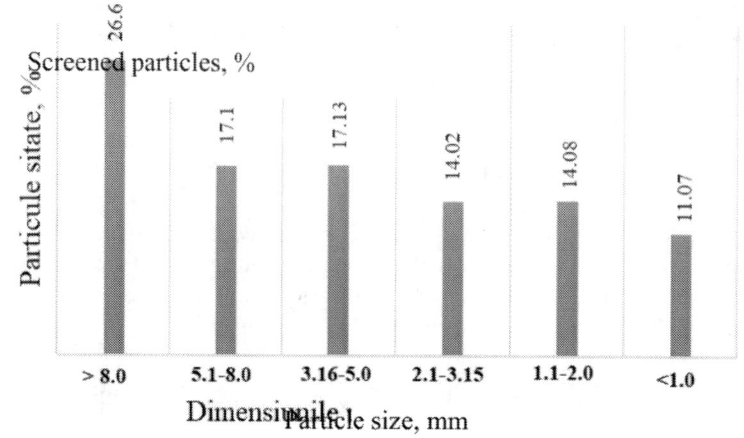

Fig. 3. Granulometric analysis of soil used in experimental research

The soil parameters during the experimental research were: pH between 5.0-6.5 and soil moisture between 62.5-77.6%.

The atmospheric conditions in the greenhouse during the growth of plants were: temperatures between 18-30 ^{0}C and air humidity between 45-80 %.

The following tables show the tracking of the evolution of the plants, in the case of soils contaminated with the three metals, namely Cu (table 3), Pb (table 4), Zn (table 5).

Table 3 - Plants subject to Cu infestation

Current no	**Plant name**	**Vegetation period, (days)**	**Harvest time (days) / Categories of concentrations (%)**				
			0	**1.5**	**3.0**	**4.5**	**6.0**
Quantity of metal in soil, mg/kg			17.6	58.9	267.2	525.1	680.8
1	Spinach	40-50	45	45	45	45	45
2	Parsley (leaf)	30-35	33	33	33	33	33
3	Parsley (root)	70-145	74	74	74	74	74
4	Radishes	30	30	30	30	30	30
5	Carrot	45	45	45	45	45	45
6	Cucumber	38-50	44	33	40	40	40
7	Tomatoes	90-130	73	73	-	93	80

Table 4 - Plants subject to Pb infestation

Current no	**Plant name**	**Vegetation period, (days)**	**Harvest time (days) / Categories of concentrations (%)**				
			0	**1.5**	**3.0**	**4.5**	**6.0**
Quantity of metal in soil, mg/kg			6.75	48.7	84.7	117.7	285.2
1	Spinach	40-50	45	45	45	45	45
2	Parsley (leaf)	30-35	33	33	33	33	33

3	Parsley (root)	70-145	74	74	74	74	74
4	Radishes	30	30	30	30	30	30
5	Carrot	45	45	45	45	45	45
6	Cucumber	38-50	44	26	40	40	26
7	Tomatoes	90-130	73	106	-	80	95

Table 5 - Plants subject to Zn infestation

Current no	**Plant name**	**Vegetation period, (days)**	**Harvest time (days) / Categories of concentrations (%)**				
			0	**1.5**	**3.0**	**4.5**	**6.0**
Quantity of metal in soil, mg/kg			39.8	202.7	534.8	921.7	1052.3
1	Spinach	40-50	45	45	45	45	45
2	Parsley (leaf)	30-35	33	33	33	33	33
3	Parsley (root)	70-145	74	74	74	74	74
4	Radishes	30	30	30	30	30	30
5	Carrot	45	45	45	45	45	45
6	Cucumber	38-50	44	26	40	26	29
7	Tomatoes	90-130	73	80	73	73	73

Considering all the cases of vegetables grown with various concentrations of heavy metal in the soil, it was observed:
- a tendency to increase the amount of heavy metal accumulated in the plant, along with the increase of the initial concentration of heavy metal in the soil.
- the curves of variation of the concentration of heavy metal in plants that are monotonously increasing, appeared more often in the case of infestations with lead and zinc (for carrots, parsley and spinach), and less with copper;
- the curves of variation of the concentration of heavy metal in oscillating plants, appeared more often in the case of infestations with copper, and less with lead and zinc;
-in radishes the curves were increasing for all three metals, while in tomatoes the curves were oscillating for all three metals;
-in cucumbers, the monotonous increasing curves appeared in the case of infestations with copper and zinc, and less with lead.

Conclusions

Following the experiments on the seven plants we found that:

- All vegetables grown on uninfested soil (0%), agricultural soil, reached plant maturity and were harvested in the optimal period.
- The spinach, parsley, radishes and carrots have reached vegetable maturity and have been harvested in the optimal period, at the same interval for all metals and concentrations.
- Cucumbers failed to reach plant maturity and were harvested earlier in the case of soil contaminated with 1.5% concentration, with Pb concentration 1.5% and 6% and Zn of concentration 1.5%; 4.5% and 6%.
- Tomatoes were the most affected plants from the ones tested; out of them, only those contaminated with 4.5% and 6% concentration and those contaminated with Pb concentration 6% were the only ones that reached plant maturity, and those that were contaminated with 3% concentration and 3% concentration Pb, they were lost drying before harvesting.

From the above mentioned, we can find that plants that have a shorter period of vegetation have survived and developed better in contaminated soil than those that have a longer vegetation period.

The research presented in the paper is important both for growers and consumers of fresh vegetables alike, but also for researchers, as this can be a starting point for future research in the field of heavy metal soil contamination and of vegetables, implicitly.

This work was supported by a grant of the Romanian Education and Research Ministry, through Programme 1 – Development of the national research-development system, subprogramme 1.2 – Institutional performance – Projects for financing excellence in RDI, contract no. 16PFE and by a grant of the Ministry of Agriculture and Rural Development, contract of sector financing, ADER type, no. 25.4.1 "*Technology for obtaining biofertilizers and/or bioinsecticides, for ecological production systems*".

References

1. N. Stan, N. Munteanu, T. Stan, Vegetables, vol III, "Ion Ionescu de la Brad" Publishing House, Iași, (2003);
2. Ruxandra Ciofu et al., Vegetable Treaty, "Ceres" Publishing House, Bucharest, (2004);
3. R. K. Sharma, M. Agrawal, F.M. Marshal, Heavy metals (Cu, Cd, Zn, Pb) contamination of vegetables in urban India: a case study in Varanasi, Environ. Poll., 154, pp. 254-263, (2008);
4. Sandeep G., Vijayalatha K.R., Anitha T. Heavy metal and its impact in vegetable crops, International Journal of Chemical Studies; vol. 7(1), pp. 1612-1621, (2019);
5. H. Zhou, W.-T. Yang, X. Zhou, L. Liu, J.-F. Gu, W.-L. Wang, J.-L. Zou, T. Tian, P.-Q. Peng B.-H. Liao, Accumulation of heavy metals in vegetable species planted in contaminated soils and the health risk assessment, Int. J. Environ. Res. Public Health 13, 289, (2016); doi:10.3390/ijerph13030289;
6. I. P. Tripathi, N. Dwivedi, P. Shukla, M. Mishra and A. Tiwari, Heavy metals in vegetations from selected market sites in Chitrakoot, Satna M.P. India, International Journal of Current Research, Vol. 7 (01), pp. 11708-11711 (2015);
7. Sandeep, G., Vijayalatha, K.R., Anitha, T., Heavy metals and its impact in vegetable crops. Int. J. Chem. Stud. 7 (1), pp. 1612–1621, (2019);
8. C. L. Jarup. Hazards of heavy metals contamination, British Med. Bull., 68, pp.167-182, (2003);
9. S. Kamran, A. Shafaqat, S. Hameed, S. Afzal, F. Samar, B. S. Muhammad, A. S. Bharwana, M. T.Hafiz, Heavy metal contamination and what are the impacts on living organisms, Greener Journal of Environmental Management and Public Safety, vol. 2 (4), pp. 172-179, (2013);
10. V. Iordache, E. Cojoc, Modeling the bioaccumulation of heavy metals in crop plants in copsha Mica area, University of Bucharest, (2019), http://www.cesec.ro/pdf/CojocEmilia2011.pdf
11. S. Bruma, Organic vegetable cultivation technologies, No. 2, "Terra Nostra" Publishing House (2004);
12. *** Order no. 756/1997 of the Ministry of Waters, Forests and Environment Protection.
13. *** FAO/WHO, Codex Alimentarius Commission, 2001.

Wastewater electrooxidation using stainless steel electrodes

Nicoleta Ungureanu[1], *Valentin* Vlăduţ[2,*], *Mario* Cristea[2] and *Dan* Cujbescu[2]

[1]University Politehnica of Bucharest, Department of Biotechnical Systems, Splaiul Independentei nr. 313, sector 6, Romania
[2]National Institute of Research – Development for Machines and Installations Designed to Agriculture and Food Industry – INMA Bucharest, Ion Ionescu de la Brad nr. 6, sector 1, Romania

Abstract. Wastewater contains essential nutrients for crops, and after proper treatment its reuse is beneficial to farmers in arid and semi-arid areas. Electrooxidation is an environmentally friendly technology used for treatment of domestic, industrial and agrozootechnical wastewater. During electrooxidation, organic compounds are oxidized at the interface between the anode and the aqueous solution, as a result of the reduction reaction at the cathode. Tests were performed on wastewater collected from the lagoon of a cattle farm, using an electrooxidation cell with stainless steel electrodes. The removal of pH, conductivity, turbidity, color, total N, total P, COD and BOD was investigated at specific voltages of 0.025 V cm^{-2}, 0.05 V cm^{-2} and 0.1 V cm^{-2} and various electrolysis times (15, 30, 60 and 120 minutes). Results show that electrooxidation is effective in reducing total N by 24 %, total P by 47 %, BOD by 47 % and COD by 82 %. After completion of tests, the suspended solids were decanted towards the anode area, due to the action of coagulant Fe_2^+/Fe_3^+ ions formed by anode dissolution, and the water was clarified. The effluent complies with the standards to be reused to irrigate crops tolerant to salinity grown in permeable soils.

1 Introduction

In 2019 water scarcity has affected over 17 % of Europe's territory and over 11 % of its population [1]. Food and Agriculture Organization of the United Nations estimates that by 2050, the global food production will have to increase by at least 50 % to feed 9 billion people [2]. Agriculture consumes about 70 % of freshwater, so the sustainable use of water resources is vital.

Wastewater reuse in irrigated agriculture became an important component in agricultural water supply in many water-scarce countries [3]. The reuse of wastewater in agriculture is part of water resource development and management that provides innovative and alternative options for agriculture, if the wastewater is properly treated to ensure that it has no harmful effects to human health or the environment.

Even in developed countries or in those not yet affected by water shortage, wastewater reuse for irrigation is a common practice. At present, over 15 million m^3 day^{-1} of treated

* Corresponding author: valentin_vladut@yahoo.com

water are used for irrigation of crops worldwide. The European countries reuse 1 billion m^3 $year^{-1}$ of treated urban wastewater, which is 2.4 % of the treated urban wastewater effluents [4]. Thus, at least 10 % of global population consumes food from crops grown by wastewater reuse [5].

Many countries have national regulations that prohibit the irrigation of agricultural crops with raw wastewater and impose limits for different physico-chemical and bacteriological parameters, including wastewater turbidity, pH, electrical conductivity, heavy metals, suspended solids, BOD, COD, nutrients, pathogenic bacteria, fecal coliforms, fecal enterococci and nematode eggs [6].

The evaluation of the possibility of reusing treated wastewater involves the qualitative analysis of the wastewater in close connection with the characteristics of soil, the type of crop and the necessary irrigation norms.

Treated wastewater from cattle farms can be used for irrigation, thus benefiting from their high content of organic substances and nutrients (nitrogen and phosphorus), but their soluble salt content requires mixing them with conventional clean water or applying low quality standards, suitable only to crops tolerant to salinity and on soils with suitable drainage.

The electrochemical treatment of wastewater is a technology that uses a constant voltage applied on a pair of metal plates (electrodes) immersed in a basin containing wastewater, to decompose partially or completely the organic pollutants. This type of treatment is classified as an advanced treatment and requires other pretreatments. On the other hand, the efficiency of electrooxidation as wastewater pretreatment depends on the anodes used to mediate the organic species oxidation present in the wastewater [7].

Electrochemical methods, including the oxidation of organic pollutants by cathodic generation of hydrogen peroxide, cathodic removal of metals, anodic oxidation using electrodes made of different materials and shapes, and electrocoagulation, are effective in the oxidative degradation of many organic pollutants in wastewater [8].

Electrooxidation is an electrochemical advanced oxidation process which involves electrolytic reactions at the electrodes surface, the formation of anodic metal and metal hydroxide cations in aqueous phase, the adsorption of soluble or colloidal pollutants on the surface of metal hydroxides, and the removal of organic pollutants by electroflotation, sedimentation and adhesion to the bubbles. Metal hydroxide precipitation is influenced by the pH of wastewater.

Different shapes, configurations and materials have been used so far for the electrodes. In terms of shape, rectangular vertical plates and pipes are widely used, connected either as monopolar-parallel (MP-P), monopolar-series (MP-S) or bipolar-series (BP-S), as shown in Figure 1 [9].

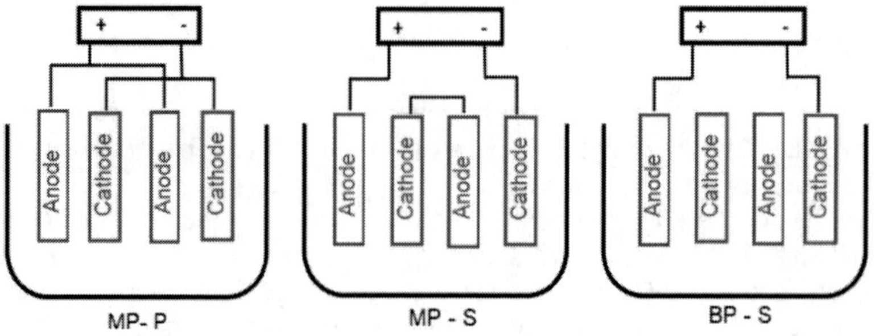

Fig. 1. Possible electrode connections in electrolysis cells for wastewater treatment [9]

Aluminum, iron, titanium, stainless steel, magnesium, graphite, boron-doped diamond for thin-film anodes, lead dioxide, tin oxide, platinum are only some of the non-hazardous, available materials for the manufacture of electrodes, and they have shown good removal rates of organic pollutants from wastewater.

The electrocatalytic properties of anodic materials have an important role in the removal efficiency of organic pollutants by electrooxidation [10].

Like all electrochemical treatments of wastewater, electrooxidation can take place with complete automation of the process, without direct addition of chemical reagents [11] because the electron itself acts as reagent which allows the agglomeration of pollutants; electrooxidation does not generate secondary pollution (which occurs in traditional chemical methods due to high concentration of chemicals), hence electrooxidation has an reduced impact on the environment.

Although electrooxidation is especially efficient in removing COD from wastewater, it is a process that generates significant quantities of sludge [12]. Another drawback of wastewater electrooxidation is the corrosion and consumption of the electrodes [13]. In time, an impermeable oxide layer is formed on the cathode, especially when Al electrodes are used. This layer slows the transmission of current within the electrolysis system, resulting in higher power consumption, higher costs and lower treatment efficiency [14]. To prevent electrodes passivation and the sediment deposits on their surface, electrodes polarity should be changed at regular intervals of time [15] or they must be replaced regularly for enhanced treatment efficiency.

To increase the efficiency of electrooxidation, wastewater containing high amounts of suspended solids must be first treated to remove the solids. For example, coupled electrocoagulation (using Al, Fe or Al/Fe electrodes) - electrooxidation processes are increasingly used. Electrocoagulation removes the finest colloidal particles in wastewater because the applied electric current fastens their collision and facilitates the coagulation.

2 Material and method

Experimental research was conducted on wastewater originated from a medium-sized cattle farm in Romania. The environmental practices according to which the farm operates involve collecting and storing the wastewater in a waterproofed lagoon provided with a mechanical stirring system (Figure 2) until its application on farmland as fertilizer.

Wastewater produced in the cattle farm is actually a mixture of wastewater from paddock scrubbing, liquid manure mixed with entrained solid residues, wastewater from milk processing on the farm's milk factory, and meteorological waters.

Fig. 2. Lagoon for wastewater storage, mechanical stirrer and wastewater sampling

The main organic pollutants in these effluents are manure and forage entrained by liquid, carbohydrates, proteins and fats originating from the farm's dairy factory. The coarse residues of organic nature represented by animal manure, forage residues and paddock litter are collected and stored on a composting platform.

Storing the wastewater in the lagoon favors the achievement of extensive local treatment, in which the wastewater is initially subjected to a physical treatment of gravitational separation of coarse impurities (clarification), and also to aerobic-anaerobic biochemical processes in which the organic substances are reduced. The stirrer increases the oxygen supply in the wastewater mass, thus contributing to the improvement of the degradation of organic substances.

During sampling, the wastewater collected in the lagoon was continuously stirred to train the solid particles deposited on the bottom. The general rules for sampling of wastewater are included in the standard SR ISO 5667-10/1994.

In electrochemical wastewater treatment, electricity is employed to conduct electrolysis processes in which the decomposition of some substances and the formation of new products at the electrode by means of electric current take place. In electrolysis, pollutants removal can be achieved by direct oxidation in the electrode process (anodic oxidation), by indirect oxidation with strongly oxidizing products resulting at the anode (O_2 and Cl_2) and by electrochemical coagulation.

The equipment used for the treatment of wastewater by electrochemical oxidation consists of an electrochemical cell (Plexiglas basin, 500 x 200 x 200 mm), in which was placed a pair of electrodes linked to a voltage source that supports a continuous current of maximum 5 A, and an apparatus for measuring the energy consumption (Figure 3). The electrodes were made of parallel stainless steel pipes (12 mm diameter, 200 mm length) with both ends welded on stainless steel plates, to form networks.

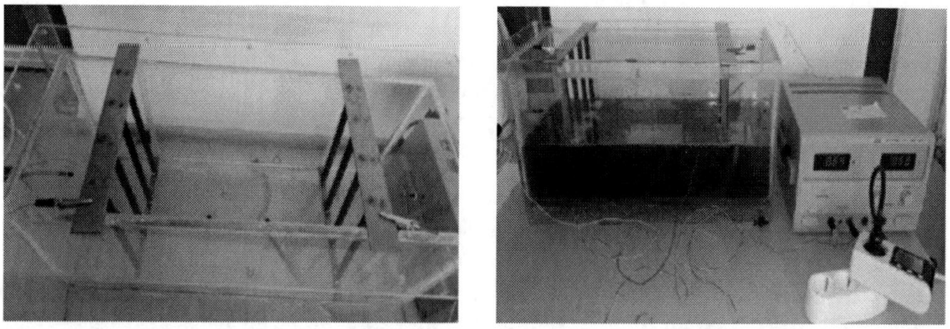

Fig. 3. Experimental setup for wastewater electrooxidation (stationary regime)

Current intensity is an important parameter that influences the efficiency of electrooxidation. As current intensity increases, the amount of anodic dissolution increases, hence an improved coagulation of pollutants will occur. In this test, different specific voltages were applied: 0.025 V cm^{-2} (test T1), 0.05 V cm^{-2} (test T2) and 0.1 V cm^{-2} (test T3) for a wastewater volume of 9 liters. During all tests it was maintained the same polarity of the electrodes, namely the positive electrode (anode) was on the right and the negative electrode (cathode) was on the left of the electrochemical cell. The evolution of wastewater treatment was observed by carrying out determinations at 15, 30, 60 and 120 minutes exposure times. Raw wastewater and samples of wastewater treated by each electrooxidation test (taken from the middle distance between the electrodes) were analyzed to determine if they comply with the existing regulations regarding the use of wastewater for irrigation of agricultural crops.

Conventional quality parameters associated with wastewater treatment include pH, conductivity, color, turbidity, total N, total P, COD and BOD, which were analyzed according to the methods mentioned by the Romanian norms corresponding to the standard in force regarding the establishment of pollutant loading limits for industrial and urban wastewater (NTPA-001/2002 and NTPA-002/2002). Among the analyzed parameters, the

Romanian STAS 9450-88 imposes limits for pH, electrical conductivity and total nitrogen, dividing the waters by quality classes. The aim of this study was to reduce pollutant loads and to assess if the effluent can be used for crop irrigation in arid and semiarid areas.

3 Results and discussion

Table 1 presents the characteristics of raw wastewater, which has high content of nitrogen and phosphorus. Raw wastewater has relatively small percentage of biodegradable organic compounds (expressed as biochemical oxygen demand, BOD) compared to the total content of organic substances (expressed as chemical oxygen demand, COD) in wastewater, due to the oxidation of the organic matter mainly by exposure to the external environmental conditions in the lagoon where the wastewater was kept for a long time.

Table 1. Initial characteristics of raw wastewater (control sample)

Parameter	Value	Parameter	Value
pH (units)	7.80	Total N (mg L^{-1})	234.2
Conductivity (mS cm^{-1})	3.84	Total P (mg L^{-1})	56.4
Color (HZ)	1092	COD (mg O_2 L^{-1})	1512.6
Turbidity (NTU)	724	BOD (mg O_2 L^{-1})	204.5

The physico-chemical parameters of farm wastewater following different electrooxidation treatment are presented in Table 2.

Table 2. Physico-chemical characteristics of wastewater after electrooxidation

Test_time	pH	Conductivity (mS cm^{-1})	Color (HZ)	Turbidity (NTU)
T1_15	7.84	3.86	1054	571
T1_30	7.84	3.86	1069	576
T1_60	7.86	3.84	1093	636
T1_120	8.14	3.81	1111	645
T2_15	7.96	3.85	980	542
T2_30	8.02	3.83	1043	576
T2_60	8.01	3.76	1134	656
T2_120	8.36	3.71	1145	736
T3_15	7.96	3.82	1139	644
T3_30	7.99	3.77	1176	708
T3_60	8.00	3.74	1167	744
T3_120	8.20	3.60	1180	770
Test_time	Total N (mg L^{-1})	Total P (mg L^{-1})	COD (mg O_2 L^{-1})	BOD (mg O_2 L^{-1})
T1_15	202.7	47.3	816.6	102
T1_30	201.3	49.3	770.2	89.9
T1_60	207.5	47.5	779.5	86.7
T1_120	200.4	44.2	705.3	74.3
T2_15	204.6	44.2	700.6	93.6
T2_30	205.5	44.5	668.2	97.5
T2_60	203.5	43.1	626.4	55.9
T2_120	178.7	38.3	626.4	36.5
T3_15	212.5	43.1	730.3	67.6
T3_30	201	45.7	663.5	57.6
T3_60	208.9	43.1	756.3	67.7
T3_120	193.1	30.0	607.8	42.8

Analyzing the data in Table 2 compared with those in Table 1, it can be observed that in all three sets of tests, the values obtained for pH, conductivity, color, turbidity, total nitrogen and total phosphorus did not change significantly. In contrast, for some samples, the chemical oxygen demand was reduced by half, and the biochemical oxygen demand was reduced by almost 5 times.

The results were plotted as a function of treatment time, as shown in Figures 4-7.

From the analysis of the pH of the treated wastewater samples (Figure 4) it can be observed that regardless of the specific voltage applied to the electrodes and the duration of treatment, when compared to the control sample, there were no significant differences in pH. All samples of wastewater treated by electrooxidation had alkaline pH, the values increasing slightly with the duration of treatment; the highest pH values were recorded for the treatment time of 120 minutes.

According to the Romanian STAS 9450-88, the samples of treated wastewater are weakly alkaline (<8.6 units) and are accepted for use as irrigation water. If the pH exceeds 8.6 units, secondary soil degradation processes would occur.

The electrical conductivity of the treated water had a decreasing tendency over time, without significant differences between the values recorded for the three specific voltages. In terms of conductivity, with respect to STAS 9450-88, in all tests, the treated water falls in salinity class C3 – it has high saline residue, so the water it can be applied on permeable soils and for irrigation of salinity tolerant crops.

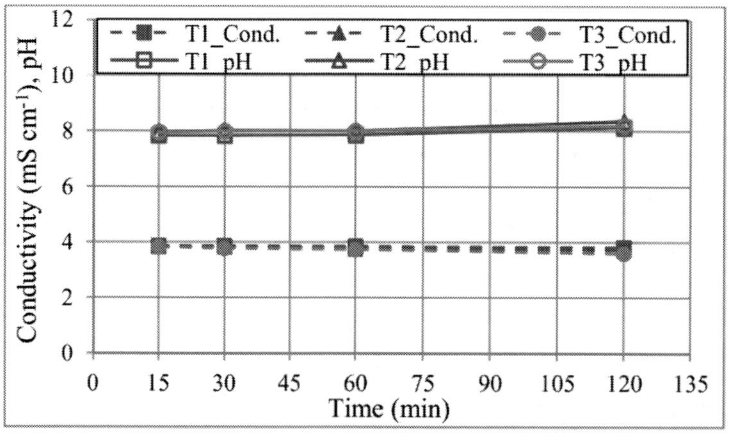

Fig. 4. Variation of pH and conductivity during testing (T1 = 0.025 V cm^{-2}; T2 = 0.05 V cm^{-2}; T3 = 0.1 V cm^{-2})

Regarding Figure 5, although it is not a standardized parameter, wastewater color influences indirectly the possibility of reuse because color is given by the concentrations of suspended solids and dissolved substances that can cause malfunctioning of the irrigation systems.

The color of treated water changed visibly between tests 2 and 3, due to air bubbles created in the electrooxidation cell, which entrained the solid matter, and to electrode dissolution which released Fe ions into the water; thus, after the last test, the formed foam was reddish. Wastewater turbidity is given by the presence of colloidal species, suspended solids and microorganisms. During each test, the turbidity increased with increasing treatment time, but in the first two tests there were lower values compared to the control sample.

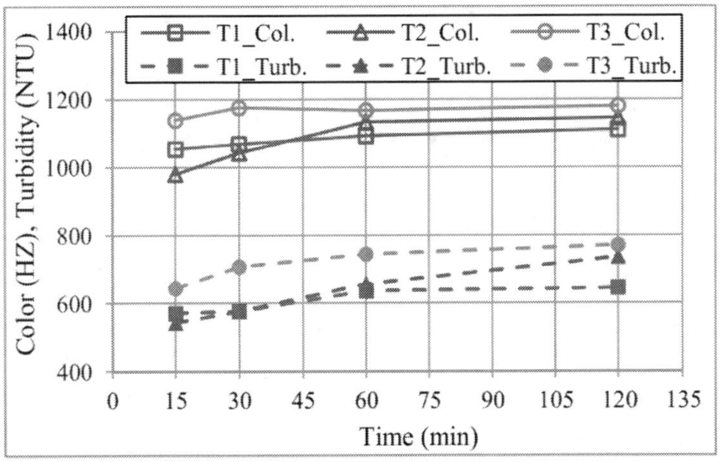

Fig. 5. Variation of color and turbidity during testing (T1 = 0.025 V cm^{-2}; T2 = 0.05 V cm^{-2}; T3 = 0.1 V cm^{-2})

Nitrogen is a common pollutant in agricultural waters as a result of the use of chemical fertilizers. Phosphorus is found in chemical fertilizers and detergents, so it is contained in domestic wastewater and in wastewater from washing the milking and dairy factory equipment within the farm. The variation in time of these nutrients is given in Figure 6. Compared to the control sample, at maximum exposure time, highest removal of total N (24 %) was achieved in test 2 (specific voltage of 0.025 V cm^{-2}), while for total P the highest removal efficiency of 47 % was obtained in test 3 (at 0.1 V cm^{-2}). According to STAS 9450-88, the values of total nitrogen fall into class C3, is usable on permeable soils and for irrigation of salinity tolerant crops.

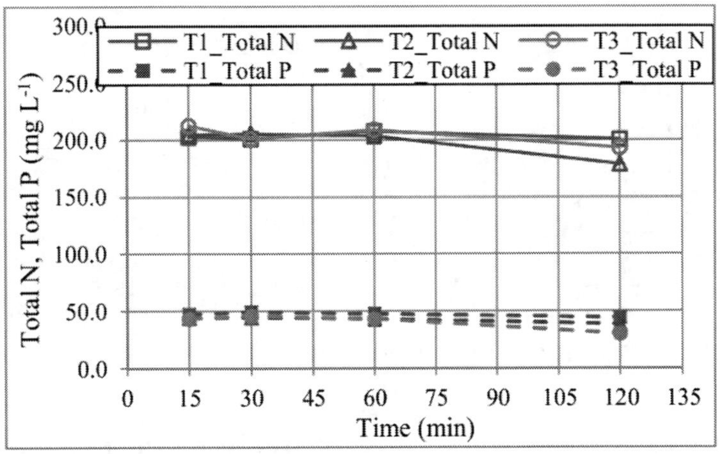

Fig. 6. Variation of total N and total P during testing (T1 = 0.025 V cm^{-2}; T2 = 0.05 V cm^{-2}; T3 = 0.1 V cm^{-2})

The concentration of organic compounds in wastewater is assessed by the chemical oxygen demand (COD) and biochemical oxygen demand (BOD), parameters by which the degrees of chemical and biochemical oxidability are estimated. Analyzing the variations in Figure 7, compared to the control sample, after 120 minutes, BOD removal rates were 64 % at specific voltage of 0.025 V cm^{-2}, 82 % at 0.05 V cm^{-2} and 79 % at 0.1 V cm^{-2}. The electrooxidation treatments have also given good results in reducing the chemical oxygen demand (COD) thus: 47 % at 0.025 V cm^{-2}, 41 % at 0.05 V cm^{-2}, respectively 40 % at 0.1 V cm^{-2}.

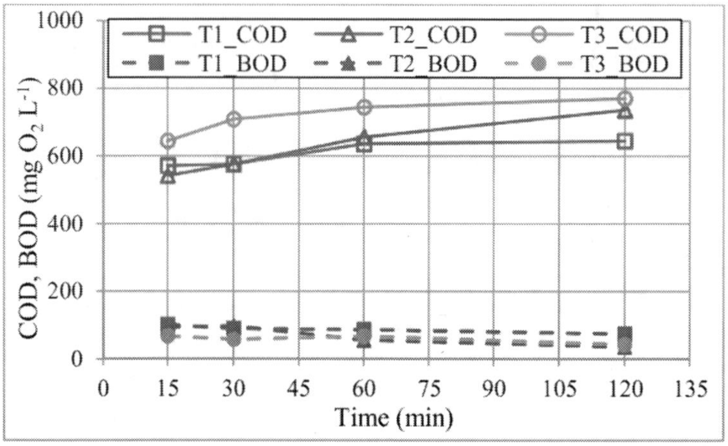

Fig. 7. Variation of COD and BOD during testing (T1 = 0.025 V cm^{-2}; T2 = 0.05 V cm^{-2}; T3 = 0.1 V cm^{-2})

Regarding the reaction of wastewater subjected to an electric field, at first there was a more intense activity in the cathode area, which towards the end of the tests moved to the anode area (Figure 8), where intense foaming occurred. Foaming shows the formation of gas (O_2 and Cl_2) at the anode. There was no important release of H_2 at the cathode because some of the electrons were transferred by the discharge of metal ions that reentered the solution, as soluble metal hydroxide, which led to a slight increase in wastewater pH. This concurrent reaction to the classical reaction of water electrolysis with the formation of H_2 has consequently resulted in slowing of H_2 emissions at the cathode.

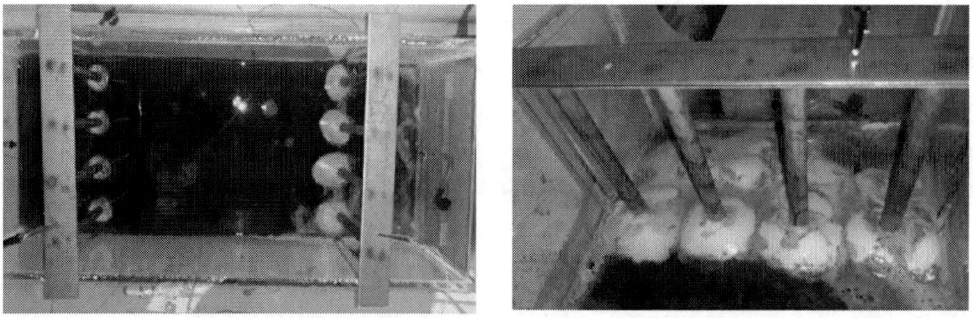

Fig. 8. Intense foaming at the anode during testing

After completion of the tests, the treated wastewater had reddish color due to the dissolution of Fe ions in the electrode material. After being left idle for 24 hours in the electrochemical cell, it was observed that the suspended solids in the water have decanted at the base of the electrochemical cell towards the anode area in a thick layer of about 2 cm, while the water cleared (Figure 9).

Sludge formation is explained by the action of Fe^{2+}/Fe^{3+} ions that are formed by anode dissolution and which act as a coagulant on the dissolved substances and the suspended matter. In addition to O_2 and Cl_2 near the anode, there are other reactive oxygen species (H_2O_2, O_3, $OH\cdot$) and reactive chlorine species (hypochlorite, $Cl\cdot$) that play a role in the oxidation of pollutants.

Fig. 9. Water after finalizing the tests, respectively clarification by decanting of suspended matter 24 hours after testing

The existence in the same cell of electrooxidation of the anodic and cathodic environment facilitates the production of secondary reactions between the metal hydroxide and Cl_2 with the formation of chlorates which in turn have an oxidizing effect.

4 Conclusions

Wastewater from cattle farm was treated by electrooxidation using different specific voltages (0.025 V cm^{-2}, 0.05 V cm^{-2} and 0.1 V cm^{-2}), aiming to reduce the loads of physico-chemical pollutants. The most important removal efficiencies were: 82 % for biochemical oxygen demand, 47 % for chemical oxygen demand, 47 % for total phosphorous and 24 % for total nitrogen.

However, close values of water quality parameters between tests prove that higher energy consumption (test 3) is not technically-economically reliable. Establishing the optimum conditions for the treatment of wastewater is mandatory to avoid the increase of the temperature in the electrooxidation cell with the increase of intensity of the electric current or to avoid the consumption of anode by its excessive dissolution.

Electrooxidation can be efficient for the treatment of wastewater from the cattle farms, and its efficiency could be increased by coupling of electrooxidation with other treatments, including disinfection, for the safe reuse as irrigation water in Romania.

This work was supported by a grant of the Romanian Ministry of Research and Innovation CCDI - UEFISCDI, Project INNOVATIVE TECHNOLOGIES FOR IRRIGATION OF AGRICULTURAL CROPS IN ARID, SEMIARID AND SUBHUMID-DRY CLIMATE, project number PN-III-P1-1.2-PCCDI-2017-0254, Contract no. 27 PCCDI / 2018, within PNCDI III.

References

1. European Commission, Environment, Water Scarcity & Droughts in the European Union (2019), https://ec.europa.eu/environment/water/quantity/scarcity_en.htm
2. Food and Agriculture Organization of the United Nations, How to Feed the World in 2050 (2019), http://www.fao.org/fileadmin/templates/wsfs/docs/expert_paper/How_to_Feed_the_World_in_2050.pdf
3. F. Pedrero, I. Kalavrouziotis, J.J. Alarcón, P. Koukoulakis, T. Asano, Use of treated municipal wastewater in irrigated agriculture - review of some practices in Spain and Greece, Agricultural Water Management **97**: 1233–1241 (2010)

4. European Commision, Environment, Water Reuse (2019), https://ec.europa.eu/environment/water/reuse.htm
5. Z. Ujang, M. Henze, Municipal wastewater management in developing countries, IWA Publishing (2006)
6. C. Becerra - Castro, A.R. Lopes, I. Vaz–Moreira, E.F. Silva, C.M. Manaia, O.C. Nunes, Wastewater reuse in irrigation: a microbiological perspective on implications in soil fertility and human and environmental health, Environment International **75**: 117–135 (2015)
7. T.E.S. Santos, R.S. Silva, C.T. Meneses, C.A. Martínez-Huitle, K.I.B. Eguiluz, G.R. Salazar-Banda, Unexpected enhancement of electrocatalytic nature of Ti/ $(RuO_2)xe(Sb_2O_5)y$ anodes prepared by ionic liquid-thermal decomposition method, Industrial & Engineering Chemistry Research **55**: 3182–3187 (2016)
8. I. Chakchouk, N. Elloumi, C. Belaid, S. Mseddi, L. Chaari, M. Kallel, A combined electrocoagulation–electrooxidation treatment for dairy wastewater, Brazilian Journal of Chemical Engineering **34** (1): 109–117 (2017)
9. D. Moussa, M. El-Naas, M. Nasser, M. Al-Marri, A comprehensive review of electrocoagulation for water treatment: potentials and challenges, Journal of Environmental Management **186** (1): 24–41 (2017)
10. S. Garcia-Segura, Ocon J.D., M.N. Chong, Electrochemical oxidation remediation of real wastewater effluents - a review, Process Safety and Environmental Protection **113**: 48–67 (2018)
11. J.M. Aquino, G.F. Pereira, R.C. Rocha-Filho, N. Bocchi, S.R. Biaggio, Electrochemical degradation of a real textile effluent using boron-doped diamond or β-PbO_2 as anode, Journal of Hazardous Materials **192** (3): 1275–1282 (2011)
12. K. Bensadok, N. El Hanafi, F. Lapicque, Electrochemical treatment of dairy effluent using combined Al and Ti/Pt electrodes system, Desalination **280**: 244–251 (2011)
13. M. Changmai, M. Pasawan, M.K. Purkait, Treatment of oily wastewater from drilling site using electrocoagulation followed by microfiltration, Separation and Purification Technology **210**: 463–472 (2019)
14. G. Hasani, A. Maleki, H. Daraei, R. Ghanbari, M. Safari, G. McKay, K. Yetilmezsoy, F. Ilhan, N. Marzban, A comparative optimization and performance analysis of four different electrocoagulation-flotation processes for humic acid removal from aqueous solutions, Process Safety and Environmental Protection **121**: 103–117 (2019)
15. A. Dimoglo, P. Sevim-Elibol, O. Dinç, K. Gökmen, H. Erdoğan, Electrocoagulation / electroflotation as a combined process for the laundry wastewater purification and reuse, Journal of Water Process Engineering **31**, 100877 (2019)

Climate changes and methods to protect vegetable crops

Adriana Muscalu[1,*], *Cătălina* Tudora[1], *Constantin* Coța[1], *Zoltan* Gyorgy[1], *Floarea* Burnichi[2], *Mariana* Bârsan[1]

[1]INMA, 6 Ion Ionescu de la Brad Blvd., Bucharest, 013811, Romania
[2]SCDL, 23 Mesteacanului Street, Buzău, 120024, Romania

Abstract. In the context of the current climate change, it has become urgently necessary to develop adequate systems to protect horticultural crops. In Romania, the losses caused to these cultures by the extreme weather phenomena, as well as the lack of active intervention measures to combat or limit them, represent the main argument for addressing this field in particular. Vegetable crops are generally sensitive to extreme weather phenomena. In the climatic conditions in our country, the extreme phenomena occur between March and September, which coincides with the vegetation period of the crops. Of these, hail causes significant damage, and in some cases, at high intensity and long-lasting, can cause the calamity of vegetable crops. Another effect can arise in the drought years, when, in the absence of rainfall, the weeds enter into a fierce competition with vegetables, as regards the specific consumption of water and nutrients. The paper presents a review regarding the current methods of vegetable crops protection against the extreme weather phenomena and of weed control in these crops, grown in organic system. Combining the available solutions, adopting integrated strategies for non-chemical weed control can be an important premise for researchers and farmers to protect vegetable crops.

1 Introduction

In recent years, an increasing importance has been given to the study of climate change, due to the negative effects generated on natural ecosystems, human health and the economy. The global warming, manifested by the increase in average and extreme temperatures of the air has been demonstrated by numerous studies carried out in different regions of the globe, during several decades [1, 2, 3, 4]. This has resulted in an increase in the humidity level of the atmosphere, increasing the intensity and frequency of negative events related to extreme precipitation, which is one of the most variable climatic elements [5].

The monitoring of extreme climatic events at the annual and seasonal scale is particularly important for the creation of protection or intervention means to reduce the losses caused by such manifestations [6]. In countries where agriculture is one of the main branches of the economy, as is the case in Romania, extreme climate phenomena

* Corresponding author: amuscalis@yahoo.com

have a profound impact [7]. Of these, hail storms are severe convective weather events that occur frequently in Central Europe during the summer, most being confined to continental Europe, where convective energy is the highest [8, 9].

In our country, about 60 % of the storms that occur between March and September are accompanied by hail falls [6]. The occurrence interval coincides with the vegetation period of most crops and the intensity of the phenomenon determines the degree of their calamity [10].

Vegetables play an important role in ensuring food and nutritional security, being one of the richest sources of vitamins and minerals. Vegetable crops are highly perishable and sensitive to unpredictable as well as extreme climate changes, which affect them in terms of quantity and quality of production, as well as the severity of environmental stress [11]. Depending on the vegetation phase of vegetable crops, hail can cause not only a drop, but even a complete loss of crops.

Also, compared to other crops, vegetables are highly sensitive to weed attack, which is a top concern among organic vegetable producers [12]. Crop protection and weed control in organic vegetable crops are major challenges for farmers.

The purpose of the paper is a presentation of the current state of protection methods against the negative effects caused by climate change for organic vegetable crops.

2. Methods of vegetable crop protection

Due to the frequency and adverse effects of hail falls, the most used protection systems are those against this extreme phenomenon.

2.1 Anti-hail protection methods

Protection methods include active and passive hail control methods.

2.1.1 Active control methods

Different methods, equipment and technologies have been used over time to reduce hail effects. Modern solutions aim at the protection of large areas, allowing the automation and mathematical modeling of processes [13]. Active control methods are based on: seeding of clouds with silver iodide or other substances, which causes freezing at higher temperatures, respectively the use of hail cannons or other types of explosive devices. These involve the generation of significant supersonic shock waves, directed to the sky, to transport the positive ions from the ground level to the cloud level, disrupting the formation of hail nuclei [13, 14, 15]. For the methods to be effective, the timing of their start-up is very important [13, 14]. These methods achieve anti-hail protection by: prevention of hail formation in advance or stopping the further growth of small ice particles in the clouds and by suppressing, stopping the growth of small and moderate hailstones, by stopping the melting of hail particles nuclei [14, 15]. The simplest method of introducing silver iodide into the clouds is to use generators, located on the ground and manually operated. The operation can also be performed by plane at the right time and place, except for heavy clouds. The method of seeding the clouds made with rocket-type devices is much more accurate and efficient, being also used in Romania. Most European countries have anti-hail systems, using the last two methods, sometimes in combination, according to national or regional strategies [13]. In order to increase the efficiency of the national anti-hail system, the research aimed at its complementary equipment, such as the anti-hail rocket launcher monitoring system [16].

2.1.2 Passive control methods

These methods refer to plant cover systems, which prevent or reduce hail physico-mechanical impact on them. The use of these systems, generally made up of the protective net and the resistance structure that supports it, requires significant investments for farmers, but in the longer term. Thus, for a culture, the structure of resistance and fixation can be used approx. 15-20 years, and the net approx. 3-4 years, or more, depending on its characteristics [14].

Agricultural crops, especially horticultural ones, are often protected by "nets" against: excessive solar radiation (shade nets), pests (insects, birds, fruit bats) or extreme weather phenomena (hail, wind, frost), depending on mesh size. These textile materials can be applied as a single coating or combined with other types of materials, as in the case of greenhouses [17]. These "nets", increasingly used for horticultural crops, are part of the so-called agrotextiles, whose generic name is used to classify the woven, nonwoven and knitted textile materials used in agriculture [18]. This sector of the economy faces many problems, as we have shown above, and these materials can offer multidimensional points of view and solutions to some of them [19]. Agrotextiles are an important segment of technical textiles, which due to their mechanical properties, high resistance to environmental factors, light processing capacity and durability can improve the yield, quality and safety of agricultural crops and products [18, 19].

In general, the advantages of using agrotextiles are: increasing agricultural production, avoiding soil drying, reducing fertilizer, pesticide and water requirements, improving product quality, increasing early maturity of non-seasonal plants, protecting against climate change and their effect [18].

Agrotextiles can be classified according to several criteria such as: the way the textile material is made (woven, knitted or non-woven), application field (agriculture, horticulture, forestry, aquaculture, animal husbandry), purpose pursued (anti-hail protection, protection against birds, insects, for shade, mulch-type mats, etc.) [18, 19]. Synthetic fibres (polyethylene, polypropylene, polyester, nylon) are widely used for their manufacture. The commonly used synthetic polymers are tape, monofilament and multifilament yarn. Natural fibres (wool, jute, cotton), although they have a much shorter lifespan compared to synthetic ones, are only used for certain purposes, when their properties related to good moisture retention, better insulation under humidity conditions, maintaining the constant temperature of the soil without generating the effect of wind tunnel, biodegradability, etc. are exploited [18, 20].

Agrotextiles are technically made, depending on their destination. Thus, those for protection against hail are lightweight, made with mesh (nets), resistant to breakage and exposure to ultraviolet radiation, to prevent crop defoliation. Usually, these nets are made of polyethylene monofilament and do not restrict plants' development [21, 22]. Textiles designed to reduce the effect of the wind can improve the quality of the plants' development by minimizing their damage and shaking. The ones used for shading are made of polypropylene monofilament slivers, in both woven and knitted form. Each plant has its own requirements for optimum development, the balance between them and the environmental conditions, can only be achieved through a correct choice of coverage [22]. In addition, agrotextiles prove their adaptability and flexibility in various locations and geographical areas [21].

The efficient application of light is essential for the growth of plants, all aspects of their development being sensitive to it. Horticulturists are studying the possibility of selective modification or activation of plants' existing features depending on the light, in order to increase the productivity and quality of the crops [23].

Following the researches, for some vegetable crops, by using photoselective nets, which have chromatic elements incorporated by an innovative technology, specific physiological and horticultural benefits were obtained, in addition to the initial protection ones, presented previously [17, 24]. In the case of sweet peppers, the use of red and yellow nets increased productivity. In addition, in their case, an increased accumulation of anti-oxidants was observed for the yellow mesh [24].

2.2 Weed control in organic vegetable crops

Weed management is a major concern for organic vegetable producers. These crops are more susceptible to weeds than the arable ones, because of their low competitive capacity, and the strategies and techniques used are established according to the crop, climate, soil technology [25, 26, 27]. For this type of crops, it is necessary to apply an integrated system of non-chemical weed control methods, and it is necessary to approach as many of them as possible in order to obtain efficient results [25, 27]. The classification of these methods can be done in the general way or by customizing according to the type of crop [25, 26, 27]. Since vegetable crops in ecological system fall into the minor crops, we present a non-chemical weed control strategy proposed for them [26].

2.2.1 Agricultural practices - preventive measures

Prevention refers to the ways in which the danger of introducing new weed species is reduced. In this case, hygiene conditions are required for the seed material or the seedlings used, for the equipment and soil. It can be disinfected with steam or steam and organic substances, such as zeolites (potassium hydroxide and carbon oxide), using the Bioflash system. Soil heating is a promising, preventive ecological method of weed control in niche crops, which bring high incomes, such as vegetables and flowers [28, 29, 30]. *Crop rotation* must be diversified by adding crop species with different life cycles. Their associated management under different environmental conditions, has an effect on the control of weed populations or causes the reduction of the weed seed bank in the soil [31, 32]. *Primary tillage and seedbed preparation* can have important effects on the dynamics of the weed communities, if the weather conditions are considered (temperature, precipitation), and facilitate the mechanical control [26, 33, 34, 35]. *Fertilization* contributes to reducing the use of herbicides and increasing crop production, but the variation in dosages, the application methods, combinations and types of fertilizers used must be adapted to maintain the biodiversity of weed communities in an agroecosystem, protecting the environment. [36, 37]. *Vegetable varieties* are chosen based on resistance to biotic and abiotic stress factors, as well as productivity and quality. Some of them due to their general size or leaf size can be competitive against the weeds, preventing the germination of their seeds [38, 39].

2.2.2 Control by physical means (mechanical or thermal)

For the mechanical control, together with the equipment fitted with classic active parts, the one equipped with new types of environmentally friendly active parts (e.g. brush weeder with horizontal or vertical axis, basket weeder, finger weeder etc.) or combinations between types of active parts can be used. Table 1 presents the characteristics of certain types of equipment intended for mechanical control between rows that can be used for any type of crop, including vegetable crops [26,40].

Table 1 Features of inter-row mechanical weed tools [26]

Inter-row mechanical weed control tools	Weeding action	Working depth	Soil texture	Inter-row distance	Optimum weed growth stage	Optimum crop growth stage	Work capacity (ha h^{-1}) / Working width
Traditional hoe (with tine cultivators)	uprooting cutting and burying	30-50 mm	both clay (with rigid shanks) and sandy (flexible shanks) soils	from 0.20-0.25 m to 0.50 m	from 4 to 6 true leaves stage	from cotyledon stage (with shanks) or from 4 true-leaves stage until the treatment is carried out without crop injury	0.5 / 1.5 m
Rotary hoe	uprooting and burying	40-50 mm	better in loose soils	from 0.40 m to 0.50-0,75m	-from cotyledon to 4 true-leaves stage -low effectiveness on grass and perennial weed species	from cotyledon stage (with shanks) or from 4 true-leaves stage until the treatment is carried out without crop injury	1.5 / 3 m
Rotary tilling cultivators (with multiple heads)	uprooting lacerating and burying	40-60 mm	both clay and sandy but not stony soils	from 0.30 m to 0.40-0.50 m	from cotyledonal 1-2 true-leaves stage to well-developed weeds	from 2 true-leaves stage until the treatment is carried out without crop injury	0.3/ 1.5 m
Horizontal brush weeder	uprooting and drying weeds on the soil surface	10-30 mm	better in loose, flat and not compacted soils	from 0.20-0.25 m to 0.40 m	from cotyledon to 4 true-leaves stage -also weeds very close to crop row	from cotyledon stage until the treatment is carried out without crop injury	0.4/ 1.5 m
Split-hoe	cutting, uprooting and drying weeds on the soil surface	10-40 mm	both clay and sandy but not stony soils	from 0.20-0.25 m to 0.40 m	from cotyledon to 6-8 true-leaves stage also weeds very close to crop row	from cotyledon stage until the treatment is carried out without crop injury	0.4/ 1.5 m

The mechanical equipment together with the thermal one for weed control are intensively used both in organic vegetable crops and in the conventional ones, because of the strict European legislation regarding the chemical substances of protection. These devices, which have been continuously developed, represent the traditional physical means used in the non-chemical weed control [41].

Regarding thermal control, it is based on the transfer of heat between equipment and plants, the contact being foliar. The specialized equipment is based on the action of: hot air, steam, hot water, saturated steam, infrared radiation, flames, laser, etc. The efficiency of the treatment depends on the amount of heat transferred, the exposure period, as well as the development stage of weeds, the younger ones being more vulnerable. Drying of plants after exposure to heat occurs because of the expansion of intracellular water, which causes the breakdown of cell membranes, young weeds being vulnerable to this [42, 43, 44]. Most used are open flame burners: atmospheric and self-vaporizer, which can have various forms (cylindrical, truncated-conic, prismatic, or truncated-pyrmidal), being powered by propane or LPG in gaseous or liquid phase [40]. For high efficiency, they have a position inclined to the vertical, the distance to the soil, respectively to the crop plant, being important [44].

Numerous researches have been made regarding the combined use of the two methods presented above, in order to increase the efficiency of the control operation [45, 46, 47]. Thus, a combined flamer-cultivator was used to control weeds in an asparagus crop, before mechanically harvesting it. The new machine achieved a satisfactory level of weed control, for a consumption of 43-87 kg ha^{-1} LPG, being especially effective on weeds in the early stages. This method can be introduced as an ecological alternative for weed control in asparagus cultivation technology, before and after harvesting it. [47].

In order to effectively control weeds, intelligent systems carry out: guiding the control devices (mechanical or thermal), detecting and identifying the weeds, eliminating them and possibly mapping them. At this moment only a few companies make and sell automatic machines for mechanical weed control: Robovator (Frank Poulsen Engineering ApS - Denmark); Robocrop (Garford Ltd. United Kingdom), IC-Cultivator (Machinefabriek Steketee BV - Netherlands), Removeed (Costruzioni Meccaniche Ferrari -Italy) [26, 40]. They have been tested in several working conditions.

The active parts of Rovobator are a pair of knives that move 1-2 cm below the ground, closing and opening according to the order received from a hydraulic servomotor connected to a camera mounted in front of it. The identification of the plants in the crop is based on colour, size and spacing [48, 49]. A computer processes the images and calculates the points at which the cutting blades need to be activated depending on the speed and location of the crop plants. For larger working widths, several such modules are mounted on a frame [48]. Thus, in a lettuce crop, Robovator removed 18 to 41% more weeds from areas with moderate to high densities and reduced the weeding time by 20 to 45 % compared to the standard cultivator, but in an onion crop it did not have a satisfactory yield [48, 50]. The Rovobator system is also made in the variant for thermal control, which uses the same plant recognition system as for mechanical control. This machine is based on the serial application of a propane-powered flame. As the flame system passes over a weed, the viewing system controls the ignition of the burners, so that a weed will be exposed to the heat of the flame several times. The viewing system extinguishes the flame when passing over the crop plants, the working speed of the equipment is 1 - 6 km h^{-1} [49].

The IC-Cultivator, has a similar construction, the blades actuated by two pneumatic cylinders, performeds a cutting capacity of 3 - 4 plants / second, at a working speed of maximum 5 km h^{-1}. Robovator and IC Cultivator were tested in lettuce thinning, achieving a level of 5.6 % because of the small size of the plants, as well as of the operating mode, based on the recognition of vegetable plants. They achieved a satisfactory weeding level, in a shorter time by 37.1 % compared to the manual one [51].

An automatic system of thermal control with hot oil used against the black nightshade and red-root weeds was studied in an early tomato crop. To a hyperspectral imaging system, based on a multispectral Bayesian classifier, a precision micro-metering system with pulsed pulses was coupled to provide organic oil at 160 ^{0}C. In the range 384-810 nm, weed species were discriminated in an average proportion of 95.9 %. The oil was sprayed over the weeds, the doses applied on the target being about 0.85 mg cm^{-2}, in 10 ms pulsed doses, during the movement with a speed of 0.04 m s^{-1}. After 15 days, the black nightshade was destroyed 95.8 % and red-root 93.8 %, while only 2.4 % of tomato plants were affected by the thermal agent. [52].

2.2.3 Mulching

It is a physical method used in the strategy of weed control in vegetable crops, consisting of the application on the soil of a layer of organic or synthetic matter, being effective especially against annual weeds. Organic mulch (cereal straw, leaves, pine needles, etc.) is most often used in organic crops because it can improve soil properties. For tomato and eggplant crops, polyethylene mulches are used, which ensure a lower soil temperature than the transparent one, preventing sunlight from entering and implicitly weed development [26].

3 Conclusions

In general, the only way to mitigate the negative impact of climate change on vegetable production, especially on their productivity, quality and yield, is to take effective measures. Crop protection against extreme weather events and weed management improvement are also among the efficient adaptation strategies, which include: the development of heat and salinity tolerant varieties, as well as flood resistant, drip irrigation, soil and moisture conservation measures, fertilizer management, etc.

Combating harmful plants in organic vegetable crops can be solved only by combining available solutions and adopting an integrated non-chemical weed management strategy. As for specialized equipment, it seems that small innovative companies will develop new weed control technologies, because based on the progress made, the future in this field belongs to automatic systems.

It is important for future research to focus on the shortcomings or weaknesses of systems and methods related to reducing the negative impact of climate change on organic vegetable crops, in order to provide farmers with viable strategies and solutions.

"The work was supported by grant of the Romanian Ministry of Research and Innovation, CCCDI - UEFISCDI project number PN-III-P1-1.2-PCCDI-2017-0659 / No.11 PCCDI/2018, within PNCDI III and a grant of the Romanian Education and Research Ministry, through Programme 1 – Development of the national research-development system, subprogramme 1.2 – Institutional performance – Projects for financing excellence in RDI, contract no. 16PFE."

References

1. A. El Kenawy, J.I. Lopez-Moreno, S. M. Vicente-Serrano, Recent trends in daily temperature extremes over northeastern Spain (1960–2006), Nat. Hazards Earth Syst. Sci., **11**, 21,(2011);
2. A. E. Croitoru, A. Piticar, Changes in daily extreme temperatures in the extra-Carpathians regions of Romania, International Journal of Climatology, **33**, 15, (2013);

3. Q. Wang, X. Fan, Z. Qin, M. Wang, Change trends of temperature and precipitation in the Loess Plateau Region of China, 1961–2010, Global and Planetary Change, **92–93**, 10, (2012);
4. A. R. Scorzini, M. Di Bacco, M. Leopardi, Recent trends in daily temperature extremes over the central Adriatic region of Italy in a Mediterranean climatic context, International Journal of Climatology, **38**, **(S1)**, 17, (2018);
5. E.B. Lupikasza, S. Hänsel, J. Matschullat, Regional and seasonal variability of extreme precipitation trends in southern Poland and central-eastern Germany 1951–2006, International Journal of Climatology **31**, 23, (2011);
6. C. Şulea-Iorgulescu, Researches on the development of an integrated computer system for monitoring the anti-hail rockets launch units, Universitaria Publishing House, Craiova, Romania, ISBN 978-606-14-0833-7 (2014);
7. A. Piticar, PhD thesis - Studies on recent climate change in Northeastern Romania, University "Babes-Bolyai" Cluj Napoca (2013);
8. S. Mohr, M. Kunz, Recent trends and variabilities of convective parameters relevant for hail events in Germany and Europe, Atmospheric Research on ScienceDirect **123,** 18, (2013);
9. H. J. Punge, K. Bedka, M. Kunz, A. Werner, A new physically based stochastic event catalog for hail in Europe, Natural Hazards **73,** 21, (2014);
10. K. A. Nicolaides , G. Photiou , K. Savvidou , A. Orphanou , S. C. Michaelides , T. S. Karakostas , D. Charalambous, C. Kannaouros, The impact of hail storms on the agricultural economy of Cyprus and their characteristics, Adv. Geosci., **17**, 5, (2009);
11. A. Pushparani Devi, Moirangthem Sanju Singh, Shubhashree Priyadarshinee Das, Jhilki Kabiraj, Effect of Climate Change on Vegetable Production- A Review, Int. J. Curr .Microbiol. App. Sci. **6(10)**, 7, (2017);
12. E. E. Braun, S. Taylor Lovell, M. Babadoost, F. Forcella, S. Clay, D. Humburg, S. E. Wortman, Abrasive grit application in organic red pepper: An opportunity for integrating nitrogen and weed management**,** HortScience 54 (9), 8, (2019);
13. D. Zugravescu, Gh. Manolea, L. Alboteanu, C. Şulea Iorgulescu, Anti-hail system - geonomic component of modern science, Proceedings of the International Conference on RISK MANAGEMENT, ASSESSMENT and MITIGATION, 5, ISSN: 1790-2769, ISBN: 978-960-474-182-3 (2010);
14. A. K. Arakelyan, Flaws and advantages of current and prospective anti-hail protection methods, stations and networks (Review), Global Advanced Research Journal of Agricultural Science (GARJAS) **7(12),** 11 (2018);
15. A. K. Arakelyan, A Way to Enhance Hail Prevention Technique and to Increase the Efficiency of Anti-Hail Protection of Unlimited Agricultural Areas, J Agron Agri Sci **2**: 013 (2019);
16. G. Manolea, I.L. Alboteanu, A. Novac, C. Sulea Iorgulescu, S. Stepan, Optimizing the Structure of the Rocket Launcher Monitoring System**,** 2019 International Conference on Electromechanical and Energy Systems (SIELMEN), Craiova, Romania, 5, (2019);
17. Y. Shahak, Photoselective Netting: The Concept, Research and Implementation in Various Crops*,* ASHS Annual Conference, 8, (2011);
18. S. K. Agrawal, Application of textile in agriculture, IJARSE **2 (7)**, 10, (2013);
19. E. Cărpuş, A Dorogan, F. Burnichi, Agrotextile Systems - Strategic Elements for Sustainable Development of the Agriculture, 7-th International Conference on Advanced Materials and Systems ICAMS, 5, (2018);

20. K. Bhavani, N. Mallikarjun, N.M. Sunilkumar, Agro textiles - Their applications in agriculture and scope for utilizing natural fibers in agro tech sector, International Journal of Applied Home Science, **4 (7&8)**, 10, (2017);
21. S. G. Annapoorani, K. Saranya, Applications of Agricultural Textiles - A Review ,International Journal for Research in Engineering Application & Management (IJREAM), **4**, ISSN: 2454-9150 Special Issue - ICDOMP'19, 4, (2019);
22. P. Marasovic, D. Kopitar, Overview and perspective of nonwoven agrotextile, TEXT LEATH REV **2 (1)**, 14, (2019);
23. M.A. Jones, Using light to improve commercial value, Horticulture Research **5**, 47 (2018);
24. Y. Shahak, Y. Kong, K. Ratner, The wonders of yellow netting, Acta Hortic. 1134, 8, (2016);
25. B. P. Baker, C. L. Mohler, Weed management by upstate New York organic farmers: strategies, techniques and research priorities, Renewable Agriculture and Food Systems, **30 (5)**, 10, (2014);
26. E. Pannaci, B. Lattanzi, F. Tei, J., Non-chemical weed management strategies in minor crops: A review, Crop Protection **96**, 15, (2017);
27. B. Melander, N. Holst, I. A. Rasmussen, P. K. Hansen, Direct control of perennial weeds between crops e Implications for organic farming, Crop Protection **40**, 7, (2012);
28. A. Peruzzi , M. Raffaelli, M. Ginanni, M. Fontanelli, C. Frasconi, An innovative self-propelled machine for soil disinfection using steam and chemicals in an exothermic reaction, J. Biosystem Engineering **110**, 9, (2011);
29. A. Peruzzi, M. Raffaelli, C. Frasconi, M. Fontanelli, P. Bàrberi, Influence of an injection system on the effect of activated soil steaming on Brassica juncea and the natural weed seedbank J. Weed Research, **52**, 13, (2012);
30. M. Raffaelli, L. Martelloni, C. Frasconi, M. Fontanelli, S. Carlesi S., A. Peruzzi, A prototype band-steaming machine: Design and field application, J. Biosystem Engineering **144**, 11, (2016);
31. A. Koocheki, M. Nassiri, L Alimoradi, R. Ghorbani, Effect of cropping systems and crop rotations on weeds, Agron. Sustain. Dev. **29,** 8 (2009);
32. D. Weisberger, V. Nichols, M. Liebman M, Does diversifying crop rotations suppress weeds? A meta-analysis, PLoS ONE **14(7):** e0219847, (2019);
33. S. Cordeau, R. G. Smith, E.R. Gallandt, B. Brown, P. Salon, A. DiTommaso, M.R. Ryan, Disentangling the Effects of Tillage Timing and Weather on Weed Community Assembly, J. Agriculture **7 (8),** 66, 18, (2017);
34. S. Gruber,W. Claupein Effect of tillage intensity on weed infestation in organic farming, J. Soil & Tillage Research **105**, 8, (2009);
35. M. H. Jorgensen Chapter 3, The Effect of Tillage on the Weed Control: An Adaptive Approach, Biological Approaches for Controlling Weed, (2018) http://dx.doi.org/10.5772/intechopen.76704;
36. L.Tang, C. Cheng, K. Wan, R. Li, D. Wang, Y. Tao, Impact of Fertilizing Pattern on the Biodiversity of a Weed Community and Wheat Growth, PLoS ONE 9(1): e84370., (2014);
37. A. A. Bajwa, Ehsanullah, S. A. Anjum, W. Nafees, M. Tanveer, H. S. Saeed, Impact of fertilizer use on weed management in conservation agriculture - A Review, Pakistan J. Agric. Res. **27(1),** 10 (2014);

38. R.K. Dhall, Breeding for Biotic Stresses Resistance in Vegetable Crops: A Review, STM Journals 2015, RRJoCST **4 (1),** 14, (2015);
39. P. Dittmar, N. Boyd, R. Kanissery, Weed Management Principles in Commercial Vegetable Production, **HS717/CV113**, 6, (2018);
40. A. Peruzzi, L. Martelloni, C. Frasconi., M. Fontanelli, M. Pirchio, M. Raffaelli, Machines for non-chemical intra-row weed control in narrow and wide-row crops: a review, Journal of Agricultural Engineering, XLVIII:**583**, 14, (2017);
41. D.L. Shaner, H.J. Beckie H.J., The Future for Weed Control and Technology, Pest Management Science **70 (9),** 11 (2014);
42. J. Banks, G. Sandral, Report on weed control using hot water / steam and herbicides in the city of Joondalup, (2007) http://api.joondalup.wa.gov.au/files/councilmeetings/2007/attach11brf210807.pdf;
43. A. M. Peerzada, B. S. Chauhan, Thermal Weed Control: History, Mechanisms, and Impacts in Jabran K., Chauhan B.S. (Eds) Non – Chemical Weed Control, Elsevier, Academic Press, 23, (2018);
44. J. Ascard, , Effect of flame weeding on weed species at different developmental stages, Weed Res. **35**, 15, (1995);
45. L. Martelloni, C. Frasconi, M. Fontanelli, M Raffaelli, A. Peruzzi, Mechanical weed control on small-size dry bean and its response to cross-flaming, SJAR **14 (1)**, 12, (2016);
46. M. Raffaelli, M. Fontanelli, L. Martelloni, C. Frasconi, A. Peruzzi, Innovative strategy and machines for physical weed control in agriculture and urban areas, International Conference RAGUSA SHWA, Italy, 7, (2012);
47. L. Martelloni, M. Fontanelli, C. Frasconi, M. Raffaelli, M. Pirchio, A. Peruzzi, A combined flamer-cultivator for weed control during the harvesting season of asparagus green spears, SJAR **15 (2),** 10, (2017);
48. B. Melander, B. Lattanzi,E. Pannacci, Intelligent versus non-intelligent mechanical intra-row weed control in transplanted onion and cabbage, J. Crop Protection **72**, 8, (2015)
49. M, Siemens, Robotic Weed Control, Proceedings of the California Weed Science Society 66, 5, (2014);
50. R. Lati, M. Siemens, J. Rachuy, S. Fennimore, Intra-Row Weed Removal in Broccoli and Transplanted Lettuce with an Intelligent Cultivator, Weed Technology **30 (3),** 9, (2016);
51. R. Smith, Evaluation of intelligent cultivators for use in lettuce production, UC Weed Science, ANR Blogs, (2016), https://ucanr.edu/blogs/blogcore/postdetail.cfm?postnum=19841&email=yes;
52. Y. Zhang, E. S. Staab, D. C. Slaughter, D. K. Giles, D. Downey, Automated weed control in organic row crops using hyperspectral species identification and thermal micro-dosing, J. Crop Protection **41**, 10 (2012);

https://doi.org/10.1051/e3sconf/202018003017

Study of the compression behaviour of grape seeds using the Finite Element Method

Carmen Vasilachi (Baltățu) [1*], *Sorin-Ştefan* Biriş [2] *and Gabriel* Gheorghe [1]

[1]National Institute of Research - Development for Machines and Installations Designed to Agriculture and Food Industry – INMA, Bucharest, Romania
[2]University Politehnica Bucharest, Faculty of Biotechnical Systems Engineering, Romania

Abstract. It is important to know all phenomena that occur during compression of grape seeds because are very complex. Studying these phenomena can help us in increasing the performance of the equipment. Also for the pressing of oilseed materials, it is helpful to understand the compression behavior of grape seeds. The purpose of this study is to find how the stress and deformation propagate in grape seeds during the compression process. Therefore, we used an advanced engineering method named finite element method that shows the propagations of stress and strain fields in grape seeds. In this paper, we made a two dimensional model of grape seed and we analyzing the dispersion of stress and strain with a dedicated software FEM. In the end, to validate the FEM model we made a compression test on grape seeds using a Hounsfield unit. This study is for highlighting the importance of the orientation of the grape seeds during the pressing process to reduce the energy consumed.

1 Introduction

Nowadays, it is very important to use alternative resources, as conventional resources are increasingly limited. That is why, in recent years, there has been a developing interest in obtaining oil from oilseeds and oil plants. Researchers have shown that these oils have nutritional elements useful for human body health and have other necessary properties which can be used in other industrial areas, for which the current technology rewards us for obtaining these resources, [1]. This category also includes grape seed oil that is obtained from grape seeds resulting from the processing of grapes in winemaking industries.

If we consider that grape seeds are a new material, it is important to know their mechanical, physical properties and behavior during pressing. At the same time, stress and strain modeling is very important because it allows us to establish a technological line for their processing and we can estimate the energy consumption needed to obtain the oil. This modeling is done with the help of the programs using the finite element method and for this we must eliminate the physical and mechanical characteristics, respectively Young's modulus and Poisson's coefficient, [2].

* Corresponding author: carmen.vasilachi@gmail.com

For a clearer picture of the physical properties of grape seeds, Avshalom Karasik and co-workers made a 3D scan two types of seeds from Hadari and Merlot grapes, as in Fig.1, where it can be noticed at point c) that the width of the seed has a slight variation of from one variety to another, [3].

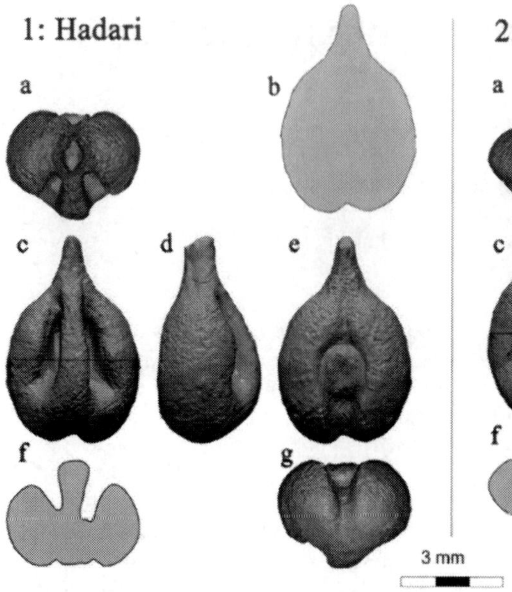

Fig. 1. 3D scan of grape seed views (a, c, d, e, g) and across- sections (b, f) from two varieties 1: Hadari (indigenous Israeli variety), 2: Merlot (international variety) [3]

In specialized literature, the physical and mechanical properties of grape seeds make subject for just small studies, these still being a topic in the early stages of development. However, L. ROLLE in his research has come to the conclusion that Young's module evolution of seeds a few weeks after harvesting the grapes tends to grow with the drying process. This is explained by the fact that Young's module is dependent on the size of the seeds, therefore it may have a variation from one seed to another or from one period to another, [4]. In this article we measured and calculated the physical and mechanical properties of grape seeds and Burgung variety.

At the same time, in specialized literature, it was concluded that the positioning of the seed at the time of compression, either vertically or horizontally, is important. In the upright position the sampler absorbs more energy to be broken than in the horizontal position. This article aims to highlight the propagation of deformations and forces inside the grape seeds during compression.

2 Material and Methods

For this paper, Burgund grape seeds were used to represent the physical and mechanical properties. The grape marc was obtained from the research station INCDBH Stefanesti harvest from October 2019. To get the seeds from grape marc, it was necessary to dry the mixture of tails, seeds, and skin, after that we separated the seeds from the rest of all. In finaly we obtain the grape seeds as in Fig.2.

https://doi.org/10.1051/e3sconf/202018003017

Fig. 2. Grape seeds of the Burgund type

The physical properties of the seeds can be observed in Table 1 and were determined as follows: the dimensions on the three axes were obtained using a digital caliper (x- length axis, y-width axis, z-thickness axis) and also the seeds were weighed to determine the unit mass but also the mass of 100 seeds.

Table 1 The physical properties of grape seeds

No.	Variety Name	Caracteristici geometrice de baza					
		Dimension [mm]			Unit mass [g]	Mass for 100 of seeds [g]	Umiditate [%]
		x - length	y - width	z - depth			
1	Burgund	6.00	3.92	2.61	0.027	33.45	4.7

We determined the mechanical properties of grape seeds using a uniaxial compression equipment, Fig.3. After the proper test, a force-displacement curve is obtained for uniaxial compression, [5]. With the help of the obtained data imported information resulted about the hardness of the seeds, the apparent modulus of elasticity, the energy consumed for breaking, drilling and moving at different times, etc.

Fig. 3. Hounsfield / Tinius Olsen mechanical test unit, model H1 KS

The grape seeds were placed between the fixed plate and the mobile plate with a speed of 1 mm / min, of Hounsfield equipment. Strength and stress were recorded and saved using the software dedicated to the machine. After performing the compression test, a force-displacement graphic was shown, Fig. 4.

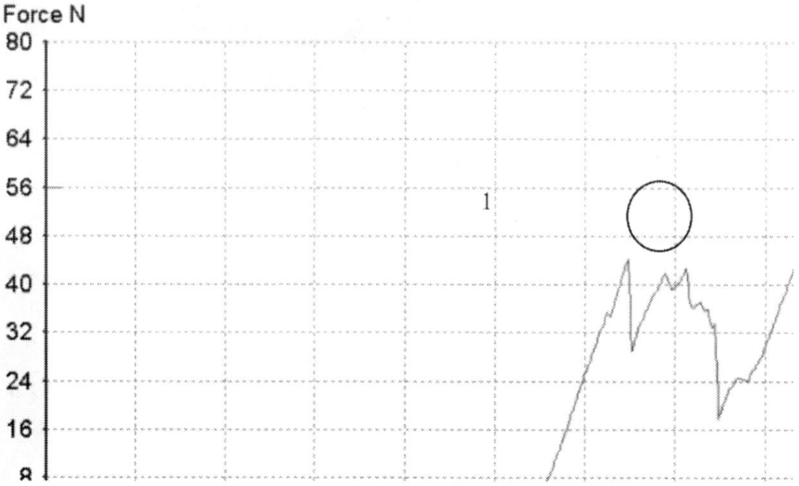

Fig. 4. Force-Extension graph for Burgund grape seeds

In the point marked with "1" on the force-displacement graph, the grape seed is broken off and appears an immediate force breakage. The maximum force applied to this seed was 56N and it was sufficient to break the seed, [6] .

The obtained results are used to create a mesh model, as in Fig. 5, which is necessary for FEM modeling of propagation of deformation stress in grape seeds. For modeling, we used a software called QuickField Elements Analysis System which created a graphic displacement for let us know what happened in the interior of the grape seeds.

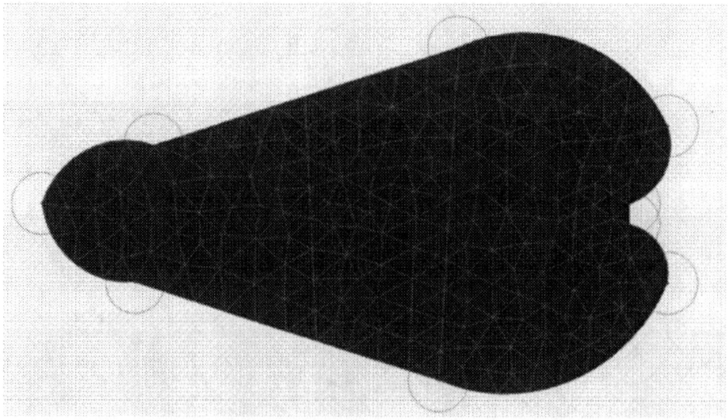

Fig. 5. Mashed model of a grape seeds

The characteristics of the material needed in seed modeling were compared with those of the specialized literature and introduced in software, such as Young's module has the value 22 MPa [5], respectively Poisson's ratio was 0.36 [6].

3 Results and discussion

The 2D drawing of the grape seed was made and the knots for the mash were made in the QuickField program, we introduced the material characteristics and applied the force on the horizontal side of the seed. As a result, we obtained the distribution of equivalent stress, Fig.6 and all the displacements within the seed, Fig.7.

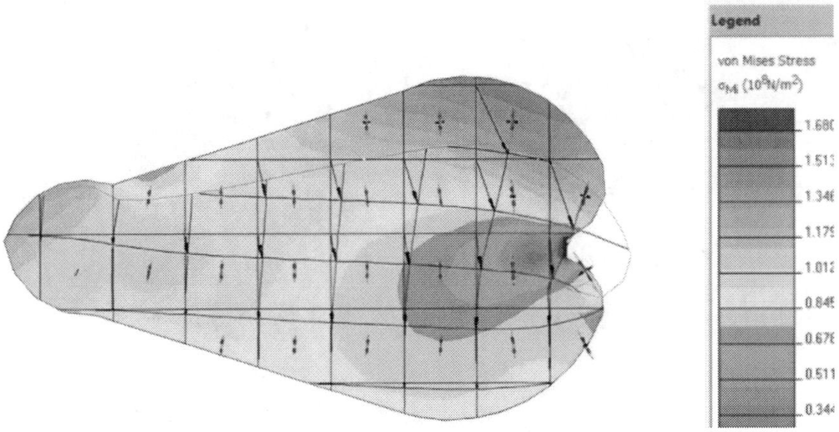

Fig. 6. Distribution of equivalent stress in the grape seeds

In Fig. 6 can be noticed that the highest stress is found in the contact area between seeds and the plate of the test equipment. It can be observed that the highest force is about 168MPa.

In Fig. 7, can be observed the distribution of displacements and how the grape seeds are deformed during the compression process. The largest displacements are found in the contact area between the moving plate and the seed, the displacements have a vertical direction.

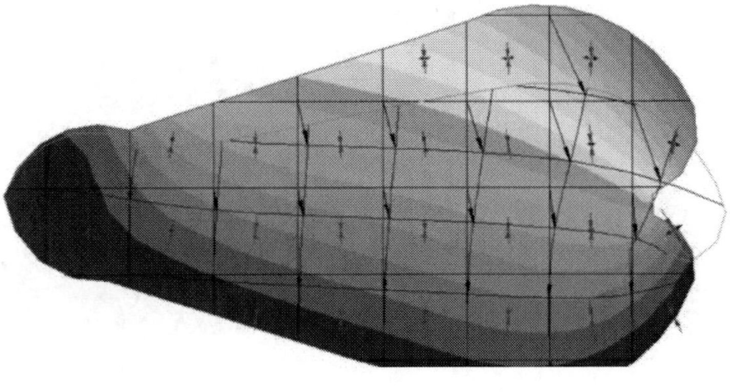

Fig. 7. Distribution of equivalent displacement in the grape seeds

In Fig. 8, the grape seed was placed upright and it is observed that the force is applied at the top of the seed. It is also noted that the contact area between the seeds and the mobile plate is smaller. The force required to compress the seed must be greater, we can conclude that in a vertical position the energy consumption of the higher seed.

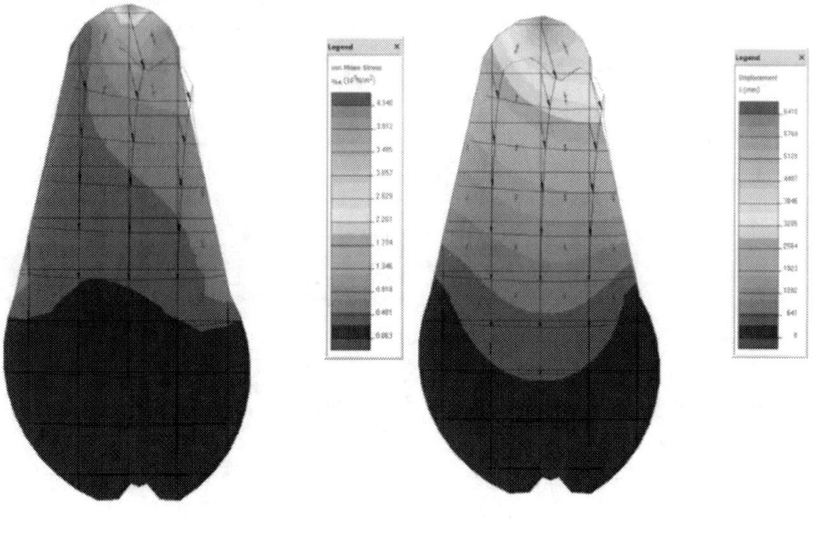

a. b.

Fig. 8. Distribution of equivalent stress(a) and displacements (b) in the grape seeds oriented vertycally

The QuickField program also created a graph of stress distribution during the simulation of compression with the force indicated by us in the system for the both positions. In Fig. 8 and Fig.9, we observe the stress - extension curve for bouth positions which has a peak of resistance after which the force has a sharp decrease which means that that is the point where the seed has broken. This graph is approximately similar to the one in Fig. 4. Different from the two graphs is that in the case of the vertical position a greater force was needed to break the grape seed.

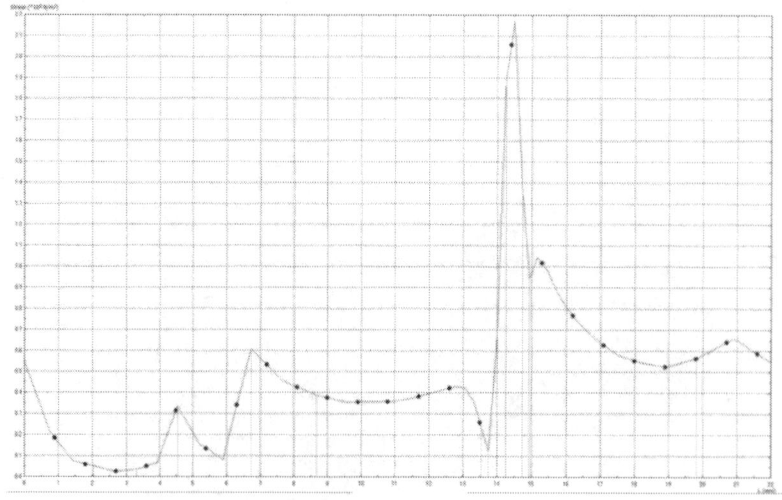

Fig. 9. Force-Extesnion graph from QuickFild when the grape seed is oriented horizontal

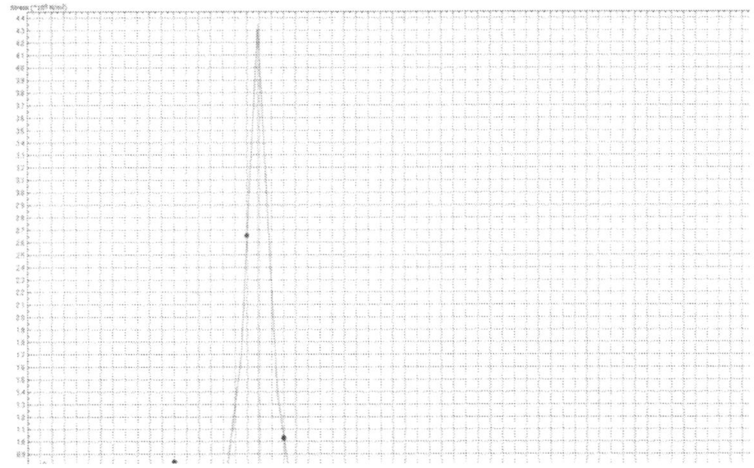

Fig. 10. Force-Extesnion graph from QuickFild when the grape seed is oriented vertically

4 Conclusions

The Finite Element Method is the advanced engineering tool for processing numerical and mathematical modeling and it was very helping in this work to demonstrate the importance of grape seeds position during the compression time.

In conclusion, it is more advantageous to positioned horizontally the grape seed during the compression test because the contact area is larger than in the vertical orientation. The displacement distribution is more constant and on a larger surface in the interior of seeds in a horizontal position. In order to optimize the pressing process by reducing the energy consumption used, a technical solution for directing the grape seeds can be found so that the horizontal parts to be in contact with the pressing screw.

This work has been funded by the European Social Fund from the Sectoral Operational Programme Human Capital 2014-2020, through the Financial Agreement with the title "Scholarships for entrepreneurial education among doctoral students and postdoctoral researchers (Be Antreprenor!)", Contract no. 51680/09.07.2019 - SMIS code: 124539.

This work was supported by the Romanian Education and Research Ministry, through through Programme 1 – Development of the national research-development system, subprogramme 1.2 – Institutional performance – Projects for financing excellence in RDI, contract no. 16PFE.

References

[1] D. J. Epasinghe, C. K.Y. Yiu, M. F. Burrow, *Mechanical properties, water sorption characteristics, and compound release of grape seed extract-incorporated resins,* JAOS Journal of Applied Oral Science, vol. **25** no.4 Bauru July/Aug. (2017)
[2] R.K. Gupta, S.K. Das, *Fracture resistance of sun¯ower seed and kernel to compressive loading,* Journal of Food Engineering, **46** (2000)

[3] A. Karasik, O. Rahimi, M. David, E. Weiss, E. Drori, *Development of a 3D seed morphological tool for grapevine variety identifcation, and its comparison with SSR analysis,* SCIENTIFC REPORTS , **8** (2018), DOI:10.1038/s41598-018-24738-9
[4] L.Rolle, F. Torchio , B.Lorrain , S. Giacosa, S. Río Segade , E. Cagnasso, V. GERBI, P.L. Teissedre, *Rapid methods for the evaluation of total phenol content and extractability*

in intact grape seeds of cabernet-sauvignon: instrumental mechanical properties and FT-NIR spectrum, J. Int. Sci. Vigne Vin, **46** (2012)
[5] C. B. Vasilachi, S. S. Biris, G. Ipate, G. Gheorghe, *Determination of the longitudinal modulus of elasticity for the different type of grape seeds*, Conference Engineering for Rural Development, (2020), Jelgava.
[6] S. S. Biris, M. Ionescu, N.E. Gheorghita, N. Ungureanu, N.V. Vladut, *Study of the compresion behavior of sunflower seeds using the Finite Element Method*, AGROFORM International Journal, vol.**4** (2019)

Study on the effects and changes of soil degradation under the influence of antibiotics

Ancuta Alexandra Petre[1,*], *Florin Nenciu*[1],*Nicoleta Alexandra* Vanghele[1], *Mariana Mădălina* Stanciu[1], *Dumitru Bogdan* Mihalache[1], *Iulia Andreea* Grigore[1] and *Laurenţiu* Vladuţoiu[1]
[1] INMA National Institute of Research - Development for Machines and Installations designed to Agriculture and Food Industry, Bucharest

Abstract: The use of antibiotics in large quantities against the combat of pests in soils, indicates an increased remanence of them, which leads to major environmental risks. After entering in the soil, antibiotics are subjected to a succession of biogeochemical processes under the action of multiple environmental factors: absorption, migration, transformation, degradation or nutritional prosperity of plants.

In order to know the current state of the environment and the effects of antibiotics it is essential to discover procedures for improving the degradation and combating the dissemination of antibiotic resistance.

Research in recent years on the extraction of antibiotics from the soil is based on complex processes, such as: Soxhlet extraction, ultrasonic extraction or accelerated solvent extraction.
Soil residue was determined using liquid chromatography coupled with mass spectrometry, being the equipment that provides the highest accuracy in the analysis of polar compounds in very low concentrations.

The purpose of the paper is to find an efficient and ecological solution for the elimination of antibiotics from soils and to highlight the evaluation of the contaminated land regarding the antibiotic residues and the impact on the composition of the bacterial community. Information about the effects of antibiotics accumulated in soil, persistence, resistance and sensitivity to chemical and microbiological substances, degradation and results related to soil fertility and crop production are highlighted.

Key words: soil, antibiotics, degradation, effect

1 Introduction

Emerging contaminants such as pharmaceuticals, personal care products, surfactants, plasticizers or pesticides are pollutants resulted from various anthropogenic activities, that are generating high levels of risk to the environment and human health.

Unlike other contaminants, they have complex structures that make them difficult to trace and remove and they cannot be monitored or decontaminated by classical waste management methods.

Antibiotics are considered by many researchers among the most important discoveries from history, due to the revolutionary contribution they have in human treatment. In

*Corresponding author : ancapetre28@gmail.com

developing countries, antibiotic use has contributed to increasing life longevity by reducing mortality caused by bacterial infections. However, antibiotics have begun to be considered an increasing threat, on the one hand because they have accumulating effects and on the other hand caused by the impossibility of evaluating in a manner the long-term effects.

These pharmaceuticals are released partially degraded or non-degraded in the environment, creating pollution with antibiotics. Bioremediation could have an important role in the treatment of these group of pollutants, being one of the few solutions that can be applied to such a complex problem [2].

Antibiotic amounts eliminated as residues in nature have too low to kill bacteria, causing microorganisms to adapt to these drugs and to gain certain levels of resistance to them.

When they reach the aquatic and terrestrial ecosystems, antibiotics as micro-contaminants show a high risk to human and animal health, due to the fact that may develop multi-resistant and very difficult to treat bacteria [1].

Numerous efforts have been made at EU level to reduce the negative effect of antibiotics ; however, it is still an important problem since antibiotic resistance in microorganisms has increased in the last years. Unfortunately, the development of new therapies and drug formulas that take into account the adaptations of bacteria to current treatments occur over long periods of time [3].

Antibiotics are classified as natural, synthetic and semi-synthetic compounds that interrupt antimicrobial activity [4]. Antibiotics have the most important role from the pharmaceutical spectrum in treating microbial infections having specific therapeutic actions on the targeted patient, as presented in Figure 1. Since 1928 when Alexander Fleming discovered penicillin by mistake, many antibiotics have been synthesized for the use on animal and plant health. Antibiotics may be used as growth promoters in livestock, for example in cattle, pigs and bird preventive treatment to improve feed efficiency [5], although they are banned in the European Union since 2006, they are still in practice in China and India [6]. The widespread use of known antimicrobials in various purposes, for treating animals and plants creates accumulations and important problems for the environment and natural ecosystem [7].

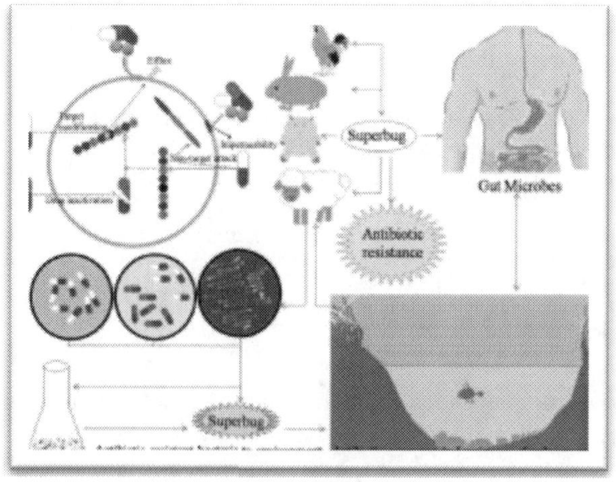

Fig. 1. Antibiotic resistance in the human body, soil and water [1]

Recent studies have been developed regarding synthesized of artificial antibiotics, considering the resistance acquired by some bacteria to main two antibiotics used on large scales, namely penicillin and streptomycin [8]. Overuse, or improper use of antibiotics determines the appearance of microbial resistance, especially in bacterial communities, main reason for that being dosing long-term administration of low antibiotic concertation [9].

Antibiotic resistance is the increase of the MIC (minimum inhibitory concentration) when treating against microbes, which prevents visible growth of a bacteria. It is a mechanism by which bacteria can survive the stress of antibiotics. Microbes can exert antibiotic resistance by passing drug targeted interaction, efflux of antibiotics from the cell (TolC efflux pump in E. coli), alteration of antibiotics and their metabolites by enzymes [10,11].

The co-substrate-assisted antibiotic-modifying enzymes, with the help of ATP being a process of rapidly measuring actively growing microorganisms through detection of adenosine triphosphate, can transfer functional groups that covalently modify antibiotics by acetylation, phosphorylation, adenylation, nucleotidylation, ribosylation, and glycosylation [12].

2 Materials and method

Biodegradation of antibiotics takes place through two biotic and abiotic processes. Biotic biodegradation requires the use of microorganisms while the abiotic biodegradation requires processes such as sorption, hydrolysis, photolysis, oxidation and reduction reactions [13]. In order to treat terrestrial and aquatic contamination, several microbial populations are analyzed, that might help in elimination of commonly used antibiotics.

In order to evaluate the concentration of residual antibiotics, there have been developed so called biomonitoring tools. These tools include Biomonitoring and Biosensors, Bioassays, biomarkers, and microbial community analyses, all these being constituted with the purpose of reducing and regulating the concentration of these difficult to quantify pollutants. In addition, an environmental monitoring and screening can be acquired, using customized sensors developed to determine some analytes and validate the used models.

The bioremediation process can be described as the metabolic potential that some microorganisms have in disintegrating some pharmaceuticals [14].

Biological methods for the removal of antibiotics are safe and efficient, because they do not employ dangerous and sophisticated treatment procedures, using mostly only bacteria, fungi, nutrients and stabilizing agents.

Another bioremediation important aspect is that can at the same time treat a wide range of other pollutants in addition to antibiotics, improving the health of the soil without having major negative effects.

Metagenomics is a novel technology made for studying genetic material, isolated from environmental samples in order to reveal the knowledge on microbial communities. These techniques could be successfully used for the identification of novel bacteria or gene clusters encoding for enzymes that degrade pollutants, and could be associated to the bioremediation processes.

Antibiotics can also have negative affect on some useful natural microbial communities. These good microbial communities are useful to ecological processes, because they are maintaining soil and water quality, being involved in the biogeochemical natural cycle of organic degradation. Antibiotics present in the environment can therefore affect both long-term and short-term microbial activities.

Indirect impact of antibiotics consist in the evolution of antibiotic resistant bacteria. Using some biodegradation processes, could remove the pollutant from the environment, when mineralized [1].

When using manure as fertilizer there is a risk of contamination with antibiotics from animal farms. Animals release 30% -90% of antibiotics that are ingesting by urine and manure.

With the application of fertilizer to provide nutrients, can be introduced into the soil and aquatic environments important antibiotic residues, such as TCs (tetracyclines) that are very persistent in the soil. The most commonly found antibiotics in fertilized soils with manure are mainly TC [15]. Antibiotics are mainly classified into different classes such as β-lactams, tetracyclines, aminoglycosides, macrolides, sulfonamides, antifungal antibiotics [16, 17].

A classification of antibiotics according to several requirements was summarized in the table 1, depending on some functional groups, activity mechanisms, molecular weight, etc. [18].

Table 1. Classification of antibiotics [33]

Class	**Antibiotic**	**Chemical formula**	**Molecular weight (g/mol)**	**Water solubility (mg/L)**	**Log KOW**	**Kd (L/kg)**	**KOC (L/kg)**
Aminoglycosides	Gentamicin	$C_{21}H_{43}N_5O_7$	477.6	100,000	−3.1	–	–
	Streptomycin	$C_{21}H_{39}N_7O_{12}$	581.6	12,800	−6.4	8–290	580–11,000
Diaminopyrimidines	Trimethoprim	$C_{14}H_{18}N_4O_3$	290.3	400	0.91	7.40	4,600
Fluoroquinolones	Ciprofloxacin	$C_{17}H_{18}FN_3O_3$	331.3	30,000	0.28	427–4,844	1,127–61,000
	Difloxacin	$C_{21}H_{19}F_2N_3O_3$	399.4	1,330	0.89	–	–
	Enrofloxacin	$C_{19}H_{22}FN_3O_3$	359.4	>53.9	0.7	0.54–5,612	39–768,740
	Norfloxacin	$C_{16}H_{18}FN_3O_3$	319.3	177,900	−1.03	591–5,791	310
	Ofloxacin	$C_{18}H_{20}FN_3O_4$	361.4	10,800	0.35	1,471–4,325	44,140
Glycopeptides	Vancomycin	$C_{66}H_{75}Cl_2N_9O_{24}$	1449.3	>1,000	−3.1	0.3–0.7	–
Ionophores	Lasalocid	$C_{34}H_{54}O_8$	590.8	750	–	9–280	2.9–4.2
	Monensin	$C_{36}H_{62}O_{11}$	670.9	0.003	5.43	0.5–65	2.1–3.8
β-Lactams	Amoxicillin	$C_{16}H_{19}N_3O_5S$	365.4	3,430	0.87	–	865.5
	Cephapirin	$C_{17}H_{17}N_3O_6S_2$	423.5	1,030	−1.15	0.21–3.83	–
	Cefuroxime	$C_{16}H_{16}N_4O_8S$	424.4	145	−0.16	–	12.4–15.5
	Penicillin G	$C_{16}H_{18}N_2O_4S$	334.4	210	1.83	–	2.68
Lincosamides	Clindamycin	$C_{18}H_{33}ClN_2O_5S$	424.9	30.6	2.16	–	70
	Lincomycin	$C_{18}H_{34}N_2O_6S$	406.5	927	0.2	–	59
Macrolides	Azithromycin	$C_{38}H_{72}N_2O_{12}$	748.9	2.37	4.02	2.18	59,900
	Clarithromycin	$C_{38}H_{69}NO_{13}$	747.9	1.7	3.16	262–400	150

Class	**Antibiotic**	**Chemical formula**	**Molecular weight (g/mol)**	**Water solubility (mg/L)**	**Log KOW**	**Kd (L/kg)**	**KOC (L/kg)**
	Erythromycin	C37H67NO13	733.9	2,000	3.06	130	10
	Tylosin	C46H77NO17	916.1	5,000	1.63	5.4–172,480	110–95,532
Sulfonamides	Sulfachloropyridazine	C10H9ClN4O2	284.7	8,200	0.31	0.90–3.5	41–170
	Sulfadiazine	C10H10N4O2S	250.3	77	−0.09	1.40–14	37–125
	Sulfadimethoxine	C12H14N4O4S	310.3	343	1.63	0.7–4.60	89–323
	Sulfadoxine	C12H14N4O4S	310.3	2700	0.7	0.6–4.9	1.8–31.3
	Sulfamethoxazole	C10H11N3O3S	253.3	610	0.89	0.6–4.9	1.2–94.9
	Sulfamethazine	C12H14N4O2S	278.3	1,500	0.89	0.23–206	60–208
	Sulfamonomethoxine	C11H12N4O3S	280.3	10,000	0.70	0.6–4.9	60–200
	Sulfapyridine	C11H11N3O2S	249.3	268	0.35	1.60–7.40	80–308
Tetracyclines	Chlortetracycline	C22H23ClN2O8	478.6	630	−0.62	1,280–2,386	794
	Doxycycline	C22H24N2O8	444.4	630	−0.02	–	–
	Oxytetracycline	C22H24N2O9	460.4	1,000	−0.9	417–1,026	2,872–93,317
	Tetracycline	C22H24N2O8	444.4	231	−1.19	417–1,026	400–93,320

Tetracyclines

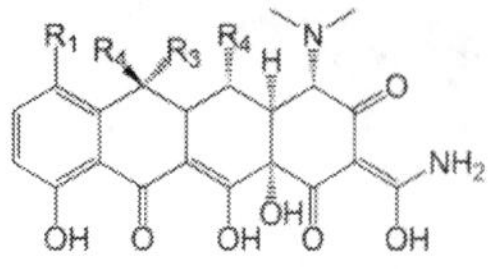

Compound	R_1	R_2	R_3	R_4
Tetracycline	H	CH_3	OH	H
Oxytetracycline	H	CH_3	OH	OH
Chlortetracycline	Cl	CH_3	OH	H
Doxycycline	H	CH_3	H	OH
Demedocycline	Cl	H	OH	H

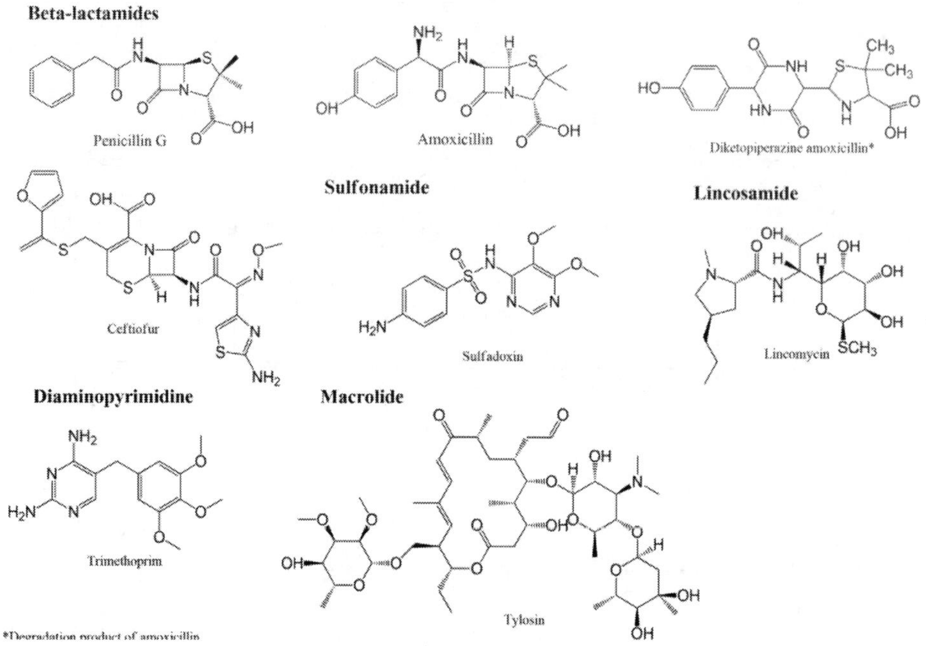

Fig. 2. Chemical structures of some antibiotic structures [33]

Kwon-Rae et al. [19] noted that sludge and manure represent the main source of antibiotics in Korea, due to the long-term use of manure containing antibiotics that accumulates into the soil. He estimated the concentrations of antibiotics total values from animal litter, using studies from the literature on excretion rate of each drug and correlating levels of antibiotic administration on farms

The concentration of macrolides coming from cattle, pigs and birds varied between 0.07–0.14, 1.05–2.1 and 0.62–1.24 mg / kg, while the sulfonamides and tetracyclines were ranged between 0.49, 8.44 and 1.39 mg / kg and 1.65, 16.56 and 15.62 mg / kg. There are also similar studies with levels of appearance of antibiotics in manure, soil, water and sediments [20-24].

The positive effects of antibiotics are severely reduced after long-term use, that is why quantifying the negative effects is currently of great research interest [25-30]. Considering that most antibiotics are not fully metabolized in the animals bodies, a high percentage of pharmaceuticals are eliminated into municipal wastewater, sewage sludge and with biosolids and reach the soils through irrigation systems, in a cycle similar to the one illustrated in Figure 3 [31,32].

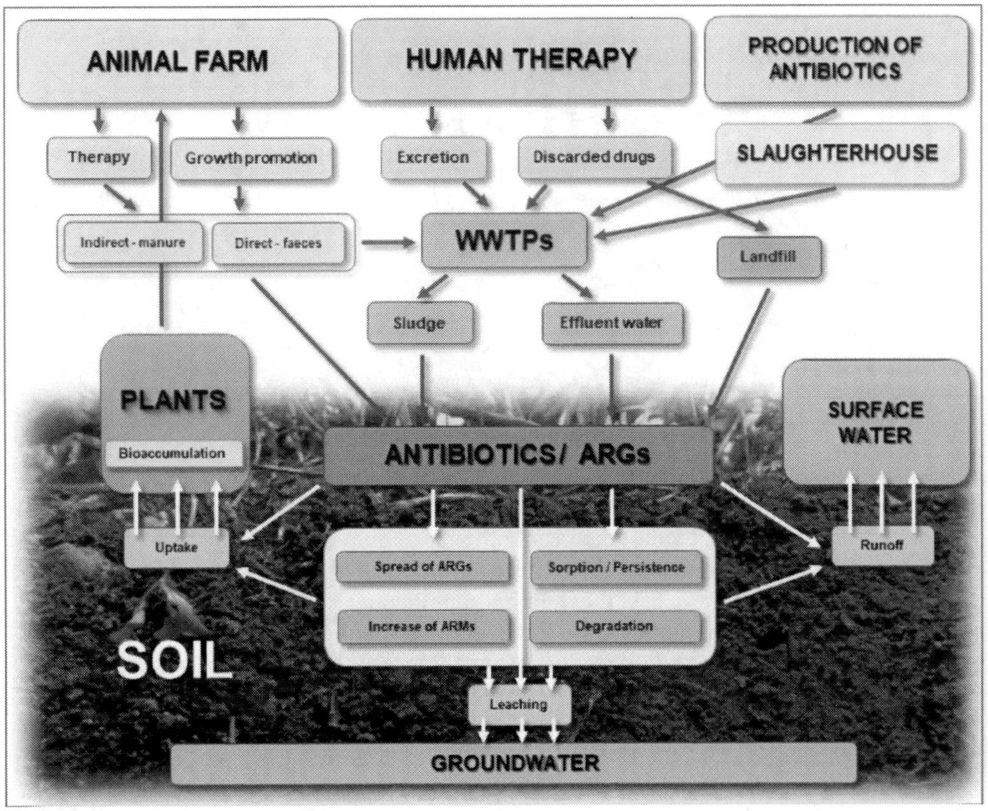

Fig. 3. The life cycle of antibiotics in the environment

3 Results and discussions

Sulfonamides are antibacterial substances that are used as antibiotics for the treatment of pigs for fattening or for various infections. Sulfonamides are eliminated then either as a parent compound or as a bioactive metabolite through urine. When fertilizing agricultural soils with manure, the antibiotic residues appear in soils in substantial quantities.

A commonly used substance from sulfonamide class is the sulfadiazine (SDZ). It is recorded that high concentrations of sulfadiazine can affect the structural and functional microbial diversity in soils and negatively affect the ecosystem. Have been assessed in various research studies the extraction of sulfadiazine from soils, using solvents as ethyl acetate, methanol, ethanol/water or acetonitrile/water. The main extraction procedures were stirring extractions, Soxhlet extraction, ultrasonic extraction and accelerated solvent extraction.

Several studies have been made to optimize extraction for residues of sulfadiazine and its two metabolites, N-acetylsulfadiazine and 4-hydroxysulfadiazine from agricultural soil.

The determination of sulfadiazine from soils have been performed using a liquid chromatography - mass spectrometry system (LC-MS), equipment for analyzing polar compounds, the system being shown in Figure 4.

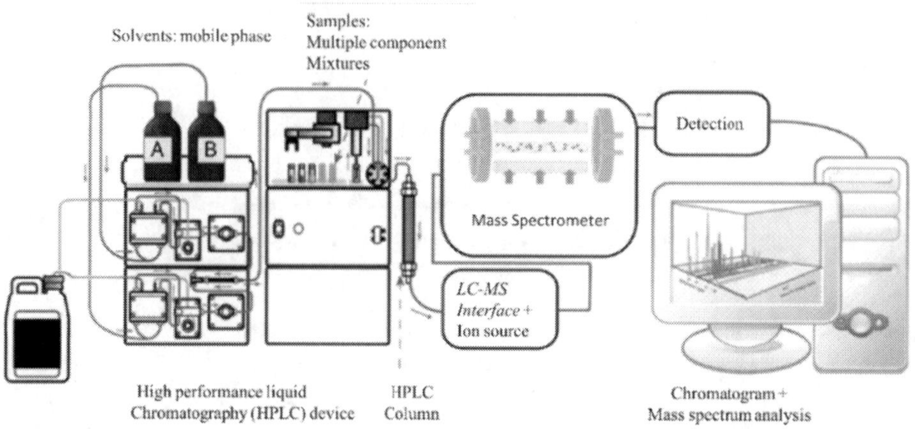

Fig. 4. Diagram of an LC-MS system [34]

Some studies identified correlations between temperatures and the levels of extraction of sulfonamides from the soil [32]. High temperatures corelated with increased pH values are increasing the degradation of the metabolites of sulfamide acetylate (N-ac-SDZ).

Extraction of sulfamide residues from soils is a difficult task, because of these compounds main characteristics: they are hydrophilic and ionizable in the typical pH range of soils. Other research studies have been conducted on investigations of the effect of solvent composition, for the extraction of solzidine residues from the soil using a basic extraction procedure.

Treatment of antibiotic residues using composting methods has recently received more interest. Besides the fact that waste is safer, the final composting products contain large amounts of organic matter and mineral nutrients, which could allow its use as an organic fertilizer.

One effective way in reducing the level of persistent organic pollutants and certain antibiotics from soils is composting process. Composting can successfully be used for sludge stabilization, by means of biological degradation of organic matter.

Among the widely used antibiotics, tetracycline family is the most is intensively studied due to high usage rates, followed by sulfonamides, macrolides and fluoroquinolones;

It was deduced from the analysis of several published articles that 82% of the research studies envisaging composting were performed on a laboratory scale using a reactor, and 18% were made in piles of compost. Composting is a useful way of reducing tetracycline, sulfonamide and macrolide families, with an elimination rate of 70-99%, while fluoroquinolone family have a clearance rate of up to 99% for enrofloxacin and norfloxacin.

In some cases, the elimination of antibiotics is correlated with several issues, such as TOC (total organic carbon), total nitrogen, phosphorus, and other trace elements [32].

4 Conclusions

Antibiotic control is an extremely important aspect to consider for the authorities and farmers, therefore strict rules are needed when selling these drugs. In addition, clear rules must be imposed on farmers regarding the safe disposal of wastes that may contain traces of antibiotic residues, as well as regarding the use of natural fertilizers in agriculture.

This bioremediation strategies have proved to be effective ways to decontaminate soils polluted with antibiotics, and bioremediation conditions have been shown to be very important in biological processes.

Bioremediation is suitable for a variety of soil pollutants, other than antibiotics, and have proved a highly potential for cleaning the anthropic ecosystem. That is why in-depth studies are needed for further developing new and improved bioremediation technologies, to address pollutants that are difficult to remedy by other methods.

This work was supported by a grant of the Romanian Education and Research Ministry, through Programme 1 – Development of the national research-development system, subprogramme 1.2 – Institutional performance – Projects for financing excellence in RDI, contract no. 16PFE and by a grant of the Ministry of Agriculture and Rural Development, contract of sector financing, ADER type, no. 25.4.1 "Technology for obtaining biofertilizers and/or bioinsecticides, for ecological production systems".

References

1. P. Grenini, *Microchem. J.* (2017)
2. M. Kumara et al., Environ Int., **124**, 448 - 461 (2019)
3. Walsh et al., *Lancet Infect Dis*,**11** ,355–362 (2011)
4. L. Catteau et al. 1–35 (2018)
5. A.J. Cowieson et al., *Anim.Feed Sci.* (2018)
6. M.G.Ronquillo et al. *Elsevier Food Control* **72**, Part B, 255-267 (2017)
7. L. N. Nielsen et al., *Environ. Pollut.*,**233** , 364-376 (2018)
8. J. Lin, Zhou et al., *Annu.Rev.Biochem.***0** (2018)
9. Wistrand-Yuen,et al., *Nat. Commun.***9 (1)**,1599 (2018)
10. V. Kumar. et al., 259–273 (2017)
11. R. Sharma et al., *J Environ. Qual*, **38**, 567–575 (2009)
12. C. Bastos,M. et al., *Land Degrad.Dev.*, **29(8)**, 2472–2484 (2018)
13. D.I. Massé, Saady et al., *Animals* **4(2)**, 146–163 (2014)
14. A.K. Dangi et al., *Crit.Rev.Biotechnol*, 1–20 (2018)
15. H. Jufer, *Pace University, Pleasantville, NY, United States* (2018)
16. Z. Shen, *Han Nan: Chinese Society of Environ. Sci.*, **4331-34** (2016)
17. J. Wang and JL J. Gu, "Journal of Livestock and poultry industry", **B 10**, 8-9 (2017)
18. K. Kümmerer, Part I, *Chemosphere*, 417-434 (2009)
19. K. Kwon-Rae, G. Owens, S. I. Kwon, K. H. So, D. B. Lee, Y. S. Ok, *Water Air Soil Pollut.*, **214**, 163– 174 (2011)
20. X. Li, Y. Xie, J. Wang, G. Christakos, J. Si, H. Zhao, Y. Ding et al., *Sci. Total Environ.*, **63–69** ,458–460, (2013)
21. G. Na, X. Fang, Y. Cai, L. Ge, H. Zong, X. Yuan, Z. Yao et al. , *Mar. Pollut. Bull.*, **69**, 233–237 (2013)
22. X. Hu, Q. Zhou, Y. Lou, *Environ. Pollut.*, **158**, 2992–2998 (2010)
23. D. R. Lapen, E. Topp, C. D. Metcalfe, H. Li, M. Edwards, N. Gottschall, P. Bolton et al.,*Sci. Total Environ.*, **399**, 50–65 (2008)
24. W. Li, Y. Shi, L. Gao, J. Liu, Y. Cai, *Chemosphere*, **89**, 1307–1315 (2012)
25. D. Fatta-Kassinos, I. K Kalavrouziotis, P. H Koukoulakis, M. I. Vasquez, *Sci. Total Environ.*, **409**, 3555–3563 (2010)
26. G. Torres-Cortés, V. Millán, H. C Ramírez-Saad., R. Nisa-Martínez, N. Toro, F. Martínez-Abarca, *Environ. Microbiol.*, **13**, 1101–1114 (2011)
27. C. Bouki, D. Venieri, E. Diamadopoulos, *Ecotoxicol. Environ. Saf.* 91 (2013)
28. K. K Brandt, O. R Sjøholm, K. A. Krogh, B. Halling-Sørensen, O. Nybroe, *Environ. Sci. Technol.*, **43**, 2963–2968 (2009)

29. X.-L. Wu, L. Xiang, Q.-Y. Yan, Y.-N. Jiang, Y.-W. Li, X.-P. Huang et al., *Sci. Total Environ.*, **487**, 399–406 (2014)
30. D. G.J. Larsson, *Ups. J. Med. Sci.,* **119**, 108–112 (2014)
31. R. Daghrir, P. Drogui , *Environ. Chem,* (2013)
32. K. Stoob., H. P. Singer, S. R Mueller, R. P. Schwarzenbach, C. H. Stamm, *Environ. Sci. Technol*. **41**, 7349–7355 (2007)
33. www.google.com
34. www.wikipedia.org

Static and modal analysis of a deep soil loosening machine

Gabriel Gheorghe[1,*], *Daniel* Lates[2], *Catalin* Persu[1] and *Marinela* Mateescu[1]

[1]National Research -Development Institute for Machines and Installations Designed to Agriculture and Food Industry, 6 Ion Ionescu de la Brad Blv., 013813 Bucharest, Sector 1, Romania
[2]IRUM, 6 Axente Sever, 545300, Reghin, Mures, Romania

Abstract. The article will present a static analysis of the equipment in order to calculate the vector field distribution of the relative - resulting displacement in the structure, the Cauchy stress tensor and tensor fields distribution of the specific deformation in the same structure. This is done in order to improve the structure and highlight the most vulnerable points within an equipment depending on the way of working. After performing the static analysis, tensometric marks were mounted in the most vulnerable points to calculate the displacement of the material and to calculate the major forces that appeared in the structure during the field experiments. At the same time, the idea of the equipment prototype was to improve the equipment with vibrating elements on the working bodies of the body type in order to increase the degree of crushing and to reduce the advancing forces, respectively to reduce the fuel consumption. Therefore, a modal analysis was performed to calculate the vibrations that appeared in the structure in order not to resonate with the frequency of the motor mounted on the working member.

1 Introduction

The development of equipment for the preparation of the germination bed is a topical topic, and in this field there are several types of equipment for the preparation of the germination bed. [5] Most of the research was performed on equipment such as chisel / subsoiler, the research in the field was performed theoretically / laboratory / field. In all these works the same essential things are determined for a tillage machine the Cauchy stress tensor, and in other works the frequencies appearing in the structure are also investigated in order to determine the lifetime of the component elements within an equipment. [6-10] But testing a plow with a vibrating element mounted on each body has not been studied.

Due to the complexity of the designed equipment (fig.1) when it was subjected to analysis due to the lack of power of the computer, the analysis could not be performed, and it was simplified to perform the analysis only on the general framework.

* Corresponding author: gabrielvalentinghe@yahoo.com

Fig.1. CAD model of agricultural reversible plow

2 Materials and methods

2.1. Boundary conditions (Structure support)

The structure is supported in three points by the system of connection to the tractor, figure 2. The connection to the tractor is done (exaggeratedly) by rigid fixing (annulment of all degrees of freedom on the contact surfaces between the tractor - plow connection elements).

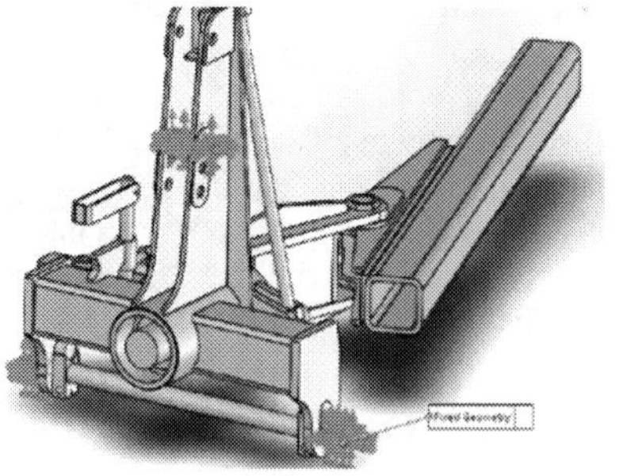

Fig.2. Way of supporting the structure

2.2. Structure loading

The loading of the structure will normally be done on the surface of the working organ in the opposite direction of advancing but taking into account that the framework is studied and not the working organs, because it is harder to replace than a working organ and much more expensive if it yields. . Therefore, a resistance force was applied resulting from the force applied on the three working organs and applied as a resistance force from the ground.

$$F_0 = ka_0b_0+\mathcal{E}\, a_0b_0v^2, \tag{1}$$

where index 0 sizes correspond to the working parts of the plow-body. In (1) F_0 is the resistance force of the soil to the action of the plow-body, a_0, b_0, are the working depth and width of the respective area of the working part, and k_0, ε_0, are the specific resistances to deformation of the soil and resistance coefficients to deformation of the soil due to the working speed.

The working speed was noted with v. In the considered example we used the values: a_0b_0=0.39 m, k= 3500 $N \cdot m^{-1}$, ε= 2200 kg/m3, v= 5 km/h. The forces were applied to the structure according to the graphical representation in figure 3; a force on the working part of about 20 000 N will result.

Also, within the analysis the mass (m=600 kg) of the rest of the structure was added in the opposite direction of the force of about 5886 N.

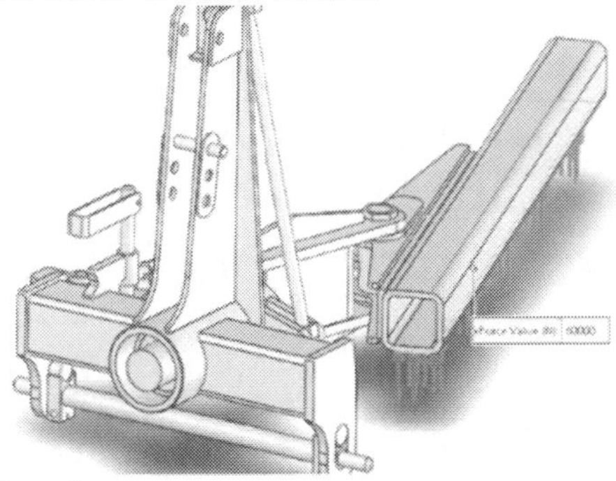

Fig.3. Way to apply loads (forces)

In addition to the force calculated using the formula, the traction force was determined also with the help of the strain gauges applied on the support for fixing the breast to the frame (Fig.4).

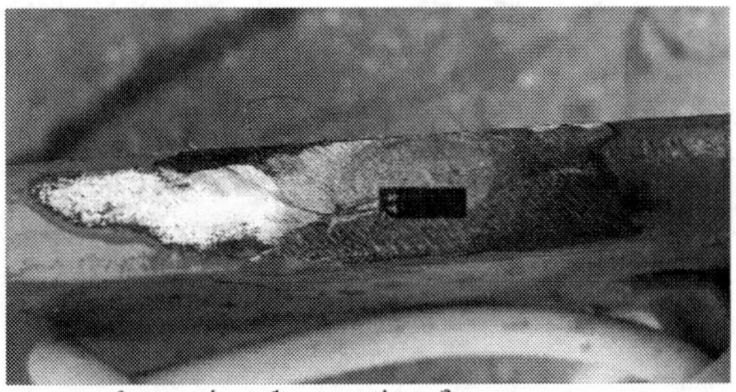

Fig.4. Attaching strain gauges to determine the traction force

This force was calculated under the field conditions and with the qualitative indices from table 1.

Table 1. Field conditions and qualitative indices

Soil compaction test-experimental plot		Soil moisture		Working width (cm)	Working depth (cm)
Depth (cm)	Penetration resistance, (kPa)	Depth, (cm)	Value (%)	*100*	*28*
0	*386*	*0-10*	15,5	*105*	*31*
2,5	*537*	*10-20*	25,7	*97*	*33*
5	*596*	20-30	28,4	*101*	*32*
7,5	*737*			*93*	*30*

https://doi.org/10.1051/e3sconf/202018003019

10	772
12,5	877
15	1123
17,5	1193
20	1404
22,5	1650
25	2071
30	2120

98	31
100	30
102	34
96	33
105	35

Figure 5 shows the diagram of the traction force resulted on the experimental plot by the aggregate AGRICULTURAL PLOW FOR 70 KW AGRICULTURAL TRACTORS + TAGRO 102 tractor.

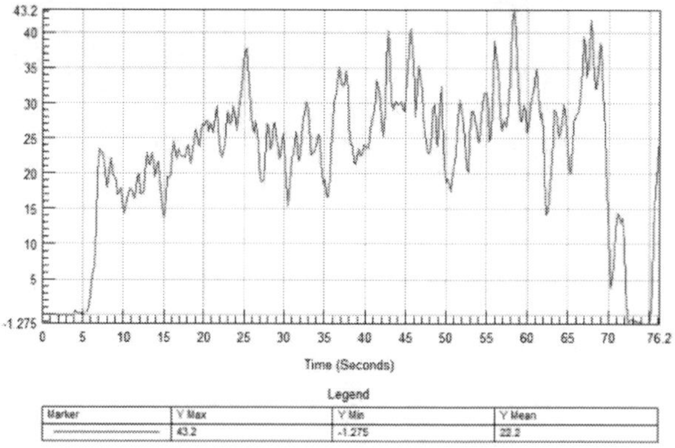

Fig. 5. Average traction force (kN)
- Traction force of reversible plow with 3 plow-bodies IRUM Reghin

For performing the linear elastic static analysis, the global contact command was applied. This condition applied by the finite element analysis operator eliminates any kind of distances, creating stress conditions corresponding to a structure more rigid than the real one. Thus, the stresses will be higher than the real ones, and the relative displacements (deformations) are expected to have lower values than the real ones. The discretization of the structure can be seen in figure 6.

Fig.6. Structure discretization: projection of the finite elements on the boundary of the structure

The material that will be used in the structure analysis is the S275JR material, which has certain properties presented in figure 7.

Property	Value	Units
Elastic modulus	2.100000031e+011	N/m^2
Poisson's ratio	0.28	N/A
Shear modulus	7.9e+010	N/m^2
Mass density	7800	kg/m^3
Tensile strength	410000000	N/m^2
Compressive Strength in X		N/m^2
Yield strength	275000000	N/m^2
Thermal expansion coefficient	1.1e-005	/K
Thermal conductivity	14	W/(m·K)
Specific heat	440	J/(kg·K)
Material Damping Ratio		N/A

Fig.7. Properties of material S275JR.

3 Results and discussion

The first result of the structural analysis are the reactions that appeared in the structure as a result of the applied force, and the other results are displacement in the structure, the Cauchy stress tensor and tensor fields distribution of the specific deformation, as well the safety factor.[1] Table 2 shows the values of the resulting forces' components, which are also found in the values of the reaction forces (in the three support zones).

Table 2. Values of the resulting forces components

Components	X	Y	Z	Resultant
Reaction force (N)	-3.01367	-54110.8	-2.17969	54110.8

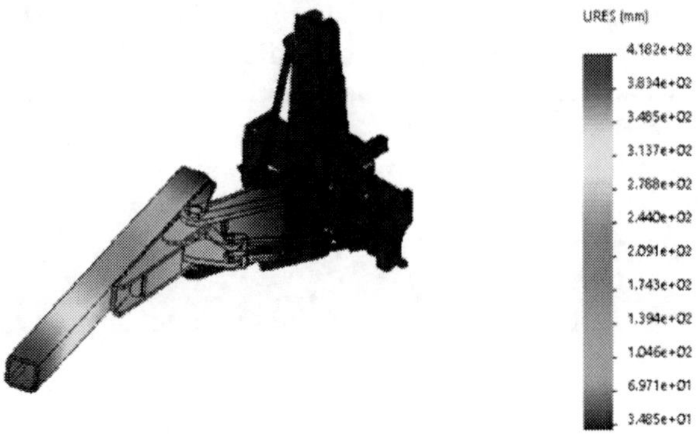

Fig.8. Distribution of the values of the relative resultant displacement field on the structure boundary

 https://doi.org/10.1051/e3sconf/202018003019

The distribution maps of the values of the relative resultant displacement field on the structure boundary are graphically shown in Figure 8. It is noted that the maximum value is located in the back of the structure. This maximum value can be exceeded if clearance of the structure and system connecting the plow and the tractor is taken into account. The increase of the displacement, under the conditions of the considered stress, with the admission of the clearances, contributes to the relaxation of the structure and implicitly the increase of the safety factor. However, exaggerated clearances generally lead to possible wear.

Fig.9. Distribution of total specific deformation field values on the structure boundary.

The distribution of the total specific deformation values is graphically represented by the colour map in Figure 9. Due to the fact that we work in the elastic-linear domain, the maximum tension will be located in the same area as the maximum specific deformation. The maximum equivalent stress is represented graphically in the same way, in figure 10, where it can be seen that very high stresses appear in the middle piece, being a very stressed piece, even if in the design and execution phase of the experimental model a reinforcement was added to eliminate these high stresses.

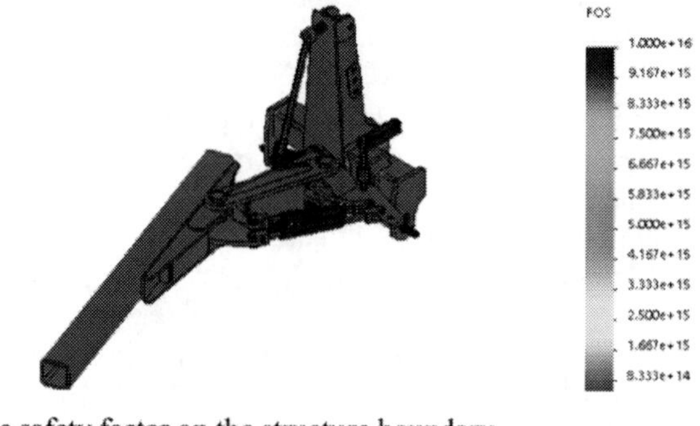

Fig.10. Distribution of the safety factor on the structure boundary

The safety factor is 8.33. For agricultural equipment that work with soil in order to prepare the germination bed, the value of the safety coefficient is between 1.8 and 2.2. The equipment can be oversized or has been designed and thinking to withstand certain accidents during work. Considering the maximum values of the equivalent stress, there is no danger for the structure material to cease.

Due to this very high coefficient, we'll perform an optimization regarding the material used; we'll run the simulation with a material that is weaker in terms of properties (fig. 11), but also cheaper, to reach a coefficient between 1.8 - 2.2.

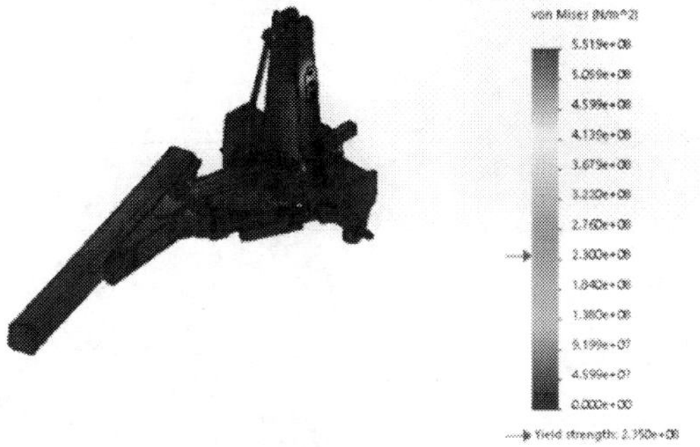

Fig.11. Representation of the equivalent stress distribution on the structure boundary for S235JR material

Because of this, you can no longer talk about observations on the safety factor since it is below 1 in that area. Due to the transformation of the experimental model into a prototype and the addition of the vibrating element to optimize the structure and avoid its resonance, a modal analysis of the basic structure of a plow was also performed. The analysed structure is a modern plow structure, with load bearing structure consisting of plates and pipes. Modal analysis or frequency analysis for this structure, aims to determine its own frequencies and to identify the components with maximum amplitude on each of the basic own frequencies.[4] The intense interaction with the tractor can induce the vibrations of the structure analysed into the resistance structure of the tractor, possibly retransmitted then through suspension to the tractor driver. The consequences can be estimated using the knowledge about vibrations effects on the human body. [2]

Starting from the findings of the modal analysis, improvements of the system can be made by small modifications of some components of the assembly or by modifying some adjustments allowed in operation or in the operating regime. Finally, we mention that frequency analysis is the first step in dynamic analysis, because the latter uses the first results to calculate the structure response to various time-varying loads. The relative displacements by directions and resultants in the structure, for each vibration mode, are given separately. Also, the colour maps of the relative displacements field are given, by component or resultant.[3] Table 3 shows, from the report of the SOLIDWORKS program, with the help of which (using the SIMULATION module) a frequency analysis was made, a list of the first five own frequencies, corresponding to five vibration modes, each with the normalized displacements on the three directions.

Table 3. First five own frequencies

Mode number	Frequency (Hertz)	Frequency (rad/s)	Direction X	Direction Y	Direction Z
1	16.128	101.33	3.6315e-005	0.00015334	4.8081e-005
2	18.625	117.03	0.021966	0.0088722	4.2245e-008
3	58.632	368.4	0.030122	0.049132	1.301e-006
4	67.257	422.59	0.00032949	0.00028952	0.00022521
5	100.68	632.59	0.00051331	0.00025398	3.3279e-005

Another way to give these results is shown in table 4, where the own frequencies and amplitude maps are given on the deformed form of the structure in the vibration modes corresponding to that frequency.

Table 4. Frequencies and amplitude maps

Mode	Frequency, Hz	Map of amplitude on the deformed form of the structure, mm
1	16.128	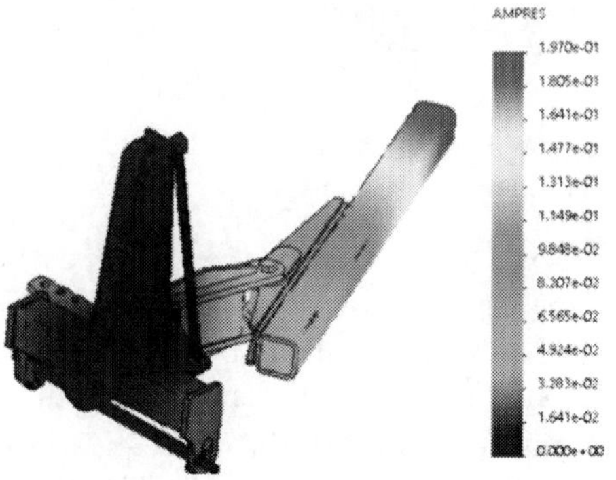
2	18.625	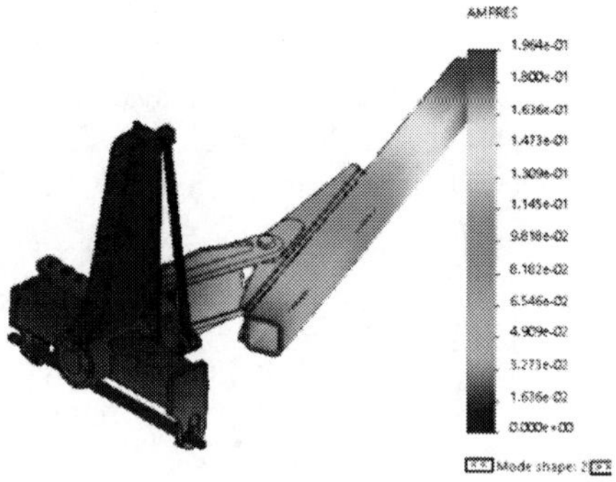
3	58.632	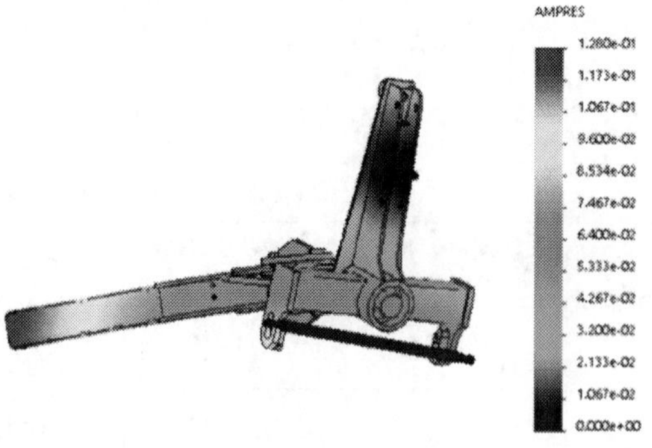

4	67.257	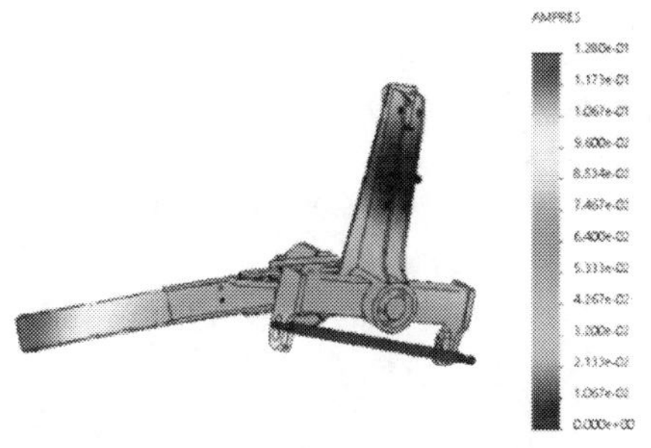
5	100.68	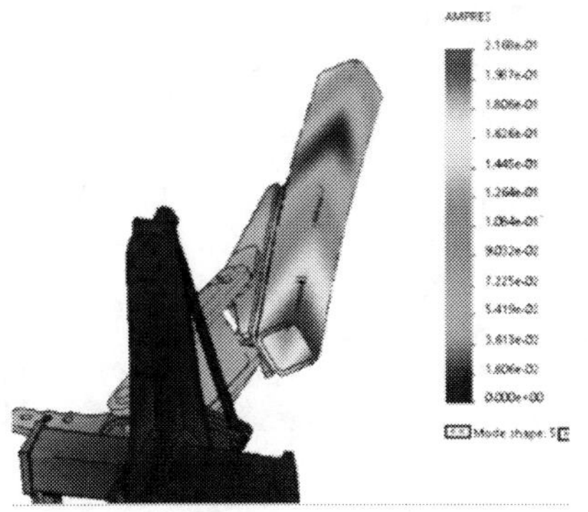

4. Conclusions

1. Finally, own spectrum is also useful in identifying the source of parasitic own oscillations. Once this oscillation is identified, the source is usually looked for in the working regime: either accidental phenomena (different deficiencies of the aggregate) or, in the existence on the analysed structure of some working devices (pumps for spraying machines, mechanical cutting devices, etc.) that work on the structure's own frequencies or on very close frequencies. In these cases, changes of the working regime of the devices are made or the isolation of these devices from the rest of the structure is improved.
2. The usefulness of this analysis is proven especially in the test phase and even in the first stages of operation, when the working regime of a product of the type analysed must be improved.
3. The frequency of the engine added for introducing vibrations into the working part is not among the five own frequencies calculated. The engine having a speed of 3000 rpm, an engine frequency of 1.21 Hz will result, which is very far from the own frequencies of the equipment structure.

This work was supported by the European Regional Development Fund under the Competitiveness Operational Program 2014-2020, Project "Rapid knowledge transfer and technical-scientific support in developing competitive products and technologies in enterprises specific to the field of bioeconomy and the production of bioresources" - contract 80/08.09.2016, MySMIS 105551, subsidiary contract 916/22.08.2017 "AGRICULTURAL PLOW FOR 70 KW AGRICULTURAL TRACTORS" („PLUG AGRICOL PENTRU TRACTOARE AGRICOLE CU PUTERE DE 70 kW") and by a grant of the Romanian Research and Innovation Ministry, through Programme 1–Development of the national research-development system, subprogramme 1.2–Institutional performance–Projects for financing excellence in RDI, contract no. 16PFE.

References

1. G. Gheorghe, C. Persu, I.Gageanu, D.Cujbescu. "Structural and modal analysis in SolidWorks of basic structure of equipment to prepare germinative bed in strips", Latvia University of Life Sciences and Technologies, 2018
2. M. Matache, Gh. Voicu, P. Cardei, V. Vladut, C. Persu, I. Voicea, Accelerated test of MAS 65 deep soil loosening machine frame, Proceedings of the 43rd international symposium on agricultural engineering, Opatija, Croatia, 2015.
3. A David., Gh. Voicu, E. Marin, M. Dutu, G. Gheorghe, Experimental researches on working qualitative indexes of a deep loosening equipment, INMATEH-Agricultural Engineering, vol 46, pp 5-12, no. 2/2015
4. M.FUTATSUKA, S. MAEDA, T. INAOKA, M. NAGANO, M. SHONO, T. MIYAKITA, Whole-Body Vibration and Health Effects in the Agricultural Machinery Drivers, Industrial Health Vol. 36 (1998) No. 2 P 127-132.
5. Croitoru Şt., Marin E., Bădescu M., Manea D., Vlăduţ V., Ungureanu N., Manea D., Boruz S., Matei Gh. - *Researches regarding the minimum work for improving soil fertility by deep loosening*, PROCEEDINGS OF THE 43 INTERNATIONAL SYMPOSIUM ON AGRICULTURAL ENGINEERING "Actual Tasks on Agricultural Engineering", pp. 165-176, 2015, Opatija – Croaţia;
6. Croitoru Şt., Vlăduţ V., Voicea I., Gheorghe Gh., Marin E., Vlăduţoiu L., Moise V., Boruz S., Pruteanu A., Andrei S., Păunescu D.- *Structural and kinematic analysis of the mechanism for arable deep soil loosening,* PROCEEDINGS OF THE 45 INTERNATIONAL SYMPOSIUM ON AGRICULTURAL ENGINEERING "Actual Tasks on Agricultural Engineering", pp. 207-216, 2017, Opatija – Croaţia;
7. Matache M., Voicu Gh., Cardei P., Vladut V., Persu C., Voicea I. - *Accelerated test of MAS 65 deep soil loosening machine frame*, PROCEEDINGS OF THE 43 INTERNATIONAL SYMPOSIUM ON AGRICULTURAL ENGINEERING "Actual Tasks on Agricultural Engineering", pp. 131-140, 2015, Opatija – Croaţia;
8. Vlăduţ V., Gheorghe G., Marin E., Biriş S.Şt., Paraschiv G., Cujbescu D., Ungureanu N., Găgeanu I., Moise V., Boruz S. - *Kinetostatic analysis of the mechanism of deep loosening system of arable soil*, PROCEEDINGS OF THE 45 INTERNATIONAL SYMPOSIUM ON AGRICULTURAL ENGINEERING "Actual Tasks on Agricultural Engineering", pp. 217-228, 2017, Opatija – Croaţia;
9. Vlăduţoiu L., Cârdei P., Vlăduţ V., Fechete L. - *Modern trends in designing and selecting the machine / equipment for soil deep tillage*, 16th International Scientific Conference "ENGINEERING FOR RURAL DEVELOPMENT", pp. 1415-1420, Jelgava, Latvia;
10. Vlăduţ V., Croitoru Şt., Constantin N., Paraschiv G., Voicu Gh., Biriş S., Bungescu S., Caba I., Ludig M., Atanasov At. - Soil tillage optimization by un-compaction and aeration of deficient deep soils, 4rd International Conference "Research People and Actual Tasks on Multidisciplinary Sciences", pp. 324-330, Lozenec, Bulgaria;

Mathematical modelling of the drying process of sugar beet pulp with macro and microelement additions in order to improve the feed quality

Elena Mihaela Nagy *[1], *Constantin* Coţa[1], *Zoltan* Gyorgy[1], and *Teodora* Deac[2]
[1]INMA Bucharest, Branch Cluj-Napoca, 400436, 59 Al. Vaida Voievod Str., Cluj Napoca, Romania
[2]Technical University of Cluj-Napoca, Faculty of Mechanical Engineering, 400641, 103-105 Muncii Bdul, Romania

Abstract. In this paper a kinetic study is performed on the drying process of beet pulp with additions of macro-elements (Ca^{2+}, Mg^{2+}, PO_4^{2+}) from hydrated lime, hydrated dolomite lime, phosphoric acid and microelements (Co^{2+}, Zn^{2+}, Fe^{2+}, Mn^{2+}), with the purpose of obtaining premixes with substantially improved value for animal feed. The experimental measurements made served to elaborate the mathematical model by means of which the speed with which the drying process is carried out is described, thus being able to calculate the stationing time of the material in the dryer. The equations for calculating the stationary time of the material in the dryer can be used for designing dryers (sizing them).

1 Introduction

The technological process of sugar production from beet processing has as a by-product a fibrous material (pulp) which represents about 5 % from the processed quantity of beet [1, 2]. This pulp could be used as valuable fodder with high energy potential and increased efficiency in feeding animals, especially ruminants: milk cows of high productivity, bulls and rams that are intensively fattened.

From the point of view of energy, the nutritional value of the sugar beet pulp is comparable with that of good quality hay with the mention that its assimilation is much easier and of high efficiency that can reach 90-95 % [3].

From a chemical point of view the pulp resulting from the technological process of sugar production is a macromolecular complex of the beet cellular membrane, composed of protopectin which under the influence of pectolytic enzymes from the digestive system is easily hydrolysed both in an acidic and alkaline medium in the intestinal tract of animals.

As a result of the process of hydrolysis the protopectins are transformed into pectinic acid and polypectins, products of low molecular mass, water soluble, that are absorbed easily and quickly [4, 5]. Another significant feature which recommends the use of pulp in the feed ration of animals is the high fiber content, 200-240 g/kg DM, material of good absorption capacity and ionic exchange, thus contributing to the adjustment of the Ph in the digestive tract of animals, [6]. On another hand, both the beet and the pulp represent major sources of protein, 150-160 g/kg DM, minerals: P-0.9 g/kg DM, K-4.8 g/kg DM, Na-

0.4 g/kg DM, Mg-2 g/kg DM, Ca-(9-10) g/kg DM and vitamins: biotin 10-50 mg/kg DM important for animal health [7-9].

The nature and chemical structure of carbohydrates and the other compounds from the composition of sugar beet pulp allows their easy transformation into useful compounds that have a positive effect on milk production and determine an increase in fat content of 8-11 % [10]. Direct metabolization of the pulp into useful compounds achieves a noticeable reduction in green-house gases (CO_2 şi CH_4) produced during digestion by the ruminants, which is a phenomenon with a positive impact on the environment. [5, 9, 11, 12].

Introduction of beet pulp or beet in animal feed is a very topical issue even if this is not a new concept. The economic efficiency of beet pulp is also due to the shorter vegetation cycle compared to that of cereals (wheat, barley etc).

The beet is a plant that adapts easily to osmotic pressure, so it can be cultivated in high alkalinity soils, which not only doesn't damage, but increases production which can reach 80 t/ha (15-17 t/ha DM) successfully replacing some cereals in the feed of ruminants and also swine [13].

Although both beet pulp and beet are highly efficient in animal feed, they cannot be stored for longer periods of time, because a higher water content determines a higher perishability. To eliminate this drawback, it is necessary to process them before storage in such a way that they will maintain their nutritional properties for longer. In practice this problem can be solved by reducing the humidity (drying) to a water content of 8-12 %, at which value the products do not spoil any more.

Another factor that requires drying the beet pulp with the purpose of long-term storage is given by the seasonal nature of operation of the sugar factories, from which, in comparatively short periods of time, large quantities of pulp are obtained that cannot be used completely at the time of production [14-17].

To increase the efficiency of beet pulp used in animal feed, it can be combined with macro-elements (Ca^{2+}, Mg^{2+}, PO_4^{2+}) and micro-elements (Co^{2+}, Zn^{2+}, Fe^{2+}, Mn^{2+}). For example, an addition of calcium is very important especially for high productivity milk cows which need a daily amount of 70-100 g of calcium, a quantity that cannot be provided only from feed [18].

It is desirable that the addition of macro-elements should be in the form of compounds that can be easily assimilated. Thus, the addition of Ca^{2+}, Mg^{2+}, PO_4^{2+} as lime and hydrated dolomitic lime to the beet pulp before drying leads to creation of organo-mineral compounds, from which the assimilation of macro-elements is superior to carbonates. Ensuring the requirements of micro-elements (Co^{2+}, Zn^{2+}, Fe^{2+}, Mn^{2+}) is achieved by adding ionic salts, respectively as sulphates ($MnSO_4$, $ZnSO_4$, $CoSO_4$, $FeSO_4$).

In this paper we conducted a kinetic study of the influence of temperature and additions of lime, hydrated dolomitic lime, H_3PO_4 and micro-elements on the speed of drying. The experimental measurements taken have been used to develop a mathematical model which can be used to describe the speed of the drying process and calculate how long the material should stay in the drier when the process takes place in the temperature interval 60-95 °C.

2 Materials and methods

The experimental study was performed with beet pulp resulted from technological process of sugar production from TEREOS-LUDUS-ROMANIA. As source for Ca^{2+} ions, hydrated lime obtained from industrial lime hydration was used. Mg^{2+} ions were added in the form of hydrated dolomitic lime prepared in laboratory by calcination at t = 850 °C of the VOLŞLOBENI dolomite followed by hydration. The source for PO_4^{3+} ions was phosphoric acid (H_3PO_4) -85% P.A. from the supplier NORDIC-CHEMICALS CLUJ. The microelements Co^{2+}, Zn^{2+}, Fe^{2+}, Mn^{2+} were added as sulphates, sulfuric acid salts of P.A. quality, from NORDIC-CHEMICALS CLUJ.

The study of drying process was carried out with the AXIS-100 moisture analyzer with 0.01% humidity accuracy and maximum working temperature 160 °C, the heat input required to evaporate the moisture is achieved by means of two halogen lamps, with which the moisture analyzer is provided. Also, this analyzer has an automatic temperature control system and a software for automatic measurement at predetermined time intervals of the sample mass. The software allows the collection and storage of measured values on a PC connected to the moisture analyzer.

The samples composition is presented in Table 1.

Table 1. Composition of samples subjected to drying in relation to the dry matter

Sample	Organic part quantity [%]	CaO	MgO	P_2O_5	Micro-element PPP
P0 – Raw pulp	18.7	-	-	-	-
P1- Raw pulp+lime+ dolomitic lime	65.63	28.5	5.85	-	-
P2 - Raw pulp+lime+ dolomitic lime + H_3PO_4+micro-elements	61.5	23.2	5.5	9.8	Co=1.2 Zn=2.1 Cu=0.15 Fe=1.5 Mn=0.25

The influence of temperature on the drying speed of the samples was studied in the range 60-90 °C for a thin layer of material

3 Results and discussions

Knowing the laws according to which the complex drying process takes place allows to find the optimal solution of the engineering problems regarding the selection and sizing of the dryer and the drying process automatization. Establishing the laws that shows the evolution in time of a certain process, physical or chemical, can be done by theoretical modeling or empirical modeling

Theoretical modeling, which implies the knowledge of the kinetic equation of studied phenomenon, is difficult to achieve for the drying process due to the large number of parameters that influence the drying speed, the multitudes and complexities of the phenomena that form the global process (internal and external diffusion of water vapor, thermal transfer, etc.) as well as due to the fact that the mechanisms by which this process is carried out may change during the course.

Considering these reasons, in the case of solids drying, empirical modeling is used, that involves experimental measurements, performed under conditions similar to those of industrial practice, whose numerical values will be correlated in a kinetic equation [19-21].

Experimentally, the drying process with constant air parameters on the surface of the material is studied. The experimental values regarding the variation in time of the sample mass are processed in the form of drying curves, which reproduce the variation in time of the material moisture content.

The analysis of results, figure 1, shows the influence of the temperature on the time required for the total humidity removal, presented in table 2.

The strong influence of temperature on the time required for total elimination of humidity is manifested for all three samples. Significant decrease, 40-44 %, of the time required for total removal of humidity at temperatures higher than 85 °C, especially in the case of P0 sample, is determined by the modification of the porous structure of the material.

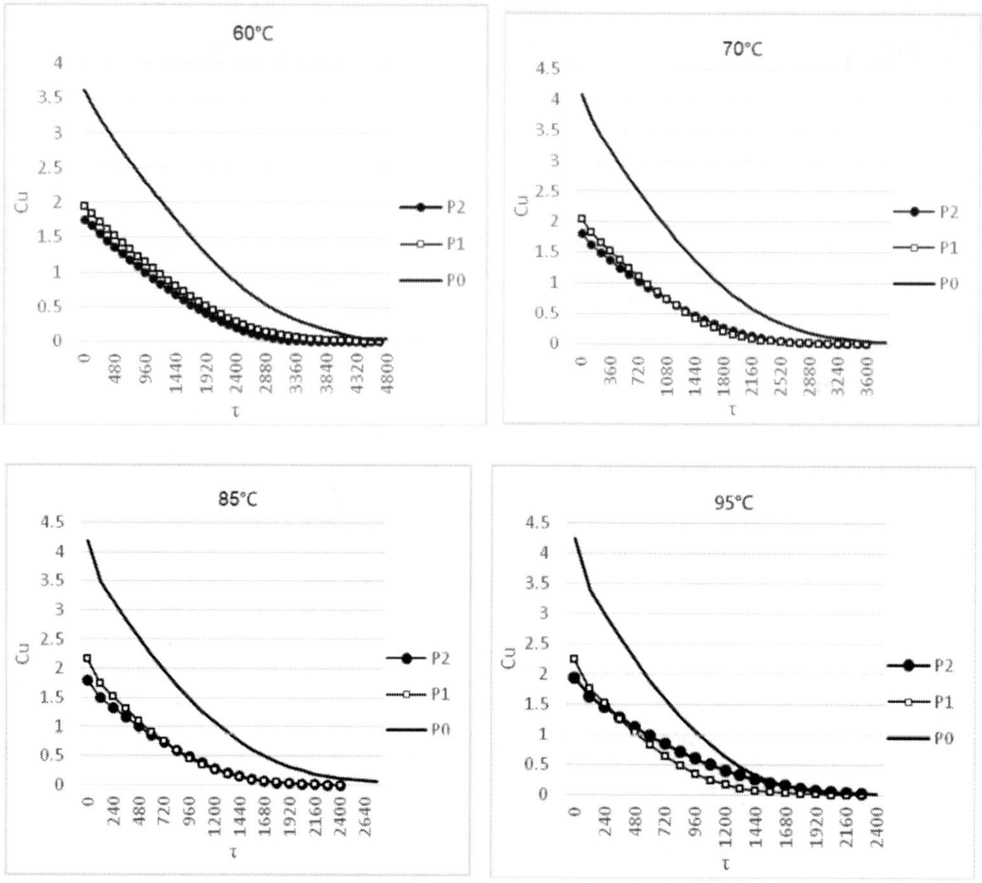

Fig.1. Variation in time of moisture content at temperatures: 60; 70; 85; 95 °C

Table 2. The time required for the total removal of humidity

Temp, [°C] / Samples	Time [s]			
	60	70	85	95
P0	5500	3700	2800	2300
P1	4000	2650	2100	2100
P2	3600	2650	2100	2100

At temperatures higher than 70 °C the cell walls of the pulp are broken and with this the micropores disappear, turning into macropores, in which the humidity is retained with lower forces, the desorption of vapors and diffusion through the crust towards the outer surface becoming easier. At low temperatures, the microporous structure of the pulp from which water vapor diffuses harder is maintained.

The results in Figure 1 and Table 2 show, on the other hand, the strong influence of micro- and macro-element additions. The additions, even at low temperatures, cause the pulp's cell walls to rupture, leading to a substantial reduction in the time required for total removal of humidity. In the case of the sample in which besides lime and dolomitic lime, H_3PO_4 and microelements were added, the breakdown of the cell walls and the modification of the porous structure occurred mainly at 60 °C, which is why the influence of the temperature is less pronounced above this temperature threshold.

https://doi.org/10.1051/e3sconf/202018003020

3.1 Mathematical modeling

Due to the reasons stated above, theoretical modeling is difficult to approach. To describe the rate of moisture removal, the empirical models presented in Table 3 will be used.

Table 3. Equations of proposed empirical models for drying speed of a thin layer of material

Model name	Model	References
Newton	$MR=exp(-kt)$	[22]
Page	$MR=exp(-kt^n)$	[22]
Modified Page	$MR=exp[-(kt)^n]$	[22]
Henderson and Pabis	$MR=a\cdot exp(-kt)$	[23]
Logarithmic	$MR=a\cdot exp(-kt)+c$	[24]
Two term	$MR=a\cdot exp(-k_0t)+b\cdot exp(-k_1t)$	[25]
Two term exponential	$MR=a\cdot exp(-k\cdot t)+(1-a)exp(-k\cdot a\cdot t)$	[26]
Wang and Singh	$MR=1+at+bt^2$	[27]

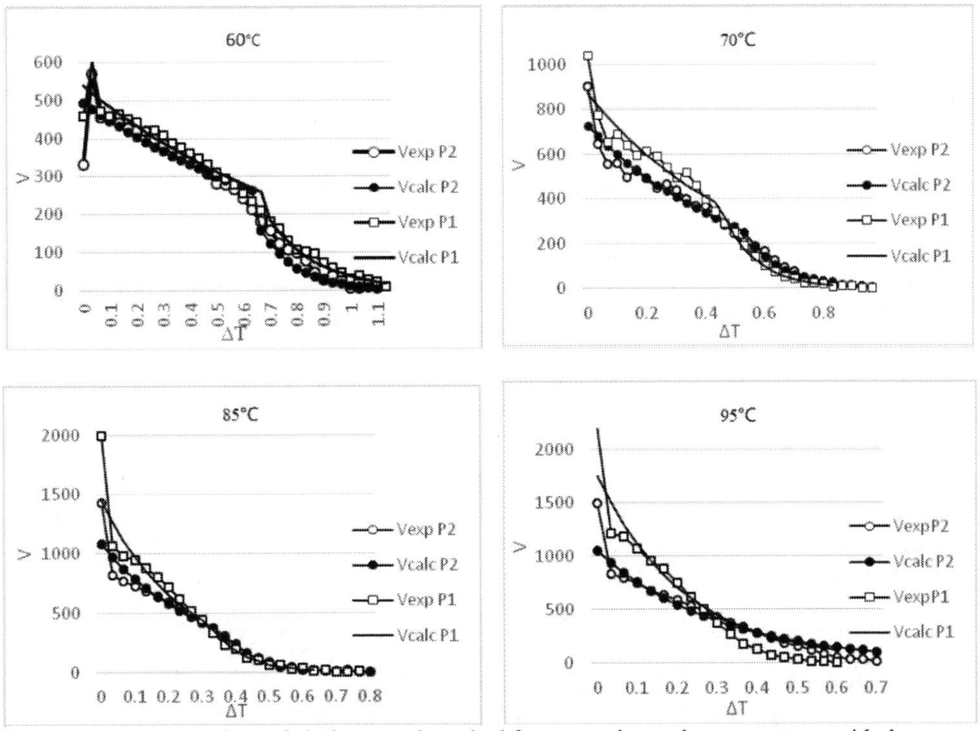

Fig. 2. Comparative values of drying speed resulted from experimental measurements with those calculated based on the proposed model

Validation of the empirical models proposed in Table 3 and their transposition into mathematical models of beet pulp drying process implies identification of the coefficients in the proposed equations.

Using the finite differences method, the experimental results, Figure 1, regarding the variation in time of humidity content were used to determine the drying speed and its variation over time.

Verification of the empirical models showed that the two-term model represents very well the experimental measurements on drying speed.

The mathematical processing of the experimental results that shows variation in time of the drying speed led to the numerical values of the constants K and A, presented in Table 4.

Table 4. Numerical values of coefficients K and A of the two-term model

Constants	T[°C] / Sample	60	70	85	95
1	P1	1.09	1.89	3.98	4.54
	P2	0.977	1.94	3.16	3.31
K_2	P1	5.36	8.85	10.07	12.3
	P2	7.45	8.41	9.81	10.3
A_1	P1	6.29	6.77	7.28	7.3
	P2	6.21	6.59	6.98	6.96
A_2	P1	8.97	9.89	9.23	9.62
	P2	10.03	10.02	9.33	9.90

The verification of the proposed model by means of the constants K and A shows that at humidity of the material greater than 90 % (± 1.5 %) there is a modification of the mechanism after which the process is carried out. Thus, two zones are distinguished: the first zone which lasts for a period of time until approx. 90 % of the moisture contained in the pulp is eliminated. During this period the drying process is carried out at a speed corresponding to the values K_1 and A_1 of the proposed model with the duration τ_1. The second zone corresponds to the elimination of approx. 10 % of the moisture of the pulp. This area is characterized by the K_2 and A_2 values of the proposed model and corresponding to a time duration τ_2. The time required to eliminate the total humidity, representing the sum of the time for the both zones.

By comparing the results of the experimental measurements of the drying speed with those calculated on the basis of the proposed model (Figure 2), a good agreement is observed throughout the studied temperature range.

Determining the time required to remove the humidity at temperatures other than those at which the experimental measurements were made implies establishing the temperature dependence of *K* coefficients of the mathematical model. The mathematical processing of the numerical values in Table 4 led to the equations in Table 5.

Table 5. Calculation equations for constants K and A as function of temperature in range 60-95 °C.

Sample	Calc. eq.of constant *K*		Calc. eq.of constant *A*	
	K_1 η=0-0.9	K_2 η=0.9-1	A_1 η=0-0.9	A_2 η=0.9-1
P1	K_1=1.09+0.098ΔT	K_2=5.36+0.2ΔT	A_1=6.29+0.029ΔT	A_2=1.09+0.098ΔT
P2	K_1=0.977+0.0661ΔT	K_2=7.45+0.074ΔT	A_1=6.21+0.0214ΔT	A_2=9.82

where: $\Delta T = T_i - 60$

T_i – working temperature [°C] in the range 60-95 °C

In order to establish the calculation formulas of the necessary time of stationing of the material in the dryer, so that its humidity decreases from the initial value X_i to the final X_f, it is based on the equation that shows the moisture content of the material, C_u, depending on the degree of its removal η

$$C_u = \frac{M_i - M_t}{m} = \frac{X_i - X_f}{1 - X_i} = \frac{X_i(A-\eta)}{1 - X_i} \quad (1)$$

where:

M_i, M_f – the initial and final humidity of the material [kgH_2O/kg DM]

η – degree of humidity removal

m- mass of dry material [kg]

C_u- humidity content of the material.

By combining Equation 1 with the corresponding ones in Table 5, we obtain the relations that allow determination of the time of stationing of the material in the dryer for temperature values in the range 60-95 °C.

$$\frac{X_i(A-\eta)}{1-X_i} = \tau_1 + \tau_2 \quad (2)$$

where: $\tau_1 = A_1 \cdot e^{-K_1\tau_1}$
$\tau_2 = A_2 \cdot e^{-K_2\tau_2}$

4 Conclusions

The kinetic study on drying process of beet pulp showed that the addition of Ca^{2+} and Mg^{2+} microelements in the form of lime and hydrated dolomitic lime determines the breakdown of the cell walls, modifying its porosity, which leads to a substantial increase of the speed of moisture removal.

The extra addition of PO_4^{3+} in the form of H_3PO_4 and microelements Co^{2+}, Zn^{2+}, Fe^{2+}, Cu^{2+} causes a more pronounced rupture of cell walls of the pulp at lower temperatures with a positive effect on the speed with which the drying takes place.

The results of the kinetic study carried out served to concretize the mathematical model by means of which the variation in time of the process is highlighted.

The concretization of the coefficients in the mathematical model allowed to establish the equations of calculation of the time of stationing the material in the dryer.

The equations for calculating the stationary time of the material in the dryer can be used when sizing the dryer.

This work was supported by a grant of the Romanian Education and Research Ministry through Programme 1 – Development of the national research-development system, subprogramme 1.2 – Institutional performance – Projects for financing excellence in RDI, contract no. 16PFE.

References

1. M. Asadi., *John Wiley&Sons, Inc.*, Hoboken, New Jersey, (2007)
2. F.N. Domsa; L. Iliescu, *Sugar Technology,* Ed. Tehnica, Bucuresti, Romania, (1973).
3. P. Melendez; P. J Pined; J. Appl. Anim. *Res*, **43**, 3, 261, (2015).
4. A. Lenart; J.M. Flink, Int. J. of Food Sci. & Tech., **19 (1)**, 45, (2017)
5. A.P.A. Monteiro; J.K. Bernard; J.R. Guo; X.S. Weng; S. Emanuele; R. Davis, G.E. Dahl; S.Tao; *J. of Dairy Sci.*, **100 (2)**, 1063, (2017).
6. A. A. M. Habeeb; A. E.Gad; A. A Taraban; I. Sanchis; L. Luján, Int. J. of Sci. Res. in Sci. and Tech., **3, 3**, 107, (2017).
7. C. M Hoffman; T. Huijbregts; N. van Swaaij; R. Jansen; Eur. J. Agro., ***30,*** 17, (2009).
8. N. Katerji; J. W. van Hoorn; A. Hamdy; M. Mastrorilli; E. Mou Karzel; Agric Water Manag. **34**, 57, (1997).
9. M. Münnich; R. Khiaosaard; F. Klevenhusen; A. Hilpold; Q. Zebeli; Animal Feed Sci. and Tech., ***224,*** 78, (2017).
10. R.E. Rauch; P.H. Robinson; L.J. Erasmus, Animal Feed Sci. and Tech., **177 (3-4),** 180, (2012).
11. A. Jokić,; Z. Zavargo; N. Lukić; B. Ikonić,; J. Marković; J. Dodić; J. Grahovac; J. on Proces. and Energy in Agri., **17 (1),** 24, (2013).
12. Ch.V. Boucque; B.G. Cottyn; F.X. Buysse, Livestock Prod. Sci., **5 2**, 171, (1978).

13. L. Mogensen; T. Kristensen , Acta. Agric. Scand. (A), **53,** 86, (2003).
14. Q. Ertl; W. Zebeli; W. Zollitsch; W. Knaus; J Dairy Sci., **99** 2, 1228, (2016)
15. E. Evans; U. Messerschmidt; J. Anim. Sci. Biotechnol., **8,** 1, (2017)
16. A. A. Wahab; A. A. Salih; J Agric Sci Technol B., **2**, 865, (2012)
17. L. Yang; Z. Hu; L. Yang; S. Xie; M. Yang; INMATEH - Agricultural Engineering, **55**, (2), 53, (2018).
18. Z. Dancea, *Nutritia animala si elemente de nutritie a omului*, Ed. Tedesco, Cluj Napoca, Romania, (2005)
19. E.K. Akpinar; S. Toraman;, Heat Mass Transfer, **8**, **2,** 110, (2015)
20. I. Doymaz; Energy Conversion and Management, **56**, 199, (2012)
21. D.G. Overhults; H.E. White; H.E. Hamilton; I.J. Ross; Transactions of the ASAE, **16**, 112, (1973)
22. A. El-Beltagy; G.R. Gamea; A.H. Amer Essa; J Food Eng, **78**, 456, (2007)
23. S.M.Henderson; S.J. Pabis; Agricultural Eng.Res., **6(3**), 169, (1969)
24. O. Yaldiz; C. Ertekin; H.B. Uzun; Energy, **26**, 457, (2001)
25. I.T. Togrul; D. Pehlivan; J Food Eng, **65**, 413, (2004)
26. E.K. Akpinar; Y. Bicer; Energy Convers. Manag., **49**, 1367, (2008)
27. A. Kaleta; K. Górnicki; Int. J. Food Sci.Technol., **45** (5), 891, (2010)

Research on methods, techniques and technologies of carbon sequestration in soil

Oana Diana Cristea[1], *Valentin* Vlăduț[1*], *Nicoleta* Ungureanu[2], *Diana Lorena* Popa[2], *Simona* Isticioaia[3], *Liliana* Dumitrescu[4], *Gheorghe* Matei[5] and *Livia* Apostol[6]

[1]The National Institute of Research – Development for Machines and Installations Designed to Agriculture and Food Industry – INMA Bucharest/ Romania
[2] University POLITEHNICA of Bucharest
[3]Agricultural Research and Development Station Secuieni, 371 Principală Street, Neamț, Romania
[4]INOE 2000-IHP, street Cutitul de Argint, no. 14, Sector 4, Bucharest, Romania
[5]University of Craiova (Romania)
[6]National Research and Development Institute for Food Bioresources (Romania)

Abstract. Soil is one of the natural reservoirs of the carbon biogeochemical cycle, incorporating approximately 6000 billion tons of carbon. Given that with the industrial development more and more carbon is emitted into the atmosphere, solutions, technologies and methods are being sought to reduce this carbon or, where appropriate, it is not eliminated into the atmosphere. The purpose of this paper is to study and identify the simplest methods to be applied in agriculture, for soil processing, by identifying the techniques, technologies and equipment to achieve this without turning the furrow, so that the carbon incorporated in the soil by plants remains sequestered there.

1 Introduction

The effects of climate change have had effects in all areas of industry. On this subject, in 2015, the "4 per 1.000 initiative: soils for food security and climate it was achievable with the aim of sequestering the organic carbon of the soil. It had the main objectives: reducing climate change, improving food security and adapting globally to climate change. Also, the Paris Agreement to request the reduction the global warming below 2°C. [1, 2].
The conclusion is that rapid decomposition is due to soil works which is also the main cause of short-term CO2 emissions. Being introduced by tillagesoil organic matter damage is determined by both soil moisture and temperature, also the dimensions of quality and quantity of organic carbon in the soil [3, 5]. Nevertheless, Research colleagues have invented higher NT emissions compared to the revised batches distributed to the decomposition of the oldest surface residues [5, 6]. Also, recently, some researchers reached the conclusion that reduction of CO2 emissions after the seeding process [7] of the no-till emission. The researchers used time emissions and exponential delimitation over time as a significance for estimating soil CO2 emissions.
There are a variety of methods which can be used in order to determine the carbon stocks in the aboveground biomass [8-10]. Even though both contact field measurements and contactless measurement use remote sensing, the Contact measurement methods are costly

[1] Corresponding author: valentin_vladut@yahoo.com

and time consuming, but the results are more accurate. Therefore, for large areas, the contact methods are almost unusable [11-14]. Conservation soil use (i.e. no-fill, minimum tillage) has had a positive impact in the areas of food security, biodiversity and water quality. the recommendation to improve the stocks of organic carbon (SOC) and total nitrogen (STN) in the soil, and even though the results were mainly reported to be encouraging with a warm climate, with rain crops and low percentage of fertilization with N, it represents an improvement in our efforts to conservate the environment [15].
For instance, in the United States, only 12 percent of total carbon dioxide emissions from the energy, transportation and industrial sectors are cloistered by the forest and croplands of the territory. Figure 1, transposed from the EPA, exemplifies processes in various forms by which trees and soils gain or lose carbon [16-18].

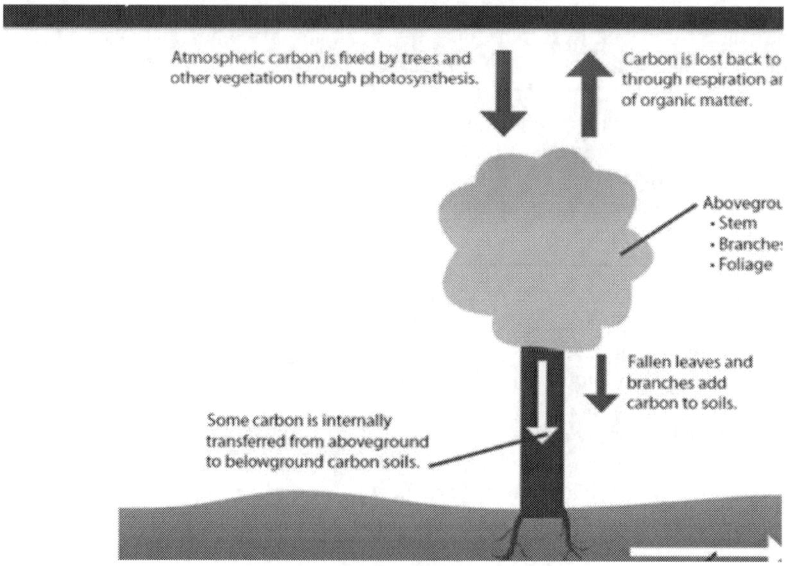

Fig. 1. Carbon pools in forestry and agriculture [18]

The already mentioned "4‰ initiative: Soils for food security and climate" from 2015 - Paris, is based on the premise that achieving food security and combating climate change are complementary. Therefore, its aim is to ensure that agriculture forwards solutions to combat the climate change [19], one of them being the increase of the soil carbon pool to offset national greenhouse gas emissions from fossil fuel combustion. The initiative also implies a voluntary action plan under the Lima Paris Agenda for Action (LPAA), and is backed up by a strong and ambitious research program [20].
Currently, all industrial carbon sequestration projects are evaluated using the Life Cycle Analysis (LCA) methodology, which takes into account the environmental impact of land use, even if carbon sequestration is not done in soil. However, carbon sequestration in the soil probably offers the chance for the greatest technological and environmental revolution ever [21]. Globally, the carbon content is about twice as high in the soil as in the atmosphere and three times as high as in the vegetation. Europe's soils are a huge reservoir of carbon, containing about 75 billion tons of organic carbon. When organic matter is formed, CO_2 is taken up from the atmosphere; reversely, when the organic matter in the soil decomposes, it releases carbon dioxide (CO_2) into the atmosphere [21].
The potential for carbon sequestration in the soil is mainly verified by pedological factors that specify the maximum physico-chemical limit for carbon storage in the soil. Amongst the major factors (Fig. 2), we can use together the texture of the soil and the clay mineralogy, the proportion of coarse fragments, the depth, the bulk density [21].

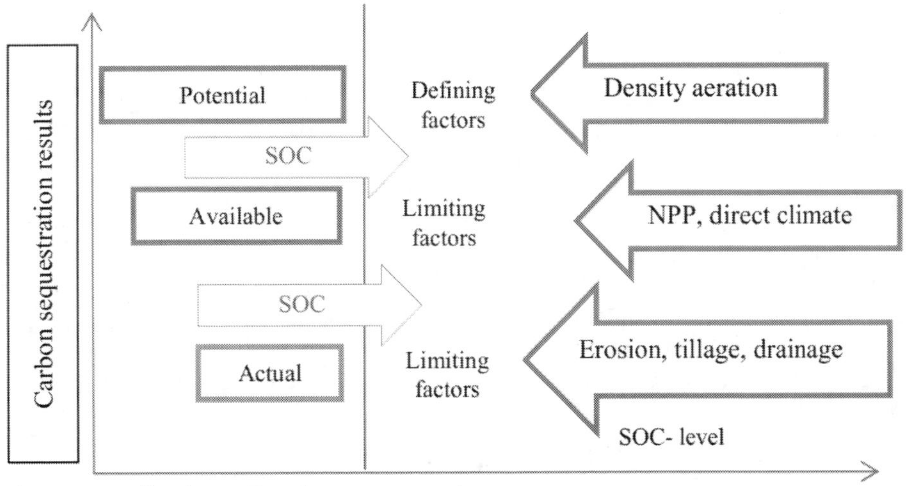

Fig. 2. Factors affecting soil carbon sequestration

The carbon sequencing obtained was defined by several factors that will determine the carbon intake in the soil system. One of the leading factor is change in photosynthesis rate minus autotrophic respiration (NPP) Practices used to manage land that increase carbon sequestration by supplementing NPPs tend to raise the level reached to be closer to the potential level, thus showing that a climate has direct and indirect effects over achievable sequestration. [23-24].

Compared to natural vegetation, agricultural soils contain less organic carbon in the soil (SOC), due to the removal of plant material as part of the crop, which itself contains carbon. As a result, agricultural soils have about 30-40% less SOC than soils with natural vegetation. Obviously, not all agricultural land could be converted back into natural vegetation because food must be produced, but changing agricultural practices could increase the amount of SOC contained in agricultural soils. This is called carbon sequestration in the soil, which signifies taking CO2 from the atmosphere and capturing it in the soil [25].

The agricultural practices that determine the increase of CO_2 emissions from the soil are varied: plowing works, deforestation, drainage of organic soils / peatlands, subsistence agriculture that determines the depletion of fertility, overgrazing etc. In the soil mineralization of organic carbon is strongly dependent on temperature, therefore global warming can increase CO_2 emissions into the atmosphere. Loosening the soil through agricultural work leads to losses of carbon and nitrogen from the soil due to increased accessibility of oxygen needed to break down organic matter and respiration, thus increasing the release of CO_2. The application of conservation works leads to an accumulation of carbon and organic matter in agricultural [26].

Recommended measures include: the transition to no-tillage management system; if an agricultural work is required, it is recommended to avoid autumn works and wait until spring; reducing the number of crossings of agricultural machinery on the soil, the use of chisel and discs only in the upper layer of the soil; the operation of agricultural equipment at lower speed [26].

The decomposition rate increases with temperature and decreases with increasingly anaerobic conditions. The dynamics of SOC pool are with a strong influence on soil management practices and which are dependent on the balance between the input carbon (C) and the output by different methodss [23-24].

2 Research Methods

The agricultural practices that determine the increase of CO2 emissions from the soil are varied: plowing works, deforestation, drainage of organic soils / peatlands, subsistence agriculture that determines the depletion of fertility, overgrazing, etc. The method of mineralization of organic carbon in the soil is strongly dependent on temperature, therefore global warming can increase CO2 emissions into the atmosphere.

Soil temperature measurement (if the no-tillage system will be applied on 76% of the arable land), is done by installing temperature sensors, the simplest of which can be made from a junction of two metals with different properties (for example a nickel-constant thermocouple). Thus, it should be noted that Figaro type oxygen sensors generate a voltage of the order of millivolts, so the accuracy of reading the datalogger must be greater than one tenth of a millivolt, and for thermocouples an accuracy of hundreds of mV is needed [27, 28].

Figure 3 shows a system with three sets of sensors, at three different depths (4, 8 and 22 cm). The number of sensor sets and their depths can be chosen depending on the purpose of the study, the specific conditions of moisture, temperature, vegetation and the limitation due to the physical dimensions of the sensors. Thus, for the strict determination of the flow at the soil surface, a set of sensors is sufficient, located at a depth between 7 and 10 cm, the concentration gradient considered in this case being the one between the surface, where the concentration of CO_2, respectively O_2, can be considered constant and measurement depth, where the concentration will be recorded by direct measurement [28].

Fig. 3. Overview and section of the monitoring system of CO_2, O_2, moisture and temperature in the soil, based on the gradient method [28]

In order to achieve higher aggregate breakdown and smaller soil aggregate sizes, three optional passes of the rotary tiller have been used and the work to be cultivated has been carried out by means of a tractor having an engine power of 77,3 kW, which had a speed of movement before 3.5 km h - 1 and a rotary tipper with 2.5 km h - 1. Prior to the works, the batches without traces of culture residues were removed. Two days before, the residues had been on the surface of the soil since the harvest period The vast majority of residues were incorporated into the soil about 0.20 m long before and 0.05 m after the soil [29].

Two efficient methods use to determine SOC have been evaluated and compared, the obtained data showing interesting results. One of the methods used is the "wet" Thurin method in modifying Nikitin (TN). At the last, "dry" measurement of the SOC was performed using the CN analyzer (EA), was performed on 95 soil samples from 17 sampling sites with a wide range of SOC (1-15%). The process of stealing samples from sites that include arable land and meadows main soil types and subtypes of Slovakia. The statistics processed revealed that the differences between the two methods are minimal in the case of a parcel of soils with the SOC content up to 3%. [30].

Percentage of C loss due to emission factors caused by soil or crop residues, could be easily summed up in accordance with the assumptions described above. Figure no. 2 presents the emission factors associated with the loss of C from crop residues in the second-year experiment. The tillage system that proved itself to cause the greatest relative loss when it comes to the additional CO2-C from the residue was MP, which led to an equivalent decrease of approximately 3.99% of C in emissions over a short period of time [29].

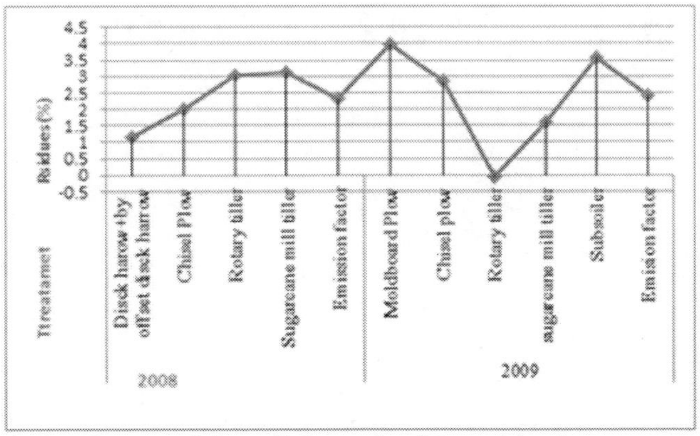

Fig.4. The main emission factors of the soil for each of the cultivation systems that were analyzed (years 2008 -2009

SOC sequestration in the relationship between percentage and induction organic C (Fig. 4) indicated that the soil under analysis was still unsaturated, even after a period of 20 years of contribution of crop residues, compost or both. The data obtained that revealed the maintenance of the stock level of SOC, is equal to zero changes, the amount subjected to the C contribution analysis in the soil was 2.04 Mg C ha^{-1} yr^{-1} (Fig. 4), which was a much lower percentage than Kong reported [31] (3.1 Mg C ha - 1 year - 1). It can be seen that after a period of 10 years of cultivation in Davis- California, in the conditions of Mediterranean climate, of Mandal [32] (3.41 Mg C ha - 1 year - 1) for a system of double-cultivated rice, of 36 years, in a subtropical climate, and Majumder [33] (3.56 Mg C ha −1 year - 1) for a 19-year-old rice wheat system in subtropical India [34].

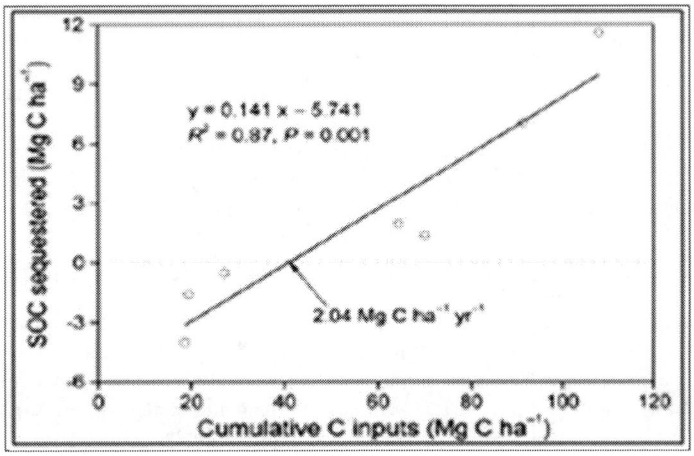

Fig. 4. SOC report after 20 years of application of inorganic fertilizers and compost [34]

3 Results and Discussion

Sequestration of carbon in the soil can be achieved, in addition to adopting measures of a managerial and administrative nature, by using appropriate methods and technologies, called conservative, tillage, the most used worldwide being presented, in the system, in table 1.

Tabel 1. Synthesis of conservative tillage technologies

Crt. No.	The technology used	The work done	Recommendations for use	Effects of use
1	Paraplow tillage technology	It cultivates the soil without turning the furrow and incorporating the vegetal remains superficially	On sloping terrain, on sandy soils, on saline terrain	Erosion reduction, water conservation, stopping structural degradation and soil dusting
2	Conservative tillage technology with CIZEL	The chisel processes the soil to a depth of max. 40 cm without overturning the furrow, causing fragmentation, crushing and loosening	Especially on sloping, saline, thin-walled and wind-eroded lands	Reducing soil erosion, water conservation in the soil, reducing structural degradation and soil dusting.
3	Soil scarifier technology	Soil tillage at working depths of 0.50 ÷ 0.60 m, without turning the furrow and a superficial crushing	Generally on heavy, heavily compacted soils	Elimination of the impermeable layer of soil (hardpan), facilitates deep water infiltration
4	Conservative tillage technology with multifunctional aggregates	Deep loosening of the soil, shredding and leveling, preparation of the germination bed, in a single pass	Spring or autumn, on stubble, on all types of soil located on sloping lands up to max. 6°	Water conservation in the soil, reduction of soil compaction, halting the decline of humidified organic matter, structural degradation and soil dusting
5	Narrow strip tillage technology	Uncultivated tillage and loosening in strips	Preparation of myrtle, for later sowing of weed crops	Water conservation in soil, reduction of soil processing, stopping the decline of humidified organic matter
6	Technology for soil compaction simultaneously with fertilization	Compaction and partial tillage at the same time as fertilization with chemical fertilizers	Stubble processing, for subsequent sowing of weed crops	Decompaction, deep aeration of soils, removal of hardpan, improvement of soil permeability
7	Technology of sowing and fertilizing weeds directly in the stubble	Partial surface processing of myrtle, sowing and fertilizing of weed crops	For sowing the second crop after straw or after weeding crops in uncultivated or insufficiently prepared soils for spring sowing	Water conservation in the soil, reduction of soil processing, conservation of organic matter

The development of technologies for this new type of agricultural exploitation could be stimulated by environmental subsidies, taking into account the fact that this results in a reduction of the environmental pollution of the global warming type [35]. The main innovative technologies for sequestering carbon in the soil that were developed by INMA are shown in figure 5.

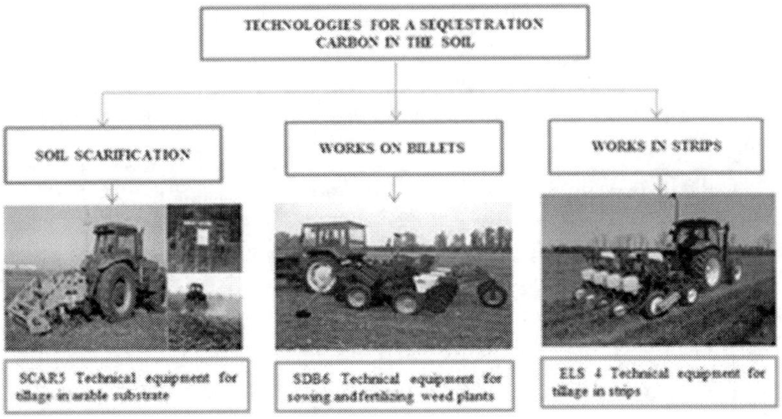

Fig. 5. Innovative technologies for carbon sequestration in the soil developed by INMA

The expansion, in practice, of soil conservation works differs from one country to another, the possibilities of mechanization are different depending on the increase of the capacity of tractors and agricultural machines and with the diversification of unloading, cultivating and sowing equipment. The technological systems of tillage have evolved a lot in the last decades, both in Romania and worldwide, evolution taking place both conceptually and in terms of extension of the conservative tillage methods.

4 Conclusions

At the global level, carbon management involves taking complex measures including actions on soil, atmosphere, water management, vegetation, resources and measures to combat climate change.
The possibility of carbon sequestration is mainly dominated by pedological factors that can establish the maximum physico-chemical delimitation when storing carbon in the soil.
these factors contain soil texture and clay mineralogy, depth, bulk density, aeration and proportion of coarse fragments.
One of the important directions approached worldwide to reduce carbon losses and implicitly the measures adopted for its sequestration in the soil, is the use of methods and technologies for carrying out conservation work on the soil, these being mainly: no tillage; strip tillage, mulch tillage, rotational tillage. 4. Worldwide, a series of conservative technologies have been developed and used, made with the appropriate technical equipment, the most used being presented in table 3 of this paper.

Acknowledgement

This paper was financed by a grant of the Romanian Education and Research Ministry, through Programme 1 – Development of the national research-development system, subprogramme 1.2 – Institutional performance – Projects for financing excellence in RDI, contract No. 16PFE.

Reference

1. J. Demenois, E. Torquebiau, H. M. Arnoult, T. Eglin, *Barriers and strategies to boost soil carbon sequestration in agriculture,* Frontiers in Sustainable Food Systems, 4:37, **14** (2020).
2. G. Barancikova, J. Makovnikova, *Comparison of two methods of soil organic carbon determination*, Polish Journal of Soil Science **48** (1), pp. 47-56, (2015).
3. H. Zhang, X. Wang, Z. Feng, J. Pang, F. Lu, Z. Ouyang, H. Zheng, W. Liu, D. Hui, *Soil temperature and moisture sensitivities of soil CO2 efflux before and after tillage in a wheat field of Loess Plateau*, China, Journal of Environmental Sciences **23**, pp 79-86 (2011).
4. O.W Heal, J.M Anderson, M.J. Swif, *Plant litter quality and decomposition: an historical overview*, Driven by nature: plant litter quality and decomposition. CAB International, Wallingford, UK, (1997).
5. K. Oorts, R. Merckx, E. Grehan, J. Labreuche, B. Nicolardot, *Determinants of annual fluxes of CO2 and N2O in longterm no-tillage and conventional tillage systems in northern France*, Soil & Tillage Research **95**, pp. 133-148 (2007).
6. B.D. Soane, B.C. Ball, J. Arvidsson, G. Basch, F. Moreno, J. Roger-Estrade, *No-till in northern, western and southwestern Europe*: a review of problems and opportunities for crop production and the environment, Soil & Tillage Research **118,** 66-87 (2012)
7. N. La Scala, A. Lopes, K. Spokas, D. Archer, D.C. Reicosky, *Short-term temporal changes of bare soil CO2 fluxes after tillage described by first-order decay models*, European Journal of Soil Science **60**, 258-264, (2009).
8. D.A. Coomes, R.B. Allen, N.A. Scott, C. Goulding, P. Beets, *Designing systems to monitor carbon stocks in forests and shrublands, For. Ecol. Manag.* 164 (1–3), pp. 89–108, (2002).
9. H.K. Gibbs, S. Brown, J.O. Niles, J.A. Foley, *Monitoring and estimating tropical forest carbon stocks: making REDD a reality*, Environ. Res. Lett. **2** (4), (2007).
10. I. Machar, J. Simon, K. Rejšek, V. Pechanec, J. Brus, H. Kilianová, *Assessment of forest management in protected areas based on multidisciplinary research*, Forests **7** (11), (2016).
11. V. Pechaneca, J. Purkytb, A. Benca, C. Nwaogua, L. Štěrbovváb, P. Cudlínb, *Modelling of the carbon sequestration and its prediction under climate change*, Ecological Informatics **47**: 50-54, (2018)
12. D.G. Goodenough, H. Chen, A. Dyk, J. Li, Multisensor data fusion for abovegroung carbon estimation, https://www.researchgate.net/publication/228338816, (2005).
13. R.A. Mandal, P. van Laake, *Carbon sequestration in community forests: an eligible issue for CDM* (a case study of Nainital, India), Banko Janakari **15** (2), pp. 53–61, (2005).
14. P. Vicharnakorn, R.P. Shrestha, M. Nagai, A.P. Salam, S. Kiratiprayoon, *Carbon stock assessment using remote sensing and forest inventory data in Savannakhet*, Lao PDR. Remote Sens. **6**: 5452–5479, (2014).
15. A. Fiorini, R. Boselli, S.C. Maris, et al., *May conservation tillage enhance soil C and N accumulation without decreasing yield in intensive irrigated croplands,* Results from an eight-year maize monoculture, Agriculture Ecosystems & Environment 296, (2020).
16. www.epa.gov/sequestration/local_scale.html

17. www.attra.ncat.org/attra-pub/PDF/carbonsequestration.pdf, Agriculture, Climate Change and Carbon Sequestration, (2009).
18. www.epa.gov/sequestration/ local_scale.html, E.P.A., Local Scale: Carbon Pools in Forestry and Agriculture, (2008).
19. B. Minasny, *Soil carbon to mitigate climate change*, http://sydney.edu.au/news/agriculture/1272.html?newsstoryid=15532, (2015).
20. E. Lin, G. Liping, J.U. Hui, *Challenges to increasing the soil carbon pool of agro-ecosystems in China*, Journal of Integrative Agriculture 17(4): 723–725, (2018).
21. C. Simota, S. Dumitru, O. Vizitiu, *Guide to good agricultural practice for mitigating the effect of climate change on agriculture*, Ghid de bune practici agricole pentru atenuarea efectului schimbărilor climatice asupra agriculturii (2014).
22. http://soco.jrc.ec.europa.eu., Sustainable agriculture and soil conservation Introduction, Correlation of soil degradation processes, non-hazardous agricultural practices and soil policies, (2009).
23. R. Lal, Carbon emission from farm operations, Environment International **30**: 981–990 (2004).
24. X. Ding, X. Han, Y. Liang, Y. Qiao, L. Li, N. Li, *Changes in soil organic carbon pools after 10 years of continuous manuring combined with chemical fertilizer in a Mollisol in China*, Soil Tillage Res. **122**, 36–41, (2012).
25. R. Teodor, *Soil conservation systems, Sisteme de lucrare pentru conservarea solului, Faculty of Agriculture*, USAMV Cluj (2019).
26. Gibbs, H.K., Brown, S., Niles, J.O., Foley, J.A., *Monitoring and estimating tropical forest carbon stocks: making REDD a reality*, Environ. Res. Lett. 2 (4), 045023, http://dx.doi.org/10.1088/1748-9326/2/4/045023 (2007)
27. R. Derpsch, K. Moriya, *Historical review of no-tillage cultivation of crops. Proceedings*, First JIRCAS Seminar on soybean research, Brazil, JIRCAS Working Report **13**, 1 – 18 (1998).
28. P.I. Moraru, T. Rusu, M. L. Sopterean, *Control of soil works and their effect on erosion, water management and carbon sequestration* , ProEnvironment **3**: 541 – 548 (2010).
29. .L.G. Teixeira, M. M. Corradi, A. Fukuda, A. R. Panosso, et al., *Soil and crop residue CO2-C emission under tillage systems in sugarcane-producing areas of southern Brazil,* Scientia Agricola, On-line version ISSN 1678-992X, Sci. agric. (Piracicaba, Braz.) 70 (5) (2013).
30. G. Barancikova, J. Makovnikova, *Comparison of two methods of soil organic carbon determination*, Polish Journal of Soil Science *48* (1): 47-56 (2015).
31. A.Y.Y. Kong, J. Six, D.C. Bryant, R.F. Denison, C. van Kessel, *The relationship between carbon input, aggregation, and soil organic carbon stabilization in sustainable, cropping systems*, Soil Sci. Soc. Am. J. **69**: 1078–1085 (2005).
32. B. Mandal, B. Majumder, T.K. Adhya, P.K. Bandyopadhyay, A. Gangopadhyay, D. Sarkar, M.C. Kundu, S.G. Choudhury, G.C. Hazra, S. Kundu, R.N. Samantaray, A.K. Misra, *Potential of double cropped rice ecology to conserve organic carbon under subtropical climate*, Glob. Chang. Biol. **14**: 2139–2151 (2008).
33. B. Majumder, B. Mandal, P.K. Bandyopadhyay, A. Gangopadhyay, P.K. Mani, A.L. Kundu, D. Mazumdar, *Organic amendments influence soil organic carbon pools and rice–wheat productivity*, Soil Sci. Soc. Am. J. **72**, 775–785 (2008).

34. J. Fan, W. Ding, J. Xiang, S. Qin, J. Zhang, N. Ziadi, *Carbon sequestration in an intensively cultivated sandy loam soil in the North China Plain as affected by compost and inorganic fertilizer application*, Geoderma 230–231: 22–28 (2014).
35. D. Cujbescu, N. Ungureanu, V. Vlăduţ, C. Persu, M.R. Oprescu, N.E. Gheorghiţă, *Field testing of compaction characteristics for farm tractor Universal 445*, INMATEH – Agricultural Engineering, **59** (3): 245-252 (2019).

Theoretical research to accomplish a system of ecological heating a greenhouse for vegetables

Eugen Marin[1,*], *Marinela* Mateescu[1], *Dragos* Manea[1], *Andrei* Dumitrascu[1], *Carmen* Vasilachi[1], *Radu* Popa[1], *Vlad* Popa[1]

[1]National Institute of Research-Development for Machines and Installations designed to Agriculture and Food Industry-INMA Bucharest 013813, Romania

Abstract. Considerating that the energy resources can be found in limited quantities and are not renewable, the most efficient and technically sustainable for heating the soil in early spring or late autumn in a greenhouse for vegetables is to use the solar energy. A heated greenhouse for vegetables is useful because the vegetation process can start more early or can continue later for can obtain several crops throughout the year. The paper presents theoretical researches accomplished for development an system of ecological heating in a greenhouse covered with a tough polyethylene wrap with triple layer. The heating system contains solar panels with heat-pipe vacuum tubes and an installation for heating the interior of the solarium using hot water hot air heaters, which adds more heat.

1 Introduction

In Romania, studies have been carried out to increase the energy independence of greenhouses with small dimensions, for example, the use of a TLUD gasification system, which uses the energy potential of agricultural residues [1]. For the use of agricultural residues in a process of energy recovery through gasification, it is necessary to grind to a size of 5 cm of the splinters, and the maximum humidity of the material should be a maximum 20% [2].

Due to the fact that the use of systems for the production of hot water is made with high costs, the most often used heating system is the one that uses flat solar collectors because it is simple and economical [3]. Lately, the researchers have emphasized the use of solar energy in the technical applications for heating greenhouses and solariums, due to the high storage performance and the efficiency of its conversion [4].

In areas of Western Europe, Central Europe, and Eastern Europe, around noon, under a perfectly clear sky and free of pollution, unit thermal energy received from the Sun can be a maximum 1000 W/m^2 (as the sum of the direct radiation and diffused), which is measured perpendicular to the direction of the sun's rays on the surface of the Earth [5].

* Corresponding author: marin_43eu@yahoo.com

In Romania, the values of the thermal flux of the diffused solar radiation in the horizontal plane are directly influenced by the position of the Sun with respect to the surface normal and the South direction having approximate values from 25 Wh/m^2 in the winter months to 62 Wh/m^2 in the summer months [6].

There are multiple possibilities for developing the use of solar energy for heating greenhouses and solariums especially in the South-East and South-Muntenia Region, according to the schematic map of the global solar radiation in Romania from Fig. 1.

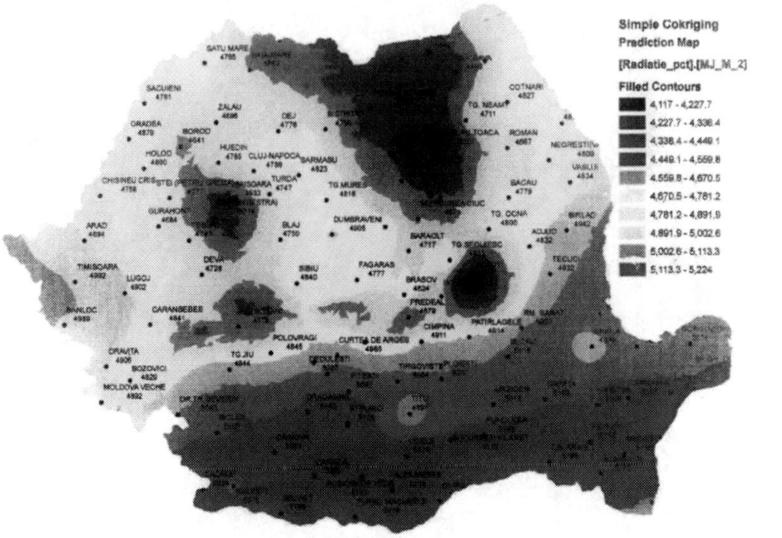

Fig. 1. Schematic map of global solar radiation in Romania [7].

The flat solar collectors (solar collector, solar thermal panel), which are designed to transform the incident solar radiation into usable heat, do not allow to following the Sun for increase the conversion efficiency, because of this the maximum efficiency (Fig. 2) is obtained only for an interval short time during a day [8].

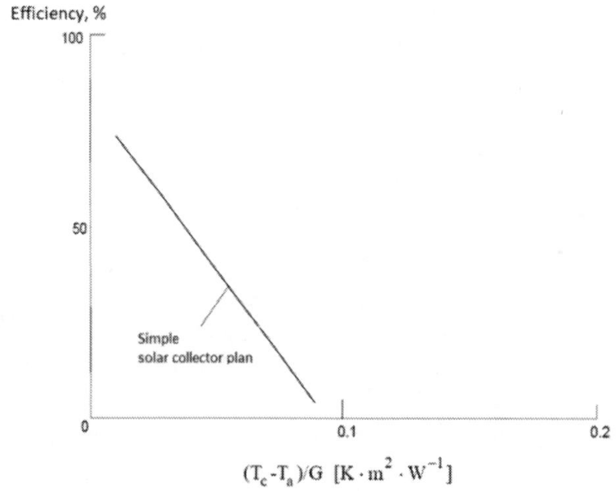

Fig. 2. Variation of the plane collector yield curvewhere, G is the incident radiation on the collector surface, in W/m^2, and T_c and T_a are the temperatures in the collector respectively the ambient temperature, in K.

There are concerns about increasing the efficiency of converting solar energy into thermal energy at solar collectors, which involves efficient capture of solar radiation.

Between a solar panel with vacuum head-pipe and a flat solar panel for domestic use, the annual yield can increase by 5 to 10 percent. [9].

Researching to account the requirements of efficient use of solar collectors that have a high percentage of solar radiation which reaches on that surface and transforms into thermal energy, INMA Bucharest has designed an experimental model of system of ecological heating needed of an agricultural farm that uses energy solar for the contribution of heat for the purpose of heating a vegetable solarium in order to grow crops during the cold time of the year (October-December and February-April).

The installation for producing heat supply uses two solar panels with 30 vacuum heat-pipe from copper of 24 mm Sontec SPA 30 58/1800, which are the most efficient in terms of capturing solar energy because the borosilicate glass layer and the vacuum inside has the role of giving a 96% efficiency (solar energy falling on the panel, compared to the energy captured by the panel). These panels can provide 70-80% of the heat input in the cold season [10].

2 Methodology

The purpose of this work was the correct sizing of the heat production installation in order to warm a vegetable greenhouse with the following characteristics [11]: width: 9 m, length: 30 m, height of the ridge: 4.05 m, the height of the pillars: 2.35 m, arch distance: 2.5 m, metallic structure: from round pipe E260 premium steel, galvanized with 275 gZn/m^2, coating: three-layered greenhouse film 200 μm that has the following properties: resistance: UV (ultraviolet), EVA (elongation): 600%, outer layer: anti-dust, transparency: 90%, diffusion: 25%, , inner layer: anti-condensation, thickness: 0.20 mm

In order to obtain the maximum yield of capturing the intensity of the direct and diffuse solar radiation, the installation of the heat supply has been dimensioned taking into account the following factors:

- the thermal energy produced by a solar collector during a day, has been observed to depend on the temperature of the thermal agent and that of the environment [12];
- the value of the intensity of solar radiation, which according to the specialized literature is influenced by: geographical region, altitude, temperature and humidity of the air, degree of air pollution, the average number of sunny hours during a day, positioning of the Sun according to the calendar month, but also by the inclination of the capture panels, the materials used in the construction, the working fluids etc;
- the volume of the greenhouse to be heated.

A special role in the design of the heat supply installation for the vegetable greenhouse had to adapt to the climatic conditions, so that during a sunny day it was considered that the entire surface area of the solar collector to be directly exposed to solar radiation for at least 4 hours during the months of October -December and February -April.

The temperature of the environment influences the thermal parameters, which is why it was taken into account the average temperature in the winter months November - March which is 2.34°C, considering the historical weather data per hour for 30 years for Bucharest provided by the National Administration of Meteorological.

In Fig. 3 is the graph for the "daily average maximum" (continuous red line) that shows the maximum average temperature of the day for each month for Bucharest [13].

Another important role in the design of the plant for the production of heat in the vegetable greenhouse also had the influence of the shadow areas and the obstacles that were estimated by a calculation method for determining the solar radiation that will be available in a sunny day during a year.

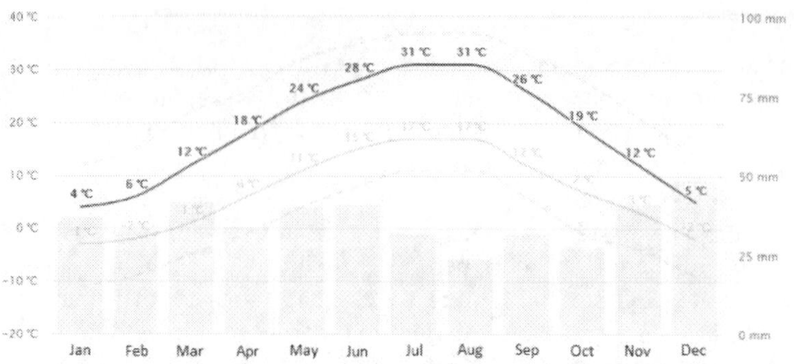

Fig. 3. Graphical representation of the average maximum temperature of one day for each month in Bucharest

According to the data provided by the National Administration of Meteorological, the ultraviolet global solar radiation, regardless of the time of year, shows an ascending daytime flow in the first part of the day, until afternoon, when the rule is reached, the maximum values. In the second part of the day, ultraviolet radiation shows a descent movement. To perform the analysis of the influence of the intensity of the solar radiation and implicitly of the heat flow received by the solar installation, the site https://www.meteoblue.com [15] offers sufficient information on the values of temperature, solar radiation, direct radiation and diffusion depending on by the geographical region of the location of the solar collectors. Based on the values offered by Meteoblue, charts were drawn regarding the values of temperature, solar radiation, direct and diffuse radiation. In Fig. 4 are represented graphically the values of temperature, solar radiation, direct and diffuse radiation realized during the period 24.03.2020-31.03.2020 from Bucharest.

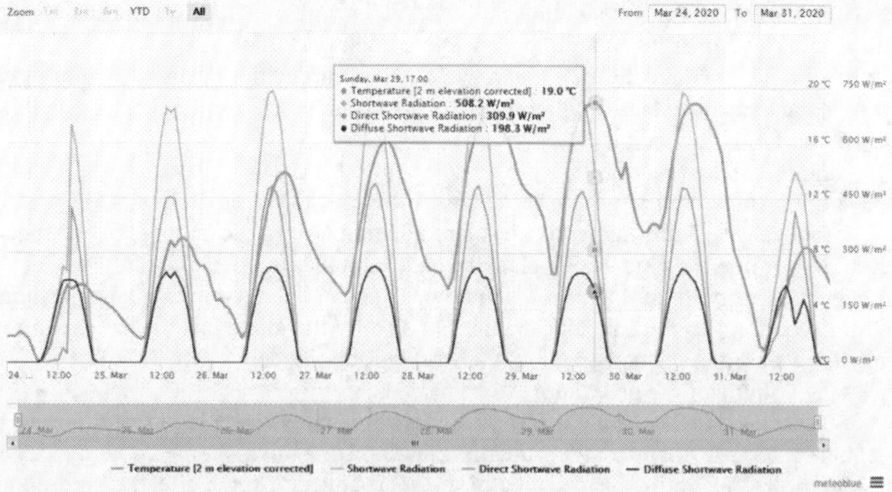

Fig. 4. Graphical representation values of the temperature, solar radiation, direct and diffuse radiation in period 24.03.2020 to 31.03.2020 from Bucharest

From the analysis of the graph presented in Fig. 4, it is observed that at 17:00 when the maximum temperature reached was 19°C, the solar radiation had a value of 508.2 W/m^2, the direct radiation had a value of 309.9 W/m^2, and the radiation diffuse represented 39.02% of the solar radiation. Accessing the database of the Meteoblue site, we could make a graphical representation of the temperature and solar radiation achieved during the same period 24.03.2020-31.03.2020 for the years 2018, 2019 and 2020 from Bucharest, as in Fig. 5.

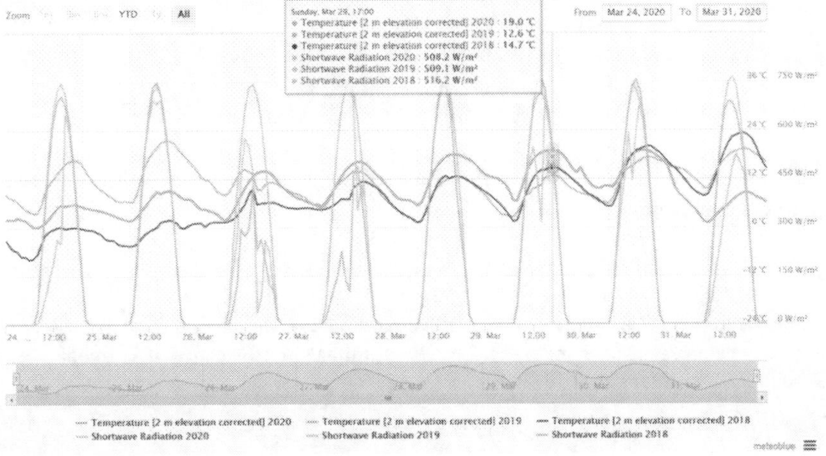

Fig. 5. Graphical representation values of temperature and solar radiation achieved between March 24-31 for the years 2018, 2019 and 2020 from Bucharest

In case of the graph presented in Fig. 5, it is observed that at 5:00 pm when the air temperature had the highest value, for the years 2018, 2019 and 2020 there were constant values of solar radiation, between March 24-31 for the years 2018, 2019 and 2020 from Bucharest, different values between 516.2 W/m^2 in 2018, 509.1 W/m^2 in 2019, respectively 508.2 W/m^2 in 2020. For choosing the solar collector from the component of the heat supply in the vegetable solar during the cold weather of the year (October-December and February-April), the power of the solar radiation from which a square meter of solar collector for the location from Bucharest (Table 1).

Table 1. Average values of solar radiation power per square meter of solar collector in Bucharest

Month	II	III	IV	X	XI	XII
	The power of solar radiation W/m^2/zi					
Clear sky	420	639	799	416	296	140
Covered sky	215	318	405	243	162	68

Analyzing the data in Table 1, it was found that the average power of solar radiation that can benefit a square meter of solar collector in cloudy weather is 235.2 W/m^2/day, of which 221.1 W/m^2/day can be captured by a solar collector of 1 m^2 that has an absorption coefficient of at least 94%. Following a technical analysis regarding the solar collectors sold in Romania, the solar collector with 30 heat-pipe drained Sontec SPA 30 58/1800 was chosen, which has the following main technical characteristics: the total area of 4.95 m^2 material of the heat pipe: copper with 99.93% purity bonded with silver alloy, capacitor size: Ø 14 x 60, tube length: 1800 mm, tube diameter: outside: 58 mm and inside: 47 mm, glass thickness: 1.6 mm, glass material: borosilicate, absorption coefficient: min. 94%, emission coefficient: max. 8% [16]. In these conditions, it was possible to design the heat production plant in greenhouse with has siz 9x30 m covered with polyethylene wrap with triple layer 200 microns, by supplementing with the following main components: two solar panels Sontec SPA 30 58/1800 with 30 vacuum heat pipes, a solar boiler with 2 500 liter Sunsystem SON 500 coils, a Grundfos Solar Oventrop 25-65 130 thermally insulated pump assembly with thermometer, flow meter, manometer, 35 liter expansion vessel, a flexible stainless steel connection for the vessel expansion, a 180°C solar vent, a 1 "thermostatic mixing valve, connecting pipes and a SR288 solar automation system with three temperature sensors.

3 Conclusions

- Theoretical studies carried out allowed to design a heat supply installation for the cold period of the year using solar collectors with vacuum tubes for heating the interior of a vegetable solar with hot water aerothermal;
- The average values of the solar radiation power per square meter of the solar collector, can easily be used in other indicative calculations for rapid predimensioning of the different types of solar collectors, being even more useful as, as a rule, such values are not indicated by manufacturing companies.

This work was supported by a grant of the Romanian Ministry of Education and Research, through Programme NUCLEU, ctr. nr. 5N / 07.02.2019, project PN 19 10 02 01-Development innovative technologies within SMART farms and also Programme 1 – Development of the national research-development system, subprogramme 1.2 – Institutional performance – Projects for financing excellence in RDI, contract no. 16PFE.

References

1. E. Maican, I. C Dutu, G. Matache, C. Dumitrescu, I.Pavel, *CFD analysis of an improved tlud based equipment for heating small greenhouses and hothouses*, INMATEH – A. E., **53**, 8, (2017)
2. E. Murad, C. Sima, T. Chereches, *Gasification of corn stover*, U.P.B. Sci. Bull., Series D, **72**, 6, (2010)
3. D. Taban, A. Pop, C. Dobre, V. Apostol, T. Prisecaru, *Establishing the optimal positioning of a three system vertical solar captors according to geographical orientation,* Rev. Term., **1**, 6, (2018)
4. Y. Tian, C.Y. Zhao, *A review of solar collectors and thermal energy storage in solar thermal applications,* App. Energ, **104**, 55 (2013)
5. https://www.reeco.ro/energie-termica-solara/
6. B. Tanase, A. Gheorghian, M. Zamfir, V. Badescu, H. Pop, *Comparative study on the positioning of solar collectors vertically or horizontally taking into account the restrictive conditions imposed by the dimensions of the renewable energy production system*, Rev. Term., **2**, 4, (2018)
7. http://energystreet.ro/fotovoltaice/harta-solara-a-romaniei/
8. F. Gáspár, V. Ros, *Comparative study of solar collectors with flat and spherical surfaces through prism solar energy conversion efficiency in thermal energy*, Rev. St. and Eng. **20**, 8, (2011)
9. http://www.calorserv.ro/articole/panouri-solare/sistemul-solar-termic-panouri-solare
10. *** Ctr. nr. 5N / 07.02.2019, PN 19 10 02 01, NUCLEU – SMART-BITECH, (2009)
11. https://www.seretransilvania.ro/ro/solarii-profesionale/solarii-gotice/solar-gotic-9m/solar-gotic-9x30m-folie-dubla-inflata.html
12. I. Ziemelis, L. Kancevica, Z. Jesko and H. Putans, *Calculation of energy produced by solar collectors,* Eng. for Rul. Dev., (2009)
13. https://www.meteoblue.com/ro/vreme/historyclimate/climatemodelled/bucure%C8%99ti_rom%C3%A2nia_683506
14. http://www.meteoromania.ro/despre-noi/meteorologie-operationala/activitati-de-baza/actinometrie/
15. https://www.meteoblue.com/ro/historyplus
16. https://www.sistemepanourisolare.ro/colector-solar-30-tuburi-vidate-heat-pipe-sontec-spa-30

Sound power determination for centrifugal pumps used for local agricultural irrigation in Romania

Mihai Chițoiu[1], *Georgiana* Moiceanu[2*], *Gigel* Paraschiv[1], *Mădălina* Ștefan[1] *and Andreea* Matache[3]

[1]University POLITEHNICA from Bucharest, The Faculty of Biotechnical Systems Engineering, 313 Splaiul Independentei, Bucharest, 060042, Bucharest, Romania
[2]University POLITEHNICA from Bucharest, Faculty of Entrepreneurship, Business Engineering and Management, 313 Splaiul Independentei, Bucharest, 060042, Bucharest, Romania
[3]National Research - Development Institute for Machines and Installations Designed to Agriculture and Food Industry, 6 Ion Ionescu de la Brad Blv., 013813 Bucharest, Sector 1, Romania

Abstract: Technological enhancements in the domain of agriculture have led to easing the burden of physical work and increased productivity. Irrigation has always been one of the most crucial elements needed for a high yielding crop, given the current uncertainty of climate change, where climate is increasing in unpredictability. Even though a wide array of methods has been developed in recent years, for the Romanian local agricultural irrigation, a popular method remains using motor-pumps. But the issue with this type of equipment is that during active functioning, it produces high sustained levels of noise pollution, and the risk of operator hearing damage is relatively unknown for long term exposures (years). In this paper, the levels of acoustic power of an irrigation motor-pump model WHT-60 were determined, taking into consideration the required standards, and directives, necessary for this activity, imposed by the European Union for all member countries.

1 Introduction

Technological advances in the domain of agriculture have increased productivity rates and have lowered physical workload for operators. Also, these technological advances, have lowered the number of personnel needed for a successful business, through acquisitions of new and advanced equipment types for specific activities. Farming is at the dawn of a major transformation not only in technology and production practices, but also in size of business, resource control and operation [1]. Constant exposure to the noise levels of these new implemented agricultural machines can cause hearing problems for operators. Exposure to noise from agricultural tools and machinery can cause partial, or even permanent hearing loss, and tinnitus. [2] Hearing loss can lead to a depreciation of life quality, and some early signs of hearing loss can include ringing or buzzing noise in the ears after work, difficulties in understanding a conversation, failing to hear background noise etc.

*Corresponding author: georgiana.moiceanu@upb.ro

Sound is a radiant energy that is transmitted through space by longitudinal pressure and it is the objective cause of hearing, and the measurement of sound is related to pressure, frequency and duration, commonly measured in decibel units [3]. Generally, the effects of noise are hidden between 30-65 dB, but exposure to sounds of over 85 dB can cause physical long-term effects, like an increase in blood pressure, decrease of heart pulse, weak muscles and retraction of blood from the skin [4].

Noise pollution in agriculture is not a new issue, its threat being recognized by the National Occupational Safety and Health, which conducted a National Occupational Exposure Survey, between 1981 and 1983, to provide descriptive data of occupational safety and health conditions, leading to the statement that workers were considered exposed to noise pollution if the noise levels were 85 dB or higher, regardless of exposure duration [5]. In a recent study, most farmers reported exposure to high noise levels for more than eight hours during agricultural activities, therefore noise is a real potential danger for the quality of life [6].

Previous studies have shown that greater risks for hearing impairment come from using medium to low-power tractors (84-101 dB), which unfortunately are frequently used in private farms, as well as combine harvesters (88-92 dB) [7]. Agricultural equipment manufacturers are constantly developing ways to reduce sound levels reached in the vicinity of operators [8].

In one study, in which 31 agricultural equipment types were tested for noise pollution at operator's ear level, the results showed the need for continued application of noise reduction techniques, in order to protect the hearing of the operator [9]. Researchers, following their researches, seem to come to a similar conclusion, in which hearing discomfort depends on magnitude, frequency, direction and duration of the exposed noise to the ear of the operator [10]

In Romania, monitoring of noise pollution is under the attributions of the Ministry of Environment, which outlined a rise in the average noise levels, from 50 dB, at the beginning of the 80's to 70 dB, in the year 1999. Romania, being part of the European Union, is applying European Directives and Standards with the sole purpose to limit the rise of noise levels, to protect the operator hearing, and to bring in new equipment that is tested and certified as within legal limits of noise emissions by equipment type [11]. When it comes to agricultural equipment, tests are realized following the instructions of Directive 2000/14/EC [12] and, of Standard SR EN ISO 3744/2011, for determining the levels of acoustic power and acoustic energy of noise sources using acoustic pressure [13]. There are studies performed using different measurement surfaces according to previously mentioned noise measurement regulations, in which different agricultural machinery were tested. In one study, the authors investigated the noise level of a motor hoe, basically having the same noise source as in our study, identified as the thermal engine, using the same testing methodology provided in Directive 2000/14/EC which will be used in our study [14]. However, the measuring surface was of hemispherical type fact which had an impact on the microphones positioning during measurements. Another study about the noise produced by a brush cutter took into consideration various working conditions, the measurement surface being also considered as hemispherical [15]. The obtained results highlighted the need for workers protection measures against the high noise produced by the equipment.

The aim of this study was to measure the acoustic power level of an irrigation motor-pump in order to establish the risks at which its operator is subjected during functioning. There were made several measurements with the water pump unit set for three output pumping heights: 10, 20 and respectively 30 meters. The measurement surface was chosen to be on a parallelepiped surface according to the noise test code imposed for this water pump unit.

2 Materials and Method

Tests were realized within the testing laboratory of National Institute for Research and Development of Agricultural Machines.
Also, given Romania's status as a member of European Union, for testing purposes we followed Directive 2000/14/EC and Standard SR EN ISO 3744/2011.
During tests we measured the noise levels of an irrigation equipment type used in Romania, a centrifugal pump type WTH 60 equipped with an internal combustion engine. These pumps are popular in Romania and are used for irrigation purposes, as well as for eliminating water from floods. It is driven by an OHV petrol motor, type 168F-1, air cooled, with a gas reservoir capacity of 3.6 litres.
Centrifugal pump WHT 60 is composed out of a petrol engine and a water pump – the pump is fixed on a frame with a shock absorbent device. The pump itself is composed out of pump body, pump lid, flow paths and the rotor. Pump aspiration is realized with a one-sided valve which does not allow liquid to penetrate the body of the pump.

Fig. 1. Centrifugal pump equipped with an internal combustion engine WHT 60.

Table 1. Main Characteristics of the centrifugal pump WHT 60

No.	Characteristics	Measuring Units	Value
1	**Motor type**	-	**OHV, petrol**
2	**Net weight**	**kg**	**30**
3	**Maximum pumping height**	**m**	**30**
4	**Maximum flow**	**m^3/h**	**60**
5	**Cylinder capacity**	**cc**	**196**
6	**Compression relation**	**HP/rpm**	**8.5:1**
7	**Maximum power**	**HP/rpm**	**6.5/3600**
8	**Reservoir capacity**	**l**	**3.6**

2.1 Testing conditions

In order to assure a high degree of certitude for the results, the following operations were realized on the centrifugal pump WHT 60:
- the lapping of the equipment was realized according to its instructions manual;
- detachable and undetachable parts were verified to ensure the relevance of the tests;
- oil levels and petrol levels were checked;
- the motor was run for 30 minutes before performing measurements.

2.2 Equipment and measuring devices used for testing

In order to accurately measure the noise levels of the WHT 60 pump, the following devices were used:

Table 2. Equipment types and devices being used for testing

No.	Name of the instrument of device	Measuring domain
1	Roulette	0÷8m
2	PULSE System with microphones	100÷10000 Hzkg
3	Thermo-hygrometer DH 50	Humidity: 5÷95% Temperature: 20÷80°C
4	Anemometer Testovent 4000	0,4-40 m/s

Fig. 2 PULSE Data acquisition board – 12 channels.

2.3 Pulse System microphone distribution

Taking into consideration the instructions from Directive 2000/14/EC and standard SR EN 3744/2011, the microphone distribution method was chosen to be on a parallelepiped surface with measurement distance d = 1 m from the reference box. The latter was a hypothetical surface defined by the smallest right parallelepiped that enclosed the noise source under test, namely the pump. In figure 3, is presented the reference box on one reflecting plane where l_1, l_2, l_3 are the length, width and height of the reference box and d_0 represents the characteristic source dimension.

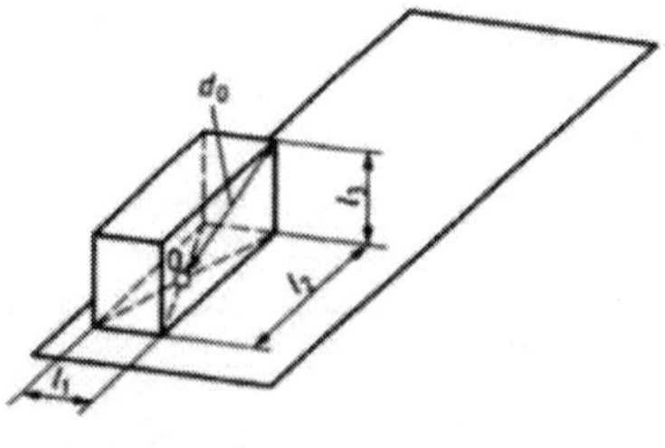

Fig. 3. Reference box on one reflecting plane [13].

The area S of the measurement surface was computed using relation (1).

$$S = 4(ab + bc + ca) \quad (1)$$

where

$a = 0.5l_1 + d$

$b = 0.5l_2 + d$

$c = l_3 + d$

2.4 Determining the acoustic power level of the WHT 60 centrifugal pump

The time-averaged sound pressure level $L_{pi(ST)}$ for the pump noise source under test was measured for each *i-th* microphone using the A scale for weighting. The measurement time interval was set at 15 seconds. There were made 3 replicas of the measurements. During the same test the time-averaged sound pressure level of the background noise, $L_{pi(B)}$ was measured before and after the 3 noise test replicas in the same conditions but with the noise source shut down. The mean time-averaged sound pressure level $L_{P(ST)}$ in decibel was then computed using relation (2).

$$L_{P(ST)} = 10lg\left[\frac{1}{N_m}\sum_{i=1}^{N_m} 10^{0.1L_{pi(ST)}}\right] \qquad (2)$$

where N_m is the number of microphones used for the measurement.

The mean time-averaged sound pressure level of the background noise, $L_{P(B)}$ was calculated using relation (3).

$$L_{P(B)} = 10lg\left[\frac{1}{N_m}\sum_{i=1}^{N_m} 10^{0.1L_{pi(B)}}\right] \qquad (3)$$

The corrections for the background noise, K_1 was computed using relation (4):

$$K_1 = -10lg(1 - 10^{-0.1\Delta L_P}) \qquad (4)$$

where $\Delta L_P = L_{P(ST)} - L_{P(B)}$.

The background noise correction condition was that if $\Delta L_P > 15dB$, then K_1 will be assumed to be zero, if $6dB \leq \Delta L_P \leq 15dB$ then the corrections will be made according to relation (4) and in all the other cases the tests will be unvalidated and renewed.

Afterward the surface time-averaged sound pressure level L_P (dB) was computed using relation (5):

$$L_P = L_{P(ST)} - K_1 \qquad (5)$$

The sound power level for the WHT 60 centrifugal pump was determined using relation (6):

$$L_W = L_P + 10lg\frac{S}{S_0} \qquad (6)$$

where $S_0 = 1\ m^2$.

Sound power level was determined using the PULSE system, equipped with 9 microphones and distributed according to the instructions from SR EN ISO 3744/2011.

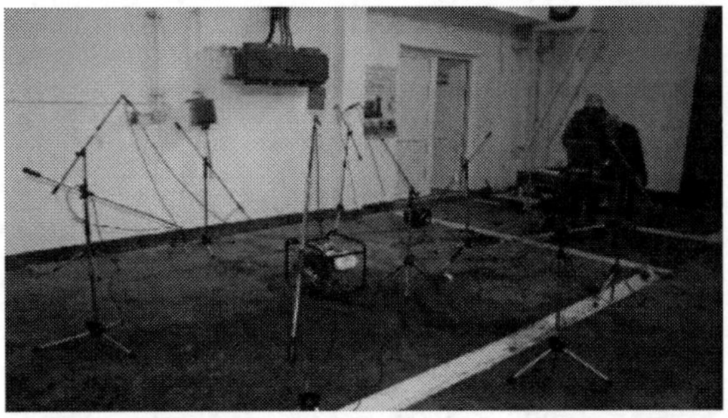

Fig. 4. Microphone distribution for determining the acoustic power of the PULSE system (9 microphones).

3 Results and discussion

The measuring surface S was calculated based on the dimensions of the noise source, and the value of 19.81 m^2 was obtained. Following the microphone distribution around the

centrifugal pump WHT 60, according to the instructions of SR EN 3744/2011, the equipment was turned on and the data acquisition was performed.
There were performed three sets of measurements on the water pump at three pumping heights, 10, 20 and respectively maximum 30 meters, in order to asses the acoustic power level in function of the load sustained by the unit.

The acoustic pressure and acoustic power results obtained for the 10 m pumping height, are presented in table 3.

Table 3. Experimental data for sound power determination for 10 m pumping height

	$L_{p1(ST)}$ (dB)	$L_{p2(ST)}$ (dB)	$L_{p3(ST)}$ (dB)	$L_{p4(ST)}$ (dB)	$L_{p5(ST)}$ (dB)	$L_{p6(ST)}$ (dB)	$L_{p7(ST)}$ (dB)	$L_{p8(ST)}$ (dB)	$L_{p9(ST)}$ (dB)	$L_{P\ (dB)}$	$L_{W\ (dB)}$
Background	38.5	38.6	39.1	39.2	39.4	39.2	39.7	39.2	38.6		
Replica 1	85.91	88.03	83.28	88.19	85.04	85.08	87.84	86.92	87.03	**86.63**	**99.6**
Replica 2	85.48	86.43	83.85	88.63	85.12	86.34	88.09	87.80	86.54	**86.70**	**99.7**
Replica 3	85.38	86.90	83.93	88.48	85.86	86.87	89.06	86.81	88.37	**87.11**	**100.1**
Background	37.8	37.2	37.2	36.9	37.4	36.8	36.4	36.6	37.3		

In table 4 are presented the results obtained for the 20 m pumping height.

Table 4. Experimental data for sound power determination for 20 m pumping height

	$L_{p1(ST)}$ (dB)	$L_{p2(ST)}$ (dB)	$L_{p3(ST)}$ (dB)	$L_{p4(ST)}$ (dB)	$L_{p5(ST)}$ (dB)	$L_{p6(ST)}$ (dB)	$L_{p7(ST)}$ (dB)	$L_{p8(ST)}$ (dB)	$L_{p9(ST)}$ (dB)	$L_{P\ (dB)}$	$L_{W\ (dB)}$
Background	38.2	38.1	38.4	38.7	39.1	39.1	39.2	38.1	38.4		
Replica 1	85.42	87.16	84.46	88.83	85.89	86.87	88.39	88.45	88.31	**87.31**	**100.3**
Replica 2	85.24	87.80	84.39	89.06	85.25	86.24	88.41	88.51	88.19	**87.30**	**100.3**
Replica 3	84.80	87.25	84.99	88.17	85.56	86.01	88.69	87.87	87.87	**87.01**	**100.0**
Background	36.2	36.8	36.5	36.1	36.7	36.5	36.9	36.9	37.2		

Final results obtained for the maximum pumping height, based on data acquisition of time-averaged sound pressure level $L_{pi(ST)}$ and the results calculated are presented in table 5.

Table 5. Experimental data for sound power determination for 30 m pumping height

	$L_{p1(ST)}$ (dB)	$L_{p2(ST)}$ (dB)	$L_{p3(ST)}$ (dB)	$L_{p4(ST)}$ (dB)	$L_{p5(ST)}$ (dB)	$L_{p6(ST)}$ (dB)	$L_{p7(ST)}$ (dB)	$L_{p8(ST)}$ (dB)	$L_{p9(ST)}$ (dB)	$L_{P\ (dB)}$	$L_{W\ (dB)}$
Background	39.5	38.4	39.4	39	39.9	39.2	39.8	39.1	39.6		
Replica 1	86.9	89.1	85.7	90.5	86.6	87.4	89.5	89.2	89	**88.46**	**101.4**
Replica 2	87.1	89.4	85.8	90.9	86.6	87.4	89.4	89.6	89.3	**88.68**	**101.6**
Replica 3	87.6	89.6	86.2	91.4	87.4	88.1	89.9	90.3	90.1	**89.24**	**102.2**
Background	35.8	35.8	36.2	36.5	38.1	36	36.9	36.6	37.1		

Analysing the obtained results for the three working conditions we conclude that the acoustic power level was maximum for the maximum pumping height. This is explained by the fact that the thermal engine auto-controller was increasing the throttle as the load was increased, thus increasing the noise level. However, the noise level observed in all three working conditions was still high, exceeding 87 dB sound pressure level and 99 dB for the acoustic power. Sounds above 85 dB are harmful for the human ear depending on the period of exposure and occurrence. The workers should wear hearing protection, such as earplugs or earmuffs.

The mean value of the sound power for the maximum pumping height $L_{W\ (dB)}$ was computed and the result of 101.9 dB was rounded at 102 dB weighted on the A scale.

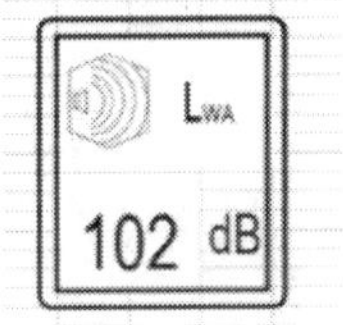

Fig.5 Sound power obtained for WHT 60 centrifugal pump.

4 Conclusions

Following the results obtained from determining acoustic power levels of a running WHT 60 centrifugal pump the following conclusions were drawn:

- Even though, according to the Directive 2000/14/EC, this device is emitting an acoustic power level within the legal limits, 102 dB is still a very high number and the worker acoustic protection is required.
- Increasing the load on the water unit pump has as result also an increase of the noise level, fact which should be considered when operating the pump.
- Long-time exposure and working in the close vicinity of this equipment type, can have negative effects on the operator's hearing health, in the long term.
- Current EU directives and standards take into consideration short to medium time exposure to noise levels generated by different equipment types.
- For effects of noise levels in regard to long time constant exposure, a multitude of data gatherings must be achieved, over a long period of time, in order to better understand these effects on the human hearing health.
- Using this equipment type for irrigation purposes during a drought would subject the operator to a higher incidence of being exposed to these high noise levels, and his hearing health could decrease;
- Further tests, on the topic of operator continuous noise level exposure must be realized, in order to raise awareness on the topic of hearing health risks, better understand how hearing is affected, and impose clearer and stricter rules.

ACKNOWLEDGEMENT

This work has been funded by the European Social Fund from the Sectoral Operational Programme Human Capital 2014-2020, through the Financial Agreement with the title "Scholarships for entrepreneurial education among doctoral students and postdoctoral researchers (Be Antreprenor!)", Contract no.51680/09.07.2019 – SMIS code: 124539.

References

[1] A. Gray, M.D. Boehlje, Department of Agricultural Economics, Working Paper 07 (2007)

[2] Government of Western Australia, Noise in agriculture – Identification, assessment and control (2004)

[3]. M.R. Dorgut, H Celen, Noise Levens of Various Agricultural Machineries, Pakistan Journal of Biological Sciences 7, pp. 895-901 (2004;)

[4] A. Sabanci, E. Uz, Ergonomy ve Agricultural Mechanization, 1st International Symposium of Ergonomy (1984)

[5] D. Pessina, M. Guerretti, Effectiveness of Hearing Protection Devices in the Hazard Reduction of Noise from Used Tractors, J. Agric. Eng. Res., pp. 73-80 (2000)

[6] D. Karamousantas, T Varzakas, A. Kanakis, Noise Levels Produced by Agricultural Machineryand Different Farming Processes, International Journal of Acoustics and Vibrations (2009)
[7] L. Solecki, Duration of exposure to noise among farmers as an important factor of occupational risk, Ann. Agric. Environ. Med., pp 89-93 (2000)
[8] C.W. Suggs, Noise Characteristics of Field Equipment, ASAE (1987)
[9] S.K. Broske, D.A. Hansen, R.L. Strand, D.T. Stueland, Hearing Loss Among High School Farm Students, Am. J. Pub. Health, pp. 619-622 (1989)
[10] A.K. Junoh, Z.M. Nopiah, W. Muhamed, A Study on the Effects of Tyre Vibration to the Noise in Passenger Car Cabin. AMO – Advanced Modelling and Optimization, **13**, pp. 53-69 (2011)
[11] CECE (COMMITTEE FOR EUROPEAN CONSTRUCTION EQUIPMENT) - Position on the future of the noise emission in the environment by equipment for use outdoors (Directive 2000/14/EC), Brussels, 18 June 2009
[12] Directive 2000/14/EC of the European Parliament and of the Council of 8 May 2000 on the approximation of the laws of the Member States relating to the noise emission in the environment by equipment for use outdoors, Official Journal L 162, 03/07/2000 P. 0001 – 0078 (2000);
[13] Standard SR EN 3744/2011, Determination of sound power levels and sound energy levels of noise sources using sound pressure – Engineering methods for an essentially free field over a reflecting plane (2011)
[14] V. Vladut, D. Manea, S. t. Biri , S. Bungescu, A. Atanasov, Test of an equipment which works in open air according to the requirements of D2000/14/EC, PROCEEDINGS OF THE 35 INTERNATIONAL SYMPOSIUM ON AGRICULTURAL ENGINEERING "Actual Tasks on Agricultural Engineering", pp 405-414, Opatija – Croaţia, (2007)
[15] E. Sorică, V. Vlăduţ, P. Cârdei, C. Sorică, C. Brăcăcescu, Comparative analysis of the noise and vibration transmitted to the operator by a brush cutter, Proceedings of the XIV-th International Conference „ACOUSTICS AND VIBRATION OF MECHANICAL STRUCTURES – AVMS-2017", published by „Springer Proceedings in Physics", pp. 165-172 (2017)

An analysis of baffles designs for limiting impact on the directional response for a partially filled tank with agro-food liquids

Gabriela Milian[1,*], *Sorin-Ştefan* Biriş[2], *George* Ipate[2], *Alexandru* Hristea[3], *Cristian* Sorica[1] and *Florin* Nenciu[1]

[1]INMA Bucharest, 6 Ion Ionescu de la Brad, 1st District, 013813, Romania
[2]University "Politehnica" Bucharest, 313, Splaiul Independentei, 060042, Romania
[3]INOE 2000-IHP, street Cutitul de Argint, no. 14, Sector 4, Bucharest, Romania

Abstract. A directional response for a partially filled tank with agro-food liquids incorporates the motion of the free surface due to roll and longitudinal accelerations into the dynamic tank model. This paper presents an analysis of efficaciousness of different designs of baffles for two different tank, to limit the impact of the response given by the liquid on the tank walls. In this paper modelling of the behaviour of the cross-section circular and ellipsoidal tank with baffles, partially filled with liquids agro-food, was developed, which performs a braking-in-a-turn manoeuver. Two different tank truck models have been used to simulate the sloshing effect of agro-food liquid cargo. Is also studied the two geometric models of the tank was done by using SolidWorks and simulating the behaviour of the tank ANSYS-CFX.

In this study, there were simulated and modeled two models of tank-type "X", respectively type "Y" used for the storage and transport of liquid agro-food products, namely water, for different speeds (v = 14m / s and v = 20m / s) and two filling levels of the tank trucks (40% and 80%). The results obtained were compared the behaviour of the tank partially filled into a braking-in-a-turn manoeuver, in which the total values of the deformations that appear as a result of the impact due to the movement of the liquid inside the tank trucks.

Keywords: design baffles, impact on the directional response, Ansys-CFX

1 Introduction

As vehicles can move with partially filled tanks, there will inevitably be an interaction between vehicle movements and fluid, which are not always perceived by the driver. Thus, situations may occur where when the driver's action coincides with the action phase of the response given by the liquid contained on the tank so that its effect on the behaviour of the vehicle is amplified and may lead to overturning.

Under the action of external excitations (different displacement, loading/unloading manoeuvres), with large amplitude or near the natural frequency of the fluid sloshing inside

* Corresponding author: gabi_isb@yahoo.com

a partially filled tank, the fluid produces violent oscillations on the tank walls and the tank structure.

When the fluid interacts with the wall, the energy exchange takes place between the two liquid-solid structures, so that the fluid can exhibit different types of movements (longitudinal, transverse, chaotic) depending on the external excitation [1].

The movement of the fluid oscillation depends on the geometry of the tank, the filling conditions and the frequency range inside, that is, the quiet time of the fluid after the sloshing effect occurs Budiansky (1960). Besides, sloshing depends on the frequency and amplitude of the forces resulting from the movement Hasheminejad and Aghabeig (2011). If the frequency produced by the sloshing is close enough to the natural frequency of the fluid-structure, resonance will occur Ibrahim (1975). Also, stability is one of the most important purposes for the design of tank vehicles that transport and storage liquid agro-food products. [2, 3].

The purpose of studying the effect of sloshing, produced inside the tank for different geometrical constructive forms is to reduce the fatigue level, produced at the level of the structure of the tank material and to improve the design of the tank, for transporting the liquid cargo under optimal conditions.

From the determinations made by theoretical and experimental research it was found that the geometry of the container and the percentage of filling of the tank with liquid can directly affect the dynamic load transfer of the liquid, as well as the sloshing duration and the amplitude of the sloshing oscillation. In order to be able to reduce and control these factors, it was concluded that by installing baffles, the sloshing of the liquid in the longitudinal direction during the acceleration and braking of the vehicle can be limited [4, 5, 6].

The results of the experimental and numerical research show that the dynamic directional responses of the vehicles with partially filled tanks obtain a special efficiency if the connection created between the sloshing and the tank vehicle is taken into account. Thus, the mounting and use of baffles of different shapes have proven to be most effective in suppressing the slosh force of the fluid under loading conditions [7-11].

2 Materials and methods

This study present an analise of the baffles design for limiting the impact regarding the steering control for a partial tank filled with agro-food liquid, more exactly of different designs of baffles for two different design tank model used to transport agro-food liquid cargo.

Two different tank truck models have been used to simulate the sloshing effect of agro-food liquid cargo. It is also explored the two geometric models of the tank using SolidWorks software, and the behaviour of the tank was simulated with ANSYS-CFX. Also, two models of tank (type "X", respectively type "Y") were simulated and modeled, for the storage and transport of water, for different speeds (v = 14m / s and v = 20m / s) and filling levels of the tank trucks (40% and 80%).

The tank model "X" is presented in Figure 1 and it consists of a 3-dimensional fluids storage, with length 3840mm and height 1300mm, with circular cross-section tank, and two baffles type (I) with height 1000mm. The tank model "Y" is presented in Figure 2, and has length 5370mm and height 1300mm, with ellipsoidal cross-section tank with two baffles type (II) height 1300mm.

The influence of liquid motion on tank was performed with the help of ANSYS software (CFD-CFX- static and transient), and the 3D modelling using the Solidworks program.

For tank model "X" with baffles type (I), CFD model is showed on Figure 3 and consist of 21673 nodes, of tetrahedral cell elements 114915. Also tank model "Y" with baffles type (II), CFD model is highlighted on Figure 4, with mesh tool 74377 nodes, of tetrahedral cell elements 36641.

Fig.1. Physical model tanker "X" and geometrical design with baffles type (I)

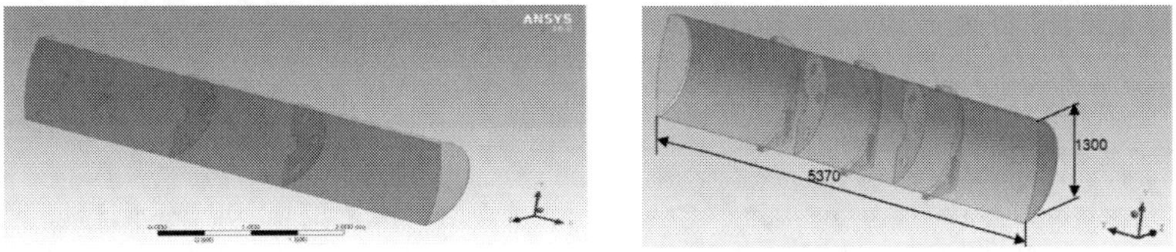

Fig.2. Physical model tanker "Y" and geometrical design with baffles type (II)

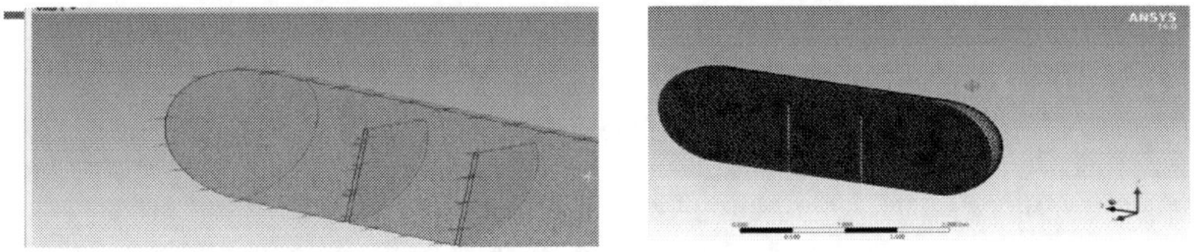

Fig.3. Mesh tanker and fluids for model tanker "X" with baffles type (I)

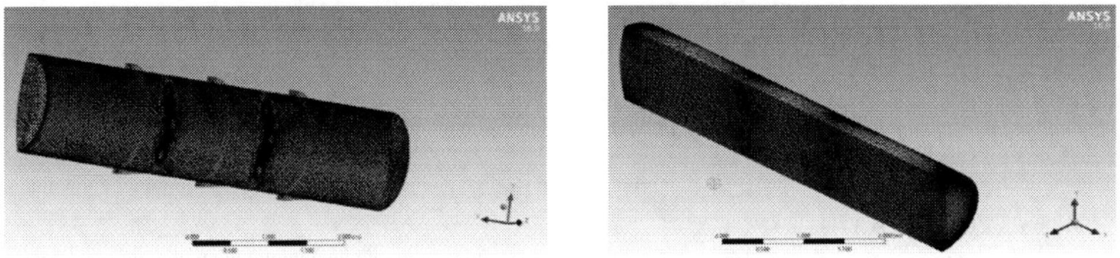

Fig.4. Mesh tanker and fluids for model tanker "Y" with baffles type (II)

The material used to make the walls of the two types of tanks "X" and Y, as well as the baffles is carbon steel with a thickness of 3mm according to SR EN 10216-2[12].

Both the simulation of the "X" tank model as well as the "Y" tank model are done consecutively, for the two filling levels 40% and 80%, respectively for the different speeds imposed by v = 14 m / s and v = 20m / s, which decrease in 10 sec. This time of 10s represents the calming time of the fluid inside the tank.

Thus, in these cases, the deceleration simulates a sudden braking and stopping of the tank during a braking maneuver.

In both cases, for the two models the acceleration is performed for 5 sec, thus reaching an acceleration of 50km / h and 80 km / h, and the deceleration during the remaining 5sec(braking) until the tank vehicle stops.

The physico-chemical properties of the fluid simulated and studied for both tank, with the two types of baffles (type (I) and type (II)) are shown in Table 1, below:

Table 1.The specifications of the fluid agro-food simulation

S.NO	NAME OF FLUID	NAME OF PROPERTY	VALUE	UNITS
1	Water	Density	998.2	Kg/m^3
		Viscosity	0001003	$Kg/m^{-1}s^{-1}$

The simulation is based on the volume fraction (VOF) formula, thus forming the free surface inside the liquid tank (water-air). Having the following expression:

$$\sum_{p=1}^{n} \alpha_p = 1 \qquad (1)$$

where : α_p - volume fraction for those two fluids(water-air),
n - no of phases present in the computational cell,
p- fluids phase.

During the simulation of the two models of the tank with the two types of baffles it was considered that the temperature inside the vessels has the value of 20°C , the pressure inside does not exceed the value of the atmospheric pressure. Thus, the models being in thermal equilibrium and in transient conditions with acceleration peak 1g in the longitudinal direction.

3 Results and discussion

Further taking into account the values of the accelerations (of the speeds imposed), as well as of the simulation of the action time and the stabilization of the liquid, the displacement manoeuvres of the tank will result in the distribution of forces and stresses on the walls of the tank, the remaining deformation of the wall of the tank, the total deformation of the tank wall structure, both in static mode and in transient mode.

Starting from all this information in the paper are analyzed the simulation for :

a)The case in which the level of filling of the tank with water is 40% and it is required a speed v = 14m / s;

b)The case in which the level of filling of the tank with water is 40% and it is required a speed v = 20m / s;

c)The case in which the level of filling of the tank with water is 80% and it is required a speed v = 14m / s;

d)The case in which the level of filling of the tank with water is 80% and it is required a speed v = 20m / s;

The analysis will be highlighted below an analysis of baffles design for limiting impact on the directional response which is based on the volume fraction (VOF) formula, how fluid moves inside the tank, thus showing the shape of the free surface, the stresses and deformations that have occurred.

In case a) the two models of the tank are simulated in order to be able to observe the behaviour and the influence of the movement of the liquid on the walls of the tank. The behaviour is shown by highlighting the stresses and deformations that can be seen in Figures 5,6.

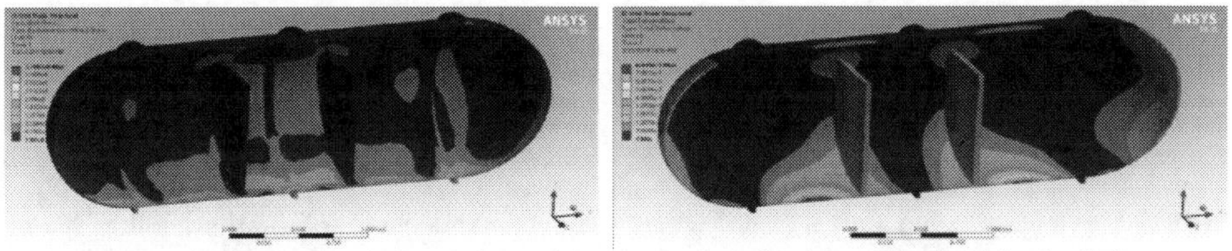

Fig.5. The result of the numerical simulation of the dynamic behavior of the fluid in the "X" type tank, with type (I) baffles for case a) for a speed v=14m/s, time 10s, fill level 40%;

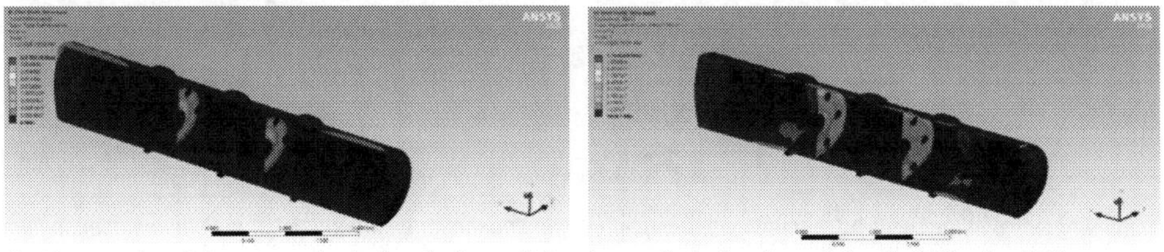

Fig.6. The result of the numerical simulation of the dynamic behavior of the fluid in the "Y" type tank, with type (II) baffles for case a) for a speed v=14m/s, time 10s, fill level 40%;

Simulations show the distributions of stresses and deformations that appear on the tank walls, both in static and dynamic (transient) conditions. The color white reproduce air, and blue the water. This is done in all cases according to the volume fraction (VOF) formula.

In the Figures of this paper, the total deformation of the tank structure is shown both in a static regime and in a transient area where the tank wall may suffer major damage, due to the very strong tension in those areas indicated by the numerical simulation by the red colour.

Further in the paper will be shown in the figures (7,8,9,10,11,12) below results obtained from the simulation of the other three cases b), c) and d) for the two models of the tank with the two forms geometric constructions of baffles.

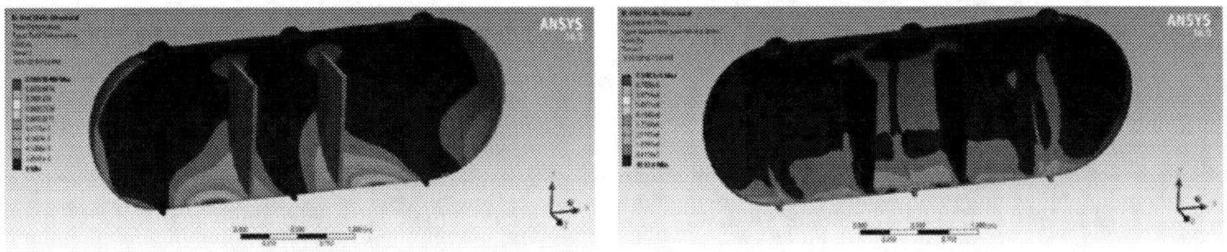

Fig.7. The result of the numerical simulation of the dynamic behavior of the fluid in the "X" type tank, with type (I) baffles for case b) for a speed v=20m/s, time 10s, fill level 40%;

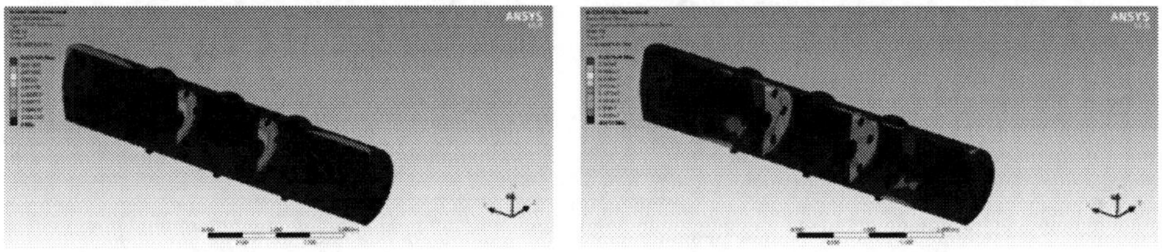

Fig.8. The result of the numerical simulation of the dynamic behavior of the fluid in the "Y" type tank, with type (II) baffles for case b) for a speed v=20m/s, time 10s, fill level 40%;

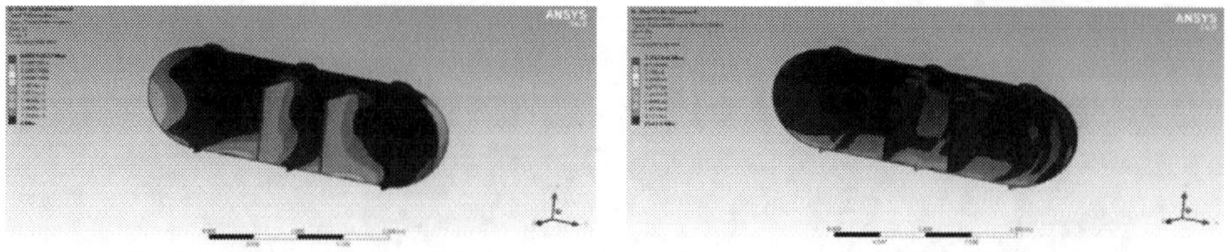

Fig.9. The result of the numerical simulation of the dynamic behavior of the fluid in the "X" type tank, with type (I) baffles for case c) for a speed v=14m/s, time 10s, fill level 80%;

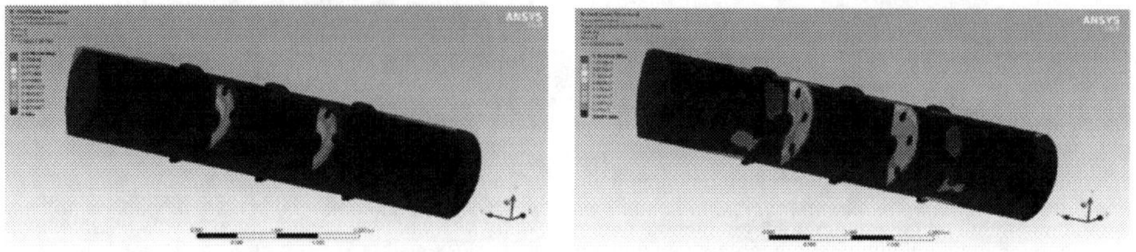

Fig.10. The result of the numerical simulation of the dynamic behavior of the fluid in the "Y" type tank, with type (II) baffles for case c) for a speed v=14m/s, time 10s, fill level 80%;

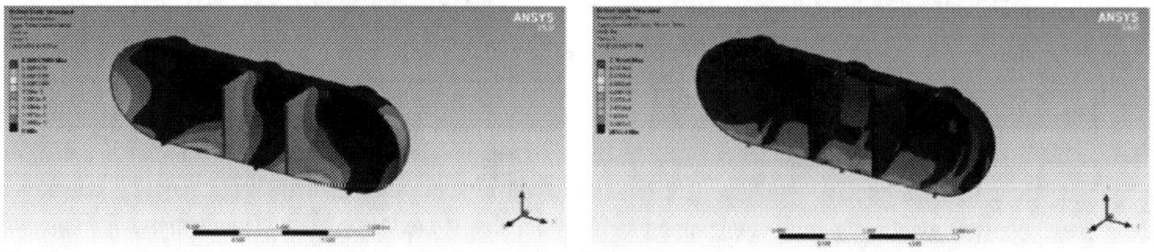

Fig.11. The result of the numerical simulation of the dynamic behavior of the fluid in the "X" type tank, with type (I) baffles for case d) for a speed v=20m/s, time 10s, fill level 80%;

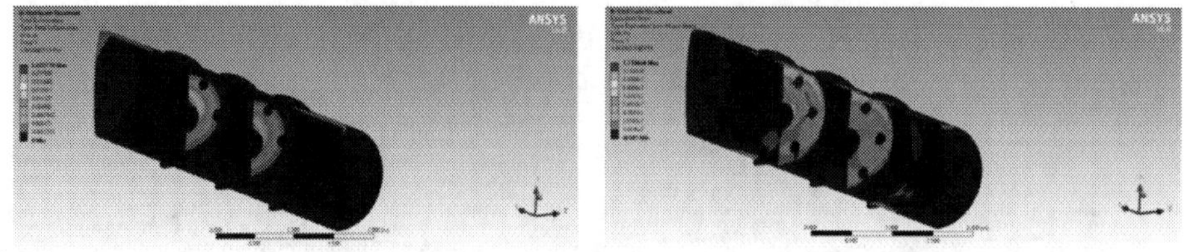

Fig.12. The result of the numerical simulation of the dynamic behavior of the fluid in the "Y" type tank, with type (II) baffles for case d) for a speed v=20m/s, time 10s, fill level 80%;

Thus, the pressure values in the two models of tank vehicle were reduced, even from the simulation, it is observed the rendering of higher pressures due to the resistance of the geometric shape of the baffles, which in time determines the damping of the voltages inside the tank vehicle as flat as seen in the figure 13.

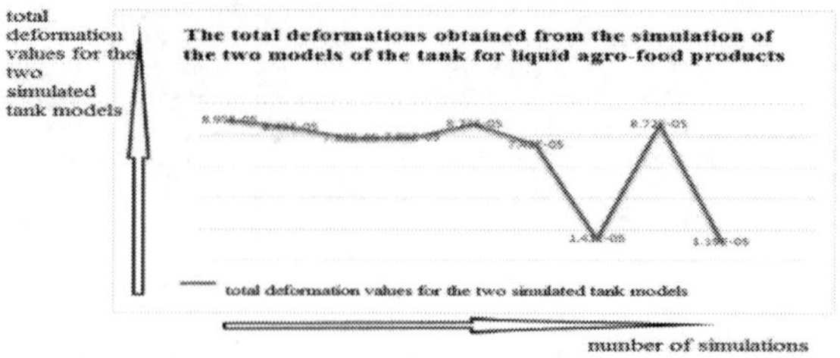

Fig.13. The result of the total deformation values for the two simulated tank models

By studying the two simulations for all four cases it was observed that the stresses on the walls of the tank tanks are higher for cases where the liquid level is lower in a case a) and b) for both tank models.

The simulations also showed that the stresses and strains have much lower values in the case of the "Y" tank model with type II baffles, compared to the other model. This fact shows from the study that it is necessary and very important to mount similar models of baffles, with different constructive geometric shapes.

4 Conclusion

In conclusion this paper presents an analysis of efficaciousness of different designs of baffles for two different design tank models, in order to limit the impact of the response given by the liquid on the tank walls.
By simulating with the two geometrical constructive forms of the baffles inside the two tank models it has been shown that the baffles play an important role and that it reduces the stresses and deformations that appear on the tank wall structure.
Comparing the results obtained by simulating with different filling levels at different speeds when performing a braking manoeuvre it can be observed that the impact of the resulting fluid can create a damaging effect and that only by mounting those baffles can the stress of a disastrous danger be alleviated.

5 References

[1] R.Suyal, *NIT Rourkela*, **n°213ME3440** (2016)
[2] R. Ranganathan,Thesis, Concordia University (1990)
[3] K. Modaressi-Terani, S.Rakheja,I Stiharu, VSD, **45** (2007)
[4] F. Sollas,Odd M. Faltisen, *IJSR*, **41** (1997)
[5] S. Rakheja , K. Modaressi-Terani , I Stiharu, *VSD*,**45** (2007)
[6] X. Zheng , X.Li, Y.Ren, Y. Wang, J. Ma, *Oc. Eng.*,**36**, (2009)
[7] S. Rakheja, A. Kolaei, M. J. Richard, *JSV*, **70**(2017)
[8] J. Zhang ,W. Wu, J.Hu, *J.Mar.*, **46**,(2016)
[9] B.Nicolsen ,L.Wang, A.Shabana, *JSV*, **405** (2017)
[10]U.A. Siddiqui ,B. Fatima , H.U. Siddiqui, M Hammad, *IJSR*, **6** (2019)
[11]R.He , E.Zhang, B.Fan, *JSAGE*, **11**, (2019)
[12] SR EN 10216-2

E3S Web of Conferences **180**, 03025 (2020) https://doi.org/10.1051/e3sconf/202018003025
TE-RE-RD 2020

Kinetostatic analysis of a scissor jack mechanism powered by a tetrad RRR-Ta-RRR

Victor Moise[1], *Ileana* Dugaesescu[1], *Lucretia* Popa[2*],
Liviu Ungureanu[1] and *Ştefan-Iulian* Moise[3]

[1] University POLITEHNICA of Bucharest, Romania
[2]INMA Bucharest, 6 Ion Ionescu de la Brad, Bucharest, Romania
[2] Etudes et Productions Schlumberger, Paris, France

Abstract. In a paper previously published by the authors, kinematic analysis of a jack mechanism driven by a motor tetrad was done. For the actual design of the mechanism, the forces and moments acting on the kinematic elements that are part of it must be determined. Also, the driving force in the hydraulic cylinder must be determined, so that the mechanism works in good conditions, for a certain technological task imposed by the design theme. The present paper makes the kinetostatic analysis of the mechanism studied, from the kinematic point of view. Therefore, the reactions in the kinematic couplings of the mechanism, as well as the driving force of the hydraulic cylinder, are determined. In order to determine the reactions of the kinematic couplings, as well as the driving force of the hydraulic cylinder, the masses of the elements, the moments of mechanical inertia, as well as the load force acting on the jack were considered. The results of the calculations were graphically represented, in reaction patterns. In order to verify the correctness of the calculations, the balance force (the hydraulic cylinder motor force) was determined by both the kinetic and the virtual power method, and the results are the same.

1 Introduction

This mechanism was patented by William P Brown and developed year by year for improving its performances [1]. There are many researches on this mechanism. Manoj R. Patil1 and S.D. Kachave studied, using Pro-E software, the construction of this mechanism intended to reduce weight of scissor jack and, at the same time, maintaining its strength and service life. The study was made using 3D modelling [2]. This mechanism is also used in the construction of the high scissor lift platform and some research were done using 3D Pro/E software [3]. Other researches about this mechanism were done by researchers from different countries, each of them trying to improve it. [4-11].

Agricultural machines have the most complex mechanisms. Therefore, it is good for the designers to know the methods for determining kinematic and kinetostatic parameters, with the help of calculation procedures on modular groups.

The jack that we studied, as shown in Figure 1, is a special jack, which is very helpful when the agricultural machine breaks down and we have nothing to raise it.

* Corresponding author: Lucreţia Popa, lucretia_popa@yahoo.com

The analysis of the mechanism was made after its synthesis, that is, after determining the dimensions of the kinematic elements. For the synthesis of the mechanism hundreds of equations are written (using the synthesis to approximate a given curve or to guide an element). At such a large number of equations the methods of numerical calculation, the choice of the initial solution of the systems of equations etc. must be mastered.

In order to determine the forces and moments acting on the elements of the mechanism, their positions are considered as known, as well as the distribution of linear and angular velocities and accelerations. At the basis of writing the force equations are used the equations of static balance, the principle of virtual work, the principle of d 'Alembert etc. For each module (structural group) in the structure of the mechanism, the system of balance equations is written, considering the forces and moments applied, the forces and moments of inertia and the forces and moments of reaction. The system of force equations is solved by an appropriate numerical method.

The kinetostatic analysis of the mechanism comprises several stages, namely:

a) kinetostatic study of each modular group;

b) the calculation program for determining the reactions in all the kinematic couplings of the mechanism, as well as the driving force of the active coupling F;

c) tracing the reaction hodographs corresponding to A, D, L and M couplings;

d) drawing the force diagram of the kinematic translational coupling F;

e) the table presentation of the reactions in A, B and D couplings for 21 positions of the mechanism elements, as well as of the balance force of the motor torques F, calculated by the kinetostatic method and by the virtual power method.

2 Material and method

The kinetostatic analysis of the mechanism was made based on the modular groups of which it is formed. The study of the mechanism as a whole would have led to a large number of equations and implicitly to greater relative and absolute errors of calculation. Regarding the exact calculation, it is known that the calculation errors depend on the numerical method chosen and on the experience of the one who solves the problem.

In this paper, we applied the scientific researches carried out from 1977 to the present. Thus, depending on the evolution of the calculation technique, software libraries were made for the kinematic, kinetostatic and dynamic calculation of the mechanisms in FORTRAN, BASIC, TURBO-PASCAL, C++, MATHCAD, MATLAB etc. [12-23]

These programs have been continuously improved. The calculation program for kinematic, kinetostatic analysis and the animation of the mechanism was made in MATLAB syntax. The program also contains the kinematic analysis of the mechanism, as this is necessary for the kinetostatic analysis.

3 Results and discussions

In figure 1, the kinematic diagram of a jack driven mechanism by a motor tetrad can be observed, with the load force acting on the element 9 being highlighted.

To make the kinetostatic analysis it is necessary to know the modules composing the jack mechanism in figure 2, where Z(0) is the base, *RRR-Ta-RRR*(1,2,3,4,5) motor tetrad and dyads RRR(6,7) and RRR(8,9). For the kinetostatic analysis of the mechanism, the applied and inertia forces in the elements' centroids are reduced; after that, the corresponding functions of each structural group are applied. It is mentioned that the kinetostatic analysis is made in the opposite direction of the kinematic analysis, namely starting with the last structural group kinematically analysed, dyad *RRR*(8,9) and ending with the motor tetrad *RRR-Ta-RRR*(1,2,3,4,5).

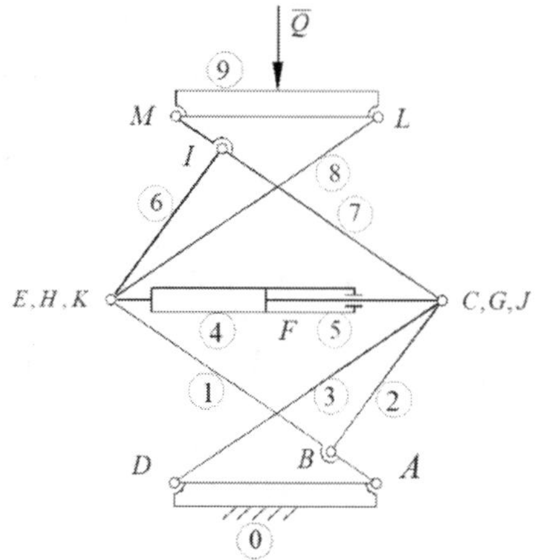

Fig. 1. The kinematic diagram of the jack mechanism

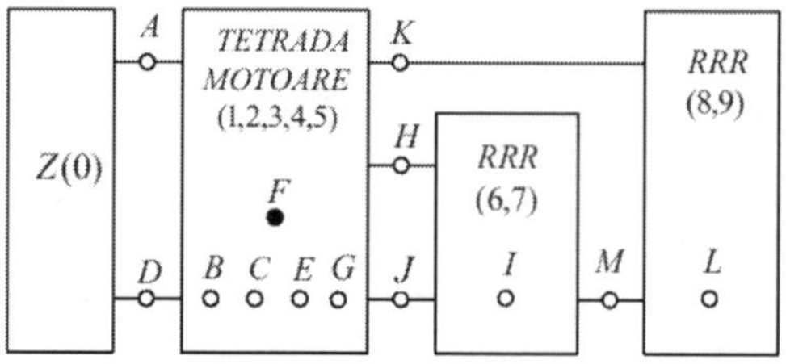

Fig. 2. The multipolar diagram of the jack mechanism

The kinetostatic analysis of the modular groups composing the mechanism

Dyad *RRR*(8,9)

On the dyad *RRR*(8,9) elements act, according to (Fig. 3,*a*)

- the load force Q;
- the weight forces: $\overline{G_8} = -m_8 \overline{g}$, $\overline{G_9} = -m_9 \overline{g}$;
- the inertia forces: $\overline{F}_{i8} = -m_8 \overline{a}_{G8}$, $\overline{F}_{i9} = -m_9 \overline{a}_{G9}$, ;
- the resulting moments of the inertia forces: $\overline{M}_{i8} = -IG_8 \cdot \overline{\varepsilon}_8$, $\overline{M}_{i9} = -IG_9 \cdot \overline{\varepsilon}_9$.

The reduction points of the force systems are considered to be in the centroids G_8 and G_9 of the kinematic elements **8** and **9.**

The accelerations of the reduction points G_8 and G_9 being known through the components on the coordinate axes, the inertia forces are:

$$\overline{F}_{i8} = F_{i8X}\,\overline{i} + F_{i8Y}\,\overline{j},\ \overline{F}_{i9} = F_{i9X}\,\overline{i} + F_{i9Y}\,\overline{j}.$$

The resulting inertia and weight forces are:

$$\overline{F}_{R8} = \overline{F}_{8X} + \overline{F}_{8Y}, \overline{F}_{R9} = \overline{F}_{9X} + \overline{F}_{9Y},$$

where:

$$\overline{F}_{8X} = -m_8 \cdot \overline{a}_{G_8X},\ \overline{F}_{8Y} = -m_8 \cdot (\overline{a}_{G_8Y} + \overline{g});$$

$$\overline{F}_{9X} = -m_9 \cdot \overline{a}_{G_9X},\ \overline{F}_{9Y} = -m_9 \cdot (\overline{a}_{G_9Y} + \overline{g}) + \overline{Q}.$$

In relation to the reduction points, the resulting moments are:

$$\overline{CM}_8 = -IG_8 \cdot \overline{\varepsilon}_8, \overline{CM}_7 = -IG_7 \cdot \overline{\varepsilon}_7.$$

The reactions in the kinematic couplings *K*, *L* and *M* are: $R_{18X}, R_{18Y}, R_{98X}, R_{98Y}$, R_{79X} and R_{79Y}. These reactions are the output data of a calculation procedure, drafted by the authors.

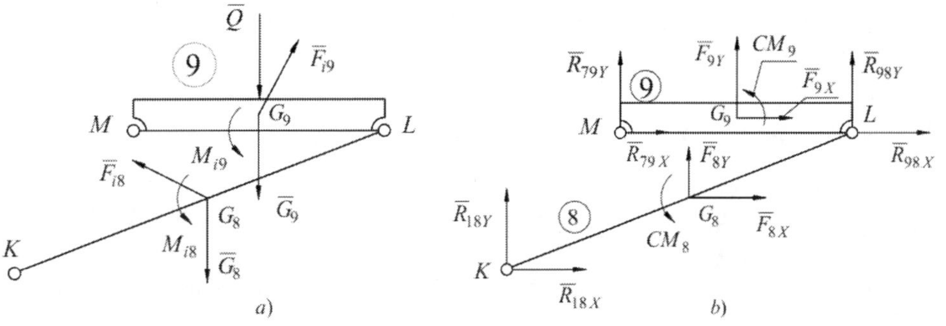

Fig.3. Dyad *RRR*(8,9)

a) highlighting the forces and moments acting on the elements of the structural group;
b) the kinetostatic diagram of dyad RRR(8,9)

Dyad *RRR*(6,7)

On the elements of dyad *RRR*(6,7), as shown in (Fig. 4,*a*) act:

- the reactions of the element 9 on the element 7 of the dyad *RRR*(6,7) namely: $\overline{R}_{97X} = -\overline{R}_{79X}$, $\overline{R}_{97Y} = -\overline{R}_{794Y}$;

- the weight forces: $\overline{G}_6 = -m_6\overline{g}$, $\overline{G}_7 = -m_7\overline{g}$;

- the inertia forces: $\overline{F}_{i6} = -m_6\overline{a}_{G_6}$, $\overline{F}_{i7} = -m_7\overline{a}_{G_7}$;

- The resulting moments of the inertia forces: $\overline{M}6 = -IG_6 \cdot \overline{\varepsilon}_6$, $\overline{M}_{i7} = -IG_7 \cdot \overline{\varepsilon}_7$.

The reduction points of the force systems are considered in the centroids G_6 and G_7 of the elements **6** and **7.**

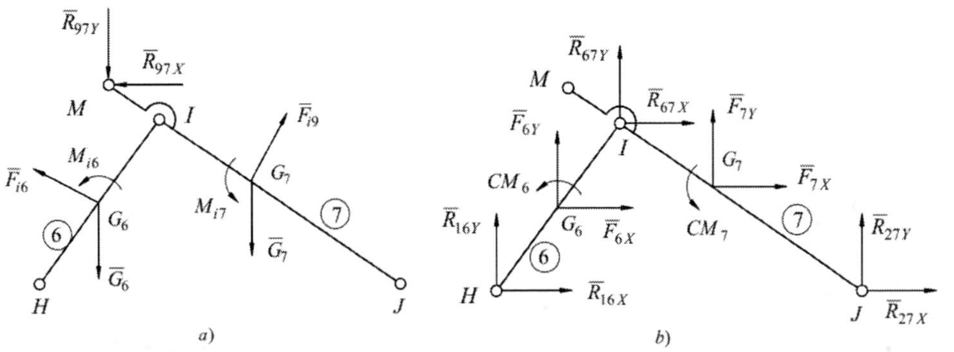

Fig.4. Dyad *RRR*(6,7)

a) highlighting the forces and moments acting on the elements of the structural group;
b) the kinetostatic diagram of dyad *RRR*(6,7)

The accelerations of the reduction points G_6 and G_7 being known through the components on the coordinate axis, the inertia forces are:

$$\overline{F}_{i6} = F_{i6X}\,\overline{i} + F_{i6Y}\,\overline{j},\ \overline{F}_{i7} = F_{i7X}\,\overline{i} + F_{i7Y}\,\overline{j}.$$

The resulting inertia and weight applied forces, are:

$$\overline{F}_{R6} = \overline{F}_{6X} + \overline{F}_{6Y},\ \overline{F}_{R7} = \overline{F}_{7X} + \overline{F}_{7Y},$$

where:

$$\overline{F}_{6X} = -m_6 \cdot \overline{a}_{G_6X},\ \overline{F}_{6Y} = -m_6 \cdot (\overline{a}_{G_6Y} + \overline{g});$$

$$\overline{F}_{7X} = -m_7 \cdot \overline{a}_{G_7X} + \overline{R}_{97X},\ \overline{F}_{7Y} = -m_7 \cdot (\overline{a}_{G_7Y} + \overline{g}) + \overline{R}_{97Y}.$$

In relation to the reduction points, the resulting moments are:

$$\overline{CM}_6 = -IG_6 \cdot \overline{\varepsilon}_6,\ \overline{CM}_7 = -IG_7 \cdot \overline{\varepsilon}_7 + \overline{G_7M} \times (\overline{R}_{97X} + \overline{R}_{97Y}).$$

The reactions in the kinematic couplings *H*, *I* and *J* are: R_{16X}, R_{16Y}, R_{67X}, R_{67Y}, R_{27X} and R_{27Y}. These reactions are the output data of a calculation procedure.

The motor tetrad *RRR-Ta-RRR*(1,2,3,4,5)

On the elements of the motor tetrad *RRR-Ta-RRR*(1,2,3,4,5) act (Fig.5):

- The reactions of the elements **6** and **8** on the element **1** of the motor tetrad *RRR-Ta-RRR*(1,2,3,4,5), are: $\overline{R}_{61X} = -\overline{R}_{16X}$, $\overline{R}_{61Y} = -\overline{R}_{16Y}$, $\overline{R}_{81X} = -\overline{R}_{18X}$, $\overline{R}_{81Y} = -\overline{R}_{18Y}$.

- the reactions of the element **7** on the element **2** of the motor tetrad *RRR-Ta-RRR*(1,2,3,4,5), are: $\overline{R}_{72X} = -\overline{R}_{27X}$, $\overline{R}_{72Y} = -\overline{R}_{27Y}$

- the weight forces: $\overline{G}_1 = -m_1\overline{g}$, $\overline{G}_2 = -m_2\overline{g}$, $\overline{G}_3 = -m_3\overline{g}$, $\overline{G}_4 = -m_4\overline{g}$, $\overline{G}_5 = -m_5\overline{g}$;

- the inertia forces: $\overline{F}_{i1} = -m_1\overline{a}_{G1}$, $\overline{F}_{i2} = -m_2\overline{a}_{G2}$, $\overline{F}_{i3} = -m_3\overline{a}_{G3}$, $\overline{F}_{i4} = -m_4\overline{a}_{G4}$, $\overline{F}_{i5} = -m_5\overline{a}_{G5}$;

- the resulting moments of the inertia forces: $\overline{M}_{i1} = -IG_1 \cdot \overline{\varepsilon}_1$, $\overline{M}_{i2} = -IG_2 \cdot \overline{\varepsilon}_2$, $\overline{M}_{i3} = -IG_3 \cdot \overline{\varepsilon}_3$, $\overline{M}_{i4} = -IG_4 \cdot \overline{\varepsilon}_4$, $\overline{M}_{i5} = -IG_5 \cdot \overline{\varepsilon}_5$ $(\overline{\varepsilon}_4 = \overline{\varepsilon}_5)$.

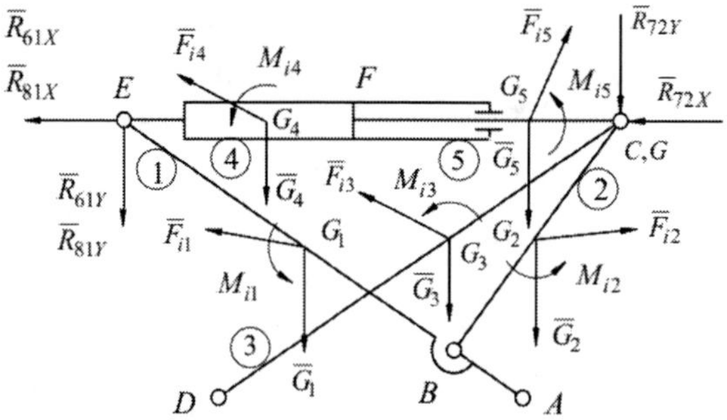

Fig. 5. The motor tetrad *RRR-Ta-RRR*(1,2,3,4,5). Highlighting the forces and moments acting on the elements of the structural group

The reduction points of the forces systems are considered in the centroids G_1, G_2, G_3 G_4 and G_5, corresponding to the elements **1, 2, 3, 4** and **5.** The kinematic parameters of the points G_1, G_2, G_3 G_4 and G_5 are calculated using a calculation procedure drafted by the authors. The resulting inertia, weight and reaction forces are as follows (Fig.6).

$\overline{F}_{R1} = \overline{F}_{1X} + \overline{F}_{1Y}$, $\overline{F}_{R2} = \overline{F}_{2X} + \overline{F}_{2Y}$, $\overline{F}_{R3} = \overline{F}_{3X} + \overline{F}_{34Y}$,

$\overline{F}_{R4} = \overline{F}_{4X} + \overline{F}_{4Y}$, $\overline{F}_{R5} = \overline{F}_{5X} + \overline{F}_{5Y}$,

where:

$$\overline{F}_{1X} = -m_1 \cdot \overline{a}_{G1X} + \overline{R}_{61X} + \overline{R}_{81X};\ \overline{F}_{1Y} = -m_1 \cdot (\overline{a}_{G1Y} + \overline{g}) + \overline{R}_{61Y} + \overline{R}_{81Y};$$

$$\overline{F}_{2X} = -m_2 \cdot \overline{a}_{G_2X} + \overline{R}_{72X};\ \overline{F}_{2Y} = -m_2 \cdot (\overline{a}_{G_2Y} + \overline{g}) + \overline{R}_{72Y};$$

$$\overline{F}_{3X} = -m_6 \cdot \overline{a}_{G3X};\ \overline{F}_{3Y} = -m_3 \cdot (\overline{a}_{G3Y} + \overline{g});$$

$$\overline{F}_{4X} = -m_4 \cdot \overline{a}_{G4X};\ \overline{F}_{4Y} = -m_4 \cdot (\overline{a}_{G4Y} + \overline{g});$$

$$\overline{F}_{5X} = -m_5 \cdot \overline{a}_{G5X};\ \overline{F}_{5Y} = -m_5 \cdot (\overline{a}_{G5Y} + \overline{g}).$$

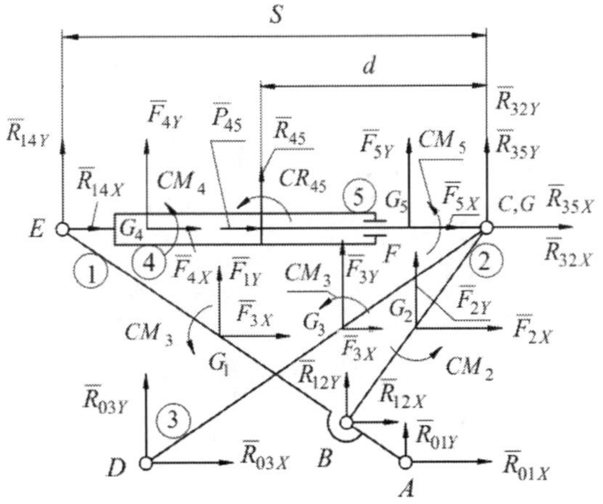

Fig.6. The motor tetrad *RRR-Ta-RRR*(1,2,3,4,5).
The kinetostatic diagram of the motor tetrad *RRR-Ta-RRR*(1,2,3,4,5)

In relation to the reduction points, the resulting moments are:

$$\overline{CM}_1 = -IG_1 \cdot \overline{\varepsilon}_1 + \overline{G_1E} \times (\overline{R}_{61X} + \overline{R}_{61Y}) + \overline{G_1E} \times (\overline{R}_{61X} + \overline{R}_{81Y});$$

$$\overline{CM}_2 = -IG_2 \cdot \overline{\varepsilon}_2 + \overline{G_2C} \times (\overline{R}_{72X} + \overline{R}_{72Y});$$

$$\overline{CM}_3 = -IG_3 \cdot \overline{\varepsilon}_3;\ \overline{CM}_4 = -IG_4 \cdot \overline{\varepsilon}_4;\ \overline{CM}_5 = -IG_5 \cdot \overline{\varepsilon}_5\ (\overline{\varepsilon}_4 = \overline{\varepsilon}_5.$$

The reactions in the kinematic couplings *A*, *B*, *C*, *D*, *E*, *F* and *G* are: R_{01X}, R_{01Y}, R_{12X}, R_{12Y}, R_{32X}, R_{32Y}, R_{03X}, R_{03Y}, R_{14X}, R_{14Y}, R_{45}, CR_{45}, R_{35X} and R_{35Y}. The driving force in the hydraulic cylinder is P_{45}. These sizes are the output data of a calculation procedure drafted by the authors.

To determine the kinematic and kinetostatic parameters of the mechanism elements, the following data were considered:

- the elements dimensions: *AB* = 0.061 m, *AE* = 0.280 m, *BC* = 0.1598 m; *CD* = 0.280 m, *BE* = *AE-AB*, *JM* = 0.280 m, *JI* = *BE*, *HI* = *BC*, *KL* = *AE*, *ML* = 0.180 m;
- the positions of the couplings adjacent to the base: *XA* = 0.0 m, *YA* = 0.0 m, *XD* = -0.180 m, *YD* = 0.0 m;
- the initial length of the hydraulic cylinder: *S0* = 0.270 m;
- the hydraulic cylinder piston stroke: = 0.098 m;
- the relative speed of the piston to the cylinder: *dS/dt* = -0.025 m· s^{-1};
- the relative acceleration of the piston to the cylinder is considered equal to zero
- the mass of the kinematic elements: m_1=1.5 kg, m_2=1.0 kg, m_3=1.5 kg, m_4= 1.5 kg, m_5=1.0 kg, m_6=1.0 kg, m_7=1.5 kg, m_8=1.5 kg, m_9=1.0 kg;
- the mechanical inertia moments: *IG*1=1.5 kgm², *IG*2=1.0 kgm², *IG*3=1.5 kgm², *IG*4= 1.5 kgm², *IG*5=1.0 kgm², *IG*6=1.0 kgm², *IG*7=1.5 kgm², *IG*8=1.5 kgm², *IG*9=1.0 kgm²;
- the load force: Q=2000 N.

Using the calculation program made by MATLAB syntax, the values of reactions in the kinematic couplings of the mechanism are obtained, as well as the driving force in the hydraulic cylinder.

In figures 7, 8, 9 and 10, the hodographs of reactions in the kinematic couplings *A*, *D*, *L* and *M* are shown.

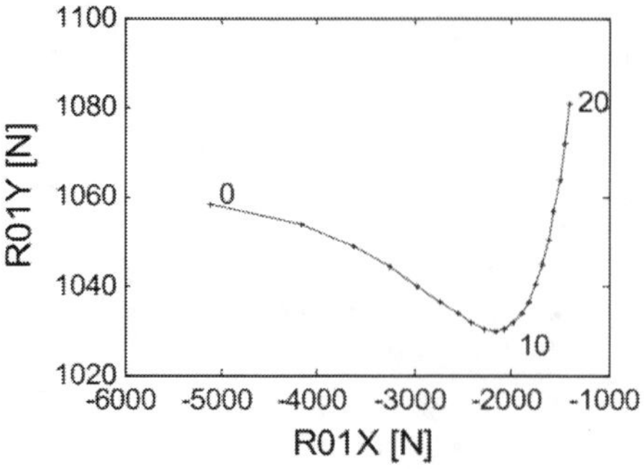

Fig. 7. The hodograph of reactions in coupling A

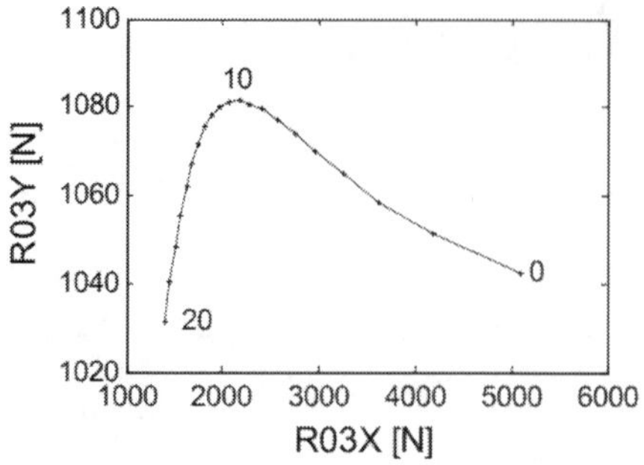

Fig. 8. The hodograph of reactions in coupling D

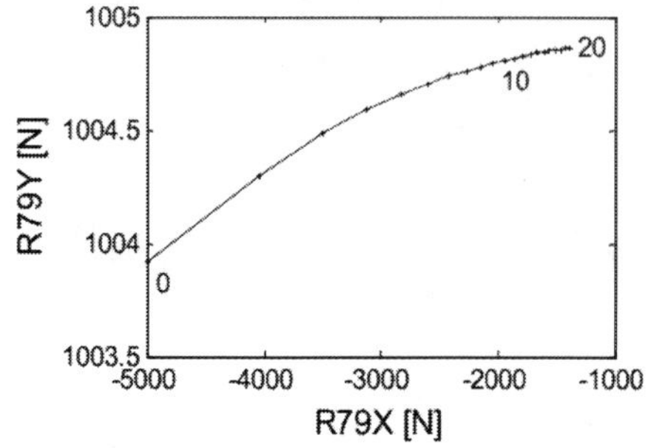

Fig. 9. The hodograph of reactions in coupling L

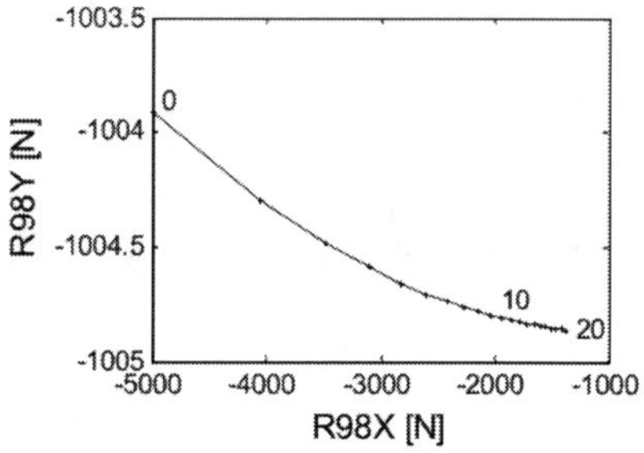

Fig. 10. The hodograph of reactions in coupling M

In Fig. 11, the diagram of the driving force in the hydraulic cylinder is presented.

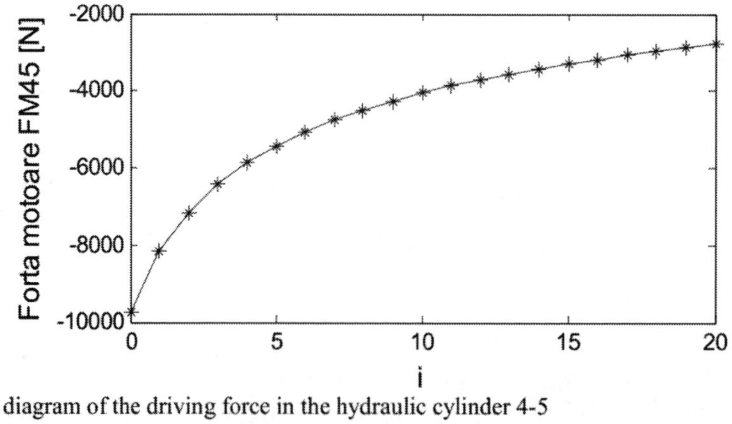

Fig. 11. The diagram of the driving force in the hydraulic cylinder 4-5

In the calculation procedure, the driving force was considered to be positive when the hydraulic cylinder is relaxing, that is when the piston rod is getting out of the cylinder.

In table 1, the components' values of the reactions in *A*, *D* and *B* couplings are presented.

Table 1 Reactions in *A*, *D* and *B* couplings

Pos.	R01X	R01Y	R03X	R03Y	R12X	R12Y
0	-5108.92	1058.11	5109.33	1042.36	-176.47	-44.19
1	-4179.34	1053.83	4179.58	1051.36	-30.55	-5.98
2	-3624.83	1048.83	3625.04	1058.70	9.01	8.53
3	-3244.66	1044.08	3244.86	1064.79	18.27	13.48
4	-2962.06	1039.90	2962.25	1069.82	17.18	13.92
5	-2740.42	1036.39	2740.60	1073.91	12.52	12.13
6	-2559.77	1033.62	2559.96	1077.09	6.99	9.28
7	-2408.21	1031.61	2408.40	1079.41	1.72	6.04
8	-2278.10	1030.39	2278.29	1080.88	-2.80	2.82
9	-2164.34	1029.96	2164.53	1081.49	-6.37	-0.10
10	-2063.32	1030.35	2063.51	1081.25	-8.94	-2.54
11	-1972.47	1031.56	1972.66	1080.16	-10.51	-4.36
12	-1889.84	1033.60	1890.04	1078.23	-11.12	-5.44
13	-1813.99	1036.48	1814.18	1075.43	-10.84	-5.72
14	-1743.76	1040.21	1743.95	1071.78	-9.72	-5.12
15	-1678.27	1044.79	1678.46	1067.26	-7.83	-3.58
16	-1616.79	1050.24	1616.99	1061.87	-5.24	-1.05
17	-1558.75	1056.55	1558.95	1055.61	-2.00	2.52
18	-1503.66	1063.74	1503.86	1048.46	1.83	7.16
19	-1451.13	1071.82	1451.33	1040.42	6.20	12.93
20	-1400.83	1080.79	1401.03	1031.48	11.04	19.87

In table 2, the values of the driving force in the hydraulic cylinder are presented, values calculated through the kinetostatic method and the virtual power method.

The small differences are due to the calculation errors.

Table 2 Driving force in the hydraulic cylinder

Pos.	P45_CINET	FE_PV45	Pos.	P45_CINET	FE_PV45
0	-9723.68	-9723.05			
1	-8175.58	-8175.30	**11**	-3861.88	-3861.86
2	-7159.70	-7159.55	**12**	-3697.68	-3697.67
3	-6428.33	-6428.24	**13**	-3548.70	-3548.69
4	-5868.95	-5868.88	**14**	-3412.60	-3412.59
5	-5422.68	-5422.63	**15**	-3287.52	-3287.51
6	-5055.45	-5055.42	**16**	-3171.93	-3171.92
7	-4746.01	-4745.98	**17**	-3064.59	-3064.59
8	-4480.32	-4480.30	**18**	-2964.48	-2964.48
9	-4248.72	-4248.70	**19**	-2870.73	-2870.73
10	-4044.27	-4044.25	**20**	-2782.62	-2782.61

4. Conclusions

Obtaining the values of reactions in the kinematic couplings of the mechanism allows passing to the next step, namely the actual design of the kinematic elements. To check the calculations, the driving force in the hydraulic cylinder 4-5 was determined also through the virtual power method and the results were identical to those in the kinetostatic calculations. Depending on the driving force of the hydraulic cylinder, as well as on the general dimensions, the hydraulic cylinder will be chosen from a provider of hydraulic elements.

This paper was financed by grant of the Romanian Education and Research Ministry, through Programme 1 – Development of the national research-development system, sub-programme 1.2 – Institutional performance – Projects for financing excellence in RDI, contract no. 16PFE.

References

1. W.P. Brown, *Scissors Jack*. US Patent Office (1946).
2. Manoj R., Patil, S.D. Kachave, Int. J. Mech. Eng. & Rob. Res., **4(1),** (2015).
3. Tian Hongyu, Zhang Ziyi, Design and Simulation Based on Pro/E for a Hydraulic Lift Platform in Scissors Type, Procedia Engineering, **16**, 772-781, (2011).
4. Islam Md Toufiqul, Yin Cheng, Jian Shengqi, Rolland Luc, Dynamic analysis of Scissor Lift mechanism through bond graph modelling, IEEE/ASME International Conference on Advanced Intelligent Mechatronics, AIM, (2016).
5. Yuantao Sun, Sanmin Wang, James K. Mills & Changjian Zhi, Kinematics and dynamics of deployable structures with scissor-like-elements based on screw theory, Chinese Journal of Mechanical Engineering, **27**, 655–662, (2014).
6. J. Hartsell, The development of a dynamic scissor lift model, West Virginia University. The Research Repository. (2010).
7. Langbecker, T. (1999). Kinematic Analysis of Deployable Scissor Structure. *International Journal of Space Structures*, **14(1),** 1–15.
8. M. Bariskan, Final Project_ Design and FEM Analysis of Scissor Jack https://www.slideshare.net/MehmetBariskan/final-project-design-and-fem-analysis-of-scissor-jack, (2014).
9. https://www.engineersedge.com/mechanics_machines/scissor-lift.htm
10. C.S. Dhamak, D.S. Bajaj, V.S. Aher, Design and Optimization of Scissor Jack, *IJAPME*, **2(1),** 29-34, (2016).
    I. Dugaesescu, L. Popa, Şt. Moise, V. Moise, V. Ştefan, Al. Rotaru. *ERDev*, (2019).
11. I.I. Artobolevski, *Theorie des mecanismes et des machines. (Theory of Mechanisms and Machines)*, (1977).
12. B. Demidovich, A.I. Maron, *Éléments de calcul numérique*, (1987).
13. N. Manolescu, Fr. Kovacs, A. Oranescu, *Teoria mecanismelor şi a maşinilor (The Theory of Mechanisms and Machines)*, (1972).
14. V. Moise, *Cinematica şi controlul mecanismelor manipulatoare cu 4 grade de mobilitate,* PhD Thesis, University Politehnica of Bucharest, (1986).
15. V. Moise, I. Simionescu, M. Ene, M. Neacşa, I.A. Tabara, *Analiza mecanismelor aplicate (Analysis of Applied Mechanisms)*, (2007).
16. V. Moise, I. Simionescu, M. Ene, Al. Rotaru, *Analysis of planar mechanisms with articulated bars. Applications in MATLAB (Analiza mecanismelor plane cu bare articulate. Aplicaţii în MATLAB).* (2015).
17. V. Moise, E. Maican, Şt Moise, *Numerical methods. Applications in MATLAB (Metode numerice. Aplicaţii în MATLAB)*. (2016).
18. Chr. Pelecudi, *Precizia mecanismelor (The precision of the mechanisms)*. (1975).
19. Chr. Pelecudi, D. Maros, V. Merticaru, N. Pandrea, I. Simionescu, *Mecanisme (Mechanisms)*, (1985).
    I. Simionescu, V. Moise, *Mecanisme (Mechanisms*). (1999).
20. V. Valcovici, St. Balan, R. Voinea, *Mecanică teoretică (Theoretical mechanics),* (1963).
21. R. Voinea, D. Voiculescu, V. Ceauşu, *Mecanică (Mechanics),* (1985).

Analysis of the process of green hemp stalks sequential harvesting

Anişoara Păun [1]*, *Gheorghe* Stroescu[1], *Alexandru* Zaica[1], *Radu* Ciupercă[1] and *Constantin* Bogdanof[1]

[1] INMA Bucharest, Romania

Abstract. HEMP (*Cannabis Sativa* - Industrial Hemp) is the industrial plant with the highest use of all these plants in the industry, since nothing is thrown away from this plant. Hemp is the plant of which everything is used, and the products obtained are of a great diversity, from the common rope to the medicinal or cosmetic substances, textiles, automotive or construction materials. It is an unpretentious plant because, apart from the fact that it wants a neutral PH and soils without excess humidity, no maintenance work need to be done and no extra care should be given. The fibre hemp is harvested at the technical maturity when the male plants shed the last pollen grains, and the stalks have a greenish-yellow colour and the leaves fell on the ground. Production management is different for an exclusive fibre production. If we look for raw materials that meet the demands of different areas of the economy with the lowest costs, we must take into account the reactivation of the hemp culture at the level of Romania, because, as we mentioned above, hemp is the plant of which everything is capitalised, and the products obtained are of a great diversity. By applying the technology for processing the stalks in green, the obtained fibre can be used with special results in the strawberry culture at the mulching operation. The paper presents the Equipment for harvesting green hemp stalks - ERCV, equipment that performs sequential harvesting of hemp stalks leaving them on the ground. At the same time, factors that influence the process of harvesting hemp stalks are presented: harvesters' cutting devices, kinematic of the cutting blades, cutting speed, cutting angles of the cutting blades, the forces acting on the blade and the power required to drive the cutting devices.

1 Introduction

Considering that raw materials that meet the needs of the different areas of the economy with the lowest costs are sought, the reactivation of the hemp culture in Romania must be taken into consideration as well, because from this plant everything is used and the products obtained are of a great diversity. Hemp is an unpretentious plant because, apart from the fact that it wants a neutral pH and soils without excess moisture, no maintenance work should be done and no extra care should be given. [1-3].

*Corresponding author: ani_paun@yahoo.com

The costs per hectare are below the costs of other well-known crops such as rape or wheat. At the same time after the cultivation of a land with this plant, the following year the soil becomes favourable for the cultivation of cereals because hemp naturally contributes to the removal of weeds, creating an inappropriate environment for their development [4].

Hemp seeds are used to make beer, oil used in the paint industry, cosmetic oil, edible oil or margarine. Medicinal teas and medicines are made from hemp leaves.

Hemp seed oil is extremely valuable for Omega 3, Omega 6, Omega 9, Vitamin E, being used by consumers to help the cardiovascular system. A lot of food can be provided from hemp. Hemp seeds are an important source of nutrients in large quantities. It is known that hemp seeds, in the form of flour, oil, cold extract, consumed raw or fried, are a healthy food and a non-medicinal remedy effective in many medical conditions. Hemp fibre has excellent moisture absorption properties, bacteriostatic, heat-resistant and antistatic properties; the hemp stalk is also a raw material for obtaining paper recently widely used for composite plates and other composite materials. At the same time, by applying the technology for processing the stalks in green, the obtained fibre can be used with special results in the strawberry culture at the mulching operation [5-6].

At this time the demand for hemp on the European market is increasing, and the value of the subsidies offered by the European Community helps the farmers to make a profit. In Romania only seeds are used, the textile industry not being interested in the hemp fibre.

Before 1989, in Romania, over 50,000 hectares were cultivated with hemp, but in 2000 the hemp disappeared from the Romanian agriculture, only after 2012 the cultivation of hemp began to get the attention of Romanian farmers. Romania offers favourable conditions for hemp cultivation over most of its territory. The success of hemp for both fibre and seed is given by its location in specialized farms well organized and technically equipped, in areas with pedoclimatic conditions. Following discussions with farmers and researchers from the research centres, they want to develop machines that allow them to harvest and process the stalks in green because there have been demands on the hemp market for the fibre, but especially for the green inflorescence part. Worldwide, there are companies that make such machines but at very high prices for Romanian beneficiaries. In this sense, within our unit was developed technical harvesting equipment that takes into account the technological requirements specific to the Romanian hemp varieties, in the case of harvesting hemp for fibre: the average cutting height from the ground does not exceed 10 cm; stalk losses are less than 5%; broken stalks are up to 6% [7-8].

The main objective of this paper is an analysis of the process of green hemp stalks sequential harvesting.

2 Methodology

The hemp for fibre is harvested at technical maturity, when the male plants shake their last pollen traces, and the stalks have a greenish-yellow colour and the leaves fallen to the ground. Production management is different for an exclusive fibre production. The selection of varieties is oriented towards the production of biomass. Harvesting hemp for fibre is done at the end of flowering, when the stalks have become golden-yellow. The main features of hemp stalks are: length (total hemp length), thickness, colour, attack of diseases and pests, mechanical injuries (hail, broken or crushed stalks), and breaking load.

The length and thickness of the stalks are influenced by many factors such as: variety, sex, day length, soil and air humidity, temperature, physical and chemical composition of the soil, applied agrotechnics, fertilizers, culture system, plant density in the field etc. In the case of hemp, the stalks with the biggest useful length, not branched and thinner, are considered to be the most valuable. They have the highest fibre content, give a high fibre yield, and the obtained fibre is more resistant.

The experimental model of Equipment for harvesting green hemp stalks, Figure 1, is intended for sequential harvesting of hemp stalks, in order to process them to obtain the fibre.

Fig. 1. Equipment for harvesting green hemp stalks ERCV: 3D

When using this equipment (description in fig 4) the cut plants remain on the ground, with the inflorescence mixed with the stalks. It should be noted that the equipment is intended for small and medium-sized farms that grow hemp.

The machine is equipped with two cutting devices, Figure 2, fixed on a mobile platform. A device is mounted on the back of the platform and cuts the stalks at a height of 100 mm from the ground in the working position; it is a double knife type with double cutting device, in which the role of the fingers and counter-cutting plates is fulfilled by the second knife moving at the same speed and in the opposite direction as the first knife.

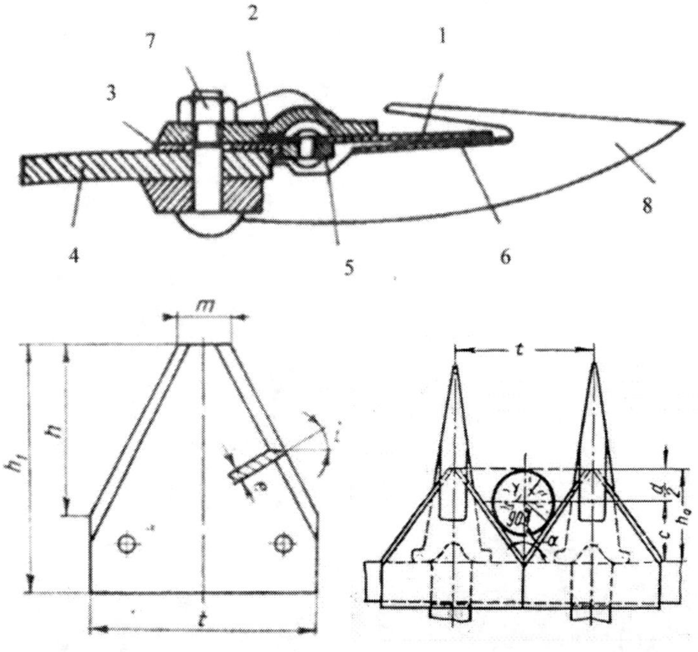

Fig. 2. Parts of the cutting device: 1 – Knife; 2 – Pressure plate; 3 – Wear plate; 4 – Cutting device support; 5 – Plate holder bar; 6 – Counter knife; 7 – Attachment screw; 8 – Peak

The simple stroke of the cutting device shown in figure 2 is determined by the relation:

$$S = t = t_0 = 76.2 \text{ mm} \quad (1)$$

Where:

S - is the knife stroke at the half turn of the crank;

t - the distance between the axes of two blades (step of the blades);

t_0 - the distance between the axes of two fingers (step of the fingers).

For actuating the knives of the cutting devices, axial crank-connecting rod mechanisms are used, at which r/l has small values, the movement of a point on the knife can be considered as a harmonic oscillation. The kinematics of the knives used on the equipment for harvesting green hemp stalks ERCV is shown in Figure 3 a and b.

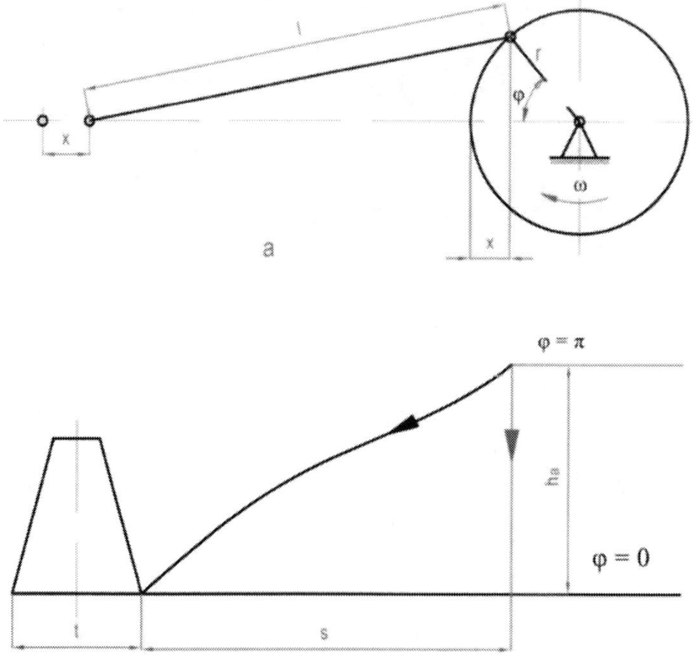

Fig. 3. The kinematics of the knife

In this case the space x will be:

$$x = r - r\cos\varphi = r(1 - \cos\cdot\omega t) \quad (2)$$

where:

ω - the angular speed of the crank, in rad/s^{-1};

t – the time the crank rotates with the angle φ ; $\varphi = \omega t$.

When the crank rotates with $\varphi = \pi$, the knife covers the space

$$x = S = r(1 - \cos\cdot\pi) = 2r \quad (3)$$

where:

S is the knife stroke

The speed of the knife is

$$v = \frac{dx}{dt} = \omega \cdot r \cdot \sin \omega t \tag{4}$$

and the acceleration of the knife is:

$$a = \frac{dv}{dt} = r \cdot \omega^2 \cos \omega t \tag{5}$$

From the expression (4) it can be noted that $v = 0$ when $\varphi = 0$ or π. When $\varphi = \pi / 2$, the speed of the knife is maximum and equal to $r \cdot \omega$.

The knife's acceleration has maximum values ($a_{max} = \pm r \cdot \omega^2$), when $\varphi = 0$ or $\varphi = \pi$

For $\varphi = \pi / 2$, $a = 0$.

During work, a point on the knife has two movements: of translation with the speed v_m, in the direction of machine movement and a harmonic oscillatory movement perpendicular to the direction of machine's advance. The trajectory described by a point on the knife is a sinusoid (3b). At a knife stroke ($\varphi = \pi$) the car covers a feed space with the speed v_m:

$$h_a = v_m t = v_m \frac{\pi}{\omega} \tag{6}$$

In general, the values of the feed space ha are 50 - 100 mm.

Cutting the plants in good conditions depends on the cutting speed v, feeding l, the fingers' step t_0, the type of cutting edges (smooth or toothed), the sharpness of the knife edge and the distance between the cutting blade and the counter-cutting plate.

The speed of a certain point of the cutting blade, which is at a distance x from the initial position, is determined by the relation:

$$v_x = S_\omega \sqrt{\frac{x}{S}\left(1 - \frac{x}{S}\right)}, \text{ [m/s]} \tag{7}$$

and the complete stroke of the knife, taking into account the offsetting i, is determined by the relation

$$S = \sqrt{(u+r)^2 - i^2} - \sqrt{(u-r)^2 - i^2} \tag{8}$$

Where : $\omega = \frac{\pi \cdot n}{30}$, is the angular speed of the crank;

u – grain length

r – radius of the crank

The relative average speed is calculated with the relation $v_c = \frac{Sn}{30}$, and the maximum one with the relation $v_{max} = \frac{S\omega}{30}$.

The absolute average speed of the knife

$$v_a = \sqrt{v_m^2 - v_c^2} \tag{9}$$

The acceleration of any point of the cutting blade that is at a distance x from the initial position is determined by the relation

$$a_x = \omega^2 \left(\frac{S}{2} - x \right), \text{ m/s}^2 \tag{10}$$

For $x = 0$ and $x = S$ the maximum accelerations are obtained $a_{max} = \frac{S\omega^2}{2}$.

The maximum inertia force of the M mass knife is

$$F_{i\max} = \frac{MS\omega^2}{2} \tag{11}$$

3 Results

When making the equipment, a number of theoretical considerations were addressed on the factors that influence the hemp harvesting process: the cutting devices of the harvesting machines; the kinematics of the cutting knife, the cutting speed; the critical angle of plant catching, the cutting angles of the cutting blades, the forces acting on the knife, in a study on which the machine design was based.

Based on the theoretical elements, the Equipment for harvesting green hemp stalks, ERCV Figure 4, was developed, which is composed of the following subassemblies:

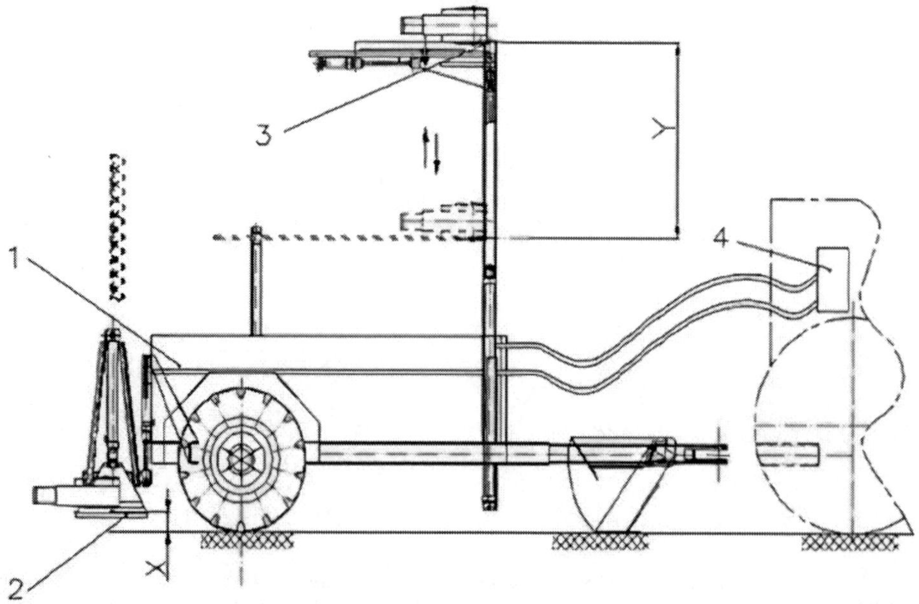

Fig. 4. Equipment for harvesting green hemp stalks ERCV: 1 – Platform; 2 – Lower cutting machine; 3 – Higher cutting machine; 4 – Hydraulic installation

Before entering the hemp crop, the tractor is started and the two cutting devices are actuated by means of the hydraulic system. They are brought from the transport position to the working position. The first cutting machine (rear platform) is set for cutting the stalks at a height of 100 mm from the ground.

Depending on the height of the hemp plant, respectively the inflorescence of the plant, the second cutting device (front platform), is adjusted vertically by means of a vertical cylinder. When the equipment operates, the cut plants remain on the ground, with the inflorescence mixed with the stalks.

In the transport position, the cutting machine is folded down and raised vertically by actuating with the help of two hydraulic cylinders working at the ends of the stroke. The second device is mounted horizontally offset to the first, in the front of the platform and performs the cutting of hemp plants' inflorescences. Because the height of the hemp varieties' inflorescences varies, this device has the possibility of adjusting the cutting height in a wide range of values with the help of a vertical cylinder. The second hydraulic cylinder works at the ends of the stroke in the working or transport position of the equipment. During transport this unit can be folded 90 degrees backwards on a support, with the help of the hydraulic system. Both knives are double-cut devices.

Each cutting device is actuated by a hydraulic motor through a distributor, operated from the hydraulic system of the equipment.

General technical characteristics of ERCV-0

- Machine type	trailed
- Energy source	65 HP tractor
- Type of cutting machine	Double-knife
- No. of cutting devices	2
- Cutting height:	
- rear cutting machine, mm	100
- front cutting machine, mm	1500-2500

4 Conclusions

Knowing the characteristics of the hemp stalks that were subject of the study allowed establishing the constructive solutions that would respond to the demands of hemp growers;

Theoretical considerations on the factors that influence the process of hemp harvesting: the cutting devices of the harvesting machines; the kinematics of the cutting knife, the cutting speed; the critical angle of plant catching, the cutting angles of the cutting blades, the forces acting on the knife, approached allowed establishing the technical and functional characteristics of the equipment for harvesting green hemp stalks.

Through the constructive solutions adopted, the aim was to create a piece of equipment that would provide the following technological requirements in the case of hemp harvesting for fibre: the average cutting height from the ground should not exceed 100 mm, stalk losses should be less than 5%, and the percentage of broken stalks should not be more than 6%, on the other hand and the realization at a lower cost price than the external ones.

This research work was supported by:

A grant of the Romanian Research and Innovation Ministry, PN 19 10 01 03-Substantion of the technology for harvesting and primary processing green hemp stalks, contract no. 5N/07.02.2019.

A grant of the Romanian Education and Research Ministry, through Programme 1 – Development of the national research-development system, sub-programme 1.2 – Institutional performance – Projects for financing excellence in RDI, contract no. 16PFE.

References

1. Carus M., Karst S., Kauffmann A. The European Hemp Industry: *Cultivation, Processing and Applications for Fibres, Shives and Seeds*. Cologne: EIHA, pp. 1–9, (2013).
2. Amaducci S., Zatta A., Pelatti F., Venturi G. *Influence of agronomic factors on yield and quality of hemp (Cannabis sativa L.) fibre and implication for an innovative production system* (Field Crops Res. 107, pp 161–169, 2008).
3. Struik P. C., Yin X., Thouminot C., Bjelková M., Stramkale V., et al. *Comparing hemp (Cannabis sativa L.) cultivars for dual-purpose production under contrasting environments* (Ind. Crops Prod. 87, pp 33–44, 2016).
4. Păun A., Bogdanof G., Ganea-Christu I., Ciupercă R., Matache M. Patent application no. A-00426/04.07.2019 *Equipment for harvesting green hemp / Echipament de recoltat cânepă verde*, OSIM, Romania, (2019).
5. Păun, A., Stroescu Ghe., Zaica Al., *Technological study on the process of green hemp harvesting*, project PN 19 10 01 03-Substantion of the technology for harvesting and primary processing green hemp stalks, INMA, Bucharest, (2019).
6. Luigi Pari a, Przemyslaw Baraniecki b, Ryszard Kaniewski b, Antonio Scarfonea, *Harvesting strategies of bast fiber crops in Europe and in China*, (Industrial Crops and Products, Volume 68, June 2015, pp 90-96).
7. Bogdanof G., Păun A., Brăcăcescu C., Stroescu Ghe. *Considerations on the hydraulic installation of the equipment for harvesting green hemp stalks*, ISSN 1454 – 8003, Proceedings of 2019 International Conference on Hydraulics and Pneumatics - HERVEX , November 13-15, Băile Govora, Romania, (2019).
8. Huang Jicheng, Shen Cheng, Li Xianwang, Tian Kunpeng, Chen Qiaomin, Zhang Bin, *Design and tests of hemp harvester*, (International Agricultural Engineering Journal Vol. 26, No. 2, June, (2017).

E3S Web of Conferences **180**, 03027 (2020) https://doi.org/10.1051/e3sconf/202018003027
TE-RE-RD 2020

Experimental testing of a helical rotor for compost distribution

Vasilica Ştefan[1*], *Ladislau* David[2], *Radu* Ciupercă[1], *Ana* Zaica[1], *Ancuţa* Nedelcu[1] and *Albert* Suvac[1]

[1]INMA Bucharest, Romania, 013811, Ion Ionescu de la Brad Bv.,Bucharest, Romania
[2] UPB Romania, 060042, Splaiul Independentei, 313, Bucharest, Romania

Abstract. It is well known that a manure spreader must be able to apply manure consistently, effectively and uniformly over the time. An over-applying doze will cause serios environmental problems (pollution) and a sub-applying doze will not contribute to plants growth and the applying of the fertilizer will be a loss. Corroborate with others problems such as maintaining a constant tractor speed, a constant rotational speed of the rotor, a constant and continuous flow of the material make the manure spreading machines an important subject for study and continuously improving . The propose of this study is to test a vertical helicoidal rotor for manure spreading in order to improve the machine distribution uniformity. The relation between the rotor position and rotation sense is a definitory issue for the machine performance. In order to achieve a high distribution uniformity a single helical rotor is tested and a series of possible situations of placing four rotors on the machine are analyzed. The optimal solution is given by two indicators: uniformity of spreading and distribution width. The test were made with an helical rotor having the following characteristics: high=1,1 m, maximum diameter of helical spiral=0.345 m, minimum diameter of the helical spiral= 0.114m, helical pitch=0.3m, inclination angle = 15°. After analyzing four possible situations it is observe that the combination of the maximum uniformity with minimum distribution with is the optimal solution. Experimental were carried out with compost and can be repeatable with others type of solid organic fertilizers, but a new calibration should be done. It is desirable that a trapezoidal, oval or triangle pattern with uniform sloping sides to be achieved. Because most of the spreaders actually produce an oval or triangle distribution pattern it is necessary subsequent passes that increase the time for fertilization operation.

1 Introduction

Manure management can have both beneficial effects, influencing the chemical and physical properties of soil, such as microbial activity, but it can also have disadvantages if application rates are not properly managed and controlled. This can be a major cause of soil quality deterioration and a source of pollution of water sources [1, 2, 3, 4]

* Corresponding author: valle_vali@yahoo.com

In order to reduce the polluting effects, efforts are made both for the production of high quality fertilizer products (compost, vermicompost) and for the production of high-performance technical equipment [5, 6, 7].

The machines for distributing solid organic fertilizers has to distribute uniformly the fertilizers on the soil, that are later incorporated into the soil according to the agrotechnical requirements of the crop plants. They must meet the following requirements [8, 9, 7, 10];

- to distribute a wide range of organic fertilizers;
- to achieve the fertilization norm imposed by the agrotechnical requirements;
- achieve uniformity of distribution according to the regulations (minimum 75%);
- be able to work on horizontal and sloping grounds;
- to allow the adjustment of the quantity of fertilizers administered per hectare (the norm);
- achieve a proper degree of shredding (particles larger than 6 cm not to exceed 25%);
- the machines are easy to operate (preparation of work and transport, filling and emptying with material, supervision of the work process);
- have a high working capacity that allows the works to be carried out during the optimum agrotechnical time);
- to maintain the initial working conditions during the work;
- have a low energy consumption;
- not to present danger of injury to the user.

The uniformity of the distribution is of great importance, because an uneven distribution makes in some areas the quantity of fertilizers to be smaller, not ensuring the expected effect, or, high concentrations of fertilizer are causing local pollution of the soil [11].

The uniformity of fertilizer distribution depends on a number of factors such as: [1, 12, 13, 14]

- properties of distributed materials: humidity, density;
- the adjustments and the dimensions of the machine: the speed of the distributing rotors, the distance between the rotors, the diameter, the pitch and the shape of the helical coil, the inclination in vertical plane of the rotors, - the mode of feeding the distribution apparatus: maintaining a constant and continuous supply flow;
- the distance from the surface of the soil;
- external conditions: wind speed, terrain slope.

A decisive factor influencing both the norm distributed per hectare and the uniformity of distribution are: inconsistent machine speed that can reach up to 25% and the inconsistent rotational speed of the rotors or inconsistent conveyor speed [9, 15, 10].

In order to achieve a high performant fertilizer machine, some experimental researches were made. A single helical rotor was tested and some recommendations were made.

2 Material and method

To perform the tests with a single rotor, the volume of the MG-5 spreader was divided into two areas by installing a partition wall, so that the material was placed to feed a single rotor (Fig.1). The vertical rotor shape and components were described in fig.2. The material used was compost.

Fig. 1. MG 5 manure spreader

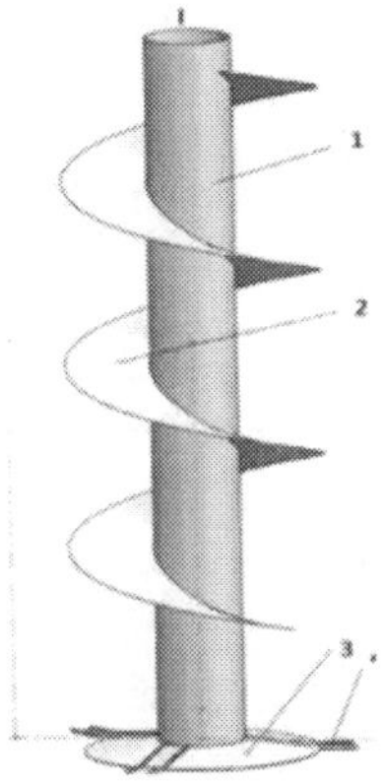

Fig.2. Helical rotor : 1) shaft; 2) helical spiral; 3)disc; 4)palette

The following initial data were considered:

- Rotor speed n = 540 rpm;
- The maximum diameter of the helical coil $Ø_{max}$= 0.345 m;
- The minimum diameter of the helical coil $Ø_{min}$ = 0.115 m;
- The pitch of the helical coil p = 0.3 m;
- High of the rotor h= 0.9 m;
- Distance between rotor base and the soil d = 1.2 m;
- The inclination of the rotor and vertical axis α = 15° (rotor inclination of 15° corresponds to the horizontal throwing of fertilizer particles leaving the helical coil at its maximum diameter 0.345 m)
- Material density: ρ = 510 kg/m^3 ;
- Dry matter u= 72.33 % °;
- Conveyor speed v_{tr} = 0.006 m/s;
- Material flow q = 1.53 kg/s.

The distribution was carried out at the stationary on a concrete platform and the collection of the material was collected from each square meter belonging to the distributed surface, placed in plastic bags and then weighed individually (Fig.3 and 4).

Fig.3. Compost distribution on concrete platform

Fig.4. Collecting the samples

3 Results and discussion

The quantities of material distributed for 25 seconds on the total area are weighed individually and presented in table 1. It is observed that the total width is 9 m (5 m to the left of the rotor axis and 4 m to the left of it) and the length is 11 m.

Table 1. The amount of material distributed (Compost) per unit area ;
n = 540 rpm, theoretical flow q = 1.53 kg / s, t = 25 s.

Amount of material distributed per unit area [kg/m²]										
		Width, m								
		-4.5	-3.5	-2.5	-1.5	-0.5	0.5	1.5	2.5	3.5
Length, m	0.5	0	0	0	0	0	0	0	0	0
	1.5	0	0	0	0	0	0	0	0	0
	2.5	0	0	0	0	1.062	0.512	0	0	0
	3.5	0	0.224	0.497	0.596	1.155	1.041	0.694	0.280	0.315
	4.5	0.041	0.240	0.954	1.295	1.170	0.775	2.309	0.470	0.170
	5.5	0.087	0.248	0.852	1.063	1.322	1.954	1.608	1.035	0.329
	6.5	0.029	0.490	0.652	0.926	1.275	1.055	1.256	1.205	0.462
	7.5	0.027	0.147	0.379	0.857	0.480	0.780	1.170	1.065	0.550
	8.5	0	0.076	0.582	0.700	0.280	0.980	0.275	0.660	0.325
	9.5	0	0.087	0.021	0.255	0.240	0.500	0.105	0.107	0.025
	10.5	0	0	0	0.025	0.005	0.102	0.047	0	0

The values were divided with 25 for finding the amount of material distributed in 1 second. The transversal distribution was determinate (table 2).

Table 2. Transverse distribution obtained with one rotor.

	Amount of material distributed per unit area									
	Width, [m]									
	-4.5	-3.5	-2.5	-1.5	-0.5	0.5	1.5	2.5	3.5	**Total**
Specific flow in 25 s, kg/ m²	0.18 5	1.51 2	3.93 7	5.71 7	6.99	7.7	7.46 6	4.82 2	2.17 6	***40.50 7***
Specific flow in 1 sec, kg/s m²	0.00 7	0.06 0	0.15 7	0.22 8	0.27 9	0.30 8	0.29 8	0.19 2	0.08 7	***1.620***

Based on the values obtained with a single rotor, the optimum variant of positioning 4 rotors is determined in order to obtain the best distribution. The rotational speeds of 540 rpm and an inclination angle of 15 ° were considered. Since the width of the machine is 2 m, the rotors are located at a distance of 0.5 m between them. Rotor 1 and rotor 2 are left-handed spiral and rotor 3 and 4 are right-Händel spiral. The rotors R1 and R2 rotate clockwise and the rotors R3 and R4 rotate trigonometrically.

Possible variants of location of the 4 rotors are shown in figure 5:

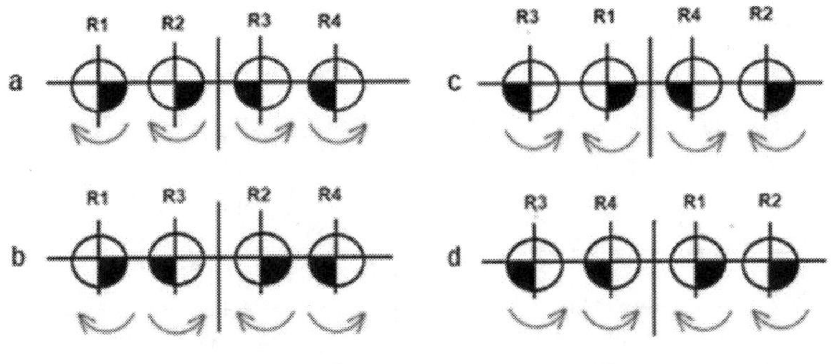

Fig.5. Variants for the location and operation of the distributor rotors
a) variant 1; b) variant 2; c) variant 3; d) variant 4

Considering that the experiments were performed with the rotor number 4 (R4), the direction of rotation and the distance between the rotors are taken into account and the amount of material distributed on each square meter is calculated similar to the previous calculations. The following results are obtained.

In variant 1 for the location of the rotors, the rotors R1 and R2 rotate clockwise and the rotors R3 and R4 rotate counter-clockwise.

It has been made the assumption that all rotors have a similar distribution, given the course of the winding direction. Considering the distribution made by 1 rotor (table 2), the cross-sectional distribution made by the machine in variant 1 of rotor placement was calculated graphically (fig.6).

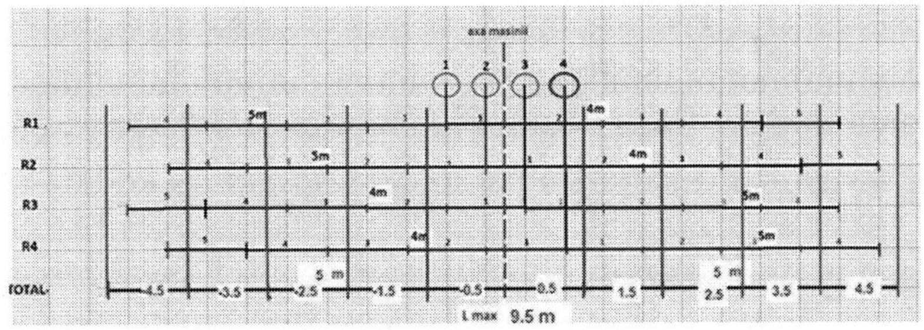

Fig.6. The overlap of the obtained distribution of 4 rotors-variant 1.

Considering the overlap of the surfaces obtained in the case of the 4 distributor rotors mounted on the machine in variant 1 with and summing the quantities distributed by each rotor on each linear meter (fig.7) the data from table 3 are obtained.

Table 3. The transverse distribution realized by 4 rotors located in - Variant 1.

	Specific flow [kg/s m^2]									
Width, [m]	-4.5	-3.5	-2.5	-1.5	-0.5	0.5	1.5	2.5	3.5	4.5
R1	0.065	0.166	0.272	0.305	0.286	0.241	0.175	0.084	0.020	0.001
R2	0.021	0.113	0.219	0.301	0.300	0.266	0.210	0.133	0.047	0.005
R3	0.005	0.047	0.133	0.210	0.266	0.300	0.301	0.219	0.113	0.021
R4	0.001	0.020	0.084	0.175	0.241	0.286	0.305	0.272	0.166	0.065
TOTAL	***0.094***	***0.347***	***0.709***	***0.992***	***1.095***	***1.095***	***0.992***	***0.709***	***0.347***	***0.094***

The distribution of the fertilizer in variant 1 has a maximum specific flow of 1.095 kg/m2·s. The obtained distribution width is 9.5 m, as it can be observed in fig.6.

Similarly, the distributions of fertilizers made by the 4 rotors in variant 2 variant 3 and variant 4 are calculated and the results are presented in table 4 and fig.7.

Table 4. The cross-sectional distribution made by 4 distributing rotors

Specific flow, [kg/s m²]												
Width [m]	**-5.5**	**-4.5**	**-3.5**	**-2.5**	**-1.5**	**-0.5**	**0.5**	**1.5**	**2.5**	**3.5**	**4.5**	**5.5**
Variant 1		0.075	0.331	0.708	0.990	1.096	1.096	0.990	0.706	0.329	0.075	
Variant 2	0.001	0.074	0.320	0.693	0.991	1.119	1.119	0.991	0.692	0.318	0.073	0.001
Variant 3	0.005	0.071	0.299	0.662	0.995	1.167	1.167	0.995	0.662	0.297	0.069	0.005
Variant 4	0.007	0.069	0.288	0.647	0.997	1.191	1.191	0.997	0.647	0.286	0.067	0.007

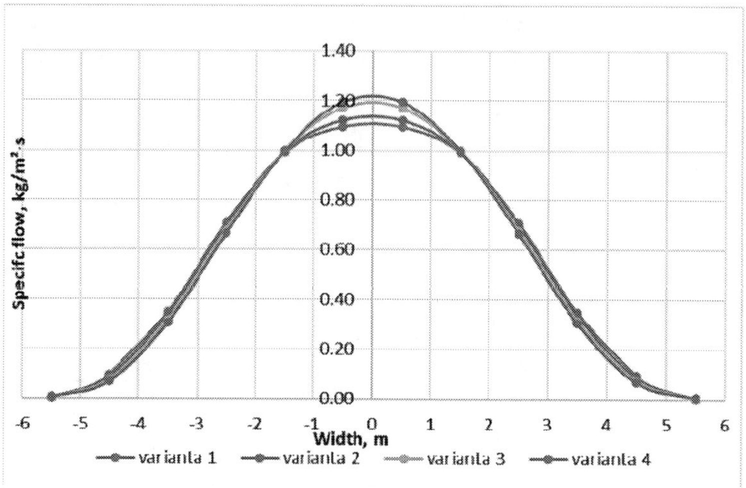

Fig.7. Comparison between all 4 variants of rotors placement

From fig.7 it is observed that in variant 2 the obtained distribution width has the value L = 11.5 m and the maximum of the specific flow reaches the value of 1.1205 kg / m²·s. For variant 3 of the location of the rotors the obtained distribution width is 11.5 m and the maximum is 1.1697 kg / m²·s. For variant 4 we obtain a width of 10.5 m and a maximum of the specific material flow of 1.1942 kg / m²·s.

The uniformity degree of the distribution (eq. 1) for each proposed variant, at a single pass of the machine (U1) was calculated. After eliminating from the calculation the extreme values, which are not representative the corrected uniformity U2 was obtained. The values obtained are presented in table 5.

$$U = \left[1 - \frac{\sqrt{\frac{\sum_{1}^{j} (m_i - m_{med})^2}{j-1}}}{m_{med}} \right] \cdot 100 \quad (1)$$

where m_i are the quantities of material collected from the soil from each square meter, m_{med} is their arithmetic mean and j is the number of collected samples.

Table 5. Uniformity of distribution for all 4 variants

Uniformity	Variant location of the rotors			
	1	2	3	4
U_1 [%]	**43.70**	17.03	14.13	12.71
U_2 [%]	**60.56**	35.49	33.41	31.55

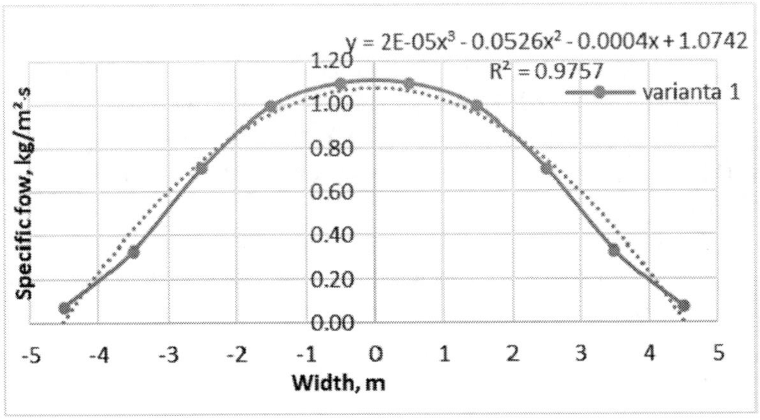

Fig.8. Optimal transverse distribution curve.

The optimal theoretical curve obtained (fig.8) is symmetrical to the axis of the machine, it has an oval/Gaussian shape and implies the need to study the optimal distance for the next steps of the machine so that the uniformity of the entire fertilizer distribution work is maximum.

This paper was financed by grant of the Romanian Education and Research Ministry, through Programme 1 – Development of the national research-development system, sub-programme 1.2 – Institutional performance – Projects for financing excellence in RDI, contract no. 16PFE.

4 Conclusions

In the experimental tests performed at the speed of the distributor rotor, n = 540 rpm the amount of material distributed in 25 seconds was 44.48 kg, which corresponds to a specific flow rate of 1.6203 kg/s m^2. In this case the width of distribution was 9 m, the largest quantity of material being in the center of the surface and towards the extremities it decreases.

When testing the machine at stationary, considering that the width of the machine is 2 m, the location of the inclined rotors with 15° is made at a distance of 0.5 m between them in 4 variants.

In version 1 (rotors R1 and R2 rotate clockwise and rotors R3 and R4 rotate counter-clockwise) the speed of the distributing rotors of 540 rpm and the specific flow of 1.2159 kg/m^2·s, the distribution width obtained is of 9.5 m and the degree of uniformity of the distribution is maximum.

At a single pass of the machine the degree of uniformity of the distribution is, along the whole width of distribution U_1 = 43.7% and after eliminating from calculation the extreme values, on which fertilizers will be distributed on the next passage, U_2 = 60.57%.

References

1. Antile D.L., Gallar L., Miller P.C.H.,Godwin R.J., "An investigation into the fertilizer particle dynamics off-the disc," *Appl. Eng. in Agr.,* vol. **31**, no. 1, pp. 49-60, 2015.
2. Landry H., Lague C., Roberge M., "Physical and rheological properties of manure products," *Appl. Eng. in Agr.,* vol. **20**, no. 3, pp. 277-288, 2004.
3. Vanotti M.B., Rashash D.M.C., Hunt P.G, " Solid-liquid separation of flushed swine manure with PAM: Effect of wastewater strenght," *Trans. of ASAE,* vol. **45**, pp. 1959-1969, 2002.
4. Ştefan (Popa) V., Ciupercă R., Popa L., Nedelcu A., Lazăr G., Petcu A.S., Zaica A., "Influenţa caracteristicilor fizice ale îngrăşămintelor organice solide asupra calităţii lucrării de împrăştiere," *INMATEH Agr. Eng.,* vol. **46**, no. 2, pp. 77-84, 2015.
5. Bohnhoff D., Converse J., "Engineering properties of separated manure solids," *Biological Wastes,* vol. **19**, pp. 91-106, 1987 .
6. Ciupercă R., Lazăr G., Popa L., Nedelcu A., Ştefan V., Zaica A., Petcu A., "Sistem ecologic de gestionare a deşeurilor vegetale şi animale," *INMATEH - Agr. Eng.,* vol. **46**, no. 2, pp. 69-76, 2015.
7. Nedelcu A., Ciupercă R., Popa L., Zaica A, Zaica Al, Anghelet A., Ştefan V, Cristescu A.C., "Aspects about organic waste composting in biocontainer," in *ISB INMA TEH*, Bucharest, 2018.
8. Cârdei P., Ştefan V., Matache M.G., "Mathematical models for calculating the qualitative parameters of organic fertilizer machines," in *Research people and actual tasks on multidisciplinary sciences*, Lozenec, Bulgaria, 2019.
9. ASABE Standards S341:4, Procedure for measuring distribution uniformity and calibrating granular broadcast spreaders, St. Joseph, Mich: ASABE, 2009.
10. Robert P. C., "Precision agriculture: a challenge for crop nutrition management," *Plant and Soil,* vol. **247**, pp. 143-149, 2002.
11. Rousselet M., Mozoyer J. , " Évaluation des performances des épandeurs de fumier: premiers results selon la norme NF EN 13080," *Ingénieries,* vol. **46**, pp. 79-92, 2006.
12. Ştefan V., Popa L., David L., Pirnă I., Ciupercă R., Petcu A., Duţu M. F., Găgeanu I , "Considerations on physical and mechanical properties of solid organic fertilizers," in *4th Int. Conf. on Therm. Equip., Ren. Energ.*, Posada, România, 2015.
13. Malgeryd J., Watterberg C, "Physical Properties of Solid and Liquid Manures and their Effect on the Performance of Spreading machines," *J. of Agr. Eng. Res.,* vol. **64**, no. 4, pp. 289-298, 1996.
14. Popa L., Pirnă I., Cojocaru I., Ciupercă R., Nedelcu A., Ştefan V., "Machine for applying manure fertilizers by means of vertical spreading apparatus," in *Sc. Conf. Dur. Agri.*, Craiova, Romania, 2008.
15. Babiciu P., Studiul interdependenţei dintre parametrii aparatelor de împrăştiere prin centrifugare şi uniformitatea de distribuţie pe sol a îngrăşămintelor minerale, Teza de doctorat, Bucureşti: Facultatea de Mecanica Agricolă - UPB, 1976.
16. Ştefan V., Cârdei P., Popa L., David L., Ciupercă R., "Influenta parametrilor regimului de lucru al masinilor de fertilizare organică, asupra performanţelor calitative ale procesului de fertilizare," *INMATEH Agr. Eng.,* vol. **58**, no. 2, pp. 115-120, 2019.

E3S Web of Conferences **180**, 04001 (2020) https://doi.org/10.1051/e3sconf/202018004001
TE-RE-RD 2020

Research of filtration and energy parameters of membrane bioreactors

Nikolay Makisha*

Moscow State University of Civil Engineering, 26, Yaroslavskoye shosse, Moscow, Russia, 129337

Abstract. Membrane bioreactors nowadays are intensively applied in wastewater treatment worldwide. The paper refers to a research, which goal was to evaluate the optimal parameters of membrane bioreactor performance. A focus was given to a part of the research that considered operation of side-stream membrane bioreactor under pressure of 0.2 MPa and variable amount of backwashes. There were three types of feed water investigated with mixed liquor suspended solids concentrations of 1.6, 2.5 and 7 g/L, respectively. The article provides results of permeate production and their processing to obtain specific energy consumption for each combination of operation parameters.

1 Introduction

Application of membrane bioreactors for wastewater treatment, which became well known in 1980s, is now a matter of a significant interest from viewpoint of research and practical implementation [1,2]. The more detailed the research are, the wider and more solid data will be obtained and the more accurate will be the operation. Real performance, of course, cannot be fully predicted during the bench-scale, pilot-scale or even full-scale research, however, all of these can significantly contribute to sustainable operation [3,4]. The experience in operation of MBR in Russian Federation is not so solid, and these blanks and gaps should be filled [5]. It considers almost every aspect of MBR – contamination removal from wastewater (with typical for Russia values of pollutants), energy efficiency of the process, fouling mitigation, operation conditions and many others [6,7]

If membrane bioreactors are taken into account submersible MBR are the most common solution in majority of cases. On the other hand, if to apply MBRs for reconstruction of small wastewater treatment plants (WWTP) the facilities' dimensions always matter and dimensions of auxiliary tanks in particular [8,9]. In this viewpoint side-stream membrane bioreactors can be focused on within a research as a probable option if to place them near-by in portable containers (and without auxiliary capital construction) [10].

* corresponding author – nmakisha@gmail.com

2 Materials and methods

The research was performed following a goal to investigate operation of side-stream MBR in case of several possible operation parameters – mixed liquor suspended solids (MLSS), operation pressure, and amount of backwashes. The objectives of the research are:

– to study suspended particles detention (of active sludge) using membrane methods at different values of operating pressure

– to increase performance of membrane bioreactors, as well as to determine the optimal range of working pressure under which it is possible to prevent clogging of membranes and breakthrough of the contaminants in the filtrate.

The bench scheme is shown in figure 1. The feed wastewater (mixed liquor after biological treatment) is supplied from the feed water tank 1 to the membrane module 3 using the pump 2. The working pressure is regulated by means of a pressure tank 4, pressure switch 5, a bypass valve 10, and a pressure gauge 12. Permeate after the membrane module was collected in the samplers 7. Ordinary capillary ultrafiltration membranes with standard characteristics were used for the bench-scale research with pore size of 0.1 micron and total surface area of $1m^2$.

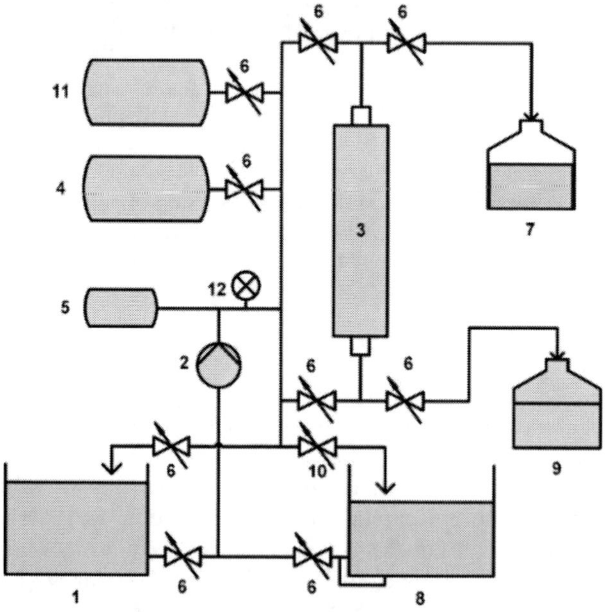

Fig. 1. Scheme of the bench-scale model: 1 – feed water tank; 2 – pump; 3 – membrane module; 4 – pressure tank; 5 – pressure switch; 6 – valve; 7 – sampler tank; 8 – washing tank; 9 – washing water sampler; 10 – bypass valve; 11 – pressure tank for washing; 12 – pressure gauge

Backwash is a mandatory option for the functioning of membrane modules in MBR. Deionized water after the reverse osmosis plant was used for backwashing and supplied from washing tank 8. Switching between feed and wash tanks was performed using valves 6. The working pump fed the wash water into the filtrate path; afterwards it was collected through the source water path to the wash water samplers 9. Switching of pipelines was carried out by means of valves 6. Pressure was controlled with pressure switch 5, a bypass valve 10 and pressure tank of washing water 11.

The whole plot of the research was divided into several sub-stages to compare obtained results in the following. The criteria of the division was operation pressure, while the MLSS value remained the same through the parts of the research. There were three concentrations of MLSS investigated within the research: MLSS=1.6 and 2.5 g/L are similar to those in conventional activated sludge systems. This approach seems reasonable, if the MBR will be applied for reconstruction of existing WWTP with current operation parameters. The third value, MLSS=7 g/L lies in a range of optimal values for submerged MBR. This paper has a focus on a part of the research with operation pressure of 0.2 MPa, as our previous study showed that higher operation pressure stimulates breakthrough of suspended solids [11].

3 Results

The first obtained results were the volumes of permeate related to variable duration of membrane filtration (Figure 2). It is predictable, that higher efficiency (greater permeate volume) was achieved at low MLSS value (1.6 g/L). Slightly higher MLSS (of 2.5 g/L) provided a slight (10-15%) decrease of permeate production, however, both sets of values may be considered as comparable. At the meantime, the permeate production dropped 3.2 times when the MLSS value was risen up to 7 g/L comparing to MLSS=1.6 g/L.

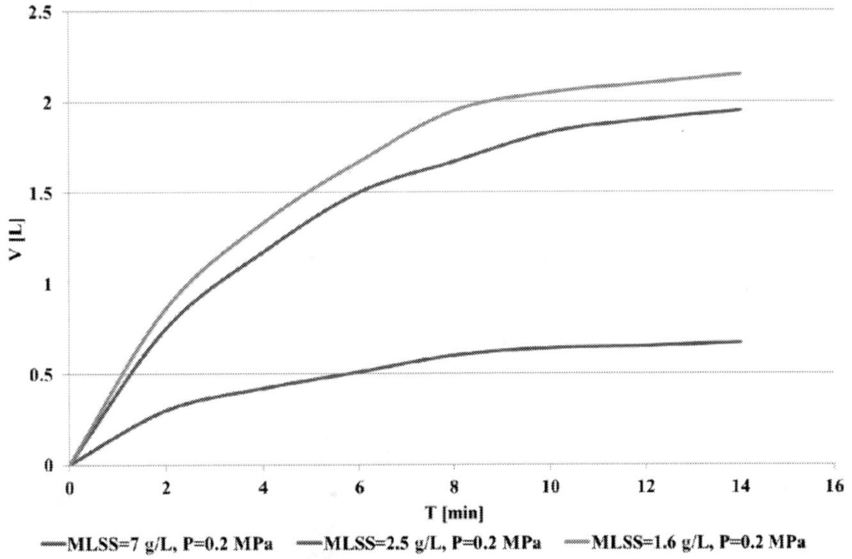

Fig. 2. Growth of permeate volume through time

That may signal that under operation pressure of 0.2 MPa the permeate production is sharply depended on the growth of MLSS. In other words, the permeate production plummets due to soar of fouling resistance cause by intensive pore clogging. The most intensive growth of permeate production is witnessed at the initial period of filtration – almost linear growth can be seen within first 3 minutes of filtration (for all three MLSS). The next part is filtration between 3 and 8 minutes, when the permeate production grows by approximately 70% under all three MLSS value. Duration of 8 minutes is a sort of checkpoint as far as the next 5 minutes period showed only 10% of permeate production growth. That means that after 8 minutes of filtration the cake layer on the membrane surface is dense enough to stop the filtration at any MLSS value. The same process can also be described through the dependence between flow and time of filtration (figure 3). The higher the MLSS value the more intensive is flux drop due to rapid pore fouling. Permeate volume in certain time can be determined if to draw straight line from abscissa axis and parallel to ordinate axis. Area of obtained figure corresponds to volume of filtered wastewater during selected time.

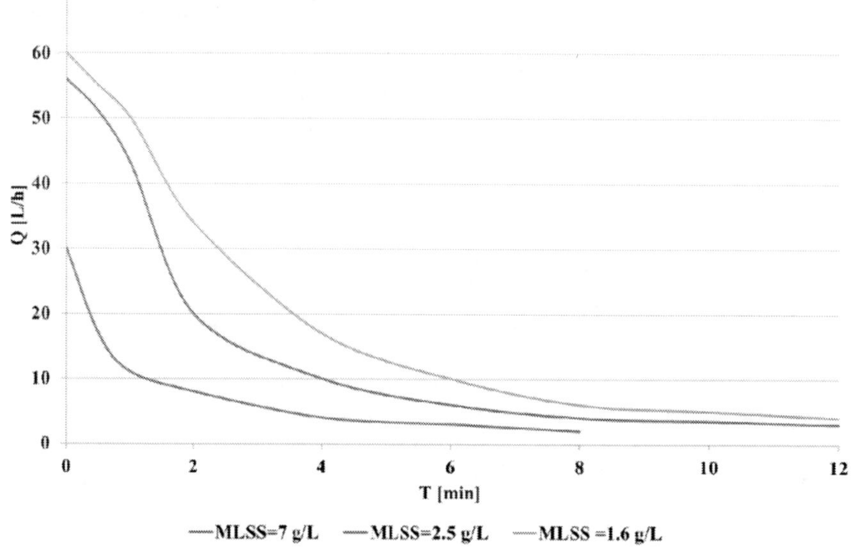

Fig. 3. Flux decrease within filtration cycle (pressure= 0.2 MPa)

For our further investigation of operation conditions, we focused at three possible operation modes – filtration cycle duration of 2, 4 and 6 minutes or 30, 15 and 10 backwashes respectively (figure 4).

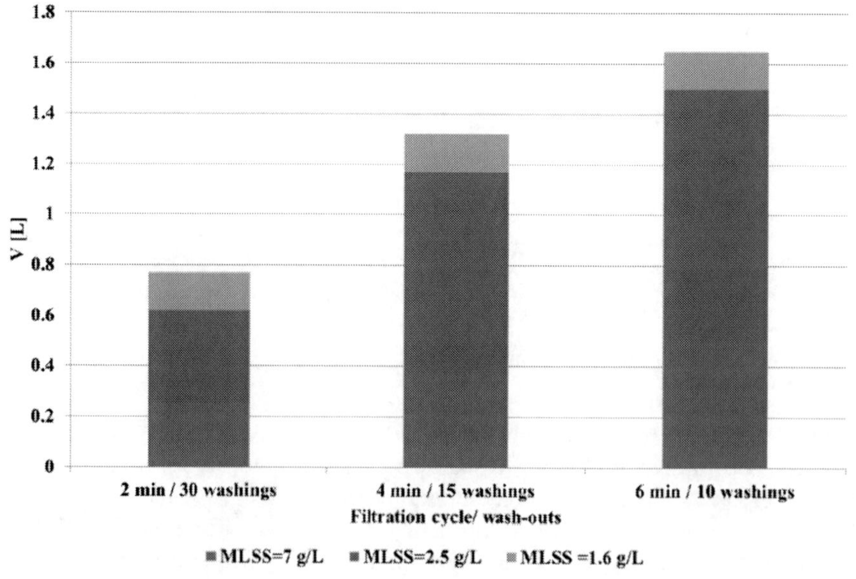

Fig. 4. Permeate volume different operation conditions

It is obvious that the shorter filtration cycle is the larger is permeating volume. On the other hand, the more often backwashes are applied, the larger is the volume of water required to implement efficient backwash and the lower is “useful” product water amount defined as a difference between total permeate amount and water used for backwashing.

https://doi.org/10.1051/e3sconf/202018004001

Backwash efficiency depends on filtration pressure value and time period of backwash. In addition, backwash efficiency can be evaluated as percentage of fouling (suspended solids) removed from membrane surface. Definition of balanced duration of filtration that corresponds to minimal operational costs was among the goals of our research.

Experimental data processing for experiments are presented on figures 5 and 6. Figure 5 shows dependencies of specific membrane flow (Lh^{-1}) to MLSS concentration and frequency of backwashes. Total permeate flow of membrane module decreases with the increase of suspended solids (and fouling, respectively) concentration.

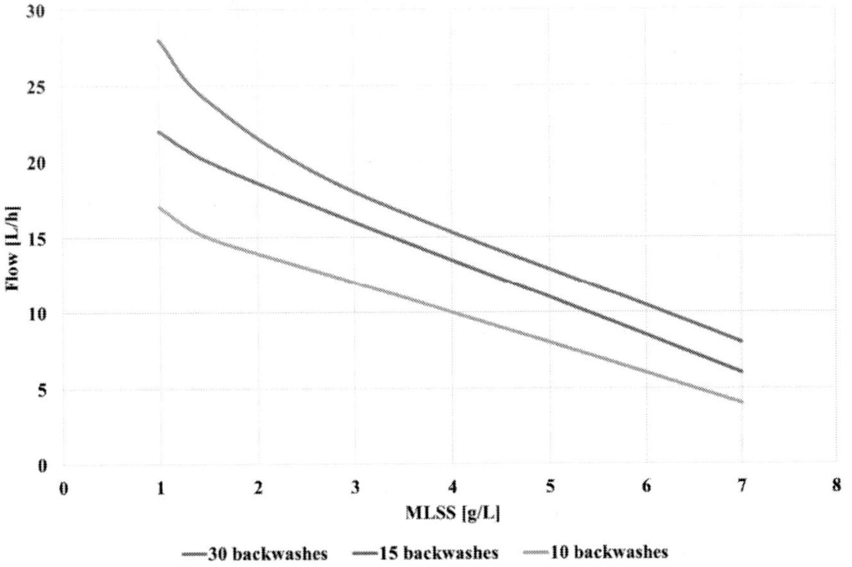

Fig.5. Flux determination in case of different backwashes amount

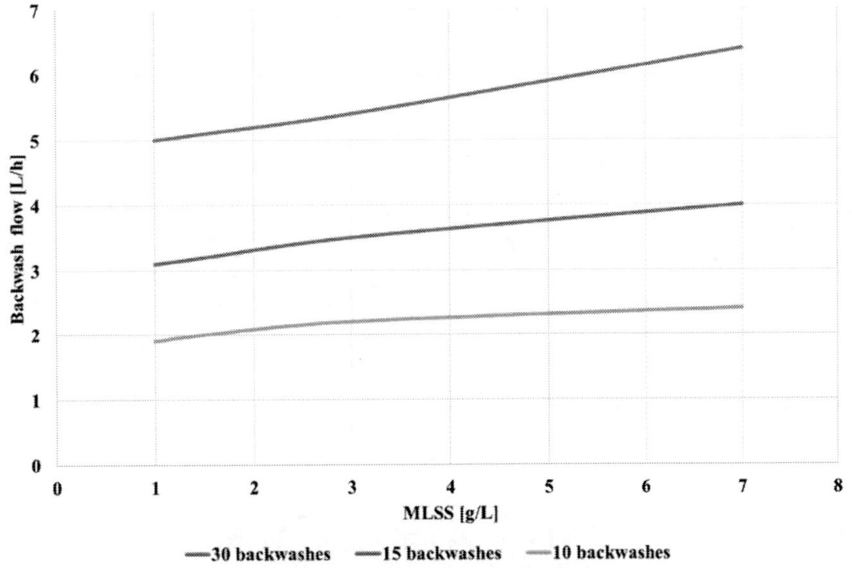

Fig.6. Backwash flow intensity in case of different backwash frequency

Figure 6 demonstrates dependencies of backwashing flow (of various backwash frequencies) to suspended solids concentration in the feed water. The increase in MLSS tends to increase in water consumption due to increase of frequency of backwashes and washing time.

Figures 7-9 gives examples of evaluation of "useful" permeate flow related to MLSS during operation of the module for various backwash frequencies performed at 0.2 MPa. Figure 7 refers to frequency of 30 backwashes, figure 8 – to 15 backwashes, and figure 9 – to 10 backwashes. There are four graphs at each of these figures presenting the following: a flow of single cycle of filtration (Q_{1f}); a flow used for backwash (Q_w); a total flow (Q_{tot}) that is a sum of Q_{1f} and Q_w; a useful flow (Q_u), which is a difference between Q_{1f} and Q_w

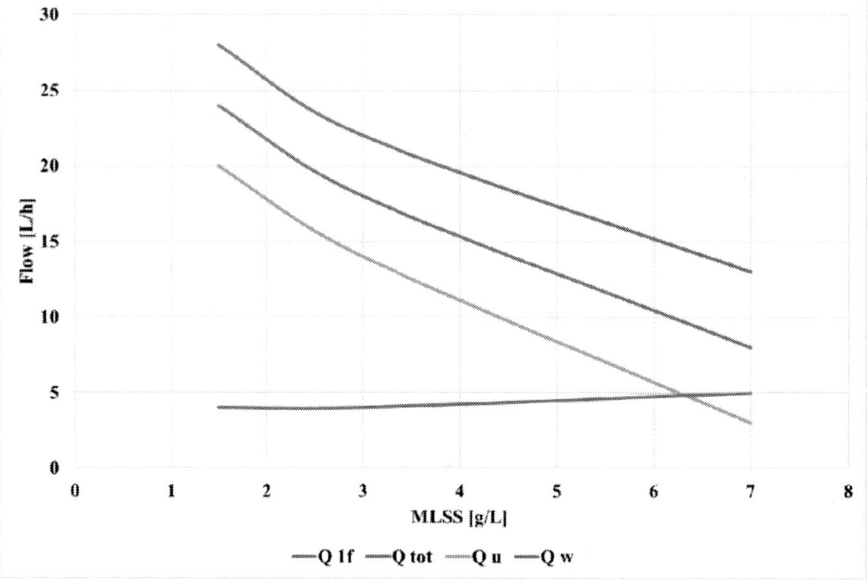

Fig. 7. Dependencies of flows to MLSS in condition of 30 backwashes

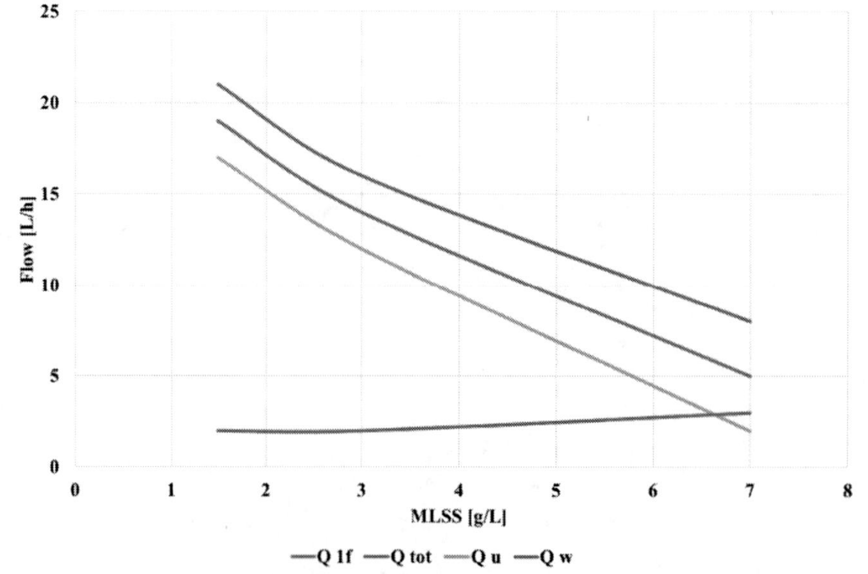

Fig. 8. Dependencies of flows to MLSS in condition of 15 backwashes

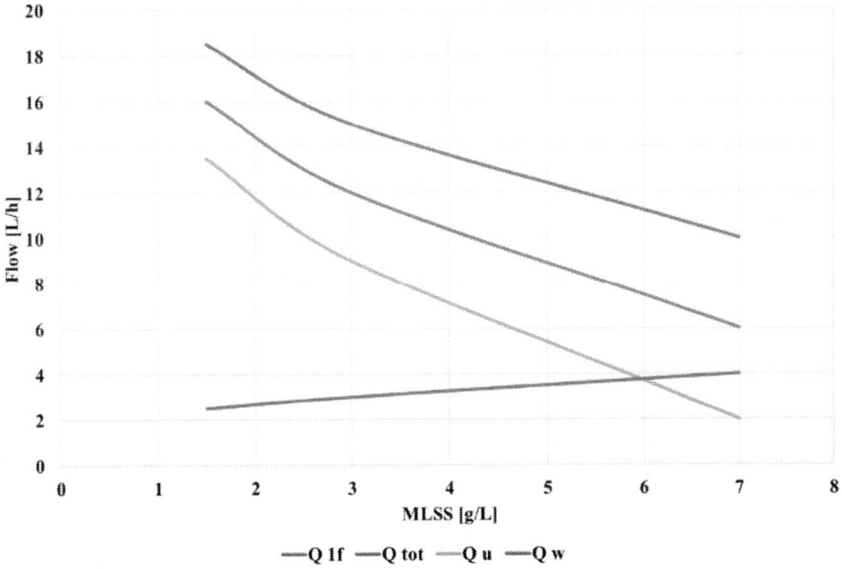

Fig. 9. Dependencies of flows to MLSS in condition of 10 backwashes

Having an optimum operation mode determined the next step of the research was to define the entire energy consumed and the specific energy consumption per flow. Within this we assumed that according to Q-H function of centrifugal pumps electric power required to pump up 1 m^3/h at 1 m of height is 0.7 kW. Figure 10 shows values of specific energy consumption calculated for the useful flow values shown on figures 7-9.

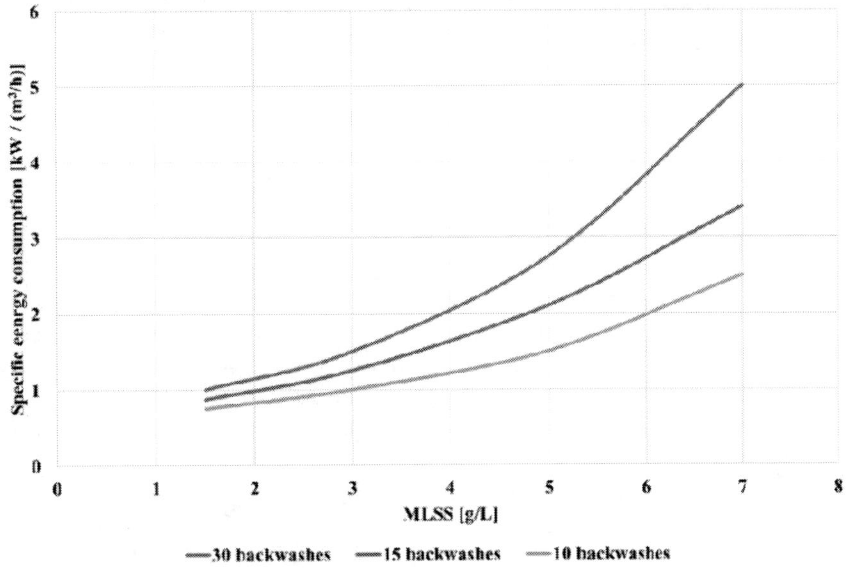

Fig.10. Specific energy consumption of the treatment process

Afterwards real energy consumption (figure 11) to provide useful flow was defined. That means that real energy consumption should sum up both power for filtration and power for backwash, so the specific energy consumption can be determined as the relation of real power consumption to useful flow.

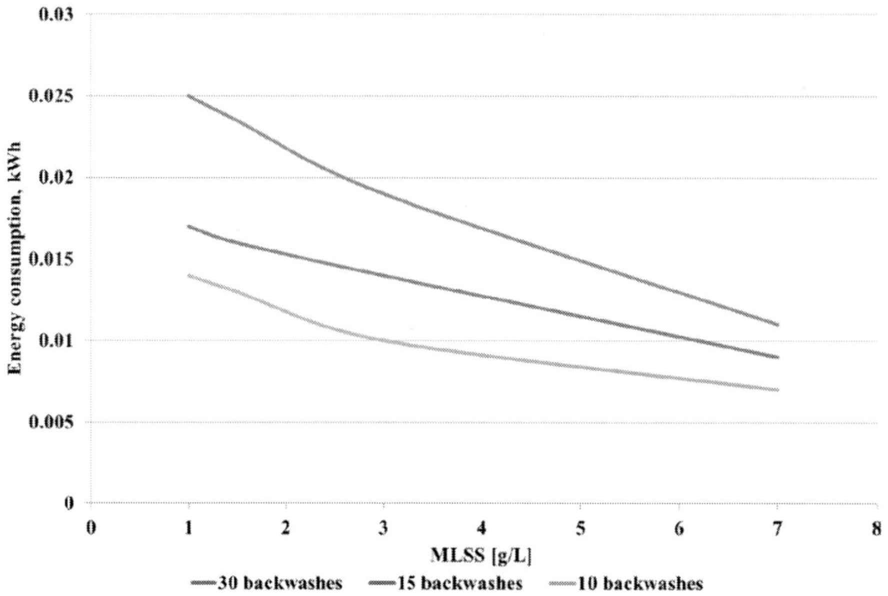

Fig. 11. Energy consumption.

It is important to note, that specific energy consumption at low MLSS value is more or less similar for different length of filtration cycle. Growth of MLSS provided a respective increase of specific energy consumption, however, the longer cycle of filtration (that means less amount of backwashing and less consumed water correspondingly) resulted in lower specific energy consumption with approx. 2 times difference if we compare 30 and 10 backwash mode respectively.

4 Conclusions

1. Permeate effluent has the most intensity within the first 3 minutes of operation, no matter what is the value of MLSS.
2. After 8 minutes of filtration, permeate production plummets due to accumulated fouling on membrane's surface (at high MLSS values in paeticular).
3. Shorter filtration cycles provides higher overall amount of filtered water, however it also means more frequent backwashes that increase water needs for that purpose.
4. Specific energy consumption has a strong relation to MLSS values and operation mode. Less frequent backwashing helps to reduce specific energy consumption by 10% at low MLSS values, and up to 50% at high MLSS values.
5. Further research is urgently needed to fit obtained results to various concentrations of pollutants in feed wastewater.

This work was financially supported by Ministry of Science and Higher Education of the Russian Federation (#MK-519.2019.8).

References

1. N. O. Yigit, G. Civelekoglu, I. Harman, H. Koseoglu, M. Kitis, Desalination, **237**, 346-356 (2009)
2. M. S. Nawaz, F. Parveen, S. J. Khan, N. P. Hankins, J. of Water Process Eng., **31**, 100861 (2019)
3. Z. Cui, J. Wang, H. Zhang, H.H. Ngo, D. Kang, Biores. Tech., **269**, 355-362 (2018)
4. K. Kimura, H. Uchida, Water Research, **1501**, 21-28 (2019)
5. N. Makisha, A. Kulakov, MATEC Web of Conferences, **178**, 09018 (2018)
6. G. Blandin, C. Gautier, M. S. Toran, H. Monclús, J. Comas, Chem. Eng. J., **3391**, 268-277 (2019)
7. I. Ruigómez, E. González, S. Guerra, L.E. Rodríguez-Gómez, L. Vera, J. of Membrane Sci., **52615,** 181-190 (2017)
8. Z. Yin, V.V. Tarabara, I. Xagoraraki, Water Research, **881**, 750-757 (2016)
9. L. Vera, E. González, O. Díaz, S. Delgado, J. of Membrane Sci., **47015**, 504-512 (2014)
10. Z. Cui, J. Wang, H. Zhang, H. Jia, J. of Membrane Sci., **53015**, 132-145 (2017)
11. N. A. Makisha, Construction: science and education, **10(1)**, 6 (2020)

The remotely controlled underwater robot system

Cristian Nita[1], *Dumitru* Deleanu[2*], and *Ionut* Voicu[3]

[1]Doctoral School of Mechanical Engineering, Electromechanical Faculty, 900663 Mircea cel Batran Street, Romania

[2] Constanta Maritime University, Department of General Engineering Sciences, 104 Mircea cel Batran street, ZC 900663, Constanta, Romania

[3] Electromechanical Faculty, Engineering Sciences in Mechanics and Environment Department, 900663 Mircea cel Batran Street, Romania

Abstract. The remotely controlled underwater robot system is used extensively by the scientific community to study the ocean, in support of diving operations, in military use, in educational outreach, in broadcast use. We built this system which has three components: 1) the surface control interface; 2) the intermediate device; 3) the robot itself. The presence of this robot replaces human activities where access is not allowed or life is endangered. The novelty of this device is the construction of component 2 (intermediate), which represents a storage frame for the underwater robot and part of the control umbilical cord. The frame descends with the robot to the depth of surveillance from where the robot leaves the storage frame and is further controlled to the research objective. The surveillance function is performed from the intermediary with the help of a video camera. The video camera is the IP type and can be accessed through the data over power network. The movement of the underwater device is accomplished by means of 8 electric motors (4 for horizontal control and 4 for vertical control). The robot has data acquisition and storage systems and piloting systems. The purpose of building this robot is educational research.

1 Introduction

The remotely controlled underwater robot system is used extensively by the scientific community to study the ocean, in support of diving operations, in military use, in educational outreach, in broadcast use (in the filming of several documentaries, coastguard services).

The remote controlled underwater robot system (RCURS) was specially designed for underwater activities that are generally deep [1]. It is connected to the ship by an umbilical cord, resistant to adverse conditions and to great depths, often used in conjunction with TMS (Tether Management System). The TMS can be a garage-type device that includes the

* Corresponding author: dumitru-deleanu@yahoo.com

https://doi.org/10.1051/e3sconf/202018004002

RCURS during its dive, or it can be a sled-like device that only carries the laying cable. The purpose of the TMS is to extend or shorten the umbilical cord, in order to dampen the power draw of the water currents.

As a use and historian, this system has been and can be used to rescue and recover objects at a great depth, assisting in the development of oil platforms where diving access is impossible, in initiating underwater constructions and repairing and maintaining them later and also they can be used extensively by the scientific community in exploring the ocean [2].

As technological variants can be designed to different sizes [3]: Micro - very small and very light (weigh less than 3kg) - used as an alternative for divers, especially in places where a diver could not have access (such as a canal, pipes or other small cavities); Mini - weigh about 15kg (they are also used as an alternative for divers); · General - usually with less than 5 horsepower (propulsion); Light Workclass - with less than 50 HP (horsepower); Heavy Workclass - usually with less than 220HP; Trenching & Burial - with more than 200 HPr, but often no more than 500 HP.

2 Recommended solution

2.1 Main system structure

The remote controlled underwater robot system is composed of three components: 1) the surface control interface; 2) the intermediate device; 3) the robot itself. The structure of the robot body is presented in the Fig. 1.

Fig. 1. The structure of the robot body

The control of the underwater robot is made from the surface with the help of specialized software installed on the computer (Fig. 1). The control signal is transmitted from the surface computer through an umbilical cord to the underwater robot. Both the electrical energy required to operate the underwater robot and the control signal are transmitted through two electrical conductors, through the data over power network system.

The 2nd component, called *intermediate*, represents a storage frame for the underwater robot and part of the control umbilical cord (Fig. 2). The frame descends with the robot to the depth of surveillance from where the robot leaves the storage frame and is further controlled to the research objective. Component number 2 is provided with a surface controlled winch that allows the robot to receive only the umbilical cord length required for movement, thus avoiding the possibility of tangling the control cord in foreign bodies.

Fig. 2. The control umbilical cord

The surveillance function is performed from the intermediary with the help of a video camera. The video camera is the IP type and can be accessed through the data over power network.

Component number 3, *the underwater robot,* hosts all motion control and navigation systems (Fig. 3).

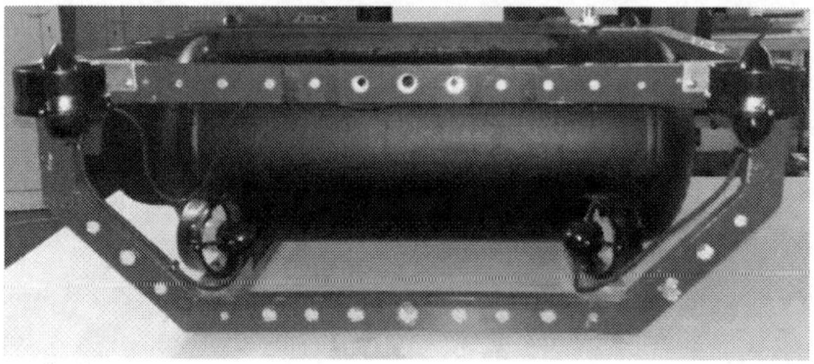

Fig. 3. The underwater robot

Motion control is carried out by means of four electric motors for horizontal control and another 4 electric motors for vertical control. Inside the robot's watertight housing are installed two video cameras, one front for navigation and one superior with interior and intermediate view, having a guiding role towards it. The electronic and electrical system installed in the underwater robot is composed of 5 power sources with an input from 220 volts AC and 12 volts DC, these have the role of supplying electric motors and auxiliary systems (Fig. 4).

Fig. 4. The electronic and electrical system

The underwater robot is powered by an alternating voltage of 220 volts from the surface.

2.2 The auxiliary structure

The auxiliary systems are the following: the data over network system, the communication system formed by a computer from the raspberry pi series, and the multi-device "*pixhawk*" pilot system. The on-board computer communicates with the *"pixhawk"* and sends the surface controls to it (Fig. 5).

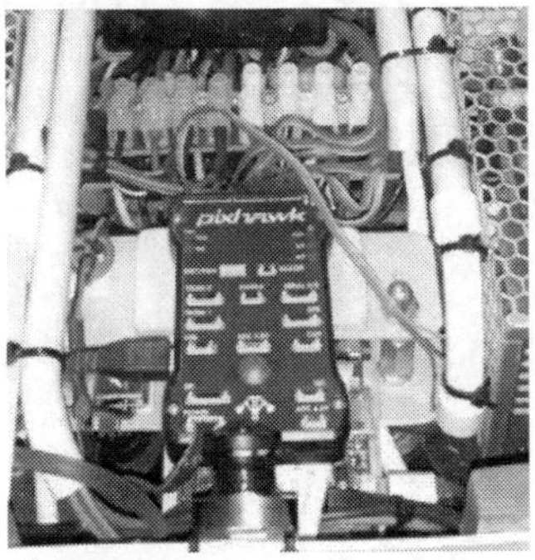

Fig. 5. The multi-device pixhawk pilot system

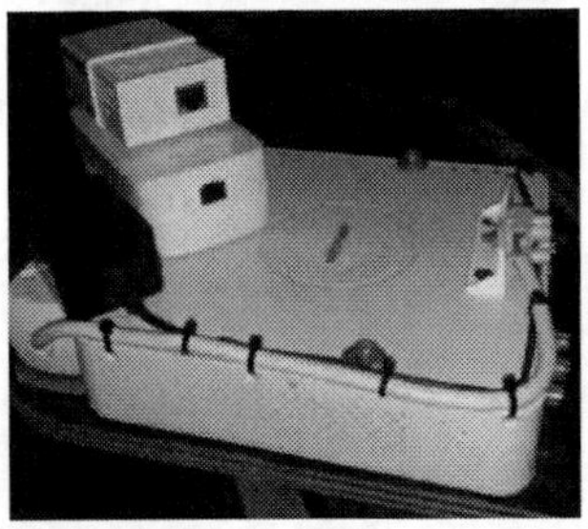

Fig. 6. The analog-to-digital signal converter

The robot is provided with indication for depth, electronic compass, temperature and inclination angle. The video signal is transmitted using an analog-to-digital signal converter, and accessed over a network over power network over IP (Fig. 6). This system is located inside the robot housing on a platform (Fig. 7).

Fig. 7. The platform with signal converter

3 Experimental researches

The general structure is fixed on the frame platform. The frame platform descends with the robot itself to a depth of surveillance from where the robot leaves the storage frame and is further controlled to the research objective (Fig. 8).

Fig. 8. The general structure of device

The component number 2 is provided with a surface controlled winch that allows the robot to receive only the length of the control cord required for movement, thus avoiding the possibility of entanglement in foreign bodies.

The robot moves in a horizontal or vertical direction with 4 electric motors on each direction (Fig. 9).

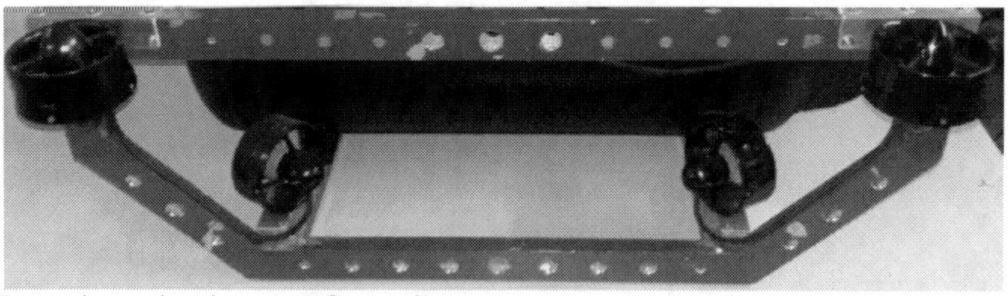

Fig. 9. Front view - electric motors for moving

The technology for making this robot has been developed for educational and research purposes. Its modular design allows configuration to meet the specific needs of a research, just as military special mission vehicles are designed [4].

4 Conclusions

Depending on the research to be carried out, this underwater robot can be equipped with various sensory and sampling devices. The purpose of this paper is to present the technological way of making such a robot for teaching and research purposes.

The novelty of this device is the construction of component 2 (intermediate), which represents a storage frame for the underwater robot and part of the control umbilical cord. The frame descends with the robot to the depth of surveillance from where the robot leaves the storage frame and is further controlled to the research objective.

Learning the technological modalities of achievement teaches students to acquire the basic skills in designing ships and submarines, and encourages them to explore the architecture and concepts of naval engineering.

This paper has financial support from the Constanta Maritime University under assistance project PN-III-P1-1.2-PCCDI-2017-0404 /31PCCDI/2018, *Holistic on the Impact of Renewable Energy Sources on Environment and Climate-HORESEC.*

References

1. D. Dumitru, C. Vlad, *Diving and underwater vehicles* (Scientific and Encyclopedic Edition, Bucharest,1986)
2. D.Dumitru, C. Vlad, *Underwater interventions* (Technical Publishing House, Bucharest, 1982)
3. R. Sundar, M. Dheepak, P. Veera Kumar, PC based remote operated underwater vehicle for marine surveillance, Int.J.of Civil Engineering and Technology, **8**, Issue 6, 716–721 (2017)
4. Pascu.E., Royal Netherlands Navy- Sea-Wasp ROV, J. Defense Romania, **20 iV** (2020)

Simulation regarding the forward part of underwater remotely operated vehicle designed for research in maritime *and riv*er districts

Andra-Teodora Nedelcu [1*], *Nicolae* Buzbuchi [2], *Liviu-Constantin* Stan [2] and *Mihail* Lucian Dumitrache [2]

[1] Mircea cel Bătrân Naval Academy, Faculty of Navigation and Naval Transport, 1 Fulgerului Street, 900218 Constanta, Romania

[2] Constanta Maritime University, Faculty of Navigation and Naval Transport, 104 Mircea cel Batran Street, 900663 Constanta, Romania

Abstract. Over the last few decades, underwater mechanical system became a necessity for ocean research and exploration. The necessity of this type of equipment provides a new type of marine platforms in different areas of oceanographic research. Until this moment, the underwater vehicles have developed in different shapes, sizes and means of propulsion. Take into consideration these characteristics, we could determinate the type and mission of the vehicle. In this paper we investigate the problem regarding the pressure exerted of remotely operated vehicle. The remotely operated vehicle is designed in order to achieve different mission regarding the surveillance in lakes, harbors and maritime environment. This mission can take place through a remote control system placed onshore or on seaborne platforms. To improve the endurance of remotely operated vehicle, we performed a simulation study during which we compare the drag forces acts on the underwater vehicle according to its shape. Using SolidWorks program to create the body simulation vehicle and Ansys CFX to realise the simulation, this paper present the result simulated for the forward part of the vehicle.

1 Introduction

A Remotely Operated Vehicle (ROV) provide an important role regarding collecting various oceano-graphic information, defense missions or applications, petrol, gas and different mineral exploration programs or biology studies. Using various integrated equipments and sensors, the vehicle could assure different assignments such as sea bottom topography, submarine detection or research the behavior of marine wildlife. Due to the fact that the remote control system is placed onshore or on seaborne platforms, an important characteristic represent the good endurance to complete different missions. Optimizing the vehicle shape it is necessary for energy utilization and endurance.

*Corresponding author: francu_andrat@yahoo.com

There was reported many studies [1]-[3] regarding the viscous fluid flow around different body shapes, for example cylinder, sphere or spheroid. The Navier-Stokes Equations explain explains and presents this behaviour regarding the viscous of fluid flow, which represent a particular forms of Newton`s Laws of Motion. For complete solution of equations, it is necessary to be solved numerically in collaboration with a computer. To reach a valid solution, the continuous flow domain must be converted in a discrete domain, represented by a numerical mesh. Using numerical discretisation scheme, we perform algebraic equation. Using a computer, the coupled algebraic equations from numerical discretisation scheme are then solved and determinate discrete values of velocity and pressure at mesh nodes along the surface body or in different points. All this process represented by the theoretical, numerical and computational techniques gather in Computational Fluid Dynamics (CFD) simulation. Also, the CFD simulation could be turbulence, thus being used for dispersion, mixing and vibrations. The turbulence play an important role regarding the momentum distribution.

There are many studies effectuated in which the authors uses CTD simulation in different submersible and underwater vehicle projects reducing costs and save a lot of time compared with experimental tests. For example in his article, Barros et al.[4] compared the analytical results obtained with numerical results. As a conclusion, the CFD method is very usefully because help to realise an economically evaluating regarding the manoeuvrability of an underwater vehicle.

2 Methodology

2.1 The geometry of body vehicle

The shape of an underwater vehicle is similar to that of a torpedo type (in other words, in a form of cylindrical body with the front part rounded). At the first design stage using SolidWorks program, the hull can be divided into four sections, namely the nose (the forward part of vehicle), middle (the part where are located the sensors and sonars equipment), tail (the after part) and the propulsion used to propel the vehicle. To model the profile of the front and after part (or we can named the bow and stern- making an analogy to the ship parts) of the vehicle body it was used Myring Equations [5]. The Myring hull profile Equations describes curves information regarding the bow and stern of the vehicle body which in the end generate the smallest possible drag coefficient along the body vehicle.

The bow is described using a modified semielliptical radius distribution described as equation (1):

$$r(x) = \frac{1}{2}d\left[1-\left(\frac{x-a}{a}\right)^2\right]^{\frac{1}{n}} \quad (1)$$

Furthermore, the stern is defined using a cubic relationship described as equation (2):

$$r(x) = \frac{1}{2}d - \left(\frac{3d}{2c^2} - \frac{tan\theta}{c}\right)(x-a-b)^2 + \left(\frac{d}{c^3} - \frac{tan\theta}{c^2}\right)(x-a-b)^3 \quad (2)$$

Where x represent the axial distance to the bow, a, b, c represent the lengths of the bow, middle and stern. The parameter d represent the middle hull diameter, n represent the index of the stern shape, and θ the stern semi-angle. All there parameters are geometric, expect of the parameter$n$. In Figure 1 are shown the schematic figure of an underwater vehicle hull and also the geometric parameters of Myring Equations.

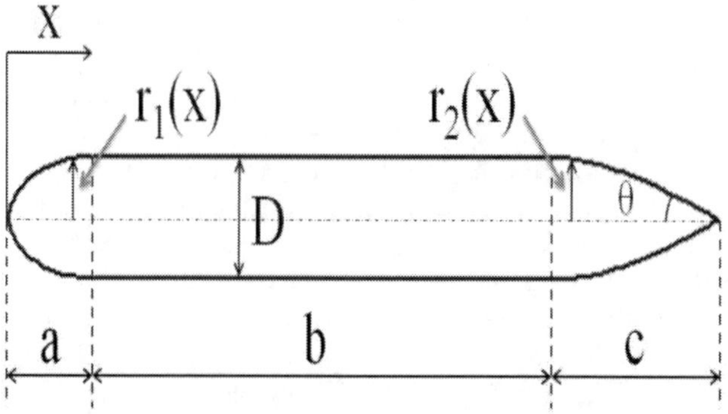

Fig. 1. Schematic figure represent the underwater vehicle hull and geometric parameters of Myring Equations.

2.2 Hull design

The concept design of the underwater vehicle is shown in Figure 2. To realise the design was used Solid Works Program.

Fig. 2. The concept design using SolidWorks Program

The steps of design the vehicle hull was as follow:

1. On the first step, was created a Sketch with vehicle body dimensions;
2. After, we doubling this Sketch created on first step in the OY direction to generate the lower part of the body vehicle along the OX axis (Figure 3);

Fig. 3. Doubling the Sketch in the OY direction

3. Using the Mirror function, we create create mirrors propulsion on both sides of the vehicle; (Figure 4)

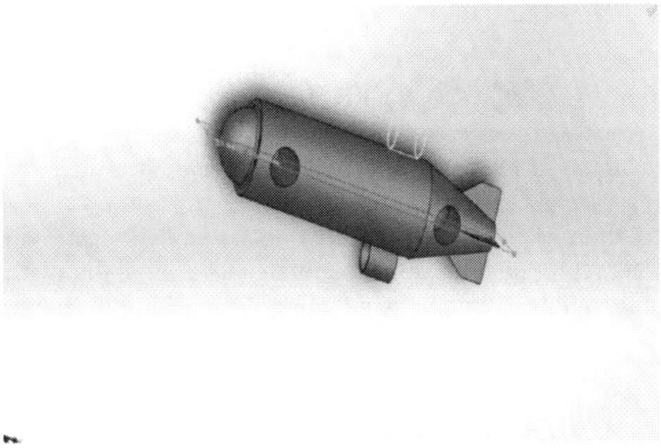

Fig. 4.Mirror propulsion on both sides

4. To realize the tail of the vehicle body, we create a new plan;
5. To create the from the tail of the vehicle was used The CirPattern function;
6. To fill the vehicle on both sides of the axle was used Revolve function;
7. Using Cut-Extrude function was create holes on the body of the vehicle;(Figure 5).

Fig. 5. The holes crated on the body vehicle

2.3 Numerical basis and turbulence model

To realize the simulation was used Ansys CFD software [7] solver.

The CFD simulation process involves three main stages:

- Pre-processing
- Solution
- Post-processing

The methodology for performing a CFD simulation includes several stages as follows:

• modelling, which involves creating the geometry step by step. First, points are created, joined by curves, then surfaces are created, and finally the body and field required to perform the analysis are generated.

• meshing (creation of the node network) also bears the name and discretization and represents a discrete representation of the geometry for each individual region which is then used to generate the volume of the node network. The meshing is made up of triangular and vertical faces.

The domain called "network" or "mesh is then discretized into a finite set of control volumes or cells. General equations of conservation of mass and momentum are discretized into algebraic equations. All equations listed are solved to create the flow field:

• calculation method;

• defining the boundary conditions;

• solution;

• post processing, which involves analyzing the results in one of two methods, graphically or alpha-numerically.

The graph shows contours, vectors, graphs, iso-surfaces, power lines and animations. Numerical are shown the integral values, calculations on the resistance to advancement, the distribution of pressure on the body surface compared to the experimental data.

The underwater vehicle is moving into an environment known as domain. In order to realise the simulation, a field considered wide enough was created to prevent the domain boundaries from affecting the flow along all the body vehicle. The dimensions of the computational domain (general parameters as length, width, depth) along the underwater vehicle are shown in Figure 6.

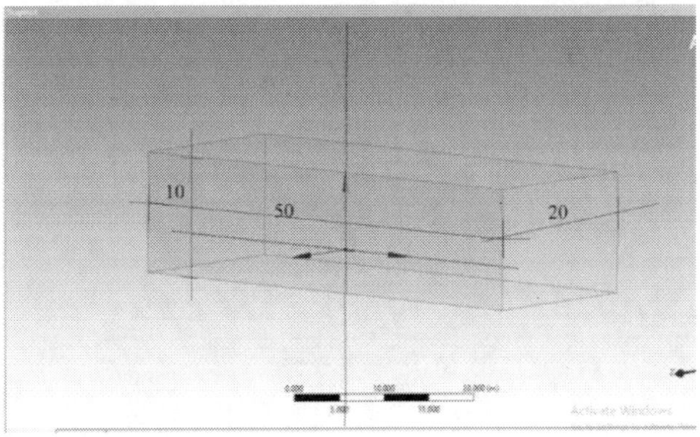

Fig. 6. The domain

For numerical determinations we consider the domain as a parallelepiped that encloses the body of the vehicle.

2.3.1 Generation of meshing

In order to numerically analyze the flow, the physical domain of fluid is discretized into network cells.[8] Flow variables are associated with each cell (element) in the network. This model uses the second-order finite element method. Compared to first-order discretization, it offers greater network convergence and greater discretization accuracy. Figure 7 shows the obtained cell network for the fluid domain in one of the studied cases.

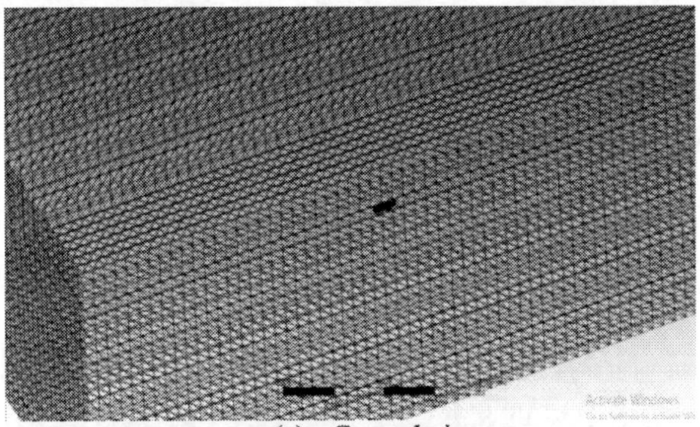

(a) General view

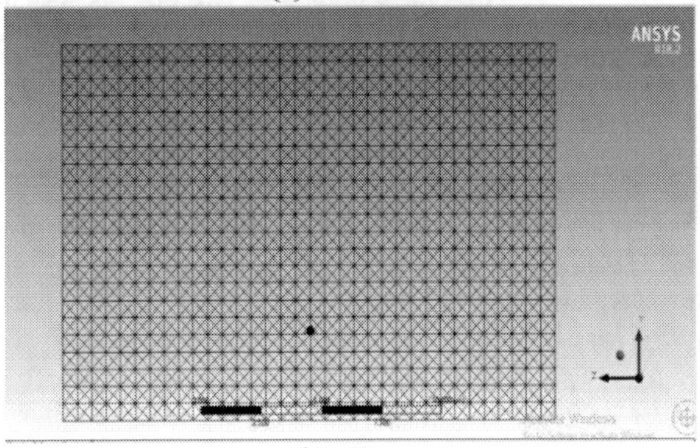

(b) Bow zone

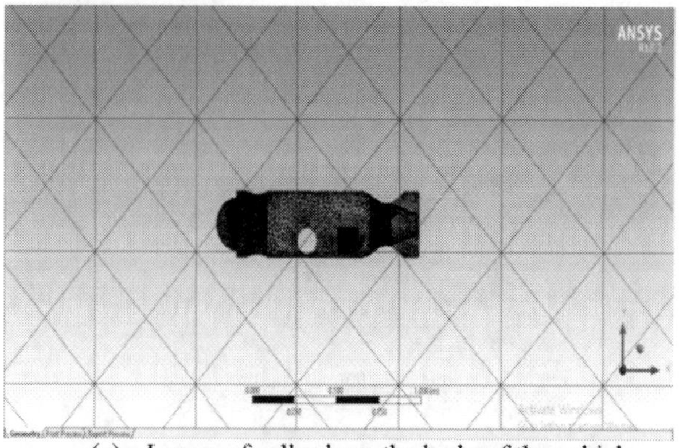

(c) Layers of cells along the body of the vehicle

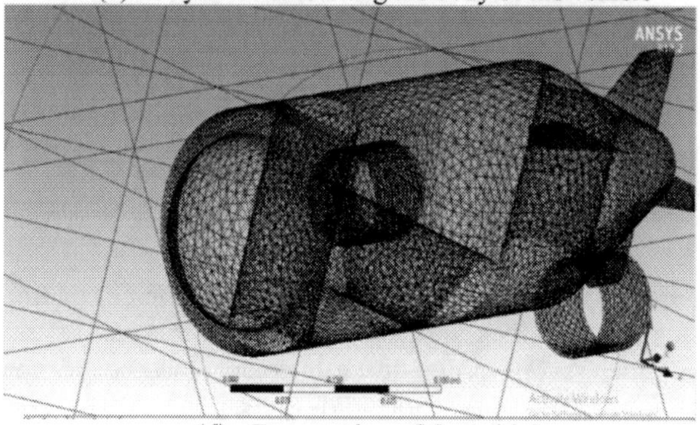

(d) Bow section of the vehicle

Fig. 7. Meshing

The conditions imposed on the mesh are: size function – curvature, maxim face size 0.605520m, mesh deseature size 3.0276e-003m, transition growth rate 1.1280, bounding box 56.7370m and minimum edge 8.0131e-006m.

To generate the phenomenon of flow during the simulations, the density of the node network was more focused in different regions of the domain. Regarding this, a fine mesh was used along the surface of the all body vehicle to resolve the equation of flow in the boundary layer.

The Skewness and Orthogonal quality functions from Mesh Statistics - Mesh Metric were used to determinate the quality of the generated network, these being the most important methods for determining the quality of the network.

The results show in final 163385 knots and 912137 elements was formed in mesh.

2.3.2 Convergence criterion

To decide if a solution has reached the desired level of convergence it is useful to monitor the residuals of the flow variables for each iteration. The residuals represent a measure of the imbalance between the left and right sides of a discretized transport equation. They are the most important measures of convergence, because they are directly related to the correct solution of the equations.

The CFX solver uses standardized residuals to determine convergence. Standard residuals are used to automatically stop when the solver reached a specified level. To observe when convergence is achieved, the user must select the residual type, MAX (maximum) or RMS (root mean square) and specify a target value. In the CFD simulations, the RMS was used residuals with a target value of 10-5, which represents according the program license a good convergence and is considered sufficient for the most part of engineering applications.

2.3.3 Boundary conditions

The condition at the output limit used was the pressure outlet, which sets the value of static pressure at the output limit, being the most suitable for CFD marine simulations. It can be assumed that the output pressure is equal to the hydrostatic pressure, if the location of the output limit is made far enough from the target object of the turbulence simulation in the flow field.

We considered the vehicle positioned at a single depth and moving with constant speed. Furthermore, we add the following assumptions and simplifications:

- consider the incompressible fluid be homogeneous flow;
- the vehicle is moving linear with constant speed;
- free surface;
- during the simulation, we not perform a dynamic mesh;
- the geometrical design of the vehicle was done on a real scale according to reality;
- fluid flow was realize in a parallelepiped area around the all body of the vehicle.

The condition imposed for the domain entry limit was of the inlet type, and the fluid velocity corresponded to the relative velocity of the body vehicle achieved by summing the velocity of the body vehicle and the velocity of the water current for each simulation. The simulation speed was 10 m/s. The flow regime was defined as subsonic, the mass and moment having normal speed values. The turbulent intensity was defined as average and the Eddy viscosity coefficient was taken into account.

The condition imposed at the output limit of the domain was of the output type, with the option of the subsonic flow regime, the mass and the moment at normal speed, and the corresponding output speed having the same value as the input.

The condition imposed on the bottom of the domain (bottom) was of the no slip wall type. Also, the roughness of the bottom of the domain was defined as a smooth wall.

The properties of the water used in the simulations are density 1000 kg / m3, the model of heat transfer being isothermal, the fluid temperature of 25 ° C. The model used for turbulence is SST (hear-stress transport)[6].

2.4. Model the flow around the body of the vehicle

The vehicle body position varies along the OY direction, the simulations was performed for different positions of the body vehicle.

The moving of the vehicle was considered stationary and subsonic. For modeling, wrte not taken into account the heat exchange and weight forces. We consider the wall roughness to be 5 μm.

During the simulations the vehicle speed and current acting along the body vehicle summed a total of 10m / s, we consider the average temperature of water to be 25 degrees. The vehicle is positioned to a maximum depth of 7 meters.

Figure 8 present the total pressure along the vehicle body. The simulation show a maximum value of pressure of 50891.5 Pa, reached in the front part of the vehicle , according to the legend where the maximum value of pressure is represented by colour red.

The minimum reached is represented by a value of -92984.5 Pa. The minimum value is located in the immediate vicinity of the vehicle's hull, besides the front part of vehicle and is represented in figure 8 using blue color.

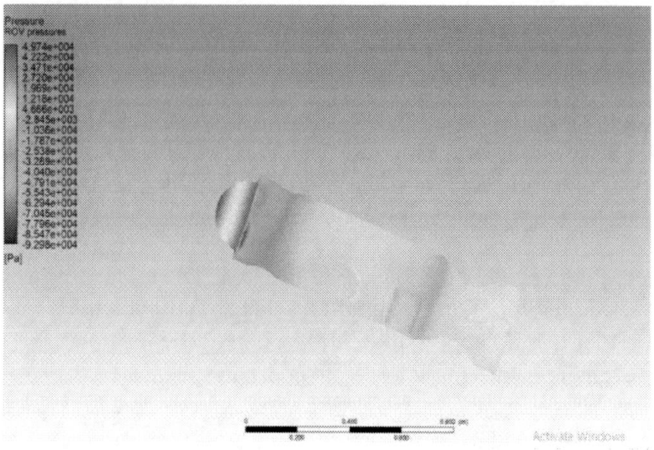

Fig. 8. Simulation regarding the pressure along the body vehicle

In Figure 9 and Figure 10 are presented the streamlines, pressure gradient lines, respectively the velocity that appear along the vehicle body.

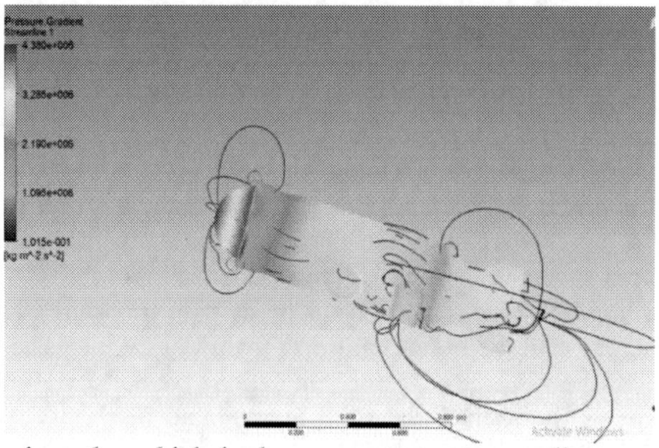

Fig. 9. Streamlines along the vehicle body - present pressure gradient

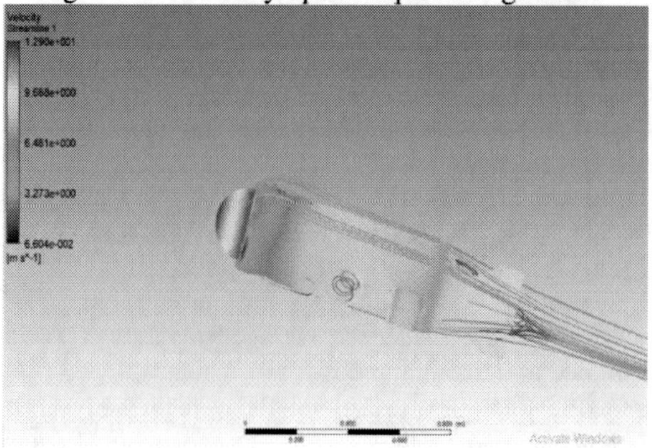

Fig. 10. Streamlines along the vehicle body – present velocity

In Figure 11, is presented the pressure distribution along the transverse surface of the body vehicle. Take into consideration the simulation, we notice a zone of maximum of 50891.4 Pa pressure in the front part of vehicle body marked with red colour, then a minimum area in the immediate vicinity of the front part of body vehicle. The average area, marked with green colour, is located near the prow, next to the propellers vertically and towards the stern of the vehicle.

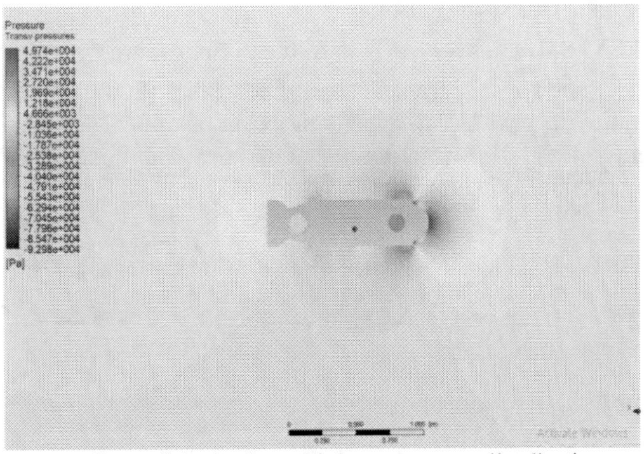

Fig. 11. The transverse surface of the body vehicle – pressure distribution

3 Conclusion

This paper presents a hydrodynamic single-pass flow distribution regarding the pressure distribution around of an underwater vehicle body.

In the future, the simulated results will be verified using the experimental tests in tank and after that in sea water. We consider that future works could improve and establish the accuracy of the CFD results comparing the numerical simulation with the underwater vehicle using a real-time adaptive identification approach in sea trial.

Vehicles can determine areas of interest for operation. The underwater vehicle has improved our ability and perspective to research the sea-floor, providing higher resolution of seafloor mapping data.

References

1. Proudman, I. and Pearson, J.R.A. (1957) Expansions at Small Reynolds Numbers for the Flow Past a Sphere and a Circular Cylinder. Journal of Fluid Mechanics, 2, 237-262. http://dx.doi.org/10.1017/S0022112057000105
2. Masliyah, J.H. and Epstein, N. (1971) Steady Symmetric Flow Past Elliptical Cylinder. Industrial & Engineering Chemistry Fundaments, 10, 293-299. http://dx.doi.org/10.1021/i160038a017
3. Rosenfeld, M., Wolfshtein, M. and Israeli, M. (1992) A Numerical Study of the Laminar Incompressible Flow Over a Prolate Spheroid at 10° Incidence. International Journal for Numerical Methods in Fluids, 15, 147-173. http://dx.doi.org/10.1002/fld.1650150203

4. Barros, E.A., Dantas, J.L.D., Pascoal, A.M. and Sá, E. (2008) Investigation of Normal Force and Moment Coefficients for an AUV at Nonlinear Angle of Attack and Sideslip Range. IEEE Journal of Oceanic Engineering, 33, 538-549. http://dx.doi.org/10.1109/JOE.2008.2004761
5. Myring, D.F. (1976) A Theoretical Study of Body Drag in Subcritical Axisymmetric Flow. The Aeronautical Quartely, 27, 186-194.
6. Menter, F. R. (August 1994). "Two-Equation Eddy-Viscosity Turbulence Models for Engineering Applications". AIAA Journal. 32 (8): 1598–1605. Bib code:1994AIAAJ .32.1598M. doi:10.2514/3.12149.
7. ANSYS CFX- Solver Theory, ANSYS, Inc., Canonsburg.
8. Juong, T., Sammut, K., He, F. and Lee, S.K. (2009) A Study on the Design Optimization of an AUV by Using Computational Fluid Dynamic Analysis. Proceedings of the 19th International Offshore and Polar Engineering Conference, Osaka, 21-26 June 2009, 696-702.

The strong connection between marine engineering and marine environmental education within a marine cluster

Mirela Iuliana Sundri[1,*] and *Feiza* Memet[1]

[1]Constanta Maritime University, Faculty of Naval Electromechanics, 104 Mircea cel Batran Street, 900663 Constanta, Romania

Abstract. Marine Environmental Education (MEE) is a vital activity, since the marine environment is impacted by industrial and technologies involved in this sector. MEE is in strong connection with marine engineering. The players from this industry have to comply with environmental protection protocols and conventions. A marine cluster is a good way of approaching issues specific to the exploitation of resources of the sea. In the present, a particularity of marine pollution is that in enclosed seas (such as Black Sea is) and on coastal areas, the pollution is higher than in the open ocean. In order to fight with this reality, it is also vital to increase the public environmental awareness throughout specific education, not only to train specialists in this respect. This paper is providing the pylons on which MEE relies. From this paper will succeed how the objectives of MEE (awareness, knowledge, attitude, skills, participation) and its specific actions (environmental education and training, involvement, bringing people together) are supported by a marine cluster. The most important result of this study can be stated as: processes developed within MEE will enforce the involvement of specialists in solving marine pollution issues and will rise the environmental consciousness of communities, in a framework provided by a cluster oriented towards MEE.

1 Introduction

A cluster is a union, an association between different entities, geographically concentrated and interrelated in a certain social field of activity, seeking to achieve a common goal [1]. According to Krugman, cluster members establish, in time, a social bond able to act in the manner of working in a positive direction [2].

Clusters are frequently used as tools of measurement of the economic specialization of a country or of a region/cross-borders region, since the competitiveness of the cluster is affected by the activity of the member entities and also by the relationship between them [3].

* Corresponding author: iuliana.sundri@cmu-edu.eu

Entities entering in a cluster are aware of the main advantages of this union, as seen in figure 1 [4].

Maritime transport is based on two pylons: maritime shipping and port activities; its performances depend on sustainability-which is achieved throughout overcoming barriers to innovation [5].

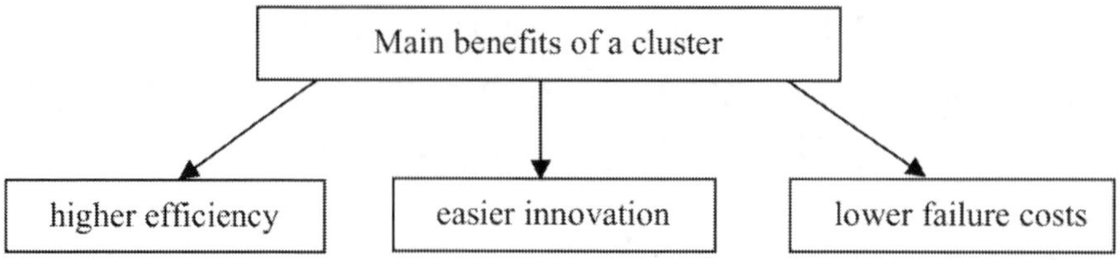

Fig. 1. Main advantages offered by a cluster to its members

It is expected that a maritime cluster will be a mean to achieve a better articulation and will strengthen the usage of synergies, while it will facilitate the construction of a sustainable and integrated perspective of the sea, of its resources and wide range of related activities (see figure 2) [6], [7].

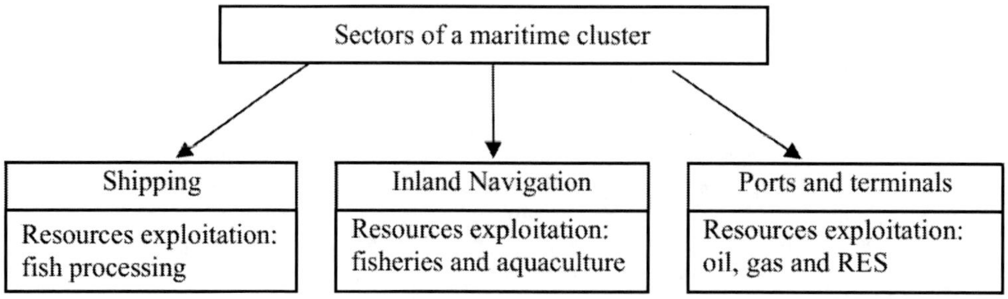

Fig. 2. Layout of sectors of a maritime cluster

In Europe, marine clusters are seen as important entities due to the fallowing realities [8]:

- approximately 40% of the world fleet belongs to our continent;
- Europe is in the top of coastal vacation destination list;
- European yacht builders lead this market, European dredging and RES companies as well;
- European maritime research, inland shipping and fishing are in the head of international hierarchy.

There are several marine clusters all over the world. Their objectives are gathering actors in order to reach desiderates such as: sustainable growth in the region, competitive maritime sector, integration of regional maritime activity in the international one. The relevance and the objective of this study consist in the integration of Marine Environmental Education activity in the classic tasks of a cluster, established in the Black Sea region, in order to form a new mentality of the people from this region and to strengthen specialist's cooperation, for solving actual pollution issues, but also to prevent future ones. In this respect, in the following we will describe the pylons on which MEE relies, and the objectives of MEE. For a successful MEE activity in the Black Sea Region, we provide the actions of MEE which must be developed within the Black Sea Marine Cluster.

2 Sources of marine pollution

The majority of coastal areas are affected by pollution, outputs of this negative aspect being found in commercial coastal and marine fisheries. This is why it is important to have in view the following brief presentation of the significant pollutants, when trying to fight against [9], [10], [11], [12]:

- *agricultural activities*: represent approximately 50% of water pollution sources in throughout agricultural input products, such as nutrients; regarding aquaculture, it is stated the need of control the effluents, since pesticides are known to be an aggressive agent acting on the aquatic ecosystem;
- *domestic and industrial activities*: domestic and industrial sewage are met in water courses, even in the ocean, resulting a diminish in the oxygen level fact that is found in a lower quality of river, sea and ocean waters;
- *oils*: oil from shipping activities is the major marine pollutant; this pollutant may end in seas and oceans by different ways (spills, daily activities, runoffs or clumping) which causes damage on marine organisms, aquatic plants and birds;
- *heavy metals and trade elements*: these pollutants (outputs of various industries) end in the seas or oceans through atmospheric or onshore effluent sources; they cause disturbs in the metabolic process and loss of intercellular constituents;
- *organic compounds*: are dangerous because of their high toxicity and long life in the surroundings and in the biological systems;
- *plastics*: this is a banned pollutant because they affect the marine environment through entangling life and damaging coral reefs;
- *suspended particulate matter*: cause difference in levels, bound in restrictions of sunlight penetration; as a result, it is registered an obstacle in the growth of plants (such as algae);
- *eutrophication and algal bloom*: have a negative impact on aquatic animals and, also, in time, it will be found an important layer of dead and decomposing organism on the bottom; moreover, they are totally unaesthetic;
- *biological pollution*: are issues of concern raised by delivery, movement and existence processes specific to non-indigenous or invasive species in waters and estuarine areas; this type of pollutants have the capacity of diminish the fitness for survival of level of biological organization.

Strictly regarding the Black Sea (which is an important sea among European Seas, since it supports economics of six countries) anthropogenic activities, eutrophication, natural variability and climatic change are topics of interest for specialists. Tendencies in the development of coastal communities lead to uncontrolled industrial and domestic wastes. In addition, it should be also considered the fact that Black Sea is of semi closed type, this is why aeration of deep water layers through lateral flows is not fast. Moreover, strong density stratification inhibits vertical mixing.

Having in view the various pollution issues specific to the Black Sea, the prosperity of the communities and the natural diversity of species, measures to increase environmental awareness have to be taken. In this respect, this paper is bringing into attention the need of a Black Sea Marine Cluster set up, dealing with Marine Environmental Education (MEE). We strongly affirm that this cluster can be used as an effective tool of educating communities and stimulating specialists through, aspect which will be discussed in the following.

3 Marine Environmental Education in the task of Black Sea Marine Cluster (BSMC)

Our opinion is that the Black Sea Marine Cluster oriented towards MEE should be developed as a geographical concentration of different bodies, from countries which border Black Sea, willing to cooperate within a busiest model that is able to bring development, innovation, competitiveness and growth in the region. These countries are: Romania, Bulgaria, Ukraine, Russia, Georgia, and Turkey. The members of BSMC should come from different fields, so that they have to be: universities, high schools, research institutes, companies, NGOs, public authorities.

Marine Environmental Education (MEE) is the link between marine engineering and marine environment. MEE is a tool at the hand of cluster members that enables involvement and learning by persons of different ages in order to review and change perspective on the preservation of clean seas and coastal areas.

We found the pylons on which MEE rely. They are given in figure 3 and we describe them below:

1. *fundamental environmental concept and elements*: should be introduced in existing disciplines delivered to students in high school and universities;
2. *new environmental chapters and modules*: included in existing courses in order to catch the detail of the environmental problem;
3. *new extra curricula programs and courses*: delivered by a different institution, which end with a certification, in order to upgrade existing knowledge and understanding;
4. *more scientific papers and research studies on projects*: must be released in order to indicate how specialists are facing the challenge;
5. *higher priority on trainers training*: in order to make them spread updated information and to enforce of tertiary education and research capabilities;
6. *intensive specific research on MEE policies and practices*: in order to get good outputs through MEE;
7. *intensive know-how exchange between users of marine environmental skills*: in order to take advantage of partners' achievements in the field;
8. *attention on adult and community marine environmental education through formal education (form of education developed in schools and universities) and non-formal education (form of education based on a strategy which do not imply student attendance)*: in order to increase the global environment literacy standards.
9.

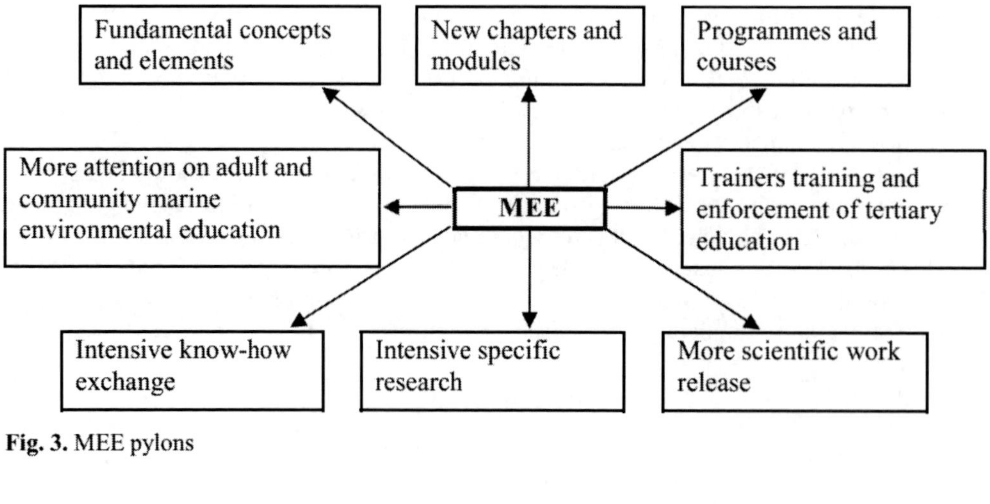

Fig. 3. MEE pylons

The aim of MEE is to make people aware of the importance of the marine and coastal environment and make them act in a manner which is in harmony with the marine environment, by gaining knowledge and skills specific to a sustainable interaction with the natural world.

The objectives of MEE, identified in this paper are as follows:

I. *awareness* – MEE has to make people awake about the problems related to the environmental debate;

II. *knowledge* – MEE has to contribute to the knowledge gaining (by individuals and social groups) related to their environment;

III. *attitude* – MEE has to determine individuals and social groups to accept modifications of behaviour in order to reach cleaner water and coasts

IV. *skills* – MEE has to make people have necessary skills needed for the identification of the marine environmental problem;

V. *participation* – MEE has to make people to feel and act responsible so that marine environmental issues are solved in the region.

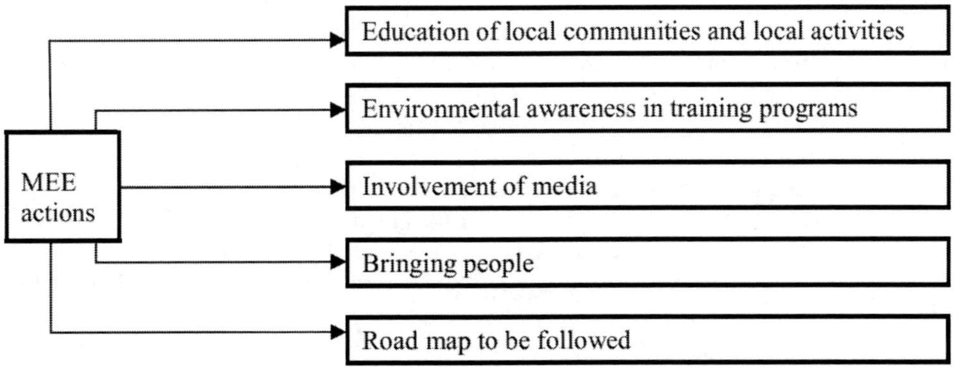

Fig. 4. MEE actions

We provide below the specific MEE actions (see figure 4):

a) *education of local communities and local representatives*: emblematic figures of the communities may decisively influence the behaviour of persons, especially in rural zones; first, an assumed attitude of this individuals towards clean seas and coastal areas has to be noticed;

b) *environmental awareness in training programs*: is a feature developed by training programs held in institutions, for working adults, in order to upgrade their knowledge, but also, marine environmental concepts might be included in student curricula;

c) *involvement of media*: in its different forms (TV, radio, newspaper) is of huge importance since media reach any individual of a community;

d) *bringing people together*: it is a vital action – BSMC has the capacity of bringing together specialists, future specialists, representatives of governmental organization and journalists, in order to effectively work within projects or networks aiming a cleaner sea and coasts;

e) *road map to be followed*: it has the following five objectives:

• management of climate change – it is about communication with enterprises, finding challenges needing further assessment and decision support, inclusion of climate change adaptation within each entity's activities;

• protection of basic resources against potential impacts related to climate change – it is about identification of potential impacts of climate change and risks determination in order to identify specific measures;

- ensure the communities against more often natural events and disasters – it is about finding the proper way of fighting against physical and economical losses; it requires adequate policies and standards, funding for specific researches and work force;
- contribution to safety, stability, security and environmental protection – it is about support actions able to serve as foundation on which safety, stability and security in the region are built;
- publicity – it is about the organization of press conferences and radio campaigns throughout banners, booklets and leaflets within an awareness raising campaign; are aimed valuable assets and workforce development opportunities, in order to reach the mentality of an important number of people living in the region.

4 Conclusions

The future of the Black Sea strongly depends on actions taken by people.

Marine and coastal pollution in the region is now subject of concern. Specialists and local communities have to change radically their mentality and life. But, as stated in this paper, this is possible throughout environmental awareness and education intensification.

We highlighted the fact that the BSMC might coordinate the actions needed to be taken in order to make communities and specialists to work together in order to establish a harmony with the sea.

This desiderate will be reached through Marine Environmental Education (MEE), managed by this cluster, as tool at the hand of the cluster members: education and research institutions, companies, NGOs and public administration entities.

The objectives and the actions of MEE, given in this paper, will enforce the involvement of specialists in solving marine pollution issues and will rise the environmental consciousness of communities, in a friendly framework provided by the cluster.

References

1. M.R. Funaric, K.P. Galic, IJHASS, **5 (5)** 650 (2011)
2. P.R. Krugman, *Geography and Trade Leuven*, 156 (1991)
3. M.H. Nijam P.W. de Langen, *ERSA 2003 Congress*, 23 (2003)
4. J.P. Rodrigue, International Transport Forum 2010, Transport and Innovation: Unleashing the Potential 22 pp. (2010)
5. P. Ripatti, *Creating a Marine Cluster. From Product Supply to System Solutions*, 70, (2014)
6. P. Monteiro, T. de Noroha, P. Neto, Sustainability **5 (9),** 4076 (2013)
7. S. Theodoropoulos, E. Tassopoulos, IJEBA, **2, 3**, 3-14 (2014)
8. European Commission, https://ec.europa.eu/maritimeaffairs/sites/maritimeaffairs/files/docs/body/report_results_en.pdf,
   Accessed in March 2020
9. M.S. Islam, M. Tanaka, Marine Pollution Bulletin, **48**, 624-649 (2004)
10. D. Xanthos D, T.R. Walker, Marine Pollution Bulletin, **118, 1-2**, 17-26 (2017)
11. M. Vikas, G.S. Dwarakish, *Aquatic Procedia* **4,** 3, 381-388 (2015)
12. L. Bat, A. Oztekin, F. Şahim, E. Arici, U. Ozsandikei, *MedFAR* **1 (2)**, 66-86 (2018)

E3S Web of Conferences **180**, 04005 (2020) https://doi.org/10.1051/e3sconf/202018004005
TE-RE-RD 2020

Mathematical modelling and numerical simulation of linear electro-hydraulic servomechanism with stepper motor

Vergil Muraru[1,*], *Constantin* Calinoiu[2], *Sebastian* Muraru[1], *Ana* Dulgheru[1] *and Cornelia* Muraru-Ionel[1]

[1]INMA Bucharest, CAE Department, 013811, Bucharest, Romania
[2]Politehnica University from Bucharest, Power Faculty, 060042, Bucharest, Romania

Abstract. The paper presents the results of research on the performance of linear electro-hydraulic servomechanisms with stepper motor, evaluated by mathematical modelling and numerical simulation with the AMESIM software package. These servomechanisms have a great potential for application, including for agricultural technical equipment. Servomechanisms of various types are used for the rapid and high precision operation of the various systems. In practice, electro-hydraulic servomechanisms have been required in the case of the systems requiring high forces or moments. These equipment are automatic hydraulic tracking systems and can be analysed with methods specific to automatic systems. The paper contains a structural description of an electro-mechanical servomechanism with stepper motor, its operation, mathematical model and its performances obtained by numerical simulation. The mathematical modelling of the electro-hydraulic servomechanism is based on the relations between the input sizes and output sizes of the equipment in its structure as well as the connection relations between these equipment. Based on the obtained results it is demonstrated that the analysis and synthesis of electro-hydraulic servomechanisms with stepper motor can be determined by numerical simulation using the AMESIM simulation software.

1 Introduction

A linear electrohydraulic servomechanism is an automatic tracking system that achieves the transformation of the input size of electrical nature into a linear displacement (output size), developing also simultaneously a high output force.
The elements from the structure of a typical linear electro-hydraulic servomechanism are: an electrohydraulic amplifier, a hydraulic cylinder, a displacement transducer whose output signal constitutes the reaction link of the servomechanism and a servo controller (electronic block) [1].

* Corresponding author: virgil.muraru@gmail.com

The performances of the electrohydraulic servomechanisms are dependent on the performances of the electrohydraulic amplifiers in their structure.
The electrohydraulic amplifier has in its structure an electromechanical converter and an electrohydraulic converter (hydraulic distributor) [2].
The electromechanical converter is an equipment that has as an input signal an electrical size, and at its output it has a displacement or a force that is proportional to the input size.
The shape, dimensions and performances of the electrohydraulic amplifier are dependent on the type of electromechanical converter from its structure [3]. In practice, the following types of electromechanical converters were imposed:

a) converter with permanent magnet and movable coil (Fig. 1);
b) converter with permanent magnets and fixed coils ("torque motor") (Fig.2);
c) proportional electromagnet (fig. 3);
d) linear motor force (fig. 4);
e) stepper motor (fig.5).

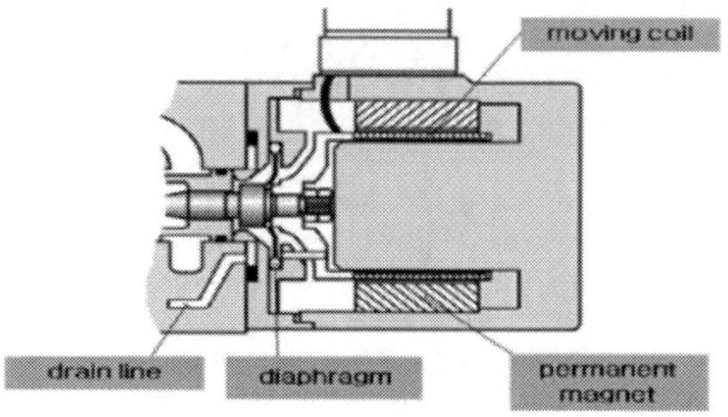

Fig. 1. Movable coil

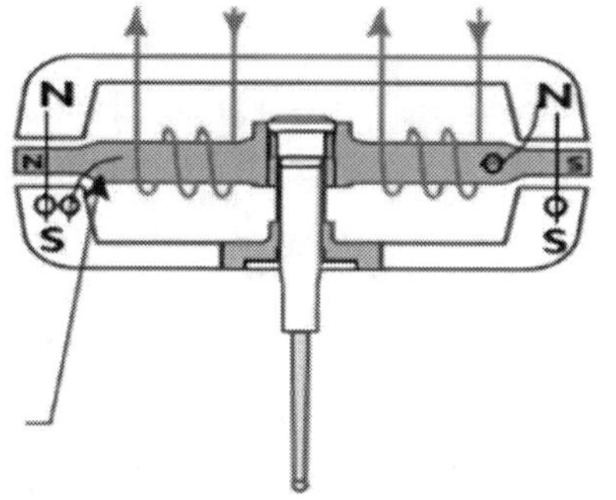

Fig. 2. Torque motor

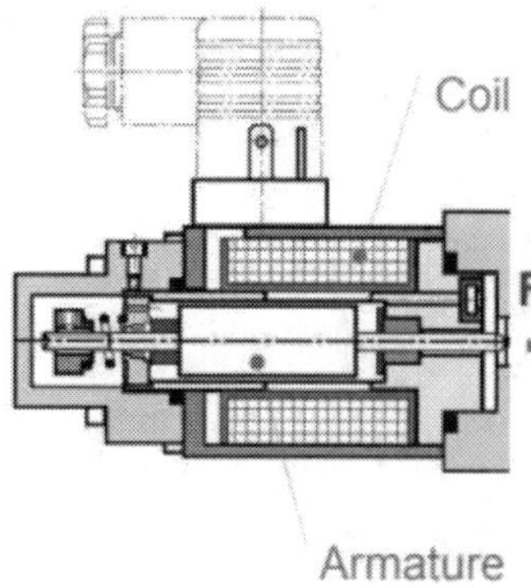

Fig. 3. Proportional magnet

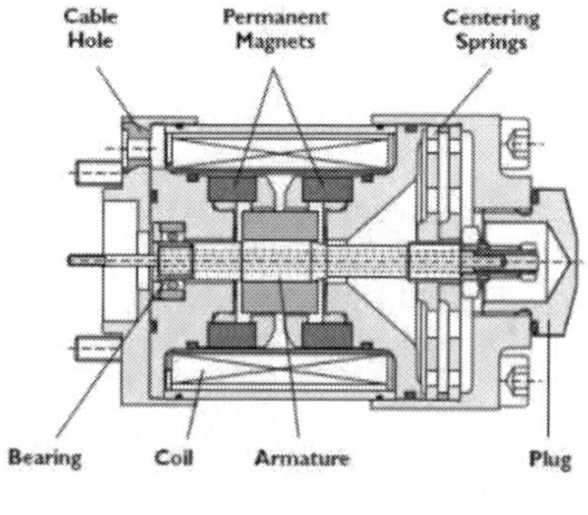

Fig. 4. Linear force motor

Fig. 5. Stepper motor

In the case of electrohydraulic servomechanisms, the Hydro-Mechanical Converter has the shape of a hydraulic distributor that transforms the input size of mechanical nature (displacement) into a hydraulic size (flow rate / debit).
The electrohydraulic servomechanisms with stepper motor do not have in their structure a displacement transducer for achieving the reaction link, their reaction size is of mechanical nature, and it is realized by means of a screw-nut mechanism.
Consequently, the servomechanism with stepper motor is a servomechanism which has the input of electric nature, the output of mechanical nature and the reaction size of mechanical nature.

2 Structure and operation

Figure 6 presents the principle diagram of an electrohydraulic servomechanism that has in its structure an electrohydraulic amplifier achieved with an electric stepper motor [4].

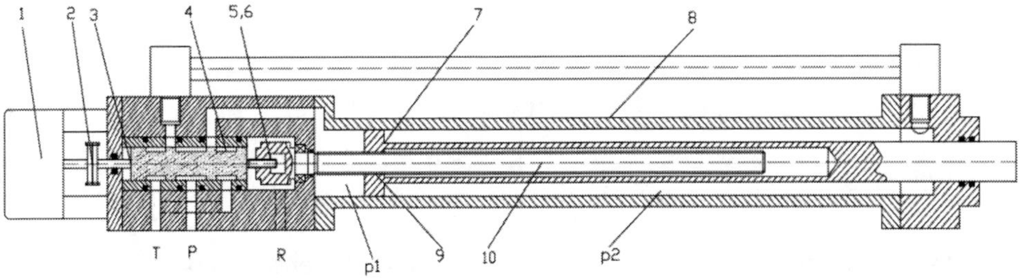

Fig. 6. Electrohydraulic servomechanism with stepper motor

The angular displacement of the shaft of the electric motor 1 is transformed into a linear displacement by means of the elastic coupling 2 and of the screw-nut assembly 5. When the motor shaft rotates with the prescribed angle $\boldsymbol{\theta}$, the screw 5 from the end of the rod 3 advances into the nut 6, which is temporarily fixed.
The axial displacement of the screw is possible because the elastic coupling 2 may deform axially, this being achieved in the shape of two membranes embedded on the contour.
The spool 4 is connected to the rod 3 through a spring (not shown on the drawing). The linear displacement of the rod is also transmitted to the spool. The spool is endowed with a rotational movement locking system, so that, it can only move linearly simultaneously with the rod.
When the motor shaft moves clockwise, the rod advances into the nut, the spool moves to the right, the distributor opens and the pressurized oil reaches the chamber from the left side of the hydraulic cylinder 8.
Simultaneously, the chamber from the right side of the hydraulic cylinder is connected to the tank. Under the action of the pressure forces, the piston 7 will move to the right with the value "z" (the cylinder rod is moving in the outward direction).
In the piston, a lead nut 9 is achieved, that has the sense of the screw thread, opposite to the direction of the nut 6.
When the lead screw moves linearly, the screw 10 is rotated and also drives the nut 6, and the screw 5, the rod 3 and implicitly the spool moves linearly in the opposite direction to the initial displacement. The displacement z of the piston stops when the distributor closes.

When the engine shaft moves counter clockwise, the pressurized oil reaches the chamber in the right side of the hydraulic cylinder and the output rod moves in the sense of its withdrawal. Thus, a linear relationship can be obtained between the input size $\boldsymbol{\theta}$ and the output size z.

3 Mathematical modelling

For the mathematical modelling of the electro-hydraulic servomechanism, it is necessary to write the relations between the input and output sizes of the equipment in its structure, as well as, the connection relations between these equipment [5-6].

3.1 The characteristic of the step by step electric motor

Angular displacement of the step by step electric motor shaft is proportional with the control voltage according to the relationship

$$\theta_y = K_m \cdot U \tag{1}$$

where:

- K_m - the transfer factor [*rad/V*];
- U – command signal (voltage) *[- (V)]*;
- θ_y – the rotation angle of the shaft [*rad*];

3.2 An equation of the linear displacement of the spool rod

The angular displacement, θ_y, of the stepper motor shaft induces a linear displacement x1 to the spool driving rod, due to the screw and nut mechanism, according to the next formula:

$$x_1 = \frac{\theta_y}{2\pi} p_{sr} \tag{2}$$

where

- x_1- spool rod displacement [*m*];
- p_{sr}- reaction screw step [*m*].

3.3 The equations of angular displacement of the screw and linear displacement of the spool rod

The lead screw has an angular displacement when the cylinder piston moves linearly, due to the advance screw and nut mechanism, according to the next formula:

$$\theta_z = \frac{2\pi}{p_{sa}} z \tag{3}$$

where:

- p_{sa}- lead screw step, [*m*];
- z – hydraulic cylinder piston displacement, [*m*].
- θ_z – screw rotation angle, [*rad*].

The lead screw also induces the nut the same angular displacement θ_z and this induces a linear displacement to the spool rod according to the other formula:

$$x_2 = \frac{\theta_z}{2\pi} p_{sr} = \frac{z}{p_{sa}} p_{sr} \tag{4}$$

where

- x_2 - spool rod displacement [m].

3.4 Comparator equation

The opening of the distributor is equal to the displacement of the spool, according to the next formulas:

$$x = x_1 - x_2 = \left(\frac{\theta_y}{2\pi} - \frac{z}{p_{sa}}\right) p_{sr} \tag{5}$$

or

$$x = x_1 - x_2 = \left(\frac{K_m \cdot U}{2\pi} - \frac{z}{p_{sa}}\right) p_{sr} \tag{6}$$

3.5 Piston displacement equation

$$A_{p1} \cdot p_1 - A_{p2} \cdot p_2 = m \cdot \ddot{z} + k_f \cdot \dot{z} + k_e \cdot z + F_v \tag{7}$$

where:

- A_{p1}, A_{p2} – the frontal areas of the piston, [m^2];
- p1, p1- the pressures in the cylinder chambers, [*Pa*];
- m - the equivalent mass of the actuated system reduced to the piston rod, [*kg*];
- z, $\dot{z}$, $\ddot{z}$ - displacement, speed and acceleration of the piston, [*m*], [*m/s*], [m/s^2];
- k_f- viscous friction coefficient [*Ns/m*];
- k_e - elastic force coefficient [*N/m*];
- F_v - disruptive force [*N*].

3.6 Continuity equation corresponding to the hydraulic cylinder

The continuity equation corresponding to the hydraulic cylinder has the next form:

$$Q_1 = A_{p1}\frac{dz}{dt} + c_{ip} \cdot (p_1 - p_2) + c_{ep} \cdot p_1 + \frac{V_1}{E}\frac{dp_1}{dt} \tag{8}$$

$$Q_2 = A_{p2}\frac{dz}{dt} + c_{ip} \cdot (p_1 - p_2) - c_{ep} \cdot p_2 - \frac{V_2}{E}\frac{dp_2}{dt} \tag{9}$$

where:

- Q_1, Q_2 - the flow rates corresponding to the cylinder chambers [m^3/s];
- c_{ip} - the internal losses coefficient [$m^3/s.Pa$];
- c_{ep} - the external losses coefficient [$m^3/s.Pa$];
- V_1,V_2 - cylinder chamber volumes [m^3].

3.7 The flow rate equation through orifices

The general equation for flowing through orifices with sharp edge is:

$$Q = c_d \cdot A(x) \cdot \sqrt{\frac{2\Delta p}{\rho}} \quad (10)$$

where:

- Q - the debit transited through the orifice [m^3/s];
- c_d – the orifice flow coefficient [-];
- A(x) – the orifice area [m^2];
- Δp- pressure drop on the orifice [Pa];
- ρ – the fluid density [kg/m^3].

In the concrete situation of the servomechanism, this equation has a particular form:
For $x \geq 0$

$$Q_1 = c_d \cdot A(x) \cdot \sqrt{\frac{2(p_s - p_1)}{\rho}} \quad (11)$$

$$Q_2 = c_d \cdot A(x) \cdot \sqrt{\frac{2p_2}{\rho}} \quad (12)$$

for $x < 0$

$$Q_1 = c_d \cdot A(x) \cdot \sqrt{\frac{2p_1}{\rho}} \quad (13)$$

$$Q_2 = c_d \cdot A(x) \cdot \sqrt{\frac{2(p_s - p_2)}{\rho}} \quad (14)$$

4 Results and discussions

4.1 Calculation of the servomechanism resolution

The servomechanism resolution is dependent on the value of the stepper motor resolution $\Delta\theta$ and the lead screw step value p_{sa} and is calculated with the next formula:

$$Rez = \frac{\Delta\theta \cdot p_{sa}}{2\pi} \; [mm]$$

The value $\Delta\theta$ is calculated with the formula

$$\Delta\theta = \frac{2\pi}{Np}$$

Where ***Np*** represents the number of steps of the motor for a complete rotation.

Table 1 shows the servo mechanical resolution values for different values of the motor resolution and different values of the lead screw step.

Table 1. Servomechanism resolution

psa [mm]	The number of motor steps per rotation		
	100	200	400
	Resolution [μm]		
4	40	20	10
6	60	30	15
8	80	40	20

4.2 Numerical simulation

Figure 7 shows the block diagram of the servomechanism

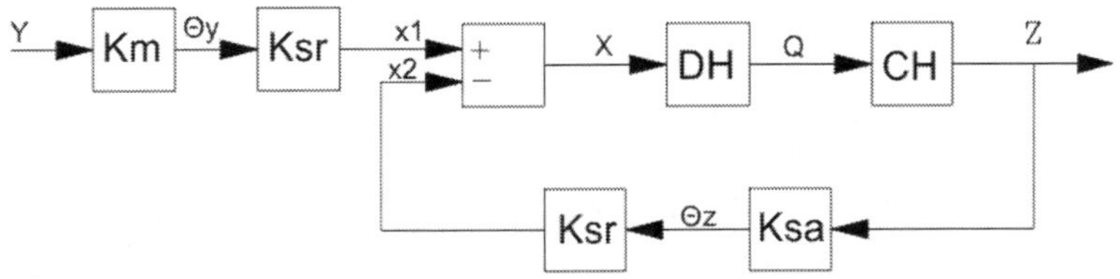

Fig. 7. Block diagram of the stepper servomechanism

For the numerical simulation of the servomechanism operation, the AMESIM software package was used [7-8]. The scheme of the simulation network is shown in Fig. 7.
The numerical simulation was performed for the following numerical values of the parameters: Piston diameter D = 50 mm, Rod diameter d = 35 mm,
The total piston stroke h = 500 mm, the supply pressure Ps = 200 bar, the reaction screw transfer factor Ksr = 2π rad / mm, the lead screw transfer factor Ksa = 4mm / 2π rad, the
For easiness of the results interpretation, the input size Y has been entered in digital form having millimeter as measuring unit.

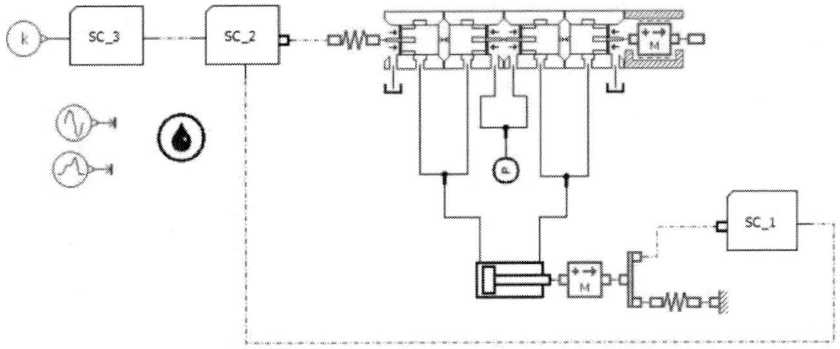

Fig. 8. The servomechanism simulation network

The results of the numerical simulations are presented synthetically in the form of diagrams.
In Figures 9, the numerically simulated command signal is presented and in Figure 10, the response of the servomechanism to the sinusoidal signal is shown.

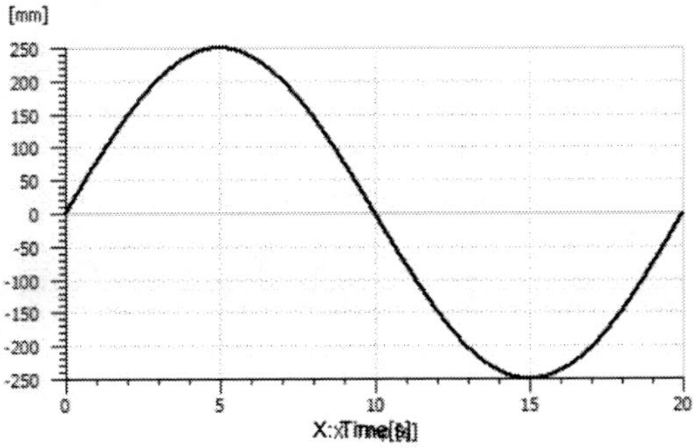

Fig. 9. The numerically simulated command signal

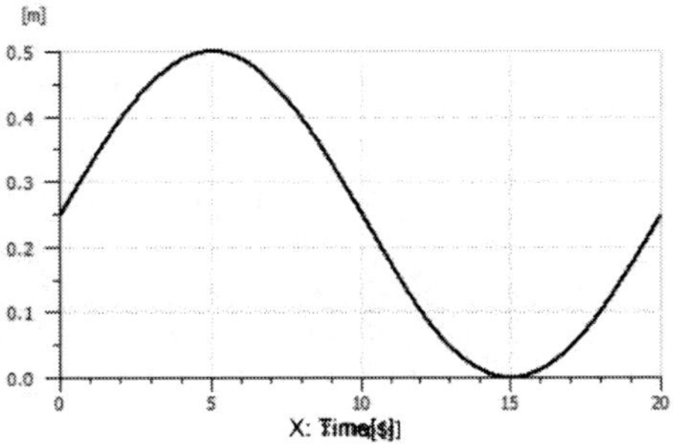

Fig. 10. Piston displacement

The servomechanism static characteristic is shown in Figure 11 and in Figure 12 is shown the variation in time of the pressures in the cylinder chambers when it is controlled with a sinusoidal signal.

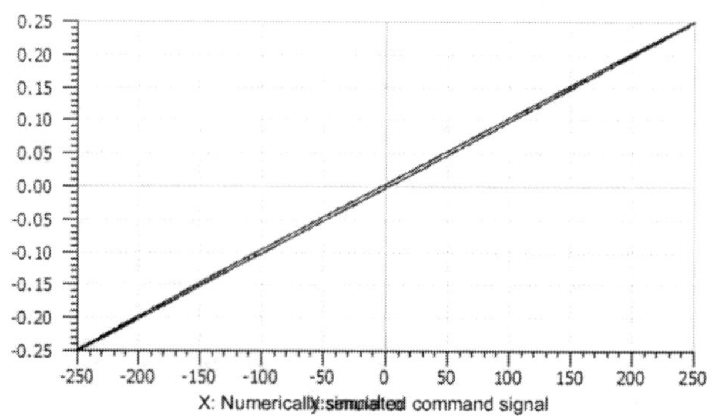

Fig. 11. The servomechanism static characteristic

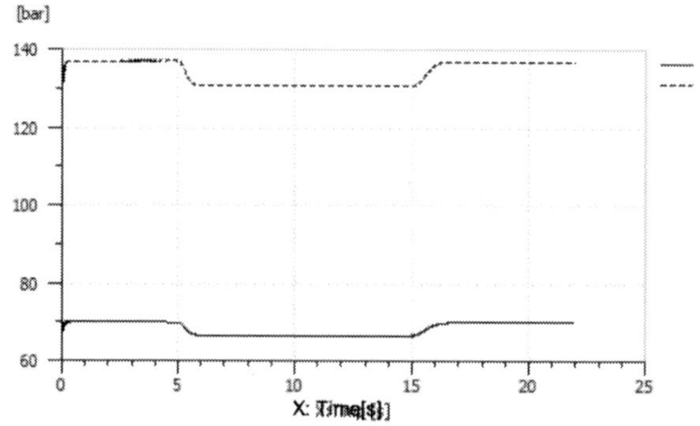

Fig. 12. Variation in time of the pressures in the servomechanism chambers. (continuous line - pressure in chamber 1, dashed line - pressure in chamber 2.

Figure 13 shows the hysteresis characteristic of the servomechanism for the situation in which the distribution windows of the distributor are equal.

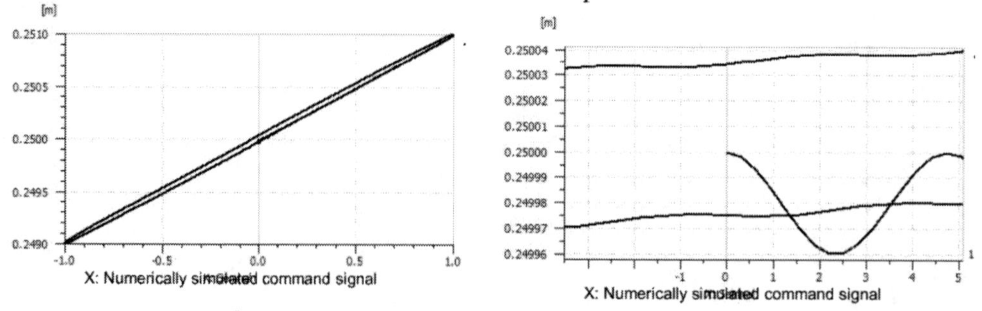

a) b)

Fig. 13 a) Hysteresis characteristic for distributor with equal distributed windows. **b)** Detail

Figure 14 shows the hysteresis characteristic of the servomechanism for the situation in which the distribution windows of the distributor are unequal.

https://doi.org/10.1051/e3sconf/202018004005

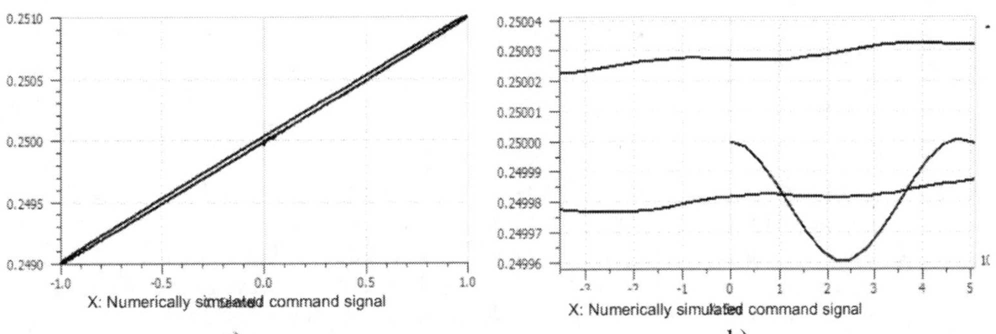

a) b)

Fig. 14. a) Hysteresis characteristic for distributor with unequal distribution windows., **b)** Detail

Figure 15 shows the response of the servomechanism to the step type signal

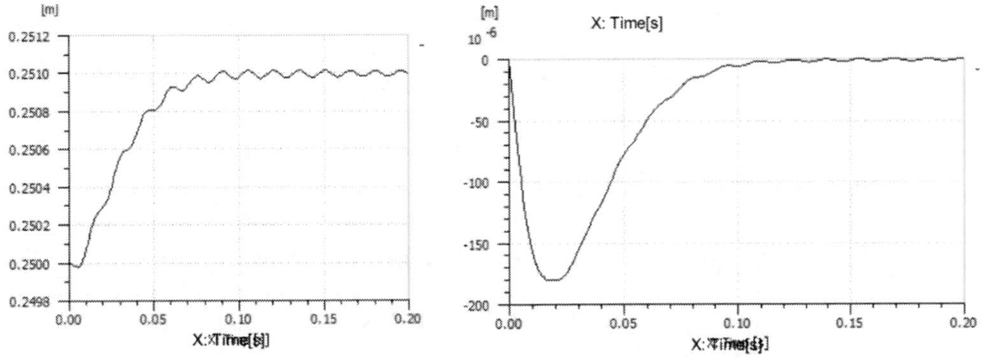

a) Piston displacement **b)** Spool displacement

Fig.15. The servomechanism response to step type signal

Figure 16 shows the evolution of the error and of the flow rates corresponding to the two chambers of the servomechanism during the transient regime, when the servomechanism was excited with a step type signal.

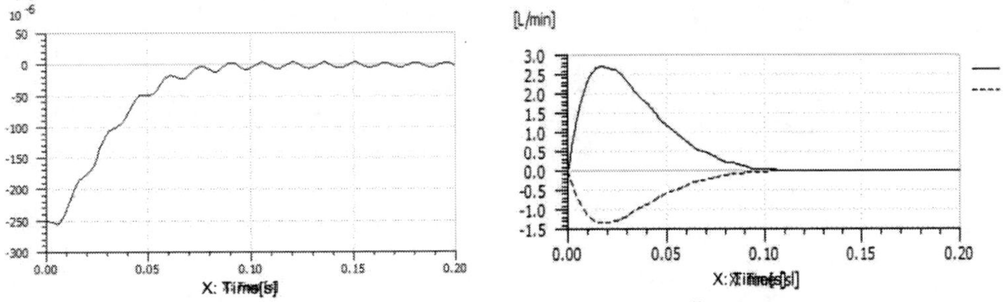

a) error **b)** flow rates

Fig.16. Variation of error and flow rates in the servomechanism chambers during the transient regime (step type excitation signal)

5 Conclusions

The analysis and the synthesis of electro-hydraulic servomechanisms with stepper motor can be determined by mathematical modelling and numerical simulation using the AMESIM simulation program.

Regarding the servomechanism performances evaluated by mathematical modelling and numerical simulation with AMESIM software, the following conclusions were established:

1) The time constant of the servomechanism is 0.03 s.
2) The resolution of the servomechanism is significantly influenced by the resolution of the stepper motor and the lead screw step.
3) The servomechanism hysteresis value is dependent on the type of distribution windows of the distributor. Because the hydraulic cylinder has the areas of the piston unequal, the hysteresis value is lower when the distribution window areas are unequal.
4) The lower value of the hysteresis was obtained for the ratio of the distributor slots areas that is equal to the piston areas ratio. For equal areas, the hysteresis is with 25 microns higher under the same operating conditions.

This work was supported by a grant of the Romanian Education and Research Ministry (RRIM), through Programme 1 – Development of the national research-development system, subprogram 1.2 – Institutional performance – Projects for financing excellence in RDI, contract no. 16PFE, and was done by "NUCLEU" Programme, developed with the support of the RRIM, project PN 19 10 01 02.

References

1. N. Vasiliu, D. Vasiliu, *Hydraulic and pneumatic drives* (Technical Publishing House, 2008)
2. H.E. Merritt, *Hydraulic Control Systems* (John Wiley and Sons Inc., New York, London, Sydney, 1967)
3. *** Technical documentation of MOOG company
4. *** Technical documentation of CURTISS –WRIGHT company
5. C. Călinoiu, N. Vasiliu, D. Vasiliu, I. Catană, *Modeling, simulation and identification of hydraulic servomechanisms* (Technical Publishing House, 1998)
6. I. Ursu, F. Ursu, *Active and semiactive control* (Romanian Academy Publishing House, 2002)
7. *** LMS INTERNATIONAL. *Advanced Modelling and Simulation Environment*, (Release **13** User Manual, Leuven, 2013)
8. N. Vasiliu, D. Vasiliu, C. Calinoiu, R. Puhalschi, *Simulaţion of fluid power Systems with LMS Imagine* (Lab Amesim CCR Press Publishing House, 2018)

Finite element analysis of the compaction plate from a garbage truck

Gheorghe Voicu[1], *Mircea* Lazea[1], *Gabriel-Alexandru* Constantin[1,*], *Elena-Madalina* Stefan[1], and *Mariana-Gabriela* Munteanu[1]

[1]Politehnica University of Bucharest, Faculty of Biotechnical Systems Engineering, Romania

Abstract. Vehicles that collect and transport household waste are equipped with complex systems that perform the loading of the waste from the stationary collection containers, take over and pre-compact of material, compaction in the body of the structure and finally the landfill of the collected waste. Of all these operations, the compaction process is by far the most mechanically demanding. In this paper, the structural analysis of the compactor plate from a garbage truck is presented. In the first stage it was carried out parametric modelling of the assembly composed of compactor plate, counter pressure plate, the back of the garbage truck and to be compacted material. After assembly, the dynamic simulation of the compaction process in the garbage truck was performed, and the compaction mechanical stresses for the compactor plate were loaded in the Simulation module of the SolidWorks 2016 program. Here the finite element analysis was performed, resulting in the value and dispersion of the equivalent stresses (calculated by von Mises criterion), displacements and relative deformations of compactor plate from the analysed garbage truck.

1 Introduction

Manufacture of component parts for garbage trucks, but also of other machines and equipments, assumes that their physical model to have followed, in the conception and design stage, several steps in which the design engineers have subjected it to modelling, simulation and analysis with specialized software operations, before being sent to execution. In this way, many possible manufacturing errors, which may occur in these component parts, can be eliminated. In the case of mechanical stresses, the most common method of analysis is finite element analysis. There are many such analyses in the literature, even in the case of municipal waste trucks, [1-4].

The purpose of this finite element 3D numerical simulation study was to simulate the behaviour of the compression plate structure of a garbage truck Valu€ Pak 1000 following the stresses that arise during the compaction process of household waste, considering that the plate is made of three different materials (Hardox 400, Hardox 450 and S355J2 steels).

The waste compaction system of such a machine consists of two basic parts, a metal plate with hinged support that takes over the material from the machine's feed area, having

[*] Corresponding author: constantin.gabriel.alex@gmail.com

 https://doi.org/10.1051/e3sconf/202018004006

a complex movement, and a counter plate disposed in the rectangular collection container (see fig. 1).

– two yoke cylinders

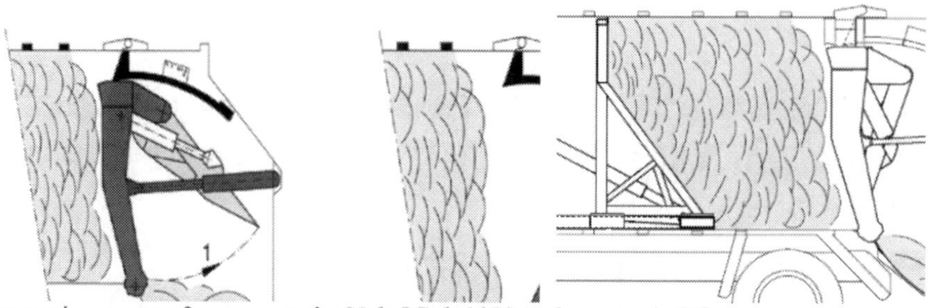

Fig. 1 Compaction system for waste at the Valu€ Pak 1000 garbage truck, [5].

In both pieces important stresses appear during the collection and compaction process, but much larger in the pushing and retrieval piece when feeding the container. At discharging, the back pressure plate has high demands because it has to push the entire amount of material from the container, the more uneven the material is more homogeneous on the cross-section of the collection container.

2 Material and method

In the first stage of this study, the three-dimensional geometric model of the components of the assembly was realized (compression plate, back pressure plate, wear part and the container). So, an ASUS K55V laptop was used, containing an Intel Core i5-3210M, 2.5GHz, NVIDIA GeForce GT 630M video card with 2GB of memory, 8GB of RAM. 3D modelling was performed with the help of the parameterized design program Solid Works Premium 2016 S.P. 0.0.

The three-dimensional modelling of the components of the assembly was performed in the module "Parts" of the design program, in fig. 2 being presented the isometric view of the compaction plate, as well as the Solid Works program interface.

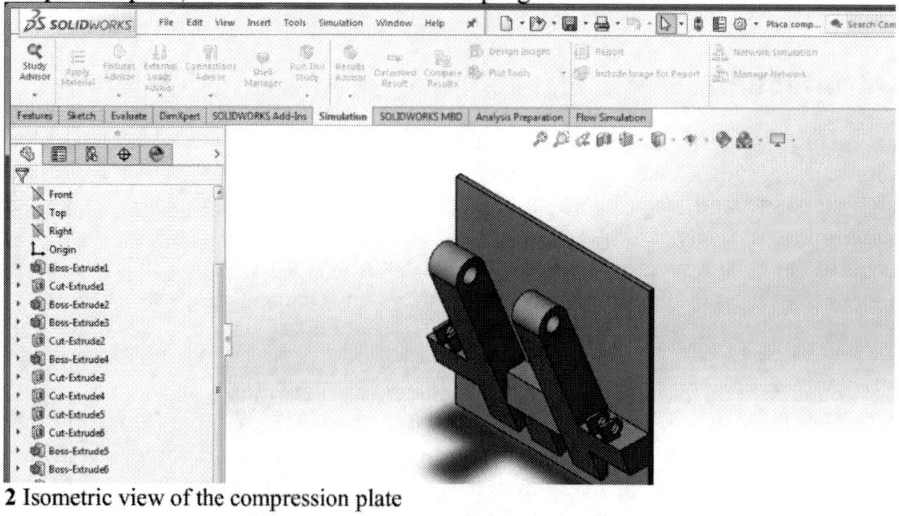

Fig. 2 Isometric view of the compression plate

The pieces were introduced in the Assembly module of the program where they were positioned by adding geometric relationships between their entities (concentricity relations, coincidence and parallelism between the edges and faces, and others). For the dynamic simulation of the compaction process of household waste, between the two plates (of compaction and counterpressure) it was modelled parameterized the material to be compacted. For this, a thickness at the bottom of about 580 mm was chosen, see fig. 3.

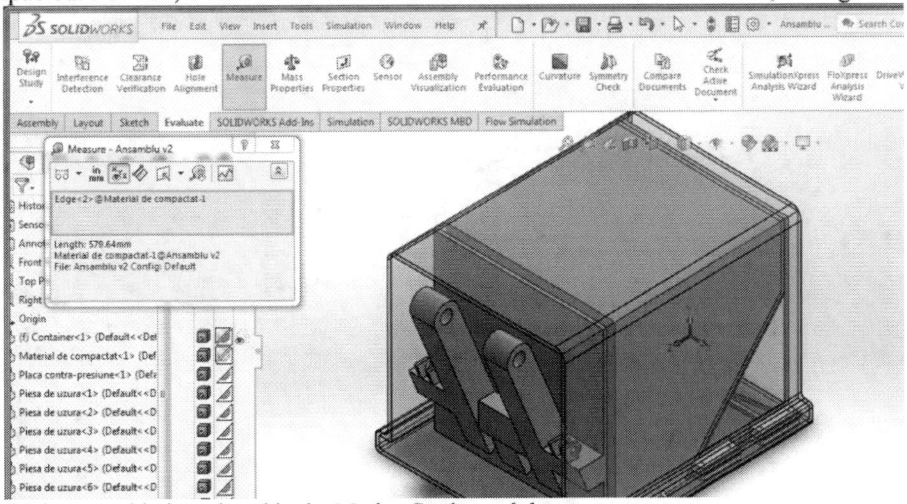

Fig. 3 The assembly introduced in the Motion Study module

The whole assembly of fig. 3 was introduced in the Motion Study module to perform the dynamic simulation of the compaction process of household waste. For this purpose it was considered that the back pressure plate is fixed, together with the compacted material, and the compactor plate moves. According to Manring N et al. [6], the speed of movement of the compaction plate without load is of maximum 150 mm/s, at the beginning of the compaction process the displacement speed is 75 mm/s and reaches the end of the compaction at 20 mm/s. Thus, a variable speed linear motor was attached to the compaction plate (see fig. 4). It should also be mentioned that the calculation period (dynamic simulation) was 12 s.

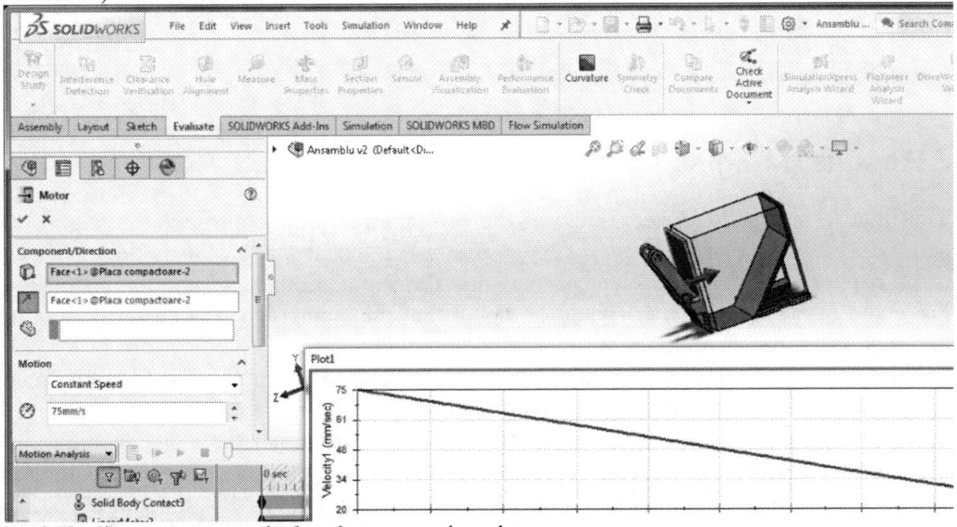

Fig. 4 The linear motor attached to the compaction plate

After completing this step, proceeded to the next step which was to import the stresses for the 3D geometric model of the compaction plate in the "Simulation" module of the Solid Works design program.

https://doi.org/10.1051/e3sconf/202018004006

According to the manufacturers specifications, the materials most often used in the manufacture of compression plates are Hardox 400, Hardox 450 and S355J2 steels. The characteristics of these three steels are presented in table 1, and in fig. 5 are shown the fatigue resistance curves of the three materials.

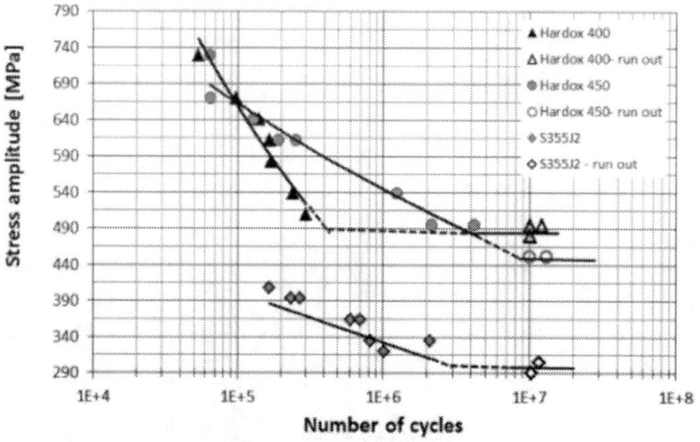

Fig. 5 Fatigue curves for Hardox 400, Hardox 450 and S355J2 steels [7-9]

Table 1. Mechanical properties of Hardox 400, Hardox 450 and S355J2 steels, [10-14]

Material / Mechanical property	Hardox 400	Hardox 450	S355J2
Elasticity modulus, N/mm^2	190000	200000	210000
Poisson's ratio, -	0.29	0.3	0.3
Shear modulus, N/mm^2	75000	78000	80000
Density, kg/m^3	8000	8100	7800
Tensile strength, N/mm^2	1250	1663	630
Yield point, N/mm^2	1020	1350	665
Coefficient of thermal expansion, 1/K	1.8e-005	1.5e-005	1.3e-005
Thermal conductivity, W/(m·K)	16	27	40
Specific heat, J/(kg·K)	500	520	470

3 Results and discussions

The finite element discretized model of the compression plate is presented in fig. 6 (there can be seen the characteristics of the mesh formed from 17904 nodes). After the discretization of the finite element network the simulation was run, its results being presented in the following. It should be noted that the simulation was performed with the "Large displacement" option selected.

Following the simulation, the design program provided the results obtained in graphical form; the geometric pattern is divided into areas of a certain colour, each area comprising the region of the geometric model in which the analysed size has the value specified in the chromatic legend on the right side of the screen. It should be noted that three simulations were run, one for each of the three chosen steels (Hardox 400, Hardox 450 and S355J2).

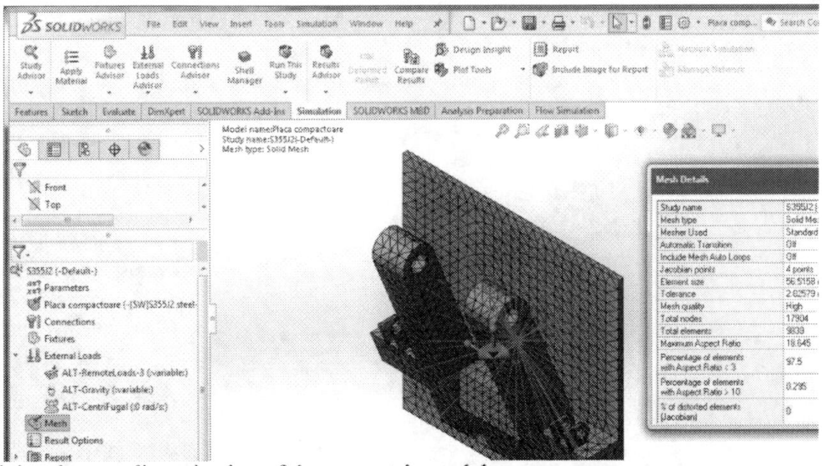

Fig.6. Finite element discretization of the geometric model

For the model of compression plate modelled and analysed, the results obtained from the simulation in Solid Works are presented below. Thus, in fig. 7 the values of the displacements that appear in the compression plate are presented during the stresses defined above.

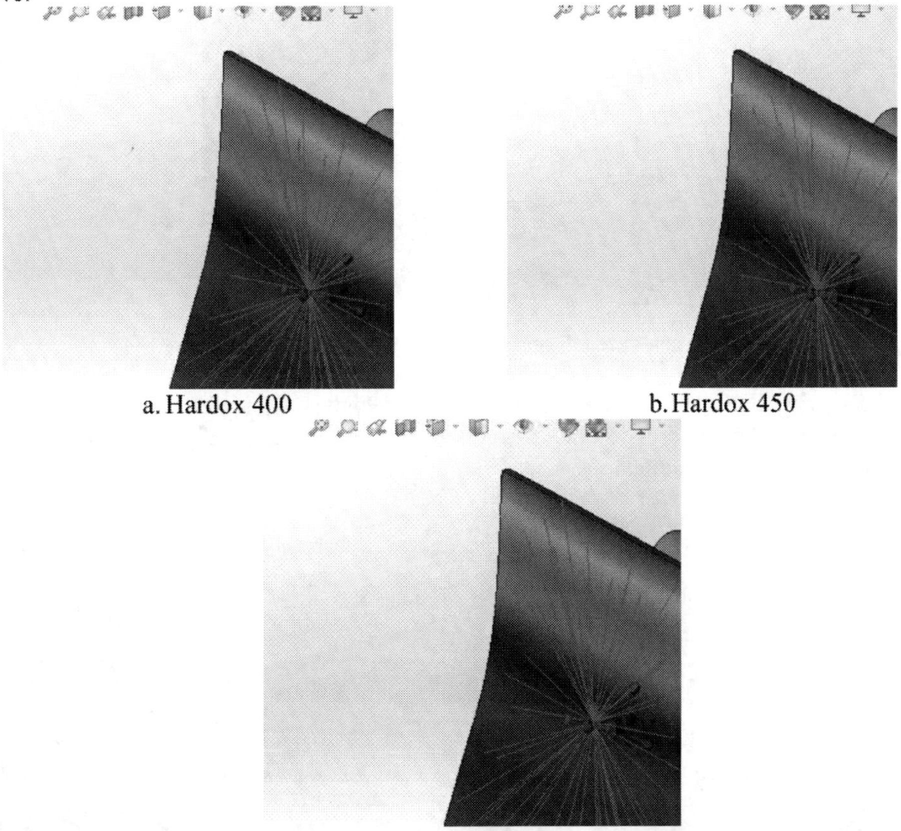

a. Hardox 400

b. Hardox 450

c. S355J2

Fig. 7. The values of displacements in the compression plate

Analysing this data, it can be seen that the values of the displacements are close (179 µm for Hardox 400, 170 µm for Hardox 450 and 162 µm for S355J2). Although S355J2 steel is not necessarily a harder material than Hardox 450, it has the least displacement during the compaction process. For all three materials, the maximum displacement appears at the top of the compaction plate.

In fig. 8 the values of the equivalent stresses in the compaction plate are presented for the three analysed cases, stresses calculated according to von Mises criterion.

a. Hardox 400

b. Hardox 450

c.S355J2

Fig.8. Equivalent stress values according to von Mises criterion

Analysing the figure it can be observed that the highest equivalent stresses appear in the middle area of the compression plate structure, and their values are well below the breaking stresses of the three materials.

a. Hardox 400

b. Hardox 450

c. S355J2

Fig. 9. The values of equivalent deformations occurring in the compression plate

Analysing fig. 9 we can observe the values of equivalent deformations that appear in the compression plate following the stresses to which it is subjected. So, the maximum equivalent deformation also emerges in the median area of the compression plate, the value of the deformations for the three materials considered being presented in the figure.

In fig. 10 the oscillation of the safety coefficient in the compression plate for the three materials considered is presented.

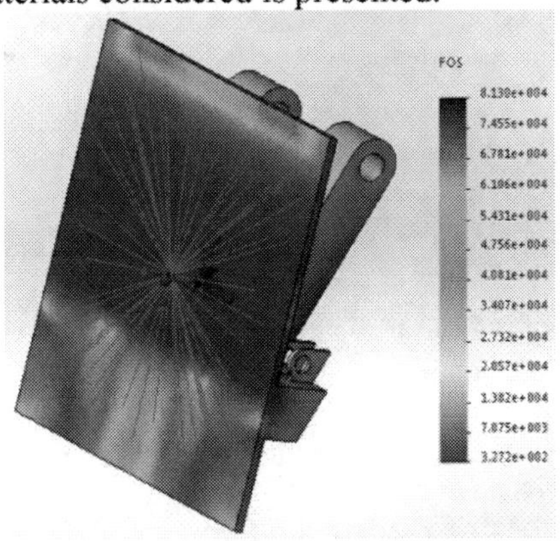

a. Hardox 400

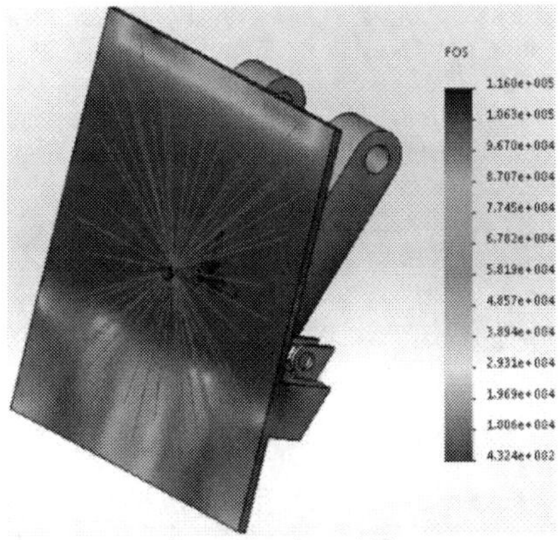

b. Hardox 450

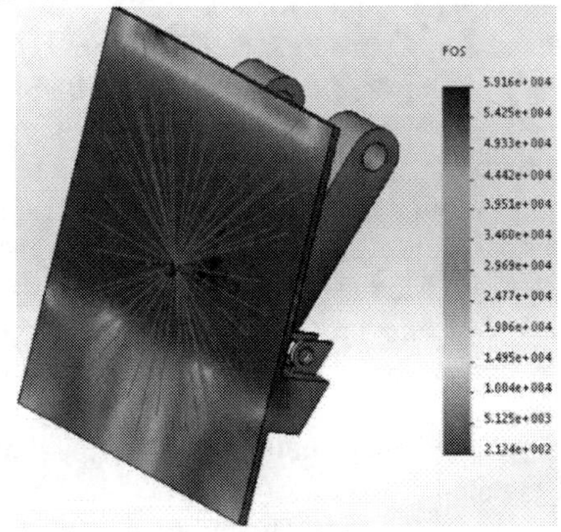

c. S355J2

Fig. 10. The variation of the safety coefficient for the three materials considered

The minimum value of the safety coefficient is reached for S355J2 steel, a steel with weaker mechanical properties compared to the other two and the maximum value is reached for Hardox 450, a steel with a rather high hardness. The minimum and maximum values are shown in the figure above. For comparison, for agricultural machines intended for ploughing, so machines that work in heavier conditions than the compression plate from the garbage truck, the coefficient of safety takes values between 1.8 and 2.2, [15].

4 Conclusions

The minimum values of the safety coefficient are 327 for Hardox400, 432 for Hardox 450 and 212 for S355J2. It can be said that this compression plate is either oversized, either it is made to work in much heavier conditions than those provided by this study, or to withstand even overload (impact with hard to compact materials, excessive friction, etc.). The analysis also showed that in the compression plate structure appear stress concentration points, located in the areas where the plate is welded to the supports that connect to the drive yoke. The maximum values of the equivalent stresses calculated with the von Mises criterion are $3.118 \cdot 10^6$ MPa for Hardox 400, $3.122 \cdot 10^6$ MPa for Hardox 450 and of $3.131 \cdot 10^6$ MPa for S355J2, values somewhere in the middle of the yield point of these materials. Taking into account the above, the next step in completing the study would be a design analysis of the compaction plate. The results presented in this paper can be useful to designers and manufacturers of household compaction machines, but not only.

The paper was funded both from the project "Sustainable development of the base of agromechanical applications in gardens, solariums, vineyards and orchards (DEZDURPRA)", CNFIS-FDI-2020-0184, from the Ministry of Education and Research through the Executive Agency for Financing Higher Education , Research, Development and Innovation, as well as from the European Social Fund within the Sectoral Operational Program Human Capital 2014-2020, through the Financial Agreement of the project “Scholarships for entrepreneurial education between doctoral students and postdoctoral researchers (Be Entrepreneur!)”, Contract no. 51680 / 09.07.2019 - SMIS Code: 124539.

References

1. P. Milosavljević, S. Jovanović, D. Jovanović, G. Radoičić, V. Blagojević, FACTA UNIVERSITATIS Series: Mechanical Engineering, **8**, 1, (2010);
2. C. Sticlaru, A. Davidescu, Facta Universitatis, Mechanical Engineering, **2**, 1, (2004);
3. M.T. Lates, *Finite elements method, aplication (Metoda elementelor finite, aplicații)*, (Ed. Universității Transilvania, Brașov, 2008);
4. S. Guanglin; C. Shuxun; L. Guangmin, International Conference on Measuring Technology and Mechatronics Automation, (2009);
5. *** (2005), Value Pack, Service manual, Geesing Norba Group, Model year 2005
6. N. Manring, R. Fales, *Hydraulic Control Systems, Second Edition*, (Wiley, 2020);
7. R. Ulewicz, P. Szataniak, F. Novy, Conference proceeding METAL (2014);
8. N. Stranghöner, D. Jungbluth, , 6th Fatigue Design conference, (2015);
9. M. Mazur, R. Ulewicz, Applied Engineering Letters, **2**, 1, (2017);
10. *** www.mtladv.com/wp-content/uploads/2016/01/151_HARDOX_400_UK.pdf
11. F. Sergejev and M. Petrov, Estonian Journal of Engineering, **18**, 3, (2012);
12. R. Ulewicz, M. Mazur, O. Bokuvka, , Transportation Engineering **41**, 2 (2013)
13. *** https://www.oblibene.com/userdata/shopimg/gamaocel/file/datasheet/hardox/Data_sheet_2062uk_Hardox_450_Sheet_2016-03-02_85_371857150_en.pdf
14. *** http://www.joostdevree.nl/bouwkunde2/jpgs/staal_23_standaarden_s355_european _standard_steel.pdf
15. G. Gheorghe, C. Persu, I. Gageanu, D. Cujbescu, 47th Int. Sym. “Actual Tasks on Agricultural Engineering”, (2019).

E3S Web of Conferences **180**, 04007 (2020) https://doi.org/10.1051/e3sconf/202018004007
TE-RE-RD 2020

Determination of granulometrical composition of the clinker by grinding in a ball mill to determine the specific consumption of additional energy

Cristian Ciobanu[1], *Paula* Tudor [3], *Gabriel-Alexandru* Constantin[2,*] *and Gabriel* Musuroi[2]

[1]CEPROCIM Bucharest, Romania
[2]University Politehnica of Bucharest, Faculty of Biotechnical Systems Engineering, Romania
[3]University Politehnica of Bucharest, Faculty of Entrepreneurship, Business Engineering and Management, Romania

Abstract. This paper presents results regarding the grinding behaviour of two batches of material (clinker) in a ball mill with its own two-step grinding process: the first one with a ball load, the second with a load of cylindrical metal bodies. The first step was performed until the residue R_{009} has reached aprox.35%, and the second stage until Blaine's specific surface was over 3800 cm^2/g. Periodically, at 10-minute intervals, the fineness of the clinker and the specific energy consumption were determined. These consumptions are quantified from the first determination and relates to the weight of the material subjected to grinding (20 kg clinker / from a source), determining the grinding ability of the material.

1 Introduction

The cement consists of dispersed solid particles whose particles have a size between 0.1 and 250 μm. Particle size, particle size distribution, but also the composition of the cement has a special influence on its behaviour during wetting, on the development of microstructures and on the properties of concrete or cement-based materials. It is important that Portland cement particle sizes fall within certain limits so that they are properly moistened and to form a material with a good subsequent behaviour. More, Excessive grinding leads to additional energy consumption (sometimes unjustified), and insufficient grinding can affect the binder properties of the cement particles and their distribution in the prepared material.

Among the properties that have been examined through experiments and computer simulation are: setting time, heat release, capillary porosity percolation, diffusivity, chemical shrinkage, autogenous shrinkage, internal relative humidity evolution, and interfacial transition zone microstructure, [1]. The authors concluded that, for lower water-cement ratio values, coarser cements can be used that can provide equivalent or even superior performance, which would reduce the production costs to the manufacturer.

[*] Corresponding author: constantin.gabriel.alex@gmail.com

Still, there are researchers who claim that cement particles larger than 45 μm are difficult to hydrate, and also the largest of 75 μm cannot be completely hydrated, when analysing the rate of reactivity and the development of stresses in the prepared concrete, [2]. Author of this paper, [2], states, however, that estimates of reactivity rates for similar cement compositions cannot be made without knowing the particle size distribution by sedimentation methods. In any case, there must be a correlation between the setting time and the fineness of the cement.

The standard method for determining particle size and particle size distribution is ASTM C115-96, with a limit of detection of a smaller size of 7.5 μm [3]. This is based on the turbidimetric method for determining the fineness of the material. The Electrical Zone Sensing (EZS) method, which is based on the Coulter principle, can operate in the particle size range between 0.2 (0.6)–800 μm, while laser diffraction is used successfully when most particles have dimensions below 1 μm, [3].

From the four techniques analysed by the authors of the previous paper: laser diffraction (LAS), detection of electrical areas (EZS), X-ray gravitational sedimentation (XRS) and scanning electron microscopy (SEM), laser diffraction (with both wet and dry dispersion), gave the best yields, this being the basis for determining the specific Blaine surface.

For cements, are two systems for determining fineness, namely, specific surface Blaine (air permeability) and turbidimetry. Blaine fineness is an indirect measure of the total surface of each cement sample and can be determined according to ASTM C204-11 "Standard Test Method for Fineness of Hydraulic Cement by Air Permeability Apparatus".

For Portland cement from the same source, the heat of hydration is affected by the fineness of the cement and the particle size distribution. Grinding to a greater fineness Blaine, in correlation with decreasing particle size, leads to an increase in the heat of hydration, experimentally confirmed [4].

The necessary fineness of the final cement is due to the milling means of the ball mills and the size distribution of the feed material. To increase the efficiency of ball mills, for optimum distribution of the final material according to the particle size, it is necessary, therefore, a correlation between the size of the balls and the size of the particles when feeding the mill, [5]. It was concluded that a ternary or at least binary mixture of balls of different sizes (50 + 20 +10 mm) leads to a final material of optimum size.

More energy efficient grinding technologies, applied in the grinding of cement, are described comparatively in the paper [6], from the point of view of operating principles, grinding efficiency, specific energy consumption, the production capacity and the quality of the cement obtained. Other papers addressing the theme of milling process with balls dedicated to grinding cement, including its control are [7-11]

The present paper presents the results regarding the grinding behaviour of two batches of material (clinker) in a ball mill with its own two-step grinding process: the first one with a ball load, the second with a load of cylindrical metal bodies.

2 Material and method

Energy consumption between the moments when the fineness of the material is determined (in our case the consumption was read directly on the electricity meter) was determined using a wattmeter. These consumptions have accumulated since the beginning of the determination and have been related to the mass of the batch (20 kg clinker/factory from 2 factories), calculating the specific energy consumption w_{1i}. The aptitude index for milling is the specific energy consumption w_1 corresponding to a reference fineness and w - the specific energy consumption of the industrial mill, [12].

$$c_1 = \frac{w}{w1} \quad (1)$$

In fig. 1 shows the scheme of a laboratory mill and the working process of a ball mill.

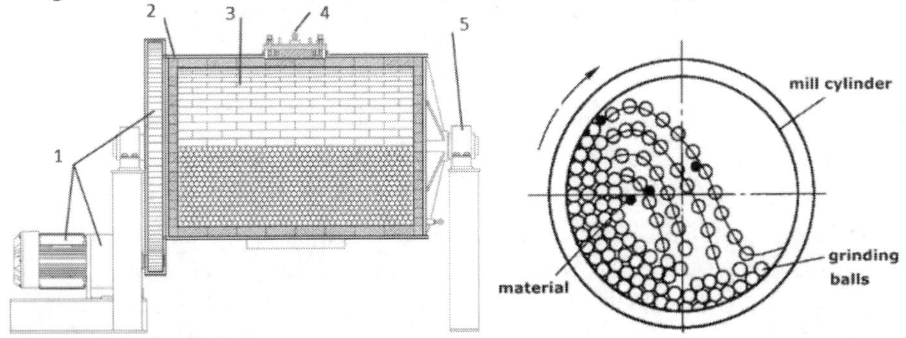

Fig.1 Schematic ball mill, [13]
a): *ball mill diagram; 1- motor, gear reducers and ring and pinion gear; 2- cylinder; 3- brick lining; 4- discharge hatch; 5- bearings;* ***b):*** *working process of ball mill*

To determine the specific surface (SSB) of the grinded material we chose the method of the Blaine permeability meter. The Blaine procedure is applicable for all cements defined in the standard EN 196-6 :2018, [14].

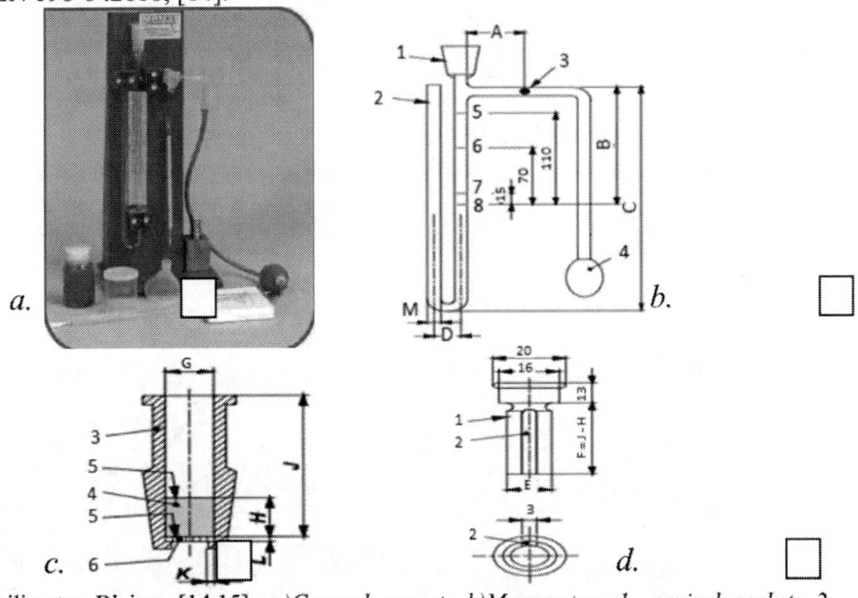

Fig.2 Permeabilimeter Blaine, [14,15]: *a)General aspect; b)Manometer: 1- conical socket; 2- manometer; 3- stopcock; 4- rubber bulb; 5-8- etched lines; c)Permeability cell: 3- cell; 4- cement bed compacted by the plunger; 5- paper filters; 6- perforated disc; d)Plunger piston: 1- piston body; 2- flat*

The Blaine permeability meter also includes: stopwatch; analytical balance; manometric liquid; apparatus for determining the density of cement (Le Chatelier 100 ml quoted balloon).

The specific surface S was calculated accordingly SR EN 196-6:2018 [16] and is conventionally expressed in $cm^2 \cdot g^{-1}$, as it is:

$$S = \frac{K}{\rho} \cdot \frac{\sqrt{e^3}}{(1-e)} \cdot \frac{\sqrt{t}}{\sqrt{10\,\eta}} \quad (2)$$

where: K - the constant of the device, e - porosity of the layer; t - time measured (s); ρ – bulk density of the material, (g·cm^{-3}), η - the viscosity of the air at the test temperature, according to table no.1, (Pa·s)

With the specified porosity e = 0.500 and the test temperature 20 ±2°C:

$$S = \frac{52.43 \cdot K \cdot \sqrt{t}}{\rho} \quad (3)$$

where: K thus determined:

$$K = 1.414 \cdot S_o \cdot \rho_o \frac{\sqrt{10\ \eta_o}}{\sqrt{t_o}} \quad (4)$$

S_0 - the specific surface of the reference cement, (cm^2·g^{-1}), ρ_0 - the bulk density of the reference cement, (g·cm^{-3}), t_0 - the average of three timed values of time, (s), η_0 - the viscosity of the air corresponding to the average of three temperatures, (Pa·s) and is given according to table 1.

Table 1. Bulk density of Hg, viscosity of the air η_0 and $\sqrt{10\ \eta_0}$ function of temperature, [16]

Temperature, °C	**Bulk density of Hg, ρ_{Hg} [g·cm^{-3}]**	**Viscosity of the air η_0, Pa·s**	$\sqrt{10 \times \eta_0}$
16	13.560	0.00001800	0.013416
17	13.560	0.00001805	0.013435
18	13.550	0.00001810	0.013454
19	13.550	0.00001815	0.013472
20	13.550	0.00001819	0.013487
21	13.540	0.00001824	0.013506
22	13.540	0.00001829	0.013524
23	13.540	0.00001834	0.013543
24	13.540	0.00001839	0.013561
Intermediate values must be obtained by linear interpolation			

For the determination of the laser particle size analysis for Clincher A and F samples, the Mastersizer 2000E (MALVERN, UK) was used, [16].

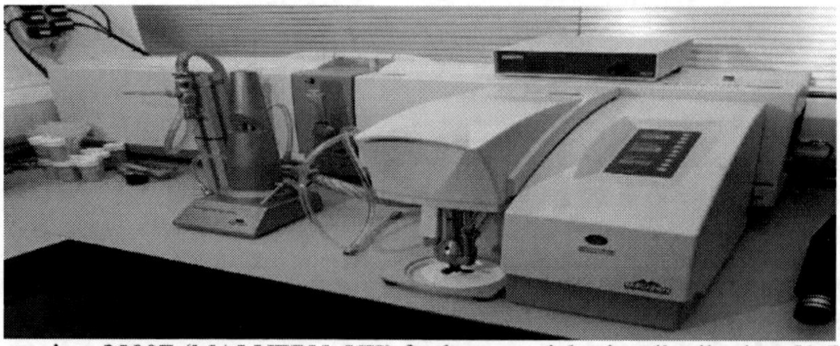

Fig.3 Mastersizer 2000E (MALVERN, UK) for laser particle size distribution, [16]

Laser particle size distribution was performed under the following measurement conditions: Wet dispersion (dispersion liquid: Ethanol ≈850ml); Refractive index dispersion liquid: 1.360; Ultrasound duration: 60 sec; Ultrasound level: 10 μm; Stirring pump speed: 2300 rpm; Laser diffraction apparatus: Mastersizer 2000E; Focal distance of the lenses:2.35; dimensional range: 0.100 μm -1000.00 μm; Optical concentration:18.45 %; Light dispersion model applied : MieAbsorbtie theory: 0.1.

3 Results and discussions

Following the determinations were recorded the following energy consumption and specific surface Blaine (SSB) for two types of clinker A and F from the 2 cement factories (Table 2).

Table 2. Specific consumption and Blaire surface area for the two types of clinker

Grinding time, [min]	No. on counter [Kwh], clinker A	No. on counter [Kwh], clinker F	Difference on counter, [Kwh], clinker A	Difference on counter, [Kwh], clinker F	Energy consumption [kwh/t] clinker A	Energy consumption [kwh/t], clinker F	Correlation coefficient c_1 clinker A	Correlation coefficient c_1 clinker F	SSB [cm²/g]-bi-cones Residue [%]-balls clinker A	SSB [cm²/g]-bi-cones Residue [%]-balls clinker F	Grinding bodies
0	7350421	7356265	0	0	-	-	-	-	-	-	Grinding balls
10	7350661	7356504	240	239	-	-	-	-	51.68 %	53.4 %	Grinding balls
20	7350882	7356729	221	225	-	-	-	-	33.6%	34.5 %	Grinding balls
30	7351096	7356946	214	217	10.7	10.85	3.08	3.03	2250	2190	Bi-cones
40	7351314	7357167	218	221	10.9	11.05	3.02	2.98	2650	2640	Bi-cones
50	7351526	7357387	212	220	10.6	11	3.11	2.99	2830	2960	Bi-cones
60	7351746	7357617	220	230	11	11.5	2.99	2.86	3180	3130	Bi-cones
70	7351968	7357837	222	220	11.1	11	2.97	2.99	3520	3290	Bi-cones
80	7352196	7358069	228	232	11.4	11.6	2.89	2.84	3590	3580	Bi-cones
90	7352417	7358290	221	221	11.05	11.05	2.98	2.98	3700	3870	Bi-cones
100	7352642	-	225	-	11.25	-	2.93	-	3870	-	Bi-cones

$$\text{Total energy consumption clinker A } = \frac{7352642-7350421}{20} = 111.05 \text{ kwh/t} \quad (5)$$

$$\text{Total energy consumption clinker F } = \frac{7358290-7356265}{20} = 101.25 \text{ kwh/t} \quad (6)$$

$W_{industrial\ mill\ consumption}$=32.92 kWh/t according to the value calculated by the cement factory.

Regarding energy consumption, Atmaca and Kanoglu found that this consumption is 24.75 kWh/ton farine, depending on the temperature and humidity of the material, [17].

Results regarding laser chemical and particle size distribution of the material subjected to grinding (clinker A and F) are presented in Table 3.

Table 3. Chemical analysis results clinker A and F

Determination	U.M.	Test code Clinker A	Test code Clinker F
Loss on calcination – L.C.	%	0.53	0.96
SiO_2	%	21.20	20.88
Al_2O_3	%	5.66	5.04
Fe_2O_3	%	2.99	3.80
CaO	%	66.34	66.78

MgO	%	1.20	0.80
SO_3	%	0.83	0.74
Na_2O	%	0.21	0.12
K_2O	%	0.50	0.34
CaO_{free}	%	0.31	0.92
$HCl - Na_2CO_3$	%	0.15	0.17

Following the measurements the following charts were obtained.

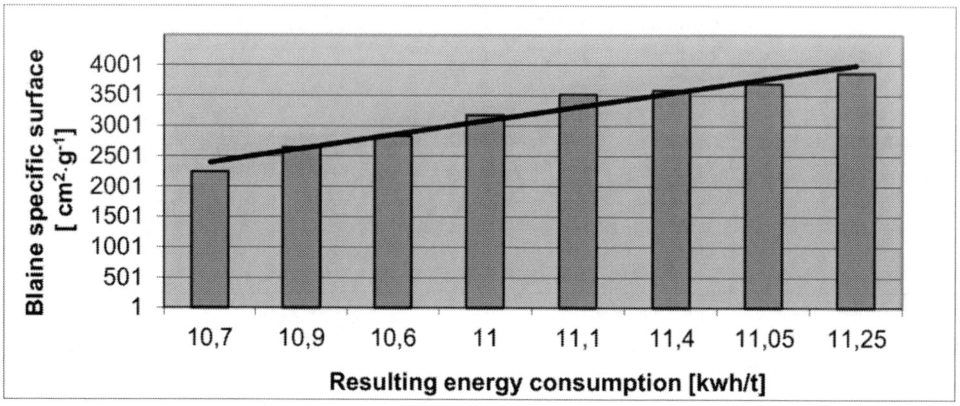

Fig.4 Grinding aptitude index clinker A

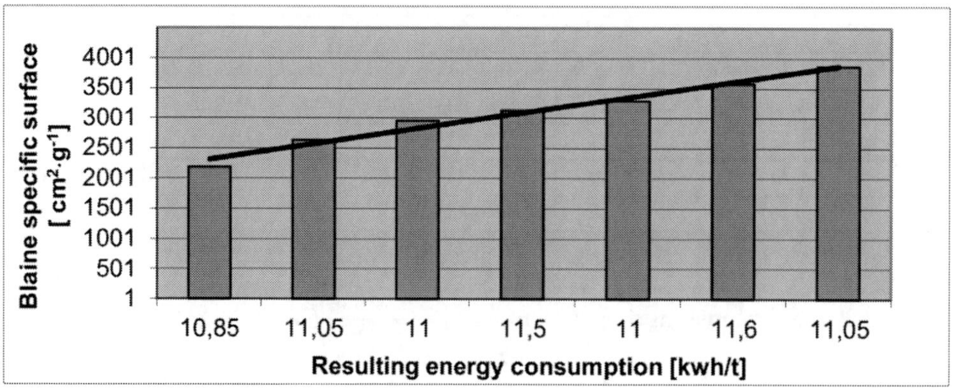

Fig.5 Grinding aptitude index clinker F

Based on the equilibrium diagrams, the basic composition of the clinkers was calculated. The calculation was made in the hypothesis of solidification at thermodynamic equilibrium and at thermodynamic nonequilibrium. In case of solidification at full equilibrium, the glass phase is completely absent. In reality, no thermodynamic balance is achieved at solidification and the mineralogical composition calculated from such a hypothesis can be considered as a potential composition, that is, possible, understanding by this that the respective chemical composition is not equal to the composition of the real minerals. Assuming the clinker as being placed in the thermal equilibrium subsystem C_3S-C_2S-C_3A-C_4AF ($MA_i > 0.64$), Bogue has established the following calculation formulas for mineralogical constituents, [12]:

$\% C_4AF = 3.04 \% Fe_2O_3$

$\% C_3A = 2.65 \% Al_2O_3 - 1.69 \% Fe_2O_3$

% C_3S =4.07 % CaO – 7.60 % SiO_2 – 6.72 % Al_2O_3 – 1.42 % Fe_2O_3

% C_2S = 8.60 % SiO_2 + 5.06 % Al_2O_3 + 1.07% Fe_2O_3 – 3.05% CaO.

Thus, from the mineralogical composition according to Bogue's calculation, the values presented in table 4 resulted.

Table 4. Mineralogical composition according to Bogue's calculation for two clinkers

Clinker A	Clinker F
C_3S = 66.48	C_3S = 71.39
C_2S = 11.48	C_2S = 6.8
C_3A = 9.95	C_3A = 6.93
C_4AF = 9.09	C_4AF = 11.55

It is observed that C_2S clinker A > C_2S clinker F (11.48 > 6.8) which results in clinker A requiring additional energy consumption in the mill because it has a much higher content of C_2S.

The results of the determinations of the laser particle size distribution for clinker A are presented in Table 5.

Table 5. Particle size distribution for clinker A

Granulometric class (µm)	*Cumulative passing (% vol)*	*Granulometric class (µm)*	*Cumulative passing (% vol)*	*Granulometric class (µm)*	*Cumulative passing (% vol)*	*Granulometric class (µm)*	*Cumulative passing (% vol)*	*Granulometric class (µm)*	*Cumulative passing (% vol)*
0.10	0.00	3.00	12.26	13.00	42.36	64.00	93.99	175.00	100.00
0.20	0.00	3.15	12.99	14.00	44.32	70.00	95.79	200.00	100.00
0.30	0.00	3.50	14.66	15.00	46.21	75.00	96.96	225.00	100.00
0.40	0.00	4.00	16.95	16.00	48.05	80.00	97.87	250.00	100.00
0.50	0.35	4.50	19.13	18.00	51.58	85.00	98.59	280.00	100.00
0.60	0.95	5.00	21.16	20.00	54.92	90.00	99.14	315.00	100.00
0.70	1.60	5.50	23.06	22.00	58.10	95.00	99.52	355.00	100.00
0.80	2.21	6.00	24.83	24.00	61.12	96.00	99.58	400.00	100.00
0.90	2.77	6.30	25.84	25.00	62.58	100.00	99.78	450.00	100.00
1.00	3.28	6.50	26.49	28.00	66.71	106.00	99.94	500.00	100.00
1.20	4.16	7.00	28.06	31.50	71.10	110.00	99.99	560.00	100.00
1.40	4.98	7.50	29.53	32.00	71.69	115.00	100.00	600.00	100.00
1.50	5.38	8.00	30.93	36.00	76.10	120.00	100.00	650.00	100.00
1.60	5.79	8.50	32.26	40.00	79.95	125.00	100.00	700.00	100.00
1.80	6.63	9.00	33.53	45.00	84.06	128.00	100.00	750.00	100.00
2.00	7.51	9.50	34.76	48.00	86.17	130.00	100.00	800.00	100.00
2.20	8.42	10.00	35.94	50.00	87.45	140.00	100.00	850.00	100.00
2.50	9.84	11.00	38.19	55.00	90.24	150.00	100.00	900.00	100.00
2.60	10.33	12.00	40.32	60.00	92.51	160.00	100.00	950.00	100.00
2.80	11.29	12.50	41.35	63.00	93.65	170.00	100.00	1000.0	100.00

The sample subjected to the laser particle size analysis was the material passed through the sieve of 1000 µm. Residue value on sieve of 1000 µm it was of R_{1000} = 0.0 %. Following the results obtained, the chart of the cumulative passage of material was drawn (% vol) depending on the granulometric class (µm) from 0.10 µm to 1000 µm (see fig. 6).

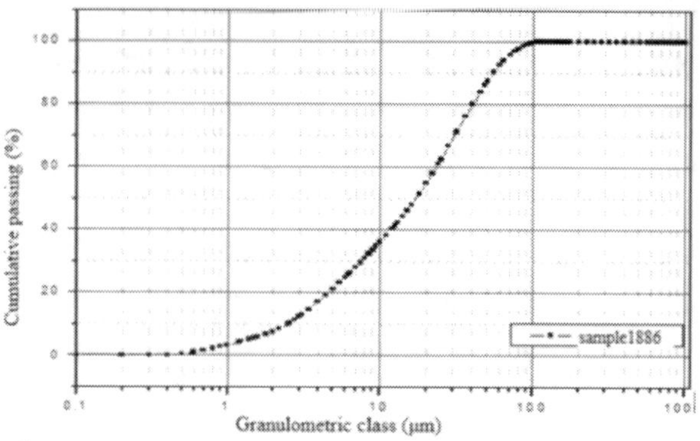

Fig.6 Chart clinker A

The results of the determinations of the laser particle size distribution for clinker F are presented in Table 6.

Table 6. Particle size distribution for clinker F

Granulometric class (µm)	*Cumulative passing (% vol)*	*Granulometric class (µm)*	*Cumulative passing (% vol)*	*Granulometric class (µm)*	*Cumulative passing (% vol)*	*Granulometric class (µm)*	*Cumulative passing (% vol)*	*Granulometric class (µm)*	*Cumulative passing (% vol)*
0.10	0.00	3.00	10.79	13.00	41.44	64.00	93.24	175.00	100.00
0.20	0.00	3.15	11.48	14.00	43.47	70.00	95.04	200.00	100.00
0.30	0.00	3.50	13.08	15.00	45.43	75.00	96.24	225.00	100.00
0.40	0.00	4.00	15.32	16.00	47.33	80.00	97.21	250.00	100.00
0.50	0.30	4.50	17.49	18.00	50.97	85.00	97.99	280.00	100.00
0.60	0.87	5.00	19.54	20.00	54.41	90.00	98.61	315.00	100.00
0.70	1.49	5.50	21.47	22.00	57.66	95.00	99.10	355.00	100.00
0.80	2.08	6.00	23.29	24.00	60.75	96.00	99.19	400.00	100.00
0.90	2.59	6.30	24.32	25.00	62.22	100.00	99.50	450.00	100.00
1.00	3.04	6.50	24.99	28.00	66.40	106.00	99.84	500.00	100.00
1.20	3.79	7.00	26.60	31.50	70.80	110.00	99.97	560.00	100.00
1.40	4.45	7.50	28.13	32.00	71.39	115.00	100.00	600.00	100.00
1.50	4.77	8.00	29.58	36.00	75.76	120.00	100.00	650.00	100.00
1.60	5.10	8.50	30.96	40.00	79.55	125.00	100.00	700.00	100.00
1.80	5.79	9.00	32.28	45.00	83.55	128.00	100.00	750.00	100.00
2.00	6.52	9.50	33.55	48.00	85.61	130.00	100.00	800.00	100.00
2.20	7.31	10.00	34.77	50.00	86.85	140.00	100.00	850.00	100.00
2.50	8.56	11.00	37.11	55.00	89.56	150.00	100.00	900.00	100.00
2.60	9.00	12.00	39.32	60.00	91.77	160.00	100.00	950.00	100.00
2.80	9.89	12.50	40.39	63.00	92.89	170.00	100.00	1000.0	100.00

The sample subjected to the laser particle size analysis was the material passed through the sieve of 1000 µm. Residue value on sieve of 1000 µm it was of R_{1000} = 0.0 %. Following the results obtained, the chart of the cumulative passage of material was drawn (% vol) depending on the granulometric class (µm) from 0.10 µm to 1000 µm (see fig.7).

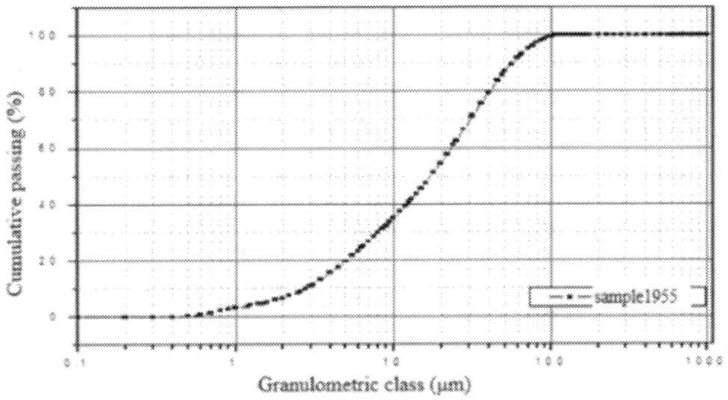

Fig.7 Chart Clinker F

4 Conclusions

Cement production requires large amounts of thermal and electrical energy. Energetic efficiency, defined as the total thermal energy consumed per ton of clinker produced and, respectively as the total electricity consumed per ton of cement produced, depends almost entirely on the technology applied in the production process.

According to the laboratory analyses it is observed that the mineralogical composition according to Bogue's calculation C_2S clinker A > C_2S clinker F (11.48 > 6.8) which results in clinker A requiring additional energy consumption in the mill because it has a much higher content of C_2S (calcium silicate) fact shown by the energy consumption to clinker A 111.05 kwh/t > clinker F 101.25 kwh/t.

Results for laser particle size distribution for clinker A and for clinker F they showed that both clinkers at granulometric class of 115 µm, has a cumulative passing (%material volume) is integral (100%). Besides reducing production costs, increasing energy efficiency contributes positively to the conservation of natural resources (fuels) as well as reducing greenhouse gas emissions (CO_2) generated both by combustion of fuels in the clinker oven (direct emissions) or in power plants for the production of electricity consumed in the cement factory (indirect emissions)

A direction of increasing energy efficiency is represented by the use of alternative fuels from waste, method with significant potential for reducing conventional fuel consumption.

Reduction of electricity consumption per tonne of product contributes to the refurbishments made in recent years at cement plants in Romania, which include:

- Replacing air compressors with some efficient ones
- Replacement of old pneumatic cement transport installations
- Resizing the ball mill load and installing systems to optimize their operation
- Installation of a high energy efficiency separator and a state-of-the-art filter
- Optimization of control of the machine flows through the Process Control System.

The paper was funded both from the project "Sustainable development of the base of agromechanical applications in gardens, solariums, vineyards and orchards (DEZDURPRA)", CNFIS-FDI-2020-0184, from the Ministry of Education and Research through the Executive Agency for Financing Higher Education , Research, Development and Innovation, as well as from the European Social Fund within the Sectoral Operational Program Human Capital 2014-2020, through the Financial Agreement of the project "Scholarships for entrepreneurial education between doctoral students and postdoctoral researchers (Be Entrepreneur!)", Contract no. 51680 / 09.07.2019 - SMIS Code: 124539.

References

1. D.P. Bentz, E.J. Garboczi, C.J, Haecker, O.M. Jensen, Cement and Concrete Research, (October 1999), 16 pages, doi: 10.1016/S0008-8846(99)00163-5;
2. D. Kim, Advances in Materials Science and Engineering, (2018), ID 1763524, 6 pages, https://doi.org/10.1155/2018/1763524;
3. C. Ferraris, V. Hackley, A. I. Avilés, Cement Concrete and Aggregates, (December 2004), 11 pages, doi: 10.1520/CCA11920;
4. A. Zayed, *Effects of portland cement particle size on heat of hydration*, Contract No. BDK84 977-13, Final Report, (December 2013), Florida Department of Transportation, Research Center;
5. N. Hlabangana, G. Danha, E. Muzenda, South African Journal of Chemical Engineering, vol.**25**, (2018), pag.79-84;
6. O. Genc, *Energy - Efficient technologies in cement grinding*, Chapter in book „High Performance Concrete Technology and Applications", (2016), doi: 10.5772/64427;
7. S. Mohanty, K. Gupta, S. Raju, Procedia Computer Science, vol.**58**, (2015), pp.217–224;
8. V. Bulgakov, S. Pascuzzi, S. Ivanovs, G. Kaletnik, V. Yanovich, Biosystems engineering, vol. **171**, (2018), pag.155- 164;
9. D.C. Tsamatsoulis, Brazilian Journal of Chemical Engineering, (2014), vol. **31**, no. 01, pp. 155-170;
10. A. Jankovic, W. Valery, E. Davis, Cement grinding optimisation, Minerals Engineering, vol. **17**, (2004), pp.1075–1081
11. J.R.G. Sarduy, J.P.M. Yanes, M.E.G. Rodríguez, J.L.Q. Ferreira and Y.M. Torres, Determining cement ball mill dosage by artificial intelligence tools aimed at reducing energy consumption and environmental impact, Ingeniería e Investigación, Vol. **33**, No. 3, (2013), pp.49-54
12. S. Opris, *Manual of engineers from cement industry*, vol. I, Ed. Tehnica (1994);
13. *** Pictures from CEPROCIM SA, B-dul Preciziei Nr.6, sector 6, Bucuresti;
14. *** SR EN 196-6:2018, *Methods for cement testing. Part 6: Fineness determination;*
15. *** SR EN 196-2:2013, *Methods for cement testing. Part 2:Chemical analysis of cements;*
16. *** https://www.n8equipment.org.uk/equipment/detail/mastersizer-2000e;
17. A. Atmaca, M. Kanoglu, Reducing energy consumption of a raw mill in cement industry, Energy, Vol. **42**, Iss. 1, 2012, pp. 261-269

E3S Web of Conferences **180**, 04008 (2020) https://doi.org/10.1051/e3sconf/202018004008
TE-RE-RD 2020

The smart transition of Power Transmission Network needed for the sustainable development of Bucharest City

Mircea Scripcariu[1*], *Cristian* Gheorghiu[2], *Ion* Tristiu[3] and *Alexandra-Gabriela* Dobrica[4]

[1]Associate Prof., Dept. of Energy Production and Use, University POLITEHNICA of Bucharest, Romania
[2]PhD. Student, Dept. of Energy Production and Use, University POLITEHNICA of Bucharest, Romania
[3]Professor, Dept. of Power Systems, University POLITEHNICA of Bucharest, Romania
[4]Master Degree Student, Dept. of Energy Production and Use, University POLITEHNICA of Bucharest, Romania

Abstract. Large cities around the world continue to grow larger. This is also the case of Bucharest, the capital city of Romania and one of the largest and most densely populated locations in Eastern Europe. In this article authors present an overview of the development of Bucharest as the largest electricity consumer amongst the Romanian municipalities and analyse possible solutions for the refurbishment of the HV (High Voltage) grid. This process needs to help the city development and its transition to a Smart City. At first authors present the strategy of local authorities for improving the living standard and decrease the environmental pollution. The impact of this strategy on the electrical energy demand is further analysed. Possible technologies are presented for substations and electrical lines starting from the current situation of the transmission network. All these can contribute to a decrease of the environmental impact and to an increase in the continuity of the power supply. Technical and economical evaluation of the refurbished grid is further presented along with a sensitivity analysis. The evaluation of the investment is made by taking into consideration the demand forecast. Lines and transformers are sized and chosen with the best available technology. For the economic analysis authors used criteria accepted by funding sources, banks especially, such as NPV and IRR. The sensitivity of the project economics is tested and discussed. As conclusions authors present the environmental benefits of gradually changing the technology used for electricity transmission in a large city such as Bucharest, mainly regarding land occupancy with switchgear and line routes and soil pollution.

*Corresponding author : mirceas1960@yahoo.com

1 Bucharest's road map towards a Smart City

In the last decades cities around the world embraced the smart development concept [1]. In 2008, the 50% threshold of persons living in cities was broken [2]. Furthermore, the United Nations Population Fund-UNFPA foresees that urban population will represent 70% of the total in 2050 [2].

By 2011, in EU-28 were living 503 million inhabitants [3], of which 42.4 % in a predominantly urban region, 35.3 % in an intermediate region and 22.3 % in a predominantly rural region [4]. According to the definition, the predominantly urban regions have less than 20 % rural population, intermediate regions between 20 % and 50 % and predominantly rural regions more than 50 % [5]. Compared to 2006, the growth of the population in the predominantly urban region is of 1.1% in the detriment of predominantly rural region [5]. At the beginning of 2020, EU-28 had 503 million inhabitants [3], rising with over 2% compared to 2011. In 2013, the world's urban areas accounted for about 64% of global primary energy use and produced 70% of the planet's carbon dioxide emissions [6].

Cities around the world are developing strategies and master plans on the road to decarbonization. In the USA, small communities are already running entirely on renewable power, including Aspen, Colorado, and Burlington, Vermont [6]. "Bigger cities have set ambitious goals, with Copenhagen, Denmark, aiming to be carbon neutral by 2025. San Diego, California, aims to be 100% powered by renewable sources by 2035, and Vancouver, Canada, by 2050" [6]. There is a remarkable trend for sustainability and cities are taking actions like joining in larger organizations with clear targets to tackle the climate change. C40 Cities Climate Leadership Group was established as a global network of cities that are developing and implementing policies to reduce emissions [6].

The European Environment Agency states that "Europe currently does not live up to its 2050 vision of 'living within the limits of our planet'. This calls for fundamental changes in lifestyles, production and consumption, knowledge and education" [7].

According to data from the National Institute of Statistics [8], Bucharest has 2.13 million inhabitants in 2019 from 2.11 million in 2015. The population growth is not spectacular, but as capital city of Romania it became a crowded area. According to Forbes, Bucharest is the main driving force of economic growth in Romania, generating 20% of the GDP. In a chart provided by Easy Park Group [9], Bucharest is ranked on the 87th position among 500 cities considered in a study from 2017, that took into consideration factors such as: Transportation and mobility, Sustainability, Administration, Business environment, Digitalization, Living standard, Experts perception regarding city development.

IMD WORLD COMPETITIVENESS CENTER ranks Bucharest on the 85th position in another smart city index. Any smart strategy brings cities economic growth and social benefits. For example, in Copenhagen, Denmark, business opportunities of over Euro 100 million were generated in 2015 [10].

To channel a large municipality such as Bucharest towards a smart city there is a need for continuous transformation of three essential pillars, the administration, the technology and the society. A smart Bucharest will be the result of the cumulated effort of local administration and public authorities, private businesses, telecoms, utilities, building owners associations, transportation companies, health organizations and education organizations (schools, universities etc.) and finally, of each citizen. The energy infrastructure in Bucharest is well developed but obsolete. The district heating system is inefficient and in poor financial situation. The electricity transmission and distribution systems are using old technologies, influencing the sustainability and the living standard. They occupy land with large substations and overhead lines and by using equipment with oil, create a hazard for soil and underground water.

Data from the residential market show that in Bucharest around 10,000 new apartments are commissioned each year. Most of these are part of residential areas developed in the outskirts of the city where there are high voltage substations and overhead lines. These are imposing restrictions to real estate developments. Along with residential developments, malls are erected. All these represent new electricity consumers that put pressure on the modernization of the electricity transmission and distribution infrastructure. The Sustainable Urban Mobility Plan shows, in the optimal scenario, that the metro system will be extended with two new mains (M5 and M6) and M2 and M3 will be lengthened [11]. New fast tramways lines are in this plan, too. One is under construction for the 2021 European Football Championship. The mobility plan shows that electrically driven systems will be developed with priority, adding on the pressure for the modernization of the electricity network. The Municipality is investing in around 300 charging stations for EVs.

Next, the authors will analyse only the smart initiatives in Bucharest city with impact on the electricity network. In section 2 authors analyse the impact of smart initiatives, of population growth and of economic development on the electricity demand forecast. Section 3 is dedicated to the presentation of the current situation of the transmission grid supplying the city of Bucharest and of the associated environmental impact. In section 4, authors present the technological solutions for the refurbishment of the transmission grid that are technically and economically analysed in section 5. The conclusions of the current research and further initiatives are presented in section 6.

2 Demand Forecast

The continuous growth in the number of inhabitants, the economic development, the development of the electrically driven public transportation and the evolution of EV market will lead the electricity demand to higher levels. Electricity networks in Bucharest need to be resized and gradually replaced to match this growth. According to E-Distributie Muntenia Sud, the electricity distribution company operating in Bucharest, the peak load was in the last years over 1,000 MW and is steadily growing. The power transmission system operator, CNTEE Transelectrica, following data from the National Commission for Strategy and Prognosis, is forecasting an electrical energy consumption growth with a yearly rate of about 1.1% [12]. The peak load demand will follow the same pattern. According to [12], 15% of the national electricity consumption is represented by Bucharest. Consequently, the electricity forecast for Bucharest is presented in Table 1.

Table 1. The annual energy consumption and the peak demand forecast for the City of Bucharest [12]

YEAR	2021	2022	2023	2024	2025	2026	2027
Wy [TWh/year]	8,625	8,700	8,790	8,895	8,985	9,075	9,165
Ppeak [MW]	1,362	1,378	1,394	1,410	1,425	1,440	1,454

The total installed capacity in (auto)transformers in the three substations supplying Bucharest is 1,900 MVA and the "n-1" criterion is not fulfilled at peak load. Beyond 2027 is expected a sharp increase in the transition from fossil fuel transportation to full-electric and hybrid. Combined with the increase in the demand of the public transportation, an increase over 10 TWh of the energy consumption can be expected beyond 2030. This forecast shows that the power transmission network needs investments in modernization and strengthening. The power growth will be ensured by wind developments in Dobrogea region (south-eastern part of Romania) and the expected development with two new nuclear units of Nuclear Power Plant (NPP) Cernavoda.

https://doi.org/10.1051/e3sconf/202018004008

3 Current Situation of the Power Transmission Grid and it's main Environmental Impact

The city of Bucharest is supplied through two 400 kV substations, Domnesti and Bucharest South, and a 220 kV substation, Fundeni. In Table 2 the loading of the power transformers in each substation at the Winter Peak Load (WPL) is presented. Substation Domnesti is equipped with transformers (T) and substations Bucharest South and Fundeni are equipped with autotransformers (AT).

Table 2. Loading Factor of the Power Transformers installed in the substations supplying Bucharest City, WPL, 2017 [13]

Equipment T/AT	Substation	Voltage ratio	Rated power (S_n)	WPL 2017 (S)	Load Factor S/S_n
-	-	-	MVA	MVA	%
AT3	Bucharest South	400/220/20	400	288.4	72.10
AT4	Bucharest South	400/220/20	400	288.4	72.10
T1	Domnesti	400/110	250	109.2	43.68
T2	Domnesti	400/110	250	98.2	39.28
AT1	Fundeni	220/110	400	193.42	48.36
AT2	Fundeni	220/110	400	189.67	47.42

Data in table 2 confirm that Bucharest South substation needs capacity strengthening. Domnesti and Fundeni substations are at the limit if considering the "n-1" criterion. The Overhead Lines (OHL) connecting these substations are loaded at over 60% of their thermal capacity. The main environmental impacts of OHLs and substations in the operation stage are: Land occupancy, Visual impact, Noise, mainly determined by the Corona phenomenon, Deforestation and Soil pollution with mineral oil and chemical substances. In the decommissioning stage, the environmental impact is determined by the huge amount of materials resulting, that need to be recycled, reused or stored.

In table 3, the typical amount of materials that result when a substation with classic Air-Insulated Switchgear (AIS) is decommissioned, is presented. Electrical apparatuses are oil insulated type and the circuit-breakers have oil for arch extinction. The example is given for a 400/220/110 kV substation with a total of 44 circuits and a Double Busbar System with a Bypass Busbar (2BB+BB) Single Line Diagram (SLD). The total oil quantity does not include auto (transformers). This data was calculated as part of a study prepared for the transmission grid operator [14]. Bucharest South substation has the same SLD (Single Line Diagram), a similar number of circuits and the same technology for the switchgear. So, when decommissioned one can expect to result the same amount of different materials.

Table 3. Typical amount of materials resulted when a substation with classic AIS (44 circuits) is decommissioned [14]

Rated Voltage	SLD	Mineral Oil	Al*	Support insulators and extinction chambers	Concrete	Steel	Other
kV	-	kg	kg	kg	m^3	kg	kg
110	2BB+BB	139,000	89,000	1,400,000	2,400	14,500	185,000
220							
400							

* Aluminium resulting only from disconnecting-switch contacts and locking coils.

When decommissioning a 400 kV OHL, only form the insulation system may result 3,200 kg/km of porcelain. From the protection wire alone may result 500 kg/km of aluminium and 700 kg/km of steel. Regarding the land occupancy, substations with AIS technology occupy important surfaces, see table 4.

Table 4. Land occupancy by the substations supplying the City of Bucharest [15]

Substation	Rated Voltage	Surface
-	kV	m^2
Domnesti	400/110	56,687
Bucharest South	400/220/110	89,155
Fundeni	220/110	59,534

The OHLs linking the three substations supplying Bucharest city need important surfaces for their routes (see table 5). The land is occupied for the entire normal operation duration of the line only by the posts. For overhead lines this duration is between 32-48 years. The OHLs crossing Bucharest exceed this operation interval.

Table 5. Routes dimensions for the OHLs connecting the substations supplying the city of Bucharest [12]

OHL	Rated Voltage	Length	Route width*	Route Surface
-	kV	km	m	m^2
Bucharest South – Domnesti	400	34.367	75	2,577.525
Bucharest South – Fundeni	220	25.030	55	1,376.650

* as per the technical norm NTE 003 for the construction of OHLs with a rated voltage of over 1,000 V [16].

According to [16], possible crossing of lines over inhabited or not inhabited buildings should be avoided as much as possible. The crossing can be accepted only with the approval of the transmission system operator and if the material of the roof is fire resistant, and if it is from metal needs to be connected to an earthing installation. The installation of posts in the proximity of intensive pedestrian-traffic areas should be avoided (such as the entrance of a building). In fig. 1 we present the situation when a post foundation for one of the above-mentioned lines is within the perimeter of two inhabited buildings. We presume that these buildings were erected after the line commissioning and are part of the development of the urban area of the city of Bucharest. Their construction may not respect the provisions of the technical norms.

Fig. 1. OHL post with foundation within the premises of an inhabited building

The transmission network supplying with electricity the capital city of Romania has a major environmental impact by occupying the land with substations and line routes, by representing a potential threat to the soil by means of oil pollution, by generating sound pollution due to the Corona effect and by having a negative visual impact.

4 The solution for improving the environmental impact and the electrical capacity of the Power Transmission Network

From the '60, electrical installation manufacturers made a technological breakthrough by using SF_6 (Sulphur hexafluoride) in circuit-breakers. The use of this gas was extended as an insulator between phases in a GIS (Gas Insulated Switchgear) installation for substations. The greatest advantage being the small surfaces occupied by substations. In Table 6 we present, for a typical bay, the approx. dimensions for AIS (Air Insulated Switchgear) and GIS technologies.

Table 6. Main Constructive features of AIS and GIS bays [12]

Rated Voltage [kV]	Technology	Bay Type	SLD	Length [m]	Width [m]	Surface [m^2]	Surface ratio (AIS/GIS)
220*	AIS	OHL	2BB	71.00	17.0	1,207.00	-
220	AIS	OHL-AT**	2BB+BB	171.10	17.0	2,908.70	-
220*	GIS	line	2BB	5.00	1.6	8.00	151
400*	AIS	OHL	2BB	89.90	22.5	2,022.75	-
400	AIS	OHL-T***	2BB+BB	209.80	22.5	4,720.50	
400*	GIS	line	2BB	7.715	2.25	17.36	117

* Selected for calculating the surface ratio (AIS/GIS),
** OHL and autotransformer (AT) bays over the same bay pitch, face to face arrangement,
*** OHL and transformer (T) bays over the same bay pitch, face to face arrangement.

SF_6 is raising concerns because it is a greenhouse gas with a contribution 23,500 times more than CO_2. For that, manufacturers guarantee high levels of safety against leakages and, at the same time, are investing in research to find a better gas from an environmental impact point of view. In the last years, large manufacturers announced the trial of GIS installations using gases with no environmental impact. At the same time keeping the same advantages regarding the surfaces needed for single bays [17 - 19].

For the connections between the substations of the transmission network supplying Bucharest, the only proven technology available is the use of high voltage cables to replace the existing OHLs. One of the most widespread technologies for high voltage cable manufacturing is to use cross-linked polyethylene (XLPE) for insulation. The conductor may be of aluminium or copper. The main advantages of these cables are flexibility, reduced weight, resistance to environmental impact and low operation costs. The mounting solution can be buried or posed in ducts. The last solution has the advantages of fast access for maintenance and increased mechanical protection. The technical and economical evaluations were done for a solution using XLPE cables posed in ducts. This mounting solution can provide also the opportunity to pose optical fibre, much needed for the development of a smart Bucharest. Choosing the optimal posing and layout of the high voltage cables should result from an in-depth study regarding the electromagnetic compatibility and the operation and maintenance procedures.

The strategy for the evolution towards a smart city for Bucharest is requesting the development and refurbishment of the transmission network. The technologies that authors consider to be the most appropriate for fulfilling this target are the use of GIS installations in all transmission substations and the use of cables for the connection between substations. The power demand forecast presented in section 2 and the load factor for the power transformers presented in section 3 were taken into consideration for the proposed profile of equipping the refurbished network with transformers. Hence, it's been proposed to replace the existing power transformers located in Bucharest South Substation with larger ones, with a rated power of 630 MVA.

The refurbishment and development of the power transmission network in Bucharest city has a number of main objectives: Reducing the visual impact and the soil pollution caused by the old technology, Increasing the power grid capacity in order to support the continuous economic development and the transition to electric transportation (EVs, the development of the underground metro network etc.). CNTEE Transelectrica SA planned measures for the development of the network, like a new substation in the centre of the city, in Grozavesti area, as a medium term measure and the development of another substation, in the eastern part of the city, but close to the city centre, as a long-term measure [12]. Both substations will be linked with cables to the existing network. These projects are very uncertain due to land issues and were not considered in our analysis.

5 The Technical and Economic Analysis of the Refurbishment Solution for the Power Transmission Network supplying the city of Bucharest

The main hypothesis for the technical calculation, were to consider that the same generation capacities exist in the power plants in Bucharest. There are no new generation units forecasted. In the power transmission network, as a node, the Brazi West substation was also included. In the first stage of the analysis the maximum three phase short-circuits currents in the main substations was calculated, in order to establish if there are any restrictions regarding the switchgear characteristics, determined by the introduction of cables, that have a specific reactance lower than that for OHLs. In table 7 the results of the calculations for the existing situation and for the proposed refurbishment are shown. The symmetrical component method [21] was used to determine the maximum three-phase short-circuit current in both the actual and the proposed scenarios, by using an equivalent-generator. The most important increase in the short-circuit current was identified in the 400 kV nodes . This value is lower than the values given for the rated breaking current by all the major GIS manufacturers. The refurbishment process needs to be planned so that the increase of currents in the fault state is carefully managed. This is caused by the introduction of cables but also by the increase in the T/AT rated power to match the conditions of the steady state.

Table 7. Three phase short-circuit current in the Power Transmission Network's nodes

Substation	Rated Voltage [kV]	Actual three phase short-circuit current intensity [kA]	Forecasted three phase short-circuit current intensity [kA]
Bucharest South	400	28.21	37.07
Bucharest South	220	23.74	24.91
Brazi West	400	26.51	26.92
Domnesti	400	28.85	37.43

For the economic analysis were used the most accepted criterions by the main funding organizations, the Net Present Value - NPV (1), Internal Rate of Return - IRR (2) and the Simple Payback Period – SPP (3).

$$NPV = \sum_{t=1}^{t_{st}} \frac{I_t - C_t}{(1+a)^t} - IC\ [EUR] \quad (1)$$

where t_{st} is the analysis time-frame, in years, I_t is the yearly income in the tth year, in EUR/year, C_t are the yearly expenditures in the tth year, in EUR/year, a is the discount rate, in %/year.

$$NPV = \sum_{t=1}^{t_{st}} \frac{I_t - C_t}{(1 + IRR)^t} = 0\ [EUR] \quad (2)$$

where the Investment Cost can be included in the yearly expenditures as a depreciation cost.

$$SPP = \frac{IC}{\frac{\sum_{i=1}^{t} I_i - C_i}{t}} [years] \quad (3)$$

where *IC* is the investment cost, in EUR.

Due to differences in the useful life of different components of the transmission network, an analysis period of 14 years, according to legal provisions, was considered [22]. The Romanian transmission operator is listed at the Bucharest stock exchange and it raises money for investments in different ways, form banks, from corporate bonds etc. [22]. Considering also the good practice in the energy sector, we used a discount rate of 10%/year. With the investment cost authors took into consideration a complete refurbishment of the substations and a possible route for HV (High Voltage) cables. Because the operation of cables generates reactive power into the network, an investment in compensation coils was added. The length and routes of cables will be clarified at the feasibility study stage. For the calculation of the operational cost with energy and power losses, the electricity price was set at 50 EUR/MWh. This value is an average price, because the transmission operator buys electricity from the spot market and from the market for bilateral contracts. For current operation, authors also considered a set of maintenance and repairing costs. The technology proposed for substations has lots of references worldwide. Transelectrica is operating GIS substations, like the 400/220 kV substation Brazi West, but no HV cable. Operation costs are significantly lower for GIS bays than for the present AIS bays. Because the continuity of the transmission service has had good indicators in the last years [12], [15], [20-23] the cost of interruptions is significantly lower than the other annual costs and was neglected.

The income is generated by the transmission activity at an average price of around 4 EUR/MWh. Because in Bucharest for the next years one can expect only a continuous increase in the electricity demand, see section 2, the income will increase annually. The value in table 9 is obtained by considering the annual increase of income as a direct effect of the annual increase in energy demand. In Table 8 the main costs and income of such a refurbishment process are shown.

Table 8. Investment, Annual Revenues and Operation and Maintenance (O&M) Costs of the Power Transmission Network Refurbishment

Financial Entry	Value	M.U.
Investment Costs (*IC*)	275,000,000	[EUR]
Annual Income (I_t)	42,600,000	[EUR/year]
Annual Cost of Power and Energy Losses (CPW)	6,900,000	[EUR/year]
Operation and Maintenance Costs (O&M)	1,200,000	[EUR/year]

Table 9. Financial Indicators for the refurbishment project

Financial Indicator	Value	M.U.
NPV	100,000	[EUR]
IRR	10.1	[%/year]
SPP	6.5	[years]

For the technology proposed for cables and substations, the operation and maintenance costs are very low compared to the existing ones. In order to obtain attractive results for the technical and economic analysis, the Transport System Operator (TSO) should raise the

 https://doi.org/10.1051/e3sconf/202018004008

transport tariff from 4 EUR/MWh to 4,2 EUR/MWh – a 8% increase, which is justifiable considering the high amount of capital invested in this project. With all these costs considered, the results of the economic analysis are presented in table 9.

The results presented in Table 9 show that the project becomes viable if the analysis period is at least 14 years, as the NPV is close to 0 and the IRR is close to the Discount Rate. Considering the average lifetime of the equipment available on the market, the actual life cycle of the project is estimated to reach at least 20 years without significant additional investment costs, which would lead to an NVP of EUR 55 million, and an IRR of 12.5%/year. With the financial indicators at limit, authors considered that a sensitivity analysis is needed. The annual cost of Power and Energy Losses is highly dependent on the electricity market price and the sensitivity of NPV with this price is shown in fig. 2. Revenues are highly dependent on the transmission tariff that is approved by ANRE-the Romanian Energy Regulatory Authority and has the sharpest gradient in fig. 2.

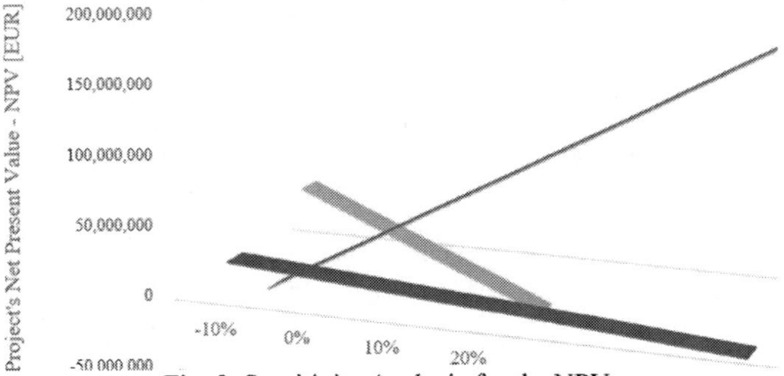

Fig. 2. Sensitivity Analysis for the NPV

6 Conclusions and further research

As presented in section 3, the equipment of the transmission network occupies around 210,000 m^2 of valuable land, meaning 0.1% of the surface occupied by the city of Bucharest. The authors of this paper propose and analyse a bold idea, the refurbishing of the transmission network supplying the electricity for approx. 2 million inhabitants.

State of the art equipment with GIS for substations and XLPE cables for the connection lines are included in this analysis. This development comes in conjunction with the strategy of Transelectrica that has as one of the objectives to build a new HV substation in the city centre, with the same technology.

From a technical point of view, the implementation of GIS technology for substations has no disadvantages and leaves a lot of space for other needs. The installation of cables for the connection between substations will increase the value of the short-circuit currents but the circuit breakers in GIS bays will cope with it. The replacement of the OHLs with cables will require compensation coils for the reactive power.

Regarding the environmental impact, the technological change in the transmission network supplying the city of Bucharest will bring: A reduction with almost 90% of land surface occupied with the installations and related constructions, the almost complete elimination of the soil pollution threat, e.g. with mineral oil; only the power transformers will remain with mineral oil insulation, but the solution used for their foundation helps avoiding soil pollution in case of incidents and an improvement in the visual impact of the electrical network; GIS can be installed in buildings.

As further research on the development of the transmission network for supplying the city of Bucharest, authors will analyse the compatibility of such refurbishment with other utility networks.

References

1. Vito Albino, Umberto Berardi and Rosa Maria Dangelico, "*Smart Cities: Definitions, Dimensions, Performance, and Initiatives,*" Journal of Urban Technology, 2015, Vol. 22, No. 1, 3 – 21, http://dx.doi.org/10.1080/10630732.2014.942092.
2. https://www.unfpa.org/ [accessed in January 2020].
3. https://ec.europa.eu/eurostat/databrowser/view/tps00001/default/table?lang=en [accessed in January 2020].
4. https://ec.europa.eu/eurostat/statistics-explained/index.php?title=File:Distribution_of_population_,_2011_-_%25,_V2.png [accessed in January 2020].
5. https://ec.europa.eu/eurostat/statistics-explained/index.php/Archive:Urban-rural_typology_update [accessed in January 2020].
6. https://www.iea.org/news/cities-are-at-the-frontline-of-the-energy-transition [accessed in February 2020].
7. European Environment Agency, "*The European environment — state and outlook 2020 Knowledge for transition to a sustainable Europe*," 2019.
8. National Statistics Institute, "*Population by residence on 1st of January 2019,*" http://www.insse.ro, Bucharest, 2019.
9. East Park, „*Easy Park Group,*" 2017. Available at: https://www.easyparkgroup.com/smart-cities-index/.
10. „Copenhagen Smart City", Smart City Report, 2015.
11. Durable Urban Mobility Plan for București – Ilfov in the 2016-2030 perspective, http://www.pmud.ro/etapele-proiectului.html.
12. Power Transmission Network Development Plan in 2018-2027 perspective, CNTEE Transelectrica SA.
13. http://www.transelectrica.ro/web/tel/consum [accessed in December 2019]
14. Nicolae Golovanov, Gheorghe Comanescu, Mircea Scripcariu and Ioan Sevastian Bitir-Istrate, "Directions and Costs for reducing the environmental impact of refurbishing overhead lines and high voltage substations of the power transmission network, in accordance with the national and European legislation," Romanian Energy and Environment Foundation, University POLITEHNICA of Bucharest, 2004.
15. https://lege5.ro/Gratuit/gqydimjugq/inventarul-bunurilor-din-domeniul-public-mijloace-fixe-concesionate-cntee-transelectrica-sa-hotarare-684-2014?dp=gy4dgojxhazds [accessed in February 2020].
16. *** NTE 003/04/00 "Technical norm for building overhead lines with rated voltages of over 1,000 V," 2000.
17. Laruelle, Elodie & Kieffel, Yannick & Ficheux, Arnaud, "g3—The Alternative to SF_6 for High-Voltage Equipment." Lecture Notes in Electrical Engineering, pp. 139-146, 2018, doi:10.1007/978-3-319-58172-9_15.
18. Kieffel, Yannick & Biquez, François & Vigouroux, Daniel & Ponchon, Philippe & Schlernitzauer, Audrey & Magous, Richard & Cros, Gérard & Owens, John, "Characteristics of g3 – an alternative to SF_6," CIRED - Open Access Proceedings Journal, pp 54-57, 2017, doi:10.1049/oap-cired.2017.0795.
19. https://www.gegridsolutions.com/press/gepress/sf6-free.htm [accessed in March 2020]
20. Romanian Parliament, "Government Decision no. 2139 from 30 November 2004 for approving the Normal Use Time and Classification of Capital Goods," 2004.
21. *** PE 134/95. "Technical norms for determining short-circuit currents in power grids with rated voltages of over 1.000 V," 1995.
22. CNTEE Transelectrica SA, 2014. [Interactiv]. Available: http://www.transelectrica.ro/web/tel/emisiune-de-obligatiuni. [accessed in March 2020].
23. Romanian Energy Regulatory Agency – A.N.R.E., "Key Performance Indicators and Technical State of the Power Transport and Distribution Networks and System Services," 2018. https://www.anre.ro/ro/energie-electrica/rapoarte/rapoarte-indicatori-performanta [accessed in December 2019]

Efficient phosphate removal by biological corrosion method

Olga Ruzhitskaya [1,*]

[1] Peoples' Friendship University of Russia, Department of Civil Engineering, 6 Miklukho-Maklaya St., Moscow, 117198, Russia

Abstract. Water quality deterioration in water bodies is directly related to the development of anthropogenic eutrophication processes. To resolve this complex issue, one needs to minimize biogenic discharge of nitrogen and phosphorus compounds into the water bodies. The article presents relevant information in the field of phosphate-removing wastewater treatment, describes the most effective biological and biological-chemical methods of phosphate-removing wastewater treatment. The article presents the results of research on phosphate-removing wastewater treatment methods using iron-bearing reinforced feed material (biological corrosion method, biogalvanic method). The placement of reinforced feed material in a standard air tank allowed to significantly increase the efficiency of organic contamination-removing biological wastewater treatment. The biological process activation ratio due to the use of reinforced feed material amounted to 1.78. The placement of reinforced feed material in the bioreactor, which operates without activated sludge return, allowed to achieve complete phosphate removal from the waste water. The maximum effect of phosphate removal (99 %) with the concentration of phosphate in treated water below the detection limit was achieved after 4 to 6 hours of wastewater treatment in the bioreactor, with the concentration of organic contamination in the incoming wastewater from 150 to 300 mgBOD/L and the concentration of phosphate upstream of the bioreactor from 4 to 6 mg PO4/L. Based on the research carried out, it was concluded that the use of biological corrosion method allows to achieve high efficiency of phosphate removal, as well as to intensify the biological process of organic contamination removal from waste water.

1 Foreword

In order to solve the issue of deterioration of the water quality of water bodies, it is necessary to minimize the discharge of nutrients (nitrogen and phosphorus compounds).

Today, many methods are used to remove nutrients, including phosphates. Which are removed by chemical, physico-chemical, biological, biological and chemical methods.

Of greatest interest are biological methods, since its use uses the biological potential of activated sludge to the maximum extent possible, but the use of the standard method of

[*] Corresponding author: ruzhitskaya-oa@rudn.ru

biological treatment achieves the removal of phosphates by only 20 - 40%, which is not enough to ensure the required cleaning effect (at least 95%).

The methods Phostrip, A / O (Anaerobic – Oxic), EASC (Extended Anaerobic Sludge Contact process) and others have been developed and are successfully used to remove phosphates.

Compounds of phosphorus and nitrogen are simultaneously removed using the methods of UCT, Bardenpho, Phoredox, etc.

The UCT method is most effective in removing nitrogen and phosphorus at the same time.

To reduce the effect of nitrates formed in the aerobic process, various modifications of the above methods are used [1].

Bardenpho is a stepwise biological treatment method with alternating aerobic and anoxic zones [2].

The organization of recirculating flows is a feature of all the schemes under consideration.

Japanese and Chinese scientists have investigated the process of intensification of nitrogen and phosphate removal using thermophilic fermentation suspension from food waste (fsfw) as an external carbon source. The efficiency of this technology in phosphorus-removing wastewater treatment exceeded 98 % [3].

Efficient phosphate removal (92.8 ±6.7) was achieved in the sequencing batch reactor (SBR) – the ASSR system [4, 5].

Stable operation and high efficiency of phosphate-removing wastewater treatment (99.1 %) was achieved by applying the biological process AAO-SBSPR [6].

Russian scientists, in their works, have carried out the estimation of potential efficiency of phosphorus compounds removal from waste water by biological method. Scientists have come to the conclusion about the possibility of phosphate removal and that certain technological parameters of biological treatment allows to meet strict requirements for phosphate removal [7].

The simultaneous removal of nitrogen (98.7 %) and phosphorus (83 %) was achieved in a study of the effectiveness of a new two-stage aerobic anoxic continuous action reactor (CFBCR) [8].

Scientists in the USA conducted the study of the reactor series (SBR) for the simultaneous removal of nitrogen and phosphorus. The reactor operating conditions for achieving high phosphate removal rate (> 99 %) were determined [9].

Modern research is aimed not only at improving recirculation processes and flows, but also at genetic modification of activated sludge bacterial composition to intensify biological phosphate removal [10, 11].

The combination of biological and chemical cleaning processes leads to the greatest cleaning effect of phosphate removal. These methods of biological and chemical cleaning differ in the place of introduction of the reagent and the composition of the reagent used.

2 Study of biological corrosion method for deep phosphate removal

We conducted research on the use of the biological corrosion method (biogalvanic method) to remove phosphorus compounds and intensify biological wastewater treatment.

2.1 Air tank – precipitation tank diagram

The placement of reinforced feed material in a standard air tank (the plant operation diagram is shown in Fig. 1) allowed to significantly increase the efficiency of organic contamination-removing biological wastewater treatment. The biological process activation ratio due to the use of reinforced feed material amounted to 1.78.

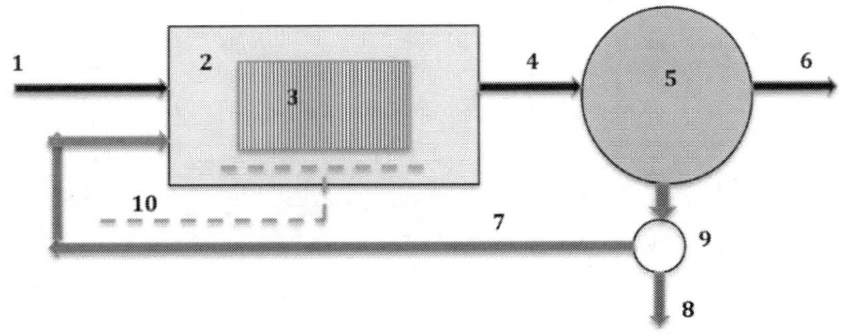

Fig. 1. Air tank with reinforced feed material operation diagram.

1 – incoming wastewater; 2 – air tank; 3 – reinforced feed material; 4 – sludge mix; 5 – secondary precipitation tank; 6 – treated wastewater; 7 – recirculated activated sludge; 8 – excess activated sludge; 9 – sludge chamber; 10 – air.

However, the main goal, which was deep phosphate removal, was not achieved, because after the reinforced feed material was placed in the air tank, the efficiency of phosphate removal rose to 99 %, but decreased to ≈ 60 % after 15 days. In the course of the experiment, the 30% decrease of the free-floating activated sludge amount resulted in 99% increase of phosphate removal. The "younger" the activated sludge, the more intensively it increases, the more phosphorus is consumed while removing it from the waste water. Since young, developing activated sludge stimulates effective phosphate removal, in order to achieve the most complete phosphate removal from the wastewater by means of reinforced feed material, it is necessary to maintain a low sludge age in the air tank by completely removing the returning activated sludge from the system maintaining sufficient amount of biomass attached to the reinforced feed material.

2.2 The bioreactor without activated sludge return diagram

On the basis of the conducted research, the diagram without activated sludge return was developed (the plant operation diagram is shown in Fig. 2).

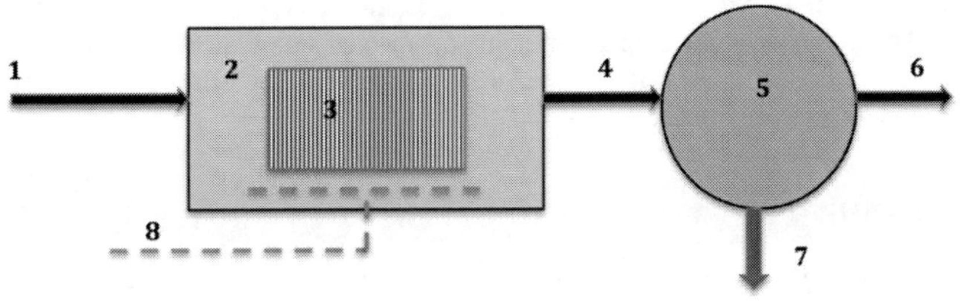

Fig. 2. Bioreactor with reinforced feed material operation diagram.

1 – incoming wastewater; 2 – bioreactor; 3 – reinforced feed material (in reference model – unreinforced); 4 – purified wastewater and biofilm mix; 5 – secondary precipitation tank; 6 – treated wastewater; 7 – excess activated sludge; 8 – air.

The maximum effect of phosphate removal (99 %) with the concentration of phosphate in treated water below the detection limit was achieved after 4 to 6 hours of wastewater treatment in the bioreactor, with the concentration of organic contamination in the incoming wastewater from 150 to 300 mgBOD/L and the concentration of phosphate upstream of the bioreactor from 4 to 6 mg PO4/L. The main data of the experiment (with aeration duration of 4–6 hours) are given in Table 1.

Table 1. Basic data from the experiment.

Parameters	Incoming drain liquids	Treated drain liquids	
		Studied	Reference
Phosphates, mg/L	4–12	BLD*–1	4–12
Iron, mg/L	BLD* – 0.3	BLD* – 0.1	-
BOD5, mg/L	100–250	3–10	10–30
Ammonium nitrogen, mg/L	15–25	8–12	15–25
Nitrites, mg/L	-	0.5–4	0.4–3
Nitrates, mg/L	-	3.5–15	BLD* – 4
Suspended substances, mg/l	60–150	BLD* – 20	15–30
Activated sludge characteristics			
Dose of biofilm attached, g/L	-	3–5	1.8–3.0
Dose of suspended sludge, g/L	-	0.1–1	0.2–0.5
Biofilm ash content	-	0.5–0.6	0.1–0.12
Sludge ash content	-	0.35	0.1
		Studied model	Reference model
Organic contamination specific oxidation rate, mgBOD/g·h		13.47–35.4	9.6–25.5
Load of organic contaminants on biomass, mgBOD/g·day		323.3–849.6	230.4–612
Phosphate specific oxidation rate, mgPO4/g·h		0.43–1.9	0–0.19
Load of phosphate on biomass, mgPO4/g·day		10.32–45.6	0–4.56

BLD* – below limit of detection

The use of the method of biological corrosion at the stage of domestic wastewater tertiary treatment showed positive results of deep phosphate removal. However, studies have shown that deep phosphate removal is mainly due to physicochemical processes of phosphate coagulation with ferrous ions, the source of which is the metal in the feed material, which is subjected to processes of electrochemical corrosion. When the biofilm loading becomes overgrown, the contact of the metal surface area with water decreases, and the purification effect decreases. After regeneration of the load, the efficiency of phosphate removal increased to its previous level.

3 Conclusions

Prospectivity of using biological and biological-chemical methods of phosphate removal from waste water has been proved. Studies have shown that the method of biological corrosion (biogalvanic method) significantly increases the efficiency of removal of phosphates and organic impurities contained in wastewater. This method can be used not only to intensify phosphate removal processes, but also for intensification of biological purification of organic contamination (BOD) in order to reduce the volume of air tanks or to intensify their operation without construction of additional sections, thus reducing the time of stay while maintaining the effect of BOD purification.

References

1. A. Kapagiannidis, I. Zafiriadis, A. Aivasidis, *Comparison between UCT type and DPAO biomass phosphorus removal efficiency under aerobic and anoxic conditions*, Water Science & Technology, 60 (2009)
2. S. Puig, Ll. Corominas, M. Balaguer, J. Colprim, *Biological nutrient removal by applying SBR technology in small wastewater treatment plants: carbon source and C/N/P ratio effects,* Water Science & Technology, 55 (2007)
3. J. Tang, X. Wang, Y. Hu, Pu Y., J. Huang, H. Ngo, Y. Li, *Nutrients removal performance and sludge properties using anaerobic fermentation slurry from food waste as an external carbon source for wastewater treatment*, Bioresource Technology, 271 (2019)
4. R. Ferrentino, M. Langone, G. Andreottola, *Progress toward full scale application of the anaerobic side-stream reactor (ASSR) process.* Bioresource Technology, 272 (2019)
5. R. Ferrentino, M. Langone, R. Villa, G. Andreottola, *Strict anaerobic side-stream reactor: Effect of the sludge interchange ratio on sludge reduction in a biological nutrient removal process.* Environmental Science and Pollution Research, 25 (2018)
6. Z. Zhu, W. Chen, T. Tao, Y. Li, *A novel AAO-SBSPR process based on phosphorus mass balance for nutrient removal and phosphorus recovery from municipal wastewater,* Water Research, 144 (2018)
7. N. Zaletova, S. Zaletov, I. Bulychev, *Potential of biological phosphorus removal*, Paper presented at the MATEC Web of Conferences, 178 (2018)
8. A. Azhdarpoor, L. Abbasi, M. Samaei, *Investigation of a new double-stage aerobic-anoxic continuous-flow cyclic baffled bioreactor efficiency for wastewater nutrient removal,* Journal of Environmental Management, 211 (2018)
9. S. Wu, J. Maskaly, *Study on the effect of total dissolved solids (TDS) on the performance of an SBR for COD and nutrients removal*, Journal of Environmental Science and Health - Part A Toxic/Hazardous Substances and Environmental Engineering, 53 (2018)
10. Y. Zhou, W. Lisowski, Y. Zhou, N. Jern, K. Huang, E. Fong, *Genetic improvement of magnetospirillum gryphiswaldense for enhanced biological removal of phosphate*, Biotechnology Letters, 39 (2017)
11. Y. Nan, G. Zhou, L. Yuan, *Function of polyphosphate kinase gene in biological phosphate removal during the wastewater treatment process.* Huanjing Kexue/Environmental Science, 38 (2017)

Elimination of the biogenic pollution effects on aquatic environments

Vladimir Sherbakov [1], *Konstantin* Chizhik [2*], *Tatyana* Shchukina [1], and *Anastasiya* Milyaeva [1]

[1]Voronezh State Technical University, Voronezh, Russia
[2]Moscow State University of Civil Engineering, Moscow, Russia

Abstract. Intensive water consumption from surface sources leads to debit imbalance and high level of pollution, especially for wastewater discharges, including urban wastewater treatment facilities. Dissolved harmful and organic substances enter a large number of water bodies, which leads to harsh conditions for the survival of the rivers flora and fauna. A demonstrative indicator of environmental troubles is the intensive development of blue-green algae, which are a nutrient medium for cyanobacteria. Their productivity can be limited by hydromechanical capture. It is proposed to use a small-sized mobile vessel with a filter unit using replaceable cartridges and a pump. As a filter material, it is recommended to use crushed reeds. This technical solution allows the environmentally safe disposal of biomass, saturated with phytoplankton and heavy metal ions that are effectively absorbed. An equation of the distribution of concentrations of blue-green algae over the distance measured from the water surface is provided. The calculated data obtained from the proposed dependence allow us to determine the required level of immersion of the suction probes. The optimal time period in days for the operation of filtering units during summer season was revealed.

1 Introduction

The urgent problem for metropolises and large cities, especially with powerful industrial centers, is the low quality of water resources. As a rule, subject to processing and transportation technologies, water supply from underground sources is satisfactory. A completely different situation occurs when using the resources of water bodies running through the territories of regions with a high population density.

The historical formation of settlements, and subsequently large cities on the main waterways, was predetermined by two main reasons: water demand for household needs (it was in abundance with such arrangement of buildings and cultivated areas) and opened up great opportunities for convenient transportation. Then, due to the active development of production activities, consumption increased significantly, which could be ensured by existing nearby rivers and lakes. At high rates of industrialization and the emergence of new high-tech industries, the construction of water bodies with huge reserves of fresh water was required. However, the effective solution of deficit issues through the creation of

*Corresponding author : irkyt-44@yandex.ru

reservoirs, the prevailing ecosystems in the occupied territories were irretrievably violated, and the negative pressure on the environment increased.

The current trend of intensive consumption of water resources, along with the imbalance introduced, causes pollution of surface sources by wastewater discharge after treatment facilities. Given the low efficiency of technological treatment cycles, wastewater entering in large volumes into rivers or reservoirs significantly worsens the quality of the habitat required to preserve the species diversity of river flora and fauna. Given that the decay time of harmful substances discharged into water bodies has a rather long period, their inevitable accumulation occurs, which poses a serious threat to the subsequent emergence of an environmental disaster. Therefore, wastewater quality indicators must be achieved at biological treatment plants so that the impurities contained in them would not cause damage to water bodies and thereby ensure the safe use of surface sources for industrial, domestic and fishery needs.

Despite rather strict legislative requirements for wastewater treatment, as well as the reduction in consumption by industrial enterprises and the population, the ecological state of water bodies in regions with large industrial centers is often of concern. Their intense pollution occurs for a number of main reasons: violations in the operating modes of treatment facilities, a mismatch with the high technical level of treatment technology and deterioration of equipment. In addition, there are other sources of pollution, which include surface runoff from residential and agricultural areas, as well as soil erosion in coastal areas. All that creates an environmental threat both for the water body itself (degradation of aquatic flora and fauna) and for the population as a whole, having a negative impact on the health of the nation.

2 Materials and methods

As a result of the pollution, biogenic elements such as nitrogen, potassium, calcium, sodium, sulfur, phosphorus, carbon, and chlorine enter water bodies. Their presence during natural water metabolism ensures the normal functioning of organisms, but significant concentrations, especially in soluble organic substances coming from sewage, are the reason for the intensive development of algae, contributing to the eutrophication of water bodies. The latter can be caused not only by anthropogenic influences, but also by the natural aging of the habitat in a water body. In prosperous ecosystems, the development of algae contributes to the formation of a richer food base for fish and other aquatic organisms, but at high growth rates caused by anthropogenic influences an additional deterioration in water quality occurs due to the so-called "flowering", turbidity and a decrease in oxygen content. At low visibility indices, the photosynthesis of algae growing on the bottom ceases and their dying off occurs, exacerbating the ecological situation. The high degree of eutrophication of water bodies achieved in this case causes the river fauna to freeze. Therefore, the intensity of growth of concentrations of blue-green algae during the "flowering" period of water is a fairly clear indicator of the ecological state of the water body. The main reasons for increasing the reproduction of cyanobacteria include the decrease in the flow rate, the formation of stagnant, well-heated shallow water and the intake of a significant amount of nutrients in the composition of dissolved organic substances. Since blue-green algae are extremely viable species of phytoplankton, they immediately respond to any environmental threat, adapting to new conditions [1-4]. They change the pace of their development depending on the hydrodynamic parameters of water bodies, weather conditions and the availability of essential nutrients. Since the vegetative period of blue-green algae is short, intensive reproduction leads to the death of a large amount of phytoplankton, which contributes to the accumulation of detritus that settles in the deep zone. The resulting biomass is a nutrient medium for destructors, which consume oxygen during their lifetime, reducing its concentration in water. In the absence of oxygen, bacteria-destructors survive due to anaerobic digestion to more auspicious conditions. As a result, the content of dissolved oxygen varies in depth, significantly decreasing in the bottom zones.

After the death of phytoplankton, the biogenic elements digested by it during the subsequent decomposition of detritus are again released at the bottom and then transported by ascending convective flows to the surface, where they are again consumed by viable cyanobacteria. Thus, the cycle of biogenic pollution is not only repeated, but also exacerbated. During intensive "blooming", water may contain up to 200 various chemical elements and compounds that pose an environmental threat [5-7]. A high concentration of blue-green algae contributes to the accumulation of acetone, butyl alcohol, acetic and butyric acids, phenols, mercaptans, amines, and indole-containing components. These processes are accompanied by a decrease in the pH of water to 5–4 and lower, as well as an increase in the temperature of the saturation zone of phytoplankton relative to the average statistical by 6–8 °C [5–7]. The latter, in turn, contributes to the intense evaporation of algal metabolites, causing air pollution and specific unpleasant odor. It should also be noted that a significant amount of dissolved organic compounds serves as a breeding ground for the preservation and development of pathogenic microflora, which exacerbates the adverse sanitary and epidemiological situation [8. 9].

If difficulties in achieving the required quality of wastewater arise, it is necessary to timely eliminate the consequences of biogenic pollution in order to prevent critical conditions in the ecological balance of water bodies. The development of eutrophication can be overcome by a fairly reliable and effective way of dealing with excess phytoplankton, based on the hydromechanical principle of filtering water resources [10, 11]. The devices of simple construction used in this case provide a directed effect on the epicenter of cyanobacterial reproduction, which helps to localize areas of increased biogenic pollution, preventing them from spreading. It is advisable to utilize collected biomass on disposable filter nozzles filled with shredded coarse aquatic vegetation, since the proteins and carbohydrates contained in phytoplankton are high quality raw materials for the production of biogas after higher calorie fats. Given the percentage ratio in terms of alternative fuel reproduction [12], under favorable anaerobic digestion conditions, up to 1300 m^3 of biogas can be obtained from 1 ton of fat, and up to 780 m^3 from proteins and carbohydrates of the same mass. This is confirmed by the results of experimental studies [13] of fermentation of raw phytoplankton in volumes of 1 dm^3 for a week at a temperature of 20-30 °C. The anaerobic process was accompanied by the formation of 700 ml or more of gas with a composition of 65% methane (CH_4), up to 30% carbon dioxide (CO_2) and an average of 1% of the following components H_2, H_2S, O_2 and N_2 [13]. According to other data, in the conversion of 1 kg of dry mass of algae [14] at a temperature of 32 ° C, biogas is produced with a volume of 0.8-1 m^3, the composition of which has practically the same indicators for the content of various impurities in methane. The amount of gas released depending on the type of utilized algae is given in the studies [15]. However, filter fillers saturated with phytoplankton during the cleaning of biogenic pollution zones do not allow evaluating the ratios of various ichthyoflora species detained in them. Therefore, in this situation, to assess the possible reproduction of biogas, one should use the averaged values obtained taking into account the actual moisture content of the raw materials sent to anaerobic digestion.

To address the issues of effective cleaning of water bodies from the effects of biogenic pollution, one should consider the distribution of phytoplankton forming in the warm season of the year in depth in zones of environmental disaster. As already noted, near the surface of the water, active photosynthesis occurs and, accordingly, favorable conditions contribute to the high reproduction of cyanobacteria, which primarily leads to their distribution in the upper layers. Therefore, the suction devices of the filtering installations should be located closer to the surface of the water and with a limitation of the depth of their immersion. It is possible to determine the appropriate range in height of the location of the probes for collecting biomass when analyzing the process of structural formation of cyanobacterial colonies in the aquatic environment. With active photosynthesis in the upper zone caused by the unhindered penetration of solar radiation, the differential equation for the distribution of the concentration of blue-green algae taking into account the hydrodynamics of the channel will look as follows:

$$\frac{dC}{dz} = \frac{m_s}{v} - \frac{C}{h}, \tag{1}$$

where C is the concentration of blue-green algae, g/m^3; z is the vertical axis directed from the surface of the water to the bottom of the reservoir, m; v - water flow rate, m/s; h is the height of the zone of intense photosynthesis, depending on the turbidity of the water, and the average taken equal to 1 m; m_s is the specific mass rate of formation of blue-green algae in 1 m^3 of the aquatic environment, g/(s·m^3). The change in the concentration of blue-green algae along the depth measured from the surface of the water primarily depends on the dynamic assimilation of the formed cyanobacteria and their accumulation in the intensive growth zone at the age corresponding to the end of the vegetative period, which is taken into account in expression (1). The solution to equation (1), including the listed components, is the dependence :

$$C = \frac{m_s h}{v} + ce^{-\frac{z}{h}}. \tag{2}$$

To determine the constant integration c as the boundary condition, we use the assumption that the concentration of blue-green algae on the surface of the water, that is, at $z = 0$, will be equal to the maximum value determined by the level of biogenic pollution of the reservoir. Then expression (2) becomes :

$$C = \frac{m_s h}{v} + \left(C_m^B - \frac{m_s h}{v} \right) e^{-\frac{z}{h}}. \tag{3}$$

The obtained equation (3) allows us to analyze the structure of the formation of colonies of cyanobacteria by the height of shallow waters at various degrees of pollution of water bodies.

3 Results

As shown by previous studies, the daily mass formation rate of blue-green algae, depending on the amount of dissolved organic substances, is in the range from 0.1 to 0.5 g/ (day·m^3), therefore, the specific mass velocity m_s, which is included in the expression (3) may have values in the range from $1.16 \cdot 10^{-6}$ to $5.79 \cdot 10^{-6}$ g/(s·m^3). The flow velocity in the expansion zones of blue-green algae, as a rule, does not exceed 0.5 m/s. Therefore, given the values of these parameters, the ratio $m_s h/v$ has a weak effect on the distribution of concentrations over the depth of the aquatic environment. Fundamental in this process is the photosynthesis of phytoplankton, which floats to the surface to absorb solar radiation and nutrients.

Consider the distribution of blue-green algae along the depth of a reservoir with varying degrees of biogenic pollution. The concentration changes shown in Fig. 1 and obtained by dependence (3) for a rather problematic environmental situation with reaching the maximum value of 0.15 g/m^3, as well as for more dangerous biogenic pollution, corresponding to 0.2 g/m^3 and to 0.25 g/m^3, show that it is the first meter zone that is subject to active saturation with phytoplankton. Therefore, to clean the reservoir, the suction devices should be located not lower than 1 m, and if possible, the depth of immersion should be rigidly fixed, they should be in the range from 0.05 to 0.5 m according to the distribution of concentrations.

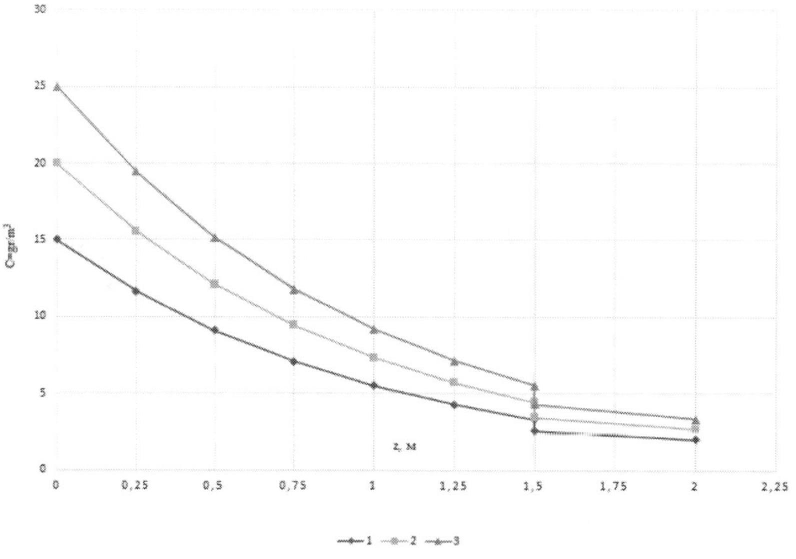

Fig. 1. Change in the concentration of blue-green algae in depth from the surface of the water: 1 - with a maximum content of blue-green algae C_m^B = 0.15 g/m^3; 2 - at C_m^B = 0.2 g/m^3; 3 - at C_m^B = 0.25 g/m^3

Given the distribution of concentrations (Fig. 1), it should be noted that after the first hydromechanical treatment of the ecological disaster zone, it is advisable to re-remove the surfaced blue-green algae after a day, thereby preventing the rapid restoration of cyanobacteria colonies. Subsequently, with a possible change in weather conditions, eutrophication can slow down significantly, which will favorably affect the habitat for ichthyofauna. For the effective organization of hydromechanical cleaning of water bodies, the operation of the devices should be limited to the period of activation of phytoplankton, namely from 10:00 to 16:00 hours. By setting immersion probes to the optimum depth and pumping water through a filter nozzle filled with shredded hard vegetation, not only the problem of removing phytoplankton is solved, but it also stimulates updating of the river flora, the aging and dying of which also contributes to eutrophication. Large hydrophytes, such as reeds or cattail during the vegetative period, are able to extract biogenic elements from water in large quantities, contributing to the natural cleaning of water bodies. For example, a dense mature reed thicket with an area of 1 ha can accumulate up to 6 tons of various mineral substances, including K - 859 kg, N - 167 kg, P - 122 kg, Na - 451 kg, S - 277 kg and Si - 3672 kg [1, 16]. Assimilating the listed elements, the reed performs cleaning functions in the reservoir, but only if its thickets will be updated every year by mowing. Cutting and removing this type of coarse algae, crushing and filling filter cartridges with it, and then through floating stations trapping phytoplankton and other impurities on them, the habitat for the river flora can be significantly improved. In addition, the sorbing properties of the reed allows the purification of water from heavy metal ions by their absorption by the porous structure of the nozzle.

The collected concentrated biomass can be sent for anaerobic digestion into the digesters of urban wastewater treatment plants or, in the absence thereof, with a minimum content of harmful substances, be added into the soil as fertilizer, since it is rich in nutrients. However, further disposal of the collected phytoplankton is preferable in a biogas plant. The reed, recommended for use for disposable nozzles, also has high methanogenesis during anaerobic digestion, which makes it possible to efficiently utilize the total biomass [17-20].

4 Conclusion

Biogenic pollution of water bodies, especially of artificial origins, at high temperature conditions in the summer seasons leads to a significant increase in the reproductive ability of cyanobacteria. In order to avoid intensive "blooming" of water in shallow water heated by solar radiation, it is advisable to periodically capture blue-green algae through floating filtering stations. The suction devices of the stations, in accordance to the obtained equation for the distribution of phytoplankton concentrations, should be located at a depth of 0.05 to 0.5 m from the surface of the water. Shredded reeds should be used as the filler for disposable filter cartridges. Such plants not only make it possible to collect excess phytoplankton, but also sorb heavy metal ions by absorbing them by the porous structure of hard aquatic vegetation filling the nozzles. Disposal of the collected biomass and cartridge filler should be carried out by anaerobic digestion, i.e. sent to the digesters of urban wastewater treatment plants or biogas plant reactors.

The proposed method for improving the environmental safety of water bodies by means of floating filtering stations allows phytoplankton capture not only to remove biogenic elements absorbed by blue-green algae, but also to additionally clean shallow water from hard ichthyoflora used to fill cartridges. All this together will cause positive dynamics in the state of water bodies under the existing threat of intense eutrophication.

References

1. L.A. Kulsky, L.A. Sirenko, Z.N. Shkavro, *Phytoplankton and water* (Kiev: Science. Dumka, 1986) 135
2. V. B. Mikhno, A. I. Dobrov, *Landscape and environmental features of reservoirs and ponds of Voronezh region* (Voronezh: Voronezh St. Pedagogical Un., 2000) 185
3. G.S. Seydaliev, I.I. Kosinova, T.V. Sokolova, K.Y.Silkin, *Ecological management of the territories of the Voronezh reservoir* (Voronezh: Origins, 2017) 186
4. A.Y. Scriabin, G.V. Popovyan, I.A. Tron, Water supp. & sanitary equip., **4** 5-8 (2017)
5. E.S. Gogina, *Removal of biogenic elements from wastewater* (M.: MGSU, 2010) 120
6. R. Song, S. Yang, H. Xu, Z. Wang, Y. Chen, Y. Wang, Int. J. Environ. Res. Public Health, **15(1)** 101 (2018)
7. L. Shang, M. Feng, X. Xu, F. Liu, F. Ke, W. Li, Toxins, **10** 26 (2018)
8. V. Shcherbakov, E. Gogina, T. Schukina, N. Kuznetsova, N. Makisha, Int. J. of Appl. Eng. Res. **10(24)** 44353-44356 (2015)
9. V.I. Shcherbakov, B.M. Kumedov, Forest J. **4** 133-137 (2015)
10. V.M. Shmandiy, V.V. Nikiforov, V.P. Alferov, E.V. Kharlamova, V.A. Pronin, Hygiene and Sanitation **6** 35-37 (2010)

11. W. Czyżewska, M. Piontek, Toxins **11** 285 (2019)

12. V. Shcherbakov, V. Pomogaeva, K. Chizhik, E. Koroleva, Advances in Intelligent Systems and Computing, **983** 361-372 (2019)

13. V.P. Alferov, V.A. Promin, V.M. Shmandy, V.V. Nikiforov, S.V. Degtyar, E.V. Kharlamova, Env. safety **3 1(9)** 9-13 (2010)

14. F. T. Shumakov, Scientific notes of Taurida Nat. Un., Ser. Geography **23(62) #2** 286-295 (2010)

15. L. Moeller, A. Bauer, H. Wedwitschka, W. Stinner, A. Zehnsdorf, Energies **11(11)** 3016 (2018)

16. R. Song, S. Yang, H. Xu, Z. Wang, Y. Chen, Y. Wang, Int. J. Environ. Res. Public Health **15** 101 (2018)

17. V. Nykyforov, M. Malovanyy, T. Kozlovs'ka, O. Novokhatko, S. Digtiar, E-Eur. J. of Enterprise Technol. **5/10(83)** 11-18 (2016)

18. C. Rodriguez, A. Alaswad, J. Mooney, T. Prescott, et al, Fuel Process. Technol. **138** 765-779 (2015)

19. M. P. Caporgno, R. Trobajo, N. Caiola, C. Ibanez, et at, Renew. Energy. **75** 374-380 (2015)

20. V.A. Milyutkin, S.A. Tolpekin, I.V. Borodulin, E.A., Collection of articles on the materials of the II All-Russian (national) scientific-practical conference 104-109 (2018)

E3S Web of Conferences **180**, 04011 (2020) https://doi.org/10.1051/e3sconf/202018004011
TE-RE-RD 2020

Development of thermal bridge numerical model, based on conjugate heat transfer and indoor and outdoor environment parameters

Martin Ivanov[1*], *Sergey* Mijorski[2]

[1]Technical University – Sofia, FPEPM, Department:" Hydroaerodynamics and Hydraulic Machines", Sofia 1000, Bulgaria

[2]SoftSim Consult Ltd., Consultant at Technical University of Sofia, FPEPM, Department:" Hydroaerodynamics and Hydraulic Machines", Sofia 1000, Bulgaria

Abstract. The presented study reveals the development of a 3D numerical model for thermal bridge assessment, based on conjugate heat transfer and CFD methods. With the developed model, thermal simulations are performed, in order to analyse the interaction between different ambient conditions and material properties. The results show that the wall boundary layer profiles are depended on the attached air flow velocity magnitude and implemented wall roughness. The parametric analysis, of the varying ambient air temperatures, confirm the linear dependence to the internal wall surface temperatures. The demonstrated correlations, in regard of the attached air flow velocity magnitude and wall roughness heights, are non-linear. The most characteristic result, achieved in the simulation study, is the impact of the wall roughness, over the internal wall temperature. The increase of the roughness leads to significant increase of the internal wall temperature. Explanation may be found in the boundary layer flow velocity magnitude near the external wall, which decreases the heat energy transfer between the solid and cold fluid medias.

1 Introduction

In today's building sector, thermal bridges calculation is addressed in the construction guidelines, in most of the European countries [1]. Throughout the building design, it is recommended to avoid thermal bridges, and this is the usual practise, prescribed in different norms and standards. [2, 3] Such technical documents give the required ambient and material parameters for thermal bridge analyses, based on long term building experience and practical studies in this area. It is very important to perform these analyses, because the linear thermal bridges may rise the heating energy requirement of the buildings with above 30%. [4] It is hard to estimate the total impact of the thermal bridges, over the energy demand of the buildings, but it may vary from few percent in old non-restored buildings, up to almost 40% in new, heavily isolated constructions, with non-covered thermal bridge areas. [5, 6] Modern facades, like double-skin, cladding systems or smart and ventilated

* Corresponding author: m_ivanov@tu-sofia.bg

façades, rise the demand for more energy efficient building envelope technologies, and hence require significant reduction in the thermal bridge appearance prerequisites. [7]

However, thermal bridges still exist worldwide, particularly in older buildings, where no energy efficiency procedures or retrofits have been applied. In such premises, the thermal bridges may have significant impact over the indoor air quality as well. In winter conditions for example, excess moisture accumulation may appear over the cold spot regions. These moisture related problems and the consequent mould growth are strongly related to important health outcomes in occupants, like advent of allergies and even asthma. Thus, analyses of the generic conditions for thermal bridge appearance is important issue for the presented area. [8, 9]

Numerical modelling of the described above phenomena is generally not an easy task, though it does help the thermal bridge avoidance in the building design process. The calculation of the steady-state temperature in the affected zones could be performed by standard finite element or temperature method, and by statically admissible flow method. [10] But, the dynamic modelling of the thermal bridges reduces the underestimation of their effect on the building performance, which is normally made by the analytical U-value method and the equivalent wall method. [11] However, the modelling methods, based on the Computational fluid dynamics (CFD), require significant modelling skills and precisely implemented initial and boundary conditions. Sometimes, the flow past complex geometries is hard to discretize, and the thermal bridges normally appear in complex junctions, between different building elements. [12]

Consequently, the development of accurate, multiparametric 3D numerical models for thermal bridge analyses, will support the buildings design practice. The presented paper suggests such primary model, based on conjugated heat transfer and CFD methods. Thermal transmittance through concrete wall is simulated, and the effect of the ambient flow conditions is analysed.

The objective of the study is to develop a 3D numerical model for thermal bridge assessment, through the means of the CFD methods and conjugate heat transfer, and to analyse different indoor, outdoor and material parameters, affecting this phenomenon.

2 Numerical model

For purpose of the study, a simplified geometry of exterior building wall is constructed, and presented on Fig. 1. The 3D model encompasses a wall section with dimensions 4 m by 4 m, including two fluid (air) and one solid regions. The test section used for the analysis is with dimensions 1 m by 1 m and wall thickness of 0.2 m. The material used for the solid region is a concrete with density of 2400 kg/m^3, specific heat of 750 J/(kg K), and thermal conductivity of 1.8 W/(m K).

Numerical discretisation is completed with snappyHexMesh meshing utility within OpenFoam®, creating a hexahedral and polyhedral mesh of 485608 control volume cells. Each of the wall surfaces within the computational domain is with 5 prism layers projecting toward inlets and outlets of the fluid regions and first layer height is 0.005 m (see Fig.1). The selected base cell size of the computational domain is based on a completed grid sensitivity study. The results of this supplementary analyses are shown on Fig. 2 and Fig. 3, in terms of surface temperatures and heat flux distribution for different base cell sizes. Three different grid sizes are tested, and four different parameters are analysed: minimum, mean, and maximum surface temperatures and resultant heat flux between the solid and internal fluid regions. Thus, an optimum mesh size of 0.05 m was considered, as there is no statistically meaningful difference for the resulting monitored parameters.

 https://doi.org/10.1051/e3sconf/202018004011

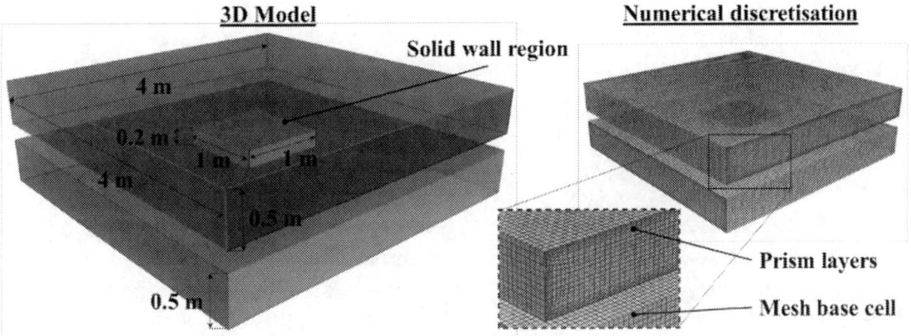

Fig. 1. 3D model and numerical discretisation

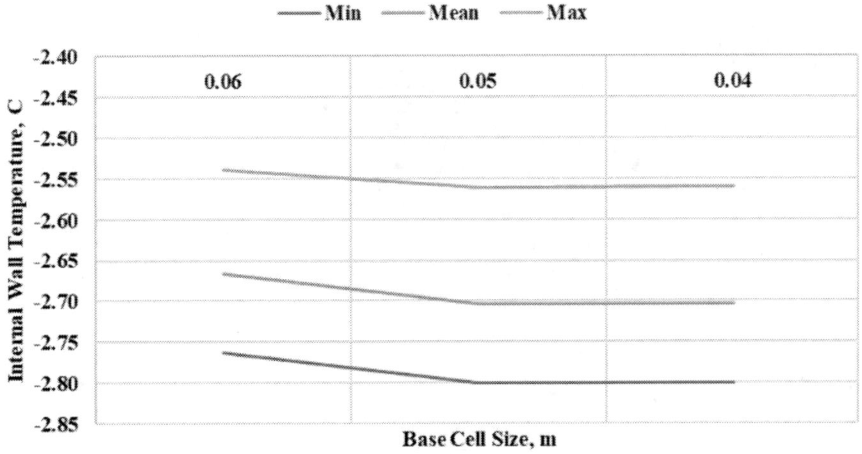

Fig. 2. Grid sensitivity - surface temperatures comparison

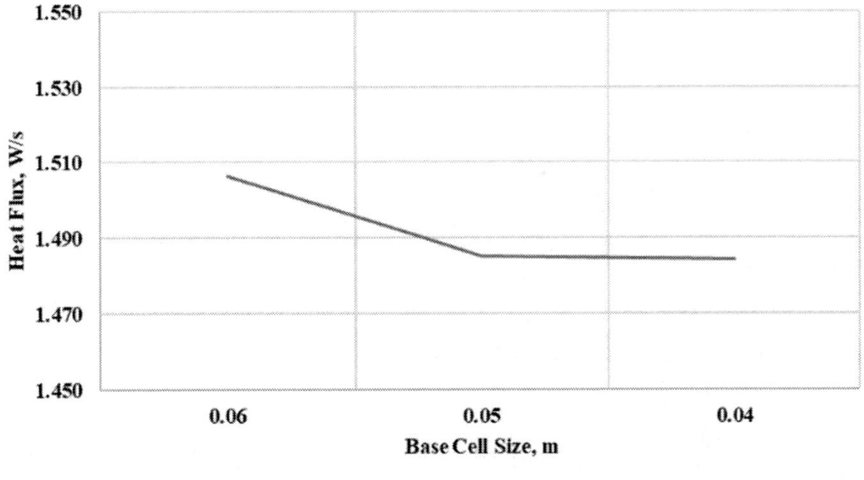

Fig. 3. Grid sensitivity - surface heat flux comparison

3 Simulations set-up and scenarios

In total, 15 steady state conjugate heat transfer simulations are performed with use of Realisable k-epsilon turbulence model and standard wall function within the CFD software Ansys Fluent 16.0. For the purposes of the analysis a constant indoor environment conditions are modelled with air temperature of 24 °C. This is accomplished by providing constant air flow along the interior of the test section with uniform velocity of 0.1 m/s, parallel to the wall. This way the impact of the different parameters (ambient conditions and wall surface and material properties) over the interior surface temperature could be compared and analysed. The outdoor environment parameters are modelled with varying air flow intensity parallel to the exterior of the test section. The configuration is tested for different external air temperatures from -20 °C to 10 °C, wall roughness heights in the range of 0.3 mm to 5.0 mm and different flow speeds – from 0.5 m/s to 3 m/s. The specified CFD model boundary conditions are outlined on Fig. 4.

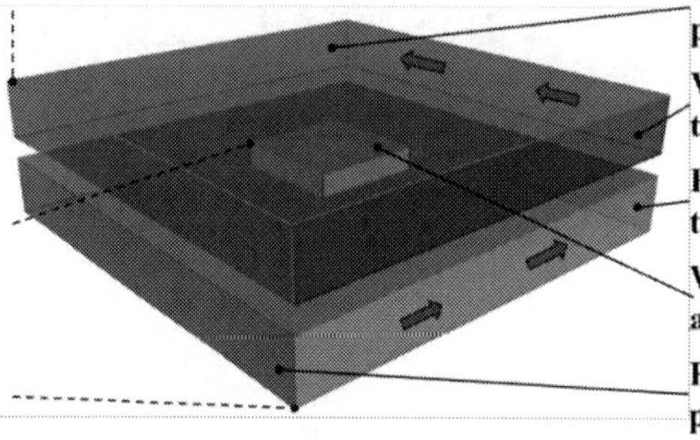

Fig. 4. CFD model boundary conditions

The standard wall function, implemented within the Fluent software, is proposed by Launder and Spalding. [13] The calculation of the momentum at the wall boundary is completed by law-of-the-wall for mean velocity as follows:

$$U^* = \frac{1}{\kappa}\ln(Ey^*) \tag{1}$$

where:

$$U^* = \frac{U_P C_\mu^{1/4} k_P^{1/2}}{\tau_w/\rho} \quad \text{and} \quad y^* = \frac{\rho C_\mu^{1/4} k_P^{1/2} y_P}{\mu} \tag{2}$$

U^* - dimensionless velocity

y^* - dimensionless distance from the wall

κ - von Kármán constant, 0.4187

$C\mu$ - empirical constant, 0.09

E - empirical constant, 9.793

τ_w - is the local wall shear stress

U_p - mean velocity of the air flow at given point

k_p - turbulence kinetic energy at given point

y_p - distance from the point to the wall surface

μ - dynamic viscosity of the air

ρ - density of air

When y^* value is greater than 11.225, the logarithmic law for mean velocity is employed, and when is less (within finer mesh regions), it is applied a laminar stress-strain relationship between mean velocity and y^*: $U^*=y^*$. While, the law-of-the-wall for energy modelling comprises two different laws:

- "linear law for the thermal conduction sublayer where conduction is important"
- "logarithmic law for the turbulent region where effects of turbulence dominate conduction".

This way, the law-of-the-wall takes the following form [14]:

$$T^* = \frac{(T_w - T_P)\rho c_p C_\mu^{1/4} k_P^{1/2}}{\dot{q}} = \begin{cases} \Pr y^* + \frac{1}{2}\rho Pr \frac{C_\mu^{1/4} k_P^{1/2}}{\dot{q}} U_P^2 & y^* < y_T^* \\ \Pr_t \left[\frac{1}{k} ln(Ey^*) + P\right] + & \\ \frac{1}{2}\rho \frac{C_\mu^{1/4} k_P^{1/2}}{\dot{q}} \{Pr_t U_P^2 + (Pr - \Pr_t) U_c^2\} & y^* > y_T^* \end{cases} \tag{3}$$

where:

$$P = 9.24 \left[\left(\frac{Pr}{Pr_t}\right)^{3/4} - 1\right] \left[1 + 0.28 e^{-0.007 Pr/Pr_t}\right] \tag{4}$$

y_T^* - non-dimensional thermal sublayer thickness
c_p - specific heat of the air
$\dot{q}$ - heat flux at the surface
T_p - temperature at given point adjacent to surface
T_w - surface temperature
Pr - molecular Prandtl number
Pr_t - turbulent Prandtl number, 0.85 at the surface
A - Van Driest constant, 26
U_c - mean velocity magnitude for $y^*=y_T^*$.

4 Results and discussion

Numerical results in form of temperature and velocity fields through the centre of the computational domain, and parallel to the outdoor air flow for one exterior condition (velocity magnitude of 0.5 m/s at -20 °C), are shown on Fig. 5 and Fig. 6. The thermal conductivity in the presented case is 1.8 W/(m K), while surface roughness is 0.3 mm.

As illustrated on the velocity and temperature field plots, the wall boundary layer flow would be sensible to the test section positioning and sizing. This way, the resulted profiles along the wall surface could not be fully developed prior to the test section. However, this depends on the wall roughness and modelled attached flow velocity magnitude. Additionally, for capturing more precisely the conjugate heat transfer process, the adjacent construction elements and corresponding thermal properties of the tested buildings should be included.

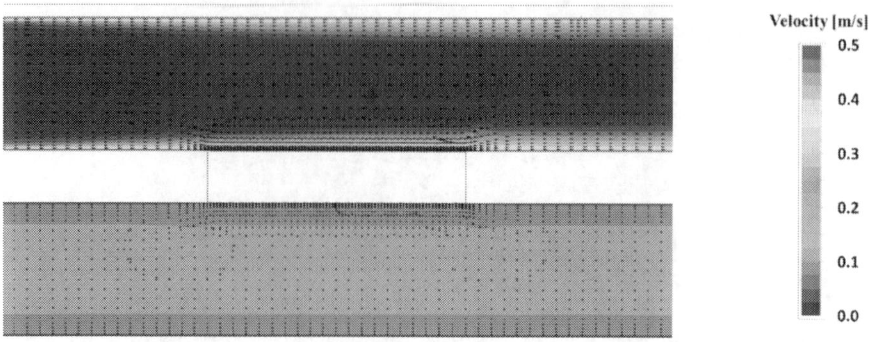

Fig. 5. Velocity field plot

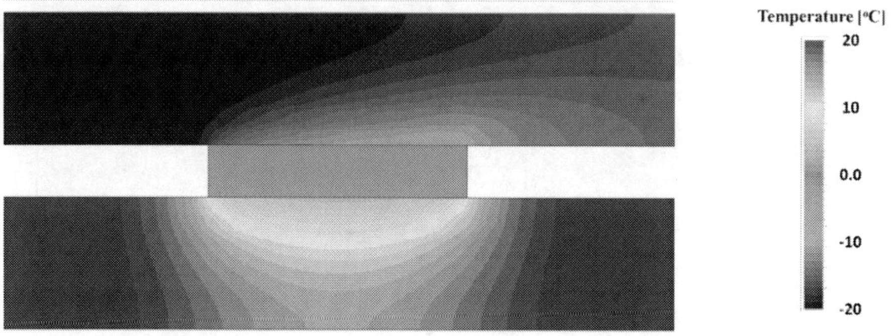

Fig. 6. Temperature field plot

Three different parametric analyses are shown by graphical representations of the mean interior wall temperature of the test section, including:

- external air temperature, presented on Fig. 7;
- external attached air flow velocity magnitude, presented on Fig. 8;
- and wall roughness heights, presented on Fig. 9.

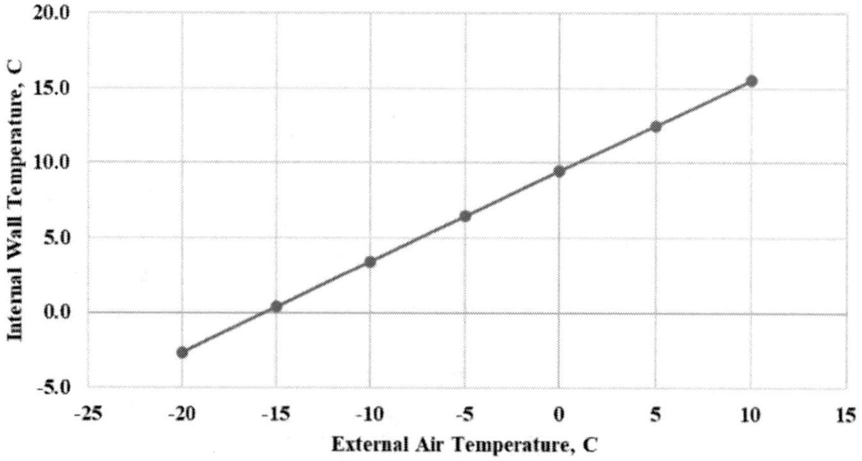

Fig. 7. External air temperature sensitivity analysis

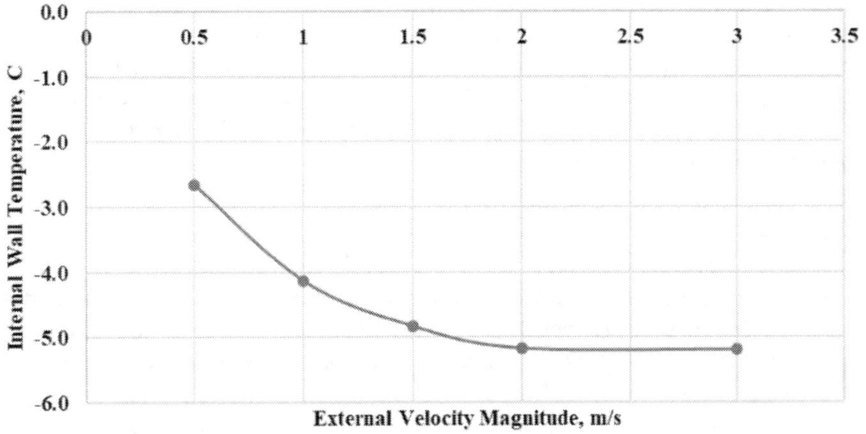

Fig. 8. Attached air flow sensitivity analysis

The parametric analysis of the varying ambient air temperatures shown on Fig. 7 illustrates a linear dependence to the internal wall surface temperatures. While, on Fig. 8 and 9 are demonstrated the non-linear dependencies in regard of the attached air flow velocity magnitude and wall roughness height.

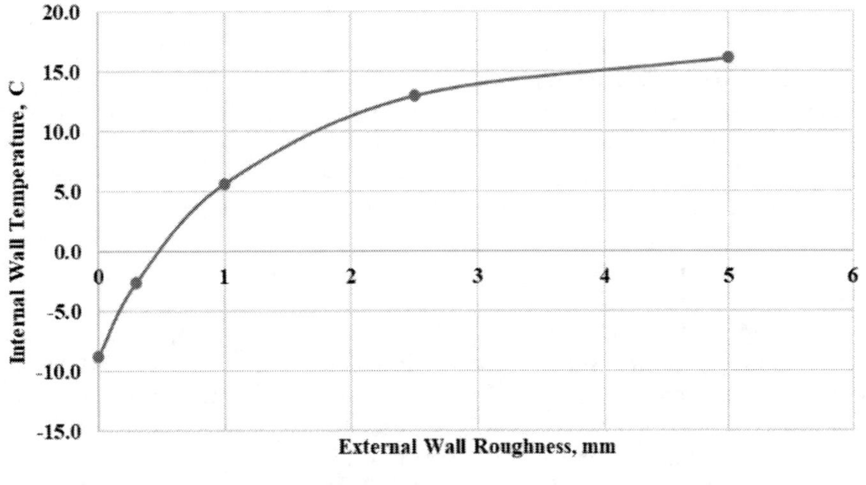

Fig. 9. External wall roughness sensitivity analysis

It is seen that, the increase of the external wall velocity magnitude from 0.5 m/s up to 3 m/s, leads to decrease of the internal wall temperature from -2.7 ^{o}C to -5.2 ^{o}C. This decrease is attained without changing the flow direction, which might have additional impact over these values. More sensible drop of the interior wall temperature is observed between 0.5 m/s and 2.0 m/s velocity magnitude. This demonstrate the need of further investigation of the means for impacting attached wind flows along building facades, thus ensuring lower velocity magnitudes within exterior walls boundary layer and lowering building heat losses.

Furthermore, the effect of the wall roughness change shown on Fig. 9 confirms the observed impact of attached flow velocity magnitude. With the increase of roughness height, more sensible is the drop of the heat loss and higher is the interior surface temperature. This way with smooth wall, the temperature drops to -8.8 °C and with coarsest material (5 mm roughness), reaches almost 16.1 °C.

This effect is explained with the boundary layer flow velocity magnitude near the external wall, which decreases the heat energy transfer between the solid and cold fluid medias. However, this phenomenon would require additional analyses of the turbulence models and applied wall functions. Also, sufficient measurement data is not yet available, in order to validate the presented numerical model.

5 Conclusions

In the presented study 3D numerical model for thermal bridge assessment is developed. The model is based on CFD methods and conjugate heat transfer. Model simulations are performed, in order to analyse the interaction between different indoor, outdoor conditions and material properties.

The results show that the wall boundary layer profiles are depended on the attached velocity magnitude and implemented wall roughness. This limits the application of the presented model to the particular length of the wall before test section and points out the need for modelling of the adjacent construction elements with the corresponding thermal properties.

The parametric analysis, of the varying ambient air temperatures, confirm the linear dependence to the internal wall surface temperatures. The demonstrated dependencies, in regard of the attached air flow velocity magnitude and wall roughness heights, are non-linear.

The results also show that the increase of the external wall velocity from 0.5 m/s up to 3 m/s, leads to decrease of the internal wall temperature from -2.7 °C to -5.2 °C, attained without changing the flow direction. Thus, a need exists for further investigation of the means for impacting attached wind flows along building facades, ensuring lower velocities within exterior walls boundary layer.

The most distinctive observed result is the impact of the wall roughness, over the internal wall temperature. The increase of the roughness leads to significant increase of the internal wall temperature. It is due to the boundary layer flow velocity magnitude near the external wall, which decreases the heat energy transfer between the solid and cold fluid medias.

Future work of the authors would include performing an experimental measurement study for validation of the proposed thermal bridge assessment CFD model. By assessing different environmental conditions and wall material properties to be proven the validity of the proposed 3D numerical method and to outline the advantages when compared with the existing simplified code thermal bridge calculations.

The presented study is supported by the National Science Fund at the Ministry of Education and Science of Bulgaria, under the activities of: "Competition for financial support for projects of junior basic researchers and postdocs – 2018", with Contract № КП-06-М27/4, entitled: "Numerical assessment of effective measures for prevention of moisture accumulation on external walls with thermal bridge, in residential rooms without organized ventilation".

References

[1] K. Kussk, J. Kurnitski and T. Kalamees, *Calculation and compliance procedures of thermal bridges in energy calculations in various European countries*, Energy Proc. J. **132**: 27-32, (2017)

[2] A. Papadopoulos, *Forty years of regulations on the thermal performance of the building envelope in Europe: achievements, perspectives and challenges*, Energy Build. **127**, (2016)

[3] ISO 10211, *Thermal Bridges in Building Construction - Heat Flows and Surface Temperatures. Detailed calculations*, Int. Org. for Standardization, (2017)

[4] H. Erhorn-Kluttig, H. Erhorn, *Impact of thermal bridges on the energy performance of buildings*, Paper P148 of the EPBD Buildings Platform, www.buildup.eu/sites/default/files/content/P_148_EN_ASIEPI_WP4_IP2.pdf., (2009)

[5] T. Theodosiou, A. Papadopoulos, *The impact of thermal bridges on the energy demand of buildings with double brick wall constructions*, J. Energy and Build. **40**: 2083–2089, Elsevier, doi:10.1016/j.enbuild.2008.06.006, (2008)

[6] K. Martin, A. Erkoreka, I. Flores, M. Odriozola and J.M. Sala, *Problems in the calculation of thermal bridges in dynamic conditions*, J. Energy and Build. **43**: 529–535, Elsevier, doi:10.1016/j.enbuild.2010.10.018, (2011)

[7] T. Theodosiou, K. Tsikaloudaki, S. Tsoka and P. Chastas, *Thermal bridging problems on advanced cladding systems and smart building facades*, J. of Cleaner Prod. **214**: 62-69, Elsevier, doi:10.1016/j.jclepro.2018.12.286, (2019)

[8] M. Ivanov, *Instantaneous field measurements of thermal bridge parameters in ground floor residential room*, Proc. "8th Int. Conf. on Thermal Equip., Ren. Energy and Rural Dev. - TE-RE-RD 2019", Web of Conf., doi: 10.1051/e3sconf/201911201016, (2019)

[9] M. Ivanov, *Dew point temperature analyses in ground floor residential room with existing thermal bridge*, Proc. "8th Int. Conf. on Thermal Equip., Ren. Energy and Rural Dev. - TE-RE-RD 2019", Web of Conf., doi: 10.1051/e3sconf/201911201017, (2019)

[10] S. Taoum, E. Lefrançois, *Dual analysis for heat exchange: Application to thermal bridges*, J. Comp. and Math. with App. **75**: 3471–3487, Elsevier, (2018)

[11] H. Ge, F. Baba, *Effect of dynamic modelling of thermal bridges on the energy performance of residential buildings with high thermal mass for cold climates*, J. Sustain. Cities and Soc. **34**: 250–263, Elsevier, (2017)

[12] A. Terziev, *Specifics in numerical modelling of flow past a square-cylinder*, Proc. of Sc. Conf. of Ruse University "Angel Kanchev, vol. **53**, book 1.2: 143-149, ISSN 1311-3321, (2014)

[13] Launder B. E., Spalding D. B., *The Numerical Computation of Turbulent Flow*, Comp. Methods in Applied Mech. and Eng., **3**:269-289, (1974)

[14] Jayatilleke C., *The Influence of Prandtl Number and Surface Roughness on the Resistance of the Laminar Sublayer to Momentum and Heat Transfer*, Prog. Heat Mass Transfer, **1**:193-321, (1969)

E3S Web of Conferences **180**, 04012 (2020) https://doi.org/10.1051/e3sconf/202018004012
TE-RE-RD 2020

Dynamic behaviour of pneumatic actuators in open-loop controlled by proportional valves

Valentin Nicolae Cococi[1], *Carmen-Anca* Safta[1*], and *Constantin* Călinoiu[1]

[1]University Politehnica of Bucharest, Power Engineering Faculty, 313 Splaiul Independentei, 060042, Bucharest, Romania

Abstract. The development of proportional and servo-mechanisms hydraulic control technique offered the premises of the development of pneumatic control, too. Despite the disadvantages of the nonlinearities appeared in the pneumatic systems (air compressibility being one of the causes), the new generation of proportional equipment made it possible to have a good quality pneumatic control system in our days. The goal of the paper is to demonstrate, by using numerical simulation technique, that the proportional valves used in pneumatic actuator drives occur at a good system dynamic at a reasonable cost as against to the directional control valve as it usually used in pneumatic actuator drives.

1 Introduction

The pneumatic system drives have been widely used in automatic production lines, machine tools, industrial robots, automotive industry, and a lot of other fields of activity where an automation equipment is needed. In our days 30% of automated processes and 43% of industrial robots are using pneumatic systems in closed-loop as positioning control systems, or in open-loop as pneumatic drives [1, 2].

Advantages of pneumatic drives to be light weight, small-sizes, simple to install and to maintain, the robust design and operation, also to be clean and cheap, to have better working temperatures (the air must not be cooled as the hydraulic oil), make them to be preferred in many automatic applications. In the last decades the development of microelectronic technology goes to improve the static characteristics and dynamic behaviour of the pneumatic systems, despite of their nonlinearities which make them difficult to control. The gas flow itself, the gas compressibility, the aerodynamic asymmetry of gas resistance, the small viscosity coefficient and the poor lubricity of the gas are some of factors which have a strong influence in the dynamic response of a pneumatic control system [1-4].

Pneumatic systems convert the compressed air energy into mechanical energy which can be a useful mechanical work or motion. The main device of this power loop is the pneumatic actuator (piston-cylinder or rotary) [3].

Pneumatic actuators are direct-driven systems because the output rod of the linear motor (or the output motor shaft of the rotary motor) is the actuator output link [5].

* Corresponding author: safta.carmenanca@gmail.com

Frequently, the pneumatic actuator is controlled by an electropneumatic control valve which can be a servo valve, a proportional valve or a solenoid valve [5]. The electropneumatic valves are designed to work with one stage, two and three stages. A single stage electropneumatic control valve is a directional operated valve. The electropneumatic control valves with two and three stages are known to have one stage to pilot and amplify the pressure on the other stage. These are known as servo valves and proportional valves. The servo valves provide a higher degree of loop control and they are used in closed loop positioning system for their positioning accuracy of ± 0.05 mm [5].

The goal of the paper is to show that a pneumatic actuator controlled by a proportional valve and working in open loop can have a good dynamic behaviour to different types of electrical signal waves. Numerical simulations are made using the simulation platform offered by Simcenter Amesim with LMS Imagine.Lab Pneumatics library [6]. The numerical results are discussed, and final conclusions are underlined.

2 Pneumatic actuator in open loop

Starting from the classical structure of a pneumatic actuator in open loop, Figure 1, controlled by a spool type direct solenoid spring return 3/2 (usually named as directional control valve, DCV), it was change the DCV with a proportional valve (Figure 2) to actuate a butterfly valve used in industrial applications.

The scheme of the pneumatic actuator controlled by the solenoid DCV will have adjustable throttles (TC1, TC2) to modify the piston velocity and two proximity sensors are attached outside of the cylinder to provide a noncontact indication of the rod piston [5, 7]. Noise dampers (or silencers, according to ISO 1219-1: 1991) are included, too. The proximity sensors and the silencers are not figured.

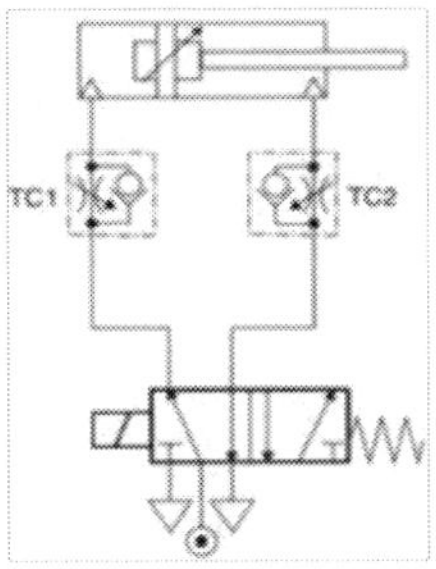

Fig. 1. Pneumatic actuator with solenoid DCV and throttle valves (TC1, TC2)

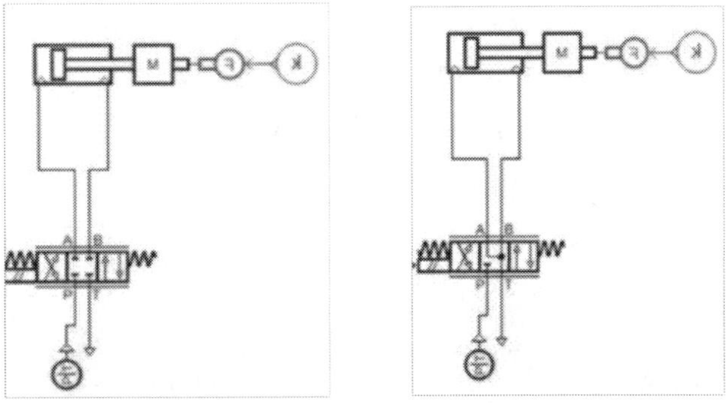

a) 4/3 closed center valve b) 4/3 ABT valve

Fig. 2. Pneumatic actuator with proportional valve, Amesim simulation scheme with M and F, K (in the mirror) symbols for mass, and constant force.

It is known that the open-loop pneumatic systems are sensitive to the initial conditions and not attenuate disturbances [5]. So, this kind of pneumatic systems are sensitive to parameter variations of the inputs. The parameter disturbance at the inputs will change the system outputs.

For this reason, it was proposed a pneumatic actuator controlled by a proportional valve with four ways and three positions of the spool, 4/3. The pneumatic drive is proposed in two configurations: with closed center (Figure 2, a) and with semi-opened center (Figure 2, b).

The pneumatic schemes were drawn in LMS Amesim.

3 Pneumatic actuator parameters and signals command

The pneumatic actuator presented in both configurations of Figure 2 has following physical characteristics of the pneumatic cylinder and proportional valve demanded by LMS Imagine.Labs Pneumatic library:

- ✓ for the double-acting pneumatic cylinder, the piston diameter is 100 mm, the rod diameter is 60 mm, the length of the stroke is 0.3 m, the dead volume in the cylinder is 60 cm^3, the total mass being moved is 50 kg, the viscous friction coefficient is 70 Ns/m, the temperature in the input and output ports of the pneumatic cylinder is 293.15 K;
- ✓ for the spool type 4/3 proportional valve, the area of control orifices is 7 mm^2, the valve natural frequency is 80 Hz, the valve rated current is 10 mA, the valve damping ratio is 0.8 and the flow coefficient is 0.72.

The pneumatic system is supplied with compressed air at a pressure of 7 barA; the pressure in the inactive cylinder chamber is 2 barA and the external temperature is the same as in the cylinder chambers. The pneumatic lines are one meter long and the inside diameter is 10 mm.

It was considered three types of signal waves command applied to the electromagnetic device of the proportional valve, with the characteristics plotted in Figure 3.

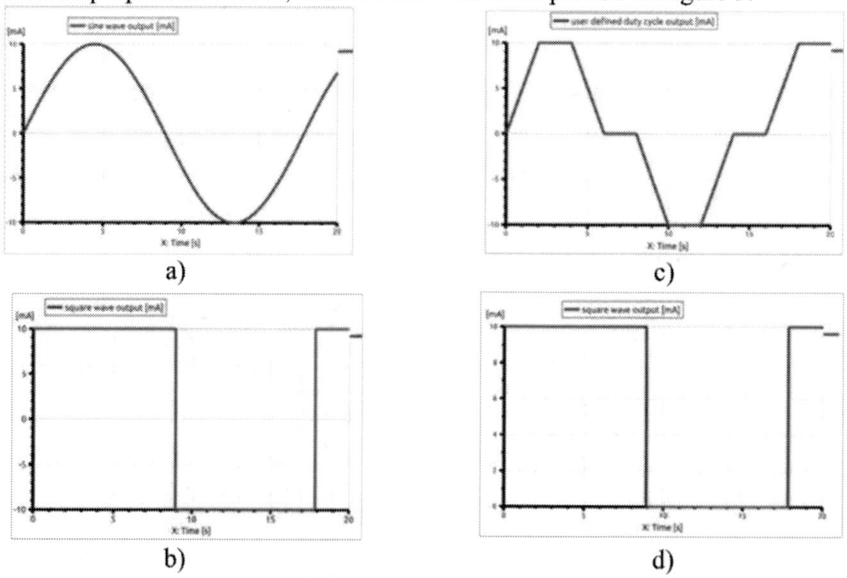

Fig. 3. Three types of signal command applied to the proportional valve: a) sine wave, b) square wave; c) trapezoidal signal; d) demi-square wave.

The sine and square wave profile signal command (Figure 3, a and b) have the amplitude ± 10mA and the wave frequency of 0.056Hz. The trapezoidal profile (Figure 3, c) has the same amplitude of ± 10mA, the same period of 18s but with the acceleration and deceleration time of 2s each and a time of 2s each for the steady state. The demi-square wave (Figure 3, d) has the amplitude +10mA and 18s period. The most usual action time diagram for a solenoid DCV has a trapezoidal profile, and this is the reason to use this kind of control signal in this study.

4 Numerical simulations of pneumatic actuator behaviour

Numerical simulations are presented using data mentioned above. In the numerical simulation model, a constant resistant force of 3846 N was considered in all studied cases. The resistant force was calculated regarding the application in water butterfly valve.

Are plotted the rod displacement against time, the velocity, acceleration and pressures evolution in cylinder chambers, for all the three signals applied in each of the pneumatic actuator scheme in Figure 2. To a better follow up of the numerical results, all six working cases are named as:

- ✓ PA-CC-PV-SINW, pneumatic actuator with closed center proportional valve and sine wave input, results were plotted in Figure 4;
- ✓ PA-CC-PV-SQRW, pneumatic actuator with closed center proportional valve and square wave input, results were plotted in Figure 6;
- ✓ PA-CC-PV-TRAP, pneumatic actuator with closed center proportional valve and trapezoidal input, results were plotted in Figure 8;
- ✓ PA-ABT-PV-SINW, pneumatic actuator with ABT proportional valve and sine wave input, results were plotted in Figure 5;
- ✓ PA-ABT-PV-SQRW, pneumatic actuator with ABT proportional valve and square wave input, results were plotted in Figure 7;
- ✓ PA-ABT-PV-TRAP, pneumatic actuator with ABT proportional valve and trapezoidal input, results were plotted in Figure 9;

Each figure from 4 to 9 has five graphical representations denoted with (a) for the numerical simulation model, (b) for rod displacement against time, (c) for rod velocity against time, (d) for rod acceleration against time and (e) for pressure of port one and two of the big and small cylinder chambers.

From the plotted results some general observations can be done.

All displacements of the rod have three specific parts: the increase part of time, when the rod extends and gets the stroke value of 0.3 m; the rest period of "rest" at the end of the stroke with zero velocities; the decrease part of time when the rod is retracted in the cylinder. In all six cases after the expand displacement, the time of rod restart (in the retract displacement) is long. The rod is no more moving, although the electromagnetic device of the proportional valve is receiving the electric signal command.

Numerical simulations highlight the compressibility of air flow. The time delay can be observed in the dynamic response of the pneumatic actuator having closed center valve control with sine wave and trapezoidal wave command (Fig. 4 and Fig. 8). In the case of pneumatic actuator with ABT valve control and the same types of command signals, the time delay is observed, too (Fig. 5 and Fig. 9).

Compared with these cases, the pneumatic actuator having square wave command will not have a delay time in the movement, not even when the pneumatic cylinder is command by proportional valve with close center, nor when it is used an ABT control valve (Figure 6 and Figure 7).

The delays are better highlighted by the velocity diagrams.

Also, it is remarked the asymmetric evolution in the extend rod displacement which has a long time of movement than the retract of the rod in the cylinder. It is interesting that this asymmetry is not available in the case of square wave signal (Fig. 6, Fig. 7).

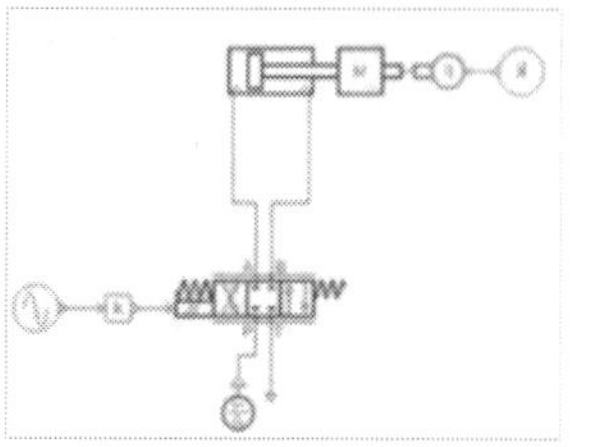

a)

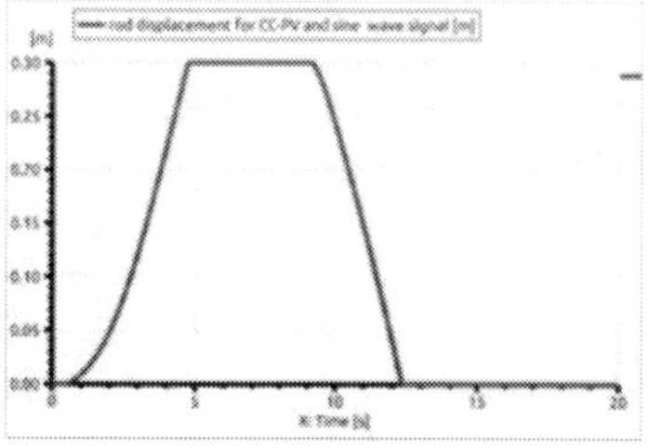

b)

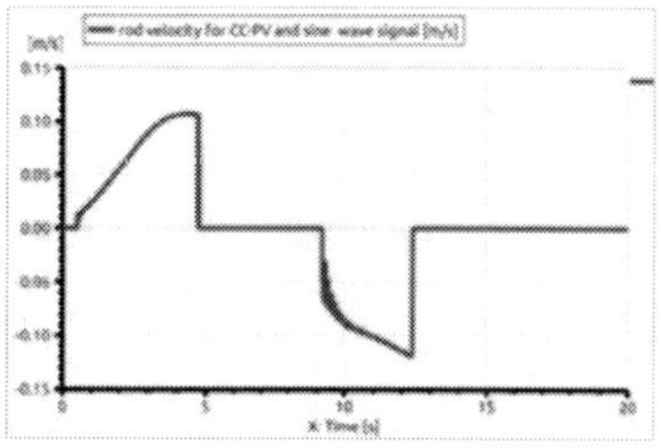

c)

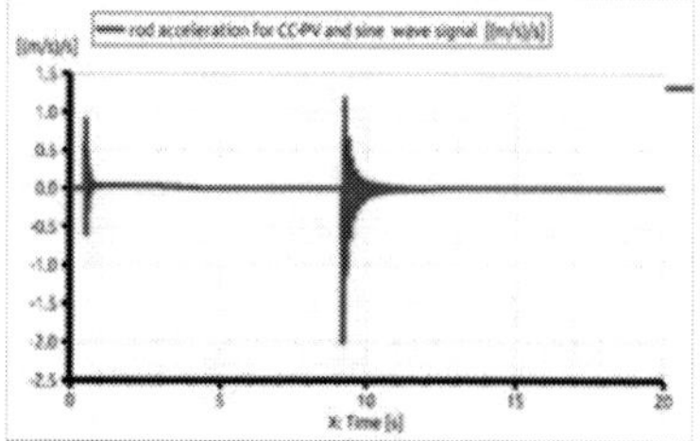

d)

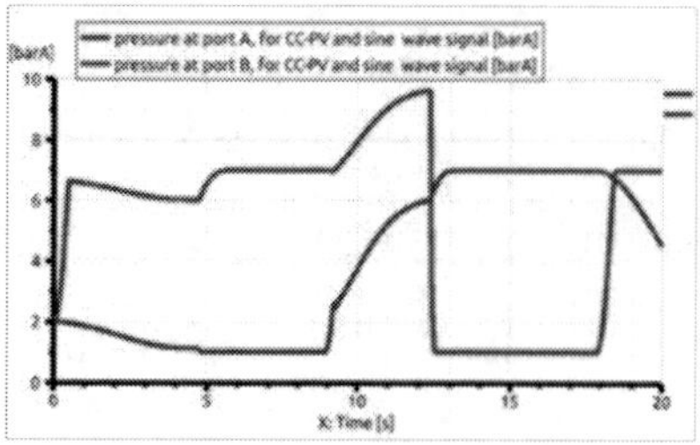

e)

4. PA-CC-PV-SINW

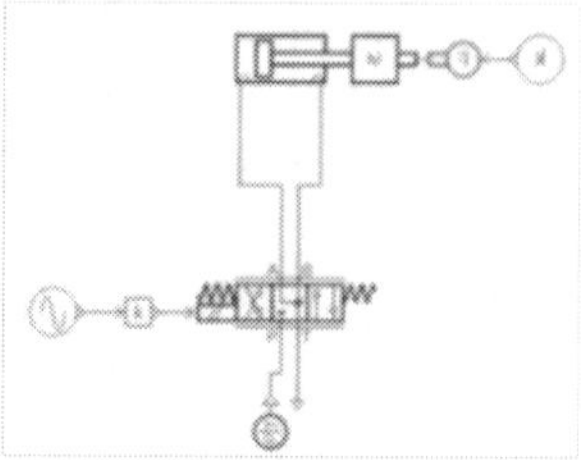

a)

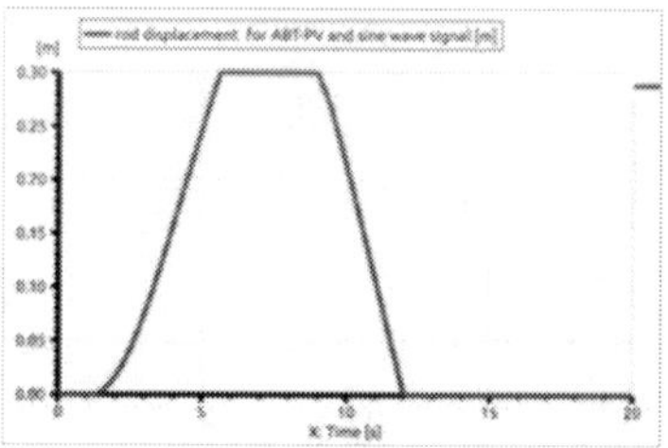

b)

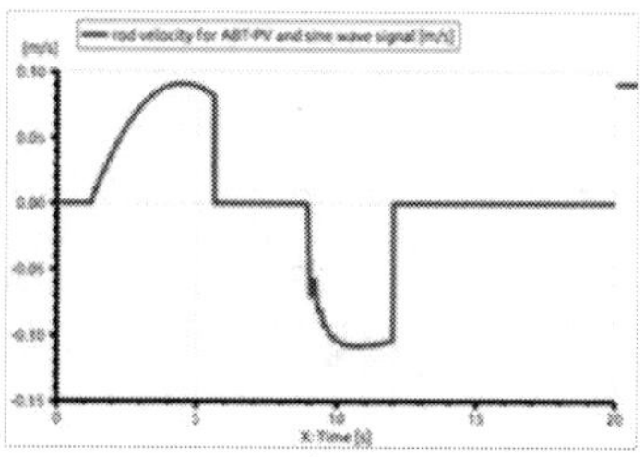

c)

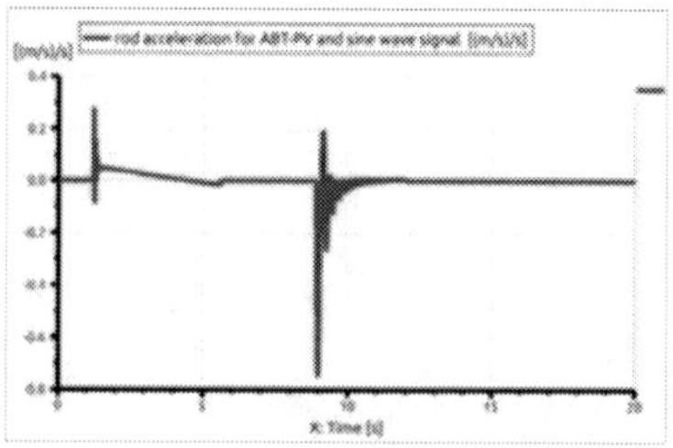

d)

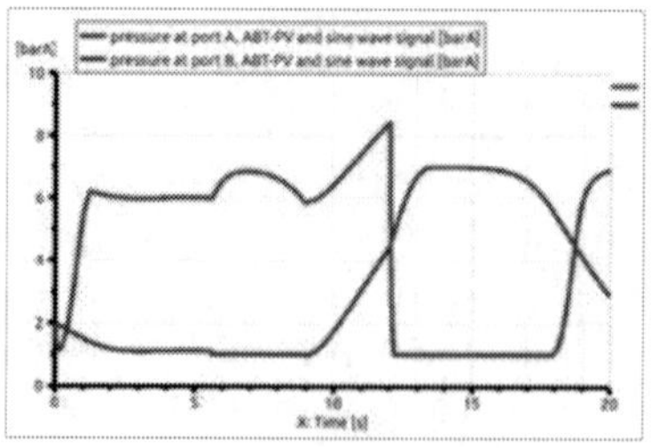

e)

Fig. 5. PA-ABT-PV-SINW

In figures 4 and 5 is observed that the time delay of the pneumatic actuator with ABT control valve is longer then in case of pneumatic actuator with closed center control valve. Also is observed that the velocity evolution is smoother in the case of ABT control valve and the velocities are almost three times smaller than the case of closed center control valve (see Fig. 4. c and Fig. 5. c). The acceleration values are smaller in the case of ABT control valve than the close center control valve.

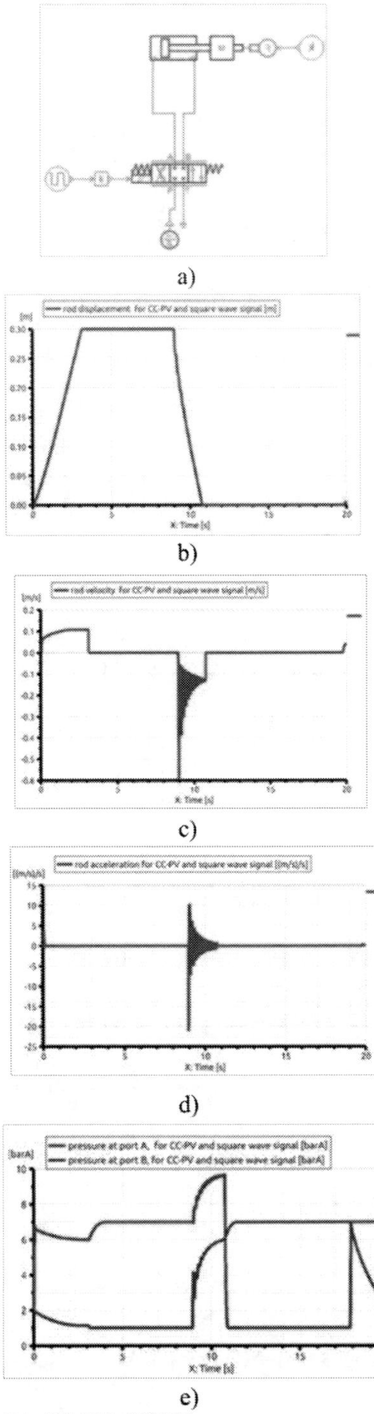

Fig. 6 PA-CC-PV-SQRW

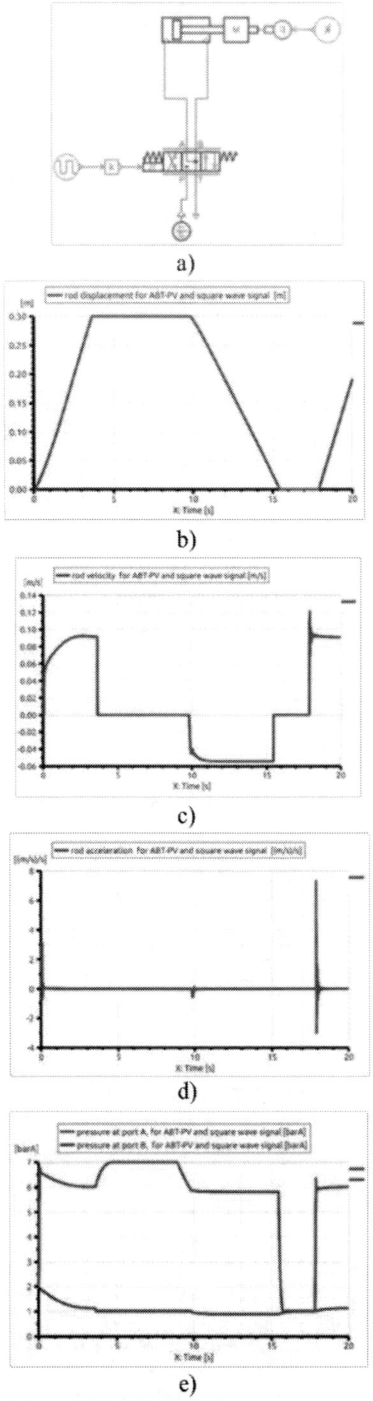

Fig. 7 PA-ABT-PV-SQRW

In figures 6 and 7 is observed the dynamic behaviour of the pneumatic actuator controlled by proportional valve at square wave signal command. In these cases, no delay time appear in the rod movement. It is observed that actuator tends to return under its own weight, Figure 7.e. To obtain this behaviour, the actuator controlled by ABT-PV was command by a semi-square wave signal (Figure 3, d).

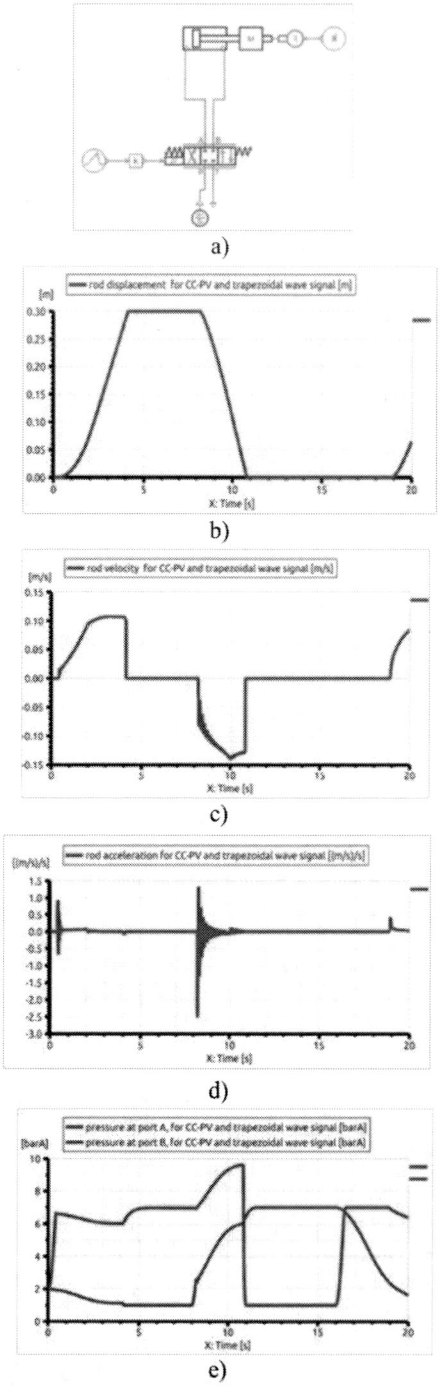

Fig. 8. PA-CC-PV-TRAP

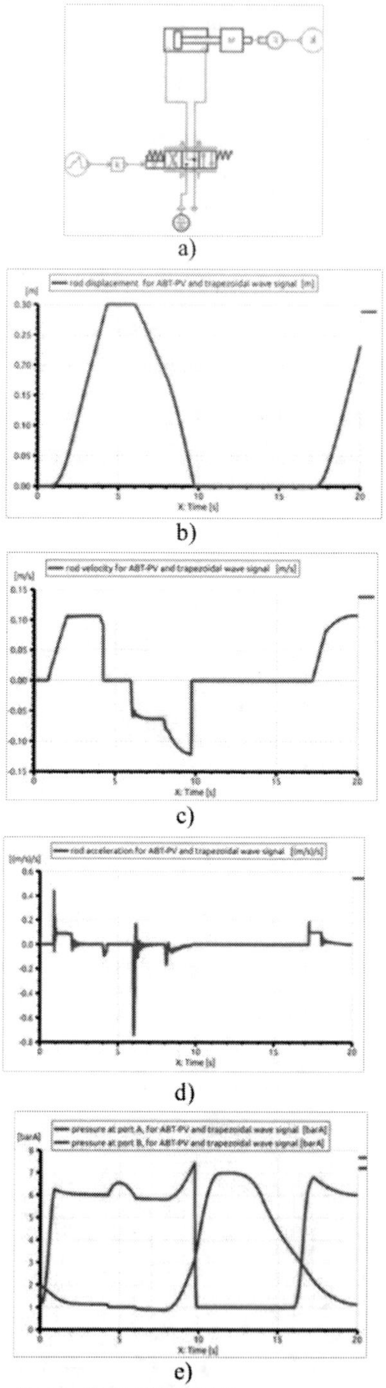

Fig. 9. PA-ABT-PV-TRAP

Regarding the pressure evolution in the cylinder chambers can be observed that the same pattern is following no matter what type of pneumatic actuator is, or type of command signal. The pressure rises quickly in the bigger cylinder chamber and when the piston is at the end of the stroke, the pressure decreases in the meantime with the increase in pressure of the smaller chamber for a complete duty cycle (Fig. 8. e and Fig. 9. e).

5 Conclusions

The goal of the paper was the study of the dynamic behaviour of a pneumatic actuator in open loop, when the control valve used is a 4/3 proportional valve in closed center and ABT configuration. Are used three types of signal waves for proportional valve command.

The numerical simulation results have been shown, that besides the nonlinearities known in the literature [8-11], the time delay of movement must be considered in the mathematical model of pneumatic actuator, too. Also, theoretically have been demonstrated that using proportional valve the pneumatic drive architecture is simplified. More, the dynamics of the pneumatic actuator in open loop can be controlled easily and parameters can be changed only from the electrical signal input of the proportional valve. Has been noticed in the case of the square wave signal that the values of acceleration of the piston which are almost three times bigger than other cases studied.

The study will be followed by experiments on pneumatic equipment to identify the parameters of the experiments with numerical simulation results.

Special thanks to OFRIM Group (https://ofrimgroup.com/) for all the support provide for this research.

References

1. Y. Yin, *High Speed Pneumatic Theory and Technology*, vol. I, Servo System, (Springer, 2019)
2. G. Chaohui, W. Chenggang, 10.5013/IJSSST.a.**17**.02.10 Int, (2016)
3. H. I. Ali, S. B. B. Mohd Noor, S. M. Bashi, M. H. Marhaban, *A Review of Pneumatic Actuators (Modeling and Control)*, Australian Journal of Basic and Applied Sciences, **3**(2): 440-454 (2009)
4. E. Richer, Y. Hurmuzlu, *A High Performance Pneumatic Force Actuator System. Part 1 - Nonlinear Mathematical Model*, ASME Journal of Dynamic Systems Measurement and Control, **122**, No.3, pp. 416-425, (2001)
5. I. Krivits, G. V. Krejnin, *Pneumatic Actuating Systems for automatic equipment. Structure and design*, (CRC Press Press Taylor and Francis Group, 2006)
6. N. Vasiliu, D. Vasiliu, C. Călinoiu, R. Puhalschi, Simulation of fluid power systems with Simcenter Amesim, (CRC Press Taylor and Francis Group, 2018)
7. M. Galal Rabie, Fluid Power Engineering, (McGraw Hill Companies, 2009)
8. Y. Tassa, T. Wu, J. Movellan, E. Todorov, *Modeling and Identification of Pneumatic Actuators, conference paper*, DOI: 10.1109/ICMA.2013.6617958, https://www.researchgate.net/publication/261479950
9. G. A. Muzy, A. S. Caporali, Positioning system of a pneumatic actuator driven by proportional pressure regulator valves, *Proceedings of the 4th Workshop on Innovative Engineering For Fluid Power* – WIEFP 2018, ABIMAQ, São Paulo, SP, Brazil, DOI: 10.3384/ecp1815611 (2018)
10. V. Constantin, G. Belforte, O. Dontu, Mihai Avram, U.P.B. Sci. Bull., Series D, **76**, Iss. 4, pp 57-68, (2014)
11. M. Sorli, L. Gastaldi, E. Codina, S. de las Heras, *Dynamic analysis of pneumatic actuators*, Simulation Practice and Theory **7**, pp. 589–602, (1999)

The influence of the vitiated air from the treatment plants on the urban and rural development

Fanel-Viorel Panaitescu[1,*], *Mariana* Panaitescu[1] and *Nicolaie* Ion[2]

[1]Constanta Maritime University, Department of Engineering Sciences in Mechanics and Environment, 900663 Mircea cel Batran street 104, Romania
[2] POLITEHNICA University, Doctoral School of Electrical Engineering, 060042 Independentei Spl. 313, Romania

Abstract. In order to reduce the emissions produced by the operational flow of a wastewater treatment plant, regardless of whether they are located in urban or rural areas, the quality of air must be analyzed. For this reason, the Constanta Nord treatment plant was chosen as a study location. This is near the Mamaia resort and does not fall within the recommended sanitary protection norms, according to NP-032-1999 (at least 300 m from the last house). Therefore, it was necessary to take measures to reduce the repellent gases generated in the neighborhood of wastewater treatment plants. The solution is in attention of all wastewater treatment plants from the world because this depend on operational flux management and health of population.The research consists of: description of the technological flow of the treatment plant and in particular of the deodorizing installation of the contaminated air, choosing a work scenario for a day with 2 shifts per hour, respectively 6 shifts per hour, the air flow needed to treat the vitiated air (the deodorization plant components) and the efficiency of the regulation system from the point of view of energy consumption. Also, possible leakage of dangerous gases into the atmospheric air was monitored (H_2S, Cl_2, CO), taking into account the fact that in the immediate vicinity there is the Mamaia resort, the student housing and rural locations Palazu and Mamaia village.

1 Introduction

Europe's air quality has been improving in recent decades. However, many people continue to be negatively affected by air pollution, especially in cities. Given its complexity, tackling air pollution requires taking coordinated action at many levels [1]. The level of emissions of pollutants released into the atmosphere may be reduced significantly through the implementation of environmental policies and strategies such as:
- greater use of renewable energy sources (wind, solar, hydro, geothermal, biomass);
- replacement of conventional fuels with alternative fuels (biodiesel, ethanol);

* Corresponding author: viopanaitescu@yahoo.ro

- the use of plants and equipment with high energy efficiency (low consumption, high yields).

"Policy makers and governments has focused their attention on air quality in the urban areas only. Air quality in rural areas remains a neglected issue so far" [2]. More than 100 substances in rural areas have been identified, like solids, vapors, gases.There are many natural sources of air pollution like:SO_2, Nox, carbonaceous particles, non-carbonaceous primary particles, ozone, volatile organic compounds (Voc; e.g. isoprene). The causes of air pollution in rural locations are natural (contaminants from animals, plants and land resources- spores, pollens, moulds, fur, feather, hair, dander, dust, grit, etc, forest fire, coal fires, dust storm, sand storm, meteorological factors, topography, air movements and climate) and anthropogenic (human activities) [3], [4], [5], [6], [7], [8].

Along with the existing causes in the rural environment adjacent to urban areas, the influences of urban pollution are added as a result of meteorological and anthropogenic factors. These caused are called trans-boundary pollution [2]. One of these sources is the pollution generated by uncontrolled emissions into the atmosphere from wastewater treatment plants. The SO_2 concentration in air would threaten physical health and even cause respiratory system diseases [4]. A study found that higher SO_2 and NO_2 particles would increase risks of suicide [5], [7].

For this reason, the Constanta Nord wastewater treatment plant (WWTP) was chosen as a study location (Figure 1). This is near the Mamaia resort and does not fall within the recommended sanitary protection norms, according to NP-032-1999 (at least 300 m from the last house). Therefore, it was necessary to take measures to reduce the repellent emissions generated in the neighbourhood of wastewater treatment plants.

Fig. 1. The sewage treatment plant in North Constanta –View after covering the objectives that gave repellent gases

The paper proposes a solution for optimizing the process of air deoderization and optimization of operating costs and energy for this solution. The solution is in attention of all wastewater treatment plants from the world because this depend on operational flux management.

There are three units of air deodorization stations in the sewage treatment plant North Constanta. These units purify the air resulting from the operation of bioreactors, secondary settler tank, distribution chambers and the connection pipes.We propose a study on the quality of the treated air, but at the same time, the study will focus on the consumption of electricity. We will also try to identify the possibilities of reducing the electricity consumption that is absorbed from the grid using oxygen detectors on bioreactors.

2 Plant description

2.1 Operational flow

The operational technological flow in Constanta North Wastewater Treatment Plant (WWTP) (Figure 2) contents: primary stage, secondary stage and tertiary stage.

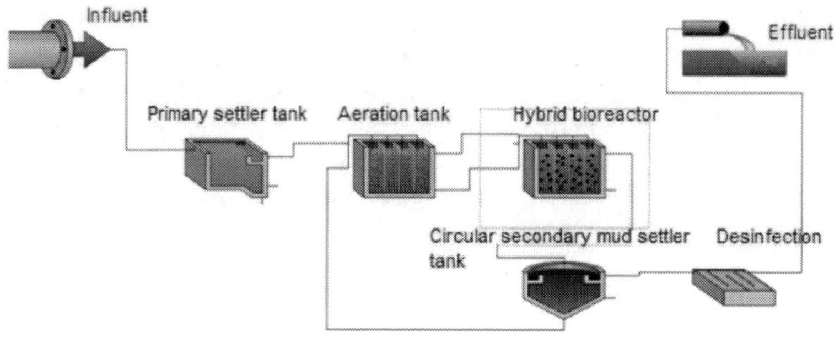

Fig. 2. Operational flow in WWTP Constanta-North.

Major processes of treatment includes [9], [10]: a) physical treatment-remove suspended large solids by settling or sedimentation and eliminate floating greases; b) biological treatment-degradation or consumption of the dissolved organic matter using the means of cultivated in activated sludge or the trickling filters; c) chemical treatment-remove other matters by the means of chemical addition or destroying pathogenic organisms through disinfection; d) advanced treatment- removing specific constituents using processes such as activated carbon, membrane separation, or ion exchange. Particular treatment processes are: a) sedimentation; b) coagulation and floculation; c) activated sludge; d) sand filters; e) membrane separation; f) disinfection.

2.2 The deodorization plant

Three deodorization units were designed using H_2S concentrations to reflect the odor levels to be removed. To treat the waste air from these surfaces and processes, we selected several activated carbon treatment systems based on the volumes corresponding to the objectives, the number of air changes required, the anticipated levels of H_2S contaminant and the imposed emission limits. Although there are several odorous substances at a treatment plant, the deodorization units were designed using H_2S concentrations to reflect the odor levels that need to be removed [9], [10].

2.2.1 Second settling deodorization plant

The deodorization units that equip each of the secondary settler tanks extract the air from the free space between the surface of the water, the walls and the roof of the settler tank through a suction tube and lead it into the air treatment compartment composed of two parallel activated carbon blankets which they absorb H_2S as well as other odorous gases as well as suspended particles. This treated air is then discharged into the atmosphere through a chimney. The deodorization units of the secondary settler tanks are installed outdoors on concrete platforms specially designed for them. Each secondary settler tank has a deodorization unit. The deodorization units of the decanters are designed to make 2 shifts of the entire volume of air per hour (Figure 3) and 6 shifts of the whole volume of air per hour (Figure 4) when an operator wishes to enter the settler tank. The air flow requiring treatment is calculated based on the number ACH - the number of air exchanges per hour, from the capsule volume. For decanters, the volume of air between the surface of the water in the decanter and the roof is 8872 m^3. The two projected situations take into account 2 ACH when the operators are not in this space and 6 ACH when they are the operators in this space.

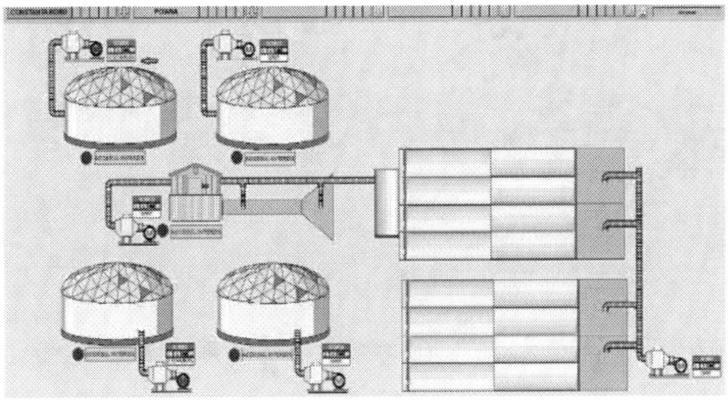

Fig. 3. Operational flow-2 shifts per hour

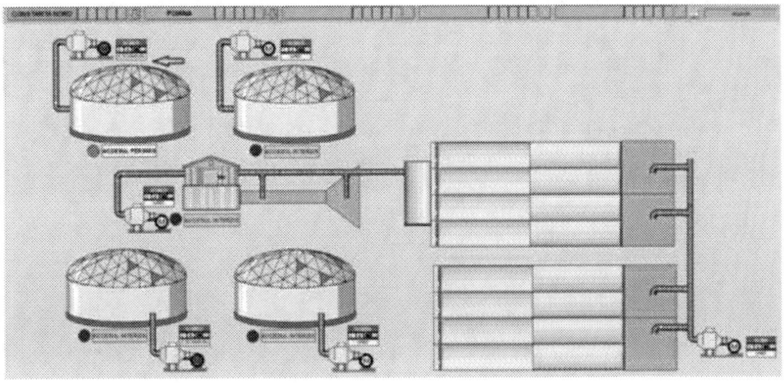

Fig. 4. Operational flow-6 shifts per hour

2.2.2 Bioreactors deodorization plant

Bioreactor deodorization unit extracts air from the free space between the water surface and roof from the four bioreactors and leads it to the air purification compartment composed of two parallel activated carbon blankets that absorb H_2S and other odorous gases as well as particles. in suspension. This treated air is then released into the atmosphere through a chimney.

The bioreactor deodorization unit is designed to operate a volume of 44,000 m^3 / h which represents the aeration rate of the bioreactor plus 10% for compensation if there is air leakage through the roof (Figure 5). The roof of the bioreactor is not designed to allow personnel to enter. The operational staff will not be able to get under the roof because the air in it does not have enough oxygen to hold the breath. The bioreactor deodorization unit is installed outdoors, on a concrete platform.

The four bioreactors are provided with roofs that form eight separate air spaces (two cells per bioreactor).The operation on the two shifts (2 shifts per hour, respectively 6 shifts per hour) implies a monitoring of the odor treatment units at the secondary settler tanks.

Fig. 5. View of the bioreactor distribution chamber

2.2.3 Deodorization plant. Distribution rooms-connection pipes

The deodorization unit of the pipes and the distribution chambers extracts the air from the free space between the surface of the water, walls and roof, leads it to the treatment compartment composed of a series of three layers of activated carbon for the absorption of H_2S and other obnoxious gases plus suspended particles. This treated air is then released into the atmosphere through a chimney. The deodorization unit of the distribution rooms and the channel is installed outdoors, on a concrete platform (Figure 6).

Fig. 6. Deodorized air defective station mechanical cleaning stage

The deodorization unit of the distribution chambers and the connection pipes is designed to provide 2 shifts of the entire air volume per hour and 6 shifts of the entire air volume per hour when an operator wishes to enter the closed spaces. The air flow requiring treatment is calculated based on the number of ACH shifts, the number of air exchanges per hour, from the capsule volume. For the distribution room, the closed volume of air is 1070 m^3.

3 Research and methods

3.1 Design scenarios

The design scenarios provide for 2 (ACH) shifts when the operators are not in the encapsulated space and 6 (ACH) shifts when they are operators in the encapsulated enclosure. The clean air comes through the ventilation slots left on the roof of the rooms. Being a very short section, the diameter of the manifold pipe is chosen: 1200 mm since, whenever it is possible and even necessary to stop and isolate one or maximum of two biorectors, it has been provided that the system of collecting pipes of the vitiated air will be equipped with elements of adjustable closing of the extracted air flow. Thus, four damps with a diameter of 600 mm were introduced into the installation, equipped with butterfly type lenses and devices for regulating the flow of extracted air. Keeping the constant velocity of 10.16 m / s of air flow through the piping system, from the central tube of 1200 mm diameter, two pipes, left / right, of 900 mm diameter, each of them ensuring the extraction of the vitiated air from two bioreactors. Further, each of these two pipes is branched into two pipes of 600 mm diameter, one for each bioreactor (table 1). Due to the production of reinforced concrete beams, four in number, in each bioreactor, with the role of creating supports for the new roofs in zone A of the bioreactors, each bioreactor can be considered to be composed of two distinct cells, cell 1 and cell 2, separated through the concrete beam to which we refer. For the extraction of the vitiated air from the two cells of each pipe of 600 mm diameter, equipped with a damper of the same diameter, another two pipes, one of 600 mm diameter each, for cell 1 and another 450 mm diameter for cell 2 are branched.

The dimensions of the pipes for extracting the foul air are calculated so that the air speed is kept below 10.16 m / s, which prevents excessive pressure losses in the pipe.For the distribution and channel chambers there are 4 types of supply pipes (Table 1).

Table 1. Pipeline distribution and construction elements.

Location	**Flow [m³/h]**	**Velocity [m/s]**	**Dimensions [mm]**
Total	7020	10.16	500
Settler tanks distribution chamber	2148	10.16	300
Settler tanks distribution chamber plus connection channel	1980	10.16	300
Bioreactors distribution chamber	4128	10.16	450
Pumps station	2310	10.16	300
Distribution chamber plus pumps station	582	10.16	200

3.2. Experimental research

The vitiated air in these enclosures passes through the deodorizing station in which there is a mixture of tree bark that retains and filters, so that at the exit of the deodorizing station a clean air is given to the atmosphere. As this deodorizer station is equipped with an extractor.

The values of parameters for the repellent gases at the inlet and outlet of the contaminated air enclosures were measured (Table 2. When the sensors detect exceedances of the repellent gases values, a red light signal illuminates which expresses the exceeding of the standard limit value.

Table 2. The inlet standard air parameters.

Gas	Threshold alarm	Danger threshold	Unit
CH_4	20	40	% LIE
O_2	19	23	% Vol.
Cl_2	0,5	1	Ppm
H_2S	5	10	Ppm
CO	30	60	ppm

During our scenario, overruns were reported for H_2S, Cl_2 and CO. During the study, operational parameters (inlet/outlet) are monitored with SCADA program, air parameters with Dräger X -am 5000 gas analyzer and GasSens system (Table 3).

The values of repellent gases were recorded for the case study:

Table 3. The recorded values of repellent gases parameters.

Location	**H_2S [% LIE]**	**Cl2 [ppm]**	**Oxygen [% vol.]**
Rare Grills	6.5	0.7	21
Desander	7.1	0.75	18
Settler tanks distribution chamber	7.3	0.75	17
Bioreactors distribution chamber+ recirculation pumps	6.7	0.6	15

GasSens is an online monitoring system for detecting dangerous gases in atmospheric air. It is designed for gas leak detection from cylinders, pipelines or gas supply equipment and for alarms, in virtually any environment from industrial installations. There are systems for monitoring different gases:

• *Sensor / transmitter* -it consists of an electrochemical gas sensor, directly coupled to a NEMA 4X transmitter. This component measures the concentration of gas and converts the measurement result into a digital signal for transmission to a receiver module. It is installed in the area where a leak or accumulation of gas is possible, either on the wall or supported by the tube for electrical cables through which the interconnecting conductors pass. Sensors / transmitters can be optionally fitted with *Auto-Test*, an electrochemical gas generator that automatically tests the sensor response daily and warns of any sensor problems. For hazardous areas there are explosion-proof sensors / transmitters.

• *Receiver module* - receives digital data from the sensor / transmitter and ensures the LED display of the gas concentration, two regulated gas alarm values, three gas alarm relays, an isolated output of 4-20 mA, as well as an alarm and a relay for damage.

• *Power supply module* - provides 12 V DC power supply. for one or two receiver modules, as well as the power supply for a siren and for charging an external battery.

• *NEMA 4X enclosures* (for alarm modules only). Four standard enclosures are available for mounting the receiver and power supply modules. The polystyrene enclosures are suitable for outdoor environments, and the digital displays and alarm indicators are very visible through a transparent polycarbonate window at the front.

• *Explosion-proof enclosures* (for gas monitors / sensors only). There are two versions that allow the use of receiver and power supply modules in hazardous areas. One version is suitable for only one receiver module, and the second version is large enough to receive both a power source and a receiver.

• *Siren* - an alarm siren, piezoelectric, of 12 V c.c. can be fitted for all NEMA 4X enclosures. The siren is connected to the receiver module and emits intermittent sounds when it reaches the set value for warning, respectively continuous sounds when it reaches the set value for alarm. The siren can be switched off by pressing the A / R switch on the front panel of the receiver.

• *Strobe lamp*- provides light status alert.

• *Backup Battery* - provides a 12 V DC, 4 AH battery and control circuits for charging, in a separate NEMA 4X enclosure. The backup battery is connected directly to the power supply module to ensure the backup power supply of the detection system.

The concentrations of gas emissions recorded at the exit from the deodorization installation are less than the allowable limit, so the impact is insignificant.

3.3 Solutions for optimizing energy consumption and monitoring oxygen levels in bioreactors

Under anoxic conditions, the denitrification process takes place, and under aerobic conditions the nitrification process takes place in the four lines of the bioreactors [11].

Therefore, in the Constanta-Nord treatment plant, four blowers are installed in operation, 1 and 2 assistance and stand-by to cover the required oxygen in the bioreactors. The electricity consumption of the turbochargers has an important weight in the general energy consumption per each stage station (table 4)[12]. The difference in consumption is due to the fact that during operation, many equipments have interruptions due to various maintenance reasons, or some have a discontinuous operating flow. Thus, each of the four THOLANDER turbochargers has a rated power of 355kW, while the FLYGT C 3400 intake pumps have a rated power of 125kW each (five pumps), and FLOTTWEG centrifuges have a rated power of 55kW each (4 centrifuges).

Table 4. Distribution of equipment by processes in Constanta WWTP.

Stage/ Power [kW]/ Energy [kWh]	**Share of equipment consumption[%]** $P_{instal}/P_{instal.total}$	**Share of equipment consumption [%]** $E_{instal}/E_{instal.total}$
Mechanical stage: P=768.83 [kW] E=4132112.4 [kWh]	25.372	20.815
Biological stage: P=1924.68 [kW] E=13704521.4 [kWh]	63.516	69.035
Mud treatment stage: P=325.86 [kW] E=2012138 [kWh]	10.754	10.136
TOTAL: P=3019.37 [kW] E=19848771.8 [kWh]	99.642	9.986

For our case study the electricity consumption for a full day of operation with 2 shifts per hour is 136 kWh (19.02 2020), and for 24 hours of operation with 6 shifts per hour is 182 kWh (20.02.2020) (Table 5).

Table 5. Operating hours index and electricity consumption index secondary deodorization stations.

Operating time	Operating hours 2 shifts/ hour [hours]	Operating hours 6 shifts/hour [hours]	The hourly energy consumption index/shifts/hour [kWh]	Energy consumption index [kWh]
18.02. 2020	4466	7642	7/2 shifts 47/6 shifts	127/2 shifts 168/6 shifts
19.02. 2020	4491	7763	6/2 shifts 46/6 shifts	136/2 shifts 175/6 shifts
20.02. 2020	4515	7862	8/2 shifts 49/6 shifts	143/2 shifts 182/6 shifts

We observe the hourly energy consumption of 2 shifts which is about 6 kWh and in the case of 6 shifts per hour it is 46 kWh. Moreover, the air flow is 18000m^3 / h, in case of operation with 2 shifts per hour, flow that does not ensure an acceptable quality for the air inside. The air flows in the two situations are 18000 m^3 / h, respectively 53232 m^3 / h.

We analyze the efficiency of the regulation system in terms of energy consumption (turbochargers have a significant share in station consumption (over 68%).

Description of the proposed solution:

□ Small bubble diaphragm diffusers absorb pressurized air. Bioreactor air supply lines are equipped with air control valves, each tank having an air control valve.

□ Two oxygen flow meters (a main probe and a reference probe) are installed at each bioreactor to adjust the volume of air intake in the bioreactor in question. The air flow at each separate bioreactor line is measured and recorded by four flow meters.

The control of the aeration systems mainly comprises a concentration of two separate control cycles.

The assessment of ambient air quality is regulated by Law 104/2011 on quality ambient air transposing Directive 2008/50 / EC of the European Parliament and of the Council on ambient air quality [13], [14]. In Constanta county, air quality is monitored by continuous measurements in 7 stations automated located in representative areas. Pollutants in the area of Constanta Nord treatment plant, Mamaia resort and Mamaia village, are monitored by CT-3 and CT-6 stations, respectively: sulfur dioxide (SO_2), nitrogen oxides (NOx / NO / NO_2), carbon monoxide (CO), ozone (O_3), benzene, particulate matter (PM10) and weather parameters (wind direction and speed, pressure, temperature, solar radiation, humidity relative, precipitation). Also, possible leakage of dangerous gases into the atmospheric air was monitored (H_2S, Cl2, CO), taking into account the fact that in the immediate vicinity there is the Mamaia resort, the student housing and rural locations Palazu and Mamaia village (Table 2, Table 3).

In order to protect the population from the possible effects of pollution due to the processes in the treatment plant and taking into account the location conditions, the hydrological and climatological conditions, it is recommended to reduce the impact of repellent gases:

-rehabilitation of the ecological balance in the area of WWTP by monitoring emissions and their impact on fauna, flora and population, as well as by planting dendrological material;

- removing sludge from the site;

- controlling the wastewater treatment and sludge treatment process and monitoring parameters of these processes;

- covered structures for sludge treatment and storage;

- avoiding crossing urban and rural areas;

- avoiding wrong operational maneuvers due to human errors;

- execution of works without causing discomfort to the inhabitants by generating noxious substances, dust, noise and vibration;

- operation at the optimal designed parameters of the technological equipment and means of transport to reduce nuisance and noise that could affect the human factor.

4 Conclusions

We have done a study on the quality of the treated air, but at the same time, the study will focus on the optimization of energy consumption and oxygene level in bioreactors. We will also try to identify the possibilities of reducing the electricity consumption that is absorbed from the grid using oxygen detectors on bioreactors.

We analyzed the reduction of energy consumption at large consumers in the station and mainly at the deodorization station by proposing work on 2 shifts, respectively 6 shifts / hour, as well as the regulation of air flow, respectively oxygen level.

The control system tests the oxygen concentration in the aerobic bioreactor and maintains its value of about 1.5%.

The quick control valve system is set to operate in the 15% -65% opening range. When the oxygen detector detects reaching the 1.5% threshold, the valves open and the air flow increases; at the same time, the power consumed by the turbocharger obviously increases.

In addition to H_2S, Cl_2 and CO are also monitored. The deodorization stations aim to purify the air from the enclosures so that the limits of the concentrations of H_2S, Cl_2 and CO are within the accepted limits. The control of gas emissions outside is done through the monitoring system of the deodorization station, and the impact of these repellent gases has proven, experimentally, to be insignificant.

This paper has financial support from the Constanta Maritime University under assistance project PN-III-P1-1.2-PCCDI-2017-0404 /31PCCDI/2018, *Holistic on the Impact of Renewable Energy Sources on Environment and Climate-HORESEC.*

References

1. Campi, M.C., Cleaner air benefits human health and climate change, Nature Journal 29-01 (2018)
2. Majra, J.P., Air quality in rural areas, J. Intechopen, **7- 27**, DOI 10.5772/16890 (2011)
3. Fang PQ., Medical and health undertakings development report in China. People's (Publishing House Press, Beijing 2014)
4. Peel J L, Tolbert P E, Klein M., Ambient air pollution and respiratory emergency department visits. Epidemiology J., **16(2)** 164– 174 (2005)
5. Bakian AV, Huber RS, Coon H, et al., Acute air pollution exposure and risk of suicide completion, American J. Epidemiology, **181(5)** 295– 303 (2015)
6. Mikael F, Raisa V, Pilvikki A, et al., Rural-urban Differences in Health and Health Behaviour: A Base-line Description of a Community Health-promotion Programme for the Elderly, Scand J. Public Health, **34** 632– 640 (2006)
7. Lilian CG, Ana CG , Ricardo TJ, et al., Air pollution and your brain: what do you need to know right now, Prim Health C. Res. Dev. J., **16(4)** 329– 345 (2014)
8. Chen S, Chen T., Air Pollution and Public Health: Evidence from Sulfur Dioxide Emission of Coal-fired Power Stations in China, Econ Res J, **8** 158– 183 (2014)
9. A. Ghiocel, M. Panaitescu, V. Panaitescu, *Methods of efficiency of treatment plants. Use of biogas obtained from the fermentation of sludge in thermal power plants (Case study: WWTP Constanța Nord)*, J.Stiinta si inginerie **7**, 28-74 (2015)
10. I.I. Panaitescu , A.A. Scupi, F.V. Panaitescu, M. Panaitescu, *Efficient use of the pumps in the wastewater treatment plant*, Proc. Of Int. Conf. Energy-Environment (CIEM), (2013).
11. E.T. Gligor, *Contributions to the energy optimization of installations and equipment within wastewater* treatment *plants* Doctorate Thesys (Romania, Oradea, 2011)
12. Robescu Dan, ş.a., *The reliability of processes, facilities and equipment for treatment and purification of water*, (Technical Ed. Bucharest, 2002)
13. Law no. 104 / 15.06.2011 on ambient air quality published in the Official Gazette of Romania, Part I, **452** - 28, (2011).
14. Directive 2008/50 / EC of the European Parliament and of the Council of 21 May 2008 on ambient air quality and cleaner air for Europe, Official Journal of the European Union (OJEU) **L 152** 11-06 (2008).

E3S Web of Conferences **180**, 04014 (2020) https://doi.org/10.1051/e3sconf/202018004014
TE-RE-RD 2020

Assessment and risk prevention of air pollution in urban sites

Mariana Panaitescu[1,*], *Fanel-Viorel* Panaitescu[1] , *Cornel* Panait[2] , *Octavia* Bardasu[3] ,and *Viorica-Ani* Merla[3]

[1]Constanta Maritime University, Department of Engineering Sciences in Mechanics and Environment, 900663 Mircea cel Batran street 104, Romania
[2] Constanta Maritime University, Department of General Engineering Sciences, 900663 Mircea cel Batran street 104, Romania
[3] Constanta City Hall, Development and European Funds Directorate, 900725 Tomis Blvd. 51, Romania

Abstract. Atmospheric pollution limitation has evolved and evolves continuously, depending on the better knowledge of the harmful effects of the different atmospheric pollutants. The main elements that have led and lead to the continuous revision of the norms are the climatic changes and acid rain associated with atmospheric pollutants. In this context, the paper presents the following aspects: two monitoring regimes aims to evaluate and prevent air pollution in urban areas; was chosen as a monitoring location Constanta city, with strategic points exposed to the risk of pollution. For the monitoring of emissions in these locations, a self-employed emission analyzer was used. Where there are risk areas, exists a risk analisys with vulnerable locations and values of emissions's concentrations. There are monitoring CO, CO_2 , NO_2 , SO_2 emissions and PM2.5 and PM10 in air of urban sites which are exposed to the risk of pollution. Finally, the recorded values are below the maximum permissible limit. The conclusion is that pollution in urban sites can also be a danger to neighboring rural settlements, if there is no strict monitoring thereof.

1 Introduction

Air pollution is a well-defined element of European environmental policy, during the last decades the policies in this field have resulted in the reduction of emissions polluting substances and the marked improvement in air quality. Air quality is determined by air emissions from stationary sources and sources mobile (road traffic), mainly in the big cities, as well as by the long distance transport atmospheric pollutants. The current European legislation in the field of air pollution is supported by the Thematic Strategy on Air Pollution 2005 (TSAP) (EC,2005) which aims to improve quality of air in 2020 compared to the situation of 2000, defining concrete objectives in this regard the impact on

* Corresponding author: panaitescumariana1@gmail.com

human health and the environment. Many publications were developed on methods and tools for assessing the health risks of air pollution at local, national and international levels [1]. The strategy establishes European legislation and the measures necessary to reach the long-term target of the Sixth Action Program for the Environment (which took place between 2002 and 2012), reaching the "quality level of air that does not endanger and does not adversely affect human health and environment" [2], [3]. In Romania, the field of "*air quality*" [4], [5] is regulated by Law 104/15.06.2011 regarding ambient air quality [6] with subsequent modifications and completions (H.G. 336/2015 for amending the annexes 4 and 5 to Law 104/2011, respectively H.G. 806/2016 for amending the Annexes 4, 5, 6 and 7 to Law 104/2011) which transposes into national law the provisions of Directive 2008/50 / EC, [7] Directive 2004/107/EC [8] and EU Directive 2015/1480. For the implementation of the law of ambient air quality, the National Air System was established Integrated Air Quality Assessment and Management (NASIAQAM) which provides the organizational framework, institutional and legal cooperation of public authorities and institutions with competences in the field for the purpose of assessing and managing the ambient air quality, in a unitary way, throughout the territory Romania, as well as for informing the population and European and international bodies on ambient air quality [9]. The NASIAQAM ensures the monitoring of the ambient air quality through the National Network Air Quality Monitoring (NNAQM), and the National Emissions Inventory System Atmospheric pollutants (NEISAP), collect and manage the information and data received from the National Network. Currently, the NASIAQAM carries out continuous measurements of sulfur dioxide (SO_2), oxides of nitrogen (NOx), carbon monoxide (CO), ozone (O_3), particulate matter (PM10 and PM2.5), benzene (C_6H_6), lead (Pb), arsenic (As), cadmium (Cd), nickel (Ni), benzo (a) pyrene. Air quality in each station it is represented by suggestive quality indices, based on the concentration values of the main atmospheric pollutants measured. The NASIAQAM comprises 41 local centers (the local Environmental Protection Agencies) which collects and transmits to the public information panels the data provided by the stations and afterwards the primary validation is sent to the Air Quality Assessment Center (AQAC) for certification within the framework of the National Agency for Environmental Protection. The municipality of Constanta falls under the management regime I of the areas in the areas and agglomerations. In this regime I of management of areas in areas and agglomerations, the municipality Constanta is recorded with excess for nitrogen dioxide and nitrogen oxides [4].

To protect human health and the environment as a whole, studies have been conducted that highlight the impact of air emissions on air quality [1], [3], [4], [10], [11], [12].

The system procedure for validating air quality data from the National Air Quality Monitoring Network takes into account the fact that air quality is determined by air emissions from stationary and mobile sources [5], [9].

1.1 Area specific data analysed

1.1.1 General information

The city of Constanța is located in the county of the same name, in the southeastern part of Romania (Fig. 1) [2]. Constanta is divided into districts: Tăbăcăria, Brotăcei, North Faleza, Coiciu, Palas, Medeea, Brătianu, Center, Peninsula, Tomis I, II, III and North, Abbey, CET, Km 4, 4-5 and 5, South Faleza, at which added new neighborhoods such as Tomis Plus and Veterans (Gate 6) (Fig. 2) [2].

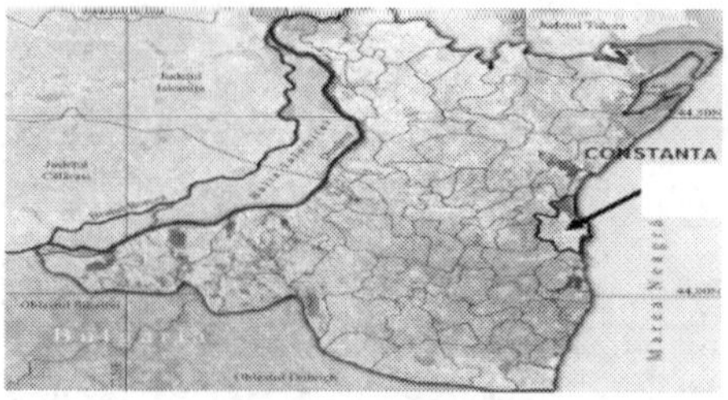

Fig. 1. Constanta city on Constanta county map.

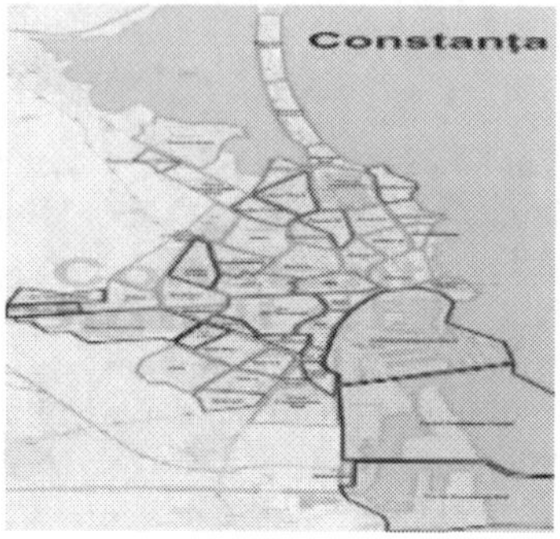

Fig. 2. The districts of Constanta municipality

1.1.2 Description of the existing air quality situation

Since 2008, air quality monitoring in Constanta has been carried out through continuous measurements, through the automatic monitoring network, component of the network national monitoring. The air quality monitoring network in Constanta consists of 3 stations automatic: *Station CT 1* – Traffic station, located in Constanta municipality – area "Casa de Cultura", evaluates the influence of traffic emissions; *Station CT 2* – urban background station, located in Constanta municipality – park area Hall; *Station CT 3* – industrial type station, located in Constanta municipality – str.Extension of the Liliac no. 6, evaluates the influence of industrial sources on air quality, having the radius of the representational area of 10 – 100 m. The measurement at fixed points of the pollutants is done by applying the reference methods.

Background pollution represents the pollution existing in the areas where the influence does not directly manifest anthropogenic pollution sources. Background pollution monitoring is a global problem, important to be able to appreciate the effects of pollutants entering the fresh air of the sphere. The monitoring of regional pollution corresponds to the monitoring of the air located relatively far from urban or industrial centers, ie between background pollution and anthropogenic polluted air.

Impact pollution represents the pollution produced in the direct impact areas of the sources anthropogenic pollution. Continuous monitoring of (local) impact pollution is required because impact pollution directly and immediately affects trophic chains and human health.

The level of annual average concentrations of air pollutants in the surrounding air at the stations the automatic air quality monitoring network in Constanta municipality is presented as follows (table 1)[2], [5]:

Table 1. The level of average annual concentrations of atmospheric pollutants in ambient air in 2017

Station	NO_2 µg/m^3	SO_2 µg/m^3	CO µg/m^3	O_3 µg/m^3	Benzen µg/m^3	PM10 µg/m^3

CT1	38. 59	5.80	0.14	*	1.88	28.84
CT2	23.11	6.86	0.1	50.53	1.93	*
CT3	19.59	7.06	0.12	47.99	*	23.95

* The indicators in question were not measured at these types of stations

The annual limit value for health protection according to Law 104/2011 is 40 µg / m^3 [6]. According to Order 589/2018 in the period 2017 – April 2018, in Constanta municipality there were exceeded NO_X, NO_2, which leads to the inclusion in regime I. The annual limit value for SO_2 emissions protection of vegetation according to Law 104/2011 is 20 µg / m^3. Maximum daily value of 8-hour averages (sliding averages) for CO emissions for health protection according to law 104/2011 is 10 mg / m^3. Target value for ozone concentrations for the protection of human health, maximum daily average value per 8 hours, according to Law 104/2011 is 120 µg / m^3. The target value for benzene for the protection of human health, according to Law 104/2011 is 5 µg / m . The annual limit value for PM10 concentrations for the protection of human health according to Law 104/2011 is 40 mg / m^3. The target value for PM2.5 concentrations according to Law 104/2011 is 25 µg / m^3. The limit value for the protection of human health for Pb, according to the law 104/2011 is 0.5 µg / m^3. The target value for As, according to Law 104/2011 is 6 ng / m^3. The target value for Cd, according to Law 104/2011 is 5 ng / m^3. The target value for Ni, according to Law 104/2011 is 20 ng / m^3.

1.2 Monitored locations from Constanta urban site

In accordance with Art. 42, section 2, of Law no. 104/2011 regarding air quality "*in order to manage the ambient air quality in each area or agglomeration delimits areas that are classified in management regimes according to the result assessment of the ambient air quality* " [2], [6]. Two management regimes are defined: *management regime I* which represents the areas in the areas and the agglomerations in which the levels for sulfur dioxide, nitrogen dioxide, nitrogen oxides, particulate matter PM10 and PM2.5,lead, benzene, carbon monoxide are greater than or equal to the limit values plus the margin tolerance; *management regime II* which represents the areas in the areas and agglomerations in which the levels for sulfur dioxide, nitrogen dioxide, nitrogen oxides, particulate matter PM10 and PM2.5, lead, benzene, carbon monoxide are lower than the limit values, respectively for arsenic, cadmium, nickel, benzo (a) pyrene, PM2.5 suspended particles are smaller than the target values [6].

The classification of areas in urban areas and agglomerations into the management regime I or II was achieved taking into account both the previous classification in management regimes and the results obtained from the local air quality assessment, which used point measurements fixed, made between 2017 and 2018, with the help of a performance measuring station [5], [10].

2 Experimental research

The paper presents the following aspects: two monitoring regimes aims to evaluate and prevent air pollution in urban areas; was chosen as a monitoring location Constanta city, with strategic points exposed to the risk of pollution: the center of the city- the City Hall, the port areas- Gate 5 – Port Constanta, the industrial area- Cora mall-Road police (Blv. I.C. Bratianu).

2.1 The center of the city- the State Theater and City Hall

- *Location: Constanta City Hall*
- *Period*: 10.10.2017 – 27.10.2017
- The values (limit, hourly, average) of parameters are presented in Table 2 (CO-Fig. 3; CO_2-Fig. 4; NO_2 –Fig. 5; Nox-Fig. 6; PM2.5-Fig. 7; PM10-Fig. 8; SO_2 – Fig. 9).

Table 2. The recorded values between in the center of city

Parameter/Unit	Limit value	Average daily value	Maximum value	Hourly Value
CO mg/m^3	< 0.10	0.11	0.919	10
CO_2 ppm	432.40	439.30	444.23	438.80
NO_2 µg/m^3/hour	60	44.7..... 57.7	< 200	193
Nox µg/m^3	> 30	80.17... 114.22	> 30	178.80....231.87
PM2.5 µg/m^3	25	25.67... 30.68	43	36.89....42.75
PM10 µg/m^3	50	57.01	50	92.11 ...111.10
SO_2 µg/m^3	350** 125 µg/m^3***	125	20*	350

* annual critical level for vegetation protection, calendar year and winter (September 1 - March 31)
** the hourly limit value for the protection of human health;
*** the daily limit value for the protection of human health.

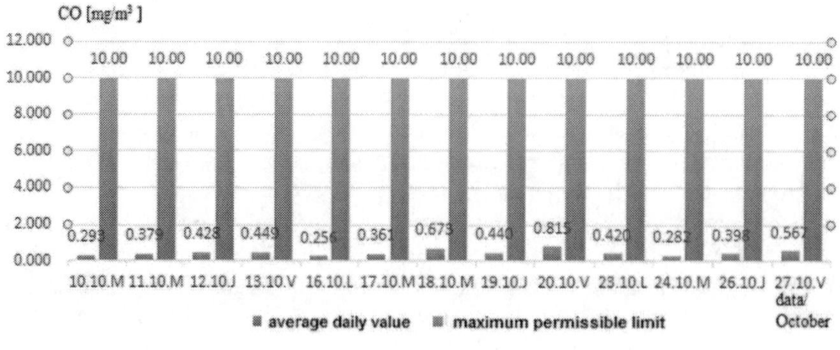

Fig. 3. CO-monitoring values in Constanta City Hall location.

CO - the values recorded are well below the allowed limit of 10 mg / m^3.

 https://doi.org/10.1051/e3sconf/202018004014

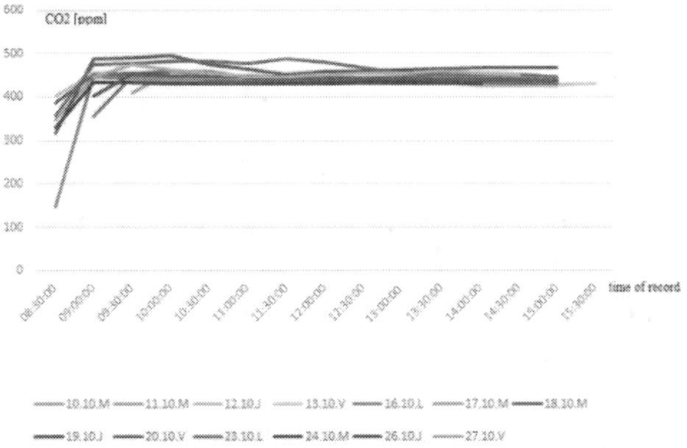

Fig. 4. CO_2-monitoring values in Constanta City Hall location.

Values are very close to all days and times. The average value per site is 439.30 ppm.

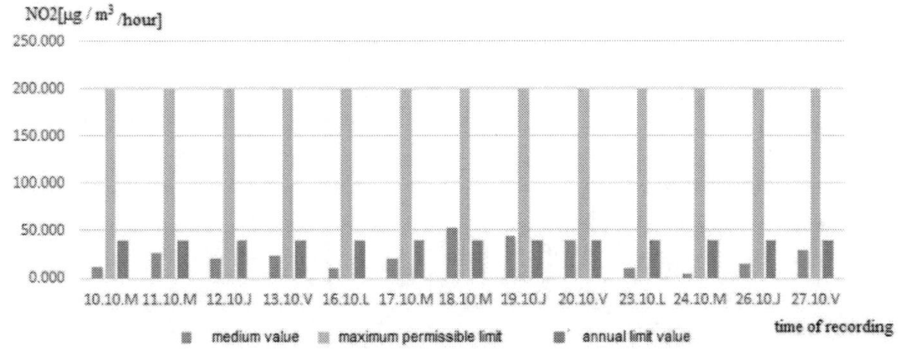

Fig. 5. NO_2-monitoring values in Constanta City Hall location-october 2017.

There were some small exceedances of 40 µg / m^3 (annual limit value for human health), respectively average daily values of 57.7 µg / m^3 and 44.7 µg / m^3.

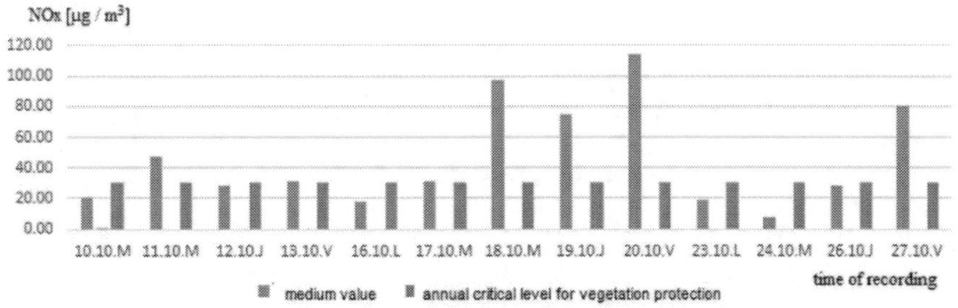

Fig. 6. NOx-monitoring values in Constanta City Hall location-october 2017.

Hourly and maximum hourly values were recorded frequently, as well as daily average values much higher than 30 µg / m^3. The maximum hours were recorded between 9.00 and 12.30.

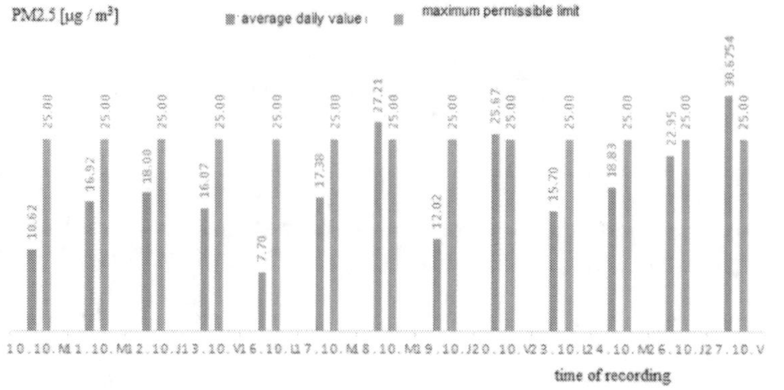

Fig. 7. PM2.5-monitoring values in Constanta City Hall location-october 2017.

Generally the hourly values and the daily average values are below the maximum allowed limit of 25 µg / m^3, however there are some exceedances to the hourly values. The maximum hours were recorded between 9.00 and 12.30.

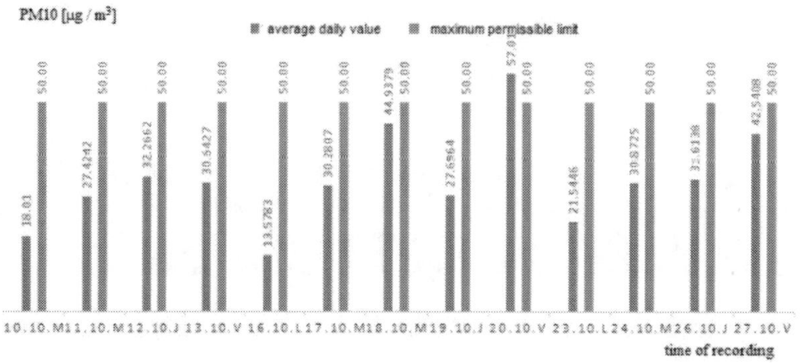

Fig. 8. PM10-monitoring values in Constanta City Hall location-october 2017.

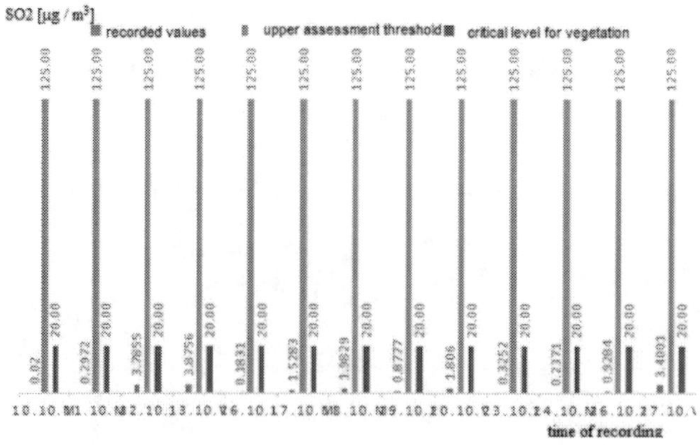

Fig. 9. SO_2-monitoring values in Constanta City Hall location-october 2017.

All recorded hourly SO_2 values and daily average SO_2 values are well below the maximum permitted limit of 350 µg / m^3 / hour. For recorded values (from 0.2371µg / m^3 / hour to 0.38756 µg / m^3 / hour) there are support possibilities decreasing up to one hour with increase in concentration. For human health, these values may cause possible irritation of the nose and eyes

2.2 The port areas- Gate 5 – Harbour Constanța

- *Location*: the port areas- Gate 5 - Harbour Constanta
- *Period:* 14.03.2018 - 31.03.2018/ 01.04.2018 - 26.04.2018.
- The values of air parameters are presented in Table 3 (CO; CO_2; NO_2; NOx; PM2.5; PM10; SO_2).

Table 3. The recorded values of air parameters in the port areaa Gate 5-Harbour Constanta

Parameter/Unit	Limit value	Average daily value	Maximum value	Hourly Value
CO mg/m^3	< 0.10	0.11	1.40 (March) 2.53 (April)	10
CO_2 ppm	459.24	459.24	459.24	459.24
NO_2 µg/m^3/hour	60	45.60 47.04 52.09	< 200	82.78 84.39 85.07 85.84 103.29
NOx µg/m^3	> 30	63.16 73.60 77.83 88.35 96.80	> 30	204.18 208.95 232.26 236.65 249.45
PM2.5 µg/m^3	25	31.92 33.00 35.75 38.30	< 25	49.12 57.2 71.61 79.82 113.40
PM10 µg/m^3	50	52.77 53.19 61.74 61.92 69.11 99.16	< 50	155.70 207.30 244.80 247.70 287.40 323.90 406.90
SO_2 µg/m^3	< 350	125	20*	350

- critical level for vegetation protection, calendar year

The registered values are below the permitted limit of 10 mg / m^3 (maximum 1.40 mg / m^3 in March and 2.53 mg/m^3 in April). The values of CO_2 are constant for all-day measurement.

2.3 The industrial area- Cora mall-Road Police (Blv. I.C. Bratianu)

- *Location*: the industrial area- Cora mall-Road police (Blv. I.C. Bratianu)
- *Period*: 27.04.2018 - 25.05.2018.
- The values of air parameters are presented in Table 4 (CO; CO_2; NO_2; NOx; PM2.5; PM10; SO_2).

Table 4. The recorded values of air parameters in the industrial area Cora mall-Road Police

Parameter/Unit	Limit value	Average daily value	Maximum value	Hourly Value
CO mg/m^3	10	0.11	0.717	0.10
CO_2 ppm	464.04	464.04	464.04	464.04
NO_2 $\mu g/m^3$/hour	40*		< 200	70.41 71.01 71.27
NOx $\mu g/m^3$	> 30**	32.91 34.25 34.92 36.71 42.57	> 30	100.41 106.67 113.198 151.50 116.85
PM2.5 $\mu g/m^3$	< 25	26.47 (April) 26.62 (4May; 18.30-0.00) 33.18 (24May; 0.30-03.00)	206.80 380.60	49.12 57.2 71.61 79.82 113.40
PM10 $\mu g/m^3$	< 50	53.27 59.89 81.55 92.62 269.11	< 50	79.14 82.26 100.50 121.60 270.30
SO_2 $\mu g/m^3$	< 350	125	20*	350

* annual limit value for human health
** annual critical level for the protection of vegetation

CO - registered values are below the permitted limit of 10 mg / m^3 (max. 1.40 mg / m^3 in March and 2.53 in April); CO_2 - average value per site is 459.24 ppm. The values of CO are constant for all-day measurement. The recorded values of NO_2 are below the maximum permissible limit of 200 µg / m^3 / hour. There were some exceedances of the hourly values compared to 40 µg / m^3 .

2.4 The risk areas

In carrying out the Air Quality Plan for the city of Constanța have data on NOx and NO_2 emission sources located in the city were interrogated Constanța and structured on the following categories:1) *fixed sources* - represented by individual or common fixed sources represented in the most of the installations of the authorized economic operators from the point of view environmental protection;2) *Surface sources* - represented by diffuse (undirected) sources of emissions released in the surrounding air;3) *Linear sources* - represent the emission sources specific to the means of road transport, railway and air, as well as non-road mobile equipment equipped with engines with internal combustion. Emissions of nitrogen oxides, nitrogen dioxide are released on the territory of the municipality of Constanța in urban areas (inhabited areas) and on industrial platforms. Once released into the air, the pollutants, due to the scattering phenomenon, they can be transported to different areas depending on the conditions meteorological conditions present. In order to manage the air quality, in each zone or agglomeration, the areas that are classified in the management regimes (I or II) were delimited according to the result of the evaluation of the ambient air quality. The management regimes are provided in art. 42 of Law no. 104/2011 [6]. Mathematical modeling of the dispersion of pollutants in the atmosphere consists in estimating concentrations of soil and height pollutants depending on the characteristics of the pollution sources, the meteorological and orographic, the processes of physical and chemical transformation that they may undergo pollutants in the atmosphere and their interaction with the soil surface [10]. Modeling the dispersion of pollutants into the atmosphere for emissions of pollutants generated by the emission sources within the municipality of Constanța were made with the AERMOD VIEW program, developed by the Canadian company Lakes Environmental [13]. The AERMOD model is a dispersion model, which allows long-term, medium and short calculation of emissions from point sources, traffic, surface sources and diffuse sources. For the municipality of Constanța, the dispersion model was run on network (13 x 13) km, with 500 m/ step. The data entered for the dispersion model are specific to 2017 [4]. The results obtained are: 1) wind rose and whether data series (Fig. 10); 2) graphic maps of the pollutant with the identification of average monthly or annual concentrations, hourly or daily concentrations, frequency of limit values according to legislative regulations; 3) data corresponding to maximum concentrations, concentrations in points network of receivers.

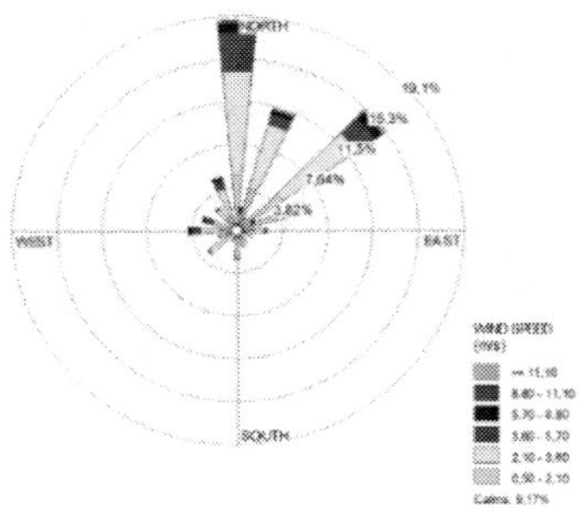

Fig. 10. Constanta wind rose and whather data series

Scenario:

-industrial sources- results of dispersion of average annual NOx concentrations in ambient air (Fig. 11, Fig. 12) and of average hourly NOx concentrations in ambient air (Fig. 13).

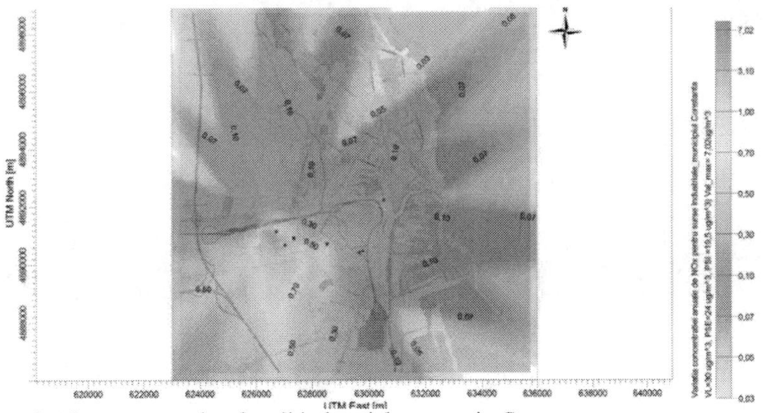

Fig. 11. Annual NOx concentration for all industrial sources in Constanta.

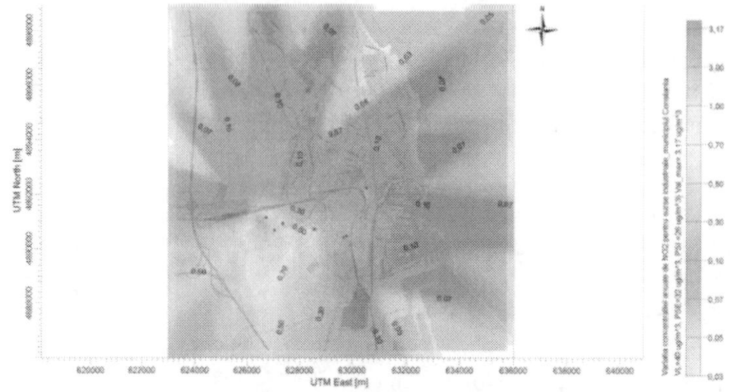

Fig. 12. Annual NO_2 concentration for industrial surface sources in Constanta.

The average annual concentration of NOx in the surrounding air for all sources, according to figure 11 is 7.02 µg / m^3, a value not exceeding the annual limit value, nor the upper threshold, but exceeds the lower threshold [2], [9].

The average annual NO_2 concentration in the surrounding air for the surface sources, according to figure 12, is 3.17 µg / m^3, a value that does not exceed the annual limit value and nor the lower and upper assessment threshold.

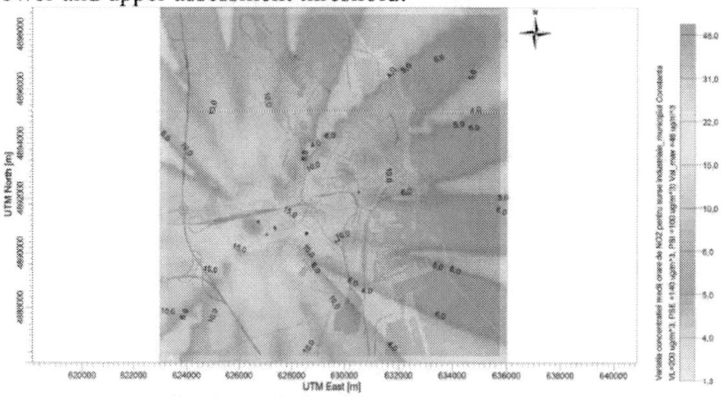

Fig. 13. Hourly NO_2 concentration for industrial sources in Constanta.

The average hourly NO_2 concentration in the ambient air for industrial sources, according to figure 13, is 48 µg / m^3, a value which does not exceed the annual limit value nor the threshold lower and upper evaluation.

- **liniar sources :**
- results of dispersion of average annual NOx concentration (Fig. 14), and of average hourly NO_2 concentrations in ambient air (Fig. 15).

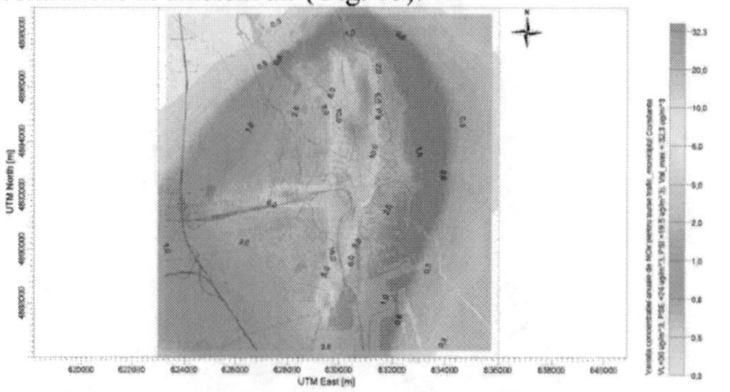

Fig. 14. Average annual NOx concentration for liniar sources in Constanta.

The average annual NOx concentration in the ambient air for linear sources, according to Figure 14, is 32.3 µg / m^3, a value that exceeds the annual limit value and the threshold. lower and upper evaluation.

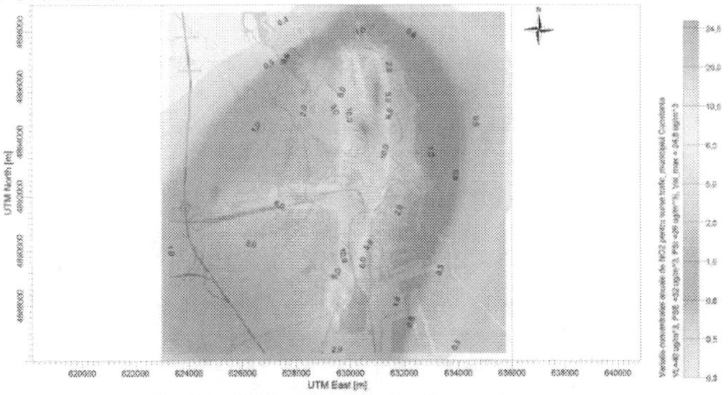

Fig. 15. Annual NO_2 concentration for linear sources in Constanta.

The average annual NO_2 concentration in the ambient air for linear sources, according to Figure 15, is 24.8 µg / m^3, a value not exceeding the annual limit value and the upper assessment threshold, but exceeds the lower assessment threshold.

- results of average hourly NO_2 concentration (Fig. 16).

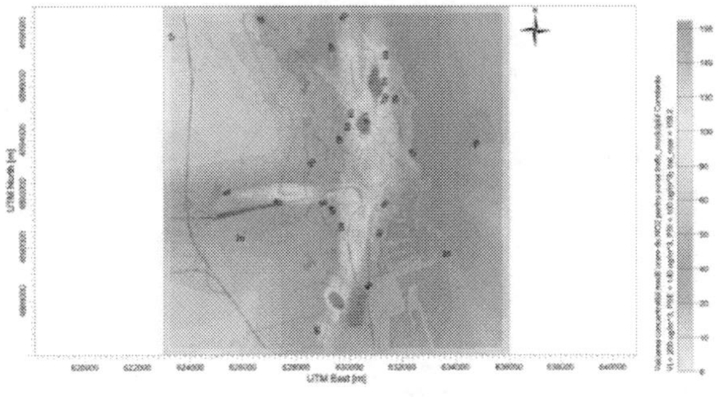

Fig. 16. Average hourly NO_2 concentration for linear sources in Constanta.

The average hourly NO_2 concentration in the ambient air for linear sources, according to Figure 16, is 158.2 µg / m^3, a value that does not exceed the annual limit value, but exceeds the lower and upper assessment threshold.

- **surface sources:**

- **r**esults of dispersion of average annual NOx concentrations in ambient air for sources of surface are presented in figure 17:

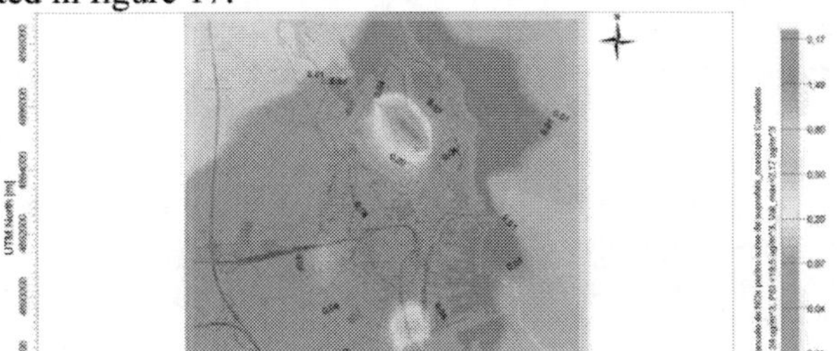

Fig. 17. Annual NOx concentration for surface sources in Constanta.

 https://doi.org/10.1051/e3sconf/202018004014

The average annual NOx concentration in the ambient air for the sources is surface, according to Figure 17, is 2.17 µg / m^3, a value that does not exceed the annual limit value and nor the lower and upper assessment threshold.

- The results of dispersion of average annual NO_2 concentrations in ambient air for sources surface are shown in Figure 18:

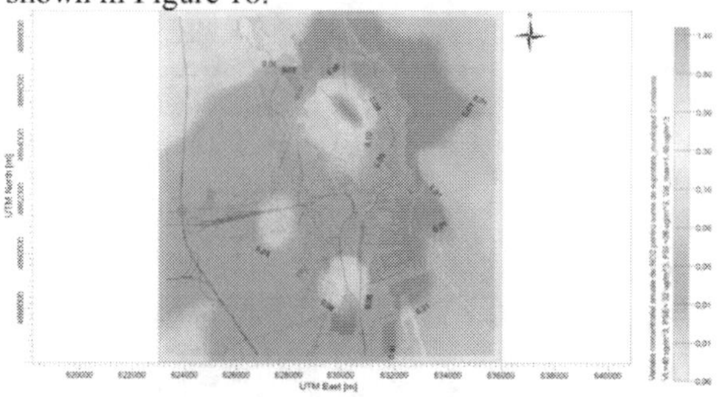

Fig. 18. Annual NO_2 concentration for surface sources in Constanta.

The average annual NO_2 concentration in the ambient air for the sources is surface, according to Figure 18, t is 1.49 µg / m^3, a value that does not exceed the annual limit value and nor the lower and upper assessment threshold.

- The results of the dispersion of the hourly average NO_2 concentrations in the ambient air for the sources of surface are shown in the Figure 19:

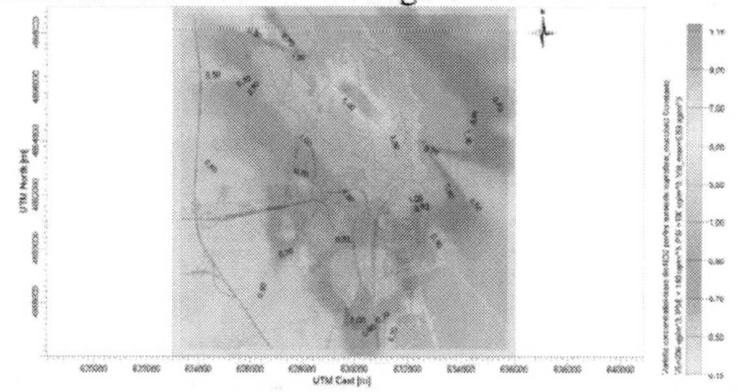

Fig. 19. Average hourly NO_2 concentration for surface sources in Constanta

The average hourly NO_2 concentration in the ambient air for the sources is surface, according to Figure 19, is 9.59 µg / m^3, a value that does not exceed the annual limit value and nor the lower and upper assessment threshold.

3 Conclusions

Air pollution has negative effects on materials and constructions, including those more representative constructions of cultural importance for Constanta. The air quality plan for the city of Constanţa is drawn up taking into account the employment regime according to Order 598/2018 by which the city of Constanta is classified in regime I with exceedances of NOx, NO_2 in the period 2017 - April 2018. the air quality level was analyzed for the following indicators: powders in suspension (PM10 and PM2.5), benzene (C6H6), sulfur dioxide (SO_2), carbon monoxide (CO), lead (Pb), arsenic (As), cadmium (Cd) and nickel (Ni).

In the Constanta urban environment the presence of nitrogen oxides is due in particular to road traffic. Once released into the atmosphere *NO* and *NO_2*, which are aerosols, are dispersed by the wind and, in areas with low speeds of air currents, depending on the density of the air, these gases will concentrate on the ground due to its higher density compared to air density. Acid rainfall on the ground can have effects different depending on the structure of the environment in which they fall, so that a calcareous land will be less affected compared to acidic soils composed mainly of silicon. Exposure of vegetation to nitrogen oxides causes plant damage through bleaching or death vegetal tissues, causing the fall of the leaves and reducing their growth rate. The short-term effects of P*M10* exposure on the respiratory tract are demonstrated, but from the point of view of mortality caused by long-term exposure, *PM2.5* presents a risk factor greater than PM10. The suspended particles accelerate the corrosion of metals, degrade paintings and sculptures and se deposit on buildings, contributing, sometimes even to the loss of heritage objects. The suspended particles negatively influence the development of flora and fauna. Above the flora acts by affecting the photosynthesis process, clogging the ostioles and disrupting the breathing, so that the plants do not grow sufficiently and the biological mass decreases. Suspended particles can absorb solar radiation by affecting or even reducing visibility. Another effect of particle pollution is to change the climate by forming clouds.

Following the modeling, it is found that for air quality implementation is necessary of measures at the emission sources that have the greatest contribution to air pollution, namely:

- for NO_2 and NOX - reduction of emissions from road transport, in particular by rehabilitation and modernization of transport infrastructure, ensuring mobility traffic, promoting the use of alternative means of transport (transport in common, bicycles, etc.);
- it is proposed to improve the bus fleet in order to reduce pollution, expand green spaces in Constanta.

This paper has financial support from the Constanta Maritime University under assistance project PN-III-P1-1.2-PCCDI-2017-0404 /31PCCDI/2018, *Holistic on the Impact of Renewable Energy Sources on Environment and Climate-HORESEC.*

References

1. Health risk assessment of air pollution – general principles. Copenhagen: WHO Regional Office for Europe, (2016)
2. J. Dragan, Air quality study and air quality plan for Constanța Municipality ANPM 1, 5-112 (2018)
3. World Health Organization Regional Office for Europe, " How far have we gotten?", *Improving environment and health in Europe XIV* 135, ISBN 978 92 890 5087 6, (2015)
4. Raport județean privind starea mediului, (2017)
5. ANPM, System procedure for validation of air quality data from the National Air Quality Monitoring Network, **PS-SCA-DCECA-02** 1-25 (2019)
6. Law no. 104 / 15.06.2011 on ambient air quality published in the Official Gazette of Romania, Part I, **452** - 28, (2011).
7. Directive 2008/50 / EC of the European Parliament and of the Council of 21 May 2008 on ambient air quality and cleaner air for Europe, published in the *Official Journal of the European Union* (OJEU) no. L 152 of 11 June 2008

8. Directive 2004/107 / EC of the European Parliament and of the Council of 15 December 2004 on arsenic, cadmium, mercury, nickel, polycyclic aromatic hydrocarbons in ambient air, published in the *Official Journal of the European Communities (OJEC)* **L 23** 25 (2005).
9. ANPM, National Air Quality Assessment and Integrated Management System (SNEGICA), **29** 5 (2020)
10. X. Lim et al., 2012, WHO, (2014)
11. F.V.Panaitescu, M.Panaitescu, C. Urda, N.A.Burcia, N.Grosan, A.Duse, Considerations about sustainable politics of rural developmen *Te-RE-RD Proc.* 367-372 (2016)
12. EEA, Air quality in Europe EEA Report **5** (2015)
13. AERMOD View™ "Gaussian Plume Air Dispersion Model", **9.9.0** (2020)

More on Smart Cultural Landscapes: technicalities of the planning processes

Alexandru Calcatinge[1*]

[1]"Ion Mincu" University of Architecture and Urbanism, Faculty of Urbanism, Department of Urban Planning and Territorial Development, 010014 Bucharest, Romania

Abstract. This paper will build upon my previous smart cultural landscapes research, by offering a technical overview of the planning processes that the policies are based on. In this respect, I will discuss, in an analytic manner, how the argumentation process should take place from a cultural landscape study. This will take into consideration the proficiencies of a planning professional, the smart characteristics of a cultural landscape policy or study and the relevant steps that should be taken to effectively solve the local rural or urban cultural landscape specific issues. The anticipated results would be to establish a well-known place of cultural landscape studies amongst the local rural and urban development processes.

1 Introduction

Various studies have shown that cultural landscapes and urban planning are closely related to the evolution of technology and communications, and thus it takes advantage of new technologies to achieve better solutions on the management side. [1] [2] [3]

In this respect, my research for the past two years has been focused on the Smart side of Cultural Landscapes. How to tackle with this concept and what it implies when it comes to the planning process at the local and regional level. In this paper I will focus on the planning and policy making processes and how this can further develop the cultural landscapes into being 'Smart'.

The urban planning process is a strange one. It is overly complex, mostly unclear, and sometimes rusty. In this respect, a major aspect to be discussed in this paper would be the preparation of two new legislative proposals in Romania: The Cultural Patrimony Code and the Territorial Development, Urbanism and Construction Code. Both are in their early stages but are bound to bring significant changes to the urban legislature. Besides those two "case studies", I will bring to the table some of the most prominent planning and cultural schools of thought, to have a complete and wide view on the technicalities of the planning processes.

Thus, having this unique opportunity to be in a time when ground-breaking changes are on the horizon, we must, as an utmost importance, see things into perspective and prepare an integral planning process in urbanism and cultural landscape studies. This will be the

*Corresponding author. E-mail address: alexandru.calcatinge@gmail.com

smart component that is currently missing. Throughout this paper you will find terms like urban and urban planning be used more than cultural landscapes, and I feel like I need to clarify this. In the process of creating urban planning projects and documentation, the landscape study is merely a component of the entire stack, with cultural landscape studies being an even smaller one. Thus, I will mostly refer to the planning process as being urban, and this will automatically include the cultural landscape documentation. What are cultural landscapes? The answer to this question will surely extend outside the scope of this paper, but I would just like to reiterate two of my findings on cultural landscapes, as a noticeably short answer to it. First, "the contemporary urban cultural landscape is a space of identity shaping, active and dynamic. Its perception is complex, of phenomenological nature, being both a cultural process and cultural product. The process of social life is directly integrated in the notion of urban landscape" [4] and "the cultural landscape is thus the mirror of the society reflecting all the attributes, features and characteristics of our existence by means of spaces created based on the knowledge acquired in time". [5]

1.1 The Problem

First and foremost, let us define the problem we are trying to solve. In a short version, the problem is > the urban (and cultural landscapes) planning process is in danger of becoming irrelevant. Is it really a planning problem, or a development problem? To define it in a definitive matter, we should first know how urban and cultural planning are relevant for the development processes. In this respect, let us take the Romanian example. In our country, urban development is a process based on the outcomes of the rules from local/regional/national urban policies. Thus, the relation is clear, and the solution lies in the way we manage to integrate all the existing material, using clear procedures, integrated and integral thinking on urbanism and cultural landscapes, and a clear way to build an urban argumentative discourse. This is what makes them smart.

1.2 The Solutions

This paper will provide one possible solution to the problem stated earlier. Thus, as I see this issue, it could be structured as follows. First, we need to define a culture of planning to determine a feasible and relevant urban practice. Then, we need to create a relevant planning theory that could be the foundation for the local urban practices and schools of thought. Thirdly we need to have an integral approach to urban and cultural landscape planning. Then we need to define the critical aspects of growth and the future of planning and development by establishing a smart way to do planning, that is the culmination of all the solutions enlisted above.

2 Methodology

The problem we are shortly discussing here is not going to receive all the possible solutions inside a short paper. It is more suitable to an extended research that could span over several years. What this means is that I find the most suitable research methodology to be a modular one. This paper would be just a module of that research. Modularity would be the foundation of a much wider solution, and each research finding will form the layers of a more complex and dynamic solution. Thus, this module would be about technicalities, the very base of every planning process.

2.1 Research strategy

The strategy followed in this paper will be built following these characteristics:

- Will follow on a clear set of principles
- Will have clear research questions
- Will tackle with all the essential issues of each module
- Will be the base for experimentation and teaching

As it seems, it is straightforward and clear, and all we must do now is to determine the principles to follow. But first, let us determine the relation between theory and methodology. Our research strategy should reveal the process that we follow to determine what we want to show with this study. According to Donald O. Case, "both theory and method must be concerned with our type of explanation (e.g., is our goal to demonstrate that one variable influences another?) and our approach to inquiry (e.g., whether we use a theory to guide observation, or whether we start making observations to build a theory)" and that "methodology is not concerned with substantive assumptions about what is being studied, which is a matter for theory (or, more specifically, for metatheory). Only theory addresses the basic assumptions that we make about the nature of reality". [6] Thus, our concern is to determine the "attempts to formulate possible factors that could be varied systematically in such a way that a corresponding variation in other variables might be observed seemed logical enough". [7] The variables we are working with are planning and culture, and a variation of one, will most certainly determine an equal variation in the other.

Our principles to follow:

- Will not look for a full coverage of the problem in such a short paper
- Will keep a modular approach throughout the paper
- Will show flexibility in dealing with complex subjects
- Will slowly build our argumentation upon layers of systematic research
- Will offer explanations on every layer of research

2.2 Methods and materials used

According to Case, "one way to conduct research that is both valid and reliable is to be found in the use of multiple methods and multiple sources of data". [8] In our case, the mostly used method will be the qualitative one, with a twist. First, we will use it to analyse and gather non-numerical data and to gain some insights about motivations and reasons behind ideas and hypotheses that we will encounter over our research.

We will not use the quantitative research method, as it does not offer us any leverage in getting a solution for planning and culture. At least not at this stage and particularly not for this module.

Regarding the study materials, we will focus on specific literature on planning theory, cultural landscape studies and technology.

3 Results

As we saw by now, our main research problem would be the irrelevance of cultural landscape and urban planning processes, and in order to get to the solution, we will have to start from the very beginning, which is the understanding of planning. This will lead us to the point where we will be able to sketch a relevant planning theory. At this point, the planning theory would give us a lead on how an integral approach on urban and cultural landscape planning should look, which nevertheless would point to a sustainable and healthy growth. At the end, the results of the research will focus on the technical side of planning and will establish a foundation for future endeavours in the fields of data science, blockchain and smart everything (cities, cultural landscapes, homes etc.).

3.1 Culture of planning

Urban planning as we know it today has its roots in the architecture and urban planning from the end of the Second World War and was closely related to the consumer culture and the way that architects and public administrations saw the need to seize the opportunity to build something new. According to Andrew Shanken, in the United States "architects embraced planning in the cultural context of the home front, when New Deal, wartime, and post-war planning overlapped". [9] Furthermore, the same Shanken stated that "As Americans lived through the transformation from New Deal to wartime to post-war planning, the city served as a site, both literally and abstractly, for thinking through potential socioeconomic, political, and physical changes". [10] At the same time, in the UK, as Nigel Taylor argues, there is a slight different approach as "after the Second World War, there was much talk of 'planning' in a general sense – that is, state intervention in, and playing a more active role in, the managing and planning of social and economic affairs generally as part of the changed political climate. (…) The prevailing view was that, with the possible exception of regional planning controls over the industry, town and country planning was concerned with the 'physical' environment and was thus most appropriately described as physical planning, as opposed to 'social' and 'economic' planning". [11]

Those two models were the blueprints for planning evolution in the western world. But on the other side of Europe, the end of the Second World War was the point when planning started to follow the Soviet model. This model failed and after the fall of the Berlin wall, the gaps between East and West started to widen even more. It was the time when changes and a lot of catching up had to be done. As "the general plan has been the object of much criticism in view of its evident failure to guide urbanisation and urban development in the required manner". [12] Romania had to follow the Soviet rules, for more than forty years, and now, thirty years since the 1989 revolution, we are still trying to impose new models for planning. The new urban general plan for the capital city started in 2013 and is still in the preliminary stages.

Anyhow, the two planning systems, the eastern and the western one, are quite different and proven hard to change and adapt to new realities, in the case of the soviet model. Those systems have already determined a culture of planning for each country or region that uses them. An history of the culture of planning is out of the scope of this paper, but the most important aspects will be shortly discussed here. One of the interesting aspects is the way architects shifted into planners, and the way Andrew Shanken sees it is important for the overall picture. He argues that "architects groused over the many threats to their profession, including old enemies, such as builders, contractors, and engineers, and new ones, especially government and corporate design bureaus". [13]

The battle was for a clear establishment of the profession into a very diffuse social, economic, and legal landscape. Shanken [14] has a short comparison between doctors, architects, and planners regarding the monopoly status of professional associations, that will establish a tight nexus between major forms of professional activities, from universities to small practices. Each profession created its own association, and between the three examples, the doctors were the firsts to impose regulatory measures for the medical practices. Then followed the architects with their plethora of professional associations that led to a failure to gage all activities. The planner's professional association had an even more difficult start and a short life until the end of the Second World War.

Thus, a new paradigm started, as "the entire western world seemed to be in the throes of a systemic shift marked by, among other things, a move toward centralized planning" [15] and it "exhibited figurative qualities, standing in for other ideas, such as the future, better living, comfort, order, and most empathically, the potential for a radical change in the social structure". [16] The eastern planning model was never something to write home about, and all the countries from the former Eastern Bloc have adopted the western model.

In this respect, the new general plan for Bucharest is designed to be a dynamic one, like a pro-active organism, a tool for monitoring, transaction and updating for local territorial beeswax. Romania took the planning model from France after 1989 and failed to improve it and build upon the good points. Nevertheless, professionals in urban planning are mainly concluding that the main general plan needs to have a reactive and pro-active component, to be able to react fast as an important tool in guiding development at different stages, thus being a dynamic regulatory instrument. [17] For the city of Bucharest, they are developing "a three-tiered planning regulation system aimed at governing developments occurring within the city" [18] and are building their conceptual and theoretical framework on three distinct theories: the theory of metropolitan functions, the theory of demand oriented urban life cycle management and science of networks. [19] The concept of the new masterplan lies, as the authors say, at the intersection of urban planning, structural field, functional field, and GIS. [20]

Fig. 1. The concept of the New Dynamic Master Plan for Bucharest. Source: T. Florescu (et. al), *The architecture of dynamic planning regulations*, Book of Proceedings, 29th Annual AESOP 2015 Congress, July 13-16, Prague, Czech Republic, pp. 2761 – 2769, p. 2764 (2015)

Therefore, what would the culture of planning be in our case? First and foremost, as architects, I see that the distinctive competence is our ability to deal with some of most complex forms and surfaces, to create a functional and liveable space. We deal with space as a metaphor of culture, an abstract entity that we can manage and transform as a tool of social creation, with cultural and economic relations. Furthermore, as urbanists (as we call the profession in Romania), we can configure spaces from micro to macro levels, from local to regional and national levels. We work with different elements such as climate, geographical data, culture, tradition, mentality, demography etcetera, which can be considered as datums; with other elements as actions, public interventions, material actions; with politics, which are the underlying actions that determine the actions, public or private; and elements like institutions, the forefront of urban/regional enactment.

Planning is a tool used to forecast and develop and has political and legal dimensions. Our culture of planning is comprised of different tools that form a planner's literacy. According to Greg Young, it has three distinct parts: a cultural literacy, an ethical literacy, and a strategic literacy. [21] The last one is the kernel of a professional planner, be it architect, urbanist, or any other profession. Nowadays strategic planning is an important component of any planner's toolkit, as it needs to create plans that integrate the objectives and resources with opportunities at all levels. The urban planner needs to be able to operate with public policies and legal instruments to fulfil his purpose. This makes the planner a good coordinator between law and decision makers, local / regional / central authorities, NGOs, and private sector actors.

The urban planner, through his documentation, needs to build relevant arguments that sustain its endeavour. Those arguments need to be regulated, and this is exactly the scope of the Cultural Patrimony Code and the Territorial Development, Urbanism and Construction Code. And for this, we need a relevant planning theory, which we will discuss in the following sub-chapter.

3.2 Relevant planning theory

Now that we know how the culture of planning has emerged, let us talk about relevant theory of planning. Dealing with all the models, concept and theories of planning is out of the scope of this paper, but we will refer to some that we personally find relevant for our contemporary practices. According to Philip Allmendinger the most important theories are systems theory, rational process theory, critical theory, neo-liberal planning theories, pragmatism, advocacy, postmodern planning and post-structuralism and communicative rationality. [22]

Besides Allmendinger, Nigel Taylor also theorized about the evolution of planning theory, about changes in planning thought and paradigms, and in the end the conclusion was simple: it all ends and starts with Faludi's procedural planning theory, as a possible singular planning theory that is constantly discussed, acclaimed or critiqued [23]. One critique comes from Michael J. Thomas, which stated that Faludi's theory is "overwhelmingly focussed on the means of planning and not the ends". [24]

In a paper from 1985, Faludi answers some of its critiques by stating that "it concerns the nature of theory and the object of planning theory as an academic discipline. For good reasons, professionals are convinced that they need both procedural and substantive knowledge". [25] Further sustaining his theories, Faludi sets two important questions to be asked by any planning professional. First is "how does planning reflect the nature of its object of concern" [26] and the second is "which aspects of the environmental demand that we should attend to the inter-relations between such interventions in the environment as o take place". [27] In order to make things simple for other researchers, Faludi answers the questions with his vision on how the environment should be seen. Thus, he proposes that the environment should be seen as a certain configuration of resources that are commonly found on land decision units, link to each other by channels of communication and protected by the land regime; and the environmental planning should be seen as the sum of operational decisions and public environmental measures. [28]

For more than thirty years, those principles were always at the forefront of planning theory, but the question that lies now is how will the planning be in the future? For this question, Allmendinger had a list of nine principles upon which planning should be built. [29] I will not remind all of them here, but I just want to focus on some of them. Thus, his second principles states that "a postmodern planning should be based on an underlying assumption that all processes and procedures are not 'closed', that is, there is no 'best' plan for every situation". [30] Further, three of the principles support the openness of the planning process, in order to keep a regularly maintained and updated framework and the last one states that "planners should take a more active and creative role in the sustenance and encouragement of fluid structures and processes". [31]

And this is not all. Besides Allmendinger, another author, Greg Young argues that culture is the salvation of planning. Thus, he said that "in order to revive, planning first needs to capture something of the true dynamism of culture, beyond the fragmented and superficial reflection of culture it usually achieves. At every planning level, culture is addressed in a conceptually fragmented, ad hoc, and frequently opportunistic fashion, that brings cultural inauthenticity to the fore". [32]

For our immediate scope, we need an integrative approach to planning, in which we need coordination between urban planning professionals and legal practitioners. This implies that the urbanists need to have a solid understanding of legal instruments, and urban concepts and jargon should be at the fingertips of any urban lawyer.

The need to codify the urban legislation is imperative, and this can only be done if urban planners, on one side, will adopt a coherent planning theory that is grounded in well-established works of theoreticians and practitioners alike, and the legal apparatus will be clarified and simplified in order to support transparency for the implementation and evaluation processes.

3.3 The smart way

Throughout my recent research, I wanted to determine the relations between cultural landscapes and open source paradigms, open innovation, and big data, as a possible smart solution for future of planning in urban / regional development. [33]

As an advocate of open source solutions for many years, I try to impose its paradigms in the current planning processes. Most important is that these paradigms are governed by two principles: accessibility and flexibility. But the future is not only open source but also a new technology that is arising and looks promising to all types of operations: blockchain. This new technology is innovative and difficult to comprehend for a non-technical mind, but some authors managed to create non-technical guides to this technology, which is extremely important for us as architects and urbanists. But what is blockchain? Well, it emerged as the technology behind crypto currencies and now is also used in proof-of-work and proof-of-stake distributed consensus mechanisms, crypto token economics, Ethereum and smart contracts, generic crypto token systems, and byzantine fault tolerance distributed consensus. [34] Nevertheless, what makes it a little bit easier to comprehend for us is our knowledge of systems theory and its relationship with planning.

For the start, let us remember a little bit about systems theory in urban planning, a concept that emerged in the 1970s and quickly gained speed into being adopted by professionals. Thus, J. Brian McLoughlin, enchanted by the evolutions in mathematics and computer processing from that time, introduced the concept of systems theory to urban planning "to provide a framework by which interested people may relate the new tools from many fields to the problems of planning cities and regions". [35] Furthermore, George Chadwick is dealing with the same concept and treats it from the same perspectives as McLoughlin, but he considers that "spatial planning, is thus seen as a more particularised branch of a general discipline, whose field may be said to impinge upon psychology, upon scientific method generally, upon general system theory, upon cybernetics, upon operations research and logistics". [36] Thus, for more than fifty years, planning was closely related to the new technologies, and scholars constantly tried to tie those two together. In this respect, Faludi considered that planning itself should be subject to planning and all boiled down to "describing a method by which decisions should be made between alternative forms of planning – or let us call them alternative types of space in which the solution to problems must be sought". [37]

Today, based on the blockchain revolution, we are getting a new perspective on systems theory, as "the purpose of the blockchain is to achieve and maintain integrity in distributed systems (...) a distributed system consists of a number of independent computers that cooperate with one another by using communication medium in order to achieve a specific objective without having any centralized element of control or coordination" [38] Blockchain is considered to be "one of the most disruptive technologies of our time. Numerous cities around the world are launching blockchain initiatives as part of the overall efforts toward shaping the urban future. However, the infancy stage of the blockchain industry leads to a severe gap between the knowledge we have, and the actions urban policy makers are taking" [39] The possibilities are endless with this technology, and most certainly it can be used in implementing urban regulations and documentations. In this respect, there are several classifications of blockchain use cases in urban planning: role based, and business model based with applications such as static registry, identity, smart contracts, dynamic registry, payments infrastructure, record keeper, digital asset market. [40] [41]

In a world dominated by blockchain technology and backed up by the open source paradigms that have been taken the world by storm [42], it would be a lot easier to enable a coherent argumentation in urban planning. And coordination, transparency, decentralization, and simplification should be the most important characteristics.

4 Conclusions

Looking back to the steps taken to determine the planning process in urban and cultural landscape studies, we can deduct that they were small and focused. The way to make urban planning relevant again is to rethink urban argumentation, to adapt it to dynamic needs of the society by defining a new culture of planning and to adapt the theory to new technologies and techniques. Besides this, it also implies a well-developed planner's literacy based on culture's principles like plenitude, connectivity, diversity, reflexivity, creativity, critical thinking, and sustainability. Thus, urban planning and cultural landscape studies are closely related, as both are based on a similar set of shared documentation structures, and the relevance of one lead automatically to the relevance of the other.

In the end, it al points to coordination and growth. Coordination at the urban planning level is not just informational exchange, it is more of a structuring of data and a coherent way to advocate it, as public administration, legal entities and urban planner all use different dialects of the same language. This should be much easier using new technologies like blockchain, to have a more transparent and decentralized process.

The future of urban planning and cultural landscapes will be closely related to the future of the cities as 'smart' dynamic entities based on smart economy, smart people, smart governance, smart mobility, smart environment, and smart living. For short, smart everything.

References

1. Ch. Garau, *From Territory to Smartphone: Smart Fruition of Cultural Heritage for Dynamic Tourism Development*, Planning Practice and Research, **29**:3, pp.238-255 (2014)
2. Ch. Garau, Smart paths for advanced management of cultural heritage, Regional Studies, Regional Science, **1**:1, pp.286-293

3. N. G. Blagovidova, N. V. Iudina, *'Smart City' system as an asset of cultural landscape development*, IOP Conference Series: Materials Science and Engineering, **775** (2020)
4. Al. Calcatinge, *The need for a Cultural Landscape theory: an architect's approach*, Urban and Spatial Planning, Volume **12**, LIT Verlag, Vienna, p.88 (2012)
5. Al. Calcatinge, *The need for a Cultural Landscape theory: an architect's approach*, Urban and Spatial Planning, Volume **12**, LIT Verlag, Vienna, p.76 (2012)
6. C. O. Case, *Looking for information. A survey of research on information seeking, needs, and behavior*, 3rd Edition, p.202 (2012)
7. L. P. Steffe, P. W. Thompson, *Teaching experiment methodology: underlying principles and essential elements*. In: R. Lesh & A. E. Kelly (Eds.), Research design in mathematics and science education (pp. 267-307). Hillsdale, NJ: Erlbaum, p. 270 (2000)
8. C. O. Case, *Looking for information. A survey of research on information seeking, needs, and behavior*, 3rd Edition, p.211 (2012)
9. A. M. Shanken, *194X: Architecture, Planning, and Consumer Culture on the American Home Front*, University of Minnesota Press, Minneapolis, p. 15 (2009)
10. A. M. Shanken, *194X: Architecture, Planning, and Consumer Culture on the American Home Front*, University of Minnesota Press, Minneapolis, p. 16 (2009)
11. N. Taylor, *Urban Planning Theory since 1945*, Sage Publishing, London, p. 5 (2009)
12. D. J. B. Shaw, *The Soviet Urban General Plan and Recent Advances in Soviet Urban Planning*, Urban Studies, volume: **20**, Issue: 4, page(s): 393-403, p. 393 (1983)
13. A. M. Shanken, *194X: Architecture, Planning, and Consumer Culture on the American Home Front*, University of Minnesota Press, Minneapolis, p. 3 (2009)
14. A. M. Shanken, *194X: Architecture, Planning, and Consumer Culture on the American Home Front*, University of Minnesota Press, Minneapolis, pp. 7-8 (2009)
15. A. M. Shanken, *194X: Architecture, Planning, and Consumer Culture on the American Home Front*, University of Minnesota Press, Minneapolis, p. 15 (2009)
16. A. M. Shanken, *194X: Architecture, Planning, and Consumer Culture on the American Home Front*, University of Minnesota Press, Minneapolis, p. 16 (2009)
17. *O decadă de planificare urbană în România*, 30 june 2014 online: https://arhitectura-1906.ro/2014/06/o-decada-de-planificare-spatiala-in-romania/ accessed on 10th aprile 2020
18. T. Florescu (et. al), *The architecture of dynamic planning regulations*, Book of Proceedings, 29th Annual AESOP 2015 Congress, July 13-16, Prague, Czech Republic, pp. 2761 – 2769, p. 2761 (2015)
19. T. Florescu (et. al), *The architecture of dynamic planning regulations*, Book of Proceedings, 29th Annual AESOP 2015 Congress, July 13-16, Prague, Czech Republic, pp. 2761 – 2769, p. 2764 (2015)
20. T. Florescu (et. al), *The architecture of dynamic planning regulations*, Book of Proceedings, 29th Annual AESOP 2015 Congress, July 13-16, Prague, Czech Republic, pp. 2761 – 2769, p. 2764 (2015)
21. G. Young, *Reshaping Planning with Culture*, Urban and Regional Planning and Development Series, Ashgate, Burlington, pp.106-110 (2008)
22. P. Allmendinger, *Planning Theory*, 2nd Edition, Palgrave Macmillan, Hampshire (2009)
23. N. Taylor, *Urban Planning Theory since 1945*, Sage Publishing, London, p. 167 (2009)

24. M. J. Thomas, *The procedural planning theory of A. Faludi*. Journal of Environmental Planning and Management, Planning Outlook, Volume: **22**, Issue: 2, pp. 72-76, p.72 (1979)
25. A. Faludi, *A decision-centred view of Environmental Planning*. Landscape Planning, Volume: **12**, pp. 239-256, p. 240 (1985)
26. A. Faludi, *A decision-centred view of Environmental Planning*. Landscape Planning, Volume: **12**, pp. 239-256, p. 240 (1985)
27. A. Faludi, *A decision-centred view of Environmental Planning*. Landscape Planning, Volume: **12**, pp. 239-256, p. 241 (1985)
28. A. Faludi, *A decision-centred view of Environmental Planning*. Landscape Planning, Volume: **12**, pp. 239-256, p. 241 (1985)
29. P. Allmendinger, *Planning Theory*, 2nd Edition, Palgrave Macmillan, Hampshire, pp.238-239 (2009)
30. P. Allmendinger, *Planning Theory*, 2nd Edition, Palgrave Macmillan, Hampshire, p.238 (2009)
31. P. Allmendinger, *Planning Theory*, 2nd Edition, Palgrave Macmillan, Hampshire, p.239 (2009)
32. G. Young, *Reshaping Planning with Culture*, Urban and Regional Planning and Development Series, Ashgate, Burlington, p.5 (2008)
33. Al. Calcatinge, *Open source paradigms in urban and rural development*, 6th International Conference on Thermal Equipment, Renewable Energy and Rural Development TE-RE-RD 2017 Proceedings, pp. 323 - 329 (2017)
34. C. Shen and F. Pena-Mora, *Blockchain for Cities – A Systematic Literature Review*, IEEE Access, Volume **6** (2018)
35. J. B. McLoughlin, *Urban and Regional Planning. A systems approach*, Faber and Faber, London, p.17 (1969)
36. G. Chadwick, *A systems view of planning. Towards a theory of the urban and regional planning process*, 2nd Edition, Pergamon Press, pp. 24-25 (1978)
37. A. Faludi, *The "Systems View" and planning theory*, Socio-Econ. Plan. Sci. Vol. **7**, pp. 67-77, p. 75 (1973)
38. D. Drescher, *Blockchain Basics: A Non-Technical Introduction in 25 Steps*, Apress, p. 17 (2017)
39. C. Shen and F. Pena-Mora, *Blockchain for Cities – A Systematic Literature Review*, IEEE Access, Volume **6**, p. 76787 (2018)
40. C. Shen and F. Pena-Mora, *Blockchain for Cities – A Systematic Literature Review*, IEEE Access, Volume **6**, p. 76810 (2018)
41. C. Shen and F. Pena-Mora, *Blockchain for Cities – A Systematic Literature Review*, IEEE Access, Volume **6**, p. 76787 (2018)
42. G. Haff, *How Open Source ate Software. Understanding the Open Source Movement and So Much More*, Apress (2018)

Study of Sequencing Batch Reactor operation with an inflow of wastewater with low concentration of organic contaminations

Elena Gogina[1,*]

[1]NRU Moscow state university of civil engineering (NRU MGSU), 129337, Jaroslavskoje sch. 26

Abstract: The paper is devoted to the study of small wastewater-treatment plants, in particular the study of wastewater treatment processes with the wastewater with low concentration of organic contaminations entering treatment plants. The primary goal of the study is to select the reactor operation mode to ensure the deep biological treatment quality of treated wastewater. The three-step experiment has been conducted in a licensed laboratory using high quality equipment. The results of the experiment and the analysis of the data obtained are presented.

1 Introduction

The environmental pollution and the contamination of water bodies in particular is the major global problem at the moment. Wastewater treatment facilities are one of the main sources of pollution, due to the fact that new wastewater treatment plants are scarcely being built, and the technical re-equipment and reconstruction of industrial and municipal wastewater treatment facilities is one of the most difficult engineering tasks [1-4].

The amount of wastewater delivered to aeration and activated sludge plants in the Russian Federation in many cases exceeds their rated capacity, resulting in deteriorating treatment quality and sanitary conditions of water bodies. In some cases, which have become more frequent recently, the pace of housing construction is constrained by the lack of reserve and unwillingness to reconstruct drainage systems and wastewater treatment facilities. The situation is further aggravated by the lack of available space for the expansion of the treatment plants complex. Therefore, it is necessary to reduce the cost of reconstruction methods as much as possible and place the reconstructed facilities in the previously allocated areas. These issues require significant innovations. Such innovations include a Sequencing Batch Reactor (SBR), which allows to combine two facilities: an aerotank and a secondary sedimentation tank. If concentration of suspended solids and organic matter in wastewater is low, there will be no need for a primary sedimentation tank. This will allow doubling the capacity of the wastewater treatment plant without using additional territory [5-8].

The use of SBR technology allows to easily regulate and, if necessary, quickly change the time of stay of treated water in the bioreactor, the activated sludge concentration, the

* Corresponding author: goginaes@mgsu.ru

sludge load, its age, the concentration of dissolved oxygen, sedimentation time, loading and unloading. All technological operations in the bioreactor can be performed either according to a given time program or according to the oxygen concentration sensor, i.e. oxygen consumption.

The aim of this paper is to study the Sequencing Batch Reactor operation mode and to select the most suitable technological mode of reactor operation with an inflow of wastewater with low concentration of organic contaminations.

The experiment consisted of studying the Sequencing Batch Reactor (SBR) operation modes, using a laboratory model, and conducting a series of sanitary and chemical analyses.

The sequencing batch reactor (SBR) technology of wastewater treatment is quite promising and widespread in Western Europe. The system is based on the Sequencing Batch Reactor. Unlike traditional methods when water flows through several tanks or chambers, in the SBR system all cyclic stages of purification take place in one tank – bioreactor. The SBR system operates on the principle of activated sludge accumulation. In the bioreactor, all contamination from the drain is absorbed by floating activated sludge microorganisms and turns into biomass.

The principle of the SBR operation is based on a cyclic change of wastewater treatment processes occurring in one reactor. Each cycle consists of certain phases, the duration and sequence of which is set according to the required biological treatment effect.

The main advantage of SBR is that all biological treatment processes (oxidation of the organics and nutrients, as well as sedimentation) take place in the same reactor, which makes it possible to regulate the duration and sequence of the treatment phases based on the concentration of wastewater, entering treatment facilities, with the use of automated control systems [9-14].

This technology allows receiving drainage water with high coefficient of flow irregularity and scarcely depend on the quality of the incoming water. The use of SBR technology allows to easily regulate and, if necessary, quickly change the time of stay of treated water in the bioreactor, the activated sludge concentration, the sludge load, its age, the concentration of dissolved oxygen, sedimentation time, loading and unloading. All technological operations in the bioreactor can be performed either according to a given time program or according to the oxygen concentration sensor, i.e. oxygen consumption.

The second peculiar feature of the SBR technology is the retention of the settled activated sludge in the bioreactor after the end of the wastewater treatment period. The settled sludge volume amounts to 35–45 % of the total volume of the bioreactor, its ash-free biomass matter concentration is in the range from 12 to 15 kg/m^3. By selecting or retaining excess sludge in the bioreactor, the concentration of activated sludge is being corrected with each new portion of the treated drainage water. Thus, the working concentration of activated sludge, its age and sludge load are regulated within the necessary limits corresponding to changes in composition or concentration of contaminations in the wastewater.

The third feature of the SBR technology is an autonomous system of sludge mix aeration in the bioreactor. Aeration is performed by mechanical turboaerators on a floating platform or by bottom hyperbolic stirrers with air supply from an individual compressor. Due to the use of the variable-frequency drive, the speed of turbines or stirrers can vary in a wide range and at very low speeds (2,050 rpm) [15,16].

The fourth feature of the SBR technology is a cyclic schedule of air supply to the bioreactor with alternating periods of intensive sludge mix aeration (organics bio-oxidation and nitrification) and periods of slow mixing without oxygen supply, denitrification, biological phosphorus removal). Depending on the concentration of contaminations in the wastewater, the sludge mix oxidation schedule can be changed rapidly, e.g. increasing the aeration time by reducing the mixing time or vice versa.

Another distinguishing feature of the SBR technology is that the same type of bioreactor can be used for wastewater treatment with different operating cycles: 8, 9, 12, 15, 16, 18, 21 or 24 hours depending on the degree of wastewater contamination and the contaminations oxidation degree [17].

Having studied the current Sequencing Batch Reactor, one can conclude that the SBR technology has not been fully explored. Nowadays, the most suitable mode of operation of the Sequencing Batch Reactor with an inflow of wastewater with low concentration of organic contaminations has not yet been developed.

2 Materials and Methods

The study consisted of three parts. The tests were carried out at the laboratory bench of the Scientific Educational Center "Water Supply and Water-Removal" of the Water Supply and Water-Removal Department of the Moscow State (National Research) University of Civil Engineering. A 30 litres plexiglass column was used in the biological reactor operation. In order to find the most suitable modes of operation, outlets were mounted in this column at different heights. A compressor, supplying air through aerators, was used to create the aeration mode, and a stirrer was used to create the mixing mode. In order to optimize the column performance as a Sequencing Batch Reactor, time relays were used, which switched on and off the aerator, stirrers in accordance with the specified time, and created conditions for the activated sludge sedimentation. The laboratory-scale plant operation diagram is shown in Fig. 1. Then there is a photo of the plant in Fig. 2.

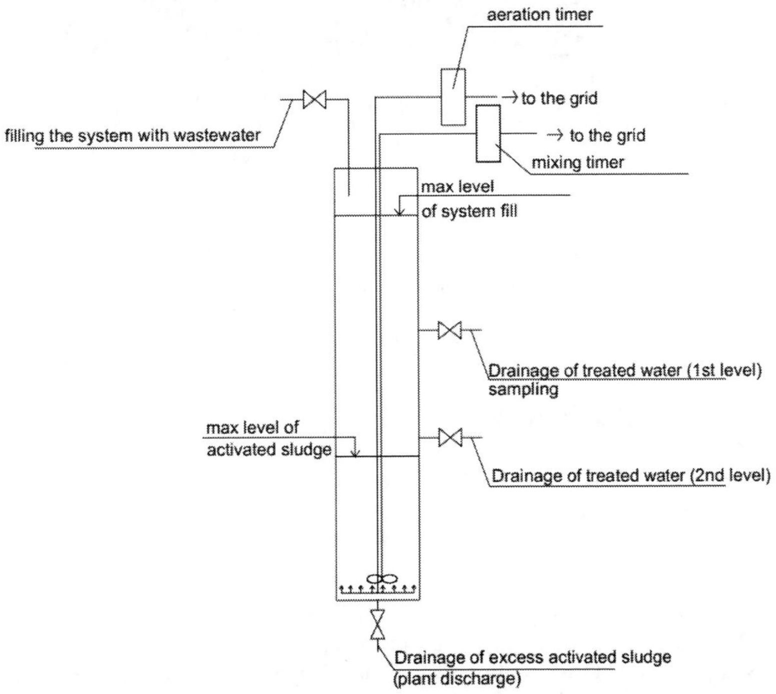

Fig. 1. The diagram of the laboratory model to study the Sequencing Batch Reactor operation.

Fig. 2. A photo of the Sequencing Batch Reactor plant.

The work was carried out with the use of artificially composed peptone-based waste liquid. The contaminations concentrations were closest to those existing in the actual waste liquid. In the course of the experiment, the concentrations of certain substances and the hydraulic load on the plant changed depending on the task assigned. The main parameters of the waste liquid entering the treatment plant are given below:

Average concentration of contaminations in wastewater

- BOD_5 – 110 mg/L
- Ammonium nitrogen (NH_4^-) – 28 mg/L
- Suspended substances – 90 mg/L.

3 Conducting an experiment

The experiment was carried out on a pilot plant in three stages, with the aim of finding the optimal mode of operation of the Sequencing Batch Reactor with wastewater with low concentration of organic contaminations. Each stage corresponded to a certain technological mode. The modes of operation differed by the operating time of the plant, as well as the duration of the phases of aeration and mixing.

First stage. During the first part of the experiment, the reactor operation mode was set at 8 hours for 15 days. The artificially composed waste liquid was as close to the real waste liquid as possible in its parameters. Wastewater consumption of the plant was 0.2 m^3/day.

The reactor operation mode included following cycles:

1. Filling the system with wastewater;
2. 2 hours – sludge mix aeration, 1 hour – mixing, then 2 hours – aeration and 1 hour – mixing;
3. Sedimentation of the purified water for 2 hours;
4. Drainage of the purified water;
5. Drainage of the excess activated sludge.

During the experiment, after each cycle, a series of sanitary-chemical analyses were conducted, which showed poor analyses results. The reactor is ineffective in this mode. BOD decreased by 90.9 % at this stage, which indicates that there is not enough time for oxidation of organic contaminations, and nitrites and nitrates indicators were above the norm. The results can be seen in the diagrams in Figures 3, 4, 5, 6.

Second stage. During this stage of the experiment, the reactor operation time increased and amounted to 16 hours. The experiment lasted 15 days. The composition of the waste liquid, the dose and age of activated sludge, the consumption of waste liquid supplied to the reactor remained the same – 0.2 m^3/day. This time, the reactor operation mode consisted of the following cycles:

1. Filling the system with wastewater;
2. 4 hours 40 minutes – sludge mix aeration, 2 hour 20 minutes – mixing, then 4 hours 40 minutes – aeration and 2 hour 20 minutes – mixing;
3. Sedimentation of the purified water for 2 hours;
4. Drainage of the purified water;
5. Drainage of the excess activated sludge.

At this stage, similarly to the first one, a series of sanitary and chemical analyses were conducted at the end of each cycle. The analyses results showed more effective purification with 91.7 % BOD removal of organic contaminations, while indicators for nitrites, nitrates and ammonia nitrogen are not much lower than those of the previous stage. This indicates that the biological purification processes were more successful than at the previous stage. To achieve the best result, it was decided to increase the plant operation time.

Third stage. During this stage of the experiment, the reactor operation mode was set at 24 hours for 15 days. Again, the waste liquid was composed artificially and was as close to the real waste liquid as possible in its parameters. The wastewater consumption of the plant was 0.20 m^3/day.

The reactor operation mode consisted of the following cycles:

1. Filling the system with wastewater;
2. 7 hours 20 minutes – sludge mix aeration, 3 hour 40 minutes – mixing, then 7 hours 40 minutes – aeration and 3 hour 40 minutes – mixing;
3. Sedimentation of the purified water for 2 hours;
4. Drainage of the purified water;
5. Drainage of the excess activated sludge.

During this stage, the plant showed the best result of the three stages. The BOD removal of organic contaminations is good and amounts to 93 %, indicators for nitrites, nitrates and ammonia nitrogen were stable at a low level, which indicates that the biological purification processes were successful. The results of the conducted analyses at all three stages can be seen in the summary diagram presented below (Fig. 3, Fig. 4, Fig. 5, Fig. 6).

 https://doi.org/10.1051/e3sconf/202018004016

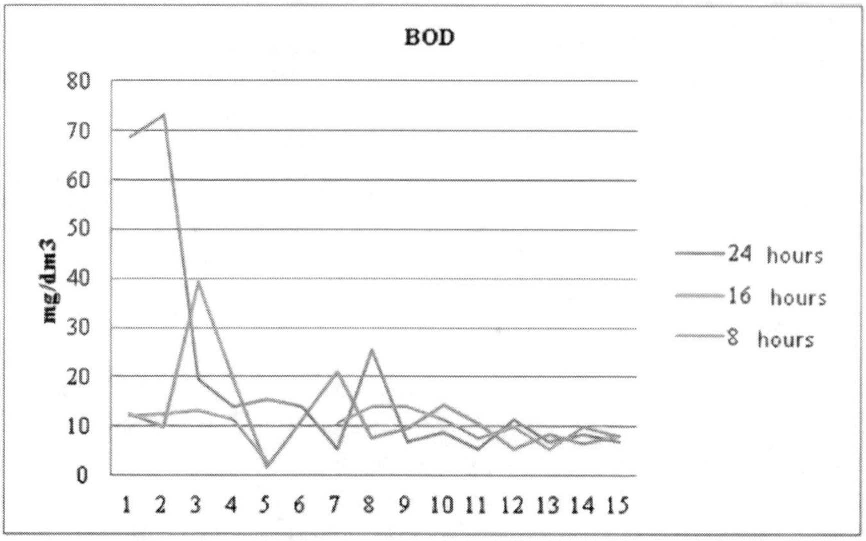

Fig. 3. BOD_5 wastewater treatment indicators summary diagram.

This diagram shows that a more or less stable process of wastewater treatment from organic contaminations occurs after 8-9 days of the plant operation in all three time cycles.

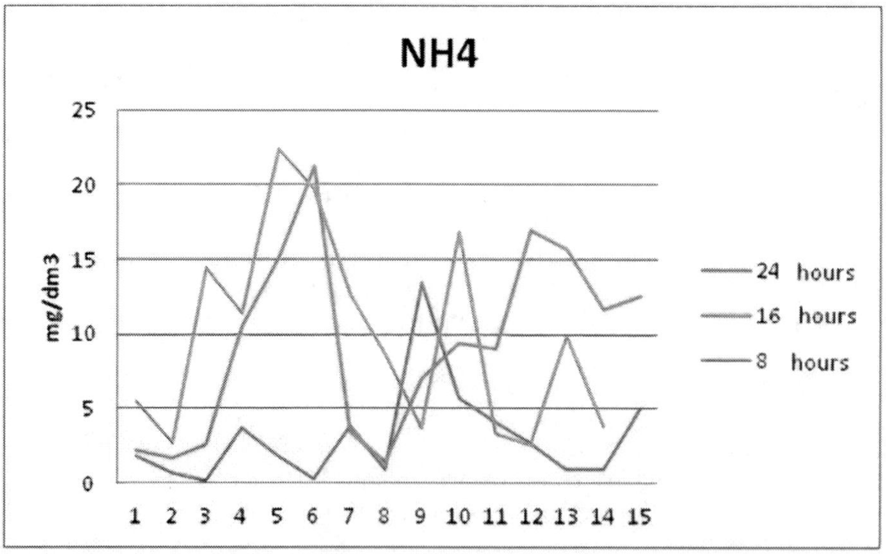

Fig. 4. Ammonia nitrogen-removing wastewater treatment indicators summary diagram.

The diagram shows that the best plant operation for the ammonia nitrogen wastewater treatment is in the 24-hour operation mode.

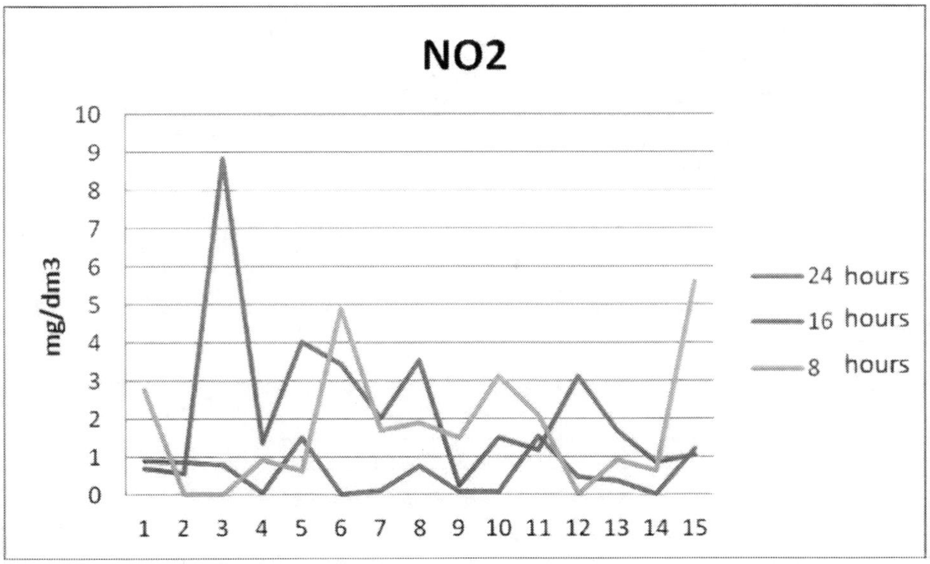

Fig. 5. Nitrites wastewater treatment indicators summary diagram.

According to this diagram it is difficult to talk about any stable plant operation, it only clearly shows that the best result of nitrites wastewater treatment can be obtained in the 24-hour plant operation mode.

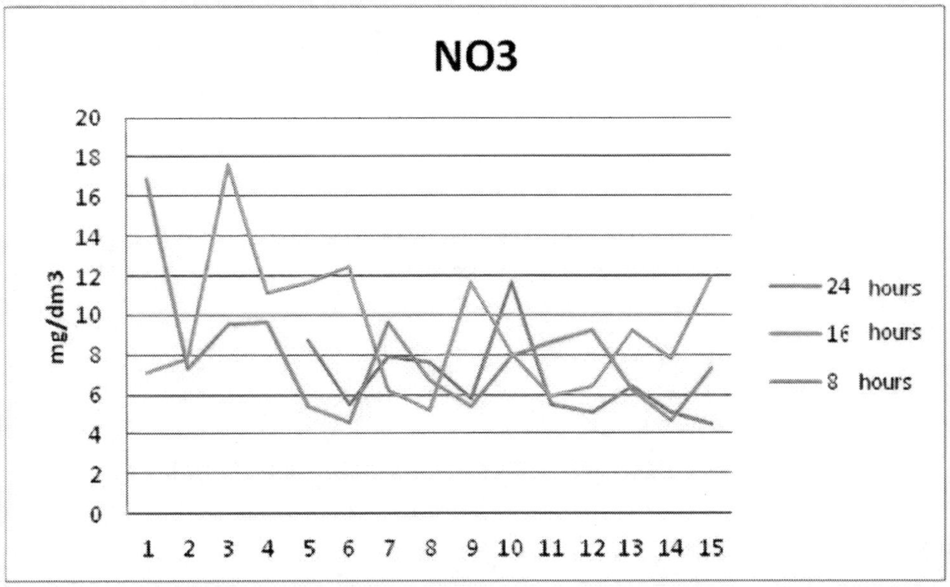

Fig. 6. Nitrates-removing wastewater treatment indicators summary diagram.

This diagram, as well as the ammonia nitrogen-removing wastewater treatment diagram, shows that more or less stable operation of the plant in the 24-hour mode starts after 9-10 days of operation, which is different from other modes. This indicates that nitrifying bacteria appeared in the treated wastewater. They are very sensitive to the slightest changes in the oxidized environment, which is associated with the changes in the operation mode of the plant.

The diagrams clearly show how the purification process is progressing and that the most acceptable analysis result is obtained in the 24-hour plant operation mode. The diagram with the results of the BOD_5 analyses is smoother, the values are approximately at the same distance, if the average values straight line is drawn.

The main data of the experiment are shown in Table 1.

Table 1. The main data of the experiment.

Parameters	Incoming waste liquid	Treated drain liquids		
		24 hours	16 hours	8 hours
OD_5, mg/L	110	7.6	9.7	10
Ammonium nitrogen, mg/L	28	1.2	7.7	9.4
Nitrites, mg/L	-	0.36	2.28	1.78
Nitrates, mg/L	-	5.2	7.9	9.3

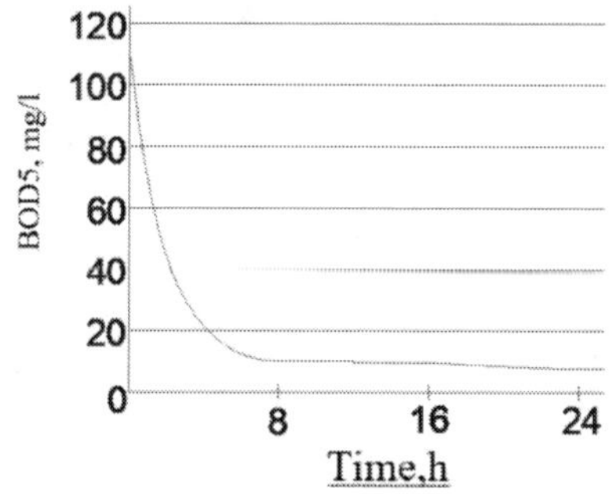

Fig. 7. Diagram of BOD_5-removing wastewater treatment dependence from the reactor operation time.

The diagram shows that with the increased time of reactor operation, BOD wastewater treatment is better. The result was obtained at the 24-hour reactor operation time.

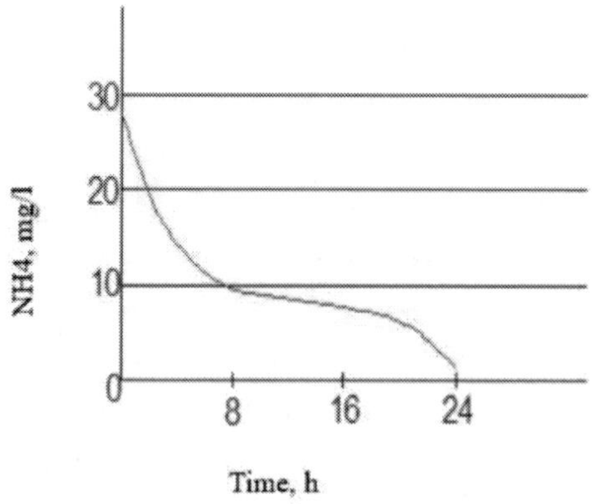

Fig. 8. Diagram of ammonium nitrogen removal dependence from the reactor operation time.

According to this diagram, almost complete removal of ammonium nitrogen occurs in the 24-hour operation mode of the reactor. The diagram shows that during the period of time from 0 to 8 hours, there is a decrease in the concentration of NH4 in water, thus it can be concluded that the nitrification process occurs in the treated water. And in the period of time from 8 to 16 hours, the diagram is smoother, which indicates that there are processes of nitrification, but there is no denitrification process yet, and after 16 hours, there is an active process of denitrification, which is displayed in the diagram with sharp decrease in ammonium nitrogen.

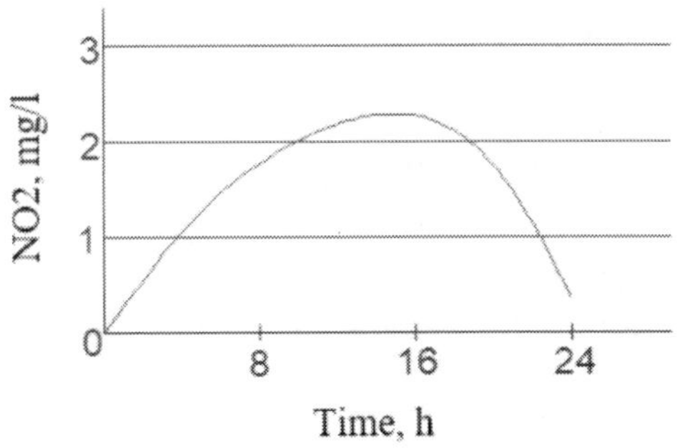

Fig. 9. Diagram of nitrites-removing wastewater treatment dependence from the reactor operation time.

From the diagram of nitrites-removing wastewater treatment dependence from the reactor operation time it can be concluded that maximum concentration of nitrites is reached in the 16-hour operation mode, then concentration of nitrites in treated wastewater decreases.

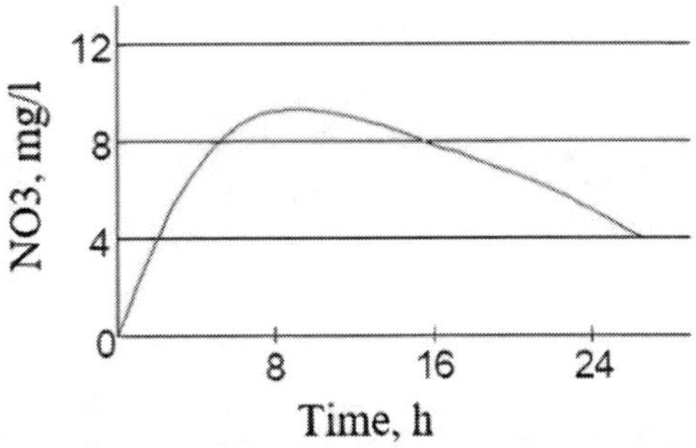

Fig. 10. Diagram of nitrates-removing wastewater treatment dependence from the reactor operation time

If simultaneously analyzing the diagrams of nitrites-removing wastewater treatment dependence from the reactor operation time and the diagram of nitrates-removing wastewater treatment dependence from the reactor operation time one can conclude that active nitrification process occurs when the reactor operation time is up to 8 hours as there are a lot of nitrites and nitrates. Then from 8 to 16 hours nitrites continue to increase and nitrates decrease, this indicates that the nitrification process has reached its peak, and the denitrification process begins. Between 16 and 24 hours, nitrites concentration decreases, nitrates also decrease, which indicates that the nitrification process is over and the denitrification process is active.

4 Conclusions

The conducted study has shown that the Sequencing Batch Reactor can potentially be used as a facility for treatment of the wastewater with low concentration of organic contaminations. The laboratory-scale plant operation showed that the 24-hour mode treatment allows to purify wastewater not only from organic contamination, but also from nitrogen compounds – ammonium nitrogen, nitrites and nitrates. The received data indicates that deep biological purification processes, which do not cause secondary contamination of the water body when discharging the purified wastewater, occurs in the reactor.

References

1. Sequencing Batch Reactor. The International Water association publishing. URL: https://www.iwapublishing.com/news/sequencing-batch-reactor
2. Sequencing Batch Reactor Design and Operational Considerations. New England Interstate Water pollution Control commission. September 2005; 1-6.
3. I. M.Panova, I. Noiberg, Ecology Production, **6**, 58-61 (2014)
4. T. George, B. Franklin, S. David, *Wastewater Engineering Treatment and Reuse* (2003)
5. T. H. Quan, N. X. Quyet, E. S. Gogina, *Construction - Formation of the environment of vital activity: International interuniversity science.-prakt. conf. students, undergraduates,* 1093-1096 (2017)
6. Summary report Sequencing Batch Reactor. United States Environmental Protection Angecy EPA/625/8-86/011. 1986.
7. N. Makisha, D. Semenova, MATEC Web of Conferences, **144**, 04016 (2018)
8. O. Ruzhitskaya, E. Gogina, V. Shmalko, (2018), MATEC Web of Conferences, 178, 09015 (2018)
9. Biological nutrient removal (BNR) Operation in Wastewater treatment plants: Manual of Practice No. 29. Water Environment Federation, 2005. Chapter 22.
10. T. H. Quan, E. Gogina, *IPB&KU International symposium on Education and Research in Global Environment Studies in Asia*, Bogor, **25** (2018)
11. C. R. Kokare, S. Chakraborty, A. N. Khopade, K. R. Mahadik, Indian J Biotechnol. **8**, 159-168 (2009)
12. Nitrification and Denitrification. Michigan Department of Environmental Quality Operator Training and Certification Unit: Online Training Course for Municipal & Industrial Wastewater.
13. U. V. Voronov, S. V. Yakovlev, *Wastewater and Wastewater treatment.* (Moscow, MGSU Publishing house ASV, 2006)
14. M. Henze, W. Gujer, T. Mino, M. Loosdrecht, IWA Publishing in its Scientific and Technical Report series (2015)
15. Standard Methods for the Examination of Water and Wastewater. 20th ed. Baltimore, American Public Health Association, 1998.
16. Unified methods for Examination of Water Quality. Part 1 - Methods of chemical analysis of water. Vol. 1 - Basic methods. Moscow, Council for Mutual Economic Assistance Publ., 1987; 1244. (rus.).
17. S. V. Yakovlev, T. A. Karuhina, *Biochemical wastewater treatment* (Moscow, Stroyizdat Publ., 1980).

E3S Web of Conferences **180**, 04017 (2020)
TE-RE-RD 2020
https://doi.org/10.1051/e3sconf/202018004017

Mathematical Description of Sequencing Batch Reactor operation of treatment of wastewater with low concentration of organic contaminations

Elena Gogina*,[1]

[1]NRU Moscow State University of Civil Engineering (NRU MGSU), 129337, Jaroslavskoje sch. 26, Moscow

Abstract: The paper is oriented towards the study of certain small wastewater-treatment plants, in particular the study of some treatment processes where the wastewater has low concentration of organic contamination. The primary goal of this paper was to obtain the characteristics of a biological treatment system which comply with the laws of enzyme kinetics. The design formulas of the Sequencing Batch Reactor with an inflow of wastewater and low concentration of organic contaminations has been obtained.

1 Introduction

Industrial development and the growing number of urban residents, continuously amplify the problems that appear in the sewerage systems in residential areas. At the same time, the regime of wastewater inflow of many cities and towns is characterized by an unsteady flow of wastewater, and a large irregularity coefficient. SBR (Sequencing Batch Reactor) technologies are used for both urban and industrial wastewater treatment, especially for lower or variable costs. Sequencing batch reactors are structures which involve the application of suspended sludge culture in wastewater treatment. Sequencing Batch Reactor treatment facilities are used in cities, residential areas, resorts, as well as in industries including dairy products, pulp and paper, tanneries, textile plants [1-5].

The advantages of using a Sequencing Batch Reactor are as follows:

• Sequencing Batch Reactor wastewater treatment plant can be placed on a limited area, compared to a standard aeration station, treatment takes place in one reactor, several biological treatment processes occur in one volume;

• The reactor operation cycle is adjusted to various required processes – aerobic, anoxic and anaerobic to intensify the process of deep wastewater treatment from biogenic elements. Thus, biological nitrification and denitrification can be performed to turn ammonia and organic nitrogen from wastewater into nitrates and then into gaseous nitrogen in one reactor;

* Corresponding author: goginaes@mgsu.ru

• In the course of reconstruction, existing facilities can be re-equip into a Sequencing Batch Reactor [6-8].

The experience in the operation of these aerobic treatment systems in Russia has revealed their clear superiority in terms of price-performance ratio over their analogues, but they were not without drawbacks. Evolution of household chemistry has led to changes in the composition of wastewater, and this generated an increase in the content of organic compounds, varying the composition and increasing the oxidation difficulty. Changes that occur in the composition of wastewaters from permanent residence single-family homes, usually led to a significant efficiency decline of any biological treatment system. If speaking about wastewater treatment plants in small towns and villages, there is usually a different situation. The low cost of drinking water leads to the fact that residents of small towns use fresh water in an improvidently manner, so the wastewater entering treatment plant is characterized by low concentrations of contaminations [9-13]. Unfortunately, there are difficulties in treating wastewater when the ratio between the concentration of organic contaminations in the wastewater and biogenic elements does not correspond to the standard biological process that prevails in most treatment plants. Therefore, the main goal of this study was to conduct a laboratory-scale experiment with the simulation of a Sequencing Batch Reactor, which was set to several modes of operation for the treatment of low concentrated wastewater [14-17].

2 Materials and Methods

The study consisted of three parts. The tests were carried out at the laboratory bench of the Scientific Educational Center "Water Supply and Water-Removal" of the Water Supply and Water-Removal Department of the Moscow State University of Civil Engineering. A 30 litres Plexiglass column was used in the biological reactor operation. In order to find the most suitable modes of operation, outlets were mounted in this column at different heights. The reactor operation mode included following cycles: filling the system with wastewater; sludge mix aeration; mixing; then aeration and mixing; sedimentation of the purified water; drainage of purified water; drainage of excess activated sludge. Wastewater treatment time varied from 8 to 24 hours. The main data of the experiment are shown in Table 1.

Table 1. The main data of the experiment

Parameters	Incoming waste liquid	Treated drain liquids		
		24 hours	16 hours	8 hours
BOD_5, mg/L	110	7.6	9.7	10
Ammonium nitrogen, mg/L	28	1.2	7.7	9.4
Nitrites, mg/L	-	0.36	2.28	1.78
Nitrates, mg/L	-	5.2	7.9	9.3

3 Interpretation

Based on the practical solution and experience in calculating such structures abroad, a formula was chosen to calculate the specific oxidation rate of organic contaminations considering activated sludge recirculation, which has number (1) in this section [18-20].

$$r = \frac{L_{en} - L_{ex}}{R * a(1-S)t} \quad (1)$$

where r – specific oxidation rate of organic contaminations, ($mgBOD_5/g_{ash\text{-}free\ basis}/h$);
Len – BOD_5 in the waste liquid entering the treatment plant ($mgBOD_5/l$);
Lex – BOD_5 in the purified liquid, ($mgBOD_5/l$);
R – Recycling coefficient calculated as Qr/Qpost ratio;
Qr – water volume, (m^3), Qpost – return sludge volume, (m^3);
a – Sludge dose which was constantly kept at 2.1 (g/l);
S – Ash content of activated sludge 0.3;
t – Sludge mix residence time in the reactor, (h).

To calculate the activated sludge load – q, ($mgBOD_5/g_{ash\text{-}free\ basis}/day$) a formula was used considering the recirculation of activated sludge.

$$q = \frac{24(Len - Lex)}{R \cdot a(1-S)t} \quad (2)$$

Len – BOD_5 in the waste liquid entering the treatment plant ($mgBOD_5/l$);
Lex – BOD_5 in the purified liquid, ($mgBOD_5/l$);
R – Recycling coefficient calculated as Qr/Qpost ratio;
Qr – water volume, (m^3), Qpost – return sludge volume, (m^3);
a – Sludge dose which was constantly kept at 2.1 (g/l);
S – Ash content of activated sludge 0.3;
t – Sludge mix residence time in the reactor, (h).

Having made the calculation, we got the following values, which are given in Table 2.

Table 2. The calculation results

Sludge mix residence time in the reactor, h.	BOD_5 specific oxidation rate, $mgBOD_5/g_{ash\text{-}free\ basis}/h$	Organic contaminations sludge load, $mgBOD_5/g_{ash\text{-}free\ basis}/day$
24	4.03	159.5
16	6.2	223.1
8	14.4	519.5

The diagrams clearly show specific oxidation rates and loads on activated sludge (Fig.1).

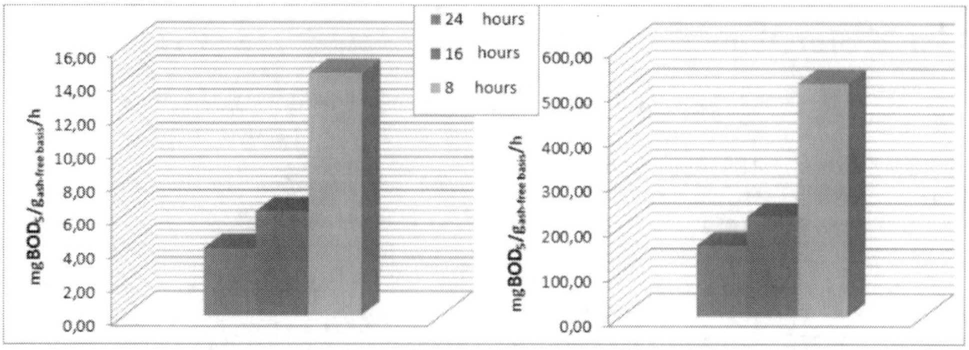

Fig. 1. Diagrams of distribution (left) of the specific rate of organic contaminations oxidation; (right) activated sludge load distribution of organic contaminations diagram.

The diagrams show that the load on activated sludge and the specific rate of oxidation of organic contaminations is minimal in the 24-hour reactor mode of operation.

The efficiency of wastewater treatment in different operating modes is presented in Table 3.

Table 3. The efficiency of wastewater treatment in different operating modes

Parameters in % to the incoming water	Treated drain liquids		
	24 hours	16 hours	8 hours
BOD_5	93	91.2	90.9
Ammonium nitrogen	95.7	72.3	66.4

The table below shows that the most efficient mode of operation of the wastewater treatment plant is the 24-hour cycle.

Figures 2, 3, 4 show the dependencies of the BOD specific oxidation rates in treated wastewater in different operation modes.

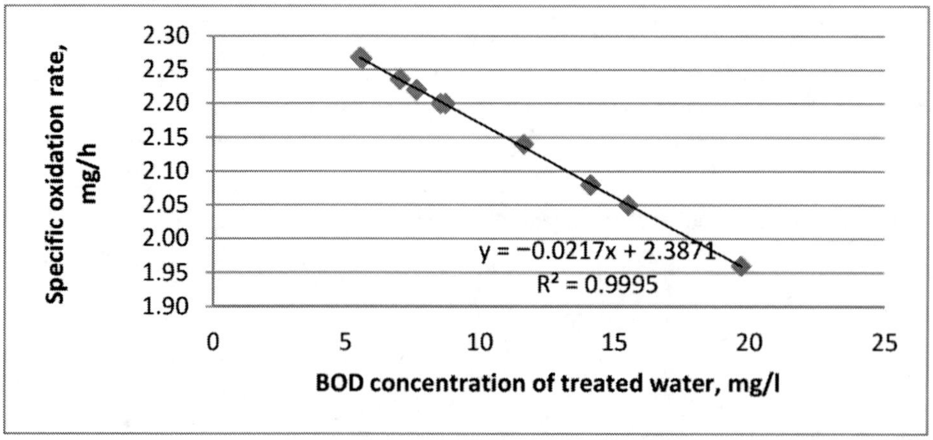

Fig. 2. Dependence of the specific oxidation rate on the BOD in treated wastewater in the 24-hour reactor operation mode.

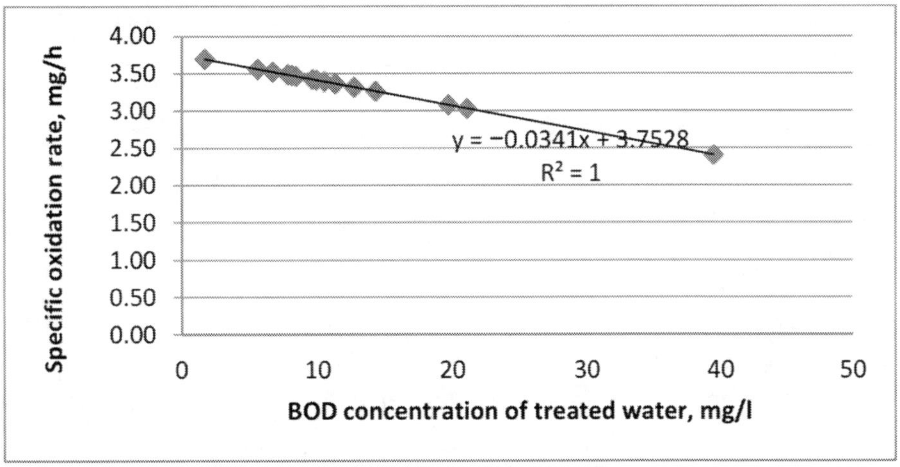

Fig. 3. Dependence of the specific oxidation rate on the BOD in treated wastewater in the 16-hour reactor operation mode.

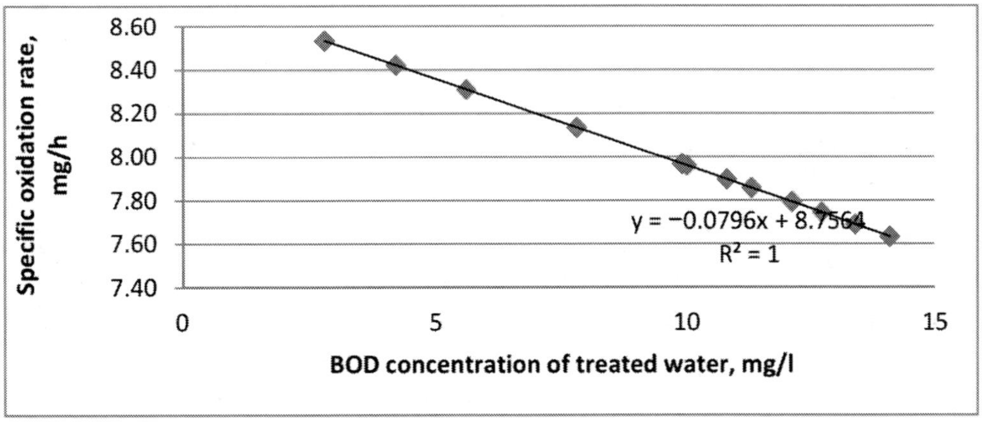

Fig. 4. Dependence of the specific oxidation rate from BOD of treated wastewater in the 8-hour reactor operation mode.

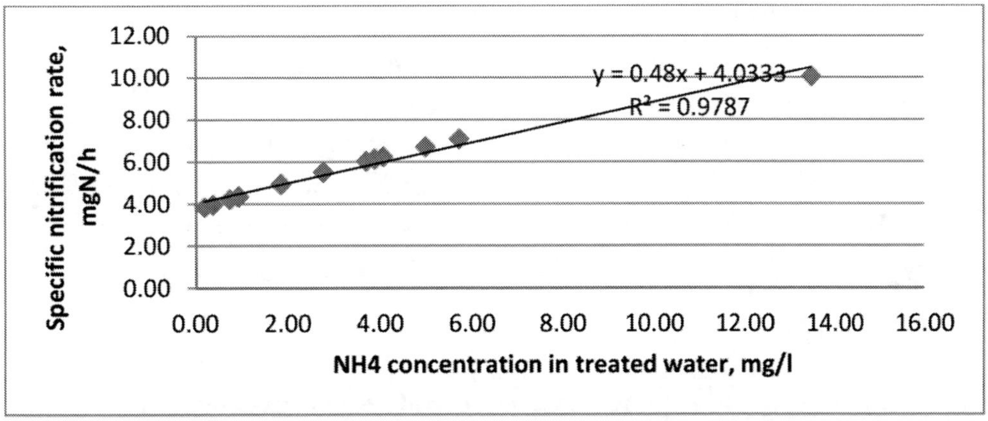

Fig. 5. Dependence of the specific nitrification rate on the BOD in treated wastewater in the 24-hour reactor operation mode.

The above diagrams show that the decrease in the BOD concentration in treated water is proportional to the decrease in specific oxidation rate, and this dependence can be described by the linear equation.

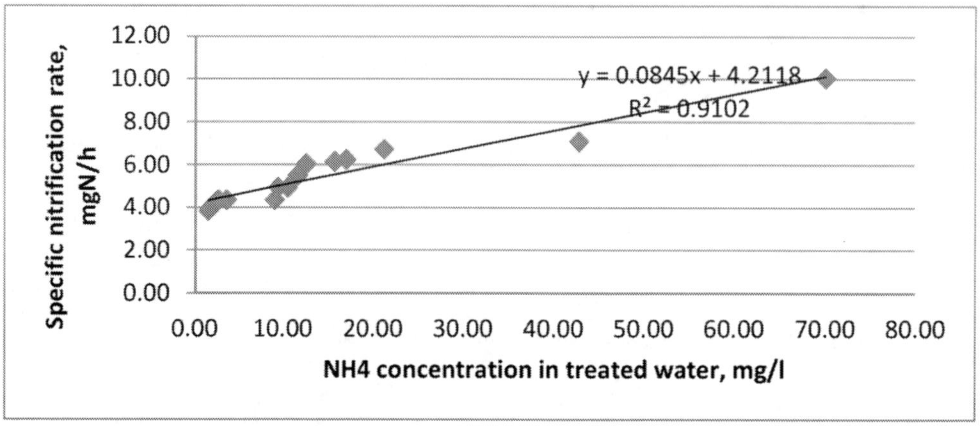

Fig. 6. Dependence of the specific oxidation rate on ammonium nitrogen concentration in treated wastewater in the 16-hour reactor operation mode.

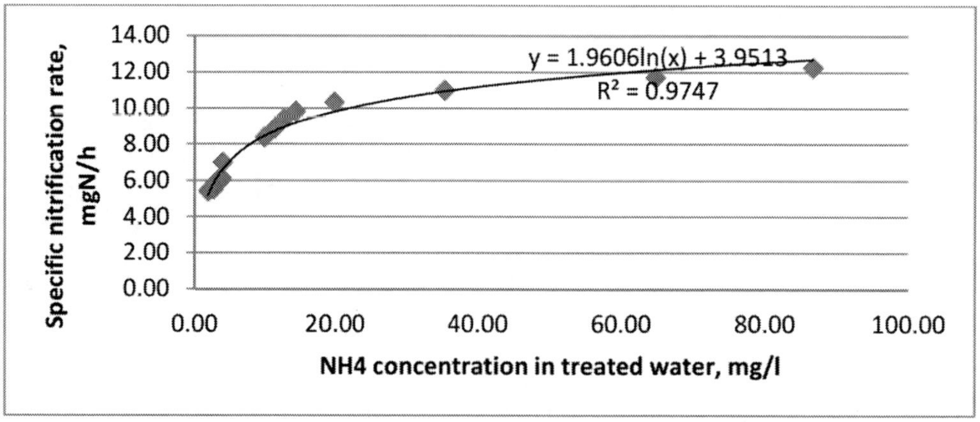

Fig. 7. Dependence of the specific oxidation rate on ammonium nitrogen concentration in treated wastewater in the 8-hour reactor operation mode.

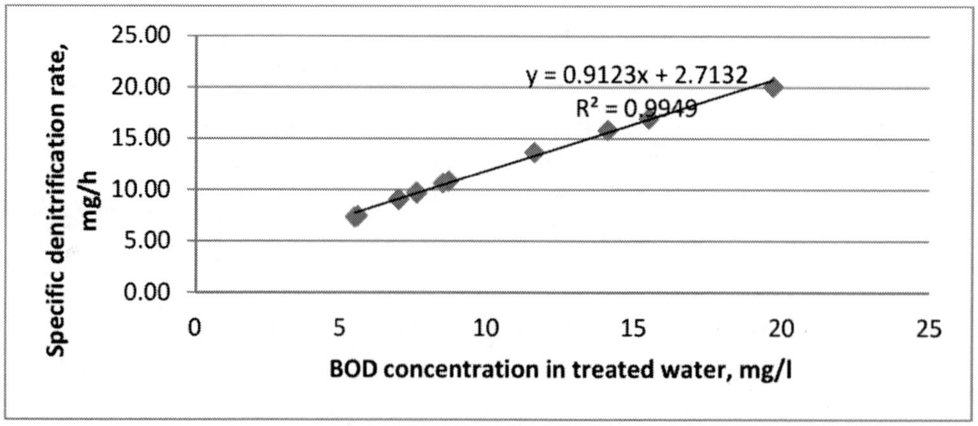

Fig. 8. Dependence of the specific oxidation rate on the BOD in treated wastewater in the 24-hour reactor operation mode.

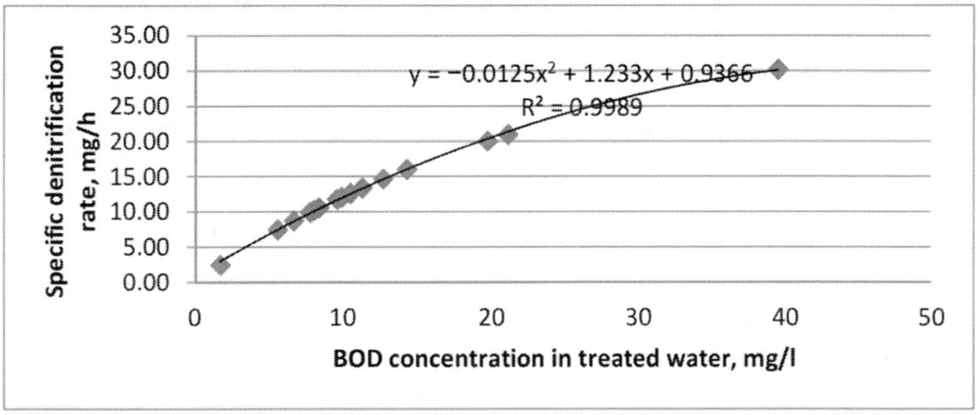

Fig. 9. Dependence of the specific oxidation rate on the BOD in treated wastewater in the 24-hour reactor operation mode.

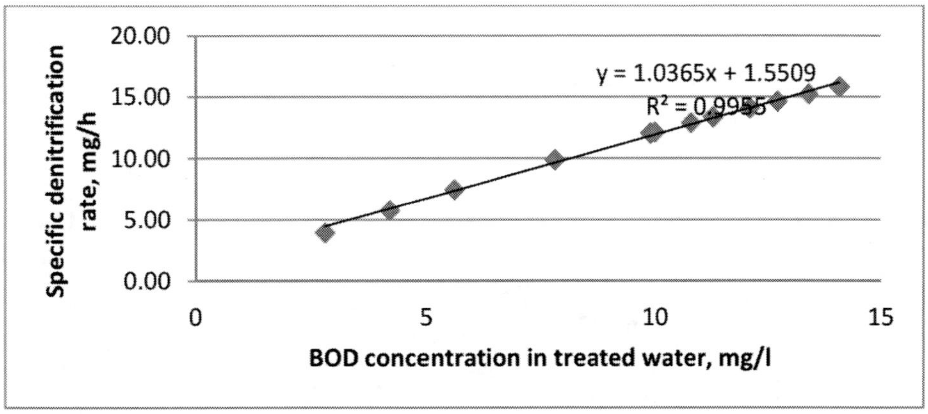

Fig. 10. Dependence of the specific oxidation rate on the BOD in treated wastewater in the 8-hour reactor operation mode.

The diagrams in Figures 8, 9, 10 show the dependence of denitrification specific rate on BOD in treated water in three modes of reactor operation, from which it is possible to conclude that the results are subordinated to linear equations. The equations show that the lower the concentration of BOD, the lower the specific denitrification rate.

The diagrams in Figures 5, 6, 7 show the dependence of the specific oxidation rate on ammonium nitrogen concentration in treated wastewater in various reactor operation mode. The higher the concentration of ammonium nitrogen in the treated water, the higher the specific nitrification rate.

 https://doi.org/10.1051/e3sconf/202018004017

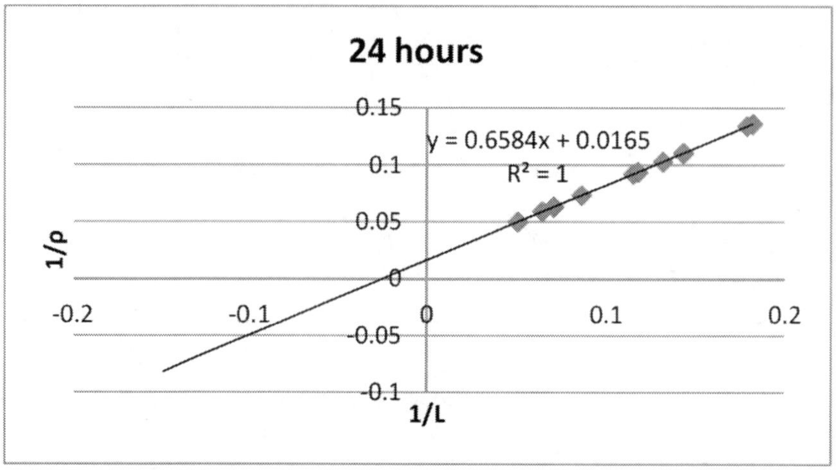

Fig. 11. The Diagram of the dependence of inverse values of organic contamination oxidation rate on inverse value of BOD5 contamination concentration

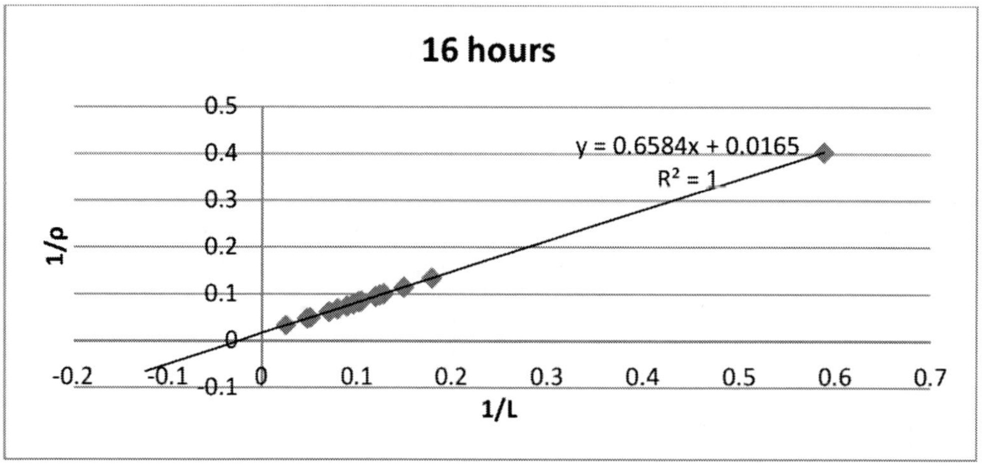

Fig. 12. The Diagram of the dependence of inverse values of organic contamination oxidation rate on inverse value of BOD5 contamination concentration

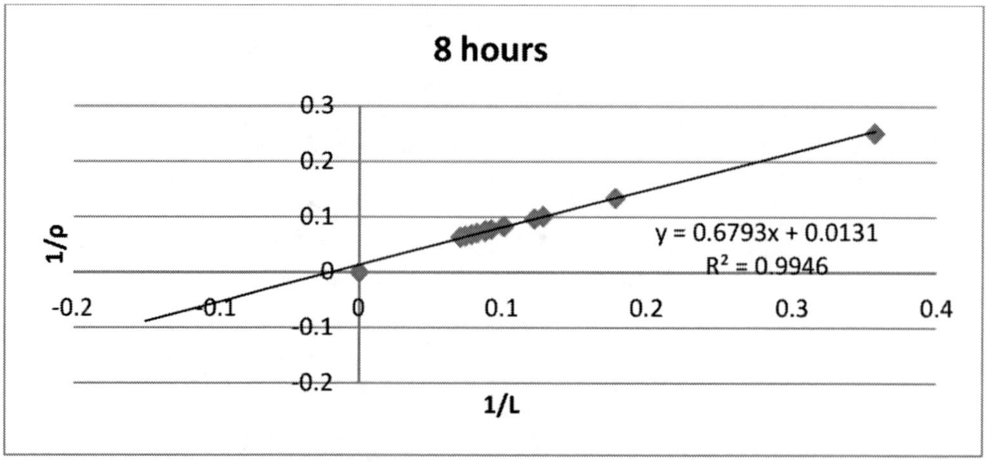

Fig. 13. The Diagram of the dependence of inverse values of organic contamination oxidation rate on inverse value of BOD5 contamination concentration

All obtained diagrams are generally subject to the law of enzyme kinetics, and this is explained by the fact that their speed is proportional to the substrate concentration: it increases to a certain level.

Based on the Michaelis-Menten provisions of enzyme kinetics, diagrams of inverse values have been constructed, which indicate that the experiment conducted and the results obtained fully comply with the laws of enzyme kinetics, despite the low concentrations of organic contaminations and nitrogen compounds in the incoming wastewater. Figures 11-13 present diagrams of the dependence of inverse values of organic contamination oxidation rate on inverse value of BOD5 contamination concentration 1/V = *f*(1/Lex). Table 4 shows the obtained values of the Michaelis-Menten constants – Vmax and km.

Table 4. The obtained values of the Michaelis-Menten constants

Stages	$V_{max,}$ mgBOD/(g·h)	k_m
24 hours	60.6	40
16 hours	60.6	40
8 hours	76.3	52.6

With the results obtained, we will calculate the oxidation rates of organic impurities using the obtained maximum values of the oxidation rate from the Michaelis-Menten constant for each process separately:

$$V_{24} = \frac{V_{max} \cdot S}{k_m + S} = \frac{60 \cdot L_{ex}}{40 + \ _{ex}} \tag{3}$$

$$V_{16} = \frac{V_{max} \cdot S}{k_m + S} = \frac{60 \cdot L_{ex}}{40 + L_{ex}} \tag{4}$$

$$V_{8} = \frac{V_{max} \cdot S}{k_m + S} = \frac{76{,}3 \cdot L_{ex}}{52{,}6 + L_{ex}} \tag{5}$$

In Figure 14 one can see that the diagrams are parallel to each other and are described by the same equation, which confirms the statement about identical kinetic dependencies in these modes of the plant operation. The diagram of the 8-hour plant operation mode differs in Vmax position, however the Michaelis-Menten constant is much higher. This allows us to say that in this system the maximum oxidation rate of contaminations is achieved faster than in the others.

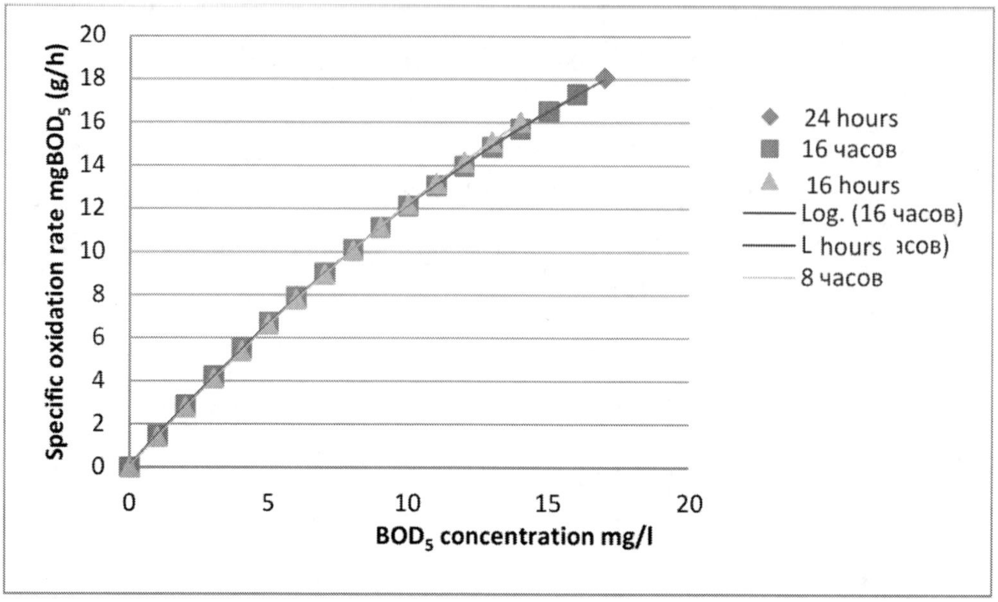

Fig. 14. Dependence of organic contaminations specific oxidation rate on BOD_5 concentration.

4 Conclusions

The conducted study of the Sequencing Batch Reactor operation has shown possibility of its application for biological treatment of wastewater with low concentration of organic contaminations. The most suitable modes of operation of the batch reactor with low concentrated wastewater with the most efficient purification have been determined. Mathematical dependences for description of wastewater treatment process in the Sequencing Batch Reactor have been received, allowing to carry out calculation of reactors when designing treatment facilities.

References

1. N. Høiby, APMIS, **125**, 272-275 (2017)
2. C. R. Kokare, S. Chakraborty, A. N. Khopade, K. R. Mahadik, Indian J Biotechnol. **8**, 159-168 (2009)
3. A. Kulakov, N. Makisha, MATEC Web of Conferences, **112**, 10019 (2017)
4. T. George, B. Franklin, S. David, *Wastewater Engineering Treatment and Reuse* (2003)
5. I. M. Panova, I. Noiberg, Ecology Production, **6**, 58-61 (2014)
6. Sequencing Batch Reactor [Electronic resource] // IWA the International water Association Publishing. URL: https://www.iwapublishing.com/news/sequencing-batch-reactor.
7. R. F. Poltak, *Sequencing Batch Reactor Design and Operational Considerations* (2005)
8. L. Hernández et al. Water, **2**, 155-169 (2010)
9. Primer for Municipal Wastewater Treatment Systems. United States Environmental Protection Agency, 2004. EPA 832-R-04-001

10. Summary report SBR. United States Environmental Protection Agency, 1986. EPA/625/8-86/011
11. E. F. Barth, *Implementation of Sequencing Batch Reactors for municipal treatment.* (1983)
12. T. Ha Quan, E. Gogina, *E3S Web Conf.*, **97**, 01017 (2019)
13. T. Ha Quan, E. Gogina, Vestnik MGSU, **5**, 589-602 (2019)
14. C. Lenore, G. Arnold, E. Andrew, *Standard Methods of the Examination of Water and Wastewater* (1999)
15. Unified methods for Examination of Water Quality. Part 1 - Methods of chemical analysis of water. Vol. 1 - Basic methods. Moscow, Council for Mutual Economic Assistance Publ., 1987; 1244. (rus.).
16. M. Henze, W. Gujer, T. Mino, M. Loosdrecht, *IWA Publishing in its Scientific and Technical Report series* (2015)
17. Biological Nutrient Removal Processes. Operation of Municipal Wastewater Treatment Plants. Water Environment Federation, 2007. Vol. 2.
18. Nitrification Denitrification. Michigan Department of Environmental Quality Operator Training and Certification Unit.
19. S. Morling, Vatten, **64**, 121-128 (2008)
20. X. Song et al. AIP Conference Proceedings, 040021 (2017)

Black sea marine litter pollution related to naval operations

Valerian Novac[1], *Luminița* Moraru[1], *Carmen* Gasparotti[1], and *Eugen* Rusu[1,*],

[1]'Dunarea de Jos' University, 47 Domneasca Street, 800 008 Galati, Romania

Abstract. The Black Sea, due to its weak exchange of waters and poor vertical circulation, has an increased sensibility to pollution. Even though this maritime basin is subject to numerous regulations at the national, regional, and international levels, the pollution issue is still actual and stringent. The present paper analyses the findings of three studies focused on marine litter issues and provide, in the concluding part, few directions which can improve the pollution prevention and containment effort. As the studies related to Black Sea litter pollution are scarce and fragmented, the authors took the opportunity and analyzed three different studies conducted in different periods and concluded that a significant part of marine debris is related to naval operations.

1 Introduction

The Black Sea is an almost landlocked basin connected to the Sea of Marmara and the Sea of Azov by the narrow Bosporus and Kerch Straits, respectively. The Black Sea is a deep basin (maximum depth of ~2200 m) with a large shelf, especially northwestern sub-basin. Its catchment area covers large parts of Europe and Asia, providing a total freshwater supply of 3 x 10^2 km^3 annually [1-3]. The important freshwater intake brought by a large number of rivers, including Danube, Dnieper, Southern Bug, Dniester, Don, and Kuban. These rivers are, also responsible for an essential introduction of pollutants including marine litter [4-6].

The region of the Black Sea is under high anthropogenic pressure from the pollution point of view, the catchment area and the coasts are well-populated with significant industrial activity, and the sea is the scene of intense shipping, tourism-related and military activities [1, 7, 8].

Accretion of marine debris in the ocean is a global issue. The condition mentioned above of the Black Sea, along with its poor vertical circulation, led to the conclusion that the marine debris is arrayed in layers corresponding with seawater disposition. Thus, the floating litter carried by surface currents, wind, waves, the densest fraction will sink to the seafloor, and a significant portion immersed amid surface and seabed. The synergic action of environmental factors will degrade the litter in time but not as fast as to avoid harmful effects on the economy, tourism, and marine flora and fauna [9, 10].

* Corresponding author: eugen.rusu@ugal.ro

Marine litter is referred to as persistent substantial matters of human origin, refined or processed, and discharged or abandoned in the marine environment or coastal area. The provenience of marine debris is associated with municipal improper waste disposal, industrial activity in the coastal area or on the rivers from the sea catchment area, fishing, tourism, military exercises, and many others [11-13].

The most common marine litter is plastic produced and used on a large scale at the global level is evident that a portion of it will end in seawater either from land or due to seaborne vectors [14].

Plastic marine debris found in the form of great stuff or macroplastic, with dimensions are easily seen and avoided by marine organisms; these larger pieces under the action of the marine environment will split into small parts (< 5 mm) called microplastic and the latest decomposed into smaller pieces (< 100 nm) – nano plastic. The latest categories can is quickly swallowed by marine animals and get further into the human food chain, though in-depth research regarding this matter is not complete [15-18].

Actual studies on Black Sea marine debris are very scarce and fragmented, but from the segments available can be easily observed that marine litter is mostly land originated, though a significant fraction owed to naval operations, merchant or military in nature. Shipping waste consists mainly, but not limited to, synthetic ropes, packaging items, pellets, food and cargo waste, fishing nets, cabin and recreational areas domestic waste. [10, 19, 20].

Due to its sensitivity to pollution matters, the Black Sea is accountable to several national, regional, and international conventions, regulations, and agreements. The most important at the regional level being the Bucharest Convention, which deals mainly with land-based pollution while, at the international level, MARPOL 73/78 regulates shipborne pollution. Following Annex V to MARPOL 73/78 Convention, Black Sea is "Special Area," thus any garbage disposal into Black Sea waters is strictly barred. At the same time, riparian states called to provide adequate port reception facilities. (PRF) [9].

2 Methods

The article analyzes the research activities conducted by some research institutions in the Romanian sector of the Black Sea. The activities conducted from 2012 through 2014 within the framework of three different projects.

2.1 Seabed litter survey facilitated by assessment of demersal fish stocks

The explorations were conducted in 2012 consisting of bottom trawling activity facilitated by demersal fish stocks assessment, such along with fishing seafloor litter samples were collected. The asset used was National Institute for Marine Research and Development "Grigore Antipa" research vessel "STEUA DE MARE 1" with its bottom trawl as fishing gear.The depths of waters hauled were between 15 m and 90 m along with the entire Romanian coastal sector, from Vama Veche to Sulina. The study investigated areas with offshore platforms (AOI 3 - Area Of Interest 3), intense navigated shipping routes and ports basins – Constanta and Midia – AOI 1 and AOI 2 and a daggers area where bottom debris is likely to be retained, is also a challenge for the trawl (Figure 1). During the demersal surveys were conducted 69 hauls in two timeframes: 05 -10.05.2012 and 13 – 16.11.2012 [12].

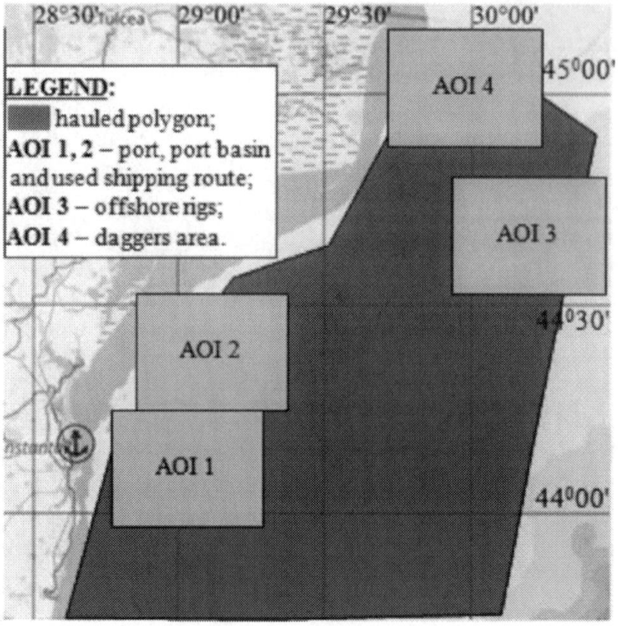

Fig. 1. Polygon hauled during demersal fish surveys – authors contribution based on [12, 21]

2.2 Black Sea marine litter survey – MISIS project

The assessment performed in the period of 22 – 31.08.2013 in the framework of MISIS Project Joint Black Sea along six polygonal surfaces, transverse of Romania, Bulgaria, and Turkey. MISIS was part of a broader program aimed to adjust the Black Sea area strategy in the marine environment protection with appropriate European policy in the field. European Council provided the funds for this project [22].

In order to have symmetry with the demersal survey (2.1) are analyzed just results related to the Romanian sector (see Figure 2 and Table 1). The researchers chose to survey one polygon in the coastal area and one in the shelf zone.

Marine debris grabbed using a beam trawl (2.5 m breadth), and the speed of trawling ship was 2.5 – 3 knots. The depths of hauls were for the coastal survey – 33 m while for shelf seabed 65 m. All trawled stuff was measured on board, categorized by type of the matter and dimension [8, 22].

Table 1. Coordinates of trawled polygons [8, 22].

Polygon	Depth [m]	Start		End		Trawled area [m^2]
		Lat	Lon	Lat	Lon	
B1	33	44^{0}10'08''N	28^{0}48'58''E	44^{0}10'07''N	28^{0}48'21''E	1250
B2	65	44^{0}10'06''N	29^{0}40'24''E	44^{0}10'06''N	29^{0}42'55''E	4600

2.3 Visual surveys of marine litter – CoCoBlas project

The observations were performed visually onboard R/V Mare Nigrum on the framework of CoCoBLAS 2014 (24 – 29.06.2014) part of broader EC Project CoCoNet. The survey area placed along the Romanian coast of the Black Sea amid the Danube delta and Constanta. The density of floating litter was estimated using the transect methodology. The same researcher conducted all observations during daylight at a low ship speed recording size, type, and position of all floating waste with dimensions above 2 cm.

Though, the authors in [4] split sighted debris in 2 main categories, for this paper considers just debris of anthropogenic origin (Anthropogenic Marine Debris - AMD) and the natural one omitted.

In the said study, were performed 30 transects, but the present paper retained just 10, the most relevant with the highest AMD values (table 2).

Table 2. Transects coordinates, surveyed area length, number of items sighted (n), and density of AMD [4].

Polygon	Length [km]	Start		End		AMD	
		Lat	Lon	Lat	Lon	n	D [items/km^2]
C1	7.75	44°12'55"N	29°01'15"E	44°15'11"N	29°06'10"E	2	23.46
C2	7.73	44°22'21"N	29°22'17"E	44°24'37"N	29°27'10"E	10	117.60
C3	7.39	44°29'59"N	29°36'11"E	44°33'35"N	29°38'35"E	3	36.93
C4	2.39	44°52'52"N	29°50'01"E	44°52'54"N	29°48'12"E	3	114.21
C5	2.01	44°52'58"N	29°43'31"E	44°52'48"N	29°42'00"E	3	135.98
C6	7.40	44°57'12"N	30°19'00"E	44°58'49"N	30°13'50"E	2	24.57
C7	4.50	45°04'56"N	30°00'55"E	45°03'32"N	30°03'44"E	4	80.75
C8	6.43	45°03'14"N	30°03'32"E	44°05'18"N	29°59'34"E	2	28.29
C9	4.50	45°05'00"N	30°00'55"E	45°03'32"N	30°03'44"E	4	80.75
C10	13.27	44°24'54"N	29°33'48"E	44°21'28"N	29°25'01"E	14	95.93

3 Results and discussions

Regarding first project presented, in 28 of 69 hauls performed was found marine litter in the trawl sack, summing 40% of total trawls conducted. Regarding the composition of debris, plastic was most frequent (containers, bags, buckets, canisters) most likely originated from ships or boats which to and fro along dedicated shipping routes. The plastic was followed by rubber, fishing gears most probably coming from illegal and unreported fishing activities which are common in assessed area [12].

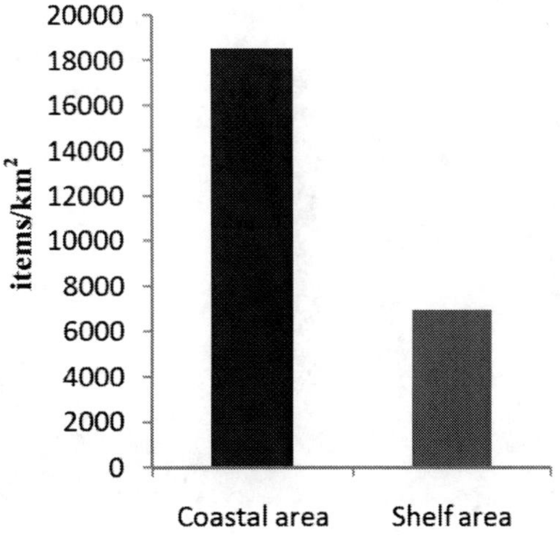

Fig. 2. Marine debris distribution in the Romanian sector [8, 22]

The second study revealed that the Romanian sector was most polluted compared with Bulgarian and Turkish ones. In the said area was registered the highest density of waste, approximately 20 000 items/km2 being more abundant near at depths below 40 m (almost a double amount) than in shelf area. The researchers who conducted the assessment concluded that mostly fishing and touristic activities were the most critical contributors to seafloor debris found. Between the main litter categories, plastic was again present in collected samples but also metal and natural products were the other debris significant constituents [8, 22].

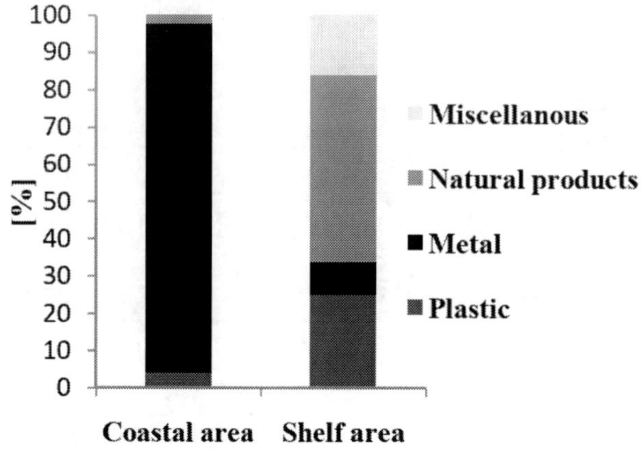

Fig. 3. Marine litter composition in the Romanian sector [8, 22]

Assessing the composition of marine debris can easily observe the abundance of metal and plastic linked with an intense shipping and fishing activity.

The last project also provides an accurate quantitative composition of debris sighted revealed that natural origin debris was more abundant (75,5%) than anthropogenic one (25,5%). The anthropogenic litter sighted consisted of plastic matters (89,1%), other man-made objects (9,1%) and styrofoam (1,8%). Composition of debris presented in the last report is in line with the previous ones regarding the likely origin of litter [4].

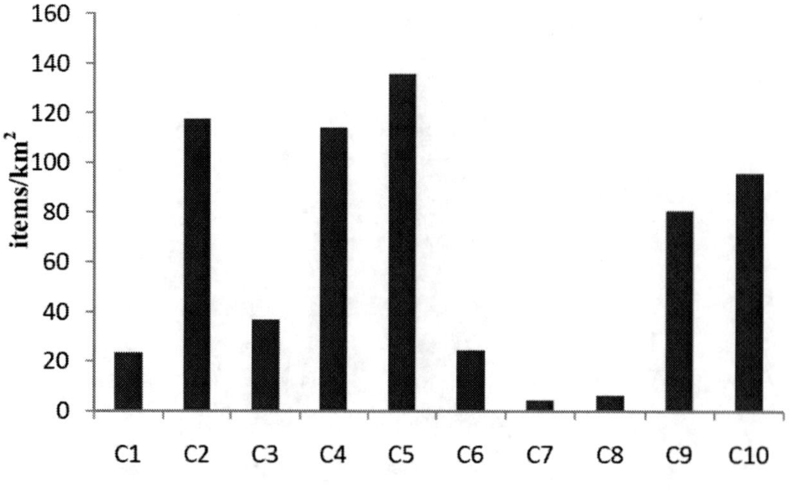

Fig. 4. Density of AMD in considered transects [4]

4 Concluding remarks

Considering significant quantities of marine debris found into Black Sea waters is evident that despite an adequate legal framework, the riparian states do not strictly adhere to provisions of international agreements signed.

More rigorous control of land discharges and onboard waste management is necessary to enforce.

Neighboring countries have to enforce surveillance from the air, surface, and subsurface – seabed employing unmanned means like Unmanned Aerial Vehicles, Unmanned Surface Vehicles, and Unmanned Underwater Vehicles, to reduce costs and eventual pollution by manned and more prominent platforms.

There is a need for multiple studies regarding pollutants type, circulation, and origins in all sensitive areas, and the effort of understanding the pollution mechanism must share at the regional level.

The most problematic remains the plastic debris, which is a threat to the environment, marine biota, and not at last for human consumptions of seafood.

As to diminish marine litter harmful effects, synergic actions shall be taken not only by naval authorities but also by local and regional responsible, and not at the end population have to be sensitized regarding pollution issues.

The paper is based on National Institute for Marine Research and Development "Grigore Antipa" research activities under the MISIS project and demersal fish stocks assessment and on the visually performed observations under the CoCoBLAS project. The authors express their gratitude for the researchers' outstanding work and for disseminating their results through referenced scientific articles.

References

1. V. Novac, E. Rusu, G. Stăvărache, Mechanical Testing and Diagnosis, **2**, 11-14, (2019)
2. V. Novac, E. Rusu, S. B. N. A., **XXI**, 607-617, (2018)
3. V. Novac, E. Rusu, *19th International Multidisciplinary Scientific Conference – SGEM, Conference Proceedings*, **19**, 683, (2019)
4. G. Suaria, M. C. Melinte-Dobrinescu, G. Ion , S. Aliani, Mar. Env. Res., **107**, 45-49, (2015)
5. E. V. Stanev, M. Ricker, Front. Mar. Sci., **6**, 1-16, (2019)
6. R. Bosneagu, I. C. Scurtu, P. Popov, R. Mateescu, L. Dumitrache, M. E. Mihailov, JEPE, **20**, 2059–2067 (2019)
7. R. Boşneagu, A.-T., Nedelcu, I.C. Scurtu, J. Phys.: Conf. Ser., **1122**, (2018)
8. S. Moncheva, K. Stefanova, A. Krastev, A. Apostolov, L. Bat, M. Sezgin, F. Sahin, F. Timofte, Tur. Journ. Fish. Aq. Sci., **16**, 213-218, (2016)
9. E. N. Topçu, A. M. Tonay, A. Dede, A. A. Öztürk, B. Öztürk, Mar. Env. Res., **85**, 21-28, (2013)
10. S. Miladinova, D. Macias, A. Stips, E. Garcia-Gorriz, Mar. Poll. Bull., **153**, 1-11, (2020)
11. A. Paiu, M. Mirea Cândea, R. M. Paiu, A. M. Gheorghe, Mar. Res. Journ., **47**, 232-239, (2017)
12. E. Anton, G. Radu, G. Ţiganov, M. Cristea, M. Nenciu, Mar. Res. Journ., **43**, 350-357, (2013)
13. C. E. Balas, A. Ergın, A. T. Williams, L. Koc, Mar. Poll. Bull., **48**, 449-457, (2004)
14. U. Aytan, A. Valente, Y. Senturk, R. Usta, F. Basak, E. Sahin, R. E. Mazlum, E. Agirbas, Mar. Env. Res., **119**, 22-30, (2016)
15. M.H. Depledge, F. Galgani, C. Panti, I. Caliani, S. Casini, M.C. Fossi, Mar. Env. Res.,

92, 279-281, (2013)
16. F. Gallo, C. Fossi, R. Weber, D. Santillo, J. Sousa, I. Ingram, A. Nadal, D. Romano, Env. Sci. Eur., **30**, 1-14, (2018)
17. J. N. Hahladakis, Env. Mon. Ass., **192**, 1-11, (2020)
18. M. Schulz, D. Neumann, D. M. Fleet, M. Matthies, Mar. Env. Res., **92**, 61-70, (2013)
19. N. Butt, Mar. Pol., **31**, 591-598, (2007)
20. M. Slišković, H. U. Boljat, I. Jelaska, G. J. Mrčelić, Res., **7**, 1-11, (2018)
21. https://map.openseamap.org accessed on 01.04.2020 10.50
22. S. Moncheva, L. Boicenco, *MISIS Joint Cruise Scientific Report, 'State of Environment Report of the Western Black Sea based on Joint MISIS cruise (SoE-WBS)'*, 401, (2014)

Determination of the minimum wind speed leading to the galloping of conductors

Muratkali Zhamanbaev[1], *Diana* Ilieva[2*], *Rakhimash* Abitaeva[3], *Bulbul* Ongar[3]

[1]Department of Power Engineering, Almaty Technological University, Tole bi Str., 100, Kazakhstan
[2]University of Telecommunications and Post – Sofia, 1 Academician Stefan Mladenov Str., 1700 Sofia, Bulgaria
[3]Department of Power Engineering KazATC after named M.Tynyshpayev, Shevchenko Str., 97, Kazakhstan

Abstract. The object of research in this paper is the split phase of overhead power lines. The study of the aeroelastic instability of the icy conductors of the split phase for a multi-span system has already been explored using the method of determining the Hurwitz stability criterion. In multi-span systems, where conductors are interconnected through a garland of insulators, the garlands themselves are involved in an oscillatory process. As a result of this, mutual influence of adjacent spans is observed - energy is transferred from one span to another. The paper investigates the aeroelastic instability of the icy conductors of the split phase in the anchor span, which is characterized by two intrinsic features: the attachment point of conductors on the supports is fixed and mutual effects between adjacent spans are not observed. The study of motion instability is carried out by the first approximation method, that is, on the basis of linearization of the nonlinear equation at the equilibrium point and further investigation of the linearized equation in the vicinity of this point. The results of the study are based on the novelty of the carried out experiments - taking into account the peculiarities of the anchor span and the findings based on the analysis of empirical data.

1 Introduction

A lot of research and papers have been devoted to the study of galloping conductors [1-11]. For example, in [1] the research is devoted to assessing the maximum excursion of single conductors and bundled galloping of conductors.

Interesting contributions may also come from the comparison between analytical/numerical results and experimental data inferred from full-scale galloping observations [2], wind tunnel tests [3,4] or on a high-voltage overhead test line [5]. Free torsional oscillation of a split conductor of an overhead transmission line is considered. The obtained calculation formulas are recommended for use in the design of new overhead transmission lines with split conductors and reconstruction of lines being already in operation [6]. Numerical approaches based on the curved cable-beam theory are considered

*Corresponding author: diana_edubg@abv.bg

promising for single cables [7]. Galloping refers to wind-induced, low-frequency large-amplitude oscillations that have been more frequently observed for a bundle conductor than for a single conductor [8]. The galloping instability analysis shows that the occurrence of negative aerodynamic damping strongly depends on the critical flow state. In all cases considered in [9], the critical Reynolds number effects are shown to include aerodynamic forces that could well be responsible for the whole range of translational galloping instabilities and the part range of torsional flutter and 3-DOF galloping instabilities for semi-elliptical cylinders.

The study of transverse galloping at low values of the Reynolds number allows the following observations to be made [10]: galloping for a square cylinder is observed only for Reynolds number larger than 140; and a sinusoidal quasi-steady model accurately reproduces amplitudes obtained from fluid–structure interaction simulations for moderately high values of the mass ratio.

2 Determination of the critical speed of wind leading to aerodynamic instability of conductors

Currently, a serious part of the analysis of the phenomena of "galloping" is determining the value of the minimum (critical) wind speed, leading to an unstable state of the icy conductor of power lines. There are few works regarding this issue [11-16], especially concerning the split-phase [14]. Below, a method for determining the critical velocity for the split-phase of the lines of the anchor section is proposed by the authors. In [17,18], a mathematical model of the galloping of conductors of a split-phase with two degrees of freedom is presented. Neglecting the generalized coordinates of torsional motion in this model, we obtain the equations of motion for galloping of conductors with one degree of freedom

$$\ddot{a}(t) + k_1\dot{a}(t) + k_2\dot{a}^3(t) + \omega_a^2 a(t) + k_3 a^2(t) + k_4 a^3(t) + k_{13} = 0 \qquad (1)$$

where the coefficients of the equation are calculated with the following expressions:

$$\ddot{a}(t) = \frac{d^2a(t)}{dt^2}; \qquad \dot{a}(t) = \frac{da(t)}{dt}; \quad k_1 = \frac{\delta\cdot\omega_a}{\pi} - \frac{2g\cdot\rho\cdot d_S\cdot(C_{D0}+C_{L0})V}{\pi\cdot P_W};$$

$$k_2 = \frac{4g\cdot\rho\cdot d_S}{3\pi\cdot P_W\cdot V}C_{L1}; \qquad k_3 = \frac{3\pi\cdot g\cdot E\cdot F}{T\cdot l^2}; \qquad k_4 = \frac{\pi^2\cdot g\cdot E\cdot F}{4\cdot P_W\cdot l^4};$$

$$k_{13} = \frac{g\cdot\rho^2\cdot d_S^2\cdot C_{D0}^2\cdot V^4}{2\cdot P_W^2}; \quad \omega_a = \frac{\pi}{l}\sqrt{\frac{g\cdot T}{P_W}\left(1 + \frac{8E\cdot F\cdot P_W^2\cdot l^2}{\pi^4\cdot T^3}\right)}..$$

Here $a(t)$ is the generalized coordinate of linear displacements, m;

l - span length, m;

g – acceleration of gravity force, m/s^2;

ρ – the density of the air stream, $daN\cdot s^2/m^4$;

d_S – the typical size of the cross section of the ice-covered conductor, m;

δ – decrement of damping;

V – wind speed, m/s;

P_0 - weight per unit length of conductor, daN/m;

P_W - weight per unit length of icy conductor, daN/m;

T – conductor tension, daN;

E – elastic modulus, daN/mm^2;
F – conductor cross-sectional area, mm^2;
C_{D0} - aerodynamic coefficient of force of frontal pressure;
C_{L0} and C_{L1} - aerodynamic lift factors;
ω_a – frequency of free transverse vibration of the conductor, rad/s.

Introducing the notation $a(t)=y_0$ and $\acute{a}(t)=y_1$, we represent (1) in the normal form

$$\begin{cases} \dot{y}_0 = y_1, \\ \dot{y}_1 = -k_1 \cdot y_1 - k_2 \cdot y_1^3 - \omega_a^2 \cdot y_0 - k_3 \cdot y_0^2 - k_4 \cdot y_0^3 - k_{13}. \end{cases} \tag{2}$$

Equating the equations in the system (2) $\dot{y}_0\ (t)=0$ and $\dot{y}_1\ (t)=0$ we obtain an expression for the determination of the equilibrium position

$$y_0^3 + \frac{k_3}{k_4} y_0^2 + \frac{\omega_a^2}{k_4} y_0 + \frac{k_{13}}{k_4} = 0. \tag{3}$$

The solution of equations (3) by the iterative method gives the following result

$$y_{01} \approx -\frac{k_{13}}{\omega_a^2}, \tag{4}$$

where y_{01} - the result of the first approximation.

We linearize system (2) in a neighborhood of the equilibrium point 0.25 point (y_{01}, 0)

$$\begin{cases} \dot{y}_0 = y_1, \\ \dot{y}_1 = \left(\frac{2k_3 \cdot k_{13}}{\omega_a^2} - \omega_a^2\right) y_0 - k_1 \cdot y_1. \end{cases} \tag{5}$$

The characteristic equation of a linearized system (5) is:

$$p^2 + k_1 \cdot p + \omega_a^2 - \frac{2k_3 \cdot k_{13}}{\omega_a^2} \approx p^2 + k_1 \cdot p + \omega_a^2 = 0. \tag{6}$$

The characteristic equation (6) has complex conjugate roots

$$p_{1,2} = -\frac{1}{2} k_1 \pm j\omega_a. \tag{7}$$

The solution of the linearized equation, taking into account the initial conditions $y_0=a\ (0) = A_0$ and $y_1=\dot{a}\ (0) = 0$ has the form

$$a(t) = A_0 e^{\frac{-k_1}{2} \cdot t} \left(cos\omega_a t + \frac{k_1}{2\omega_a} sin\omega_a t\right), \tag{8}$$

where

$$k_1 = \frac{\delta \cdot \omega_a - 2b_3 \cdot (C_{D0} + C_{L0}) \cdot V}{\pi}.$$

As follows from formula (8), under the condition $k_1=0$, the split phase performs undamped periodic oscillations. Equating the coefficient k_1 to zero, we find the minimum wind speed at which the split phase performs undamped oscillations with a constant amplitude A_0 and frequency ω_a (the value of coefficient b_3 is taken into account during the transformation)

$$V_{KR}^{H} = \frac{P_W \cdot \delta \cdot \omega_a}{2g \cdot \rho \cdot d_S \cdot (C_{D0} + C_{L0})}. \tag{9}$$

It follows from formula (9) that the lower critical speed is influenced by the damping decrement [20], the weight of the ice sediment, the frequency of natural transverse vibrations of the conductors (therefore, the characteristics of power lines), air density, geometric dimensions, and stationary aerodynamic coefficients of ice deposition. A similar study was carried out in [19], only with respect to a multi-span system, when the conductors are interconnected through garlands of insulators. In [19], processes for a multi-span system are considered whereas this paper considers an anchor span. The novelty of the present work consists in experiments and conclusions related to various features of the anchor span. The analysis of experiments has shown that the value of the critical wind speed at which the galloping of conductors is excited does not depend on the method of attaching the conductor to the support (rigid attachment or through garlands of insulators that give some mobility). If in article [19] the calculation formula is obtained by applying Hurwitz stability criterion, the novelty in the present elaboration consists in researching the stability of nonlinear systems which has been conducted in the first approximation, i.e. based on the linearization of nonlinear equations at the equilibrium point and on the further study of the linearized equation in the neighbourhood of these points.

Therefore, the value of the lower critical speed is not affected by the way the conductor is fixed to the support, that is, rigid anchor fastening or movable - through the garlands of insulators.

3 The modeling process

The verification of the obtained results has been carried out by modeling of the initial nonlinear system (2) in the vicinity of the critical velocity (9). Transients are obtained using Mathcad and applying the Runge-Kutta method with a fixed step which is shown in Fig.1,2. The calculations have been performed for a split phase of 3 conductors AC-300/39.

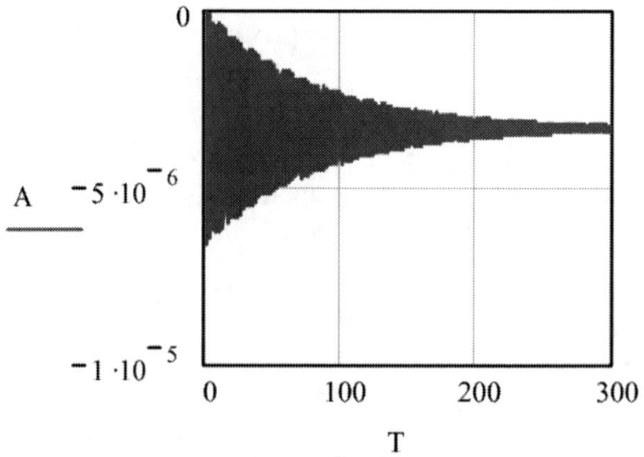

Fig. 1. Transient processes in $l = 200$ m and 10 daN/mm^2 where $V = 1m/s < V_{KR}$

In the figure - the ordinate: A, m - amplitude galloping of conductors; on the abscissa: T, s - the current time.

As follows from Fig.1, if the wind speed is less than the critical speed determined by the formula (9) ($V < V_{KR}$), then the oscillation process (galloping of conductor) is not supported by the system and fades over time.

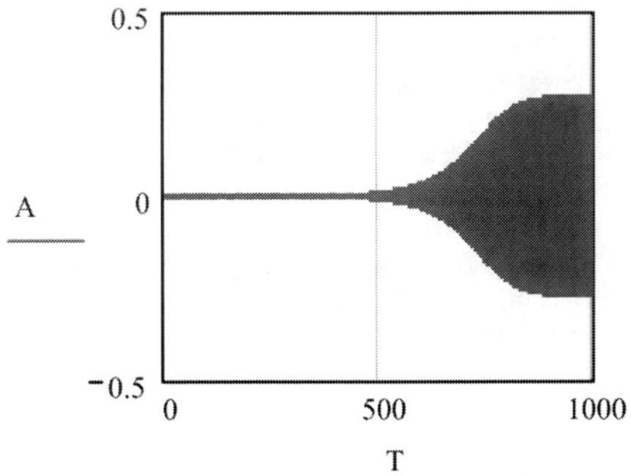

Fig. 2. Transient processes in $l = 200$ m and 10 daN/mm^2 where $V = 1{,}6\ m/s > V_{KR}$

If the wind speed is greater than the critical speed determined by the formula (9) ($V > V_{KR}$), then the oscillation process is supported by the system and after the end of the transition process, a stable oscillation process with an amplitude of 0,25 m is observed (Fig.2). It should also be noted, that usually the frequency of the galloping coincides with the frequency of the free oscillation of the conductor.

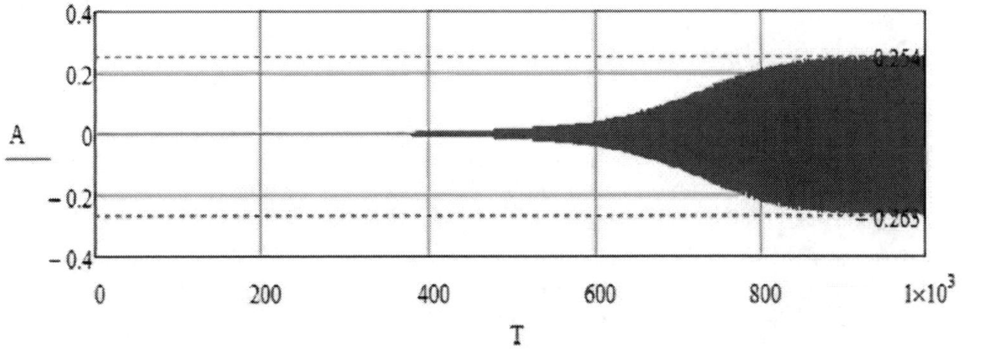

Fig. 3. Transient processes in $l = 200$ m and 10 daN/mm^2 where $V = 2{,}1\ m/s > V_{KR}$

Fig. 3 shows the simulation result for the wind speed *V=2,1 m/s* and for the obtained critical speed *V_{KR}=1,72 m/s*. As can be seen from Fig.3, after the end of the transition process, the duration of which has been approximately 900 s, there is a stable oscillatory process with amplitudes of 0,254 m and 0,263 m, respectively. It should be noted, that the amplitudes are not the same when the phenomenon of conductor galloping occurs. The movement of the conductor from the static balance to the upper position is slightly higher compared to the lower position. This has been repeatedly confirmed by numerous observations of the galloping of conductors on existing power lines and on experimental installations.

4 Conclusions

The proposed calculation formula for the critical wind speed allows for an analysis of the characteristics of the air line, stationary aerodynamic coefficients of ice deposition and damping decrement on the stability of icy split-phase conductors in the wind flow.

Based on this analysis, the following conclusions are drawn:

1. The value of the lower critical wind speed at which the excited galloping of the conductors is significantly affected by the frequency of the free transverse vibrations of the conductors of the split phase, the damping decrement, and the aerodynamic characteristics of the profile of the icy conductors;
2. The configuration of the split phase (the number of conductors in the phase) practically does not affect the critical wind speed;
3. The lower critical wind speed does not depend on the method of attaching the conductor to the support (rigid or having a certain degree of freedom), that is, the galloping of the icy conductors of the split phase in the anchor span and multi-span system is excited under the same conditions.

Along with the novelty presented in the work, another significant advantage of the study is the developed calculation formula of the critical speed which can be used as an engineering method for estimating the critical (threshold) wind speed, which will allow dividing the boundary of a stable galloping from an unstable one and thus make it possible to construct areas of occurrence of the galloping of conductors for a specific air line at the specified parameters, conductor grade, aerodynamic characteristics and the decrement of oscillation. The scientific and practical results of this research work can also be used for further studies of conductor galloping, for compiling zoning maps of the territories according to the frequency of occurrence of galloping, taking into account climatic and geographical features, and for developing measures for protecting overhead power lines from galloping.

References

1 J.L. Lilien, D.G. Havard. Galloping Data Base on single and bundle conductors. Prediction of Maximum Amplitudes. IEEE Trans. power delivery. **15(2)**, pp.670-674 (2000)

2 C.B. Gurung, H. Yamaguchi, T. Yukino. Identification of large amplitude wind-induced vibration of ice-accreted transmission lines based on field observed data. J.Engineering Structures, **24** pp. 179-188 (2002)

3 O. Chabart, J.L. Lilien. Galloping of electrical lines in wind tunnel facilities. J.Wind engineering and industrial aerodynamics **74-76**, pp. 967-976 (1998)

4 P. V. Dyke, Andre Laneville. Galloping of a single conductor covered with a D-section on a high-voltage overhead test line. J. Wind engineering and industrial aerodynamics **96,** pp.1141-1151(2008)

5 L. Zhou, B. Yan, L. Zhang, S. Zhou. Study on galloping behavior of iced eight bundle conductor transmission lines. J. Sound and vibration **362**, pp.85-110(2016)

6 Y.N. Bocharov, V.V. Titkov, R. Abitaeva, A.B. Bekbaev, M.A. Djamambaev. Torsion vibrations of splitted conductor of overhead transmissions. St. Petersburg polytechnic university. Engineering sciences and technology, **23**(1), pp.84-91(2017)

7 Y. Zhitao, L. Zh., S. Eric, L. William E. Galloping of a single iced conductor based on curved-beam theory. J. Wind eng. ind. aerodyn. **123**, pp.77-87, (2013)

8 Y. Zhitao, S. Eric, L. Zh., L. William E. Galloping of iced quad-conductors bundles based on curved beam theory. J. Sound and vibration **333**, pp.1657–1670. (2014)

9 W.Y. Ma, Q.K. Liu, X.Q. Du, Y.Y. Wei. Effect of the Reynolds number on the aerodynamic forces and galloping instability of a cylinder with semi-elliptical cross sections. J. Wind eng. ind. aerodyn. **146**, pp.71–80, (2015)

10 A. Joly, S. Etienne, D. Pelletier Galloping of square cylinders in cross-flow at low Reynolds numbers. J. Fluids and structures **28**, pp.232–243, (2012)

11 P. Augustin. Fluttering of the "electrical line galloping" type for conductors of continuous mass,- Buletinul institutului politehnic "Gheorghe Gheorghiu-dej" Bucuresti. TVomul. XXIX, **3**, pp. 103-107, (1967)

12 V.I. Vanko, I.K. Marchevsky. Pitching conductors of PLW - Lyapunov instability. - Energy. News of higher educational institutions and energy associations of the CIS. **6,** pp. 14-23, (2014) [in Russian]

13 L.K. Marchevsky. Mathematical modeling of the flow around a profile and research of its stability in a stream according to Lyapunov / dissertation abstract for the degree of candidate of physical and mathematical sciences. //. - M.: N.E. Bauman MSTU, (2008), [in Russian]

14 V.Ya. Gorin, N.N. Davidson, E.A. Marasina. Methodology for determining the critical wind speed when galloping conductors of overhead power lines. - // Science.Pratsi DonNTU, seriya "Electrical Engineering and Energy", **7** (128) .- Donetsk: DVNZ DonNTU, pp. 52-57, (2009), [in Russian]

15 F.N. Shklyarchuk, A.N. Danilin. Nonlinear vibrations and galloping of conductors with icing. - Bulletin of TulSU. Technical science. Energy, electricity, electric drive. **11** (188), pp. 188-197, (2013), [in Russian]

16 V.A. Feldstein, S.V. Kolosov, S.V. Ryzhov. VL conductor gallope model and calculation of protective equipment. - M .: Pub.house IAC "Energy". pp. 145-156. (2010), [in Russian]

17 M.A. Dzhamanbaev, N. Tokenov Mathematical model of the dance of the split phase of overhead power lines (anchor span) / proceedings of the II - international scientific conference "High Technologies - the Key to Sustainable Development", /. - Almaty, KazNTU named after K.I. Satbaev. vol. **II**, p. 5. (2013), [in Russian]

18 N. Tokenov, M. Dzhamanbayev, A. Bekbayev, D. Eskendirova, O. Baimuratov. Mathematical Model for Calculating Aerodynamic Characteristics of Overhead Transmission Lines. - J. Applied Mechanics and Materials. **610,** pp 52-59. (2014)

19 M.A. Dzhamanbaev, K.S. Chakeeva, J.E. Karataeva, Z.A. Dzhumabekova. The method of calculating critical wind speeds in determining the area of aerodynamic instability of the conductors of the split phase of a multi-span power line // J. Measuring equipment, **6,** pp. 25-30. (2019), [in Russian]

20. Terziev A., Analysis of the terrain specifics and roughness factor on the wind shear over complex terrains, IOP Conference Series Materials Science and Engineering 595:012043, September 2019, DOI: 10.1088/1757-899X/595/1/012043

Mechano-hydraulic pumping solutions for reduction off energy losses in agricultural mobile machinery

Alexandru Hristea[1*], *Bogdan* Tudor[1], *Stefan* Sefu[1] and *Gabriela* Milian[2]

[1]INOE 2000-IHP, street Cutitul de Argint, no. 14, Sector 4, Bucharest, Romania

[2]INMA, street Ion Ionescu de la Brad, no. 6, Sector 1, Bucharest, Romania

Abstract. Hydraulic digital power consists in a hydraulic plant having one or more discreet valued components actively with which we are controlling the system output flow. In this paper authors study a solution of 4 fixed pumps parallel connected. This complex pumping system can achieve 15 flows. The pumping system consists in an electrical single speed motor, 4 fixed pumps, the repartition will achieve with the help of 4 electrohydraulic distributors and an electronic micro-controller. The hydraulic plant was made in our Research Institute in the digital hydraulic laboratory for conducting tests and experiments.

1 Introduction

The evolution of technology was one of the biggest developments for human society, and within it a very important role it has played by the development of mechanical, electrical, hydraulic and pneumatic drives. Further will be analyzed how the hydraulic drive past from movements required in order to achieve the role of agricultural machinery to technical means of making movements with energy losses as low as possible. This shift towards energy saving was made in time, in several stages, each stage being represented by a specific technology.

In the first stage, with fixed pump systems, initial achievements were oriented toward movements' realization and only secondarily to reducing energy consumption by reducing losses [1].

In the second stage, adjustable stage pump systems, was pursued to modify the displacement of the pumps (power consumption) during operation, by means of mechano-hydraulic means like controllers.

*Corresponding author: hristea.ihp@fluidas.ro

 https://doi.org/10.1051/e3sconf/202018004020

The third stage, the energy consumption approach one (from the primary source) to the requirements of each phase of the cycle of operation, is dominated by equipping adjustable pumps with devices controlled with servo equipment or load sensing [2].

In the fourth stage, the digital hydraulics, was passed from the regulation of the pump, in order to reduce losses, to the use of several units of the system to close the primary energy consumption to energy needs of the driven machinery.

Great novelty of the 90s was the transfer of digital hydraulics from theory to real world practice. This thing was manageable due to specialists such as Scheidl, Linjama, Vilenius, Bishop, Wadsley, Merrill, Kudzma, Ploeckinger, etc., who had the first notable practice results.[3]

Digital Fluid Power means hydraulic and pneumatic systems having discrete valued component(s) actively controlling system output. Very important for digital hydraulics is the "intelligent control". In Europe there is accepted that digital hydraulics has a branch that was based on parallel connections, with a center located in Tampere, and a branch based on the switching technology in Linz. It should be specified that digital hydraulics does not mean traditional hydraulics, classic or modern with analog element which we drive with digital electronics. Digital hydraulics means hydraulics or pneumatics systems which have in their structure at least one hydraulic element to ensure at the exit specifically discrete values and active control. This thing involves beside the digital hydraulic elements a proper electronic command, so a good computerization of the system.

2 The advantages of digital hydraulics

The advantages of digital hydraulics are extremely clear at the theoretical level, but less visible in industrial practice. The hopes related to speed, robustness, simplicity and equipment reliability is not yet materialized in industrial products used on the large scale or in some specific areas. Comparative with the standard on/off hydraulics, digital hydraulics has a response time much smaller, but it keeps the durability and work capacity in dirty mediums of the classic hydraulics drives. Specialists from Tampere estimate that there is the possibility to save a percentage of fuel between 20 and 40%, machines becoming simpler and light-weight without a price rising. Digitization of all mobile machines can reduce consumption on road with a percentage between 5 and 10%. Even the general architecture of the hydraulically operated vehicle can be simplified by using digital hydraulics, as in figures 1, 2, as indicated by L. Wadsley.[4]

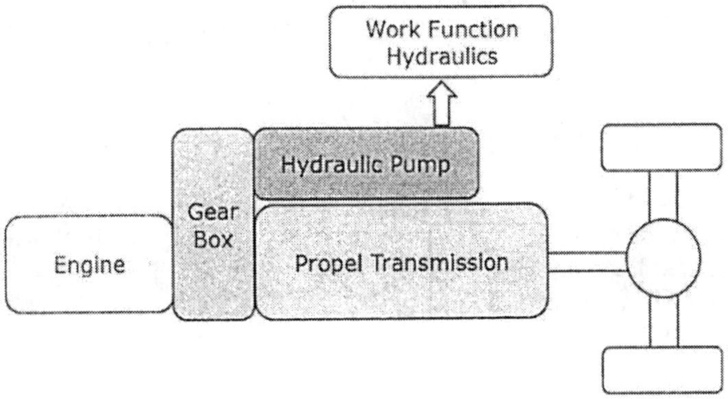

Fig. 1. Typical mobile machine layout [4]

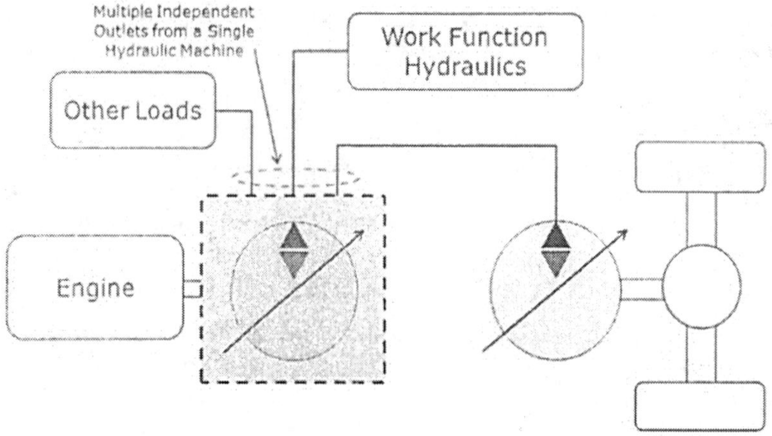

Fig. 2. Alternative mobile machine layout [4]

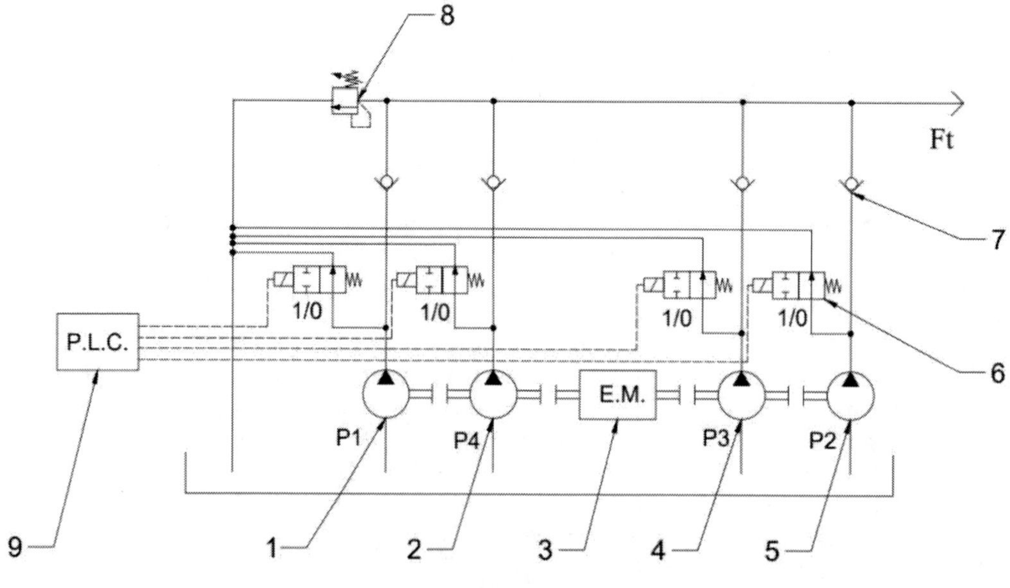

Fig. 3. Hydraulic system with 4 fixed displacement pumps consists in: (1,2,4,5) fixed displacement pumps, (3) biaxial electric motor, (6) 2/2 directional valve, (7) one-way valve, (8) pressure relies valve

3 The 4 fix displacement pumps solution with 15 selectable flows

Hydraulic systems which used pumping units with 4 coaxial pumps (Figure 3) or 4 pumps driven by independent electric/thermic motor were actually the first steps towards digital hydraulics. Each pump has its own flow (Q1, Q2, Q3, Q4) of which we can select for the system one, two or even all for, getting the flow rates with the help of a programable logic controller (P.L.C.):

Table.1 Depending on flow requirements, one or more pumps will be selected

Nr.crt.	Pump combination	Total displacement $[cm^3/rot]$	System Flow [L/min]	Power N [kW]
1	P1	4	6	2
2	P2	8	12	4
3	P1+P2	12	18	6
4	P3	16	24	8
5	P3+P1	20	30	10
6	P3+P2	24	36	12
7	P3+P2+P1	28	42	14
8	P4	32	48	16
9	P4+P1	36	54	18
10	P4+P2	40	60	20
11	P4+P2+P1	44	66	22
12	P4+P3	48	72	24
13	P4+P3+P1	52	78	26
14	P4+P3+P2	56	84	28
15	P4+P3+P2+P1	60	90	30

P1=First pump with the flow displacement of 4 cm^3/rot

P2= Second pump with the flow displacement of 8 cm^3/rot

P3= Third pump with the flow displacement of 16 cm^3/rot

P4= Forth pump with the flow displacement of 32 cm^3/rot

The flow rate from the column four was calculated at the nominal speed of 1500 rpm

Using the electric automation system there is selected one of the 15 variants of flow, which can provide one of the 15 different speeds for the hydraulic motor E.M.

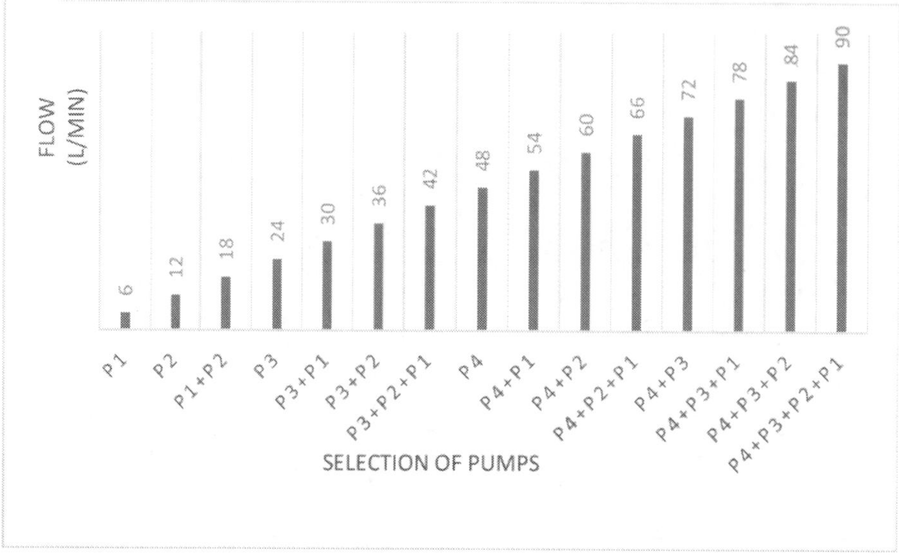

Fig 4. Flow chart

Output power is:

$$N \cong \frac{Ps \cdot (Q1+Q2+Q3+Q4)}{600\eta}, \text{ where } Q_2=2Q_1;\ Q_3 = 2Q_2;\ Q_4=2Q_3 \quad (1)$$

N [kW]= Total output power, Ps [bar]= System pressure, Q [l/min] = system flow composes of the 4 pumps, η= total system efficiency

For each of the 15 phases, besides output power N there is also power installed in excess, that will not be consumed; there will be lost only the part of this power which ensures oil recirculation through the pumps that are not actively involved in that particular phase, plus normal losses caused by operation of the pumping system. The amount of power that is lost is small compared to the amount of useful power consumption. The great advantage of this solution is that besides economical use of power the system allows simple automation, flow accuracy up to 12.5% and also functional structure quite simple and inexpensive.

4 Conclusions

This material refers to the hydraulic systems with 4 fixed displacement pumps, and it highlights certain aspects such as the variable flow pumps equipped with various regulators have high energy efficiency because flow and pressure automatically adapt to customer needs. Reducing energy losses which cause overheating of hydraulic oil leads to simplification of hydraulic system by:

• Although fixed displacement pump hydraulic systems have wasteful behavior in terms of energy, there are a lot of cases in which they can be kept within reasonable limits;

• Complex high-power systems lead to extremely significant energy losses, and therefore hydraulic technology has made considerable progress by using adjustable pumps and more recently with the digital hydraulic witch is an affordable alternative to the complex pump systems.

• Possibility to mount a smaller capacity tank and to remove the oil cooler, pressure reduction valve or electrical decoupling valves.

This paper has been developed in INOE 2000-IHP and funded by a grant of the Romanian Ministry of Education and Research under the National Research Programme NUCLEU, Financial agreement no. 18N/08.02.2019, Additional Act no. 9/2020, Project acronym: OPTRONICA VI, Research theme PN 19-18.01.02, titled: "Advanced research on developing synergic border architectures used in solving global challenges and improving knowledge-based competitiveness", Phase no. 6, titled: 'Methods, procedures and equipment for experimental validation of mathematical models and numerical simulations, used for modernizing hydraulic equipment in order to reduce energy losses'.

References

1. G. Matache, St. Alexandrescu, A. G. Pantiru, Gh. Sovaiala, M. Petrache: The analysis of flow losses through dynamic seals of hydraulic cylinders, *HIDRAULICA* no. 1, pp. **52-60**, ISSN 1453 – 7303, (2013)
2. C. Cristescu, L. M. Micu, I. C. Dumitrescu, P. Krevey, "Using Load Sensing control systems to increase energy efficiency of hydrostatic transmissions", *HIDRAULICA* no. 4, pp. **71-77**, ISSN 1453 – 7303, (2015)
3. C. Cristescu, P. Drumea, D. Ion Guta, C. Dumitrescu, "Theoretical research and laboratory experimental tests regarding the dynamic behavior of hydraulic system for energy recovery at the braking of motor vehicle", in: Proc. of *The 8-th International Fluid Power Conference-IFK 2012 Fluid Power Drives!*, pp. **435 - 446**, (www.ifk2012.com), (Dresden, Germany, March 26-28, 2012)
4. PhD.Eng. Heinrich THEISSEN "Fluid Power For Sustainability", *International Conference of Hydraulics and Pneumatics-HERVEX*, ISSN 1454- 8003, pp. **137-157** , (Calimanesti-Caciulata, Romania, November 7-9, 2012)
5. Rudolf SCHEIDL, Helmut KOGLER" Hydraulic Switching Control - State of the Art and Challenges" (2012)
6. Andreas PLÖCKINGER, Bernd WINKLER, Rudolf SCHEIDL," Development and Prototyping of a Compact, Fast 3/2 Way Switching Valve with Integrated Onboard Electronics" *The 11th Scandinavian International Conference of Fluid Power, SICFP*, ISBN 978-91-7393-589-0, pp. **66-67 (**Linkőping, Sweden, 09 June 2-4 2009)
7. Luke WADSLEY," Optimal System Solutions Enabled by Digital Pumps" *The 52nd National Conference on Fluid Power,* **pp.7-13** (Las Vegas, March 23-25, 2011)

AUTHOR INDEX

AUTHOR INDEX

AUTHOR INDEX